# Innovative Materials and Systems for Energy Harvesting Applications

Luciano Mescia
*Politecnico di Bari, Italy*

Onofrio Losito
*Politecnico di Bari, Italy*

Francesco Prudenzano
*Politecnico di Bari, Italy*

A volume in the Advances in Environmental Engineering and Green Technologies (AEEGT) Book Series

ENGINEERING SCIENCE REFERENCE
An Imprint of IGI Global

| | |
|---|---|
| Managing Director: | Lindsay Johnston |
| Managing Editor: | Austin DeMarco |
| Director of Intellectual Property & Contracts: | Jan Travers |
| Acquisitions Editor: | Kayla Wolfe |
| Production Editor: | Christina Henning |
| Typesetter: | Lisandro Gonzalez |
| Cover Design: | Jason Mull |

Published in the United States of America by
Engineering Science Reference (an imprint of IGI Global)
701 E. Chocolate Avenue
Hershey PA, USA 17033
Tel: 717-533-8845
Fax: 717-533-8661
E-mail: cust@igi-global.com
Web site: http://www.igi-global.com



Library of Congress Cataloging-in-Publication Data

Innovative materials and systems for energy harvesting applications / Luciano Mescia, Onofrio Losito, and Francesco Prudenzano, editors.

pages cm

Includes bibliographical references and index.

Summary: "This book reports on some of the best tools and technologies available for powering humanity's growing thirst for electronic devices, including piezoelectric, solar, thermoelectric, and electromagnetic energies"-- Provided by publisher.

ISBN 978-1-4666-8254-2 (hardcover) -- ISBN 978-1-4666-8255-9 (ebook) 1. Energy harvesting--Equipment and supplies. 2. Transducers. I. Mescia, Luciano, 1973- II. Losito, Onofrio, 1969- III. Prudenzano, Franceso, 1964-

TK2897.I56 2015
621.31--dc23

2015003787

This book is published in the IGI Global book series Advances in Environmental Engineering and Green Technologies (AEEGT) (ISSN: 2326-9162; eISSN: 2326-9170)

British Cataloguing in Publication Data
A Cataloguing in Publication record for this book is available from the British Library.

# Advances in Environmental Engineering and Green Technologies (AEEGT) Book Series

ISSN: 2326-9162
EISSN: 2326-9170

Mission

Growing awareness and an increased focus on environmental issues such as climate change, energy use, and loss of non-renewable resources have brought about a greater need for research that provides potential solutions to these problems. Research in environmental science and engineering continues to play a vital role in uncovering new opportunities for a "green" future.

The **Advances in Environmental Engineering and Green Technologies (AEEGT)** book series is a mouthpiece for research in all aspects of environmental science, earth science, and green initiatives. This series supports the ongoing research in this field through publishing books that discuss topics within environmental engineering or that deal with the interdisciplinary field of green technologies.

Coverage

- Air Quality
- Electric Vehicles
- Waste Management
- Renewable energy
- Contaminated Site Remediation
- Alternative Power Sources
- Policies Involving Green Technologies and Environmental Engineering
- Water Supply and Treatment
- Green Transportation
- Cleantech

IGI Global is currently accepting manuscripts for publication within this series. To submit a proposal for a volume in this series, please contact our Acquisition Editors at Acquisitions@igi-global.com or visit: http://www.igi-global.com/publish/.

The Advances in Environmental Engineering and Green Technologies (AEEGT) Book Series (ISSN 2326-9162) is published by IGI Global, 701 E. Chocolate Avenue, Hershey, PA 17033-1240, USA, www.igi-global.com. This series is composed of titles available for purchase individually; each title is edited to be contextually exclusive from any other title within the series. For pricing and ordering information please visit http://www.igi-global.com/book-series/advances-environmental-engineering-green-technologies/73679. Postmaster: Send all address changes to above address. 

## Titles in this Series

*For a list of additional titles in this series, please visit: www.igi-global.com*

*Green Services Engineering, Optimization, and Modeling in the Technological Age*
Xiaodong Liu (Edinburgh Napier University, UK) and Yang Li (British Telecom, UK)
Information Science Reference • copyright 2015 • 315pp • H/C (ISBN: 9781466684478) • US $225.00 (our price)

*Handbook of Research on Social, Economic, and Environmental Sustainability in the Development of Smart Cities*
Andrea Vesco (Istituto Superiore Mario Boella, Italy) and Francesco Ferrero (Istituto Superiore Mario Boella, Italy)
Information Science Reference • copyright 2015 • 491pp • H/C (ISBN: 9781466682825) • US $325.00 (our price)

*Progressive Engineering Practices in Marine Resource Management*
Ivelina Zlateva (Independent Researcher, Bulgaria) Violin Raykov (Institute of Oceanology, Bulgaria) and Nikola Nikolov (Technical University of Varna, Bulgaria)
Engineering Science Reference • copyright 2015 • 456pp • H/C (ISBN: 9781466683334) • US $245.00 (our price)

*Promoting Sustainable Practices through Energy Engineering and Asset Management*
Vicente González-Prida (University of Seville, Spain) and Anthony Raman (NTEC Tertiary Group, New Zealand)
Engineering Science Reference • copyright 2015 • 402pp • H/C (ISBN: 9781466682221) • US $235.00 (our price)

*Handbook of Research on Advancements in Environmental Engineering*
Nediljka Gaurina-Medjimurec (University of Zagreb, Croatia)
Engineering Science Reference • copyright 2015 • 660pp • H/C (ISBN: 9781466673366) • US $345.00 (our price)

*Soft Computing Applications for Renewable Energy and Energy Efficiency*
Maria del Socorro García Cascales (Technical University of Cartagena, Spain) Juan Miguel Sánchez Lozano (University Centre of Defence at the Spanish Air Force Academy, Technical University of Cartagena, Spain) Antonio David Masegosa Arredondo (University of Granada, Spain) and Carlos Cruz Corona (University of Granada, Spain)
Information Science Reference • copyright 2015 • 408pp • H/C (ISBN: 9781466666313) • US $235.00 (our price)

*Optimum Design of Renewable Energy Systems Microgrid and Nature Grid Methods*
Shin'ya Obara (Kitami Institute of Technology, Japan)
Engineering Science Reference • copyright 2014 • 430pp • H/C (ISBN: 9781466657960) • US $235.00 (our price)

*Nuclear Power Plant Instrumentation and Control Systems for Safety and Security*
Michael Yastrebenetsky (State Scientific and Technical Centre for Nuclear and Radiation Safety, Ukraine) and Vyacheslav Kharchenko (National Aerospace University- KhAI, Ukraine, and Centre for Safety Infrastructure-Oriented Research and Analysis, Ukraine)
Engineering Science Reference • copyright 2014 • 470pp • H/C (ISBN: 9781466651333) • US $265.00 (our price)

IGI GLOBAL
DISSEMINATOR OF KNOWLEDGE
www.igi-global.com

701 E. Chocolate Ave., Hershey, PA 17033
Order online at www.igi-global.com or call 717-533-8845 x100
To place a standing order for titles released in this series, contact: cust@igi-global.com
Mon-Fri 8:00 am - 5:00 pm (est) or fax 24 hours a day 717-533-8661

# Editorial Advisory Board

# Table of Contents

## Section 2
## Piezoelectric Systems for Energy Harvesting



## Section 3
## Thermal and Solar Systems for Energy Harvesting

# Detailed Table of Contents



This chapter summarizes the physical properties of THz antennas, provides a summary of some of the most important recent developments in the field of energy harvesting of Earth long-wave infrared radiation, discusses the potential applications and identifies the future challenges and opportunities. In particular, a THz antenna is designed in order to transform the thermal energy, provided by the Sun and re-emitted from the Earth, in electricity. The proposed antenna is a square spiral of gold printed on a low cost dielectric substrate. Simulations have been conducted in order to investigate the behavior of the antenna illuminated by a circularly polarized plane wave with an amplitude chosen according to the Stefan-Boltzmann radiation law. Moreover, these THz antennas could be coupled with other components to obtain direct rectification of T radiation. As a consequence, these structures further optimized could be a promising alternative to the conventional photovoltaic solar cells.



The solar energy is able to supply humanity energy for almost another 1 billion years. Optical nano-antennas (ONAs) are an attractive technology for high efficiency, and low-cost solar cells. These devices can be classified to semiconductor nano-wires and metallic nano-antenna. Extensive studies have been carried out on ONAs to investigate their ability to harvest solar energy. Inspired by these studies, the scope of the chapter is to highlight the latest designs of the two main types of ONAs. The metallic nano-antennas are discussed based on the following points: plasmon, modeling, and performance of antenna designs using different configurations and materials. Moreover, the semiconductor nano-wires are studied thoroughly in terms of photonic crystals, antenna design with different patterns, nano-wire forms and materials. Also, the applications of ONAs and their fabrication aspects such as diode challenges are presented in detail. Finally, three novel designs of ONAs are presented and numerically simulated to maximize the harvesting efficiency.



This chapter proposes an overview of microwave energy harvesting with focuses on the design of high efficiency low power rectifying circuits. A background survey of RF energy harvesting techniques is presented first. Then, the performances of conventional rectifier topologies are analyzed and discussed. A review of the most efficient rectenna designs, from the state of the art, is also presented. Design considerations for low power rectifier operations are detailed and new high efficient rectifying circuits are designed and evaluated in both GSM and ISM bands under low power constraints.



Energy harvesting (EH) which scavenges electric power from ambient, otherwise wasted, energy sources has been explored to develop self-powered portable electronic devices. Vibration energy, a widely available ambient energy source, can be converted into electric power using a piezoelectric energy harvester that generates electric potential in response to applied mechanical strains. As a compact and durable design paradigm, a piezoelectric energy harvesting skin (PEH skin) which can be directly attached onto the surface of a vibrating engineered system has been proposed to scavenge electric power from vibration energy. The goal of this chapter is to describe the core technologies for the realization of the PEH skin from a system integration perspective as four parts: (a) modeling, (b) design, (c) manufacturing, and (d) demonstration. The readers will be able to learn the entire procedure of developing the PEH skin and applying it to self-powered wireless sensor network (WSN) through this chapter.

This chapter provides a detailed study on the harvest of the energy contained in raindrops by means of piezoelectric transducers. The energy harvester has the role of an electric source, able to recharge storage devices of small electronic components, such wireless sensors, by using the vibrational energy released by the drops hitting the transducer, reducing in such a way the chemical waste of conventional batteries. In technical literature, diverse studies agree on the level of suitable generated voltage on the electrodes of a piezoelectric transducer subjected to rainfall, but a complete characterization on the supplied power is still missing. This work, also to limit optimistic forecasts, takes into account the behavior of the transducers in different scenarios: subjected to real and artificial rainfall, standalone or in parallel configuration, in conventional geometries, due to the commercial format or in customized shape, free to move or with an imposed optimal deflection.



After a brief introduction of piezoelectric materials, this chapter focuses on the characterization of vibrating freestanding piezoelectric AlN devices forced by different external forces acting simultaneously. The analyzed vibrating forces are applied mainly to piezoelectric freestanding structures stimulated by irregular vibration phenomena. Particular kinds of theoretical noise signals are commented. The goal of the chapter is to analyze the effect of the noise in order to model the chaotic vibrating system and to predict the output current signals. Moreover, the author also shows a possible alternative way to detect different vibrating force directions in the three dimensional space by means of curved piezoelectric layouts.



There exist numerous low-frequency excitation sources, such as walking, breathing, and ocean waves, capable of providing viable amounts of mechanical energy to power many critical devices, including pacemakers, cell phones, MEMS devices, wireless sensors, and actuators. Harvesting significant energy levels from such sources can only be achieved through the design of devices capable of performing effective energy transfer mechanisms over low frequencies. In this chapter, two concepts of efficient low-frequency piezoelectric energy harvesters are presented, namely, variable-shaped piezoelectric energy harvesters and piezomagnetoelastic energy harvesters. Linear and nonlinear electromechanical models are developed and validated in this chapter. The results show that the quadratic shape can yield up to two times the energy harvested by a rectangular one. It is also demonstrated that depending on the available excitation frequency, an enhanced energy harvester can be tuned and optimized by changing the length of the piezoelectric material or by changing the distance between the two tip magnets.

## Section 3
## Thermal and Solar Systems for Energy Harvesting



The authors propose in this chapter an original, self-sustainable, power supply system for wireless monitoring applications that is powered from an energy harvesting device based on thermoelectric generators (TEGs). The energy harvesting system's purpose is to gather the waste heat from low temperature sources (<90°C), convert it to electrical energy and store it into rechargeable batteries. The energy harvesting system must be able to power a so-called condition monitoring system (CMS) that is used for the monitoring of heat dissipation equipment. The setup used for measurements (including mechanical details) and the experiments are described along with all the essential results of the research. The electronic system design is emphasized and various options are discussed.



This chapter presents a study of thermoelectric energy harvesting with nano-sized thermopiles (nTE) in a planar 65 nm silicon CMOS process. These devices generated power from a 5C temperature difference at a density comparable to commercially available thermoelectric generators, following a metric used in the research literature (Hudak, 2008). By analyzing these devices as a thermoelectric harvesting system, the authors explore the impact of additional performance metrics such as heat source/sink thermal impedance, available heat flow density, and voltage stacking, providing a more comprehensive set of criteria for evaluating the suitability of a thermal harvesting technology. The authors use their thermoelectric system theory to consider the prospects for several thermoelectric energy harvesting applications.



The use of solar energy as a renewable source is one of the most promising resources to generate electricity. The viability of concentrating solar power (CSP) systems depends on the development of highly reflective materials that are inexpensive and maintain their optical properties for extended lifetimes under outdoor environments. In this chapter, the implementation of flexible polymer substrates plated with silver by the Dynamic Chemical Plating technique (DCP) is proposed because of its low cost, and easy and rapid deposition, in addition to the high speed at which the deposit is made. However, the deposits made under this technique have certain nanoscale imperfections, which begin to exist certain permeability of substances that can stain the silver over time, so a study of this feature is performed, to help assess their durability.

## Section 4
## Non-Conventional Systems and Materials for Energy Harvesting



Portable electronic systems and wearable sensor networks are offering increasing opportunities in fields like healthcare, medicine, sport, human-machine interfacing and data sharing. The technological research is looking for innovative design solutions able to improve performances and portability of wearable systems. The power supply strategy is crucial to improve lifetime, reduce maintenance, preserve the environment and reduce costs of smart distributed electronic systems applied to the body. The conversion of biomechanical energy of limbs and joints to electricity has the potential to solve much of the actual limitations. The design and building of wearable energy harvesters for wearable applications require different approaches respect to traditional vibratory energy harvesters. This chapter focuses on transduction materials, modeling strategies, experimental setups, and data analysis for the design of biomechanical energy harvesters; a case study based on system integration and miniaturization is also described for applications in the field of human-machines interfacing.



The concept of harvesting energy from flow-induced vibrations has received a great deal of attention in the last few years. This technology would help in the replacement of small batteries that require expensive and time consuming maintenance and development of self-powered electronic devices, such as health monitoring sensors, medical implants, data transmitters, wireless sensors, and cameras. In this chapter, a particular focus is paid to the concept of harvesting energy from aeroelastic instabilities, such as flutter in airfoil sections, vortex-induced vibrations in circular cylinders, and galloping in prismatic structures. Nonlinear electroaeroelastic models for these energy harvesters are derived and validated with experimental measurements. It is shown how linear and nonlinear analyses can be used to breach traditional barriers in the design and performance enhancement of these aeroelastic energy harvesters, characterization of their behaviors, and identification of the contribution of different types of nonlinearities.

Energy is one of the most important issues in the world owing to the steady growth of energy demands, serious concern about fossil fuel exhaustion, and environmental problems. Reverse electrodialysis (RED), which is a technique to capture the useful energy from mixing the solutions of different concentration, is recently recognized as an efficient method to alleviate the current issue associated with the energy. This chapter reviews the RED studies using nanoporous membranes (named as nanofluidic reverse electordialysis, NRED) in place of conventional polymer membranes having no physical pore. This chapter is made up six parts: the mechanism, the fabrication of nanostructures, the influence of operating parameters (concentration gradient and pore size), the materials, the efforts on enhancing performance (surface modification technique and large area platform), the numerical modeling, and, lastly, the feasible applications.



The energy harvesting based on electrostrictive polymers has great potential for remote applications such as in vivo sensors, embedded micro-electro-mechanical systems devices. The harvested energy and action are controlled by the permittivity, the Young's modulus and their dependence on frequency and level of stress. One should use a model which takes into account mechanical losses in order to obtain efficient devices. This chapter provides a brief overview of the methods for harvesting mechanical to electrical energy using elestrostrictive polymer. The second paragraph presents two new techniques which enable the improvement of the electromechanical performance of electrostrictive polymers in order to demonstrate their potential for the vibrational energy recovery. Based on the strong relationship between the frequency of operation and the mechanical strain from one could improve the electromechanical conversion. The development of a new prototype based on electrostrictive polymers for increasing the conversion AC–DC power is discussed.

# Foreword

Energy harvesting (also known as scavenging) is the use of ambient energy to provide electrical power for electronic and electrical devices (in general small) making them self-sufficient, with or without the use of additional batteries.

The term "energy harvesting" is quite new; however, the concept has been known for a long time: we have been wearing automatic wrist watches (powered by our arm's movement) for decades.

Research in the field is bringing many different approaches and the energy sources exploited with success by energy harvesters are thermoelectric, biomechanical, vibrations, waste water, infra-red emission, ambient fluid motion, and solar (just to cite the ones developed in the book). Another problem is transferring the power to the device that should use such power: it cannot be always done by a simple cabled connection, and research in the topic include the design of special antennas and systems, commonly referred to as wireless power transfer systems. As clearly evidenced in the book, the two problems (harvesting energy and correctly delivering it to the device to be powered) are frequently related to each other and the engineering approach to face both of them is crucial, as highlighted in this book.

A potential huge market for energy harvesting systems is awaiting in the next years, and every engineer should be aware of the opportunities and challenges this topic offers. This book is a unique opportunity for the reader to develop a basic, yet accurate, knowledge of energy harvesting sources, technologies and their applications.

Each chapter covers a specific topic and it is written by well-known researchers, mainly from the academic world, with the aim of not only giving readers the fundamentals of each topic but also letting them see future trends and possibilities. It is the perfect starting point for an engineer who wants to enter into this fascinating world.

The editors of the book are also well known researchers and are currently working in the Electromagnetic Fields Group of the Dipartimento di Ingegneria Elettrica e dell'Informazione at Politecnico di Bari. Their research interests include theoretical aspects for the development of fiber lasers and amplifiers as well as the design of antenna system, shielding, EMC test, microwave and optical devices; therefore they coordinated the development of the book making it an exciting reading experience "despite" its highly technical content.

*Sami Barmada*
*University of Pisa, Italy*

**Sami Barmada** *was born in Livorno, Italy, 18 November 1970. He received the M. Sc. degree in June 1995 in Electrical Engineering from the University of Pisa. In November 1995 he joined ABB Teknologi AS (now ABB Corporate Research) in Oslo, Norway, working in the Power Systems and Components Department. In 1997 he joined the Ph.D. courses in Applied Electromagnetics, at the University of Pisa. Prof. Barmada received the Ph.D. degree discussing the thesis "Wavelet Expansion Methods for the Numerical Analysis of Transmission Lines". During the Ph.D. he has been visiting student at the University of Akron (OH, USA), Electrical and Computer Engineering Department. Between years 2001 and 2003 he has been a Post Doctoral Student at the Department of Electric Systems and Automation, where he currently is Associate Professor. He is author of several publications in refereed Journals and Conference Proceedings; he has served as session chairman (PIERS and ACES conferences), instructor of short courses and tutorials at international conferences, member of editorial boards and serves as reviewer for international journals. He has been General Chairman of the ACES 2007 Conference held in Verona, Italy, in 2007, and has been Technical Program Chairman for the conference PIERS, held in Pisa in 2004. He has been recipient of the John F. Alcock Prize (2004), awarded annually to a paper published by the Institution of Mechanical Engineers which concentrates on technical innovation in the railway traction field. Prof. Barmada is currently a member of the Board of Directors of the Applied Computational Electromagnetics Society (ACES) and member of the Institute of Electrical and Electronic Engineers (IEEE).*

# Preface

The significant increasing in the demand of wearable electronics, wireless sensors and passively powered devices is attracting considerable research interest, from both the academic and the industrial communities, in the area of energy harvesting. In fact, energy harvesting technology makes possible the utilization of some of the available and otherwise wasted energy in the environment (electromagnetic waves, sunlight, mechanical vibration, thermal gradients, convection flows, chemical, biological and mechanical motion or other forms of harvestable energy) in a very innovative and useful way.

After a first and necessary development phase, energy harvesting technology has reached maturity, as shown by its application to several fields. This is due to many different aspects, but probably the most relevant one is that energy harvesting devices are attractive to significantly prolong overall battery life and in some passively low powered wireless electronic devices as replacement of the batteries. Considering that batteries are impractical, costly or dangerous to replace, and that the power provided by energy harvesting devices is clean, infinite and perpetually renewable it is clear the significant environmental benefit. Therefore, now this technology represents an important alternative to eliminate the need of batteries in our household electronics providing devices that could be considered as efficient energy eco-systems. Within this context, the synthesis, modeling and characterization of innovative materials, the development of innovative power management techniques and technologies as well as the design and optimization of innovative components and functional devices, have huge potentiality in enabling the future energy harvesting applications.

This book is aimed at showing this aspect in the fields of engineering. In other words, the book reports neither all the possible applications of a specific engineering problem nor the most recent solutions of each considered problem evaluating the multitude of applications of energy harvesting devices to the different problems in engineering. Therefore, the main goal is collecting examples of the employment of these devices in problems which are fundamental in engineering from a conceptual point of view, providing the main ideas of their application to practical cases.

On the basis of these considerations, the proposed contributions face the implementation of thermoelectric, biomechanical, piezoelectric, aeroelastic, and rainfall energy harvester, the design and optimization of innovative antenna systems to harvest energy, the modelling and synthesis of novel materials as well as the development of novel techniques to enhance the overall system performance.

It is shown that the concepts of energy harvesting are applicable to cases characterized by circuit architectures requiring particular performance and devices whose behaviour is described by complex mathematical models. Thus, efficient methods and circuits to manage and gather the energy are a must. Therefore, optimization and accurate numerical modelling are needed for device performance prediction and to improve the capability of the harvester in collecting and transforming the energy wasted in the environment.

This book reports technical reviews, discussions, and basic knowledge regarding the most important energy harvesting sources and technologies including piezoelectric, solar, thermoelectric, and electromagnetic. Therefore, it also illustrates research activities regarding non-conventional energy harvesting sources, devices and materials. The topics analysed in the chapters will allow readers to conceptualize the promise of the energy harvesting-based technologies. Moreover, the authors of the chapters are researchers of academics, industry, and national laboratories.

The book scheme reproduces the organization of the book chapters, whose considered topics are affine. All the chapters have a practical spin-off in terms of real world application, otherwise all the problems could constitute a theoretic exercise which would not be significant in engineering.

The first part of the book shows how energy harvesting technology can be applied to the development of electromagnetic harvesters in different fields, like wireless energy transfer, terahertz and optical antennas. In particular, in Chapter 1 the problem of the design and optimization of terahertz antenna are dealt with. In this chapter, some design strategies and guidelines are illustrated with the target to enhance the capability of THz antenna to collects the longwave radiation emitted from the Earth and to convert it into electricity. The chapter summarizes the physical properties of THz antennas, provides a summary of some of the most important recent developments in the field of energy harvesting of Earth long-wave infrared radiation, discusses the potential applications and identifies the future challenges and opportunities. A detailed numerical investigation is illustrated with the aim to study the effects of metal properties, ground plane, substrate and metal layers thickness as well as gap size on the electric field enhancement at the antenna feeding point. This kind of technology, could be used to overcome the main drawback of the current solar technology, i.e. the operation only during the morning. The same technology could also be used in most industrial process where there is waste heat that could be gathered to produce electricity. Moreover, this kind of antennas may also be used to remove the heat and to cool electronic devices and, if ad hoc designed, should be a good candidate to improve the building thermal insulation and to reduce the energy consumption due to the operation of air-conditioners and fans. Finally, due to the employed materials, this technology should be more eco sustainable with respect to the conventional ones. It is straightforward to observe that the researchers and designers working in the specific considered field can benefit of the proposals of this chapter. Moreover, engineers and scientists who want to become familiar with the rectennas and the methods of their optimization are advised to study this chapter.

As well as the terahertz antenna, the optical nano-antennas (ONAs) could be a very interesting devices that collects the solar energy in visible and near infrared wavelength range. A detailed description of the science, concepts of design, modelling key issues and the history of the ONAs are provided in Chapter 2. This chapter highlights the latest designs of the ONAs taking into account the following points: plasmon and electromagnetic, modelling and performance of antenna designs using different configurations and materials. Three novel designs of flower shape metallic nano-antenna, decagonal semiconductor nanowires with semiconductor and hybrid cores are presented and analysed using 3D full vectorial finite difference time domain method. The performance of the suggested ONAs with different materials and configurations has been also investigated based on the absorption efficiency spectra, absorbed power profiles and emission patterns. The applications of ONAs and their fabrication aspects such as diode challenges are presented in detail, too. The intended audience of this chapter is constituted by students, scientists and engineers who might use or be involved with the illustrated devices. Moreover, the chapter topic fits in new emerging area as photodetection, light emission, sensing, heat transfer and spectroscopy.

The problem of RF-DC conversion circuits dedicated to microwave energy harvesting is also faced in Chapter 3. In detail, the chapter presents an overview of wireless power transfer with focus on the design of high efficiency low power rectifying circuits. A background survey of RF energy harvesting techniques has been presented. Moreover, the low power rectenna topologies as well as the performance of conventional rectifier topologies have been analysed and discussed considering that the rectification circuits have be optimized and conveniently designed to reduce the minimum power-threshold for the system operation. Solutions for low power rectifier operations has been detailed and new high efficient low power rectifiers has been designed and measured in both the GSM and ISM bands. Several guidelines on the optimization of the rectenna performance in terms of power conversion efficiency and DC output voltage have been provided. In particular, the design of 1) low power and high efficiency rectenna in the 900 MHz band, 2) dual bands lower power rectifier, and 3) medium power rectifier is presented. However, the chapter is intended for high frequency circuit designers and high speed system engineers.

The second part of the book reports some examples of piezoelectric energy harvesting systems. An innovative design for a piezoelectric energy harvester is discussed and proved in Chapter 4, where an alternative design paradigm of the cantilever beam is proposed. A piezoelectric energy harvesting (PEH) skin directly attached onto the surface of a vibrating engineered system has been proposed to scavenge electric power from vibration energy. The chapter aims at comprehensively describing the sequentially executed steps for the realization of the PEH skin from a system integration perspective as four part: modelling, design, manufacturing, and demonstration followed by the output power enhancement. The electromechanically-coupled analytical model and multi-physics finite element model are used for analyse the PEH skin. Moreover, the design methodology for segmentation of the PEH skin is carried out in order to perform the design optimization for maximizing harvestable electric power in a cost-effective manner. The manufacturing process of the PEH skin is explained with the standard operating procedure (SOP). Moreover, the experimental results demonstrates that the amount of the output power generated by the PEH skin is sufficient to supply energy to wireless sensors in real-time.

The application of piezoelectric transducers to harvest the energy contained in raindrops is also investigated in Chapter 5. A comprehensive and detailed explanation of piezoelectric energy harvesting of rainfall and its limits is provided. To better define the amount of power extracted from a single drop of rain, different systems consisting of two types of commercial transducers are compared. Moreover, the chapter illustrates the behaviour of the transducers in different scenarios: subjected to real and artificial rainfall, standalone or in parallel configuration, in conventional geometries, due to the commercial format or in customized shape, free to move or with an imposed optimal deflection. Finally, in order to improve the performance of the harvester, some solutions including the use of non-conventional geometries and the optimization of electromechanical efficiency imposing an optimized deflection are suggested. On the basis of the above mentioned topics, the chapter could be considered very interesting by scientists and engineers working in the field of the piezoelectric materials and devoted to the development of wireless sensors for monitoring of environmental conditions.

Chapter 6 deals with vibrational energy harvesting systems, employing piezoelectric effect. After an introduction illustrating the piezoelectric materials and structures, the characterization of vibrating freestanding piezoelectric devices forced by different external forces acting simultaneously is illustrated. Some examples of output current trends are discussed focusing the attention on one, two and three harmonics as vibrating sources. Very irregular generated currents are observed thus highlighting the problem to collect efficiently the produced energy in a vibrating turbulence. Moreover, the effects of the noise are analysed in details to model the chaotic vibrating system and to predict the output current signals.

The use of curved layouts is also discussed. This solution allows to collect better at low frequencies the vibration energy in the three dimensional space. Finally, some useful aspects for laboratory testing processes are discussed suggesting some preliminary experimental procedures.

Variable-shaped piezoelectric energy harvesters and piezomagnetoelastic energy harvesters are the main topics of the subsequent Chapter 7. In particular, linear and nonlinear electromechanical models are developed and validated with experimental measurements. Motivated by the need of developing accurate models for the design of low-frequency piezoelectric energy harvesters, a parametric study is performed to investigate the effects of both the length of the piezoelectric material and the spacing distance between the two magnets on the fundamental natural frequency of the harvester and its performance. In particular, it is demonstrated that depending on the available excitation frequency, an enhanced energy harvester can be tuned and optimized by changing the length of the piezoelectric material or by changing the distance between the two tip magnets.

The third part of the book presents some recent experimental and theoretical advances in the field of thermoelectric energy conversion and harvesting.

In detail, Chapter 8 introduces an original, self-sustainable, power supply system for wireless monitoring applications that is powered from an energy harvesting device based on thermoelectric generators. The energy harvesting system purpose is to gather unused heat, convert it to electrical energy and store it into rechargeable batteries. The chapter provide a detailed description of the mechanical, thermal and electronic part of the proposed energy harvesting system. The device is completely autonomous and employs a two stage energy storage system. Moreover, the device can be used for monitoring various parameters such as vibrations, temperatures, pressures, pollution, gases etc. with the purpose of minimizing the repair costs and preventing equipment damage. On the basis of the above mentioned topics, the chapter could be considered very interesting by the end user of the thermoelectric device that designs and optimizes energy harvesting systems.

Chapter 9 illustrates a study of thermoelectric energy harvesting with nano-sized thermopiles in a planar 65 nm silicon CMOS process. In particular, a thermoelectric system theory is used for evaluating the prospects for several thermoelectric energy harvesting applications. The impact of additional performance metrics such as heat source/sink thermal impedance, available heat flow density, voltage stacking, material and device figures of merit are explored for providing a more comprehensive set of criteria for evaluating the suitability of a thermal harvesting technology. This is relevant for many industrial applications.

Finally, Chapter 10 proposes organic materials and engineering applications to protect the highly reflective metal surfaces of solar concentrators. In this chapter, different configurations are considered, using aluminium as substrate, or the silver plating on polymer obtained by Dynamic Chemical Plating technique with a protective coating. In addition, several coatings, prepared by sol-gel technique, are illustrated and their optical properties are studied as well as feasibility for their use in solar technology.

The last part of the book reports researches dealing with the application of non-conventional systems and materials for energy harvesting. To this purpose, Chapter 11 describes a work on wearable energy harvesters for wearable applications. The chapter focuses on transduction materials, modelling strategies, experimental setups, and data analysis for the design of biomechanical energy harvesters. Moreover, a "case study" based on system integration and miniaturization is also described for applications in the field of human-machines interfacing. In the first part of the chapter, the state of the art is critically analysed, by means of applicative-oriented approach and harvesting systems feasibility. The main problems of portability, comfort and integration are analysed with particular focus on innovative materials. The description of test benches design and building, experiments as well as the prototypes fabrication is the second part of the chapter, where the main focus is given to the practical issues to be solved in the verification process of performance and reliability. The main requirements of testing repeatability and experimental confidence with real operative conditions are described, too. Testing on the human body is then showed by presenting some results from biomechanical energy harvesters applied to the human body. This work can be particularly interesting for industrial and academic researchers addressed to design and to optimize the technologies related to portable and wireless electronics. Moreover, the chapter topic fits to the ever increasing need of portable devices for applications in health care, medicine, bioengineering, human-machine interface (HMI), insurance, military, sport and entertainment.

Chapter 12 shows the concept of harvesting energy from aeroelastic instabilities, such as flutter in airfoil sections, vortex-induced vibrations in circular cylinders, and galloping in prismatic structures. The chapter presents nonlinear electroaeroelastic modeling and design of these aeroelastic energy harvesters. To determine the accuracy of the developed mathematical models, the obtained numerical results are compared with experimental measurements. Linear and nonlinear analyses are performed to investigate the effects of the system parameters on the onset of instability and the performance of these harvesters. The nonlinear normal form is derived and used to characterize the type of instability and to determine the importance of different structural or/and aerodynamic nonlinearities in the harvester response. Finally, qualitative and quantitative comparisons are presented and discussed. The topic of the chapter resides in the field of non-linear dynamic and aerodynamic instability phenomena. Thus it is very beneficial and can be used to breach traditional barriers in the design and performance enhancement of these aeroelastic energy harvesters, characterization of their behaviours, and identification of the contribution of different types of nonlinearities.

Chapter 13 reviews the reverse electrodialysis (RED) studies, one of the strong candidates in concentration gradient energy, using nanoporous membranes in place of conventional polymer membranes having no physical pore. In the first part of the chapter, the basic principle for the RED system and the effect of various parameters on the RED performance are illustrated. Then it follows a section where existing studies on the nanoslit array, anodic aluminum oxide membrane, track-etch membrane, and nanocrystal structure are summarized. The numerical modeling, the efforts on performance enhancing, the fabrication techniques and the used materials are also explained in the last part of the chapter. It follows that researchers and scientists engrossed in the design of implantable biomedical devices, e microelectromechanical systems, portable personal electronics, and nanofluidic devices could be very interested in the reading of this chapter.

Chapter 14 deals with topics in the field of the recovery of human mechanical energy. The electro-active polymers are studied as a transducer for converting mechanical energy provided by human movement into electrical energy. Four different themes complementary and necessary were treated. After a state of the art materials used in MEMS and performance, two distinct objectives were defined. At begin, the study is focused on the synthesis process materials and types of filler to be added to polymers to improve properties in operation and energy recovery. Moreover, a test bench with the measurement protocols is illustrated too. In the last section of the chapter, a mathematical model is established for maximizing energy. Finally, thanks to the specific circuits for energy management, a system with frequency doubler and other circuits is illustrated and the energy generation of complete system is tested.

The previously reported information related to the contents of each chapter shows the growing and recent interest in energy harvesting technology. In particular, the attention has been focused on electromagnetic, thermal and piezoelectric harvester without neglecting the non-conventional energy harvesting system and materials. In particular, approaches based on this technology have often proved to be suited to face applicative and technological challenges. Therefore, this book should be intended to be a space in which to find knowledge about theoretical and applicative advances, new experimental discoveries and novel technological improvements of energy harvesting. The resulting work constitutes a collection of high quality chapters which are representative of the existing research trends.

The target audience of the book is mainly those with an interest in applying energy harvesting technology in practical applications. Academicians, researchers, advanced-level students, technology developers, application engineers and producers of energy harvesting materials and/or systems will find this book useful in furthering their research activities and knowledge in the field of the energy harvesting. End-user organizations could find interest in this book to understand how likely the technology is able to address their needs. The researchers can find the methods and the most prominent concepts of energy harvesting technology in a unique book. Moreover, this volume can provide insights and support executives, concerned with the management of expertise, knowledge, information and development in different types of work communities and environments.

*Luciano Mescia*
*Politecnico di Bari, Italy*

*Onofrio Losito*
*Politecnico di Bari, Italy*

*Francesco Prudenzano*
*Politecnico di Bari, Italy*

# Section 1
# Electromagnetic Harvesters

# Chapter 1
# Earth Long-Wave Infrared Emission, New Ways to Harvest Energy

**Luciano Mescia**
*Politecnico di Bari, Italy*

**Pietro Bia**
*Politecnico di Bari, Italy*

**Onofrio Losito**
*Politecnico di Bari, Italy*

## ABSTRACT

*This chapter summarizes the physical properties of THz antennas, provides a summary of some of the most important recent developments in the field of energy harvesting of Earth long-wave infrared radiation, discusses the potential applications and identifies the future challenges and opportunities. In particular, a THz antenna is designed in order to transform the thermal energy, provided by the Sun and re-emitted from the Earth, in electricity. The proposed antenna is a square spiral of gold printed on a low cost dielectric substrate. Simulations have been conducted in order to investigate the behavior of the antenna illuminated by a circularly polarized plane wave with an amplitude chosen according to the Stefan-Boltzmann radiation law. Moreover, these THz antennas could be coupled with other components to obtain direct rectification of T radiation. As a consequence, these structures further optimized could be a promising alternative to the conventional photovoltaic solar cells.*

## INTRODUCTION

In the last two decades, the worldwide energy demand has increased by 40% but the main power generated nowadays is still produced using fossil fuels emitting tons of carbon dioxide and other pollution every second. As result, the harmful effects of hydrocarbon-based power sources as global warning, air pollution, acid precipitation, ozone depletion, forest destruction are increasingly apparent. In order to

DOI: 10.4018/978-1-4666-8254-2.ch001

limit these drawbacks, suitable actions aimed to reduce the dependence on the fossil fuels are needed, and the search for clean and renewable alternative energy resources is one of the most urgent challenges to the sustainable development of human civilization (Manzano-Agugliaro, Alcayde, Montoya, Zapata-Sierra, & Gil, 2013; Arigliano, Caricato, Grieco, & Guerriero, 2014).

The sun is the most powerful source of energy for the Earth. It provide a continuous stream of energy and for this reason it can potentially play a very important role in providing most of the heating, cooling and electricity needs of the world. As result, several approaches and technologies to directly or indirectly harvest energy from the sun have been successfully proposed and implemented. In particular, beside the conventional energy resources such as petroleum, coal, and nuclear plants, renewable ones, such as wind, solar, hydropower, geothermal, hydrogen and biomass/biofuel have been positively used to give a strong contribution to power generation without increasing environmental pollution (Khaligh & Onar, 2010).

Among the several kinds of solar techniques, the photovoltaics (PV) one is the most mature technologies. According to the National Renewable Energy Laboratory (NREL), the worldwide demand for photovoltaics (PV) is increasing every year and industry estimates suggest as much as 18 billion watts per year could ship by 2020. So, to meet the increased demand for solar-conversion technologies, dramatic improvements are required in terms of PV efficiency and cost/complexity reduction (Tiwari & Dubey, 2010). To this aim, conversion efficiency of PV technologies has been increasingly evolved during the last 40 years.

The energy created by the fusion reaction in the Sun is converted in thermal radiation and transferred in the form of electromagnetic waves into the free space. Solar radiation occurs over a wide range of wavelengths, nevertheless the main range of this radiation includes ultraviolet ( $\lambda < 0.4\ \mu m$ ) of which the content is less than 9%, visible (light, $0.4\ \mu m < \lambda < 0.7\ \mu m$) where the content is approximately 39%, and the remaining 52% consists of infrared radiation ( $0.7\ \mu m < \lambda < 100\ \mu m$ ). Approximately 30% of the solar radiation is scattered and reflected back to the space from the atmosphere, and about 70% is absorbed by the atmosphere and by the surface of the Earth. By absorbing the incoming solar radiation, the Earth temperature rises and, as an heated object, mainly reemits electromagnetic radiation in the wavelength range from 8 μm to 14 μm with a peak wavelength of about 10 μm. Due to the different spectral properties of the Sun and Earth emission, they are classified as short-wave and long-wave infrared (LWIR) radiation, respectively. The reemitted LWIR radiation energy is under-utilized by current technology. Considering that solar energy falls on the surface of the earth at a rate of about 120 petawatts and that PV technologies at present have low efficiency, it is clear that the Earth LWIR could be a potential renewable energy source and could have wide application and deep impact on our society (Byrnes, Blanchard, & Capasso, 2014).

A potential system able to collects the longwave radiation emitted from the Earth and to convert it into electricity is based on rectifying antenna (rectenna) technology. It typically consists of a THz antenna, intercepting the electromagnetic waves within a specific frequency band, and a diode which rectify the ac voltage due to the electromagnetic at the antenna terminals (Ma & Vandenbosch, 2013; Sabaawi, Tsimenidis, & Sharif, 2013; Shrestha, 2013). Despite some research activities are devoted to experimentally demonstrate the feasibility of rectenna to harvest the Earth LWIR radiation (Byrnes, Blanchard, & Capasso, 2014; Kotter, Novack, Slafer, & Pinhero, 2010), accurate numerical modeling is needed for device performance prediction, design and refinement as well as to obtain peculiar properties that may be exploited in the fabrication of more complex systems as antenna array. In fact, the optimization of antenna geometry and materials is crucial to improve the rectenna performance in terms of matching efficiency, electric field enhancement at the feeding point of the antenna, and RF-to-DC conversion efficiency. The antenna shape influences the polarization and direction of both emission and reception, the bandwidth and impedance matching. Moreover, the frequency-dependent dielectric constant of the metals strongly effects the guideline used at RF frequencies for the antenna design (Biagioni, Huang Jer-Shing, & Hecht, 2012; Bharadwaj, Deutsch, & Novotny, 2009; Novotny, 2007). In fact, at THz frequencies the electromagnetic penetration depth is comparable with the dimensions of the antenna leading a volume currents. The research efforts in metallic nanostructures have opened novel opportunities for the enhancement of light-harvesting capability (Novotny & van Hulst, 2011; He, Cui, Ye, Zhang, & Jin, 2009; Gadalla, Abdel-Rahman, & Shamim, 2014). These structures, exhibit potential advantages in terms of polarization, tuneability, and rapid time response. In fact, they have 1) a very small detection area, 2) allow the electromagnetic field localization beyond the diffraction limit, 3) they very efficiently release radiation from localized sources into the far field, 4) they make possible the tailoring of the interaction of electromagnetic field at the nanoscale, 5) they can be tuned to a specific wavelength. Moreover, the nanoscale antenna dimensions combined with the high electric field enhancement in the antenna gap enable a small device footprint making it compact enough to be monolithically integrated with electronics and auxiliary optics. Finally, these structures can provide a means to transduce free-propagating electromagnetic waves into localized surface plasmon resonance modes resulting in the reduced effective wavelength of wire waves.

## MAIN FOCUS OF THE CHAPTER

During the last few years, the use of nano-antennas have gained a great interest for solar energy harvesting (Gallo, Mescia, Losito, Bozzetti, & Prudenzano, 2012; Sabaawi et al., 2013; Alda, Rico-Garcia, Lopez-Alonso, & Boreman, 2005; Bharadwaj et al., 2009; Gonzalez & Boreman, 2005; Kocakarin & Yegin, 2013). These antennas, couple electromagnetic radiation at very high frequencies, THz and IR regimes, in the same way that RF antennas do at the corresponding wavelengths. As a consequence, several studies are currently focused in translating the concepts of RF antennas into the THz and IR frequency regime.

Because the size of optical antennas is in the range from a few hundred nanometres to a few microns, the technological limits did not allow their realization until a few years ago. However, thanks to the development of electron beam lithography and similar techniques the required level of miniaturization for the realization and demonstration of nano-antennas has been obtained (Bareiß et al., 2011; Kotter et al., 2010; Gritz, Puscasu, Spencer, & Boreman, 2003).

The guideline for the nanoantenna design are quite similar with those used at RF frequencies, but crucial differences in their physical properties and scaling behavior occur. In fact, in contrast to perfectly conducting concept used at RF frequencies, at optical frequencies metals no longer behave as perfect conductors and their interaction with electromagnetic field is determined by the frequency dependent complex dielectric function (Biagioni et al., 2012).

The radiation efficiency, directivity and bandwidth of the antenna are critical parameters to take into account. Moreover, the design of the best antenna for a given application is a problem not easy to solve since there are trade-off between antenna parameters (Kocakarin & Yegin, 2013; Ladan, Ghassemi, Ghiotto, & Wu, 2013). In fact, a strong directivity, a large bandwidth, a small size and a large radiation resistance, need to be combined. Furthermore, these limitations are an important consideration when designing an antenna for specific applications and in addition the typical design strategies of the radiowave antenna engineering cannot be completely used without careful considerations. In fact, at the THz and IR frequencies the metal losses became a constraint that antenna and circuit engineering have to take into account. Rather a lot of the energy in the surface modes are carried in the dielectric above the antenna. Compared to RF regime, the large losses and the finite skin depth generate consequences as reduced radiation efficiency, lower quality factor of the resonances, deviating radiation patterns and current distribution. Finally, the well know impedance matching circuits based on passive stub-like resonator structures have to be carefully designed since the metal losses strongly reduce the overall radiation efficiency (Biagioni et al., 2012; Bharadwaj et al., 2009).

One important aspect that a rectenna has to verify is that it should be able to concentrate the propagating free space LWIR plane waves having a wide spectral bandwidth and incoming from a range of directions of incidence. As result, the design of the isolated antenna plays an important role for the overall rectenna efficiency. Among various type of antenna, planar antennas are gaining popularity owing to their low profile, light weight, and simple coupling with rectifying element. In addition, they offer versatility in terms of resonant frequency, polarization, radiation pattern, and impedance. They are supported by a substrate and considering that it is electrically thicker at THz frequencies a decreased efficiency occurs respect their RF counterparts. In order to overcome this drawback, printed antenna having grounded substrates are generally preferred. In fact, due to the image dipole generated by the ground interface the antenna impedance is modified and substrate thickness can be reduced to increase the efficiency. Moreover, the presence of the ground allows the radiation in only one direction. On the contrary, the radiation properties of these antennas become sensitive to substrate losses, especially when the substrate thickness increases, and the substrate permittivity, acting as a parasitic impedance, causes a red shift of the resonant frequency. As result, for a given substrate permittivity there is a particular substrate thickness maximizing the performance of the printed antennas. Until now, dipole (Ma & Vandenbosch, 2013; Locatelli, 2010; Gonzalez & Boreman, 2005; Sabaawi et al., 2013), crossed dipole (Stokes et al., 2014; Kocakarin & Yegin, 2013), bowtie (Briones, Alda, & González, 2013; Gonzalez & Boreman, 2005; Krenz et al., 2012), log-periodic (Gonzalez & Boreman, 2005; Semenov et al., 2007), square-spiral (Gallo et al. 2012; Kotter et al., 2010), and Archimedean spiral (Cherednichenko et al., 2011; Gonzalez & Boreman, 2005) geometries have been proposed for IR and THz antennas.

In this chapter, some design strategies and guidelines will be illustrated with the target to enhance the uncapability of THz antenna to collects the longwave radiation emitted from the Earth and to convert it into electricity. A spiral antennas made of a thin film of gold printed on a $SiO_2$ substrate will be considered. Due to their broad bandwidth, spiral antennas have been proposed to collect solar energy (Gallo et al., 2012; Kotter et al., 2010). They allow to concentrate the electric field in the gap between two metallic arms which constitutes an appropriate point to transport energy needed to supply other circuitry. These antennas are good resonators and it is expected to capture a large electric field at resonance. Moreover, the gain performance of the spiral antenna can be easily improved by increasing the number of arms. Round spiral antennas are generally designed by using Archimedean spiral geometries which have linear growth rates and frequency independent radiative characteristics. Moreover, the frequency independency is limited to a wavelength band determined by the antenna size. Spiral antennas can be constructed as planar structures and they can radiate linearly or circularly polarized waves. However, the main drawback with this type of antenna is the difficulty in configuring an array. Despite that, equiangular spiral can be chosen as the array element since it allows 1) convenient connection of DC lines at the tips of the spiral arms, 2) possible dual polarization, 3) convenient feed point for diode connection.

The feasibility of spiral antenna to harvest Earth thermal radiation has been investigated through simulations by illuminating the antenna with a circularly polarized plane wave having the amplitude and a frequency of the actual Earth radiation. The amplitude of the circularly polarized plane wave emitted by the Earth has been properly calculated through the Stefan-Bolzmann theory. Moreover, the Lorentz-Drude model has been taken into account in order to model the variation of the metal permittivity versus the frequency within the infrared frequency region. The physical properties of spiral antennas, the antenna parameters, computational considerations as well as important aspects pertaining the radiation efficiency, directivity, bandwidth and polarization are illustrated. Furthermore, to improve the coupling efficiency of the free-space radiation into the antenna, a suitable choice of the antenna material as well as an accurate design of the antenna geometry has been performed. In particular, the substrate material has been carefully designed to achieve desirable electrical and physical parameters. Considering that the local field enhancement is a critical design parameter for antenna performance we discuss the optimal conditions for the maximum field enhancement. In particular, the effects of the metal and substrate thickness, gap and antenna shape and dimensions are investigated.

## RECTENNA AND PHOTOVOLTAIC TECHNOLOGY ASPECTS

Photovoltaic conversion is the direct conversion of sunlight into electrical energy without any heat engine to interfere. The conventional photovoltaic devices are cheap (<1$/W), rugged and simple in design, have long life, require very little maintenance, have integrated large scale manufacturability, and their construction as stand-alone systems provides outputs from microwatts to megawatts. Moreover, silicon, the major material of photovoltaic panel, is abundant and environmentally safe. However, the application of PV technologies can be dangerous under several conditions. In fact, multi-junction PV cells using materials as gallium arsenide and cadmium telluride, can cause harm to the environment if leaked. Finally, in concentrated solar power techniques, coolant and lubricant could be a problem.

Solar cells convert sunlight directly into electricity. They are made of semiconducting materials that can absorb the photons from sunlight. The working principle of these devices is based on the photovoltaic effect. In particular, when a solar photon interacts with an electron of the semiconductor valence band, the energy of the photon can be absorbed by the electron which can move to the conduction band, leaving a hole in the valence band. Electrons and holes are accelerated by the drifting electric field in the depletion zone of the PN junction and can produce photocurrent.

The photovoltaic effect depends on interaction of photons, with energy equal to, or more than the band-gap of the photovoltaic materials. As a result, in the semiconductor solar cells only sunlight of certain energies can efficiently create electricity. So, the efficiency of PV is fundamentally limited and a lot of research activities are in progress to improve it. The first generation of solar cells, used in 90% of today's cells, use a single p-n junction to extract energy from sunlight photons and are manufactured from silicon semiconductors. They have about 30% efficiency, but result in a price too high to compete with fossil fuels (Tyagi, A.A. Rahim, N.A. Rahim, & Selvaraj, 2013). The second generation of solar cells, called thin-film solar cells, are made from amorphous silicon or non-silicon materials and exhibit low production costs but result in much lower efficiency rates (Tyagi et al. 2013; Becker et al. 2013). Third-generation of solar cells are being made from variety of new materials including solar inks, solar dyes, and conductive plastics. Some new solar cells use plastic lenses or mirrors to concentrate sunlight onto a very small piece of high efficiency PV materials (Tyagi et al. 2013; Abdina et al. 2013). However, the PV material is more expensive, and because the lenses must be pointed at the sun, the use of concentrating collectors is limited to the sunniest parts of the country. Moreover, more complex solar cells that can harvest energy from a wider range of the electromagnetic spectrum with higher efficiency have been proposed, but they are too expensive for widespread use. Finally, another drawback of PV-based technologies is the fact of being strongly depended on daylight, which in turn makes them sensitive to the weather conditions (Goetzberger & Hoffmann 2005).

Since the incoming LWIR is an electromagnetic wave radiation at terahertz wavelengths, it can be collected by tuning an antenna in such a way that it is resonant at THz and IR frequencies. This can be achieved by shrinking the dimensions of the antenna to the scale of the wavelength. The main advantage of the antenna approach is that conversion does not be band-gap limited as in PV cells. These antennas can enhance the interaction of IR waves with nanoscale matter providing an high electric field at the feeding point of the antennas (Kotter et al. 2010). Nano-antennas exhibit potential advantages in terms of polarization, tuneability, and rapid time response. In fact, they have 1) a very small detection area, 2) allow the electromagnetic field localization beyond the diffraction limit, 3) they very efficiently release radiation from localized sources into the far field, 4) they make possible the tailoring of the interaction of electromagnetic field at the nanoscale, 5) they can be tuned to a specific wavelength. Finally, the nanoscale antenna dimensions combined with the high electric field enhancement in the antenna gap enable a small device footprint making it compact enough to be monolithically integrated with electronics and auxiliary optics.

The fabrication of THz and IR antenna requires reliable and reproducible structuring techniques able to accurately define critical antenna dimensions such as gap size and arm length. Various top-down and bottom-up nanofabrication approaches have been applied to experimentally realize these kind of antennas. In particular, top-down approaches such electron-beam lithography (EBL) (van Dorp & Hagen, 2008; Weber-Bargioni et al., 2010) and focused-ion beam (FIB) milling (Orlo, Utlaut M., & Swanson, 2002) are capable of fabricating large arrays of nearly identical nanostructures with defined orientation and distances. On the other hand, bottom-up approaches take advantage of chemical synthesis and self-assembly of nanoparticles in solution but they often require precise size selection and nanopositioning as well as assembly strategies to create non-trivial structures. Moreover, the design of these novel antennas by using well known printing techniques allowing costs reduction and a quick prototyping approach is another important aspect to consider.

Since the Earth mainly emits its thermal energy within the infrared band during the night, this kind of technology could be used to overcome the main drawback of the current solar technology, i.e. the operation only during the morning. The same technology could also be used in most industrial process where there are waste heat that could be gathered to produce electricity. Because the temperature affects the received wavelength band, the antenna should be ad hoc designed to operate and collect around the frequency range where the industrial process radiates most of its thermal energy. Moreover, this kind of antenna may also be used to remove heat and to cool electronic devices by substituting, for example, fans mounted in every personal computer, and at the same time it could continuously charge the batteries. Ad hoc designed array antenna should be a good candidate to improve the building thermal insulation and to reduce the energy consumption due to the operation of air-conditioners and fans. In fact, by coating the roofs, windows and walls such system could cool down buildings without the external power sources. Due to the employed materials, this technology should be more eco sustainable with respect to the conventional ones.

## THEORY

The design of the antenna able to collect the electromagnetic radiation emitted by the Earth has been performed by illuminating the system with a circularly polarized plane wave. As result, some theoretical considerations are needed to set in the simulations the right value of the electric field of the incoming plane wave.

The Sun is mainly composed of hydrogen and helium. The extreme pressure and temperature at its core fuse together nuclei of hydrogen atoms to form the nuclei of helium releasing huge amounts of energy. The energy created by the fusion reaction is converted in thermal radiation and transferred in the form of electromagnetic waves into the free space. Solar radiation is a polychromatic electromagnetic wave and its frequency is extended over a broad spectrum. Considering that the coherence area of a source can be calculated by

$$A_{co} = \frac{\lambda^2}{\Omega}$$

and that the sun solid angle and average wavelength are about $\Omega \approx 6.8 \times 10^{-5}$ str and $\lambda = 0.5\ \mu m$, respectively, the coherence area of solar light is about $3.7 \times 10^{-3}\ \text{mm}^2$, i.e. a circle with a radius of 0.03 mm. Part of the solar radiation is scattered and reflected by clouds and part is absorbed by the atmosphere and by the surface of the Earth. By absorbing the incoming solar radiation, the Earth warms up and its temperature rises.

As an objects with a temperature above absolute zero, the Earth has the capability to radiate electromagnetic waves. In particular, it emits thermal radiation at much lower intensity with respect to the Sun because it is cooler. To study the thermal radiation emitted by the Earth, it can be approximated as a black body that absorbs all the incoming radiation and emits a temperature-dependent radiation spectrum. According to the Planck's law (Planck, 1901), the radiation emitted by a black body at the temperature *T* per unit of area and wavelength (spectral irradiance) is given by

$$I(\lambda, T) = \frac{2\pi c^2 h}{\lambda^5} \frac{1}{\exp(hc / \lambda \kappa T) - 1}$$

where *c* is the speed of light, *T* is the temperature [K], $h = 6.63 \times 10^{-34}$ [J s], $\kappa = 1.388 \times 10^{-23}$ [J K$^{-1}$] are the Plank and Boltzmann constant, respectively, *I* is the spectral irradiance [W m$^2$ μm$^{-1}$]. The integration of (2) over all wavelengths gives the expression for the power, *R*, emitted per unit area on the surface of a black body at temperature *T*

$$R = \int_0^\infty I(\lambda, T)\, d\lambda$$

The solution of (3) is the Stefan-Boltzmann's relation in which the amount of radiation emitted by a blackbody is proportional to the fourth power of its temperature

$$R = \sigma T^4$$

where $\sigma = 5.670 \times 10^{-8}$ [W m$^{-2}$ K$^{-4}$] is the Stefan-Boltzmann constant. The black body spectral irradiance curve (2) exhibits a peak depending on the temperature and corresponding to a wavelength $\lambda_{max}$. A steeply and slowly spectral irradiance decreasing occurs at shorter and longer wavelengths, respectively. In particular, considering a mean Earth temperature $T$=287 K (Jones, New, Parker, Martin, & Rigor, 1999) the maximum of the spectral irradiance occurs at $\lambda_{max} = 10.6\ \mu m$, corresponding to a frequency $f_{max} = 28.3$ THz.

An accurate estimation of the Earth thermal radiation requires a different approach because it reflects part of the incoming energy. The emissivity, $e$, of the Earth, defined as the ratio between the energy radiated by a material and the energy radiated by a black body at the same temperature, is less than 1. For this kind of bodies, named grey body, the equation (3) becomes:

$$R = \int_0^{\infty} I\left(\lambda, T\right) d\lambda = e\sigma T^4$$

Generally, the emissivity value of the Earth depends on gases density (water vapor and carbon dioxide), but a mean value of 0.9 can be considered (Trenberth, Fasullo, & Kiehl, 2009).

Since an antenna has a limited bandwidth, a fraction of the total power density of the solar spectrum can be captured by a single element antenna. In particular, if $f_1$ is the lower frequency, corresponding to the upper wavelength $\lambda_U$, and $f_2$ is the upper frequency, corresponding to the lower wavelength $\lambda_L$, the fractional power density available in the antenna bandwidth can be calculated by

$$S_E = \int_{\lambda_L}^{\lambda_U} \frac{2\pi c^2 h}{\lambda^5} \frac{1}{\exp\left(hc / \lambda \kappa T\right) - 1} d\lambda$$

By considering the plane wave approximation, the power density emitted by the Earth, $S_E$, is related to the incident electric field $E_i$ by

$$S_E = \frac{E_i^2}{2Z_0}$$

where $Z_0 = 377\ \Omega$ is the intrinsic impedance of the free space. As result, the incident electric field is given by

$$E_i = \sqrt{2Z_0 S_E}$$

The antenna operating in receiving mode captures the incident electromagnetic waves coming from the free space and stores the electromagnetic energy in its near electric and magnetic field. The antenna reactance corresponds to the stored energy and the resistance contributes to the power dissipation. As result, the receiving antenna acts as a RLC resonator. Due to the absence of explicit voltage source the equivalent circuit of receiving antenna is more complicated than the transmitting one. In fact, in the receiving mode the antenna is driven by the induced current due to the incident electromagnetic field. However, when operated at its resonant frequency, the receiving antenna can be modeled by a voltage source, $V_O$, and an impedance in series, $Z_A = R_A + jX_A$. In particular, $V_O$ is the open circuit voltage occurring at the end of the antenna when no load is connected and $Z_A$ is the antenna impedance, where $X_A$ is the antenna reactance and $R_A$ is the antenna resistance which is a combination of the radiation resistance, modelling the radiated power, in series with the loss resistance, modelling the conductive and dielectric losses.

The voltage $V_O$ can be expressed as

$$V_O = 2E_i\sqrt{\frac{R_A A_{eff}}{Z_0}}$$

where

$$A_{eff} = \frac{\lambda^2 G}{4\pi}$$

is the effective area of the antenna, and $G$ is the antenna gain. Using (8)-(10) the final expression of the open circuit voltage is

$$V_O = \sqrt{\frac{2R_A\lambda^2 G}{\pi}S_E}$$

When the incident electromagnetic field illuminates the metallic structures of the antenna, it excites the collective oscillation of conduction electron in the metal propagating at the boundary of the metal surface. At THz and IR frequencies, metals with a finite thickness cannot be considered as perfect conductors and their interaction with electromagnetic field is determined by the frequency dependent complex dielectric function. In fact, at these frequencies the conductor loss could be very significant and the finite skin depth can generate consequences as reduced radiation efficiency, lower quality factor of the resonances, deviating radiation patterns and current distribution.

The more simple theory used to investigate the behavior of metals is based on the Drude-Sommerfeld model of free-electron gas. Unfortunately, this model fails to reproduce the experimental spectrum at the range of visible frequency. This is because the Drude-Sommerfeld includes only electromagnetic response of nearly-free electrons and ignores the response of bound electrons. To overcome this drawback, the Lorentz-Drude (LD) model is typically adopted. In this model, the processes of interband transitions (excitation of bound electrons to conduction electrons) of electrons in metals is taken into account and in particular the equation of motion for bound electrons is given by

$$m_e \frac{d^2\mathbf{r}}{dt^2} + \gamma \frac{d\mathbf{r}}{dt} + \alpha \mathbf{r} = e\mathbf{E}e^{-j\omega t}$$

where $m_e$, $\Gamma$, $e$ and $\alpha$ are the effective mass, the damping coefficient, the charge and the spring constant of the bound electron, respectively. Assuming that the macroscopic polarization can be expressed as

$$\mathbf{P} = Ne\mathbf{r}$$

where $N$ is the number of the electrons per unit volume, eq. (12) can be rewritten as

$$\frac{d^2\mathbf{P}}{dt^2} + \Gamma \frac{d\mathbf{P}}{dt} + \omega_0^2 \mathbf{P} = \frac{Ne^2}{m_e} \mathbf{E}e^{-j\omega t}$$

where

$$\Gamma = \frac{\gamma}{m_e}$$

and

$$\omega_0 = \sqrt{\frac{\alpha}{m_e}}$$

is the natural frequency of the bound electrons.

The harmonic solution of (14) is

$$\mathbf{P}(\omega) = \frac{Ne^2}{m_e} \frac{\mathbf{E}}{\left(\omega_0^2 - \omega^2\right) + j\Gamma\omega}$$

and considering that the macroscopic polarization can be also formulated in term of electric field

$$\mathbf{P} = \varepsilon_0 \chi \mathbf{E}$$

where $\chi$ is the electric susceptibility, the following relation is obtained

$$\chi(\omega) = \frac{\omega_p^2}{\left(\omega_0^2 - \omega^2\right) + j\Gamma\omega}$$

where

$$\omega_p = \sqrt{\frac{Ne^2}{\varepsilon_0 m_e}}$$

is the plasma frequency. Finally, the relative permittivity is given by

$$\varepsilon_r\left(\omega\right) = 1 + \chi = 1 + \frac{\omega_p^2}{\left(\omega_0^2 - \omega^2\right) + j\Gamma\omega}$$

The DL equation (21) can be easily extended for multi-band transitions by considering the following equation

$$\varepsilon_r\left(\omega\right) = 1 + \sum_{i=0}^{k} \frac{f_i\omega_p^2}{\left(\omega_i^2 - \omega^2\right) + j\Gamma_i\omega}$$

where $k$ is the number of oscillators with natural frequency $\omega_i$ and $f_i$ is the oscillation strength.

For free electrons the spring constant $\alpha$ can be neglected and eq. (22) becomes

$$\varepsilon_r^f\left(\omega\right) = 1 - \frac{\Omega_p^2}{\omega(\omega - i\Gamma_0)}$$

As result, if the material has both free and bound electrons eq. (22) can be rewritten as

$$\varepsilon_r\left(\omega\right) = \underbrace{1 - \frac{\Omega_p^2}{\omega(\omega - i\Gamma_0)}}_{\text{intra-band contribution}} + \underbrace{\sum_{i=1}^{k} \frac{f_i\omega_p^2}{\left(\omega_i^2 - \omega^2\right) + j\Gamma_i\omega}}_{\text{inter-band contribution}} = \varepsilon_r^f(\omega) + \varepsilon_r^b(\omega)$$

where $1/\Gamma_i$ is the lifetime, $\Omega_p$ is the plasma frequency of the intra-band transitions with oscillator strength $f_0$ and damping constant $\Gamma_0$, and

$$\varepsilon_r^b\left(\omega\right) = \sum_{i=1}^{k} \frac{f_i\omega_p^2}{\left(\omega_i^2 - \omega^2\right) + j\Gamma_i\omega}$$

is the inter-band contribution to the material permittivity due to the bound electrons. Figure 1 illustrates the plot of eq. (24) for the gold. The negative values of the real part of the permittivity give a small pen-

*Figure 1. Real part (blue curve), and imaginary part (green curve) of the gold permittivity calculated by using the Lorentz-Drude model*

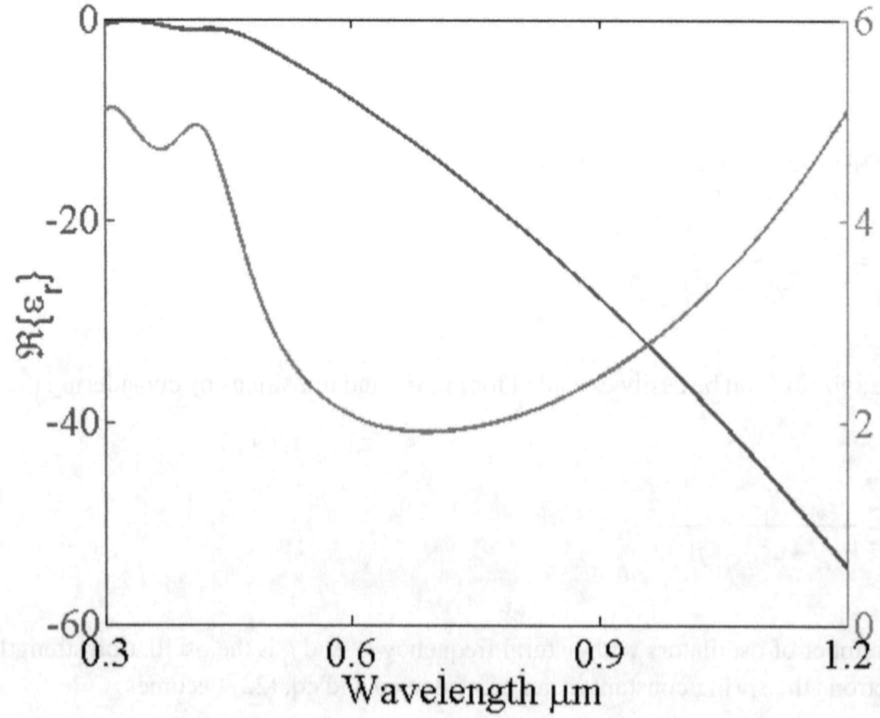


etration depth of electromagnetic field. On the other hand, the imaginary part is positive and contributes to the dissipation of the energy in the metal. As result, the availability of this model allows the calculation of electrical properties of the gold, such as the permittivity, the skin depth and losses.

## NUMERICAL RESULTS

A spiral antennas made of a thin film of gold printed on a $SiO_2$ substrate is considered. A square spiral geometry is taken into account since its advantages in terms of size with respect to a circular one. In fact, considering that the optimal reception of a spiral antenna occurs when the spiral arm length equals approximately one wavelength, which correspond to a diameter of $D = \lambda / \pi$, for the circular spiral, and a side length $W = \lambda / 4$ for the square spiral, the last one allows to obtain a comparable antenna gain when the width of the square spiral is approximately 75% of the diameter of the circular spiral antenna. A commercial full-wave 3D electromagnetic solver, CST Microwave Studio Suite 2013, has been used to perform simulations. In particular, the main solver parameters as cell number, computational domain size, frequency band have been optimized to reduce the computational efforts and obtain at the same time accurate numerical results. The substrate has permittivity $\varepsilon_{SiO_2} = 4.86$ and thickness $t_{SiO_2}$. The ground plane and antenna arms are constituted by a gold thin film having thickness $t_{Au}$ and dispersive properties defined by eq. (24) with five (*k*=5) oscillator terms (Rakic, Djurisic, Elazar, &

Majewski, 1998). The two antenna arms are separated by a gap $t_{gap}$ . Figure 2 shows the geometry of the antenna. A substrate material of dimensions 20×20 μm$^2$ and a squared spiral antenna of dimension 4×4 μm$^2$ are considered in the simulations. The total length of each arm is about 16 μm. The antenna materials as well as the length of each linear filament composing both the arm have been designed to expand the radiation patter and to improve the coupling efficiency of the free-space radiation into the antenna.

Most applications of rectenna are based on their capability to concentrate the incident electromagnetic field at the feeding point. As result, high gain and efficiency are key design parameters to take into account. In order to evaluate the efficiency of the antenna in collecting the thermal radiation, a number of simulations were performed by illuminating the antenna with a circularly polarized plane wave. The amplitude of this plane wave depends on the actual thermal power emitted by the Earth. Because the radiation emitted from the Earth has a maximum at the wavelength close to 10.6 μm, the electromagnetic analysis was performed from 7.5 μm (40 THz) to 15 μm (20 THz). In particular, by considering $T$=287 K and $e$=0.9 as the mean Earth temperature and emissivity, respectively, the power density emitted from the Earth can be calculated by integrating the equation (6) in the considered wavelength range. The obtained value is $S_E$=152.5 W/m$^2$ and using eq. (8) the plane wave amplitude considered in simulation was $E_i$=239 V/m.

The antenna size mostly depends on the frequency band of operation. Considering that the perfectly conducting concept used at RF frequencies cannot be used at THz and IR frequencies, the resonant length of the antenna not exactly scales linearly with the incident frequency. In fact, the effect of current waves propagating along the metal is quite different at THz and IR frequencies from those arising at microwave

*Figure 2. Geometrical configuration of the square spiral antenna*

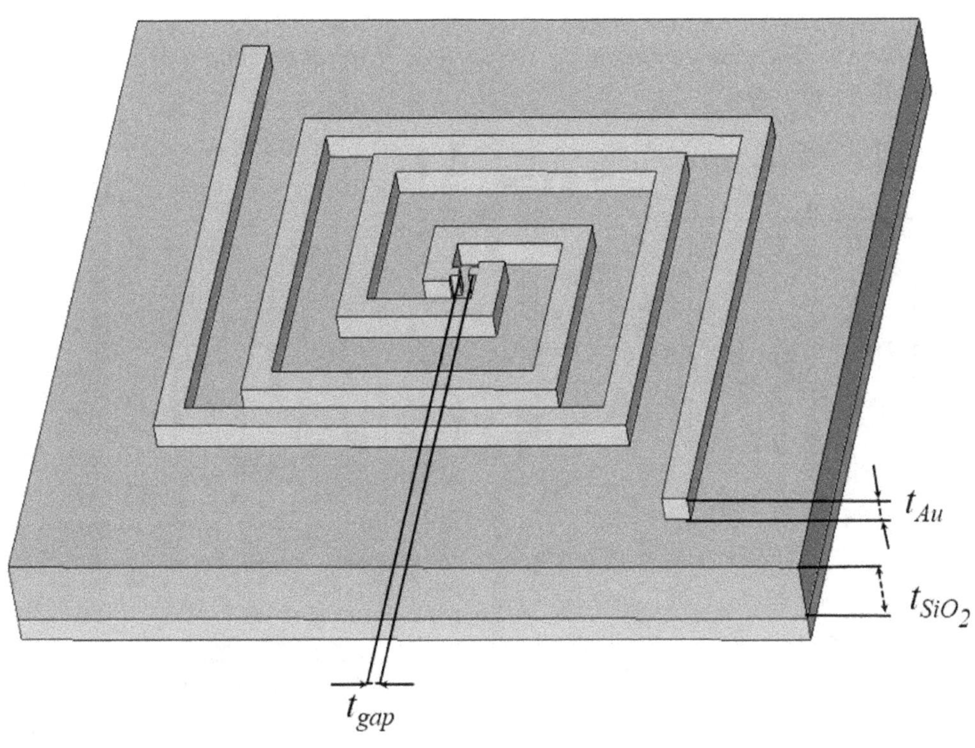

and lower frequencies. Moreover, the material of the arms and ground plane affects the efficiency of the antenna. As result, the simulation has to be performed taking into account the gold dispersion instead of perfect electric conductor (PEC) boundary condition for the antenna arms and substrate and, as a consequence, more computational efforts are required. This occurrence is well illustrated in Figure 3 when the frequency of the exciting plane wave is 28.3 THz. In fact, when the dispersive properties of the gold are considered (Figure 3(a)) the electric field distribution on the top surface of the spiral antenna exhibits a clear field enhancement at the feeding point of the antenna respect the PEC case (Figure 3(b)).

The lower conducting layer acts as a perfectly reflecting ground plane, bouncing energy back through the substrate and into the free-space. In particular, the ground interface allowing the radiation in only one direction makes possible an improvement of the antenna performance. Moreover, the ground plane can be used to control the radiation quality factor and to match the absorption quality factor of the antenna. Figure 4 shows the electric field distribution on the top surface of spiral antenna with and without the gold ground plane when the frequency of the exciting plane wave is 28.3 THz. It is worthwhile to note that the presence of ground plane strongly improves the electric field. In fact, the maximum electric field of about $2.1 \times 10^4$ V/m, calculated using the ground plane (Figure 4(a)), is higher than 721 V/m calculated neglecting the ground plane (Figure 4(b)). This occurrence indicates that the ground plane plays a major role in the calculation of the antenna efficiency.

There are several factors that contribute to deciding the dimension of the antenna and its behavior such as the substrate material used and its thickness. In fact, the type of substrate plays a major role in the calculation of the antenna dimensions, and its thickness is a considerable parameter to take into account during the design procedure. The most desirable substrate to design antenna having a large bandwidth and high efficiency are the ones that are thick with low dielectric constant. However, this comes at expense of a large volume antenna and a surface waves formation. On the other hand, thin substrates

*Figure 3. 2D electric field distribution on the top surface of spiral antenna for (a) dispersive gold and (b) PEC arms and substrate*

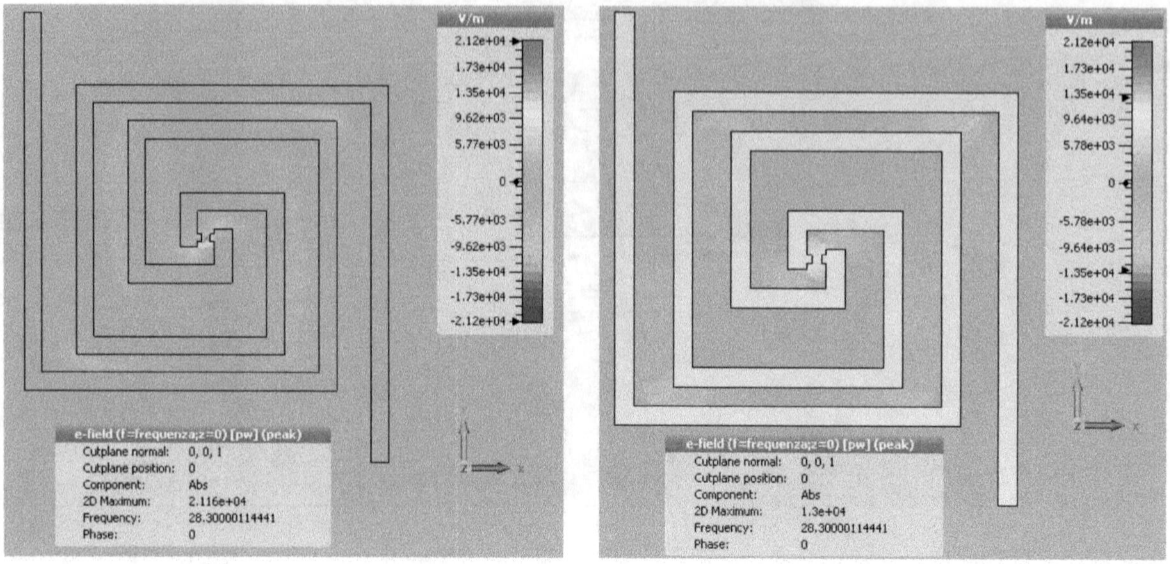


(a) (b)

*Figure 4. 2D electric field distribution on the top surface of spiral antenna (a) with gold ground plane and (b) without ground plane*

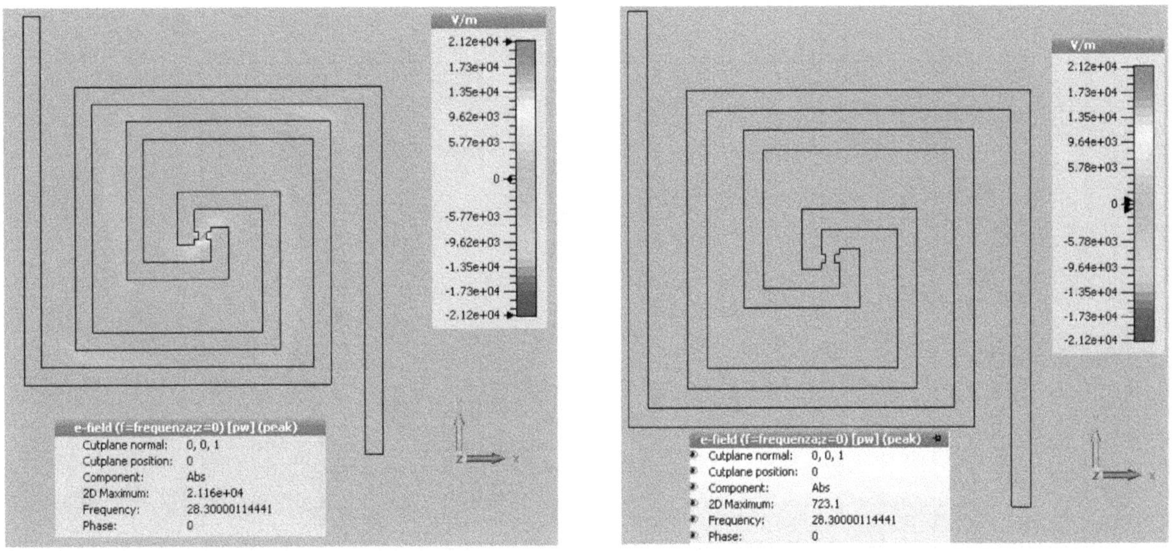


(a) (b)

with high dielectric constants reduce the overall size of the antenna, since the fringing fields are tightly bound to the substrate. However, because of the relatively higher loss tangents, they are less efficient and have relatively smaller bandwidth (Singh, 2010). Moreover, considering that the radiation quality factor of the antenna can be engineered tuning the distance between the coated gold ground plane and the spiral arms it is clear the need to optimize the substrate thickness. To this aim, the effect of the substrate thickness on the induced gap voltage $V_O$ are shown in Figure 5 for different thickness of gold layers. Also in this case, the frequency of the exciting plane wave is 28.3 THz. The calculation are performed as the thickness of $SiO_2$ is changed from 0.5 μm to 1.5 μm. As shown in Figure 5 the induced gap voltage increases by increasing the $SiO_2$ thickness exhibiting a peak around 1.1 μm. It is important to note that the optimum substrate thickness is close to the quarter-wavelength (λ/4) thickness. In fact, at λ/4 the radiation and absorption quality factor of the antenna are matched and a maximum electric field enhancement occurs. For $SiO_2$ thickness lower than the optimal one the real and image antenna radiation cancel each other more.

The geometry of the antenna strongly influences its properties and the gap between the two arms represents an essential feature of the antenna structure. In fact, the gap size plays an important role in the field confinement and enhancement and determines the spatial size of the concentrated light spot. To this aim, the effects of the gap size on the value of the induced gap voltage have been investigated. Simulations results are shown in Figure 6 for different gold thickness. Also in this case, the induced gap voltage increases by increasing the gap size showing a peak close 0.11 μm. Gap size higher than the optimal one results in a significant decreasing in the electric field intensity, but a very slight variation in the resonance frequency occurs. Field enhancement increases with the decreasing of the gap size due

*Figure 5. Gap voltage between the two antenna arms versus the substrate thickness for different thickness of gold: $t_{AU} = 0.9\ \mu m$ (blue curve), $t_{AU} = 1.2\ \mu m$ (red curve), $t_{AU} = 1.5\ \mu m$ (green curve).*

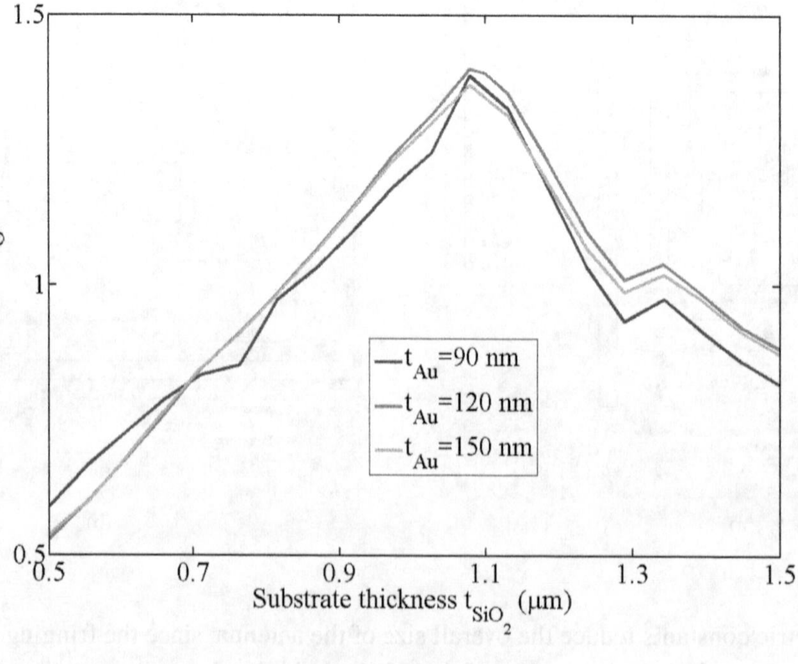


*Figure 6. Gap voltage between the two antenna arms versus the gap size for different gold thickness: $t_{AU} = 0.9\ \mu m$ (blue curve), $t_{AU} = 1.2\ \mu m$ (red curve), $t_{AU} = 1.5\ \mu m$ (green curve).*

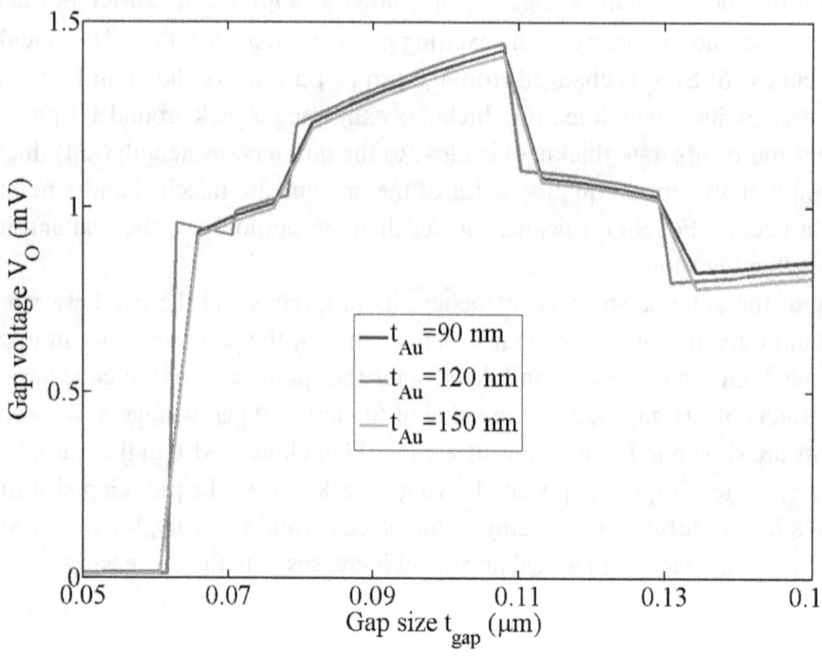

to the increase of the charge concentration near the gap, but for very small gap the coupling between the spiral arms is very strong and generates a shift of the antenna resonance. Thus the antenna is not resonating at 28.3 THz and as a consequence the induced voltage strongly decreases.

Simulations have also been performed by varying the gold thickness to verify how this parameter affects the capability of the antenna in collecting the thermal radiation. In fact, in these antenna structures the incident Earth thermal radiation excites surface waves at the metal-dielectric interface which are channeled toward the gap. In particular, the incident electromagnetic field excites collective oscillations of free electrons of the gold flowing along the antenna structure. This excited modes show a resonant characteristic which depend on the metal type and geometric properties of the metal layers.

Figure 7 illustrates the obtained results pertaining the induced gap voltage versus the gold thickness. It worthwhile to note that the gold thickness does not strongly effects the induced gap voltage. However, all the three curves highlight an optimal gold thickness maximizing the gold voltage. Moreover, as illustrated in Figure 5, the effect of substrate thickness on the electric field enhancement is apparent.

With the help of the reciprocity theorem, the same antenna can be used for the transmission and reception purpose. As result, the radiation characteristic of the proposed receiving antenna can be simulated by considering it as transmitting one. To this aim, a lumped port connecting the two antenna arms have been considered as a feed of the antenna. The simulated far-zone radiation patterns for the directivity at the frequency 28.3 THz are illustrated in Figure 8. In particular, Figure 8(a) and Figure 8(c) refer to the simulations in which gold is used for the antenna arms and substrate, while Figure 8(b) and Figure 8(d) refer to the simulations where the PEC boundary condition is used for the antenna arms and substrate

*Figure 7. Gap voltage between the two antenna arms versus the gold thickness for different substrate thickness:* $t_{SiO_2} = 0.7\ \mu m$ *(blue curve),* $t_{SiO_2} = 1.1\ \mu m$ *(red curve),* $t_{SiO_2} = 1.5\ \mu m$ *(green curve).*

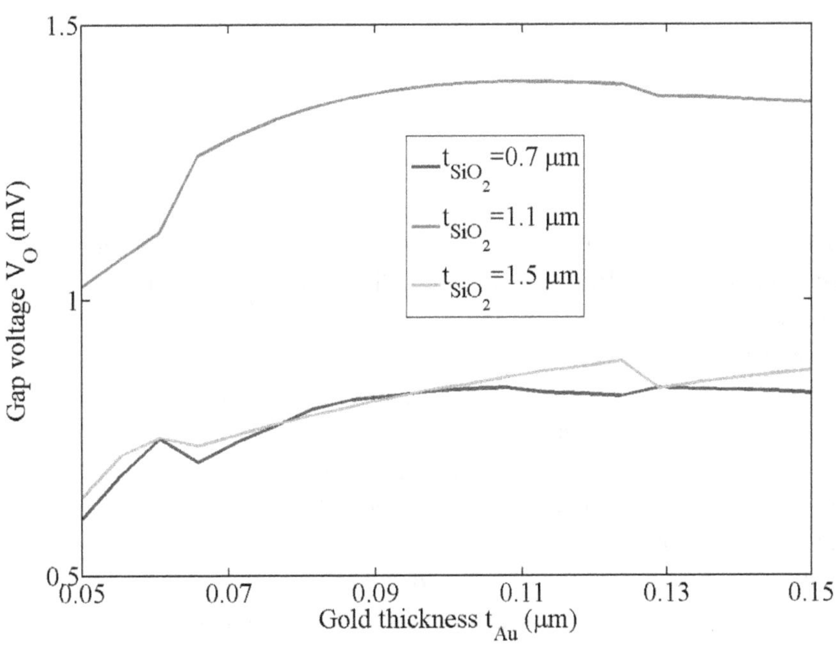

*Figure 8. Far field radiation pattern of the directivity in E ($\phi$=0) and H ($\phi$=90°) plane, (a) polar and (c) rectangular plots for dispersive gold, (b) polar and (d) rectangular plots for PEC boundary condition*

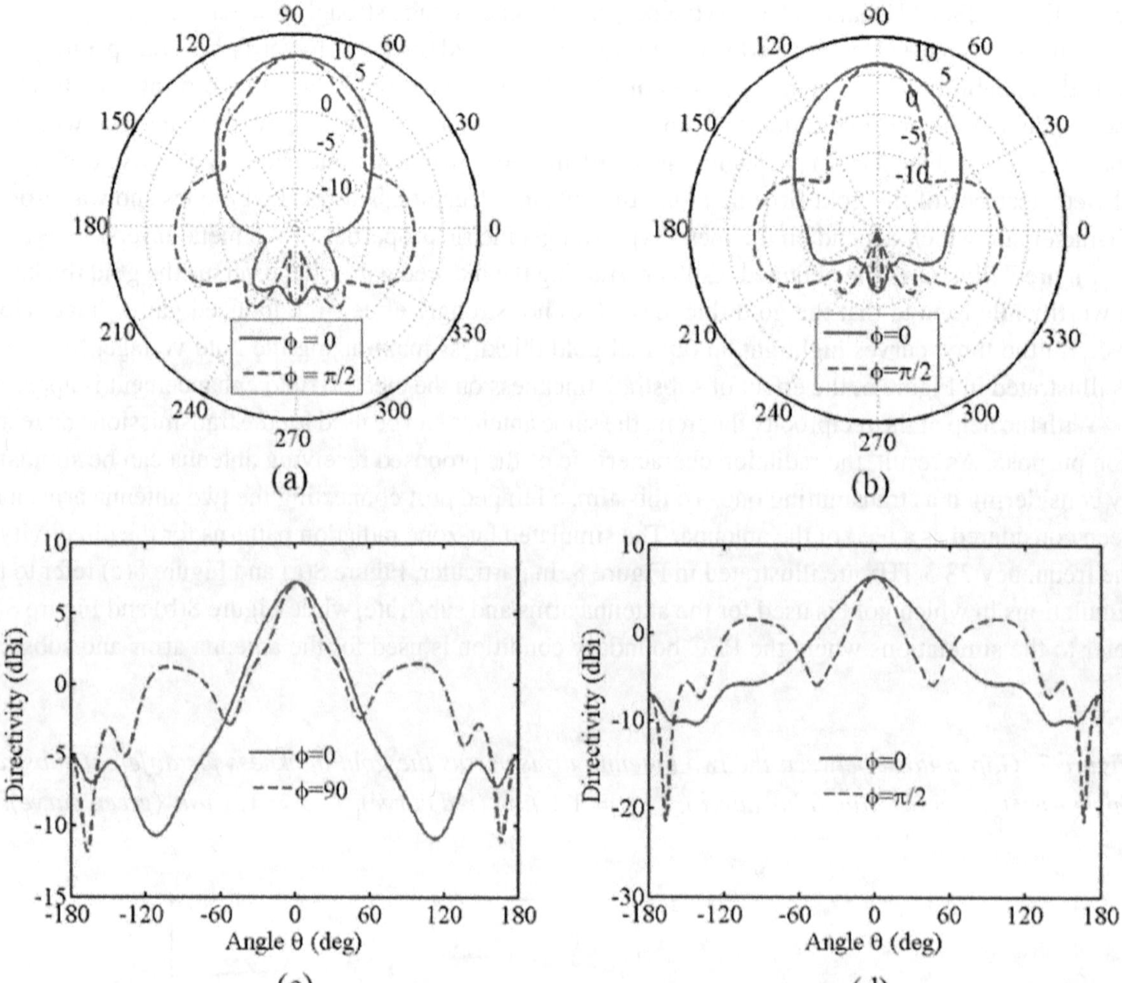


The radiation patterns in both E and H plane have been plotted, too. Figure 9 shows the radiation solid regarding (a) the gold and (b) the PEC boundary condition. The obtained numerical results highlight that the dispersive properties of the gold introduce a not negligible differences respect to the PEC case. In fact the radiation pattern exhibits a maximum directivity value of about 7.2 dBi and 6.4 dBi for gold and PEC cases, respectively. Moreover, strong differences in beam shape are apparent. In particular, in the E-plane the same – 3 dB angular aperture of about 50 degree has been calculated for the gold and PEC case, but quite different values of – 3 dB angular aperture in H-plane occurs. By an inspection of Figure 9 it can be observed that the radiation solid for the gold case exhibits side lobes level lower than that of the main lobe, while for the PEC case there are a number of side lobes having levels comparable with the main lobe level. Moreover, due to the gold losses the radiation efficiency is lower than that calculated for the PEC case. However, a half power beam large enough to allow the receiving of thermal energy from different directions have been calculated.

*Figure 9. Radiation solid for (a) dispersive gold and (b) PEC boundary condition*

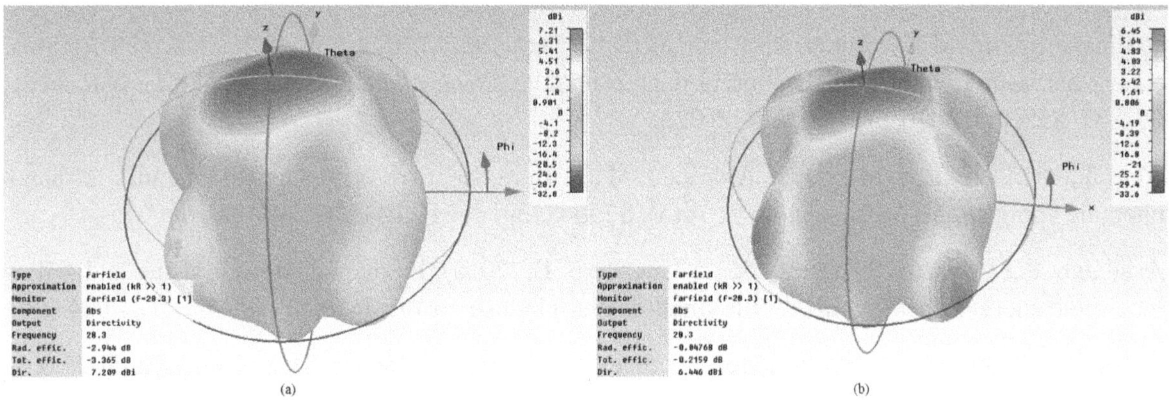


## CONCLUSION

A square spiral antenna able to collect the longwave radiation emitted from the Earth has been numerically investigated. In the chapter, the basic formulas regarding the antenna performance parameters have been provided. A brief review of the material properties of the metals are illustrated, too. We explored the different features of antenna in receiving and transmitting operation. In receiving mode, the antenna collects the electromagnetic field coming from the free space and focuses the energy with strong local electric field in the small gap region. The radiation characteristic have been investigated by feeding the antenna with a lumped port connecting the two antenna arms, and the receiving properties by illuminating the antenna with a circularly polarized plane wave. The magnitude of this plane wave has been calculated using the Stefan-Boltzmann law and considering the Earth as a grey body having a mean temperature of 287 K. Moreover, in the analysis the dispersive characteristics of the gold have been taken into account through the use of a detailed Lorentz-Drude model. The accurate design of the antenna is a key topic to improve the harvesting capability and the system efficiency for electricity generation. To this aim, a detailed numerical study has been performed with the aim to identify optimal geometric parameters as gold thickness, substrate thickness and gap size ensuring local field or gap voltage enhancement. The obtained numerical results show that with the right value of gap size, shape and materials as well as substrate and gold thickness the proposed antenna can efficiently collect energy at THz frequencies becoming a valid alternative to the traditional solar cells.

Until now, the technology is not sufficient to make an efficient, cost-effective system able to collects the longwave radiation emitted from the Earth. Further research efforts have to be fulfilled to identify the suitable materials and technology for the design and fabrication of efficient THz antennas. Although many antenna configurations are now being realized at THz and IR frequencies, the study of this kind of antennas is still in its initial stage and extensive research needs to be performed on energy conversion method and antenna structures to efficiently generate overall system electricity. We hope that the scientific community will provide vigorous research effort and will consider this technology to create a role for energy harvesting in the renewable energy landscape.

## REFERENCES

Abdina, Z., Alim, M. A., Saidur, R., Islam, M. R., Rashmi, W., Mekhilef, S., & Wadi, A. (2013). Solar energy harvesting with the application of nanotechnology. *Renewable & Sustainable Energy Reviews*, *26*, 837–852. doi:10.1016/j.rser.2013.06.023

Alda, J., Rico-Garcia, J. M., Lopez-Alonso, J. M., & Boreman, G. (2005). Optical antennas for nano-photonic applications. *Nanotechnology*, *16*(5), S230–S234. doi:10.1088/0957-4484/16/5/017

Arigliano, A., Caricato, P., Grieco, A., & Guerriero, E. (2014). Producing, storing, using and selling renewable energy: The best mix for the small medium industry. *Computers in Industry*, *65*, 408–418.

Bareiß, M., Tiwari, B. N., Hochmeister, A., Jegert, G., Zschieschang, U., Klauk, H., & Lugli, P. et al. (2011). Nano antenna array for terahertz detection. *IEEE Transactions on Microwave Theory and Techniques*, *59*(10), 2751–2757. doi:10.1109/TMTT.2011.2160200

Becker, C., Amkreutz, D., Sontheimer, T., Preidel, V., Lockau, D., Haschke, J., & Rech, B. et al. (2013). Polycrystalline silicon thin-film solar cells: Status and perspectives. *Solar Energy Materials and Solar Cells*, *119*, 112–123. doi:10.1016/j.solmat.2013.05.043

Bharadwaj, P., Deutsch, B., & Novotny, L. (2009). Optical antennas. *Advances Optics and Photonics*, *1*(3), 438–483. doi:10.1364/AOP.1.000438

Biagioni, P., Huang, J.-S., & Hecht, B. (2012). Nanoantennas for visible and infrared radiation. *Reports on Progress in Physics*, *75*(2), 024402. doi:10.1088/0034-4885/75/2/024402 PMID:22790344

Briones, E., Alda, J., & González, F. J. (2013). Conversion efficiency of broad-band rectennas for solar energy harvesting applications. *Optics Express*, *14*(S3), A412–A418. doi:10.1364/OE.21.00A412 PMID:24104428

Byrnes S. J., Blanchard R., & Capasso F. (2014). Harvesting renewable energy from earth's mid-infrared emissions. *PNAS Early Edition*, 1-6.

Cherednichenko, S., Hammar, A., Bevilacqua, S., Drakinskiy, V., Stake, J., & Kalabukhov, A. (2011). A room temperature bolometer for terahertz coherent and incoherent detection. *IEEE Transaction Terahertz Science Technology*, *1*(2), 395–402. doi:10.1109/TTHZ.2011.2164654

Gadalla, M. N., Abdel-Rahman, M., & Shamim, A. (2014). Design, optimization and fabrication of a 28.3 THz nano-rectenna for infrared detection and rectification. *Scientific Reports*, *4*, 1–9. doi:10.1038/srep04270 PMID:24599374

Gallo, M., Mescia, L., Losito, O., Bozzetti, M., & Prudenzano, F. (2012). Design of optical antenna for solar energy collection. *Energy*, *39*(1), 27–32. doi:10.1016/j.energy.2011.02.026

Goetzberger, A., & Hoffmann, V. U. (2005). *Photovoltaic solar energy generation*. Berlin, Heidelberg: Springer-Verlag.

Gonzalez, F. J., & Boreman, G. D. (2005). Comparison of dipole, bowtie, spiral and log-periodic IR antennas. *Infrared Physics & Technology*, *46*(5), 418–428. doi:10.1016/j.infrared.2004.09.002

Gritz, M. A., Puscasu, I., Spencer, D., & Boreman, G. D. (2003). Fabrication of an infrared antenna-coupled microbolometer linear array using chrome as a mask. *Journal of Vacuum Science & Technology B Microelectronics and Nanometer Structures*, *21*(6), 2608–2611. doi:10.1116/1.1629714

He, S., Cui, Y., Ye, Y., Zhang, P., & Jin, Y. (2009). Optical nano-antennas and metamaterials. *Materialtoday*, *12*, 16–24.

Khaligh, A., & Onar, O. (2010). *Energy harvesting: solar, wind, and ocean energy conversion systems*. CRC Press.

Kocakarin, I., & Yegin, K. (2013). Glass superstrate nanoantennas for infrared energy harvesting applications. *International Journal of Antennas and Propagation*, *2013*, 1–7. doi:10.1155/2013/245960

Kotter D. K., Novack S. D., Slafer W. D., & Pinhero P. J. (2010). Theory and manufacturing processes of solar nanoantenna electromagnetic collectors. *Journal of Solar Energy Engineering*, *132*(1).

Krenz, P. M., Tiwari, B., Szakmany, G. P., Orlov, A. O., González, F. J., Boreman, G. D., & Porod, W. (2012). Response increase of IR antenna-coupled thermocouple using impedance matching. *IEEE Journal of Quantum Electronics*, *48*(5), 659–664. doi:10.1109/JQE.2012.2189758

Ladan, S., Ghassemi, N., Ghiotto, A., & Wu, K. (2013). Highly efficient compact rectenna for wireless energy harvesting Application. *IEEE Microwave Magazine*, *14*(1), 117–122. doi:10.1109/MMM.2012.2226629

Locatelli, A. (2010). Analysis of the optical properties of wire antennas with displaced terminals. *Optics Express*, *18*(9), 9504–9510. doi:10.1364/OE.18.009504 PMID:20588796

Ma, Z., & Vandenbosch, G. A. E. (2013). Optimal solar energy harvesting efficiency of nano-rectenna systems. *Solar Energy*, *88*, 163–174. doi:10.1016/j.solener.2012.11.023

Manzano-Agugliaro, F., Alcayde, A., Montoya, F. G., Zapata-Sierra, A., & Gil, C. (2013). Scientific production of renewable energies worldwide: An overview. *Renewable & Sustainable Energy Reviews*, *18*, 134–143. doi:10.1016/j.rser.2012.10.020

Novotny, L. (2007). Effective wavelength scaling for optical antennas. *Physical Review Letters*, *98*(26), 266802-1, 266802–266804. doi:10.1103/PhysRevLett.98.266802 PMID:17678117

Novotny, L., & van Hulst, N. (2011). Antennas for light. *Nature Photonics*, *5*(2), 83–90. doi:10.1038/nphoton.2010.237

Orlo, J., Utlaut, M., & Swanson, L. (2002). *High resolution focused ion beams: FIB and applications*. Kluwer Academic/Plenum Publishers.

Planck, M. (1901). Über das Gesetz der Energieverteilung im Normalspektrum. *Annalen der Physik*, *4*(3), 553–558. doi:10.1002/andp.19013090310

Rakic, A. D., Djurisic, A. B., Elazar, J. M., & Majewski, M. L. (1998). Optical properties of metallic films for vertical-cavity optoelectronic devices. *Applied Optics*, *37*(22), 5271–5283. doi:10.1364/AO.37.005271 PMID:18286006

Sabaawi, A. M. A., Tsimenidis, C. C., & Sharif, B. S. (2013). Analysis and modeling of infrared solar rectennas. *IEEE Journal of Selected Topics in Quantum Electronics*, *19*(3), 9000208. doi:10.1109/JSTQE.2012.2227686

Semenov, A. D., Richter, H., Hübers, H. W., Günther, B., Smirnov, A., Il'in, K. S., & Karamarkovic, J. P. et al. (2007). Terahertz performance of integrated lens antennas with a hot-electron bolometer. *IEEE Transactions on Microwave Theory and Techniques*, *55*(2), 239–247. doi:10.1109/TMTT.2006.889153

Shrestha, S., Noh, S. K., & Choi, D. Y. (2013). Comparative study of antenna designs for RF energy harvesting. *International Journal of Antennas and Propagation*, *2013*, 1–7. doi:10.1155/2013/385260

Stokes, J. L., Yu, Y., Yuan, Z. H., Pugh, J. R., Lopez-Garcia, M., Ahmad, N., & Cryan, M. J. (2014). Analysis and design of a cross dipole nanoantenna for fluorescence-sensing applications. *Journal of the Optical Society of America. B, Optical Physics*, *31*(2), 302–310. doi:10.1364/JOSAB.31.000302

Tiwari, G. N., & Dubey, S. (2010). *Fundamentals of photovoltaic modules and their applications*. RSC Publishing.

Trenberth, K. E., Fasullo, J., & Kiehl, J. (2009). Earth's global energy budget. *Bulletin of the American Meteorological Society*, *90*(3), 311–323. doi:10.1175/2008BAMS2634.1

Tyagi, V. V., Rahim, A. A., Rahim, N. A., & Selvaraj, L. (2013). Progress in solar PV technology: Research and achievement. *Renewable & Sustainable Energy Reviews*, *20*, 443–461. doi:10.1016/j.rser.2012.09.028

van Dorp, W. F., & Hagen, C. W. (2008). A critical literature review of focused electron beam induced deposition. *Journal of Applied Physics*, *104*(8), 081301. doi:10.1063/1.2977587

Weber-Bargioni, A., Schwartzberg, A., Schmidt, M., Harteneck, B., Ogletree, D. F., Schuck, P. J., & Cabrini, S. (2010). Functional plasmonic antenna scanning probes fabricated by induced-deposition mask lithography. *Nanotechnology*, *21*(6), 065306. doi:10.1088/0957-4484/21/6/065306 PMID:20061594

## ADDITIONAL READING

Alvarado, U., Juanicorena, A., Adin, I., Sedano, B., Gutiérrez, I., & de Nó, J. (2012). Energy harvesting technologies for low-power electronics. *Transactions on Emerging Telecommunications Technologies*, *23*(8), 728–741. doi:10.1002/ett.2529

Bailey, R. L. (1972). A proposed new concept for a solar energy convertor. *Journal of Engineering for Power*, *94*(2), 73–77. doi:10.1115/1.3445660

Balocco, C., Kasjoo, S. R., Zhang, L. Q., Alimi, Y., & Song, A. M. (2011). Low-frequency noise of unipolar nanorectifiers. *Applied Physics Letters*, *99*(11), 113511. doi:10.1063/1.3636437

Bia, P., Caratelli, D., Mescia, L., & Gielis, J. (2013). Electromagnetic characterization of supershaped lens antennas for high-frequency applications. *Proceedings of the 43rd European Microwave Conference*, 1679-1682.

Bozzetti, M., de Candia, G., Gallo, M., Losito, O., Mescia, L., & Prudenzano, F. (2010). Analysis and design of a solar rectenna. *Proceedings IEEE International Symposium on Industrial Electronics*, 2001–2004. doi:10.1109/ISIE.2010.5637486

Choo, H., Kim, M.-K., Staffaroni, M., Seok, T. J., Bokor, J., Cabrini, S., & Yablonovitch, E. et al. (2012). Nanofocusing in a metal-insulator-metal gap plasmon waveguide with a three-dimensional linear taper. *Nature Photonics*, *6*(12), 838–844. doi:10.1038/nphoton.2012.277

Chou, S. Y., Krauss, P. R., & Renstrom, P. J. (1996). Nanoimprint lithography. *Journal of Vacuum Science & Technology B Microelectronics and Nanometer Structures*, *14*(6), 4129–4133. doi:10.1116/1.588605

Collado, A., & Georgiadis, A. (2013). Conformal hybrid solar and electromagnetic (EM) energy harvesting rectenna. *IEEE Transactions on Circuits and Systems. I, Regular Papers*, *60*(8), 2225–2234. doi:10.1109/TCSI.2013.2239154

Costanzo, A., Romania, A., Masotti, D., Arbizzani, N., & Rizzoli, V. (2012). RF/baseband co-design of switching receivers for multiband microwave energy harvesting. *Sensors and Actuators. A, Physical*, *179*, 158–168. doi:10.1016/j.sna.2012.02.005

Dragoman, D., & Dragoman, M. (2013). Geometrically induced rectification in two dimensional ballistic nanodevices. *Journal of Physics. D, Applied Physics*, *46*(5), 055306. doi:10.1088/0022-3727/46/5/055306

Hammar, A., Cherednichenko, S., Bevilacqua, S., Drakinskiy, V., & Stake, J. (2011). Terahertz direct detection in YBa2Cu3O7 microbolometers. *IEEE Transactions Terahertz Science Technology*, *1*(2), 390–394. doi:10.1109/TTHZ.2011.2161050

Hardin, B., Hoke, E. T., Armstrong, P. B., Yum, J.-H., Comte, P., Torres, T., & McGehee, M. D. et al. (2009). Increased light harvesting in dye-sensitized solar cells with energy relay dyes. *Nature Photonics*, *3*(7), 406–411. doi:10.1038/nphoton.2009.96

Jain, K., Klosner, M., Zemel, M., & Raghunandan, S. (2005). Flexible electronics and displays: High-resolution, roll-to-roll, projection lithography and photoablation processing technologies for high-throughput production. *Proceedings of the IEEE*, *93*(8), 1500–1510. doi:10.1109/JPROC.2005.851505

Jones, D., New, M., Parker, D. E., Martin, S., & Rigor, I. G. (1999). Surface air temperature and its changes over the past 150 years. *Reviews of Geophysics*, *37*(2), 173–199. doi:10.1029/1999RG900002

Karasik, B. S., Sergeev, A. V., & Prober, D. E. (2011). Nanobolometers for THz photon detection. *IEEE Transactions. Terahertz Science Technology*, *1*(1), 97–111. doi:10.1109/TTHZ.2011.2159560

Krishnan, S., Goswami, Y., & Stefanakos, E. (2012). Nanoscaale rectenna for thermal energy conversion to electricity. *Technology & Innovation*, *14*(2), 103–113. doi:10.3727/194982412X13462021397651

Lui, K. W., Vilches, A., & Toumazou, C. (2010). Ultra-efficient microwave harvesting system for battery-less micropower microcontroller platform. *IET Microwave Antennas & Propagations*, *5*(7), 811–817. doi:10.1049/iet-map.2010.0250

Ma, Z., & Vandenbosch, G. A. E. (2014). Wideband harmonic rejection filtenna for wireless power transfer. *IEEE Transactions on Antennas and Propagation*, *62*, 1–7. doi:10.1155/2014/802815

Majewski, L. A., Balocco, C., King, R., Whitelegg, S., & Song, A. M. (2008). Fast polymer nanorectifiers for inductively coupled RFID tags. *Materials Science and Engineering B*, *147*(2-3), 289–292. doi:10.1016/j.mseb.2007.08.031

Marian, V., Allard, B., Vollaire, C., & Verdier, J. (2012). Strategy for microwave energy harvesting from ambient field or a feeding source. *IEEE Transactions on Power Electronics*, *27*(11), 4481–4491. doi:10.1109/TPEL.2012.2185249

Massaro, A., Spano, F., Missori, M., Malvindi, M. A., Cazzato, P., Cingolani, R., & Athanassiou, A. (2014). Flexible nanocomposites with all-optical tactile sensing capability. *RSC Advances*, *4*(6), 2820–2825. doi:10.1039/C3RA45678A

Masuch, J., Delgado-Restituto, M., Milosevic, D., & Baltus, P. (2013). Co-Integration of an RF energy harvester into a 2.4 GHz transceiver. *IEEE Journal of Solid-State Circuits*, *48*(7), 1565–1574. doi:10.1109/JSSC.2013.2253394

Moddel, G. (2013). Will rectenna solar cells be practical? In Garret, Grover, Sachit (Eds.), Rectenna solar cells model, (pp. 3-24). Springer. doi:10.1007/978-1-4614-3716-1_1

Olgun, U., Chen, C., & Volakis, J. L. (2011). Investigation of rectenna array configurations for enhanced RF power harvesting. *IEEE Antennas and Wireless Propagation Letters*, *10*, 262–265. doi:10.1109/LAWP.2011.2136371

Peter, T., Rahman, T. A., Cheung, S. W., Nilavalan, R., Abutarboush, H. F., & Vilches, A. (2014). A novel transparent UWB antenna for photovoltaic solar panel integration and RF energy harvesting. *IEEE Transactions on Antennas and Propagation*, *62*(4), 1844–1853. doi:10.1109/TAP.2014.2298044

Qiang, R., Chen, R. L., & Chen, J. (2004). Modeling electrical properties of gold films at infrared frequency using FDTD method. *International Journal of Infrared and Millimeter Waves*, *25*(8), 12163–1270. doi:10.1023/B:IJIM.0000042759.67055.7a

Rephaeli, E., Raman, A., & Fan, S. (2013). Ultrabroadband photonic structures to achieve high-performance daytime radiative cooling. *Nano Letters*, *13*, 1457–1461. PMID:23461597

Schnell, M., Alonso-Gonzalez, P., Arzubiaga, L., Casanova, F., Hueso, L. E., Chuvilin, A., & Hillenbrand, R. (2011). Nanofocusing of mid-infrared energy with tapered transmission lines. *Nature Photonics*, *5*(5), 283–287. doi:10.1038/nphoton.2011.33

Singh, G. (2010). Design considerations for rectangular microstrip patch antenna on electromagnetic crystal substrate at terahertz frequency. *Infrared Physics & Technology*, *53*(1), 17–22. doi:10.1016/j.infrared.2009.08.002

Soumya, S., Mohamed, A. P., Paul, L., Mohan, K., & Ananthakumar, S. (2014). Near IR reflectance characteristics of PMMA/ZnO nanocomposites for solar thermal control interface films. *Solar Energy Materials and Solar Cells*, *125*, 102–112. doi:10.1016/j.solmat.2014.02.033

Takacs, A., Aubert, H., Fredon, S., Despoisse, L., & Blondeaux, H. (2014). Microwave power harvesting for satellite health monitoring. *IEEE Transactions on Microwave Theory and Techniques*, *62*(4), 1090–1098. doi:10.1109/TMTT.2014.2303425

Wang, R., Ye, D., Dong, S., Peng, Z., Salamin, Y., Shen, F., & Ran, L. et al. (2014). Optimal matched rectifying surface for space solar power satellite applications. *IEEE Transactions on Microwave Theory and Techniques*, *62*(4), 1080–1089. doi:10.1109/TMTT.2014.2300451

Yu, M., Long, Y.-Z., Sun, B., & Fan, Z. (2012). Recent advances in solar cells based on one-dimensional nanostructure arrays. *Nanoscale*, *4*(9), 2783–2796. doi:10.1039/c2nr30437f PMID:22473573

Zhu, Z., Joshi, S., Grover, S., & Moddel, G. (2013). Graphene geometric diodes for terahertz rectennas. *Journal of Physics. D, Applied Physics*, *46*(18), 185101. doi:10.1088/0022-3727/46/18/185101

## KEY TERMS AND DEFINITIONS

**Energy Harvesting:** Process by which environmental energy is captured and converted into usable electrical energy.

**Infrared Emission:** Electromagnetic radiation with wavelengths extending from 700 nm to 1 mm.

**Lorentz-Drude Model:** Function modelling the frequency dependence of the dielectric permittivity.

**Planar Array Antenna:** Antennas collection in which all of the elements are in the same plane.

**Radio Frequency:** Electromagnetic radiation with frequency extending from 3 kHz to 300 GHz.

**Rectenna:** Special type of antenna converting electromagnetic energy into direct current electricity.

**Spiral Antenna:** Frequency independent antennas shaped as a two or more-arms spiral.

# Chapter 2
# Optical Nano-Antennas for Energy Harvesting

**Salah Obayya**
*Zewail City of Science and Technology, Egypt*

**Nihal Fayez Fahmy Areed**
*Zewail City of Science and Technology, Egypt & Mansoura University, Egypt*

**Mohamed Farhat O. Hameed**
*Zewail City of Science and Technology, Egypt & Mansoura University, Egypt*

**Mohamed Hussein Abdelrazik**
*Zewail City of Science and Technology, Egypt & Ain Shams University, Egypt*

## ABSTRACT

*The solar energy is able to supply humanity energy for almost another 1 billion years. Optical nano-antennas (ONAs) are an attractive technology for high efficiency, and low-cost solar cells. These devices can be classified to semiconductor nano-wires and metallic nano-antenna. Extensive studies have been carried out on ONAs to investigate their ability to harvest solar energy. Inspired by these studies, the scope of the chapter is to highlight the latest designs of the two main types of ONAs. The metallic nano-antennas are discussed based on the following points: plasmon, modeling, and performance of antenna designs using different configurations and materials. Moreover, the semiconductor nano-wires are studied thoroughly in terms of photonic crystals, antenna design with different patterns, nano-wire forms and materials. Also, the applications of ONAs and their fabrication aspects such as diode challenges are presented in detail. Finally, three novel designs of ONAs are presented and numerically simulated to maximize the harvesting efficiency.*

## INTRODUCTION

The search for alternative energy sources in the 21st century is a growing academic and industrial pursuit. Rising costs of carbon-based fuels coupled with increased emissions has placed a greater demand on the clean energy sector. Recently, renewable energy gives strong contributions to power generation without increasing environmental pollutions. A promising energy source is photovoltaics (PVs), which has traditionally been made from high purity, expensive crystalline silicon (c-Si). In addition, the PV solar cell

DOI: 10.4018/978-1-4666-8254-2.ch002

devices are designed to absorb solar energy in the visible region (400 nm-700 nm) which constitutes 46% of the solar spectrum (Malinovska, 2010). Therefore, the search for cheaper, high-performance materials for solar cells based applications is mandatory.

One of the alternative approaches to PV technology is the optical antennas that can harvest more solar energy by extending the absorbed range to the infrared (IR) region (Malinovska, 2010). The optical antenna can be fabricated by using metallic nano-particles or semiconductor nano-wires. Therefore, the optical antenna can be classified to metallic nano-antenna (rectenna) and semiconductor nano-antenna. The use of rectenna for energy harvesting depends on the fact that when an electromagnetic wave is incident on a nano-antenna, a time varying current will be induced on the antenna surface, and hence a voltage will be generated at the feeding point of the antenna. The generated wave will oscillate at the frequency of the incident wave. Consequently, in order to obtain Direct Current (DC) power, a suitable rectifier should be embedded at the feed point of antenna. These types of energy harvesting systems are called “rectennas”; which basically consist of antennas connected to a rectifier that converts the received signal to DC power and produces electricity (Malinovska 2010; Berland *et al.* 2001). An infrared rectenna structures based on metal–insulator–metal (MIM) diodes between dipoles (Berland *et al.*, 2003) has been designed for operation at the operating wavelength $\lambda$=10 μm. However the efficiency of this system was less than 1% (Berland *et al.*, 2003). A spiral nano-antenna for solar energy harvesting has been designed and fabricated at mid-infrared region by Kotter *et al.* (2010). In addition, Midrio *et al.* (2011) have designed and analyzed mono pole antenna based on nickel for the reception of thermal radiation. Vandenbosch *et al.* (2012) have introduced upper bounds for the solar energy harvesting efficiency of metallic nano-antennas. In this regard the silver exhibited the highest efficiencies, both in free space and on top of glass ($SiO_2$) substrate, with radiation efficiencies near or slightly above 90% (Vandenbosch *et al.*, 2012). Although, the metallic nano-antenna offers outstanding harvesting efficiency over a wide frequency range, no suitable rectifier diodes are described in the literature to meet the requirements for solar energy harvesting.

Semiconductor nano-antennas have recently gained tremendous interest due to their light absorption (Hong *et al.*, 2013). Light absorption in standing semiconductor nanowires is a complex phenomenon, with strong dependence on nano-wire dimensions, lattice arrangement and absorption coefficient of the raw materials (Hong *et al.*, 2013). Ordered semiconductor nano-antenna structure can be fabricated through a few unattractive methods, such as lithography and polystyrene ball assembly, due to their high cost and complexity (Hong *et al.*, 2013).

Following this introduction, a detailed description of the science, concepts of design, modeling key issues and the history of optical nano-antenna (ONA) solar cells will be provided. In addition, the latest designs and fabrication challenges of the two types of optical nano-antennas are discussed thoroughly. On top of that, three novel designs of flower shaped metallic nano-antenna, decagonal semiconductor nano-wires with semiconductor and hybrid cores are presented and analyzed using 3D full vectorial finite difference method (FDTD) (Taflove & Hagness, 2005). The performance of the suggested ONAs with different materials and configurations has been investigated based on the absorption efficiency spectra, absorbed power profiles and emission patterns. The proposed flower shaped design exhibits higher efficiency over the conventional dipole antenna by at least 15%. In addition, the total harvesting efficiency is enhanced by 32.7%. Moreover, the suggested nano-antenna offers large bandwidth in the wavelength range from 450 nm to 1400 nm. Additionally, the flower-shaped dipole nano-antenna has better efficiency of 74.6% compared to the conventional solar cells at $\lambda$= 500 nm at which the sun irradiance is maximum. Furthermore, the calculated ultimate efficiency of the proposed D-SiNWs with solid semiconductor core is 19.3%. This ultimate efficiency can be maximized either by tuning the geometric parameters of the proposed D-SiNWs or using hybrid core. The reported structure with hybrid core offers an ultimate efficiency of 32.62% while the suggested tuned D-SiNWs design offers an ultimate efficiency of 38%.

## 1. Metallic Nano-Antenna

Solar energy with solar spectrum shown in Figure 1(a) is considered as the most significant source of green renewable energy. It should be noted from this figure that, the solar energy oscillates over a wide range of wavelengths. Most of this energy is concentrated in the visible, whereas small amount is distributed over the ultraviolet and infrared. Photovoltaic (PV) solar cells absorption performance is mainly limited by problem of semiconductor band gap. Figure 1(b) shows the absorption coefficients and absorption bands of major semiconductor materials (Foll *et. al.*, 2010). It can be noticed from this figure that the silicon as an example cannot absorb light energy more than 1100 nm due to the silicon band gap of 1.1 ev at room temperature. In spite of the developed industry of the PV solar cells, the solar cells conversion efficiency doesn't exceed 30% (Gadalla, 2013).

One of the alternative approaches to PV solar cells with higher efficiency for energy conversion is rectenna (Gadalla, 2013). Figure 2 shows a schematic block diagram of "rectenna" system with load. As illustrated in Figure 2, a rectenna system includes a metallic nano-antenna, a rectifier diode and DC pass filter all in parallel. In addition, the load must be included in the scheme because it may affect the output voltage and then the rectenna efficiency. The rectifier diode is an ultra-high speed diode that converts the AC power to DC power. The conversion efficiency of the diode is a key factor of the rectenna performance. The DC pass filter is present to smooth the signal before delivery to the load.

Metallic nano-antenna is a light coupling device that collects the solar energy in the optical frequencies band for generating electricity with high efficiency conversion. It consists of nanometer scale metallic particles mounted on substrate as shown in Figure3 (a). Currently investigated optical nano-antennas include various designs in terms of different material constitutions, configurations, and arrangements.

*Figure 1. (a) Solar energy spectrum of the sun, (b) Bandgap of common semiconductor materials (Foll et. al., 2010).*

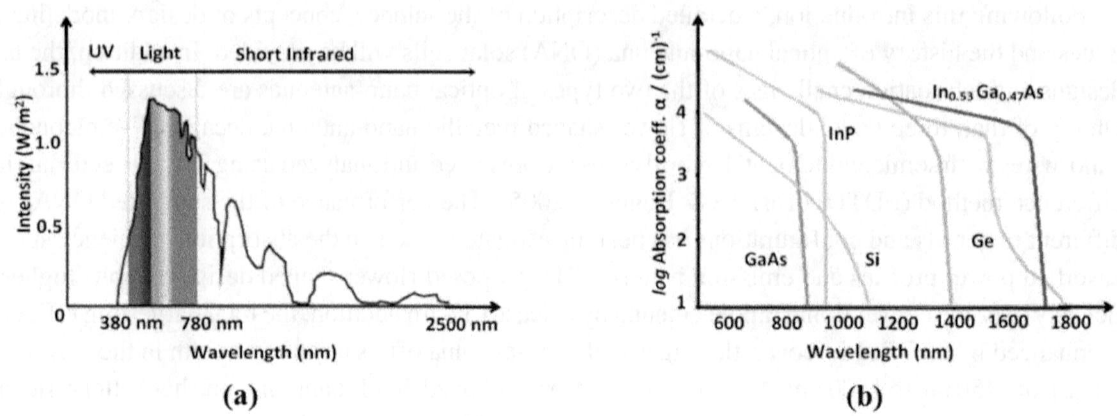

*Figure 2. Schematic diagram of the rectenna system (Gadalla, 2013)*

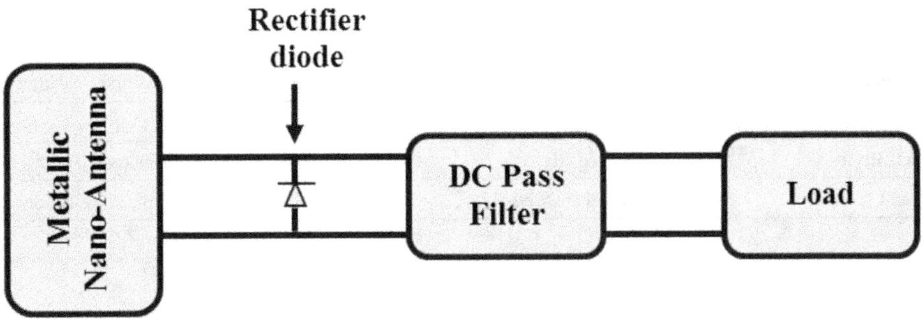


Figure 3 (b) shows three common configurations for the metallic nano-antenna such as rectangular, bowtie and elliptical shapes. The interaction between surface plasmons of metallic nano-particles induced by light is capable of focusing and confining visible and near infrared (NIR) lights into nanometer scale dimensions and this will be explained in details in the following subsections. Nano-antennas based on surface plasmons can be used for many applications, such as antennas for nano scale imaging and spectroscopy (Hartschuh, 2004) and for improving solar cells efficiency (Hagglund *et al.*, 2008). The scope of our research is to use the nano-antennas for energy harvesting applications.

Rectenna in principle has no efficiency limitations since the major problem of semiconductor band gap does not apply here. Efficient collection by the antenna, perfect matching between the diode and antenna and efficient rectification nominates rectenna to have a theoretical conversion efficiency of 100%. Beside to the advantage of higher efficiency for energy conversion of rectennas, Table 1 shows many other better aspects as compared to the traditional solar cells. As indicated by the table, rectennas can operate during the whole day and independently of weather conditions such as humidity and cloud cover compared with traditional solar cells that operate during day time only and require clear and dry weather. On top of that, unlike solar cells operation, rectennas radiation, being isotropic in nature, does not require a certain orientation for energy collection. Further, rectennas can be fabricated on substrates much cheaper than silicon. In addition, rectennas could be used to cool buildings and other structures, transforming the once wasted heat into electrical energy (Gadalla, 2013).

Several technical challenges have faced the THz rectenna devices such as: design, simulation and fabrication of nano-antennas (Gadalla, 2014). Due to very small geometric features, the simulations of the nano-antennas require great computational resources (Gadalla, 2014). Nano-antennas materials should also be selected with proper optical and electrical characteristics in THz range (Kotter *et al.*, 2010).In

*Figure 3. Metallic nano-antenna: (a) Side view, (b) Top views for common antenna configurations*

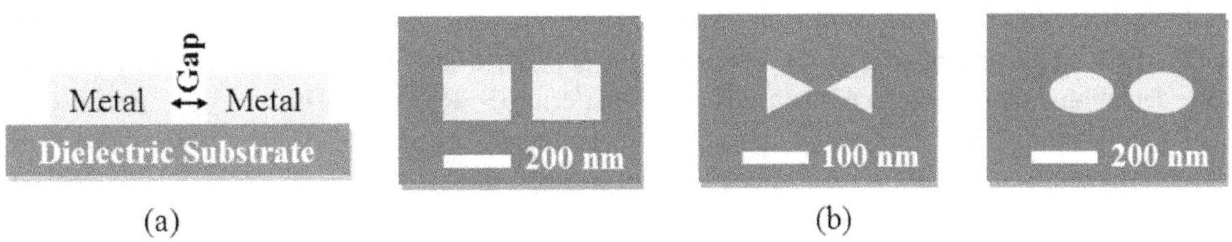

*Table 1. Comparison between the traditional solar cells and rectennas (Gadalla, 2013)*

| | Solar cell | Rectenna |
|---|---|---|
| **Operation time** | Day time only | Whole day |
| **Weather conditions** | Requires clear and dry weather | Independent of weather conditions |
| **Theoretical efficiency of conversion** | Low | High |
| **Orientation sensitivity** | Very sensitive | Less sensitive |
| **Absorption band** | Limited absorption | Wide range absorption |
| **Cost** | High | Low |

addition, nano-antennas fabrication process needs advanced machines like electron beam lithography (EBL) (Rai-Choudhury, 1997). The design and fabrications of THz diode with high conversion efficiency from AC to DC (Dagenais *et al.*, 2010) are also challenge problems for THz rectennas. The growth of nanometer oxide layer needs accurate and precise fabrication process. Furthermore, the MIM diode has very high input impedance with high mismatching between the diode and nano-antennas (Abdel-Rahman *et al.*, 2004).Finally, the MIM diode exhibits a considerable parasitic capacitance (Gadalla, 2014; Dagenais *et al.*, 2010).

## 2.1 Metals, Plasmonics, and Electromagnetics

Paul Drude in 1900 (Fox, 2012) supposed that the behavior of electrons in metals looks like a sea of continuously moving electrons that collide between themselves and oscillating into and out of heavy positively charged ions as well. The collisions lead to loss in the electrons kinetic energy and damping of electric field takes place. Owing to the chaotic thermal motion of the conduction electrons in metals, a thin double electrical layer of electron clouds and positively charged ions exist near metal surface. This layer has an electric field that prevents the electrons of flying far away from the metal surface. The interaction of metals with electromagnetic (EM) fields varies according to the frequency of the interacting electromagnetic wave (Kostiučenko, 2011). Metals are so reflective; hence do not allow the electromagnetic waves propagate through them for the Microwaves or Far Infrared (FIR) frequency regimes. At higher frequencies of Near Infrared (NIR) and Visible (VIS), metals behave somewhat different and become less resistant to electromagnetic waves penetration. At the Ultraviolet (UV) frequencies, metals gain much of dielectric attitude, enabling the penetration and propagation of electromagnetic waves with varying attenuation levels, depending on the specifics of the electronic band structure. Gold for example shows a strong absorption in UV frequency regime due to transitions between electronic bands (Kostiučenko, 2011).

Plasmonics is a novel quickly developing nano-photonic challenging branch that studies light-metal interaction with the purpose of enhancing or localizing electromagnetic fields at extremely tiny structures on the order of several hundreds of nanometer (nm). Plasmonics find diverse applications in different science fields, like: optical sensors, biological surface plasmon microscopy, and plasmon couplers, molecule detection/emission, surface enhanced Raman scattering (SERS), optical microscopy, spectroscopy, high-density optical data storage and optical circuits and many others(Gadalla, 2013; Hartschuh *et al.*, 2004; Hagglund *et al.*, 2008; Kostiučenko, 2011).

Plasmons are defined as quasi particles located on a metal surface due to the oscillations of free electrons with respect to fixed positively charged ions in metals. The oscillatory electrons through a thin metal that is distinguished by its negative real or small positive imaginary dielectric constant are stimulated either by an electron or a photon. At the interface between metal and dielectric, surface plasmons will couple with the photons and result in a coherent electron charge oscillations in the visible and NIR portion which is called surface plasmon polarition (SPP). In this case, the surface electromagnetic waves propagate parallel to metal/dielectric interface direction for distances in the order of tens to hundreds of microns, and decay evanescently in perpendicular direction of interface with 1/e decay lengths on the order of 200 nm (Gadalla, 2013). Localized surface plasmons (LSP) exist when light interacts with particles much smaller than the incident wavelength. This condition is met in the IR and visible wavelength region for metal/dielectric interfaces. Over few years, scientific community introduced the optical metallic nano antenna that comprises two nano-scale metal particles with a gap as shown in Figure4. The optical properties of the metallic nano-antenna show that, when exciting the antenna's surface by a visible or infrared light, it generates surface plasmon oscillations that drive current towards the feed point of the antenna creating a hot spot at which the field intensity is enhanced as shown in Figure4(c).

## 2.2 Modeling and Characterization of Metallic Nano-Antennas

Figure 5 describes the circuit model of the receiving optical antenna. The optical antenna in a receiving mode can be modeled as impedance in series with voltage source (Gadalla, 2013; Ma *et al.*, 2013). In this model, we take into considerations the effect of air gap between dipole arms which acts as a capacitor with capacitance $C_{gap}$. The input impedance of antenna impedance is defined as

$$Z_{ant} = R_{ant} + jX_{ant} \tag{1}$$

where $R_{ant}$ and $X_{ant}$ are antenna resistance and reactance, respectively. In general, antenna resistance consists of two parts radiation resistance ( $R_{rad}$ ) and loss resistance ( $R_{losss}$ )

*Figure 4. Schematic diagrams illustrating (a) surface plasmon polariton (Kostiučenko et al., 2011) (b) localized surface (Kostiučenko et al., 2011) plasmons, (c) field enhancement in the gap between two nano-metal particles*

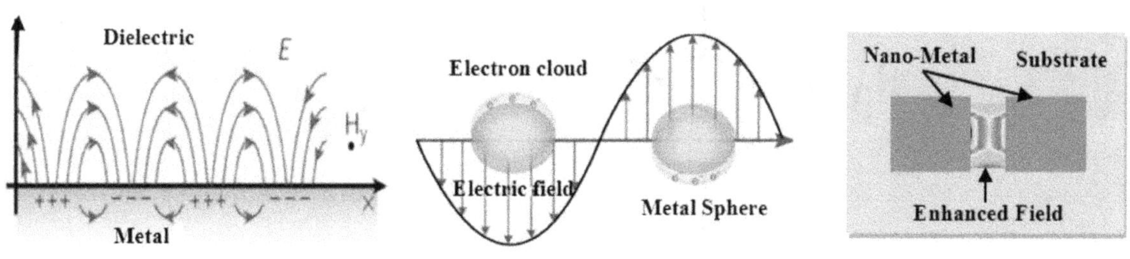


(a) (b) (c)

*Figure 5. Equivalent circuit of the receiving optical antennas*

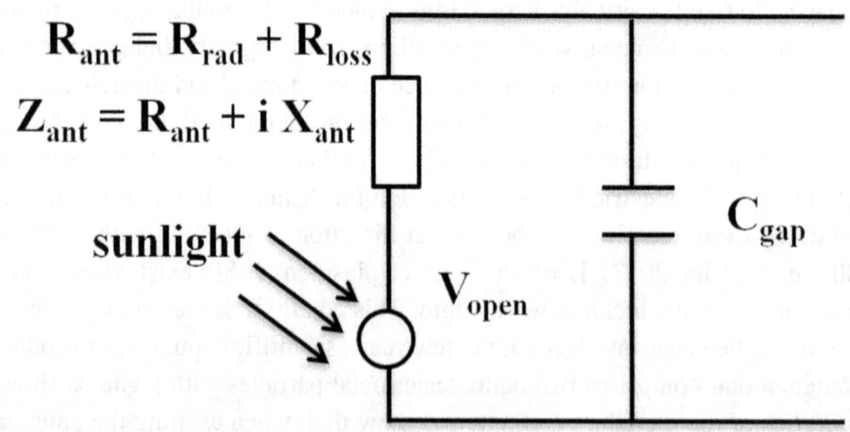


$$R_{ant} = R_{rad} + R_{losss} \tag{2}$$

The $R_{rad}$ and $R_{loss}$ represent the modeling of the radiated power and dielectric loss, respectively (Gadalla, 2013; Ma *et al.*, 2013).

Metal insulator metal (MIM) tunneling diode is most widely rectifier in terahertz range (Grover & Model, 2011). The MIM diode can be represented as a capacitor ($C_D$) parallel with nonlinear resistance ($R_D$) as shown in Figure 6. The total impedance of the MIM diode is given by

$$Z_D = R_D + jX_D \; Z_D = R_D + jX_D \tag{3}$$

where $X_D$ is the diode reactance.

The cutoff $f_c$ frequency can be expressed in terms of $R_D$ and $C_D$

*Figure 6. Equivalent circuit of the MIM diode antennas*

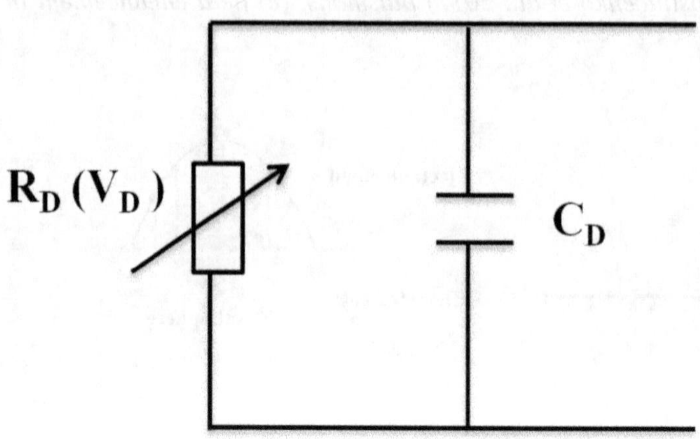

$$f_c = \frac{1}{2\pi R_D C_D} \tag{4}$$

The circuit model of solar rectenna system consists of optical antenna connected parallel with MIM diode as shown in Figure 7. The input impedance of the rectenna is given by (Garret & Sachit, 2013)

$$\mathrm{Z_{rec} = Z_{ant}Z_D = \frac{(R_D R_{ant} - X_D X_{ant}) - j\left(R_D X_{ant} - R_{ant} X_D\right)}{\left(R_D + R_{ant}\right) - j\left(X_D + X_{ant}\right)}}$$

$$Z_{rec} = Z_{ant} // Z_D = \frac{(R_D R_{ant} - X_D X_{ant}) - j(R_D X_{ant} - R_{ant} X_D)}{\left(R_D + R_{ant}\right) - j(X_D + X_{ant})} \tag{5}$$

The power received by impedance of the rectifier is given by (Vandenbosch *et al.*, 2012)

$$P^{rec} = \frac{1}{2}\left|V_{open}\right|^2 \frac{R_{rec}}{\left(Z_{rec} + Z_{ant}\right)^2} \tag{6}$$

At matching condition $Z_{rec} = Z^*_{ant}$ $\mathrm{Z_{rec} = Z^*_{ant}}$ and the maximum power received is equal to

$$P^{rec} = \frac{\left|V_{open}\right|^2}{8R^2} \tag{7}$$

The performance of the nano-rectennas depends on the efficiency of light capturing by the nano-antenna as well as transferring to its terminals and the capability of the rectifier to transform the captured

*Figure 7. Equivalent circuit of solar rectenna diode system*

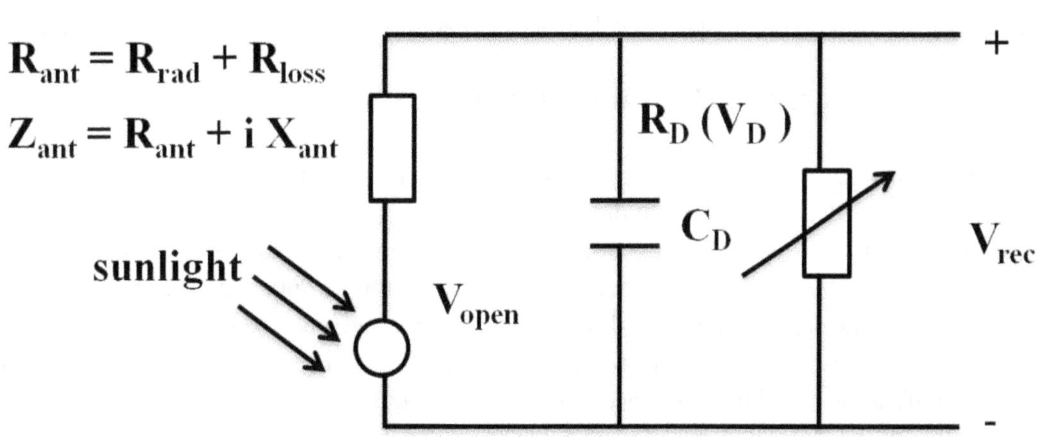

light into low frequency power. Based upon the reciprocity theory (Vandenbosch *et al.*, 2012), it is deduced that the efficiency at receptions is similar to that of the antenna in converting input power, given at its terminals, into radiation, which is called the radiation efficiency $\eta^{rad}$ of the antenna (Dagenais *et al.*, 2010; Ma *et al.*, 2013; Biagioni *et al.*, 2012). In addition, at small scale, no quantum effects have to be taken into account, and then the antenna is able to transmit and receive electromagnetic waves rather than particles. In effect, the coupling between electromagnetic waves and the nano-antenna is the same as at microwave frequencies. From the values of the extinction and scattering cross-sections, based upon Mie theory (Biagioni *et al.*, 2012), the optical radiation efficiency of nano particles can be expressed as

$$\eta^{rad} = \frac{C_{sca}}{C_{ext}} = \frac{C_{sca}}{C_{sca} + C_{abs}} \tag{8}$$

where $C_{sca}$, $C_{ext}$, and $C_{abs}$ are scattering, extinction, and absorption cross-sections, respectively that can be calculated by series expansion of the internal and scattered fields into a set of partial waves described by vector harmonics (Dmitruk *et al.*, 2010). For energy harvesting applications, and to characterize the antenna, namely, the total harvesting efficiency is used which is given by (Vandenbosch *et al.*, 2012)

$$\eta^{tot} = \frac{\int_0^\infty P(\lambda,T) \times \eta^{rad}(\lambda)\, d\lambda}{\int_0^\infty P(\lambda,T)\, d\lambda} \tag{9}$$

where λ is the wavelength and P is Planck's law for black body radiation defined by

$$P(\lambda,T) = \left(\frac{2\pi hc^2}{\lambda^5} \times \frac{1}{e^{hc/\lambda kT} - 1}\right) \tag{10}$$

where T is the absolute temperature of the blackbody (in K), h is Planck's constant ($6.626\times 10^{-34}$ Js), c is the speed of light in vacuum ($3\times10^8$ m/s), and k is the Boltzmann constant ($1.38\times10^{-23}$ J/K). In the case of solar energy harvesting, the temperature T is the temperature of the sun surface (Gallo *et al.*, 2012).

## 2.3 History of Optical Metallic Nano-Antennas

Although the idea of harvesting solar energy using nano-rectennas was first introduced four decades ago, significant research with fabrication technologies have been conducted recently. The idea was originally proposed by Bailey (1972) and the first patent on solar rectenna was issued to Marks (1984).

In 1996, Lin *et al.* (1996) reported the first empirical design for the optical nano antenna that operates in the visible light. The design comprises dipole antenna array and a p-type doped boron single crystal wafer with 3 nm oxide layer for rectification. The power of Lin's results was owing to its validation for electric rectification and light resonance in the visible frequencies. After four years, Kottman *et al.* (2001) introduced and simulated the idea of using irregular nano-particles to maximize the incident field enhancement compared to the field enhancement due to a regular one. Kottman (2001) utilized the finite element method (FEM) to validate his idea and showed that an irregular triangle can enhance the incident field by five orders of magnitude at resonance wavelength of 470 nm compared to four orders of magnitude using an equilateral triangle of side length 27nm at resonance wavelength of 329nm. In 2005 Sundaranurthy *et al.* (2005) of Stanford University were able to fabricate two opposing gold 60° bowties with only 16nm tip to tip separation and operating at optical regime. The thickness, length, and radius of curvature of the bowtie were taken as 20nm, 88nm and 12nm, respectively. The antenna was fabricated from Au and mounted on 4nm adhesion layer of chromium. Sundaranurthy used Indium tin oxide layer over a fused silica substrate to improve lithography using electron beam. Sundaranurthy (2005) utilized the finite difference time domain (FDTD) analysis to detect electric field enhancement around the bowtie tip. For a linearly polarized incident plane wave with substrate illumination, a factor of enhancement of 1654 at 856nm for 16nm gap was obtained. The power of this study was the ability to visualize the current distribution along the antenna surface and the current behavior at the gap. The numerical analysis showed that there is a strong displacement current flowing through the gap which is highly uniform and continuous with the displacement current on the antenna's surface. The displacement current and hence the electric field enhancement are maximized as we get closer to the tip. One major problem with Sundaranurthy device was the imperfect fabrication of the nano antenna, especially the sharpness of the bowtie's edges. In 2009 McMahon *et.al* proposed the nano-antenna design of two gold dimers with a diameter of 90 nm. McMahon *et.al* (2009) used the FEM to study the gap effect between the gold dimers on the electric field intensity enhancement. The numerical analysis show that, gap of 5nm, 1nm, 0.5nm and 0.25 nm enhanced the field by 2, 3, 3 and 5 orders of magnitude, respectively with a resonance wavelength of 550nm, 600nm, 700nm and 900nm, respectively. Further, McMahon *et.al* (2010) reported the designs of cylindrical nano-wires and bowtie antennas using nonlocal dielectric constants. McMahon showed that taking non-local dielectric constant into consideration, will result in the decline of the near field intensity enhancement by one order of magnitude compared to using local dielectric functions. Further, Kotter *et al.* (2010) designed and fabricated a spiral nano-antenna for solar energy harvesting at mid-infrared region .In 2011, Midrio *et al.* (2011) have designed and analyzed mono pole antenna based on nickel for the reception of thermal radiation. Moreover, a trial to increase the frequency of operation of the 60° bowtie nano-antenna to IR regime was in the beginning 2012 by Chen *et al.* (2012). Chen fabricated a 60° bowtie nano-antenna with 200 nm arm length and a gap of 30nm with carbon nano tube photo detector. The 60° bowtie resulted in a relative intensity enhancement of 220 times at 0.77 μm. In August 2012, Thorsten Feichtner (2012) used the evolutionary algorithm along with FDTD numerical analysis to optimize the dimensions of hybrid design of split ring and two wires for improving the field enhancement at the hot spot. The hybrid design of Feichtner resulted in relative enhancement of 3500 at the middle of 10 nm gap. Most of the reported nano antennas are based on detecting the field enhancement in the gap by several techniques. On top of that, Vandenbosch *et al.* (2012) introduced upper bounds for the solar energy harvesting efficiency of nano-antennas. This efficiency was investigated in terms of the dimensions of the nano-antenna and the metal (gold, silver, copper, aluminum, and chromium) used as a conductor. These results set upper bounds for any possible process transforming the light into electrical energy. In this regard the silver exhibited the highest efficiencies, both in free space and on top of a glass ($SiO_2$) substrate, with radiation efficiencies near or slightly above 90% (Vandenbosch *et al.*, 2012).Recently in 2014, Gadella *et al.* (2014) have proposed a THz bowtie antenna integrated with maximum near field intensity enhancement at the infrared energy regime (Gadella *et al.*, 2014). In addition, Gadella *et al.* (2014) concentrated on designing and fabricating THz copper oxide rectifier with high AC to DC conversion efficiency. The thin film diode offers low zero bias resistance of 500 Ω, thus achieving a good impedance match between the antenna and the rectifier.

## 2.4 Novel Design of Flower Shape Dipole Nano-Antenna

In this section, a novel design of nano-antenna for energy harvesting is proposed and analyzed using 3D FDTD. The new design consists of three elements nano-antenna with elliptical shape and with air gap. The numerical simulations are investigated for improving the harvesting efficiency of the nano-antennas within the wavelength range from 400 nm to 1400 nm. The suggested design has high efficiency of 74.6% at 500 nm where the irradiance of the sun is maximum. The proposed nano-antenna shows an improvement in the harvesting and the total harvesting efficiency, over the conventional dipole antenna by 15% and 32.7%, respectively.

### 2.4.1 Simulation Results

In order to validate the simulation results, initially the conventional dipole antenna (Vandenbosch *et al.*, 2012) shown in Figure 8(a) has been considered. The conventional dipole antenna (Vandenbosch *et al.*, 2012) consists of two elements of equal length L, and height H separated by a distance G. In this investigation, the two elements are made from gold with dipole length L=250 nm, width W=40 nm, height H= 40 nm, and gap G= 10 nm. In this evaluation, the radiation efficiency calculated by the FDTD method is compared to that of the method of moment (MOM) technique (Vandenbosch *et al.*, 2012). Figure 8(b) shows the variation of the wavelength dependent radiation efficiency using MOM techniques (Vandenbosch *et al.*, 2012) and FDTD method at different mesh grid size $\Delta x = \Delta y = \Delta z = \Delta$. It is revealed from this figure that a good agreement between MOM and FDTD results at $\Delta$=10 nm. Therefore, mesh grid size $\Delta$ will be fixed to 10 nm in all simulations throughout this study. Figures 8(c) and (d) show the scattering and the absorbing fields along XY plane of the conventional dipole using air substrate. It can be noted from the figures that the fields are concentrated in $E_z$ component and this indicates that the antenna can harvest linearly polarized light. Figure 8(e) shows the induced current distribution along the metal surface of the conventional dipole

It may be noted form figure 8(c) that, the majority of the field is located along the edge. Therefore, the conventional rectangular dipole design is modified in this study to elliptical shape with elliptical air gap. In addition, the dimension of the air gap is equal to the region of minimum field distribution as shown in Figure9 (a). Figure 9(b) shows the top view of the proposed structure. Figures 9(c) and (d) show the scattering and the absorbing fields along XY plane of the proposed elliptical dipole using air substrate.

The reported design as shown in Figure 9(a) (Hussein *et al.*, 2014 a) consists of two elliptical shape elements with air gap separated by a distance G. Each element of height H and length L has minor and major radii $a_1$, and $b_1$, respectively. However, the elliptical air gap has major and minor radii $a_2$ and $b_2$, respectively. The structure geometrical parameters can affect the harvesting efficiency (Vandenbosch *et al.*, 2012) of the suggested design. Therefore, the effect of the antenna dimensions and gap size on the performance of the reported nano-antenna is investigated thoroughly.

The impact of the major $a_2$ and minor $b_2$ radii of the air gap is first investigated. However the other parameters are fixed at G = 10 nm, H= 40 nm. In addition, the major $a_1$ and minor axes $b_1$ of the metals are taken as 125 nm, 40 nm, respectively. First the effect of the minor radius of the air gap is studied

*Figure 8. (a) Conventional dipole structure (Vandenbosch et al., 2012), (b) variations of the wavelength dependent radiation efficiency using MOM (Vandenbosch et al., 2012) and FDTD techniques at different mesh grid size$\Delta$, (c) Scattered Ez field magnitude along XY plane (d) Absorbed Ez field magnitude along XY plane (e) Current distribution along XY plane. (©2014, IET. Used with permission.)*

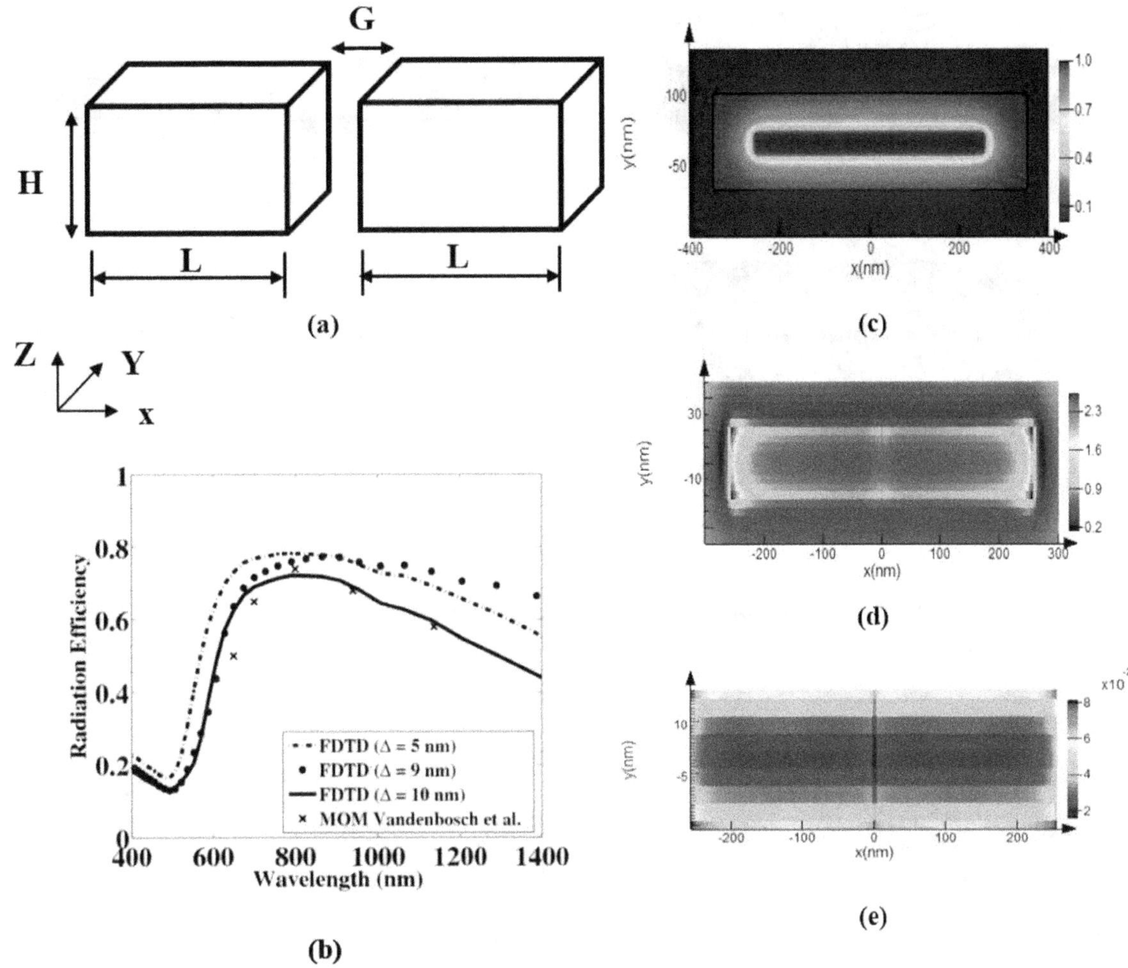


while the major radius is fixed at 70 nm. Figure 10 (a) shows the wavelength dependent radiation efficiency at different radii of the minor radius of the air gap. It is evident from this figure that the optimum value for the minor radius $b_2$ is equal to 20 nm which corresponds to the highest efficiency and broadest bandwidth. Next the effect of the major radius $a_2$ of the air gap is reported while the minor radius $b_2$ is taken as 20 nm. The numerical results reveal that the behavior is almost the same for different major radii of the air gap as shown in Figure 10(b). In addition, maximum harvesting efficiency is obtained at $a_2$= 80 nm, with a slight change in the bandwidth of the proposed antenna.

In microwave regime, the electrical length should be equal to an integer multiple of half the wavelength in order to obtain the resonance case which is not the case with nano-antennas. In this regard, the resonance condition is affected by the length L of each element, radii of the metal ellipse $a_1$ and $b_1$ and the air gap radii $a_2$, and $b_2$. Figure 11 shows the wavelength dependent radiation efficiency at different

*Figure 9. The proposed elliptical dipole nano-antenna, (b) Top view of the proposed elliptical dipole nano-antenna (c) Scattered Ez field component along XY plane (d) Absorbed Ez field component along XY plane (©2014, IET. Used with permission.)*

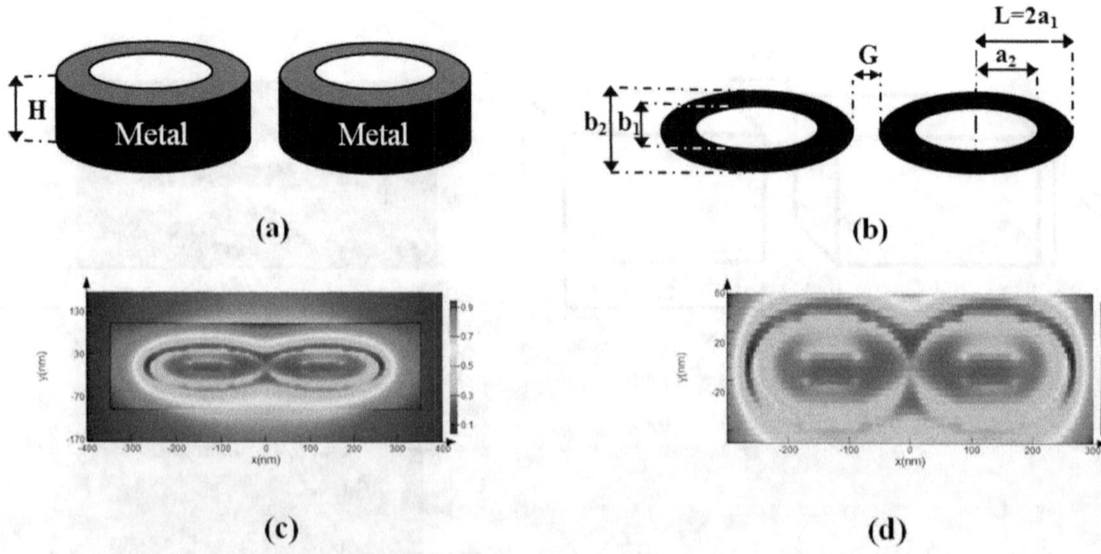


*Figure 10. Variation of the radiation efficiency of the proposed elliptical dipole with the wavelength at different (a) minor radii of the air gap b2, (b) major radii of the air gap a2 (©2014, IET. Used with permission.)*

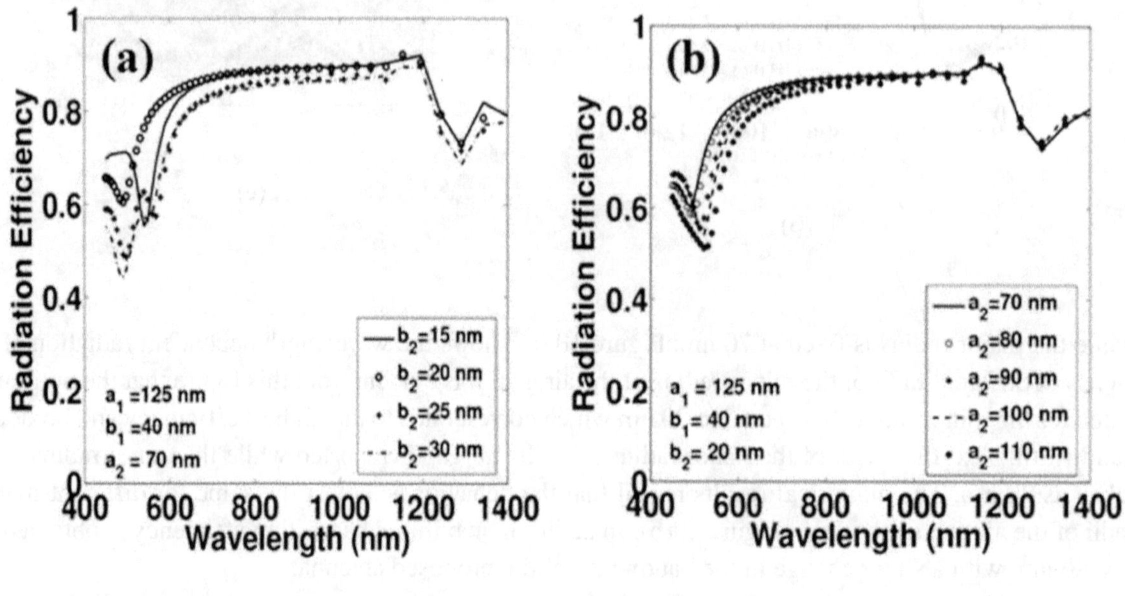

*Figure 11. Variation of the radiation efficiency of the proposed elliptical dipole with the wavelengths at different values of the length L of each element (©2014, IET. Used with permission.)*

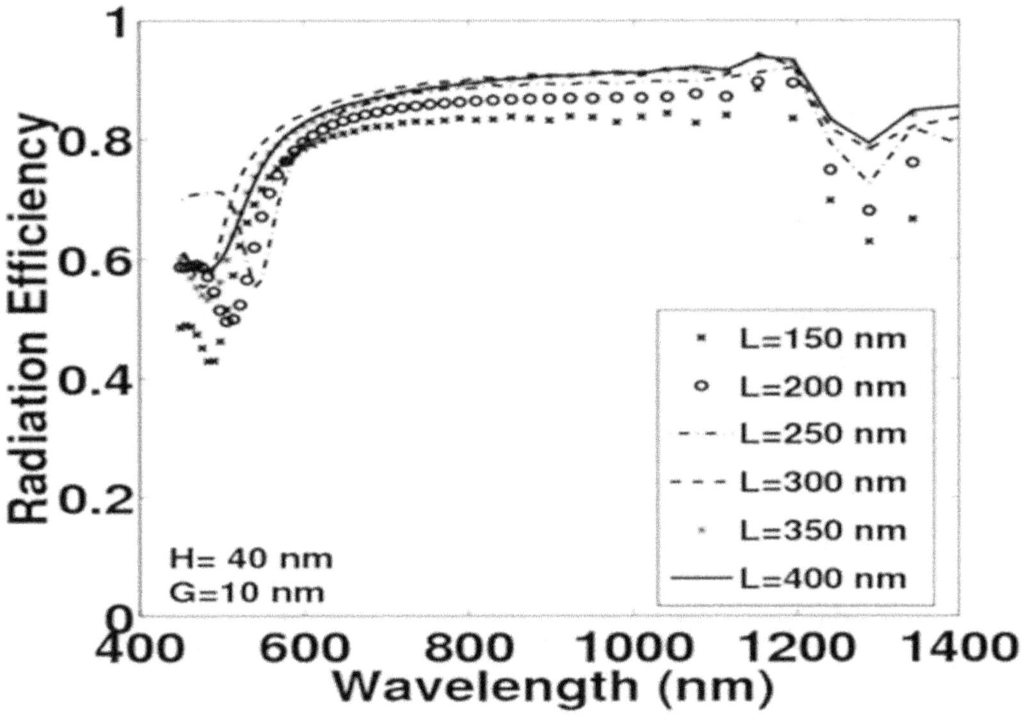


values of the length of each element L. It should be noted that for the proposed elliptical nano-antenna, the length L is equal to the major diameter $2a_1$. In this study, the height H and gap G are fixed to 40 nm, 10 nm, respectively. In addition, different cases are studied; L=150 nm ($a_1$= 75 nm, $b_1$=24 nm, $a_2$= 42nm, $b_2$= 9 nm), L=200 nm ($a_1$= 100 nm, $b_1$=32 nm, $a_2$= 56 nm, $b_2$= 12 nm, L=250 nm ($a_1$= 125 nm $b_1$=40 nm, $a_2$= 70 nm, $b_2$= 15 nm), L=300 nm ($a_1$= 150 nm $b_1$=48 nm, $a_2$= 100 nm, $b_2$= 18 nm), L =350 nm ($a_1$= 175 nm $b_1$=56 nm, $a_2$= 133 nm, $b_2$= 21 nm) and L=400 nm ($a_1$= 200 nm, $b_1$=64 nm, $a_2$= 177 nm, $b_2$= 24 nm). It is found that the harvesting efficiency and bandwidth increase with increasing the length of each element L as shown in Figure11.

Next, the effect of the height H on the nano-antenna performance is studied. In this evaluation, the height of the antenna element is changed from 40 to 80 nm with a step of 10 nm. However, the other parameters are fixed at $a_1$=125 nm, $b_1$= 40 nm, G=10 nm, $a_2$=100 nm, and $b_2$=20 nm. Figure 12 shows the wavelength dependent radiation efficiency at different H values. It is evident from this figure that the radiation efficiency increases with increasing the height H up to certain limit where the absorption losses will cancel the enhancement of the increase of the density wave of electrons. In addition, the harvesting efficiency shows an effective improvement by increasing the antenna height.

The impact of the gap distance G between the two elliptical elements is also reported while the other parameters are fixed at H= 40 nm, $a_1$=125 nm, $b_1$= 40 nm, $a_2$=100 nm, and $b_2$=20 nm. Figure 13 shows the variation of radiation efficiency of the proposed elliptical dipole nan-antenna with the wavelengths at different gap distances G. It is evident from this figure that the radiation efficiency decreases with increasing the gap distance between the two elements.

*Figure 12. Variation of the radiation efficiency of the proposed elliptical dipole with the wavelengths at different dipole height H values (©2014, IET. Used with permission.)*

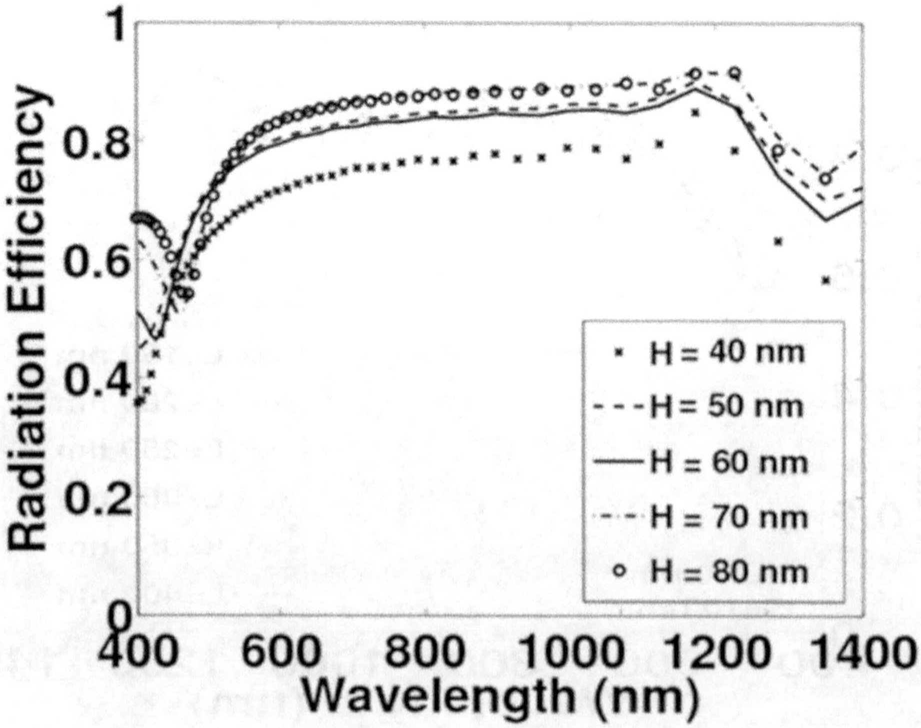


*Figure 13. Variation of the radiation efficiency of the proposed elliptical dipole with the wavelengths at different gap distances (©2014, IET. Used with permission.)*

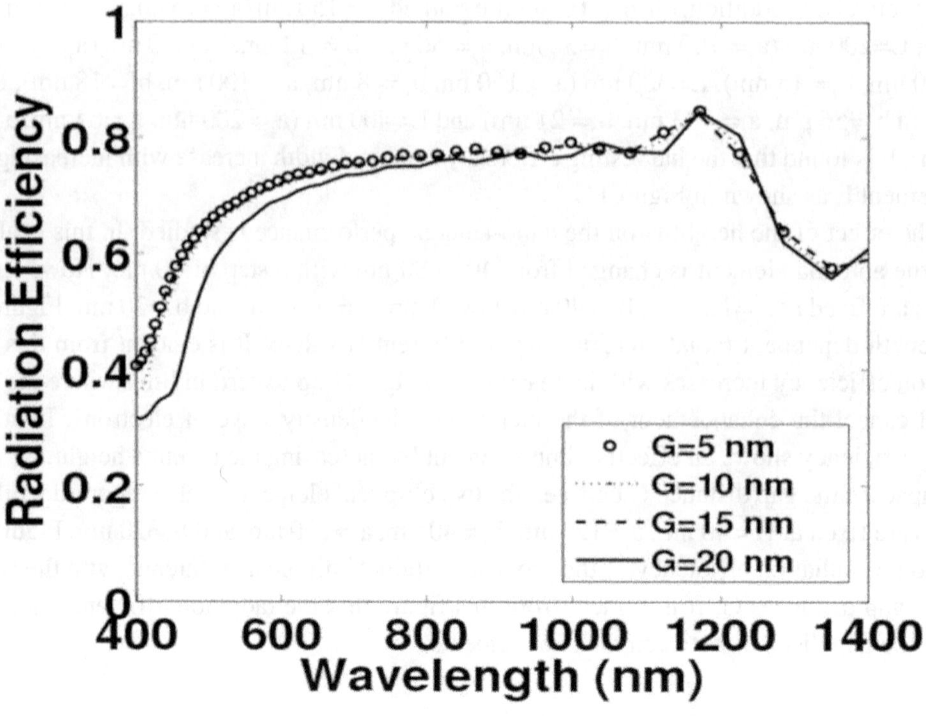

A comparison between the traditional dipole nano-antenna and the proposed elliptical dipole nano-antenna with air gap, in terms of the radiation efficiency and total harvesting efficiency, is also demonstrated. In this study the dimensions of the traditional dipole nano antennas are taken as W=40 nm, H=40 nm and L=250 nm. However, the dimension of the proposed design are fixed to $a_1$=125 nm, $b_1$= 40 nm, $a_2$= 100 nm $b_2$= 20 nm and H=40 nm. Figure14 shows the variation of radiation efficiency with the wavelength for the conventional dipole and the proposed elliptical shape nano-antenna. It is evident from this figure that the harvesting efficiency of the suggested elliptical nano-antenna is greater than that of the conventional dipole antenna. In addition, the investigated nano-antenna has harvesting efficiency of 74.6% at 500 nm where the maximum irradiance of the sun. Moreover, the obtained harvesting efficiency is more than twice that of the traditional crystalline silicon wafer based solar cells. This is due to the metal thickness and the air gap that acts as a cavity which harvests more energy. Therefore, the proposed structure, improves the radiation efficiency and the total harvesting efficiency by 10% and 17.7%, respectively. In addition, the new design reduces the size of the metal used in the nano-antenna fabrication. Moreover, the suggested antenna has a bandwidth from 450 nm to 1200 nm with harvesting efficiency of approximately 80%.

Next elliptical nano-antenna with two elliptical shapes in each element is considered. In this type, each element consists of two ellipses with the same size and with rotation angle θ between their major axes as shown in Figure 15 (a). In order to detect the effect of the rotation angle θ on the antenna perfor-

*Figure 14. Variation of the radiation efficiency with the wavelength for the conventional dipole (Vandenbch et al., 2012) and the proposed elliptical nano- antenna (©2014, IET. Used with permission.)*

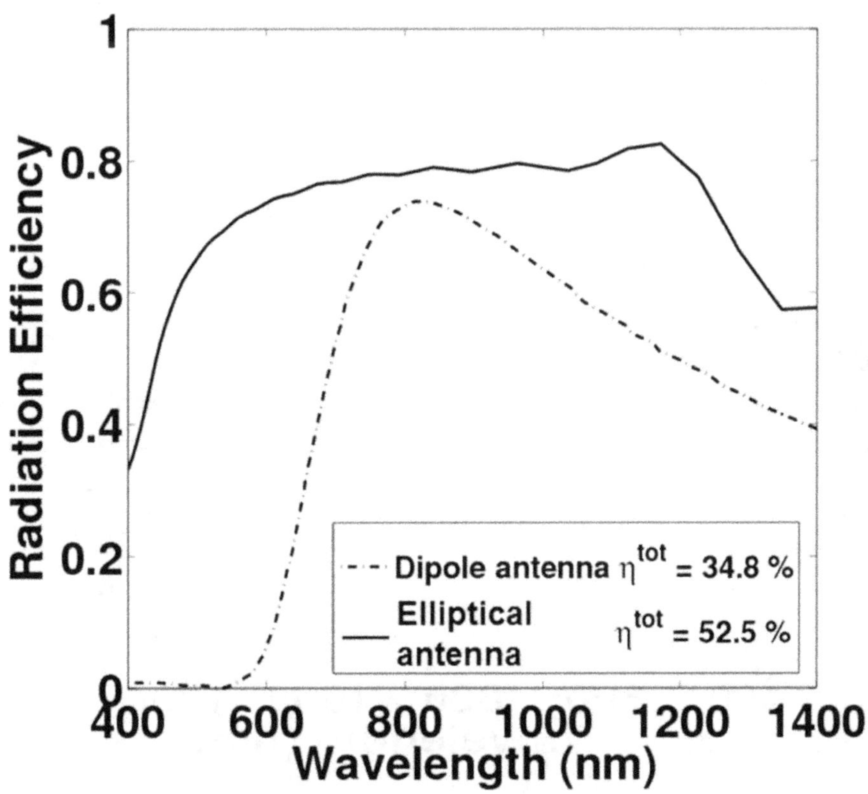

*Figure 15. Schematic diagram of (a) elliptical dipole nano-elements with two ellipses in each element (b) flower shape dipole nano-antenna (©2014, IET. Used with permission.)*

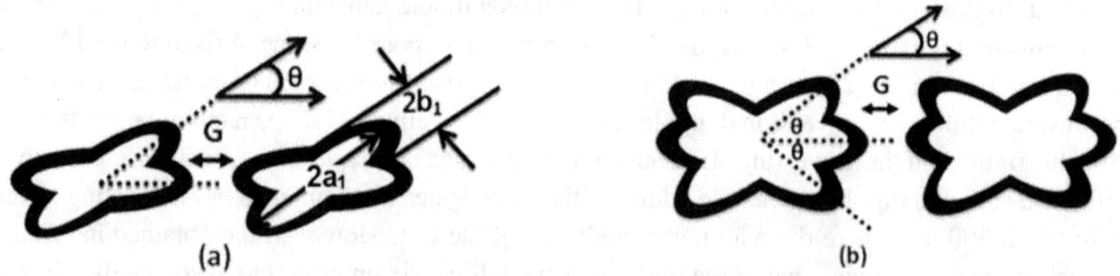


mance, the other parameters are fixed to $a_1$=125 nm, $b_1$= 40 nm, $a_2$= 100 nm, $b_2$=20 nm, G=10 nm and H=40 nm. Figure 16 shows the wavelength dependent radiation efficiency at different rotation angles, 5°, 10°, 15°, 20°, 30° and 45°. It is found that the radiation efficiency is slightly affected by the rotation angle θ as shown in Figure 16.To increase the capability of energy harvesting, a flower shaped nano-antenna, shown in Figure15 (b), is suggested. The proposed flower nano-antenna dipole is composed of three ellipses in each element, with rotation angle θ=10° to harvest energy within the wavelength range from 400 nm to 1400 nm. Figure 17 shows the calculated radiation efficiency of the conventional dipole antenna (Vandenbosch *et al.*, 2012) and proposed elliptical dipole antennas with one ellipse (type one),

*Figure 16. Variation of the radiation efficiency of the proposed two elliptical elements dipole antenna with the wavelengths at different rotation angle between two ellipses (©2014, IET. Used with permission.)*

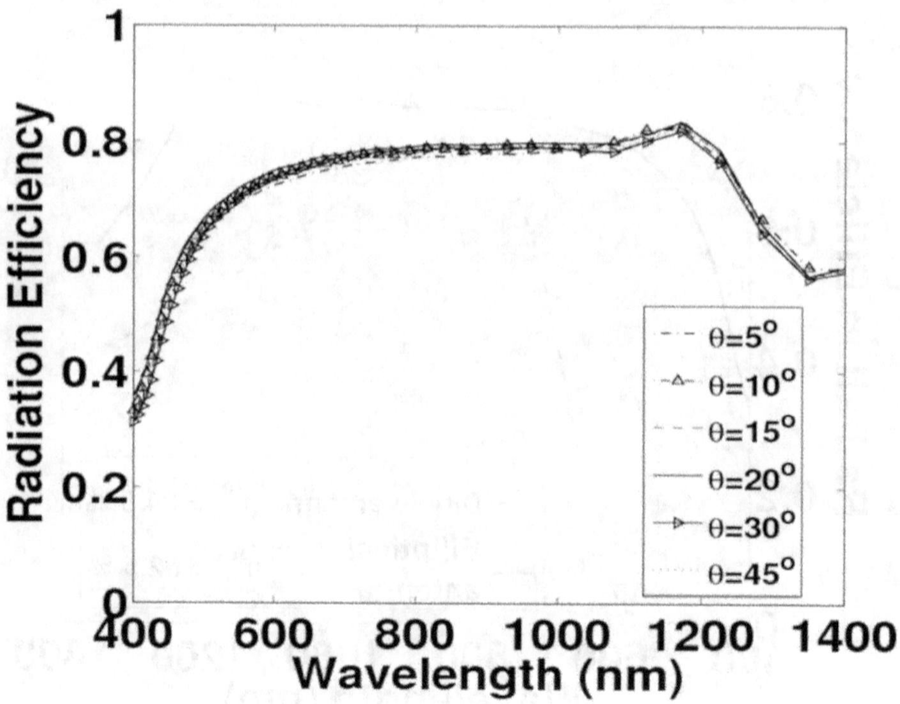

*Figure 17. Variation of the radiation of the conventional dipole nano-antenna (Vandenbosch et al., 2012) and the proposed elliptical nano-antennas (©2014, IET. Used with permission.)*

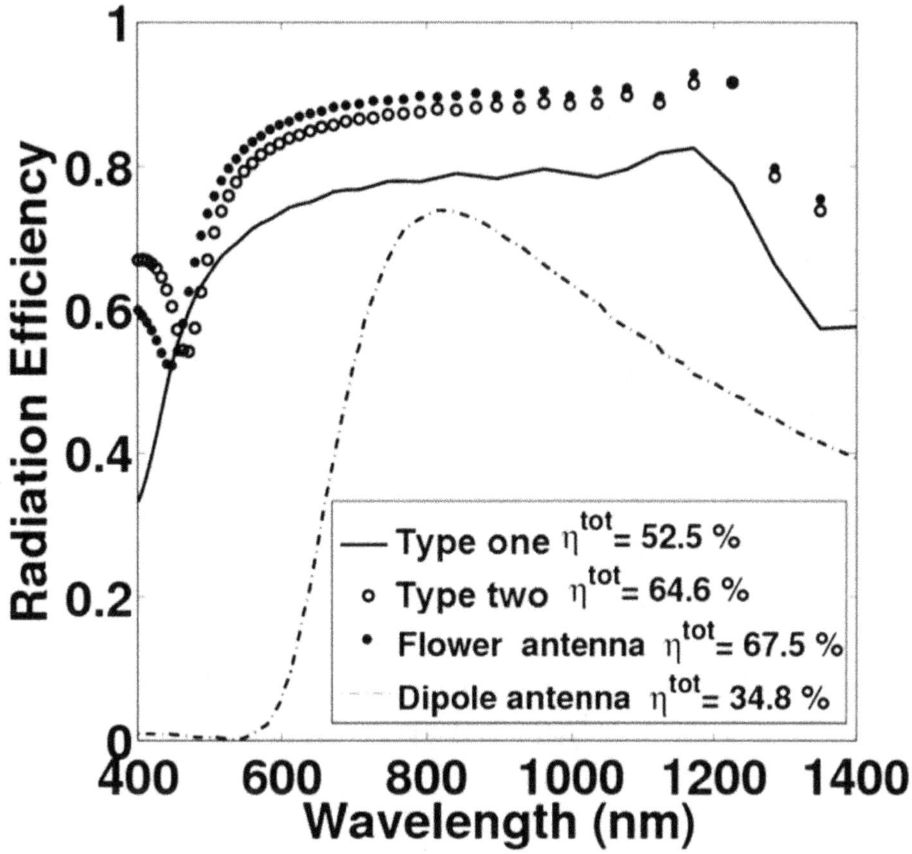


two ellipses (type two) and the three ellipses (flower shaped) in each element. In this investigation, the dimensions of the traditional dipole nano-antennas are taken as W=40 nm, H=40 nm and L=250 nm. However, the dimension of the proposed designs are $\theta=10^{o}$, H= 40 nm, G=10 nm, $a_1$=125 nm, $b_1$= 40 nm, $a_2$=100 nm, and $b_2$=20 nm. It is evident from this figure that the flower shaped dipole shows an increase in the radiation efficiency and the total harvesting efficiency by 15% and 32.7%, respectively over the conventional dipole antenna. These results considerably agree with the upper bound declared by Vadanbosch (Vandenbosch *et al.*, 2012) for nano-antennas.

In an attempt to find a metal that is more abundant and hence considerably less expensive to employ as antenna, different metals are tested such as Cr, Al, Au and Ag. Figure 18 shows the wavelength dependent radiation efficiency of the suggested flower shape dipole at different metals. In this study the other parameters are fixed to $\theta=10^{o}$, H= 40 nm, G=10 nm, $a_1$=125 nm, $b_1$= 40 nm, $a_2$=100 nm, and $b_2$=20 nm. It is evident from Figure18 that the radiation efficiency depends essentially on the metal conductivity. In this regards, the silver possesses the highest efficiency from 450 nm to 800 nm; whereas gold has obvious high efficiency from 800 nm to 1200 nm. This is in a good agreement with the results for dipole antenna studied by Vadanbosch (Vandenbosch *et al.*, 2012). It is also evident form Figure18 that metals with high electrical conductivity such as gold and silver have much better performance than metals with low electrical conductivity such as Al and Cr in the wavelength range of interest.

*Figure 18. Variation of wavelength dependent radiation efficiency of the proposed flower shape dipole at different metals (©2014, IET. Used with permission.)*

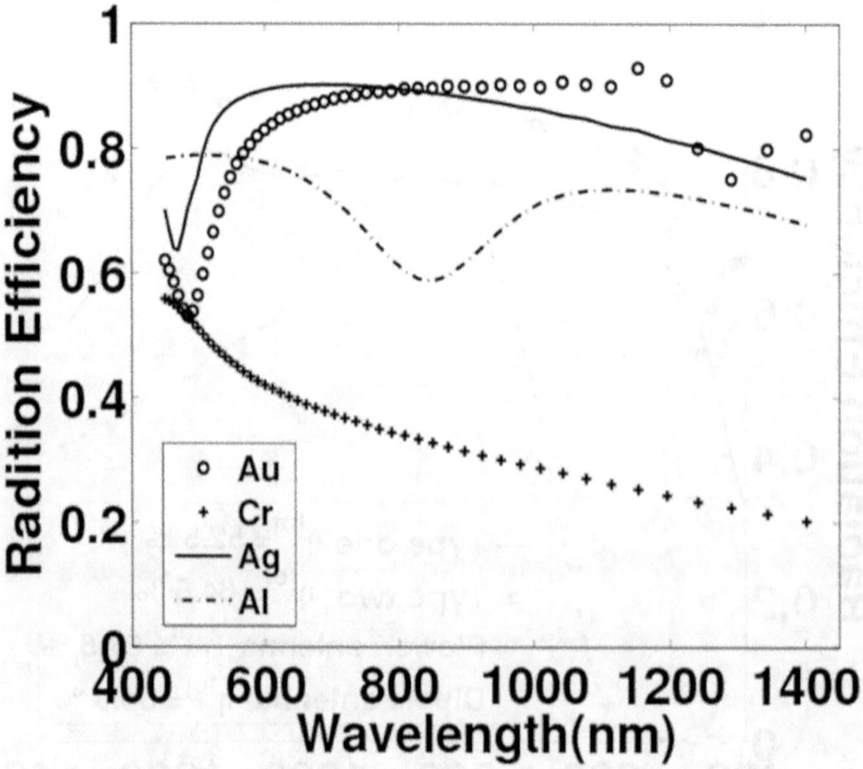


In recent practical applications, especially in biophysics and sensors, it is essential to take into account the influence of the surrounding medium on the optical properties of the nanoparticles (Biagioni *et al.*, 2012). As the proposed design is intended for different applications, the dependence of energy harvesting on the refractive index of the surrounding medium is essential in this scenario. Figure 19 (a) shows variation of the wavelength dependent radiation efficiency of the proposed flower shape dipole antenna and the conventional dipole antenna using different surrounding mediums, air ($n_m = 1$), water ($n_m = 1.33$) and fused silica ($n_m = 1.46$). However, the other parameters are kept constant at $\theta$=10° H= 40 nm, G=10 nm, $a_1$=125 nm, $b_1$= 40 nm, $a_2$= 100 nm, and $b_2$=20 nm. In addition, the conventional dipole parameters are taken as W=40 nm, H=40 nm and L=250 nm. It is revealed from Figure19 that the radiation efficiency for the flower shaped dipole antenna decreases with increasing the refractive index of the surrounding medium.

Figure 19 (a) also shows that the index of the surrounding medium has slight effect on the radiation efficiency of the suggested nano-antenna at higher wavelengths. In addition, the behavior of the radiation efficiency for the suggested flower shaped nano-antenna follows the same trend as the conventional dipole antenna. At 500 nm, the radiation efficiency of the flower shaped antenna surrounded by air ($n_m$=1) shows 40% increase over the conventional dipole antenna. Additionally, the suggested nano-antenna can harvest a linearly polarized light over a broad bandwidth extending from 500 nm to 1200 nm. It is also evident from Figure 19(a) that 15% and 20% improvements in the radiation efficiency of the proposed

*Figure 19. (a) Variation of wavelength dependent radiation efficiency of the proposed flower shape dipole and conventional dipole antenna at different surrounding mediums, (b) Absorbed electric field $E_z$ along XY plane for flower shape antenna using air substrate, (c) Absorbed electric field $E_z$ along XY plane for flower shape antenna using fused silica substrate, (©2014, IET. Used with permission).*

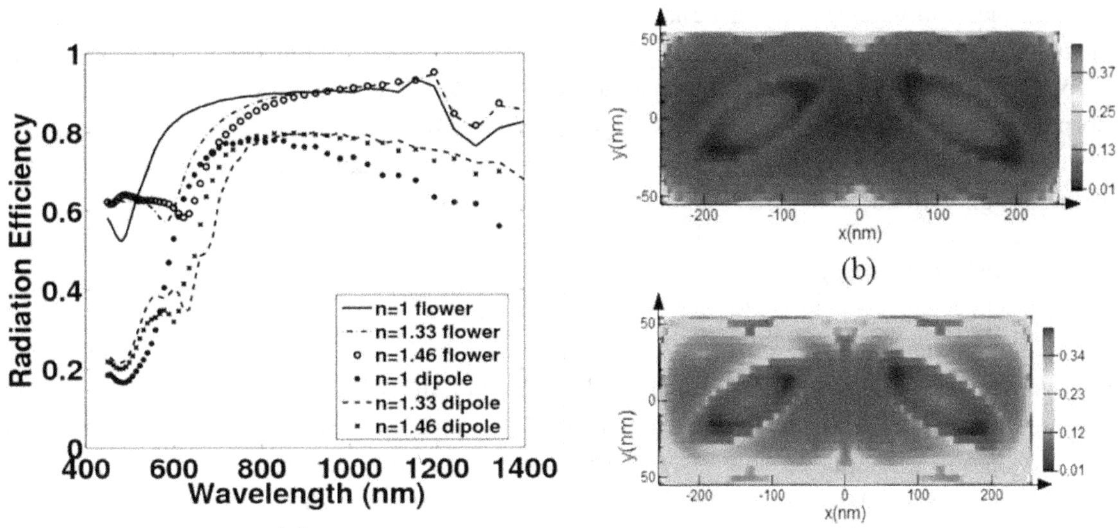


(a) (c)

nano-antenna are obtained with surrounding mediums of refractive indices 1.33 and 1.46, respectively in the wavelength range from 800 to 1200 nm, over the conventional dipole antenna. Figures 19 (b) and (c) show the captured absorbed fields with air and fused silica mediums at 550 nm, respectively. It can be observed from these figures that use of fused silica substrate induces surface waves which minimize the radiation efficiency by about 20% compared with air substrate.

## 3. SEMICONDUCTOR NANO-ANTENNA

### 3.1 Introduction

Recently, there is a dramatic increase in energy supply requirements facing humanity with negative environmental impacts of fossil fuel combustion. Therefore, the search for renewable energy and energy efficiency technologies become mandatory. Among the available renewable energy sources, the solar cell attracts the interest of a lot of researchers. However, the implementation of solar energy technologies suffers from their high costs and resources availability. As a result, conversion and storage of solar energy with high efficiency and low cost have recently gained tremendous interest.

In the previous sections, the efficiency of the solar cell has been improved by using rectenna. In addition, the metal is used instead of semiconductor with its bandgap constrains. However, the rectenna depends on diode in the THz range which is difficult for fabrication which affects the total performance of the rectenna. Therefore, nano-structured devices have been used in recent years instead of bulk materials due to significant improvements in fabrication techniques for nano-materials. The semiconductor nano-antennas (NAs) are of particular interest in solar energy conversion devices. The semiconductor NAs can decrease the reflection and improve light scattering and trapping with less material than bulk and pure structures. This will reduce the fabrication cost of the solar cells with abundant and low-cost elements. The nano-wires (NWs) can be arranged in different configuration like periodic photonic crystal structure to increase the overall efficiency. The Light absorption in standing semiconductor nanowires is a complex phenomenon, with strong dependence on nano-wire dimensions, lattice arrangement and absorption coefficient of the raw materials (Hong *et al.*, 2013).

In this section, the semiconductor nano-antennas are studied thoroughly in terms of photonic crystal, antenna design with different patterns, nano-wire forms and materials.

## 3.2 Photonic Crystals

The photonic crystals (PCs) (Russell, 2003) are periodically structured electromagnetic media. There are one-, two-, or three-dimensional PC structures depending on the periodicity of the dielectric materials as shown in Figure 20. The one dimensional (1D) PC has a periodic modulation of the permittivity in one direction only while the structure is uniform in the other two directions as shown in Figure 20(a). The Bragg grating is an example of the 1D PC that is widely used as distributed reflector in vertical cavity surface emitting lasers. In addition, the Bragg grating PC can be used as anti-reflecting coatings to decrease the reflectance from the surface and hence improve the efficiency of lenses, prisms and other optical components.

The two dimensional (2D) periodic PCs as shown in Figure 20(b) are periodic along two directions. Therefore, the 2D PC can have more design flexibility than 1D PC due to the periodicity of the permittivity along two directions, with uniform medium in the third direction. The 2D PCs can be made by drilling or etching holes in a dielectric material which is very promising for integrated photonic circuits. In addition, the photonic crystal fiber (PCF), also known as holey fiber (HF) is an example of the use of 2D PC structures.

The 3D PC has a periodicity along all three directions. Therefore, the 3D PC has more flexibility than 1D and 2D PCs to confine the light in any direction. In addition, the number of possible PC configura-

*Figure 20. Photonic crystal types (a) 1D, (b) 2D, and (c) 3D (Joannopoulos et al., 2008)*

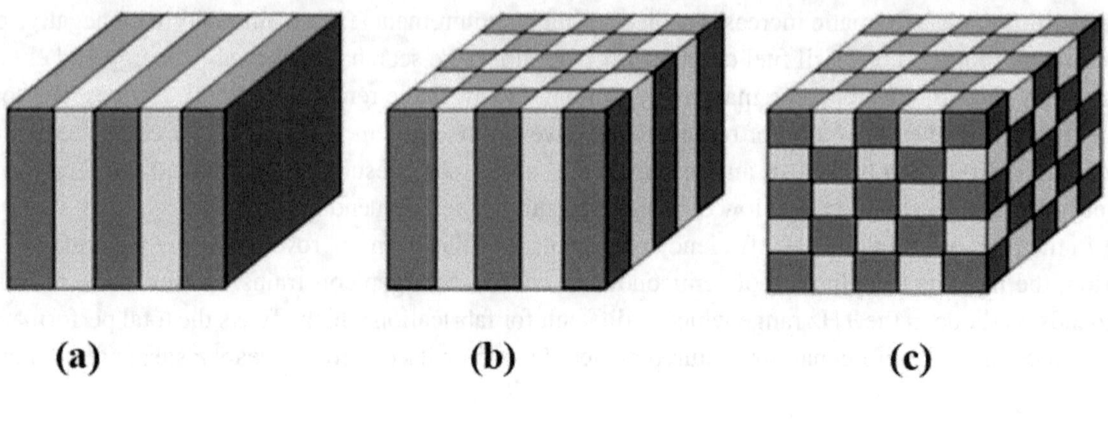

tions is much larger than that of 1D or 2D PC. However, the 3D PC fabrication is complex and more costly. Consequently, the 2D-periodic PCs are the more widely used structures. In this study, the 2D planar PC structures are used to increase the ultimate efficiency of the semiconductor nano-antennas.

The periodic PCs rely on the photonic band-gap (PBG) defined as ranges of frequency in which light cannot propagate through the structure. Therefore, certain frequencies can pass through the structure, while a select frequency is reflected. The PCs have the ability to confine the light effectively. Therefore, the planar PCs can be used to create all-photonic integrated circuits (ICs). The frequencies within the PBG cannot propagate through infinite PC structure. Therefore, a defect in the periodic structure can be made to confine the forbidden frequency through the defect. The defect can be represented by missing of some elements or group of elements of the periodic structure, shift or variation of the PC parameters. This property can be used for a wide range of applications such as PC waveguides, micro resonators, spontaneous radiation management, optical insulators, nonlinear elements and PCFs. For example, the micro resonators operation depends on the localization of the radiation inside the defect area of the periodic structure. In addition, the PC waveguides can be obtained by removing a raw of the PCF structure which is called line defect. Such line defect possesses wave guiding properties within quite a wide range of wavelengths.

It should be noted that the conductivity of the semiconductor NWAs based solar cells is greater than that of bulk silicon solar cells. This is due to surface structure defects that produce reflection and scattering through the NWAs and hence improve the absorption efficiency.

## 3.3 Latest Designs of Semiconductor Nano-Antenna

The current crystalline Si cells (Nelson, 2003) are widely used in the Photo voltaic (PV) market with efficiency of around 25%. However, they are expensive due to the high material costs. Therefore, the use of cheaper materials such as polycrystalline thin films or organic semiconductors can decrease the material cost. In addition, the amount of required material to absorb the light can be decreased further by increasing the concentration of the light through a small area. Moreover, a dramatic improvement in the fabrication techniques can further minimize loss mechanism and hence increase the solar cell efficiency. The use of semiconductor nano-antennas (NAs) or nano-wires (NWs) can fulfill these requirements. In this regard, the use of nanowire solar cells with structures of diameter of a few nanometers (Tsakalakos, 2008; Kandala *et al,* 2008) has been considered for photovoltaic power generation. The semiconductor NWs has advantages in terms of optical absorption enhancement (Garnett &Yang 2010; Kelzenberg *et al.*, 2010; Tian, 2007; Hu & Chen, 2007). This is due to the small reflectivity of the NW array (NWA) with large open area on the frontal surface. In addition, the light is trapped by multiple total internal reflections (Cao *et al.*, 2010) through each nanoscale cylindrical resonator NW. Moreover, the NW diameter, period and wavelength have the same order. Therefore, strong light scattering and hence further light trapping (Muskens *et al.*, 2008; Bao *et al.*, 2010) occurs. The behavior of the semiconductor NWs depends mainly on the structure geometrical parameters of the NWA. Therefore, theoretical studies of optical absorption are needed to maximize the ultimate efficiency of the NWAs based solar cells. The NWAs with optimized parameters can offer higher broadband absorption than a solid film (Lin & Povinelli, 2009).

Initially, NWAs with uniform NW diameter and length have been studied and experimentally achieved (Lin *et al.*, 2009; Li *et al.*, 2009). In addition, NWAs with random orientation, length, diameter, and position as well have been experimentally investigated in (Muskens *et al.* 2008 ; Street *et al.* 2009).

In these studies (Muskens *et al.* 2008; Street *et al.*, 2009), reflection or backscattering is noticed with slight effects on absorption. However, significant absorption enhancement is achieved using random one-dimensional multilayer and two-dimensional cylinders due to photon localization in random media (Ruan & Kaviany, 2005; Ruan & Kaviany, 2006). In addition, the optical properties for vertically aligned silicon NWAs with three types of structural randomness, i.e., random position, diameter, and length are numerically investigated (Bao & Ruan, 2010). The NWAs with random position show slight absorption enhancement, while those with random diameter or length show significant absorption enhancement. This is due to the stronger optical scattering in a random structure. Therefore, the structural randomness in vertical nanowire arrays will not destroy but rather can further enhance optical absorption compared to ordered nanowire arrays.

The reflectance and absorption characteristics of slanting silicon NWAs have also been reported (Hong *et al.*, 2013) for its application in solar cell. The slanting NWAs can offer better suppression of light reflection and light harvesting ability than the vertical structure. Therefore, the slanted structures can further improve the performance of Si NWs based solar cells. The slanted NWAs offer an ultimate efficiency of 33.45% which is greater than 28.36% of optimum vertical Si NWs (Hong *et al.*, 2013). In addition, solar cell device based on the slanting Si NW structure has been successfully fabricated by Fang *et al.* (2008).

The optical properties of the solar cells based on large-lattice-constant structures should be studied carefully to optimize the solar cell parameters. Therefore, it is important to understand the dependence of the absorption enhancement (Chutinan & John 2008; Tumbleston *et al.*, 2009; Duché *et al.*, 2008; Cao *et al.*, 2009&Mallick *et al.*, 2010) on the optical guided modes. For this purpose, the guided resonance modes, also known as slow Bloch modes (Duché *et al.*, 2008), have been extensively studied in photonic-crystal slabs (Fan & Joannopoulos, 2002; Tikhodeev *et al.*, 2002; Prasad *et al.*, 2008). The guided resonance modes have high electromagnetic field intensity within the photonic-crystal slab. This is due to the coupling between the incidents light from air to the leaky modes of the two dimension (2D) periodic structure (Fan & Joannopoulos, 2002; Tikhodeev *et al.*, 2002 ; Prasad *et al.*, 2008). Therefore, the optical absorption will be strongly enhanced when the guided resonance modes are excited by incident sunlight.

The ordered Si NW structure can be fabricated using different methods such as lithography (Mein *et al.*, 2011), and polystyrene ball assembly (Huang *et al.*, 2007). However, these methods suffer from high cost and complexity. Recently, low cost maskless method (Huang *et al.*, 2011; Peng *et al.*, 2006) with electroless metal catalyst wet etching and metal nano-particles catalyst has been used for Si NW formulation. The etching direction of this approach is strongly dependent on the Si crystallography. In addition, the etching speed is much faster along the (100) orientation (Kuiqing *et al.*, 2007; Li, 2012; Ming-Liang *et al.*, 2008). Moreover, Si NWs with different slanting angles have been fabricated using silicon wafers with different crystallographic orientations. Furthermore, the slanting SiNWs can be fabricated by controlling the etching solution concentration (Huang *et al.*, 2010).

## 3.4 Novel Design of Decagonal SiNWs

The 3D FDTD is used to study and analyze the optical properties of a novel decagonal semiconductor nanowire arrays (Si NWAs) for solar cell applications. The new design consists of vertical Si semiconductor NWs arranged in a decagonal shape with a large Si NW embedded in the core region. Several optimized parameters such as the central core radius, nano pillar radius as well as the distance between adjacent Si NW pillars are obtained through schematic methodology. The numerical simulations are carried out for improving the ultimate efficiency of the nanowire within the wavelength range from 300 nm to 1100 nm. The proposed semiconductor NWs design exhibits an efficiency of 38%; thus enhancing the ultimate efficiency over the slanting Si NWAs by about 7%.

### 3.4.1 Design Consideration

Figure 21 (a) shows 3D view of vertically aligned two layers of SiNWs arranged on decagonal lattice (D-SiNWs) reported by (Hussein *et al.*, 2014b). Figure 21(b) and (c) illustrate both the schematic top view of the simulated SiNWs and the SiNW unit isosceles triangle, respectively. In Figure 21(b), the proposed SiNWs are arranged along the circumference of two circular layers. The isosceles cell of the proposed structure comprises five identical pillars with diameter d and core pillar of diameter ($2a$), as indicated in Figure 21(c).The radial distance between the two layers is $\Lambda$; while the distance between the SiNW centers along the same layer is equal to $2\Lambda sin(\Theta/2)$. The numbers of SiNWs in the first and second layers are equal to $360/\Theta=10$ and $2\times(360/\Theta)$ **=20**, respectively; where $\Theta$ is the vertex angle between the long arms of the isosceles triangle cell shown in Figure 21(c).

### 3.4.2 Simulation Methodology

In this study, the D-SiNWs are vertically aligned as shown in Figure 21(a). The D-SiNWs design is illuminated from the top by plane wave normal to x-y plane with the electric field polarized along the $x$-axis. The 3D-FDTD method is employed for investigating the optical properties of the SiNWs over the main part of the solar spectrum ranging from $\lambda$ = 300 nm to $\lambda$ = 1100 nm. The unit cell is bounded by periodic boundary conditions in $x$ and $y$ directions and perfect matched layers in $z$ direction. The refractive indices of the D-SiNWs are taken from the literature (Palik, 1997). The absorption is calculated as $A(E)=1-R(E)-T(E)$, where $R(E)$ and $T(E)$are energy-dependent reflections and transmission, respectively. The ultimate efficiency ( $\eta$ ) is used to evaluate the absorption performance of the decagonal SiNWs. The ultimate efficiency is calculated from the following equation

$$\eta = \frac{\int_{E_g}^{\infty} P(E) A(E) \frac{\mathrm{E}g}{\mathrm{E}} dE}{\int_{0}^{\infty} P(E) dE} \tag{11}$$

where $E_g$ is the bandgap of Si, $P(E)$ is the standard AM 1.5 direct and circumsolar intensity spectrum[20], $E$ is the energy of incident sun light and $A(E)$ is the absorption efficiency. In ultimate efficiency calculations, it is assumed that each photon absorbed with energy higher than $E_g$ produces one electron-hole pair and all generated carrier are collected without recombination losses. In this study, $Eg$ is equal to 1.12 ev which is the energy corresponding to bandgap wavelength 1127 nm of the Si. The integrals are evaluated using the trapezoid rule with 801 sampling point.

*Figure 21. (a) Schematic 3D diagram of the proposed D-SiNWs structure (b) top view for the proposed two layers SiNWs (c) top view for unit isosceles triangle SiNWs (©2014, Optical Engineering. Used with permission.)*

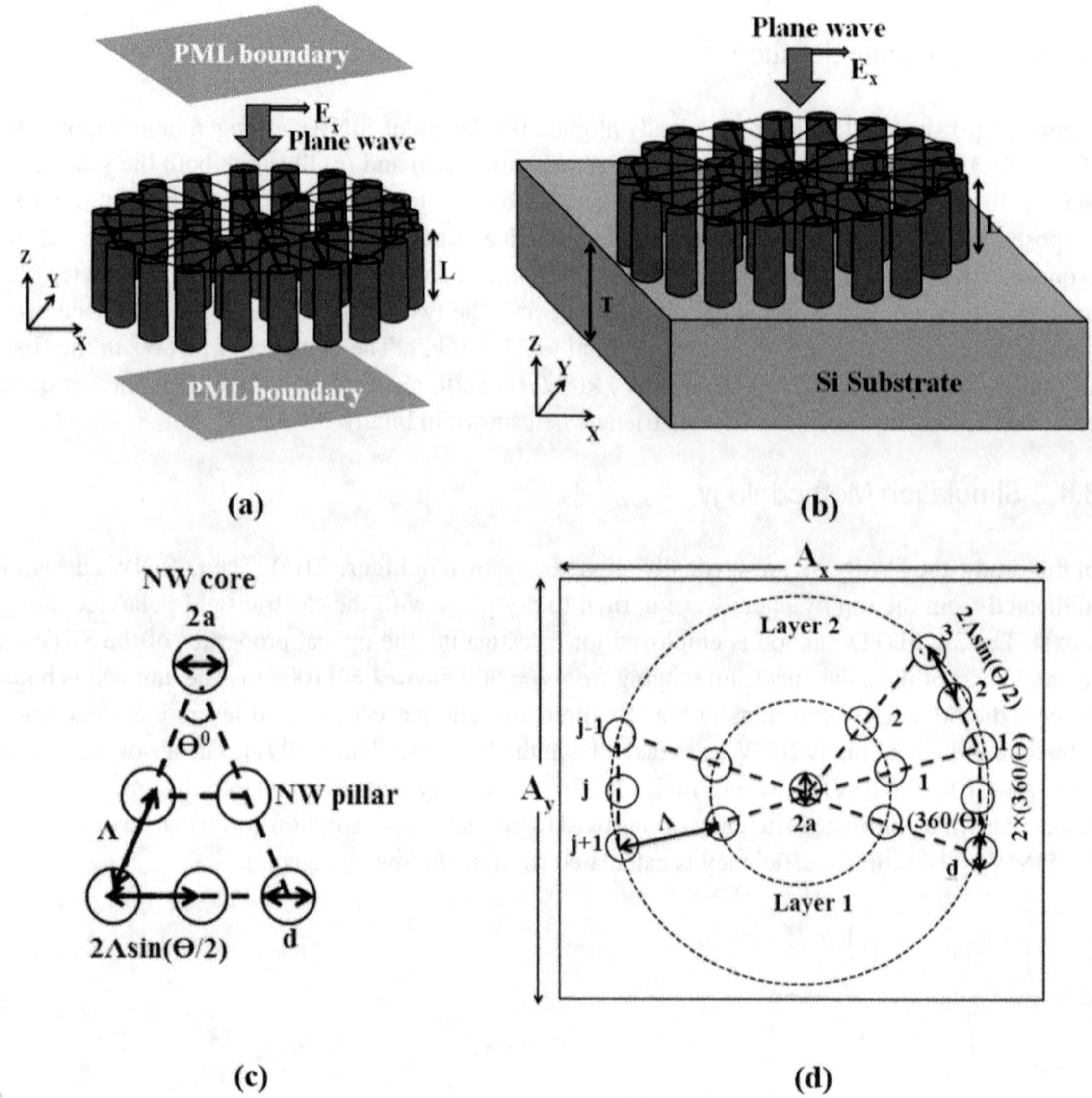


### 3.4.3 Simulation Results

In order to validate the simulation results, the square nanowire arrays are initially simulated using 3D-FDTD method and compared with published results carried out by Lu Hu *et al.*(2007). The Si NW has radius r, length L and periodicity P. In this investigation, r = 25 nm, L =2.33 μm nm and P = 100 nm. Figure 22 shows the variation of the absorption versus energy of the Si NWS; the results obtained by

*Figure 22. Variations of the calculated energy absorption spectrum of Si NWAs compared with TMM results published by (Hu & Chen 2007). (©2014, Optical Engineering. Used with permission.).*

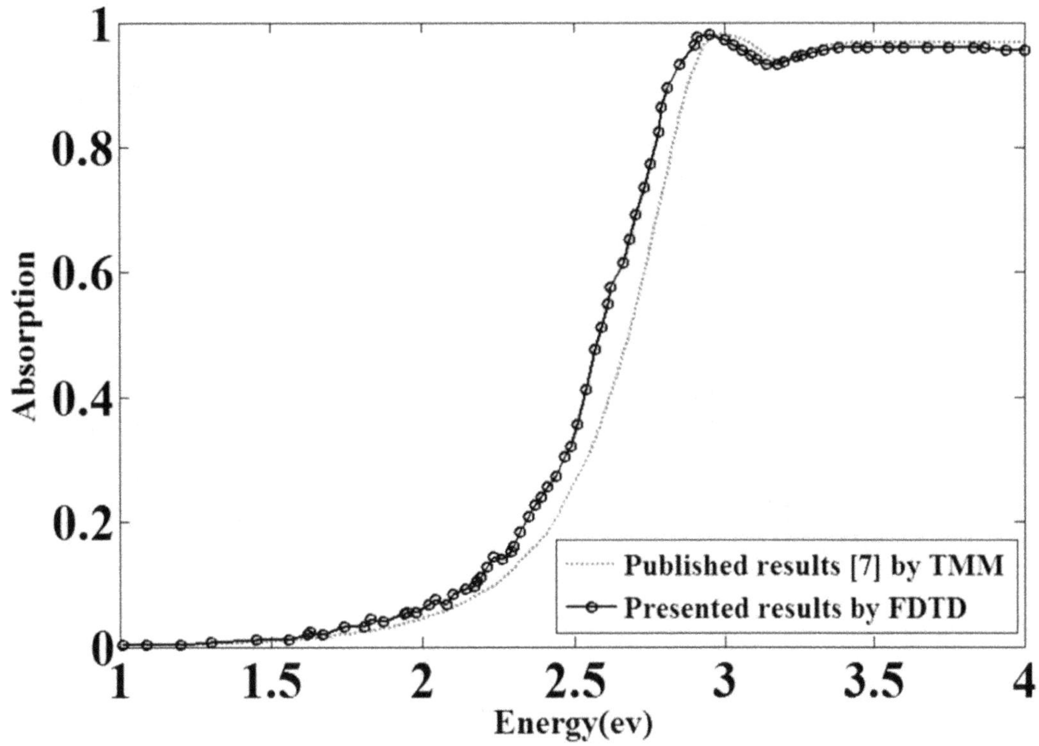


the 3D-FDTD method and those obtained by the transfer matrix method (TMM) (Hu &Chen, 2007). A good agreement is revealed from this figure between the 3D-FDTD results and the published results (Hu & Chen, 2007).

In the aim of achieving the optimum ultimate efficiency of the proposed D-SiNWs, different geometric parameters are studied and analyzed. First, the optical effects of cell size on the absorption are investigated as shown in Figure23. In this evaluation, the diameter of all pillars and the hole pitch $\Lambda$ are changed at fixed (d/Λ) ratio of 0.5 and L= 2.33 µm. The figure shows that the optical absorption increases by increasing the SiNWs diameter due to the shift of the absorption edge toward low energy region as shown in Figure23. As a result, the ultimate efficiency increases by increasing the SiNW diameter at fixed *(d/Λ)* ratio of 0.5 as shown in Figure23.The highest ultimate efficiency of 32% is obtained at SiNW diameter of 240 nm. This agrees with the theoretical interpretation (Hu &Chen, 2007; Lin & Povinelli, 2009; Li *et al.*, 2009), where the SiNWs diameter has direct effect on the ultimate efficiency. Next, the effect of the radial distance $\Lambda$ between SiNWs layers is also reported, whereas the other parameters are fixed at $d$ = 240 nm and $L$= 2.33 µm. Figure 24 shows the energy dependent optical characteristics of the D-SiNWs at different values of Λ. It is revealed from Figs. 24 that maximum ultimate efficiency of 33.1% occurs at $\Lambda$=500 nm which meets the maximum irradiance of the sun.

Further optimization is carried out on the central core radius to enhance the absorbed power and hence the ultimate efficiency. Figure 25 shows the relation between the ultimate efficiency and the

*Figure 23. Optical characteristics of D-SiNWs at different d values with fixed (d/Λ) ratio of 0.5. (©2014, Optical Engineering. Used with permission.).*

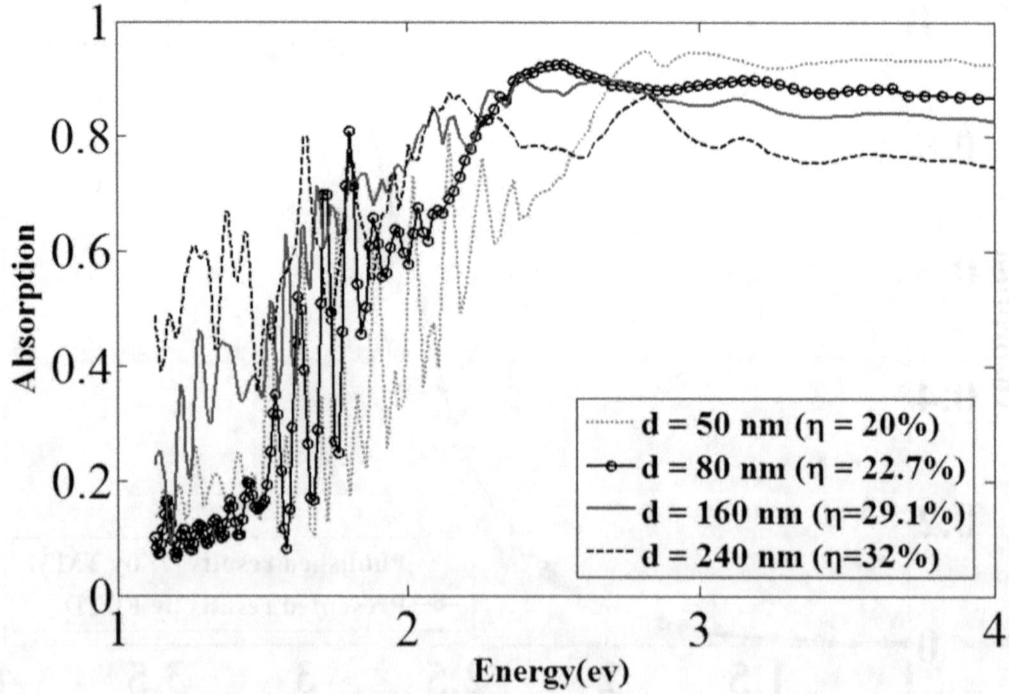


*Figure 24. Optical characteristics of D-SiNWs at different Λ values with fixed d = 220 nm. (©2014, Optical Engineering. Used with permission.).*

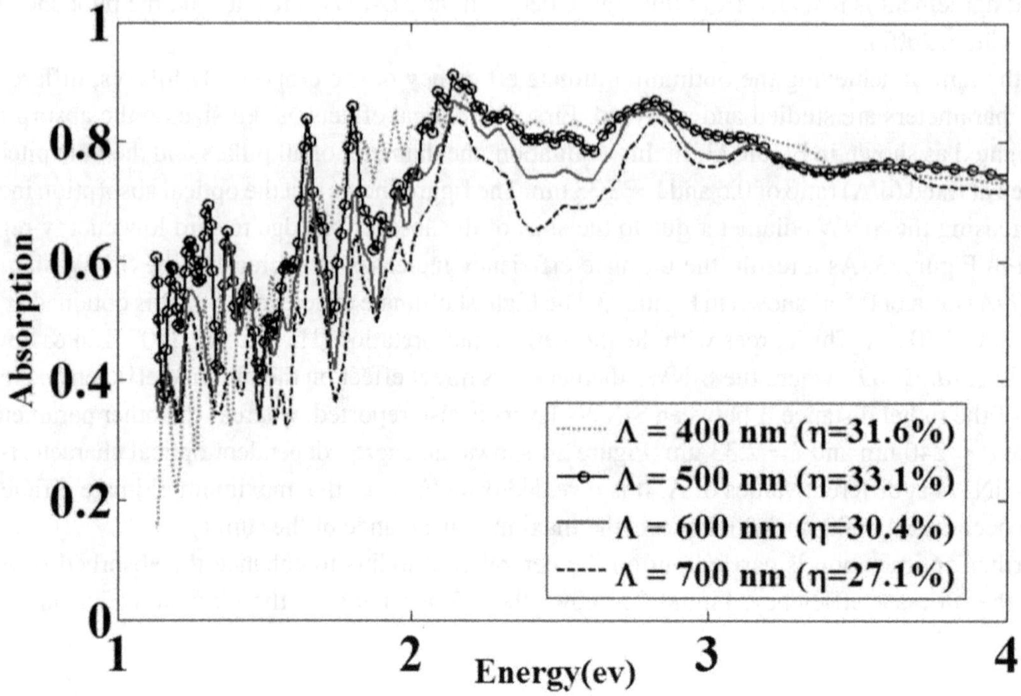

*Figure 25. Absorption efficiencies of D-SiNWs as a function of core radius. (©2014, Optical Engineering. Used with permission.).*

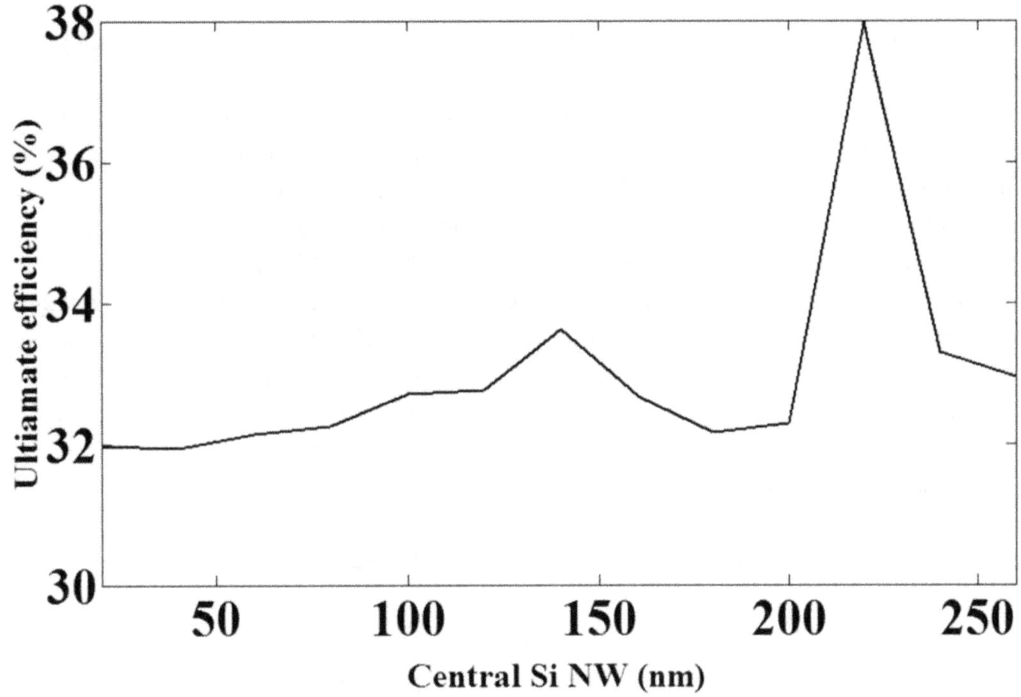


*Figure 26. Variation of the absorption spectra of square lattice and the proposed D-SiNWs at different core radius.*

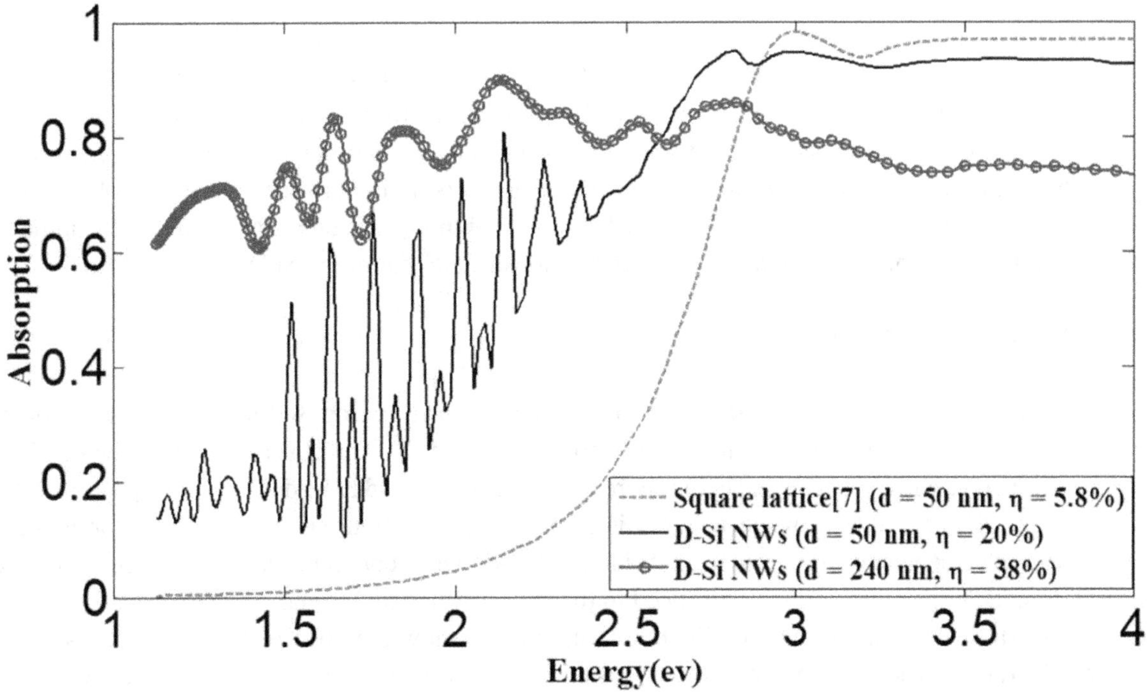

central SiNW radius. In this investigation, the other parameters is taken as $d = 240$ nm, $\Lambda$=500 and $L$= 2.33 µm. It should be noted that the absorption in SiNWs is dependent on the NWs diameter. Therefore, the SiNWs diameter in the first and second layers as well as the central NW diameter will lead to two different resonance frequencies. The combined effects of the two radii will accordingly broaden the bandwidth of the absorbed power and consequently the ultimate efficiency. It is thus found that the optimum efficiency of 38% is obtained at central core diameter $2a = 420$ nm, radial distance between the two layers $\Lambda$=500 nm, and SiNWs diameter $d$ =240 nm. Figure 26 shows the relation between the optical absorption and energy (ev) for Si NWs square lattice, and the proposed D-Si NWs. The simulation is executed for the square lattice with d=50 nm, and the D-Si NWs with diameters d = 50 nm, d=240 nm. The novel design apparently has the advantage of higher filling factor compared to the square lattice Si NWs for the same unit cell. The filling factors for the square and D-Si NWs are equal to 0.2 and 0.3 respectively. The square lattice has an efficiency of 5.8%, while the proposed structure shows an ultimate efficiency of 20%. Moreover, Figure 26 shows the absorption spectra of the D-Si NWs at d= 240 nm as related to the energy (ev). The proposed design exhibits an ultimate efficiency of 38%, greater than that reported previously structures (*Hong et al.*, 2013) by 7.5%. This is due to increasing the filling factor from 0.2 to 0 .39.

## 3.5 Novel Design of Decagonal SiNW with Hybrid Core

The 3D FDTD method is used to study the semiconductor absorption of decagonal SiNW design with hybrid core. The proposed structure improves the light absorption and reduces the reflection due to Fabry-Perot modes, plasmonic modes, and hybrid modes. The reported structure with hybrid core offers an ultimate efficiency of 32.62%.

### 3.5.1 Design Considerations

Figure 27 shows a schematic diagram of the suggested decagonal semiconductor NWs with central hybrid core (Hussein *et al.*, 2014c). In addition, three different configurations of the hybrid core are shown in Figure 27 (c). The gold/silicon core of diameter 120 nm is used to increase the ultimate efficiency of the proposed solar cell. The cladding NWs are arranged in a decagonal shape with diameter 50 nm and height h=2.33µm. The distance between two neighboring NWs in the same ring is equal to $\Lambda_1$=90 nm, while the distance between two neighboring NWs in two neighboring rings is $\Lambda_2$ = 45 nm. It should be noted that the height of the metal layer of the core is lower than that of the Si layer by 250 nm.

### 3.5.2 Simulation Results

Figure 28 (a) shows the semiconductor absorption spectra for the proposed structures with solid Si core, hybrid core 1, hybrid core 2, and hybrid core 3, respectively. The absorption efficiencies of the reported structure with solid Si core, hybrid core 1, hybrid core 2, and hybrid core 3 are equal to 19.7%, 23.97%, 19.26%, and 32.62%, respectively. It is evident from this figure that a significant broad band absorption enhancement occurs by using hybrid core 1 and hybrid core 3 instead of using solid Si core. In addition, the higher absorption is enhanced in the long wavelength regime with excellent absorption (>95%) in short wavelength regime. Figures 28(b) and (c) show the absorbed power profiles of the proposed structure with solid Si core, hybrid core 1 at the wavelength $\lambda$ = 940 nm, respectively. The

*Figure 27. Schematic of decagonal SiNW with central metal NW (a) Top View (b) Perspective view of one NW, (c) Top views of the hybrid Cores, (©2014, IEEE, NUSOD' 2014 Proc. Used with permission.).*

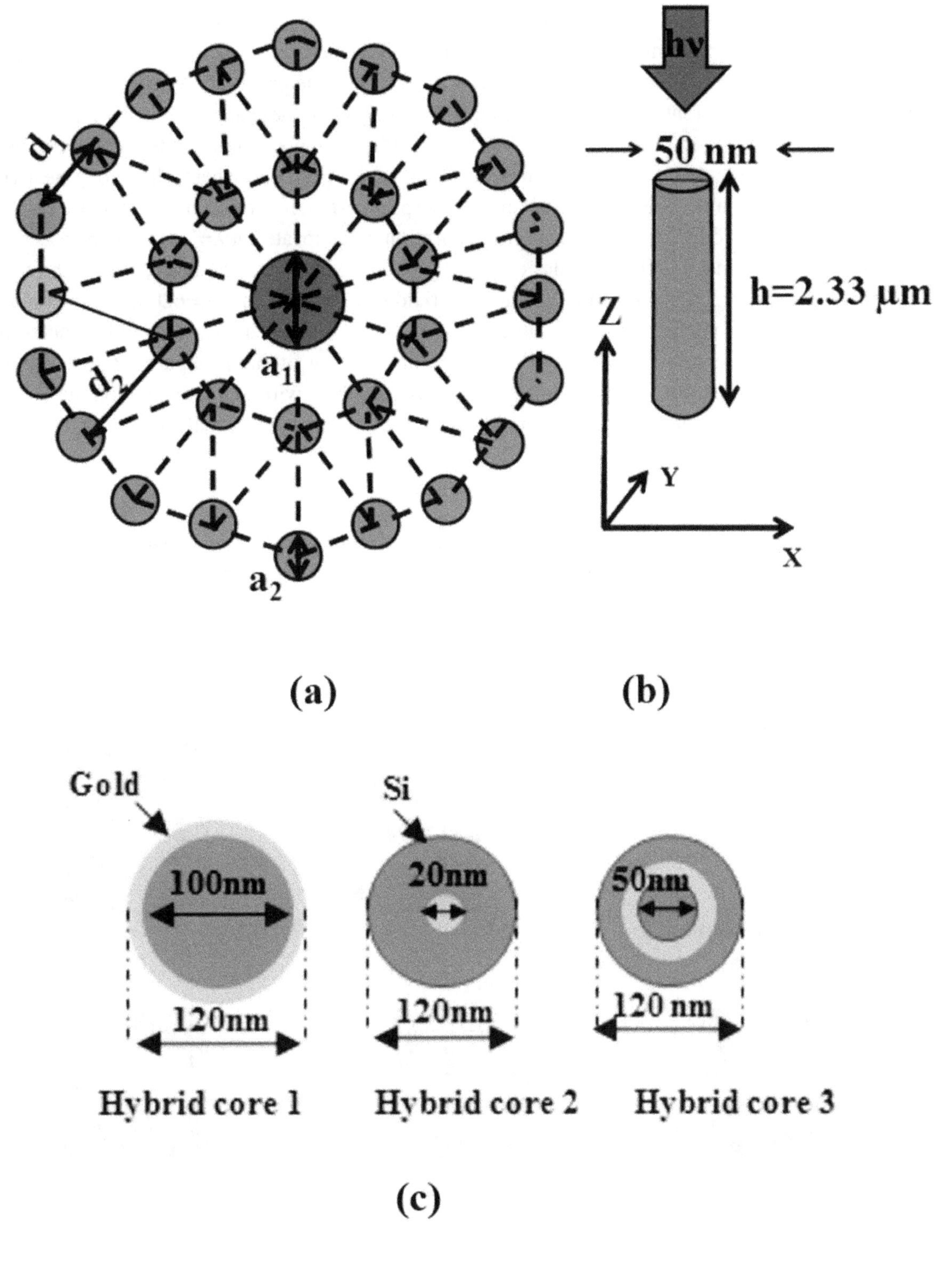

figures show that the maximum value of the absorbed power per unit volume using in the decagonal SiNW using a hybrid core 1 is improved by around 50% compared to that with solid Si core. It should be noted that these visualized profiles are compatible with the calculations of the absorption efficiency previously shown in Figure28 (a).The improvement of the absorption efficiency is due to the combined effect of different absorption modes that increase the semiconductor absorption band. Figures 28 (d), (e) and (f) show the normalized absorption mode profiles inside the proposed decagonal SiNWs with hybrid core 1at three different wavelengths. Based on the analysis of these absorption features, the optical mode scan be divided into three types, Fabry Perot (FP mode), plasmonic mode and hybrid mode. The absorption mode located at 404 nm can be ascribed to the FP mode where, the strong near-field enhancement appears in the SiNWs. The absolute absorption efficiency of such FP mode is relatively higher compared to the other optical modes. The absorption mode located at $\lambda$=820 nm can be ascribed to plasmonic modes, since the field profiles exhibit a pronounced plasmonic feature, that is, the strong near-field enhancement in the gold core. Because the field profiles located at $\lambda$=620 nm show both the FP-resonant and plasmonic features, it is categorized as hybrid mode. Finally, it should be noted that most field profiles of these absorption are hybrid, indicating that few of them are pure FP modes or pure plasmonic modes. These three types of absorption modes contribute with each other to obtain the broad band absorption enhancement shown in Figure28 (a).The obtained results pave the road to low cost and high efficient SiNW solar cell.

*Figure 28. (a) Semiconductor absorption spectra, Absorbed power profiles at $\lambda$=940 nm for SiNWs in decagonal lattice with: (b) hybrid core (c) solid core, Typical absorption mode profiles in the hybrid core: (d) Fabry Perot mode excited at 404 nm, (e) Hybrid mode 620 nm, (f) Plasmonic Mode at 820 nm, (Adapted from, Figure 2,IEEE, NUSOD'2014 Proc.)*

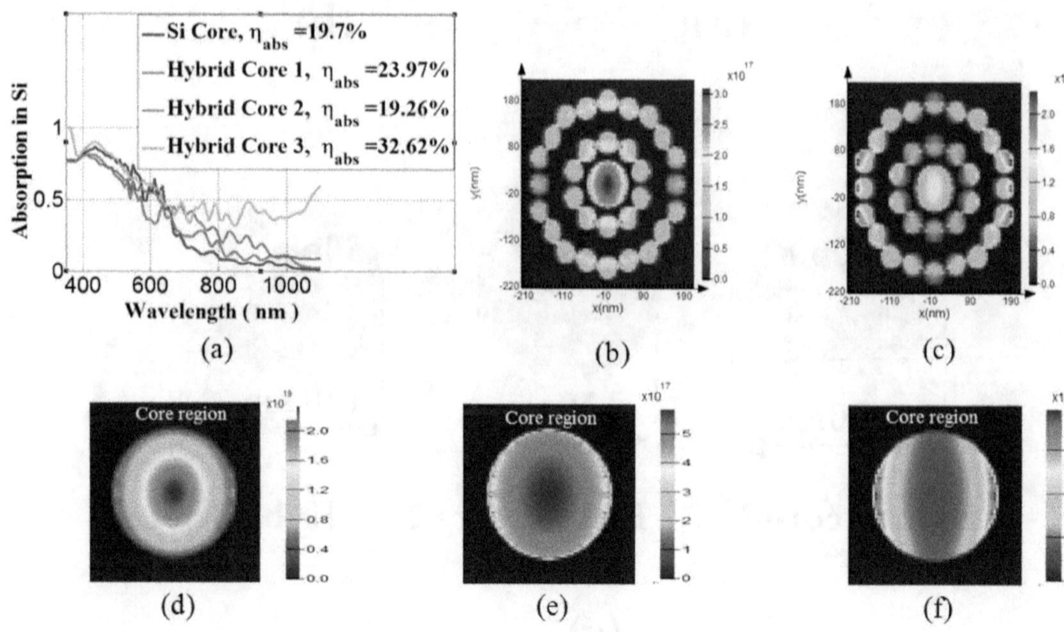

## REFERENCES

Abdel-Rahman, M., González, F., & Boreman, G. (2004). Antenna-coupled metal-oxide-metal diodes for dual-band detection at 92.5 GHz and 28 THz. *Journal of Electronics Letters*, *40*(2), 116–118. doi:10.1049/el:20040105

Bailey, R. L. (1972). A proposed new concepts for solar energy converter. *Journal of Engineering for Power*, *94*(2), 73–77. doi:10.1115/1.3445660

Bao, H., & Ruan, X. (2010). Optical absorption enhancement in disordered vertical silicon nanowire arrays for photovoltaic applications. *Optics Letters*, *35*(20), 3378–3380. doi:10.1364/OL.35.003378 PMID:20967072

Berland, B. (2003). *Photovoltaic technologies beyond the horizon: Optical rectenna solar cell.* Retrieved from http://www.nrel.gov/docs/fy03osti/33263.pdf

Berland, B., Simpson, L., Nuebel, G., Collins, T., & Lanning, B. (2001). Optical rectenna for direct conversion of sunlight to electricity. In *Proceedings of the Conference National Center for Photovoltaics Program Review Meeting (NREL).* Shaffer Parkway Littleton, Colorado.

Biagioni, P., Huang, J., & Hecht, B. (2012). Nanoantennas for visible and infrared radiation. *Journal of Progress in Physics*, *75*, 1–10. PMID:22790344

Cao, L., White, J. S., Park, J. S., Schuller, J. A., Clemens, B. M., & Brongersma, M. L. (2009). Engineering light absorption in semiconductor nanowire devices. *Journal of Nature Material*, *8*(8), 643–647. doi:10.1038/nmat2477 PMID:19578337

Cao, L. Y., Fan, P. Y., Vasudev, A. P., White, J. S., Yu, Z., Cai, W., & Brongersma, M. L. et al. (2010). Semiconductor nanowire optical antenna solar absorbers. *Journal of Nano Letters*, *10*(2), 439–445. doi:10.1021/nl9036627 PMID:20078065

Chen, H., Xi, N., Lai, K. W. C., Chen, L., Fung, C. K. M., & Lou, J. (2012). Plasmonic-resonant bowtie antenna for carbon nanotube photodetectors. *International Journal of Optics*, *2012*, 1–9. doi:10.1155/2012/318104

Chutinan, A., & John, S. (2008). Light trapping and absorption optimization in certain thin-film photonic crystal architectures. *Journal of Physical Review A*, *78*(2), 1–15. doi:10.1103/PhysRevA.78.023825

Dagenais, M., Choi, K., Yesilkoy, F., Chryssis, A. N., & Peckerar, M. C. (2010). Solar spectrum rectification using nano-antennas and tunneling diodes. In *Proceedings of the conference on Optoelectronic Integrated Circuits XII.* San Francisco, California: SPIE.

Dmitruk, N. L., Malynych, S. Z., & Moroz, I. E., & Kurlyak, V. Y. (2010). Optical efficiency of Ag and Au nanoparticles. *Journal of Semiconductor Physics, Quantum Electronics & Optoelectronics*, *13*, 369–373.

Duché, D., Escoubas, L., Simon, J. J., Torchio, P., Vervisch, W., & Flory, F. (2008). Slow bloch modes for enhancing the absorption of light in thin films for photovoltaic cells. *Journal of Applied Physics Letter*, *92*(19), 193310–193313. doi:10.1063/1.2929747

Fan, S. H., & Joannopoulos, J. D. (2002). Analysis of guided resonances in photonic crystal slabs. *Journal of Physical Review B*, *65*(235112), 1–8.

Fang, H., Li, X., Song, S., Xu, Y., & Zhu, J. (2008). Fabrication of slantingly aligned silicon nanowire arrays for solar cell applications. *Journal of Nanotechnology*, *19*(255703), 1–10. PMID:21828663

Feichtner, T., Selig, O., Kiunke, M., & Hecht, B. (2012). Evolutionary optimization of optical antennas. *Journal of Physical Review Letters*, *109*(127701), 1–8. PMID:23005987

Föll H., Leisner M., Cojocaru A. & Carstensen J., (2010), Macroporous semiconductors. *Journal of Materials, 3*, 3006-3076.

Fox, M. (2012). *Optical Properties of Solids*. Oxford: Oxford University Press.

Gadalla, M. N. (2013). *Nano antenna integrated diode (Rectenna) for infrared energy harvesting*. (Master dissertation). Retrieved from Kaust Repository.

Gadalla, M. N., Abdel-Rahman, M., & Shamim, A. (2014). Design, optimization and fabrication of a 28.3 THz nano-rectenna for infrared detection and rectification. *Journal of Scientific Reports*, *4*(270), 1–9. doi:10.1038/srep04270 PMID:24599374

Gallo, M., Mesciaa, L., Lositoa, O., Bozzettia, M., & Prudenzanob, F. (2012). Design of optical antenna for solar energy collection. *Journal of Energy*, *39*(1), 27–32. doi:10.1016/j.energy.2011.02.026

Garnett, E. C., & Yang, P. (2010). Light trapping in silicon nanowire solar cells. *Journal of Nano*, *3*, 1082–1087. PMID:20108969

Grover, S., & Moddel, G. (2011). Applicability of metal/insulator/metal (MIM) diodes to solar rectennas. *IEEE Journal of Photovoltic*, *1*(1), 78–83. doi:10.1109/JPHOTOV.2011.2160489

Hagglund, C., Zach, M., Petersson, G., & Kasemo, B. (2008). Electromagnetic coupling of light into a silicon solar cell by nano disk plasmons. *Journal of Applied Physics Letters, 92*, 053110 1-3.

Hartschuh, A., Beversluis, M. R., Bouhelier, A., & Novotny, L. (2004). Tip-enhanced optical spectroscopy. *Journal of Philosophical Transactions of the Royal Society of London, Series A: Mathematical. Physical and Engineering Sciences*, *362*(1817), 807–819. doi:10.1098/rsta.2003.1348

Hong, L., Rusli, Wang, X., Zheng H., Wang, H., & Yu, H. (2013). Design guideline of Si Nanohole/PEDOT:PSS hybrid structure for solar cell application. *Journal of Nanotechnology*, *24*, 1–6. PMID:23940111

Hong, L., Wang, R. X., Zheng, H., & Wang, H. (2013). Design guidelines for slanting silicon nanowire arrays for solar cell application. *Journal of Applied Physics*, *114*(8), 084301–0843016. doi:10.1063/1.4819175

Hu, L., & Chen, G. (2007). Analysis of optical absorption in silicon nanowire arrays for photovoltaic applications. *Journal of Nano Letters*, *7*(11), 3249–3252. doi:10.1021/nl071018b PMID:17927257

Huang, Z., Fang, H., & Zhu, J. (2007). Fabrication of silicon nanowire arrays with controlled diameter, length, and density. *Journal of Advanced Materials*, *19*(5), 744–748. doi:10.1002/adma.200600892

Huang, Z., Geyer, N., Werner, P., Boor, J. D., & Gosele, U. (2011). Metal assisted chemical etching of silicon: A review. *Journal of Advanced Materials*, *23*(2), 285–308. doi:10.1002/adma.201001784 PMID:20859941

Huang, Z., Shimizu, T., Senz, S., Zhang, Z., Geyer, N., & Gosele, U. (2010). Oxidation rate effect on the direction of metal-assisted chemical and electrochemical etching of silicon. *The Journal of Physical Chemistry C*, *114*(24), 10683–10690. doi:10.1021/jp911121q

Hussein, M., Areed, N. F. F., Hameed, M. F. O., & Obayya, S. S. A. (2013). Flower-shaped dipole based nano-antenna for energy harvesting. In *Proceedings of the Conference on Progress in electromagnetics research symposium (PIERS)*. Stockholm, Sweden: PIER.

Hussein, M., Areed, N. F. F., Hameed, M. F. O., & Obayya, S. S. A. (2014a). Design of flower-shaped dipole nano-antenna for energy harvesting. *Journal of IET*, *8*, 167–173.

Hussein, M., Areed, N. F. F., Hameed, M. F. O., & Obayya, S. S. A. (2014c). Hybrid core semiconductor nanowires for solar cell applications. *In the proceeding of the 14th International Conference on Numerical Simulation of Optoelectronic Devices (NUSOD)*. Alma de Mallorca, Spain.

Hussein M., Hameed, M. F. O., Areed, N. F. F., & Obayya, S. S. A. (2014b). Ultra high efficient solar cell based on decagonal arrays of silicon nanowires. *Journal of Optical Engineering, 53*, 117105 1-7.

Joannopoulos, J. D., Meade, R. D., & Winn, J. N. (2008). *Photonic crystals: Molding the flow of light*. New Jersey: Princeton University Press.

Kandala, A., Betti, T., & Fontcubertai, A. M. (2008). General theoretical considerations on nanowire solar cell designs. *Journal of Physica Status Solidi*, *206*(1), 173–178. doi:10.1002/pssa.200723672

Kelzenberg, M. D., Boettcher, S. W., Petykiewicz, J. A., Turner, D. B., Putnam, M. C, & Warren, E. L. Spurgeon, J. M., Briggs, R. M., Lewis, N. S., & Atwater, H. A. (2010). Enhanced absorption and carrier collection in Si wire arrays for photovoltaic applications. *Journal of Nature Materials*, *9*, 239–244. PMID:20154692

Kostiučenko, O. (2011). *Nano-structure-induced field enhanced nonlinear optics of plasmonic substrates*. (Master dissertation). University of Southern, Denmark.

Kotter, D. K., Novack, S. D., Slafer, W. D., & Pinhero, P. J. (2008). Nanatenna electromagnetic collector. In the Proceedings of Conference on Energy Sustainability (ASME). Florida, USA.

Kottmann, J. O., Martin, D. S., & Schultz, S. (2001). Non-regularly shaped plasmon resonant nanoparticle as localized light source for near-field microscopy. *Journal of Microscopy*, *202*(1), 60–65. doi:10.1046/j.1365-2818.2001.00866.x PMID:11298871

Kuiqing, P., Mingliang, Z., Aijiang, L., Ning, W., Ruiqin, Z., & Shuit, L. (2007). Ordered silicon nanowire arrays via nanosphere lithography and metal-induced etching. *Journal of Applied Physics Letter*, *90*(16), 163123. doi:10.1063/1.2724897

Li, J., Yu, H., Wong, S. M., Li, X., Zhang, G., & Lo, P. G. (2009). Design guidelines of periodic Si nanowire arrays for solar cell application. *Journal of Applied Physics Letter*, *95*, 243113-1–4.

Li, X. (2012). Metal assisted chemical etching for high aspect ratio nanostructures: A review of characteristics and applications in photovoltaics. *Journal of Current Opinion in Solid State and Materials Science*, *16*(2), 71–81. doi:10.1016/j.cossms.2011.11.002

Lin, C., & Povinelli, M. L. (2010). Optical absorption enhancement in silicon nanowire and nanohole arrays for photovoltaic applications. In the Proceedings of Conference on Next Generation (Nano) Photonic and Cell Technologies for Solar Energy Conversion, (SPIE). California.

Lin, G. H., Abdu, R., & Bockris, J. O. M. (1996). Investigation of resonance light absorption and rectification by sub nanostructures. *Journal of Applied Physics*, *80*(1), 565–568. doi:10.1063/1.362762

Ma, Z., & Vandenbosch, G. A. E. (2013). Optimal solar energy harvesting efficiency of nano- rectenna systems. *Journal of Solar Energy*, *88*, 163–174. doi:10.1016/j.solener.2012.11.023

Malinovska, D. D. (2010). The state-of-the-art and future development of the photovoltaic technologies–the route from crystalline to nanostructured and new emerging materials. *Journal of Physics: Conference Series*, 253.

Mallick, S. B., Agrawal, M., & Peumans, P. (2010). Optimal light trapping in ultra-thin photonic crystal crystalline silicon solar cells. *Journal of Optics Express*, *18*(6), 5691–5706. doi:10.1364/OE.18.005691 PMID:20389585

Marks, A. M. (1984). *Device for conversion of light power to electrical power.* U.S. patent 4445050.

McMahon, J. M., Gray, S. K., & Schatz, G. C. (2010). Optical properties of nanowire dimers with a spatially nonlocal dielectric function. *Journal of Nano Letters*, *10*(9), 3473–3481. doi:10.1021/nl101606j PMID:20715807

McMahon, J. M., Henry, A. I., Wustholz, K. L., Natan, M. J., Freeman, R. G., & Van Duyne, R. P. (2009). Gold nanoparticle dimer plasmonics: Finite element method calculations of the electromagnetic enhancement to surface-enhanced Raman spectroscopy. *Journal of Analytical and Bioanalytical Chemistry*, *394*, 1819–1825.

Mein, W. S., Yu, Y. H., Yali, L., Junshuai, L., Wei, S. X., Singh, N., Lo, P. G. Q., & Dim, K. (2011). Boosting short-circuit current with rationally designed periodic Si nanopillar surface texturing for solar cells. *Journal of IEEE Transaction on Electron Devices*, *58*, 3224–3229.

Midrio, M., Romagnoli, M., Boscolo, S., Angelis, C., Locatelli, A., Modotto, D., & Capobianco, A. (2011). Flared monopole antennas for 10 μm radiation. *Journal of Quantum Electronics*, *47*(1), 84–91. doi:10.1109/JQE.2010.2057501

Ming-Liang, Z., Kui-Qing, P., Xia, F., Jian-Sheng, J., Rui-Qin, Z., Shuit-Tong, L., & Ning-Bew, W. (2008). Preparation of large-area uniform silicon nanowires arrays through metal-assisted chemical etching. *Journal of Physical Chemistry*, *112*, 4444–4450. PMID:18433167

Moddel, G., & Grover, S. (2013). *Rectenna solar cells*. New York: Springer. doi:10.1007/978-1-4614-3716-1

Muskens, O. L., Rivas, J. G., Algra, R. E., Bakkers, E. P. A. A., & Lagendijk, A. (2008). Design of light scattering in nanowire materials for photovoltaic applications. *Journal of Nano Letter, 8*, 2638–2642.

Nelson, J. (2003). *The physics of solar cells*. London: Imperial College. doi:10.1142/p276

Palik, E. D. (1997). *Handbook of optical constants of solids*. San Diego: Academic Press.

Peng, K., Hu, J., Yan, Y., Wu, Y., Fang, H., Xu, Y., & Zhu, J. et al. (2006). Fabrication of single-crystalline silicon nanowires by scratching a silicon surface with catalytic metal particles. *Journal of Advanced Functional Materials, 16*(3), 387–394. doi:10.1002/adfm.200500392

Prasad, T., Colvin, V. L., & Mittleman, D. M. (2008). Dependence of guided resonances on the structural parameters of terahertz photonic crystal slabs. *Journal of Applied Physics, 25*, 633–644.

Rai-Choudhury, P. (1997). *SPIE handbook of microlithography, micromachining and microfabrication*. Washington: SPIE Press.

Ruan, X. L., & Kaviany, M. (2005). Enhanced nonradiative relaxation and photoluminescence quenching in random, doped nanocrystalline powders. *Journal of Applied Physics, 97*(10), 104331–104335. doi:10.1063/1.1900937

Ruan, X. L., & Kaviany, M. (2006). Enhanced laser cooling of rare-earth-ion-doped nanocrystalline powders. *Journal of Physical Review B, 73*(15), 155422-1–15. doi:10.1103/PhysRevB.73.155422

Russell, P. (2003). Photonic crystal fibers. *Journal of Science, 299*(5605), 358–362. http://www.fgglass.com/light_and_thermal_performance.html doi:10.1126/science.1079280 PMID:12532007

Sabaawi, A. M., Tsimenidis, C., & Sharif, B. S. (2013). *Overview of nanoantennas for solar rectennas*. Retrieved from http://link.springer.com/chapter/10.1007%2F978-1-4614-3716-1_11

Street, R. A., Wong, W. S., & Paulson, C. (2009). Analytic model for diffuse reflectivity of silicon nanowire. *Journal of Nano Letters, 9*(10), 3494–3497. doi:10.1021/nl901683y PMID:19588905

Sundaramurthy, A., Crozier, K., Kino, G., Fromm, D., Schuck, P., & Moerner, W., (2005). Field enhancement and gap-dependent resonance in a system of two opposing tip-to-tip Au nano triangles. *Journal of Physical Review B, 72*, 165409 1-6.

Taflove, A., & Hagness, S. (2005). *Computational Electrodynamics: The Finite Difference Time-Domain Method* (3rd ed.). London: Artech House.

Tian, B., Zheng, X., Kempa, T. J., Fang, Y., Yu, N., Yu, G., & Lieber, C. M. et al. (2007). Coaxial silicon nanowires as solar cells and nanoelectronic power sources. *Journal of Nature, 449*(7164), 885–889. doi:10.1038/nature06181 PMID:17943126

Tikhodeev, S. G., Yablonskii, A. L., Muljarov, E. A., Gippius, N. A., & Ishihara, T. (2002). Quasi guided modes and optical properties of photonic crystal slabs. *Journal of Physical Review, 66.*

Tsakalakos, L. (2008). Nanostructures for photovoltaics. *Journal of Materials Science and Engineering, 62*, 175–189.

Tumbleston, J. R., Ko, D. H., Samulski, E. T., & Lopez, R. (2009). Absorption and quasi guided mode analysis of organic solar cells with photonic crystal photoactive layers. *Journal of Optics Express*, *17*(9), 7670–7681. doi:10.1364/OE.17.007670 PMID:19399146

Vandenbosch, G. A. E., & Ma, Z. (2012). Upper bounds for the solar energy harvesting efficiency of nano-antennas. *Journal of Nano Energy*, *1*(3), 494–502. doi:10.1016/j.nanoen.2012.03.002

## KEY TERMS AND DEFINITIONS

**Finite Difference Time-Domain (FDTD) Method:** One of the popular computational techniques for the solution of electromagnetic problems.

**Optical Antennas:** Devices for converting propagating radiation into confined/enhanced fields at nanoscale.

**Photonic Crystals:** Periodic structures that exhibit a periodic variation in the permittivity with a lattice period in the same order of magnitude as the wavelength of light.

**Radiation Efficiency:** The ratio between the radiated powers to the input power of the antenna.

**Rectenna:** A device combining an antenna and rectifier to convert the electromagnetic radiation received by the antenna to a direct current signal by the diode.

**Surface Plasmon (SP):** A collective oscillation of surface free electrons at metal dielectric interface.

**Tunneling Diode (Metal Insulator Metal):** A nonlinear device that can be used for performing a very fast operation in the optical region.

# Chapter 3
# Wireless Energy Transfer:
## On the Efficiency of Low Power Rectenna Topologies

**Aya Mabrouki**
*ESEO / IETR, France*

**Mohamed Latrach**
*ESEO / IETR, France*

## ABSTRACT

*This chapter proposes an overview of microwave energy harvesting with focuses on the design of high efficiency low power rectifying circuits. A background survey of RF energy harvesting techniques is presented first. Then, the performances of conventional rectifier topologies are analyzed and discussed. A review of the most efficient rectenna designs, from the state of the art, is also presented. Design considerations for low power rectifier operations are detailed and new high efficient rectifying circuits are designed and evaluated in both GSM and ISM bands under low power constraints.*

## INTRODUCTION

During the last decades, the increasing demand for wireless communication systems, used in several applications ranging from smartphones to low power electronic devices, has led to a trend involving constant system miniaturization for a better integration in the environment (Al Agha et al., 2009; Gungor & Hancke, 2009). However, there are still some constraining points that should be addressed. One of the most confounding problems is the use of batteries as power supply. This induces more limitations, not only in terms of miniaturization, but also in terms of portable devices cost and weight reduction due to the battery occupied volume. Hence, the portability and autonomy of the wireless devices will be also affected depending on their batteries lifetime. Batteries used in many wireless sensors need periodical recharging or replacement, which may be hard, expensive or even impossible in some applications. This problem affects a large spectrum of applications such as habitat and environmental monitoring (Mainwaring et al., 2002), structural monitoring (Lee et al., 2006; Walsh et al., 2005), and health monitoring

DOI: 10.4018/978-1-4666-8254-2.ch003

of patients (Paing et al., 2007). Increasing the battery capacity is not a reliable solution since it comes at the expense of size, weight and cost. Moreover, used battery induces environmental pollution that constrains the "being green" today's trend evolution. On account of these limitations, research on innovative wireless power supply should offer more availability for portable electronic devices. Energy harvesting from sources in the sensors environment is becoming an emergent research area that could be a long lasting and inexpensive solution making wireless devices truly independent (Papovic, 2006).

## WIRELESS POWER TRANSMISSION: BACKGROUND SURVEY

The concept of wireless power transmission ages more than 100 years and has been firstly presented by Nicolas TESLA in his United States patent No. 685,954, dated November 5, 1901, (Glenn, 1994), where he describes the process of electromagnetic waves transfer as transmitting electrical signal "disturbances" through the natural media. In the claim 11 of his patent, he stated the principles of wireless energy transmission, the storage of the energy in a capacitor and energy management over the time. Although Tesla wireless power transmission systems implementation was unsuccessful, he did transmit power from his oscillators that operated up to 100MV at 150 kHz. Research efforts continued later in the 1930s to promote wireless power transmission by the invention of magnetron and klystron.

Rectification of microwave signals for supplying dc power through wireless transmission has been proposed and researched in the context of higher power beaming since 1950s. This era witnessed also the advancement of high-power microwave tubes by Raytheon Company, Waltham, MA (Brown, 1984) and a 15-kW average power S-band cross-field amplifying tube was developed with 81% overall DC to RF efficiency by 1958. A good review of the early history of microwave converters is given in (Edwards, 1947). In the early 1960s, the first application employing microwave rectification of dc power extraction was performed. Innovative researches at Raytheon have been carried out and the term "rectenna", referring to a receiving antenna followed by a rectifier built around balanced bridge or single semiconductor diode, has been introduced. Practical demonstrations of wireless power transmission began to appear in military and space-related research programs. The Solar Power Satellite (SPS) project, (Glaser, 1968) proposed by Peter Glaser in 1968, is considered as one of the most ambitious project dealing with high power WPT. It consists of a gigantic satellite designed as an electric power plant orbiting in the Geostationary Earth Orbit (GEO) that capture permanently the solar power and send it towards ground stations through microwave beam. An illustration is given in Figure 1, (Falkenstein, 2011). Despite the 100% efficiency that could be reached by the SPS, the project was never put into practice due to the large antenna sizes and costs. For 5GW of transmitted power, the satellite would need to be 5.2 km wide and 10.4 km long, with an emitter antenna of around 1km in diameter (Marian et al., 2012).

Conversion efficiency of the rectenna is defined as the amount of microwave power that is converted into dc power. Efforts to increase rectenna performances continued latterly and in 1977, Brown, Raytheon Company, recorded the greatest conversion efficiency of 90.6% (Raytheon, 1977). The rectenna element was built around GaAs-Pt Schottky barrier diode and used aluminium bars to construct the dipole and transmission line. In microwave power transmission, antennas have well-defined polarizations, and high conversion efficiency is enabled by single frequency high microwave power densities incident on an array of antennas and rectifying circuits. Several applications for this type of WPT have been proposed for helicopter powering (Sinohara & Matsumoto, 1998), inter-satellite power transmission (Yoo & Chang, 1992) including utility power satellites, mechanical actuators for space-based telescopes (Epp

*Figure 1. Microwave power experiment in space on a small satellite and on the International Space Station (Falkenstein, 2011)*

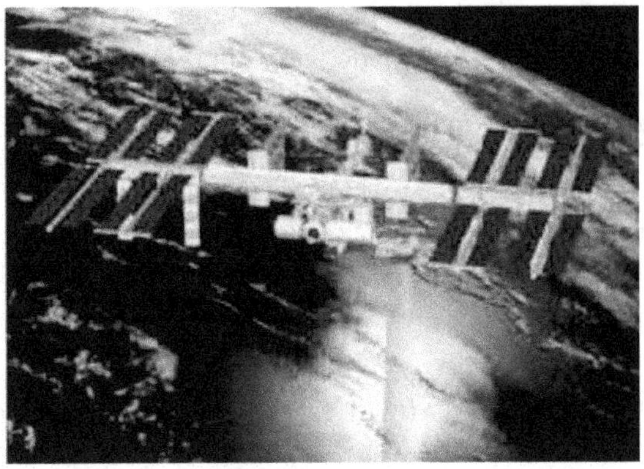

et al., 2000), and small dc motor driving (Fujino et al. 1994)… Rectifying antennas with linear (Brown, 1991), dual linear (Epp et al., 2000) and circular polarization (McSpadden & Chang, 1994) have been developed and the achieved efficiencies ranged from around 85-90% at lower microwave frequencies to around 60% at X-band and around 40% at Ka-band (Yoo & Chang, 1992).

WPT applications include also RFID (Radio Frequency Identification) technology that emerged in 1980 and which is widely used for supply chain management and inventory control. RFIDs are powered by the interrogating radio-frequency wave typically in the UHF range (Falkenstein, 2011). Active RFIDs tags, developed in recent years, are the most concerned of wireless powering to ensure a more reliable service.

## ENERGY HARVESTING PRINCIPLES

Nowadays, several electronic devices powering solutions employ energy harvesting techniques, using either different single energy sources or a combination of two or more sources. Power density is one of the main differences between these sources, for example RF (0.01-0.1uW/cm$^2$), vibration (4-100uW/cm$^2$), photovoltaic (10uW/cm$^2$- 10mW/cm$^2$), and thermal (20uW/cm$^2$-10mW/cm$^2$) (Mikeka & Arai, 2011). Wireless energy transfer can be realized using mainly two different approaches. The first one is based on magnetic resonant coupling between tow structures and is dedicated to relatively low frequency applications. The major limitation of this technique is that energy can only be transferred over low distances, generally in the same order of magnitude as emitter and receiver size. The second approach consists in radiative high frequency energy transfer based on electromagnetic waves (generally above 1GHz). This technique allows both short and long range wireless energy transfer with relatively compact emitter and receiver sizes. RF energy, though considered as a limited source on account of incident power density, is preferred to feed low power devices like industrial sensors or sensor networks over distances of several meters. Wireless energy transfer using electromagnetic waves has been mostly applied for UHF applications in the GSM and ISM bands. It is realized through three processes. First, the high frequency power signal is generated and sent by the emitter antenna. Power is then transported under the form of an electromagnetic wave beam towards the receiver, where it is captured by the receiver antenna and transformed into DC power by the rectifier. The term rectenna refers to the association of antenna and rectifier. The principle of the whole power transmission process is described in Figure 2. This chapter will focus on the design of receiving portion of energy transfer system and more specifically the conversion of electromagnetic waves into DC power.

## Electromagnetic Environment

The receiving portion of a wireless power transfer system is illustrated in Figure 3.

*Figure 2. Schematic of wireless power transfer process*

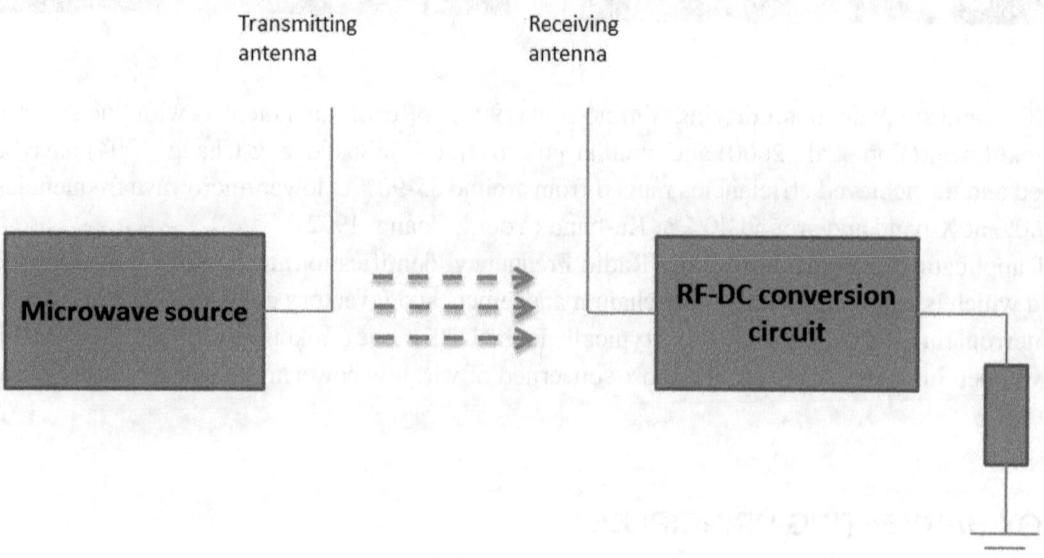


*Figure 3. Schematic of a rectenna element*

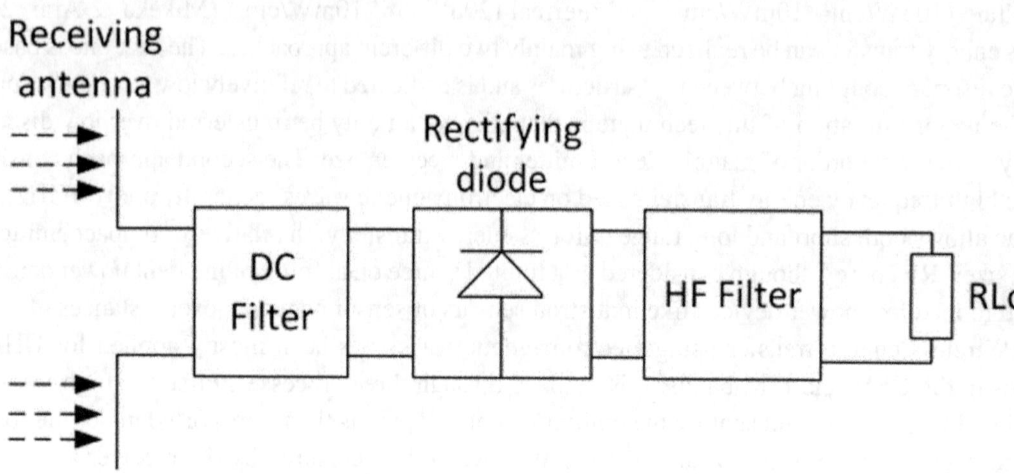

The incident RF power is captured by the antenna and transformed into DC power by the diode-based converter. The main conditions that must be met to insure high RF-to-DC conversion efficiency rectifier are: (a) Minimize reflections at the fundamental frequency in order to maximize power transfer to the diode; (b) Harmonic filtering to reduce voltage and current overlap; (c) Choose the appropriate DC load to maximize the DC power. In order to preserve the antenna re- irradiating the high order harmonics generated by the rectifier, a band-pass filter is inserted between the rectifier and the antenna. At the output a DC filter (low pass filter) is used to minimize the harmonics. Rectenna designs depend strongly on the incident waves that carry energy. Several cases of wireless powering can be considered (Papovic, 2006):

- One or more high directivity narrowband transmitters with well-known and fixed polarization and well-known power levels. In such case, the efficiency can be very high.
- One or more medium power semi-directional transmitter that illuminate a range in space, with multipath present. The incident power density is known approximately but there are multipath effects that change polarization and special distribution of power density.
- Unknown transmitter over a range of frequencies, power levels, generally unpolarized, with varying low level spatial power densities.

In this chapter we will focus on the two last power transfer scenarios.

## RF-DC Conversion Efficiency

The RF-DC conversion efficiency of a rectenna, given by (1), is defined by the ratio of the DC power, $P_{DC}$, measured at the load to the received power, $P_{RF}$, injected at the input of the converter. It depends on several factors: the impedance match between the antenna and the rectifier, antenna efficiency and the DC load impedance.

$$\eta = \frac{P_{DC}}{P_{RF}} \tag{1}$$

The RF power collected at the converter level depends on the characteristics of the transmitting and receiving antennas and can be evaluated by the Friis equation (Marian et al., 2012):

$$P_{RF} = P_t.G_t.G_r.\left(\frac{\lambda}{4\pi D}\right)^2 \tag{2}$$

where $G_t$ and $G_r$ are the gain of the transmitting and receiving antennas respectively. $P_t$ is the transmitted power. $\lambda$ is the wavelength and $D$ is the distance between emitter and receiver. $R_{load}$ is the load resistance and $V_{DC}$ is the DC output voltage.

Higher conversion efficiency of rectenna require high gain antenna to ensure high DC levels at the load. However, high gain comes at the expense of antenna size, which is not suitable for electronic devices miniaturization trend. Antenna polarization is also an important factor that should be chosen depending on the application. Despite the fact that circular polarized antenna offers the possibility to keep the DC voltage constant even if there is a rotation of the emitter or receiver, the corresponding gain is usually less than 3dB compared to that of linearly polarized antenna. Rectenna designs need to achieve several specifications in terms of high conversion efficiency, low cost, high reliability, and low mass at the transmitting and receiving ends of the system.

## LOW POWER RECTENNA TOPOLOGIES

High power density levels are not available everywhere. Microwave energy can be strongly attenuated due to multipath effects. Moreover, stringent restrictions on power densities are required in respect to the health standard. According to the ICNURO (International commission on Non-Ionizing Radiation Protection) exposure guidelines, power densities are limited to 0.45mW/cm$^2$ (41V/m) at 900MHz and 0.9mW/cm$^2$ (58V/m) at 1.8 GHz for mobile phone base station frequencies, and 1mW/cm$^2$ (6V/m) at the microwave frequency 2.45GHz (ICNURO, 1998) (IEEE Std, 1999). Announced by WHO (World Health Organization), typical maximum public exposure levels are 10uW/cm$^2$(6V/m) for TV and radio transmitters and for mobile phone base stations, 20uW/cm$^2$ for radars, and 50uW/cm$^2$ for microwave ovens (WHO, 1999). The investigation of electric field in practical cases is between 0.14V/m (5.2nW/cm$^2$) and 3V/m (2.4uW/cm$^2$) as shown in the CPIC newsletters and articles (Retrieved from http://www.developpement-durable.gouv.fr/IMG/pdf/rapport_COPIC_31_juillet_2013) (Zhou, 2013).

Rectenna, aside antenna, are built around diodes or diode mounted transistors. For microwave rectifiers at low input power levels (a few mW) the full-wave rectifier approach is not efficient and a single-diode rectifier should be considered. The main factor in high efficiency rectifier design is the diode threshold voltage. In fact, under high power levels, diode threshold is not an issue since it still small compared to incident high frequency voltage amplitude. However, it becomes critical in the case of low incident power levels (below 1 mW) due to losses. Generally, zero bias schottky diodes with low threshold voltage ($V_j$) of 150mVand low junction capacitance $C_{jo}$ of around 0.18 pF are used in rectenna designs. Input matching circuits are also an important factor to ensure maximum power transfer between the receiving antenna and the rectifier. The optimal power transfer can be achieved when the antenna input impedance is equal to the conjugate of the impedance of the rectifying circuit:

$$Z_{antenna} = Z^{*}_{rectifier} = Z_0 \tag{3}$$

The power injected to the rectifier and the effective value of the voltage delivered by the antenna can then be calculated, respectively, by:

$$P_r = \frac{V_{eff}^{\ 2}}{4.Z_0} \tag{4}$$

$$V_{eff} = \sqrt{4.Z_0.P_r} = \sqrt{4.Z_0.\frac{|E|^2.G.\lambda^2}{4.\pi.Z_{air}}} = |E|.\lambda.\sqrt{\frac{Z_0.G}{\pi.Z_{air}}} \tag{5}$$

where $Z_{air}$ is the air characteristic impedance, $G_r$ is the gain of the receiving antenna and E is the electric field efficient value at the receiver position. For an input impedance of 50Ω, $V_{eff}$ can be estimated to 450mV and 140mV for input power of 1mW and 100uW respectively at the rectifier levels (Marian et al., 2012). Hence, for very low input power, the voltage level decreases and becomes very inferior to the diode threshold which implies a degradation of the RF-DC conversion efficiency. Given these considerations, the major focus of this chapter is to develop high efficiency low power rectennas that satisfy today system miniaturization trend requirements. Rectifying circuits will be designed in the GSM and ISM band which offer the best compromise between free-space attenuation and antenna size reduction.

## Conventional Schottky Diode Based Rectifier Topologies

In this section we review the conventional rectifier topologies built around either schottky diodes or transistor mounted diodes. These structures depend on the position and number of diodes used for rectification.

### Series Mounted Diode Rectifier

The schematic of this topology is given in Figure 4. The rectifier is based on a single diode used in series with the signal path. The diode $D_1$ is blocked during the negative half wave of the input signal then becomes conductive during the positive one. Hence, the energy flows from the source and input filter towards the output filter which blocks high frequency harmonics from reaching the load.

#### *Shunt Mounted Diode Rectifier*

In this rectifying structure, Figure 5, the diode $D_1$ is used in parallel between the input and output filters. $D_1$ is directly polarized by the DC voltage it generates. During the negative half wave, $D_1$ is conductive and the input filter is charged. Then, in the positive half wave, $D_1$ is blocked and energy flows from the source and input filter towards the output filter.

*Figure 4. Series mounted diode rectifier topology*

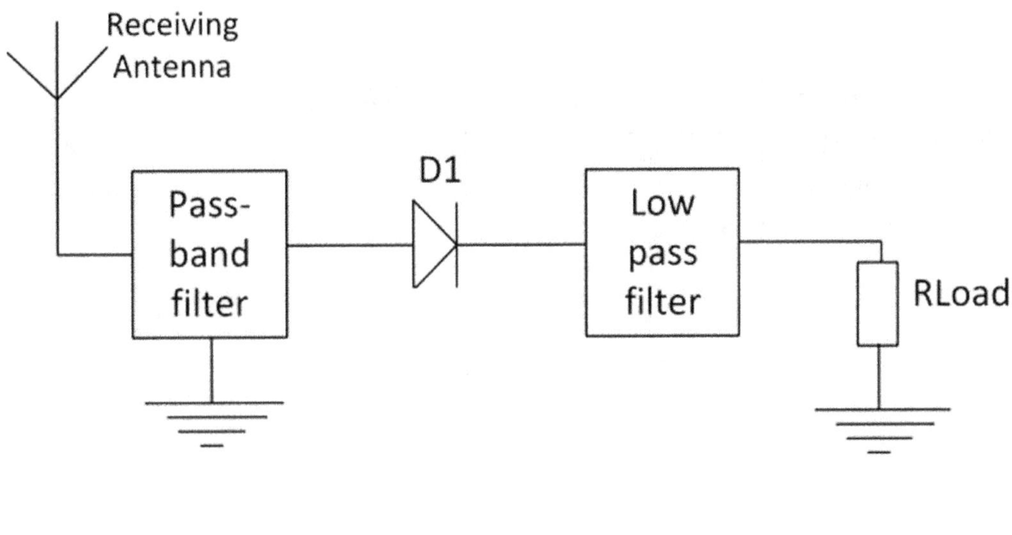

*Figure 5. Shunt mounted diode rectifier topology*

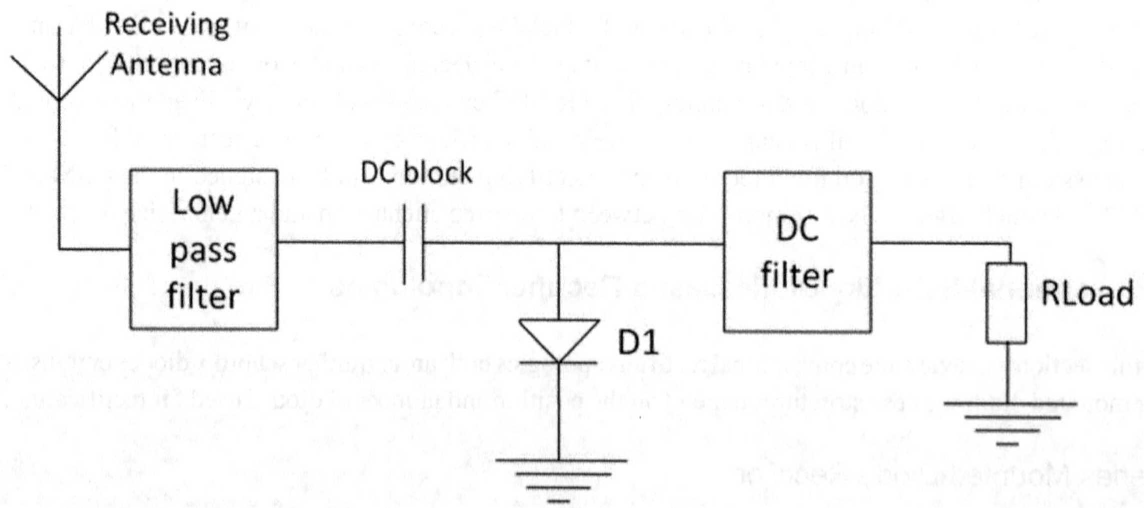


### *Voltage Doubler Rectifier*

This topology is based on the association of the series and shunt mounted diode structures, Figure 6. $D_1$ is conductive during the negative half wave and the input filter gets charged. During the positive half wave, the energy coming from the antenna and the energy stored in the input filter are transferred towards the output filter and load, through $D_2$ which becomes conductive. This structure offers higher DC output voltage levels with less conversion efficiency in comparison with the single diode rectifier structure.

*Figure 6. Voltage doubler rectifier topology*

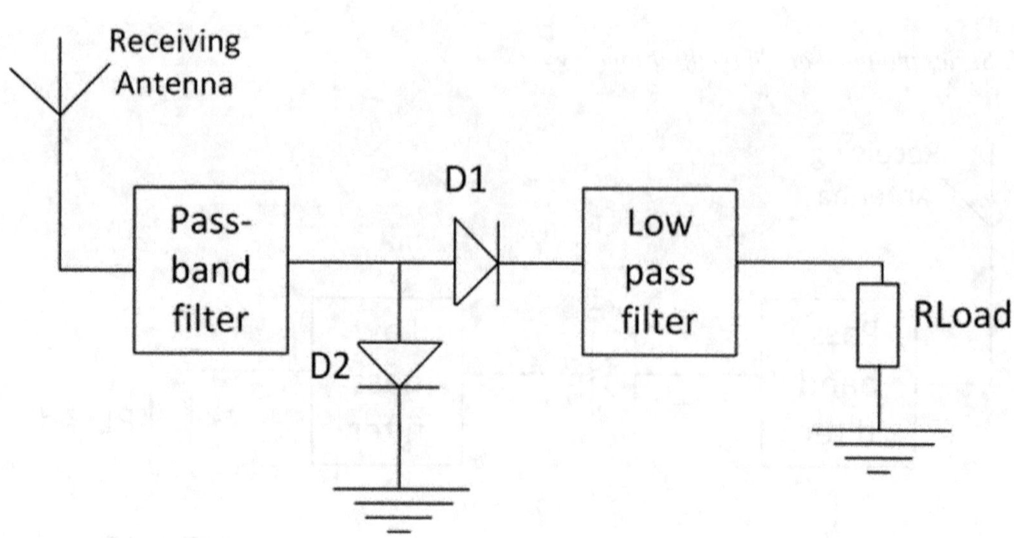

This rectifier can be cascaded several times to obtain an N stage voltage multiplier, as described in Figure 7.

The output DC voltage can be estimated by:

$$V_{outN} = N.V_{out1} \tag{6}$$

where, $V_{out1}$ is the output DC voltage of the single diode rectifier for the same input power.

### *Diode Bridge Rectifier*

The diode bridge rectifier is widely used in low frequency rectification, Figure 8. The principle of full wave rectification consists in the restitution of the entire incident wave at load level compared to the half wave rectification where the load is disconnected from the source for half the time. Diodes $D_1$ and $D_4$ are conductive during the positive wave length, while $D_2$ and $D_3$ are blocked. Diodes conduction states are then inversed during the negative half wave and the current flow direction is the same as in previous cases. This topology offers high power handling capabilities. As two diodes are conductive simultane-

*Figure 7. N stage voltage doubler rectifier*

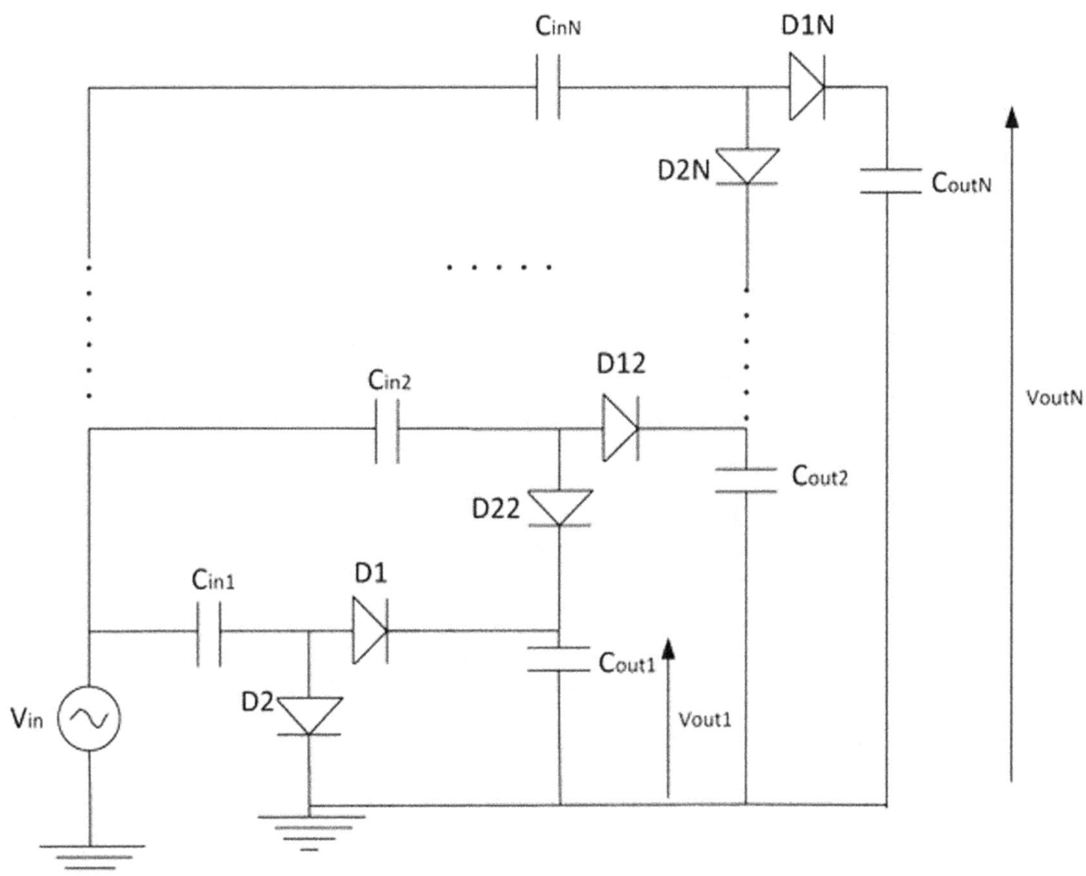

ously, the critical power is reduced by the increase of the resistance per diode. However, losses increase in the case of lower power incidences since the signal has to overcome tow threshold voltages, which makes the topology not suitable for low power applications.

RF designers should find a trade-off between high output voltage and good power conversion efficiency when choosing the rectifier structure. A comparison between the performances of conventional rectifier topologies, mentioned above, as a function of input power, has been widely discussed in several works. In (Marian et al. 2011), a Rectenna Figure of Merit (*RFoM*), expressed in (7), has been evaluated at 2.45GHz to compare the performances of four traditional rectifying topologies: series and shunt mounted diode configuration as well as a single stage and two stage voltage-doubler rectifiers.

$$RFoM(P_{in}) = V_{DCopencircuit} . \eta_{optimalLoad} \tag{7}$$

It reported that voltage doubler rectifier has the lowest *RFoM* for low input power due to their RF-DC conversion efficiency. In the other hand, the two stage voltage doubler circuit has higher *RFoM* for high input power due to its high DC output voltage. The shunt and series diode rectifier has almost the same performances for input power levels around 30μW. However, series mounted diode rectifier topology has the highest *RFoM* for low input power levels (below -5dBm) and hence, it offers the best compromise between DC output voltage and power conversion efficiency.

*Figure 8. Diode bridge rectifier topology*

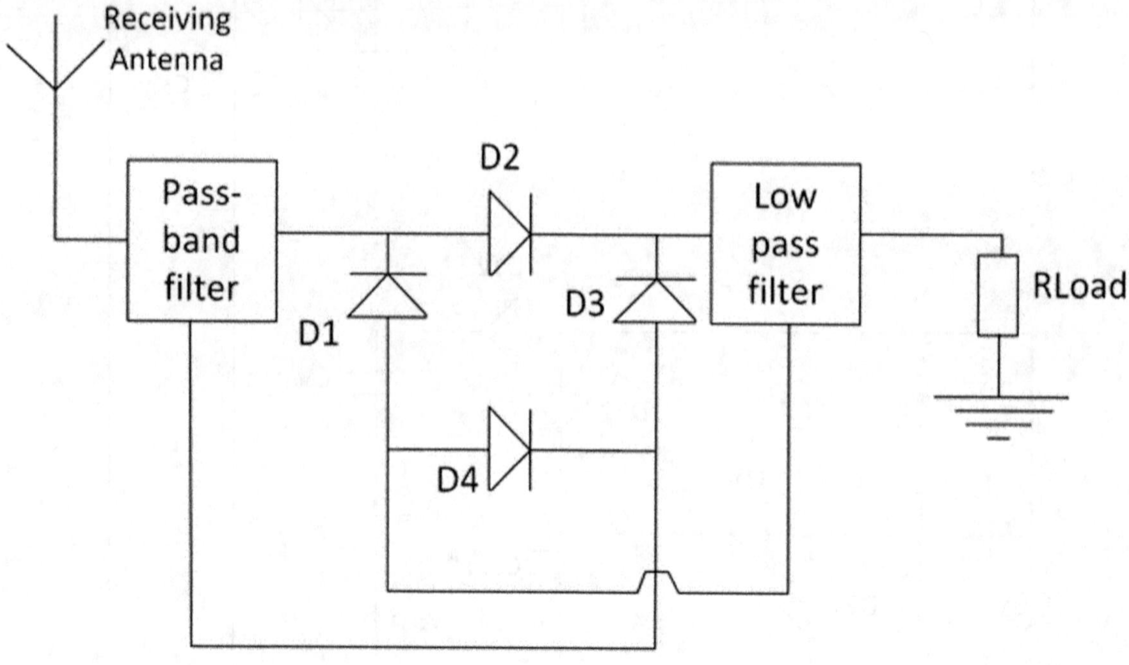

## Conventional Transistor Based Rectifier Topologies

Diode connected transistors are also widely used in RF-DC conversion circuits. Power conversion efficiency (PCE) of a diode connected transistor rectifier is generally worse than that of the schottky diode due to its large threshold voltage ($V_{th}$). Alternative solutions should be proposed to circumvent or diminish the "dead-zone" in voltage rectification and otherwise reduce or "cancel" the effective threshold voltage in standard CMOS rectifier designs. Many $V_{th}$ cancellation techniques proposed in the literature, have been discussed and applied for very low voltage rectifier designs, (Kotani & Ito, 2007; Nakamoto et al., 2007; Umeda et al., 2006). In this section we review two basic MOS transistor based rectifiers, widely described in the state of the art (Karolak et al., 2013). These structures are derived from the topologies described above. The traditional single stage half-wave rectifier is presented in Figure 9.

During the negative half wave, when $V_{D1M} > V_{in}+V_{th}$, the transistor M1 turns on while the transistor M2 turns off as $V_{CG} < V_{out}$. The DC voltage at $V_{CG}$ node is equal to $V_{in}+V_{D1M}-V_{th}$. In the positive half wave, transistor M1 turns off while M2 turns on and the charge is transferred from $C_{in}$ to $C_{out}$. The energy is then stored in the storage capacitor $C_{in}$ at the end of the cycle. The available DC output voltage is determined by:

$$V_{out} = 2(V_{in} - V_{th}) \tag{8}$$

where $V_{in}$ is the input RF signal amplitude and $V_{th}$ is the threshold voltage of MOS transistor. As can be seen, the achievable DC voltage is limited by the threshold voltage of MOS transistor. To increase the output voltage, N stages multiplier structure can be cascaded as depicted in Figure 10. As a result, higher DC voltage can be obtained due to the accumulation of stored charges transferred from the output of one stage to the input of the following one.

*Figure 9. Half-wave MOS transistor based rectifier topology*

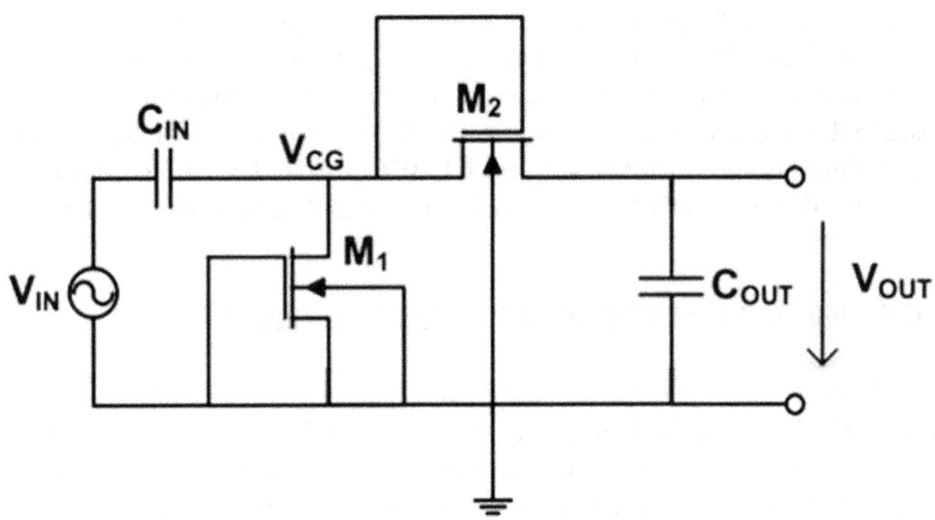

*Figure 10. N stage voltage multiplier structure*

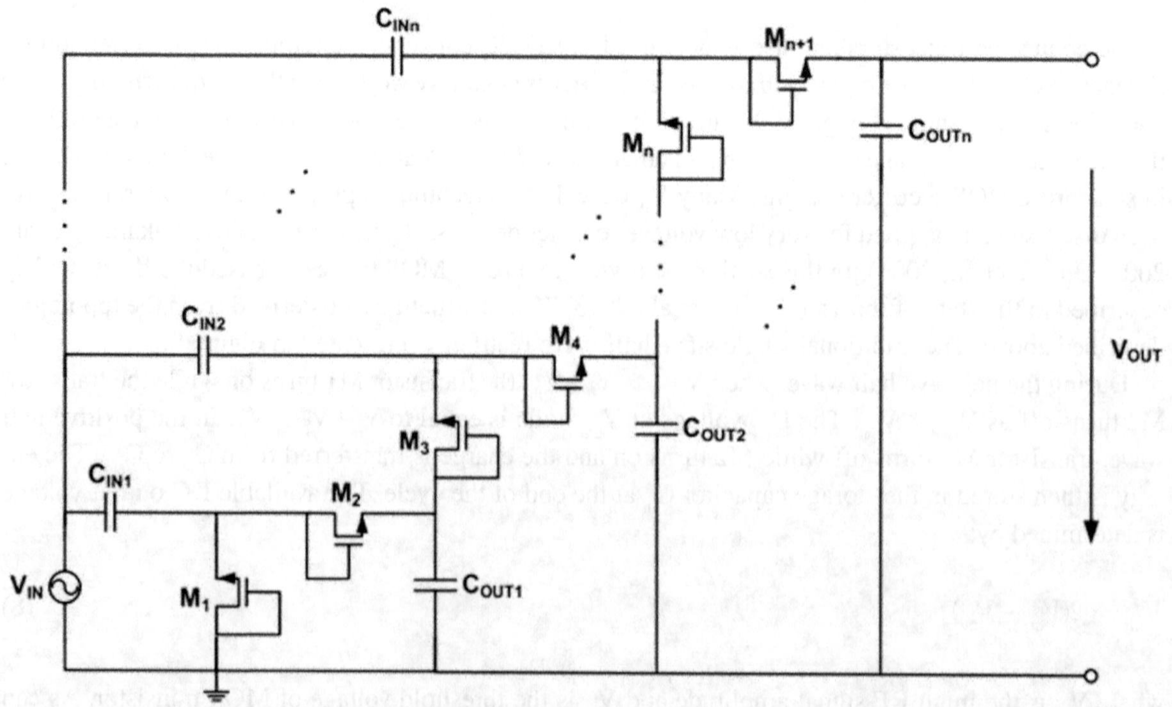


The output voltage at $C_{out}$ of N stage cascaded multiplier depends on the number N and can be expressed by (Karolak et al., 2013):

$$V_{out} = 2.N.(V_{in} - V_{th}) \tag{9}$$

According to (8) and (9), the threshold voltage of MOS transistor has a significant impact on the achievable DC voltage. Rectifier performances optimization technique implies the reduction of threshold voltage. In addition, increasing the number of multiplier stage leads to an increase of power dissipation and thus power conversion efficiency will decrease for low power application. Many solutions have been proposed in the literature to resolve this problem, for example in (Kotani, Sazaki, & Ito, 2009) a high efficiency differential drive CMOS rectifier for UHF RFIDs has been designed in 0.18µm CMOS technology with 67.5% measured PCE at 935MHz for -12.5dBm RF input power and 10KΩ output load.

## DESIGN OF HIGH EFFICIENCY LOW POWER RECTENNAS

According to (2) the power at the receiver position drops off at the rate of $1/D^2$ as distance from the source is increased. Hence, the available power to the receiver decreases by 6dB for every doubling of distance from the transmitter. The decrease rate of power densities becomes much faster than $1/D^2$ in the case of multi-path fading. As a consequence, design of power conversion circuits at very low power incidences and longer operating distances, becomes a critical task. Rectification circuits must be optimized to improve on the minimum power-threshold it takes for the system to operate, (Karthaus & Fisher, 2003; Le, Mayaram, & Fiez, 2008). Several diode-based rectifiers have been developed, many integrated with antennas, with a good comparison presented in (Falkenstein, Roberg, & Papovic, 2012) and in recent research works focusing on low power applications, (Falkenstein, 2011) (Hagerty et al., 2004). The main step in improving rectenna conversion efficiency is the suppression of the harmonic components of the rectifying devices composed by diode or transistor. In (Imai et al., 2011; Mabrouki, Latrach, & Sayegh, 2013; Takhedmit et al., 2010), output and input harmonic filters have been used to increase the conversion efficiency. Harmonic termination is an efficient method that have been widely used for RF PAs efficiency improvement and employed in a single stage shunt diode rectifier design optimization (Roberg et al., 2012). In (Ungan et al., 2009), a high Q crystal quartz resonator achieved 1V DC output voltage and more than 22% conversion efficiency for -30dBm power level but at a medium wave frequency of 24MHz. A spiral rectenna has been designed in (Bouchouicha et al., 2010) and experiments have shown 0.7% conversion efficiency for 3.55nW/cm$^2$ power density at 1.85GHz. The study of antenna impedance effect on power conversion efficiency optimization has been also investigated in (Marian et al., 2011) and the designed rectifier achieved 300mV DC output voltage for -15dBm input power and 6KΩ output load at 2.45GHz. In (Georgiadis, Andia, & Collado, 2010), a rectenna based on a square aperture-coupled patch antenna has been optimized for low power application at 2.43GHz using both electromagnetic (EM) and harmonic balance simulations (HB). The circuit exhibits 15.3% and 11.3% conversion efficiencies at 0.15μW/cm$^2$ power density (-19.2dBm) for the vertical and horizontal polarizations respectively.

1. Rectenna modeling

Modeling and analyses of rectenna element should take into account the behavior and interaction between several parameters such as non-linear components, radiating components, and transmission lines. Generally, the design and optimization of each sub-part of the circuit separately from the one other will induce discrepancies between simulation and experimental results. Among the approaches used for HF devices analyses, there is the electromagnetic technique that takes into account the electromagnetic phenomena in the device and which is based on Maxwell equations solutions. Analytic modeling of circuit is also another way to design high efficient devices, however its accuracy is still constrained by the model simplifications and approximations and also it is only valid in a well-defined domain. Rectenna optimization should offer a good compromise between power conversion efficiency and DC output voltage. To do so, design simulations should take into account all the interactions and coupling between the different parts of the circuit and in the same environment. Several electromagnetic tools can be used for the designs; they are usually adapted in the case of simulated devices comparable in size to the wavelength. The main difference between them consists in the method used for solving Maxwell's laws. This can be done either in time domain or in frequency domain. Time domain analysis takes into account non-linear component behavior and can do a wideband analysis in a single simulation. However the computing time it takes can be very long. Finite Difference Time Domain (FTFD) and the Transmission Line Method (TLM) are the most commonly used time domain method. Modeling non-linear elements in the frequency domain, in the other side, seems to be more difficult and the analysis is generally faster than the one in time domain. The Method of Moments (MOM) and Finite Elements Method (FEM) are the most common techniques used in frequency domain analyses. Developing accurate models that predict non-linear behavior of diode or transistor based rectenna is generally a challenging task for RF designers. In (Akkermans et al., 2005) an analytical model for low power rectenna has been proposed for the series mounted diode rectifier topology.

2. Schottky diode's choice

Schottky diode characteristics have a significant role in rectenna conversion efficiency optimization. Specifications on schottky diodes choice for RF-Dc conversion circuits are mainly:

- Larger operation frequency (many GHz).
- Good sensitivity for low input power densities.
- High conversion efficiency.
- Low cost.

The equivalent circuit model of schottky diode is given in Figure 11. The resistance $R_s$ is the parasitic series resistance of the diode, the sum of the bond-wire and lead-frame resistance, the resistance of the bulk layer of silicon, etc. RF energy coupled into $R_s$ is lost as heat. Thus, it does not contribute to the rectified output of the diode. $C_d$ is the parasitic junction capacitance of the diode, controlled by the thickness of the epitaxial layer and the diameter of the Schottky contact. Schottky diode performances mainly depend on two elements of its equivalent circuit model, the junction resistor $R_j$ and the junction capacitor $C_d$. $I_d$ is the current passing through the junction resistance $R_j$ and $V_d$ is the voltage at its two ends. $L_p$ is the parasitic inductance. $C_p$ is the parasitic capacitance.

The junction capacitance expression is given by (Kilkowski, 1994):

$$C_d = C_{j0} \cdot \left(1 - \frac{V_d}{V_j}\right)^{-M} \quad (10)$$

where $C_{j0}$ is the zero-bias junction capacitance (0.7pF for HSMS-2820, 0.18pF for HSMS-2850 and HSMS-2860).

$V_j$ is the junction potential (0.65V for HSMS2820 and HSMS2860, 0.35V for HSMS-2850). M is the grading coefficient (1/2 for an abrupt junction, 1/3 for a linearly graded junction).

Implemented on metal-semiconductor junction rather than P-N junction, Schottky diodes are characterized by fast switching time, low reverse-recovery time, low forward voltage drop, and low junction capacitance. As shown in Figure 11, the junction capacitor is connected in parallel with the junction resistor. Thus, the distribution of electricity depends on the impedances of the junction capacitor and resistor. Diode with small junction capacitance converts better the RF energy and hence, has good rectifying performances.

For a single diode configuration, the conduction loss can be computed, in terms of the current and the series resistance, by (11) (Merabet et al., 2009):

*Figure 11. Equivalent circuit model of schottky diode*

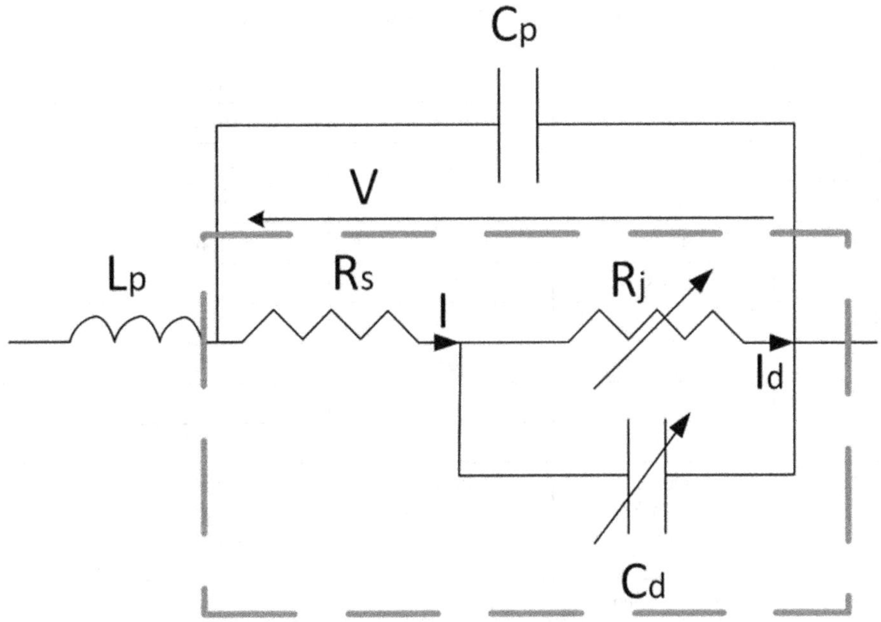


$$P_{loss} = \frac{1}{T}\int_{\frac{T}{2}+t_0}^{t_0} R_s . i_d\left(t\right)^2 dt \tag{11}$$

where T is a rectifying period.

As can be concluded from 11, higher value of the junction resistance $R_s$ will induce the increase of conduction losses. In RF-DC conversion circuits, fast switching time equates to high speed capability and lower forward voltage drop equates to less power dissipation when conducting. Schottky diodes with low forward voltage and small junction capacitance are the well suited for RF energy harvesting applications.

The impact of schottky diode HSM-28xx parameters on power conversion efficiency has been studied in (Zhou, 2013) for a single diode rectifier configuration and a good comparison has been provided. This study can be resumed by analyzing the typical parameters of schottky diode HSM-28xx given in Table 1.

For low power incidences levels, diodes HSM-2850 is the best candidate for low power conversion efficiency circuits thanks to is very low forward voltage drop $V_f$ (150mV) and its small junction capacitance (0.18pF), however, it is limited for application below 1.5GHz. HSMS-2860 is also well suited for low power rectifying circuits in both GSM and ISM bands. In the other hand diodes HSM-2810 and HSM-2820 are suited for high power applications due to their small series resistances and large breakdown voltages. In (Roberg et al., 2012), the SMS7630 schottky diode has been used for high efficiency rectenna design at 2.45GHz for medium to high power conversion (beyond 0dBm). This diode is also well suited to low power conversion thanks to its very low forward voltage drop $V_f$ (60mV-240mV) and its small junction capacitance (0.14pF). In this chapter, diodes HSM-2850 and HSM-2860 will be used to design low to medium power rectifiers in th GSM and ISM bands.

3. Low power rectenna designs.

*Table 1. Typical parameters of schottky diode HSMS-28xx [27]*

| Diode Type | HSMS-2810 | HSMS-2820 | HSMS-2850 | HSMS-2860 |
|---|---|---|---|---|
| BV(V) | 25 | 15 | 3.8 | 7 |
| $I_{BV}$(A) | 10E-5 | 10E-4 | 10E-4 | 10E-5 |
| $C_{j0}$(pF) | 1.1 | 0.7 | 0.18 | 0.18 |
| $E_G$(eV) | 0.69 | 0.69 | 0.69 | 0.69 |
| $I_s$(A) | 4.8E-9 | 2.2E-8 | 3E-6 | 5E-8 |
| N | 1.08 | 1.08 | 1.06 | 1.08 |
| $R_s$(Ω) | 10 | 6 | 25 | 5 |
| $V_j$(V) | 0.65 | 0.65 | 0.35 | 0.65 |
| $X_{TI}$ | 2 | 2 | 2 | 2 |
| M | 0.5 | 0.5 | 0.5 | 0.5 |
| $V_f$ | 400mV ($I_f$=1mA) 1V ($I_f$=35mA) | 340mV ($I_f$=1mA) 0.5V ($I_f$=10mA) 0.7V ($I_f$=30mA) | 150mV ($I_f$=0.1mA) 250mV ($I_f$=1mA) | 350mV ($I_f$=1mA) 0.6V ($I_f$=30mA) |
| Power level | Low Flicker noise | >-20dBm | <-20dBm | $P_{in}$<-20dBm (freq<1.5GHz) $P_{in}$>-20dBm (freq>4GHz) |
| Frequency Band | RF | RF | <1.5GHz | 915MHz-5.8GHz |

a. Low power high efficiency rectenna design in the 900MHz band.

As explained previously, the main challenges in rectenna design is to achieve high conversion efficiency for low power incidences while respecting today's system miniaturization trend. Most of the rectenna design optimizations have been carried out in the ISM band at 2.45GHz. Rectenna miniaturization becomes more difficult for frequencies in the GSM 900MHz band due to the increasing size of the receiving antenna. In this section, we propose a miniaturized high efficiency rectifier dedicated to low power operations (below -10dBm). The rectifier topology we propose is depicted in Figure 12 (a), (Mabrouki et al., 2014). It is based on a voltage booster formed by a voltage doubler type « Latour ». The circuit includes two capacitors $C_1$ and $C_2$ and two rectifier diodes $D_1$ and $D_2$. The voltage source supplies two separate branches in parallel. A first branch is constituted by the series connection of the diode $D_1$ and the capacitor $C_1$, and the second branch is constituted by the diode $D_2$ and the capacitor $C_2$ in series, one terminal of the capacitor $C_1$ being connected to one terminal of the capacitor $C_2$. One capacitor is charged at the positive half wave and the other at the negative half wave. The output voltage $U_{out}$ is collected at the terminals of the circuit formed by the two capacitors $C_1$ and $C_2$. For open circuit case, the output voltage is twice the input peak voltage. The circuit architecture including passive components and microstrip lines is described in Figure 12(b). According to the analysis presented in Table 1, schottky diode HSMS-2852 is the well suited candidate for this rectifying circuit thanks to its low threshold voltage of 150mV and its low junction capacitance $C_{j0}$ of 0.18pF. An input inductance and a shunt capacitor of 3 pF have been added to improve the input matching. The input inductance has a

*Figure 12. (a) "Latour" doubler structure; (b) simulated schematic circuit; (c) photograph of the rectifier*

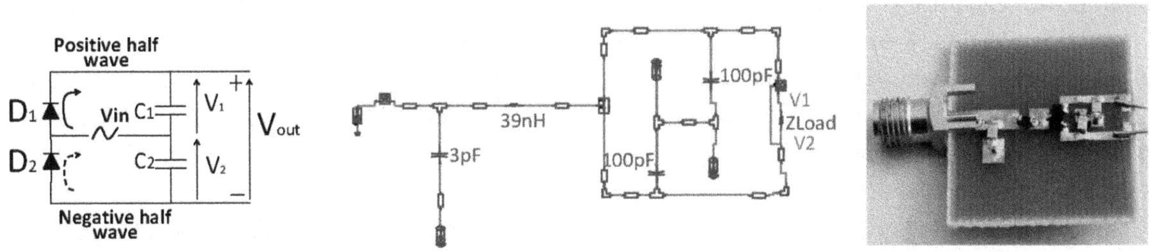


value of 39nH and a Q factor of 35 at 800MHz. The output capacitors $C_1$ and $C_2$ are equals to 100pF. The proposed structure is designed on 1.6mm thickness FR4 substrate (εr=4.5 and tanδ=0.025), Figure 12(c). The circuit has been simulated under Harmonic balance (HB) routine from Agilent ADS. Large Signal S-parameters (LSSP) simulations have been carried out to optimize the input matching network of the rectifier to the standard 50Ω input impedance at 915MHz (for input power below -10dBm). The use of CMS components for matching helps reducing the area of the circuit in comparison with the use of microstrip lines. However, attention should be paid to the choice of the Q factor of the passive component in order to reduce the effect of dispersion on the sensitivity of the circuit. The implemented circuit occupies an area of 2cm*2.3*cm.

The measurement setup, used to characterize the circuit, includes a directional coupler that monitors the second harmonic power level and a power meter to measure the reflected power. The DC output voltage $V_{out}$ is obtained by subtracting the voltage $V_-$ from voltage $V_+$ . In Figure 13 (a) and 13(b), the measured output DC voltage and conversion efficiency of the circuit are plotted for three different input power levels (-20dBm, -15dBm, -10dBm) for 10 KΩ resistor load as a function of frequency. Maximum DC voltages of 144mV and 610mV are obtained at 850MHz for -20dBm and -10dBm respectively. The circuit exhibits good correlation between simulated and experimental results. A 21% maximum power conversion efficiency is measured at 850MHz for -20dBm input power and 10KΩ resistor load. The results are well situated in the state of the art, in comparison with the few works dealing with low power rectifition designs using zero bias schottky diodes at 2.45GHz (Georgiadis et al. 2010; Marian et al. 2012; Zhou 2013).

*Figure 13. (a) Measured output voltage; (b) Measured conversion efficiency versus frequency for different input power levels*

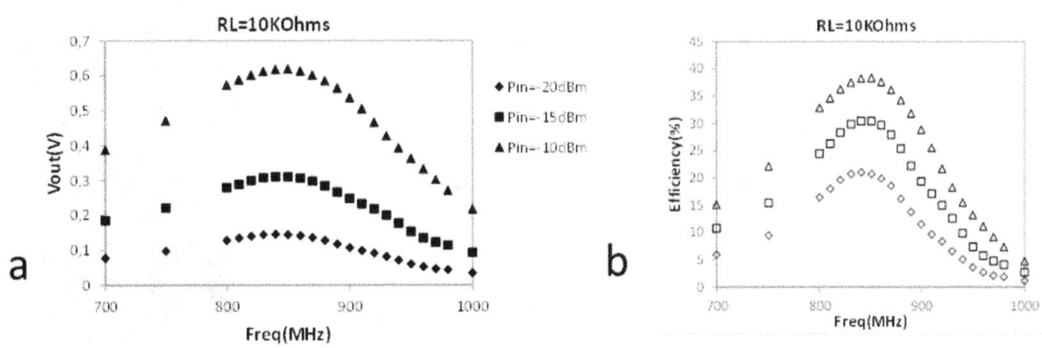

It is also important when characterizing a rectifier circuit to analyze its response as a function of input power even if it is dedicated to low power applications. Hence, the measured output voltage and conversion efficiency versus input power at 850MHz have been carried out and plotted in Figure 14(a) and 14(b) respectively. 610mV and 1.17V DC output voltages are measured respectively at -10dBm and -5dBm input powers and 46% maximum RF-DC conversion efficiency is achieved at 0dBm input power. It can be seen that efficiency drops off beyond 0dBm input power due to the diode characteristics suitable for low power operations.

We designed another alternative of the circuit presented in Figure 12(b). The idea is to analyze the impact of input matching circuit on the performances of the rectifier and also to reduce its area. The architecture is slightly different from the first one and different CMS component values are used to optimize the circuit. The schematic of the rectifier and its photograph are given in Figure 15(a) and 15(b). The implemented circuit is more compact than the first presented structure with 35% occupied area's reduction.

*Figure 14. (a) Measured output voltage; (b) Measured conversion efficiency versus input power at 850MHz*

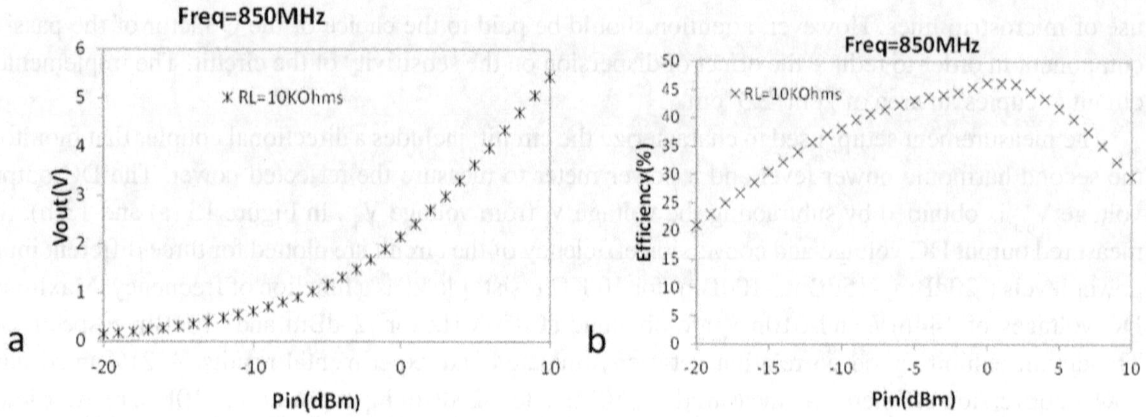


*Figure 15. (a) Simulated schematic circuit; (b) photograph of the rectifier*

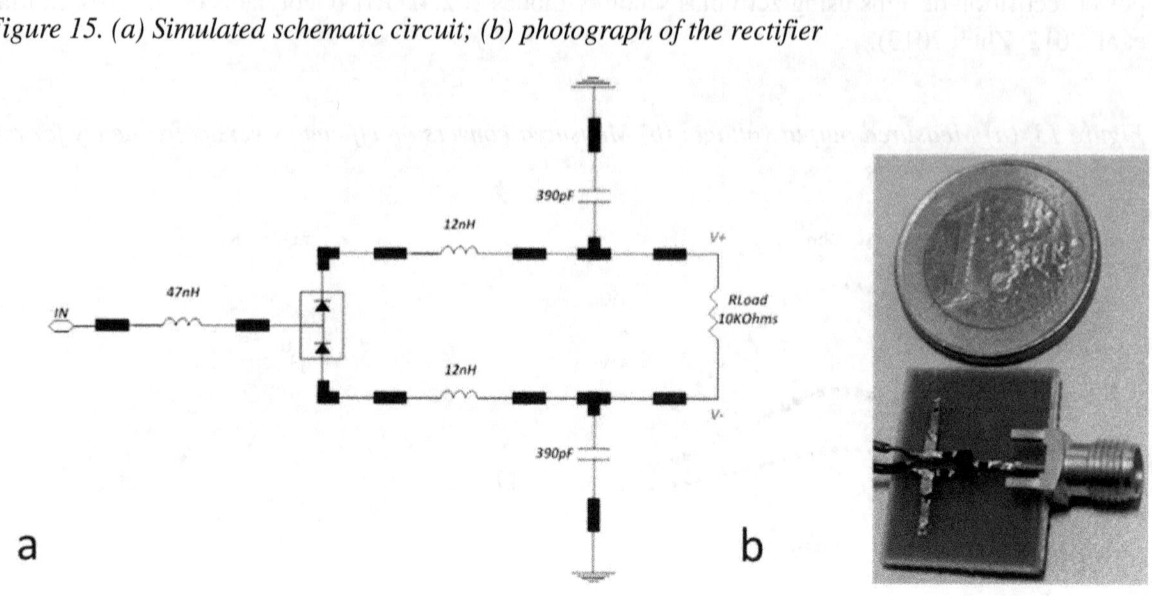

The experimental analyses, carried out previously, have been also repeated for this rectifier and the different curves of measured DC output voltage and conversion efficiency versus frequency and versus input power are plotted respectively in Figure 16(a) and 16(b) for 10KΩ resistor load. The circuit is well matched at 875MHz and exhibit almost 4% conversion efficiency optimization in comparison with the first one. In Figure 17(a) and 17(b), the impact of resistor load value on the circuit performances is illustrated. As can be observed, for low input power (below-10dBm), reducing the resistor load by 32% has a slight effect on conversion efficiency. Higher values of load are not generally practical because they are hard to emulate with a boost converter (Dolgov, 2010; Falkenstein, 2011). This structure is more efficient and compact than the first one.

A miniaturized rectenna using the mentioned above rectifier circuit, Figure 18(a), has been characterized at 875MHz. It uses a half loop antenna with one SRR cell with 10cm*10cm ground plane (Ramanandraibe et al., 2013). The addition of SRR helps miniaturizing the antenna by reducing the

*Figure 16. (a) Measured output voltage; (b) Measured conversion efficiency versus frequency for different input power levels*

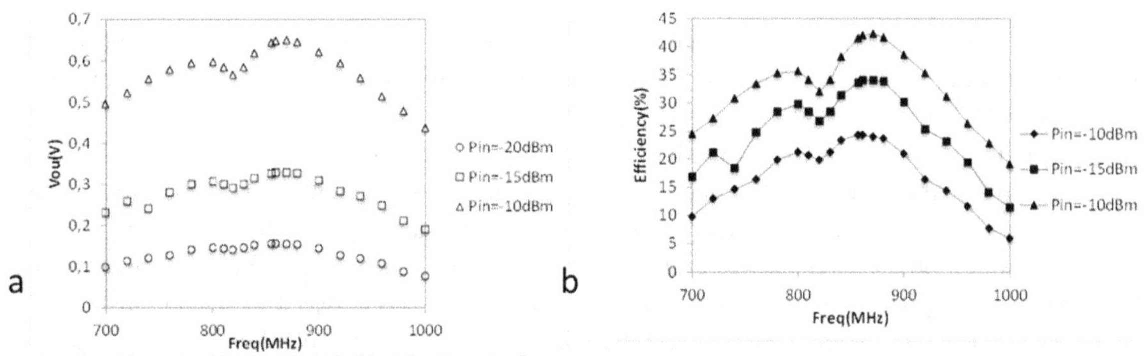


*Figure 17. (a) Measured output voltage; (b) Measured conversion efficiency versus input power at 875MHz for different resistor load values*

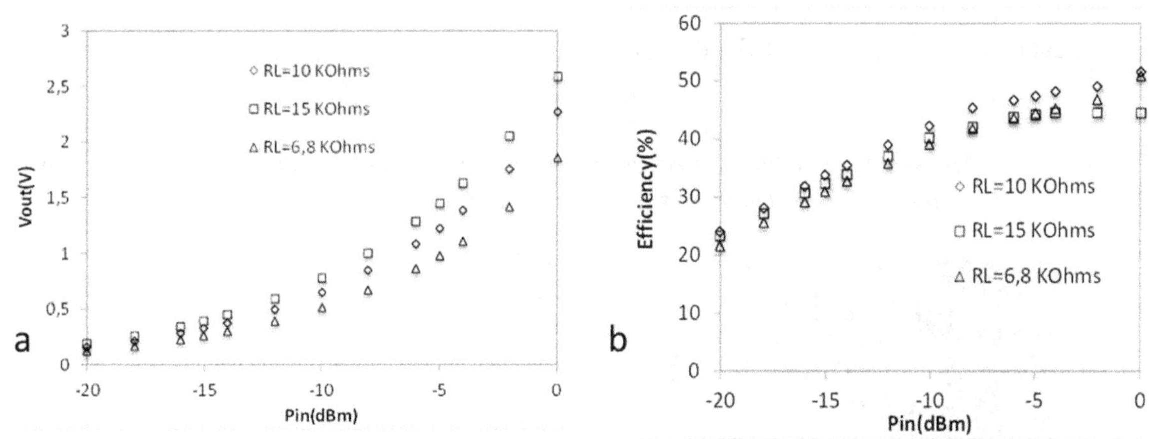

*Figure 18. (a) Photograph of the miniaturized rectenna; (b) simulated and measured gain of the antenna*

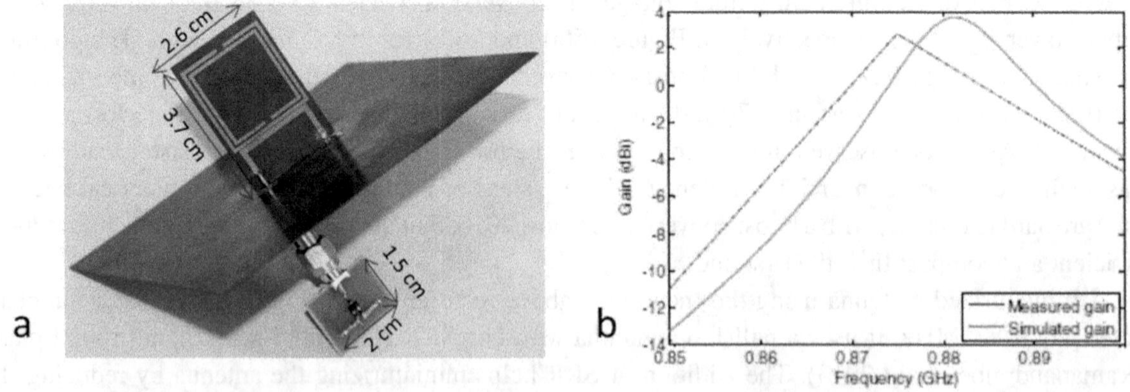


resonance frequency (70% of frequency shit). It contributes also to the impedance matching of the antenna and hence it increases the antenna efficiency. The antenna has a gain of 2dB at 875MHz with 71% efficiency, Figure 18(b).

The measurement setup of the rectenna in anechoic chamber is presented in Figure 19(a). A horn antenna with 4dBi gain is used as transmitting antenna placed at a distance of 1m from the rectenna. The evaluation of DC output voltage as well as conversion efficiency as a function of th received power $P_r$, at the input of the rectifier, calculated according to (2), are presented in Figure 19(b) and 19(c) respectively.

The rectenna exhibits good performances with 19% conversion efficiency at -20dBm input power for 10KΩ DC load. The 5% efficiency degradation is due to the matching losses between the rectifier and the antenna since the two circuits were designed separately. This highlights the importance of the whole rectenna modelling which should take into account the interactions between the different sub-parts of the circuits. The experimental results of the designed antenna are still well situated in the state of the art in comparison with the rectenna performances developed in (Georgiadis et al., 2010) which show 15.3% and 11.3% conversion efficiencies at 0.15μW/cm$^2$ power density (-19.2dBm) for the vertical and horizontal polarizations respectively. The rectenna size reduction is also an important result especially for low power applications in the 900MHz GSM band.

b. Dual bands lower power rectifier design.

*Figure 19. (a) Measurement setup in anechoic chamber; (b) measured output voltage; (c) measured efficiency of the miniaturized rectenna versus the received power*

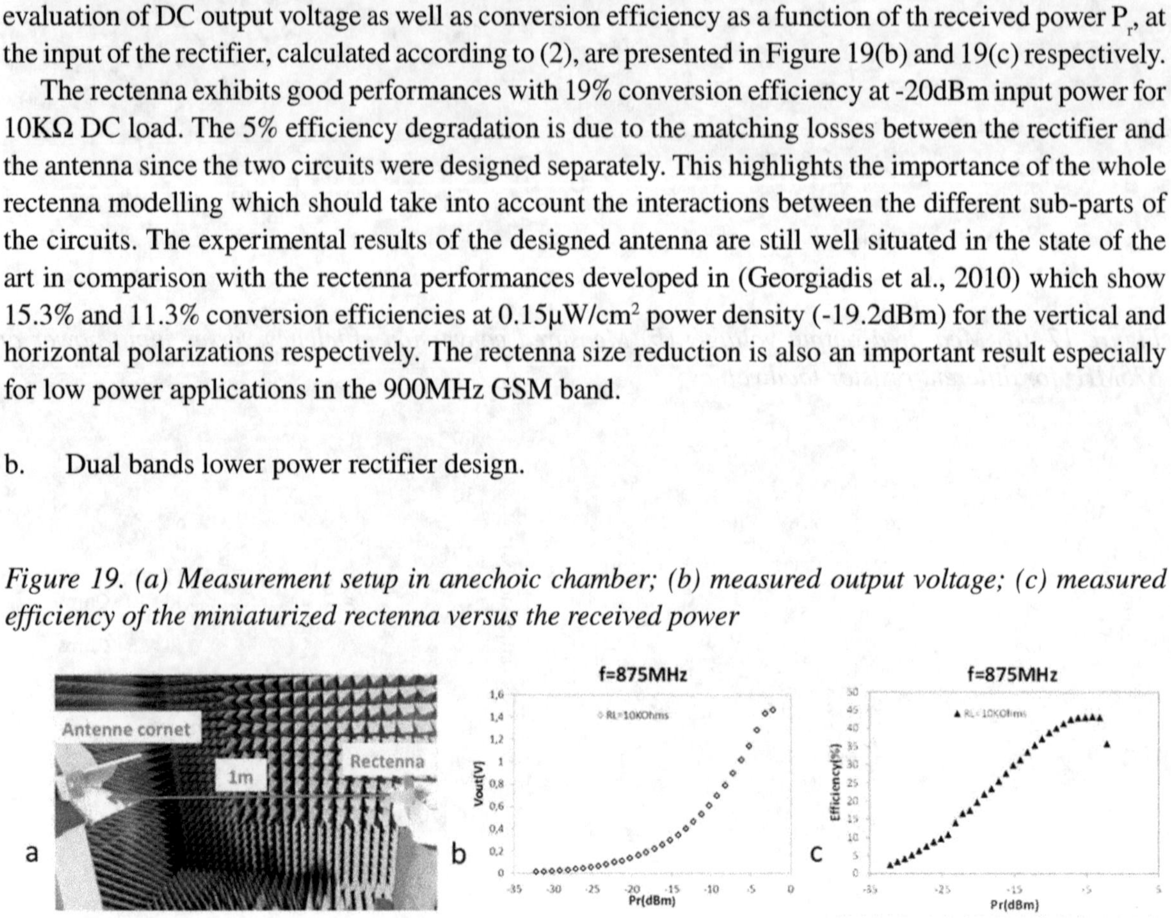

Inspired from the previously presented circuits, we developed a dual band rectifier at 900MHz and 1800MHz GSM bands. The circuit is based on tow voltage doubler topologies optimized at the two operating frequencies; Figure 20(a) and 20(b). The uses of CMS components for impedance matching helps miniaturizing the circuit but it still a critical point that affects the circuit sensitivity and especially induces frequency shift depending on the component Q factor. The HSMS-2852 schottky diode is also selected for the low power operations.

The optimal measured characteristics of the rectifier are obtained at 830MHz and 1.735GHz, as illustrated in Figure 21(a) and 21(b). 15% and 35% conversion efficiencies are achieved at 830MHz for -20dBm and -10dBm input power respectively. The measured efficiencies at 1.735GHz for the same input power are 25% less than the ones at 830MHz. This is due to the limited operation frequency of the HSMS-2852 schottky diode as explained in Table 1. One alternative to optimize the rectifier at 1800MHz can be the use of the HSMS-2862 diode which is also suitable for low power incidences (beyond -10dBm) and for higher frequency operations.

The impact of load resistor values has been also studied for this rectifying structure and measurement results are depicted in Figure 22(a) and 22(b). The same conclusion can be cited here for the uses of DC load value between 6KΩ and 10KΩ for which the rectifier has approximately the same behavior in terms of conversion efficiency.

Future optimization of these low power rectifiers implies the uses of high Q CMS components to ensure a better sensitivity of the circuits. Modelling the whole rectenna element in the same environment should also be processed to achieve higher conversion efficiencies with less discrepancy between simulated and experimental results.

4. Medium power rectifier designs.

In this section, we will present the design of two rectifiers for medium to high power applications (0dBm-20dBm). According to Table 1, diode HSMS-2862 can be selected for this power range operation. The first rectifying structure is based a single series mounted diode topolofy. The input band-pass filter is described in Figure 23(a) (Douyere, Alicalapa, & Lan Sun Luk, 2008; Mabrouki et al., 2013). The structure includes a microstrip quarter wave length stub connected to the ground. The filter has been optimized to prevent reflections at the fundamental frequency. The output DC filter, Figure 23(b),

*Figure 20. (a) Photograph of the dual band rectifier; (b) simulated schematic circuit*

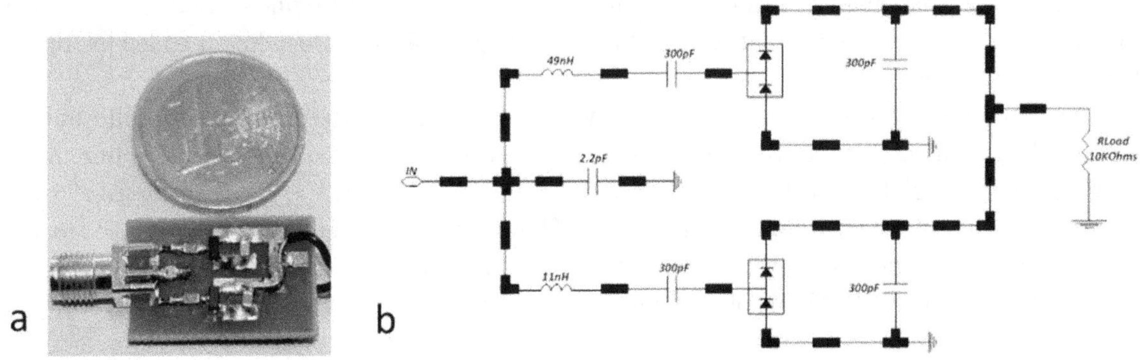

*Figure 21. (a) Measured output voltage; (b) measured conversion efficiency versus frequency for different input power levels*

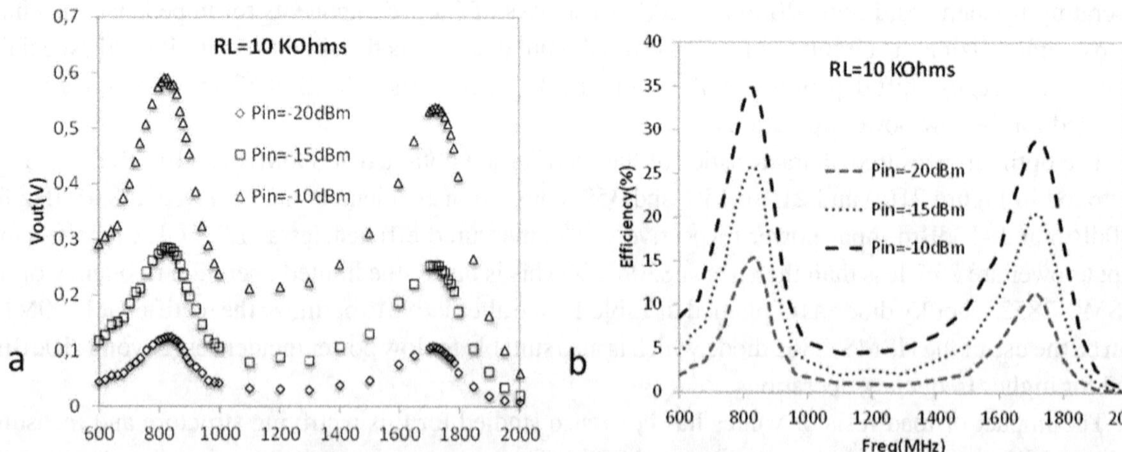


*Figure 22. (a) Measured output voltage; (b) measured conversion efficiency versus input power for different resistor load values*

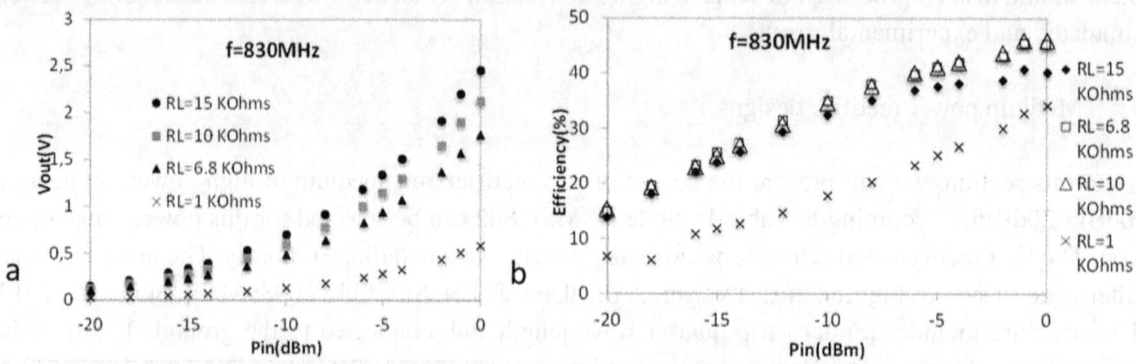


is composed by two open circuit shunt stubs. A quarter wave length stub (λ/4) is used to present a short circuit termination at the fundamental and odd harmonics (3f, 5f…). The second (λ/8) stub presents a short circuit termination at the second harmonic. A large (λ/4) microstrip line is used as a capacitance to minimize the ripple in the output voltage. Finally, A large (λ/4) microstrip line is added to eliminate the forth harmonics.

The rectifier is well matched at 2.45GHz with a measured S11 of -16dB. Figure 24(a) illustrates the measured DC voltage ($V_{out}$) for two values of $R_{load}$. Increasing the DC load beyond the optimum value (1kOhms) will certainly increase $V_{out}$. In fact, an increase of 200mV of $V_{out}$ can be achieved at 10dBm input power when $R_L$ is varied from 1kΩ to 1.68 KΩ However, the conversion efficiency will be reduced. This is depicted in Figure 24(b), where a maximum efficiency of 54% is obtained at 12dBm input power for 1KΩ DC load and only 42% is obtained for 1.68 KΩ at the same input power.

*Figure 23. (a) Input filter, (b) output filter, (c) photograph of the rectifier*

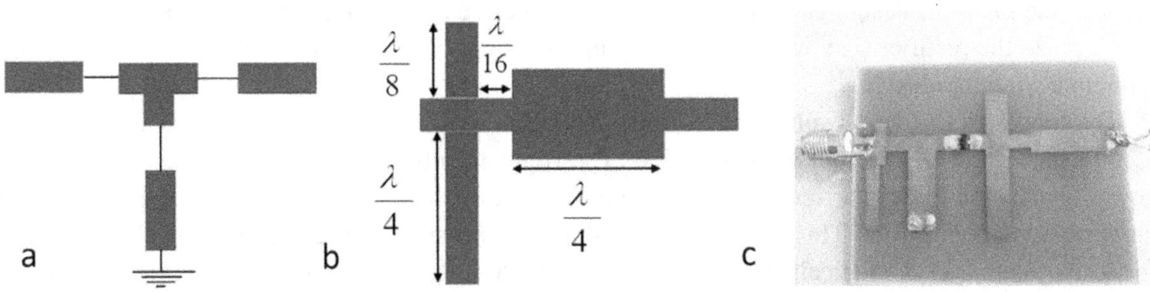


*Figure 24. (a) Measured output voltage for two resistor load values, (b) Power conversion efficiency as a function of input power*

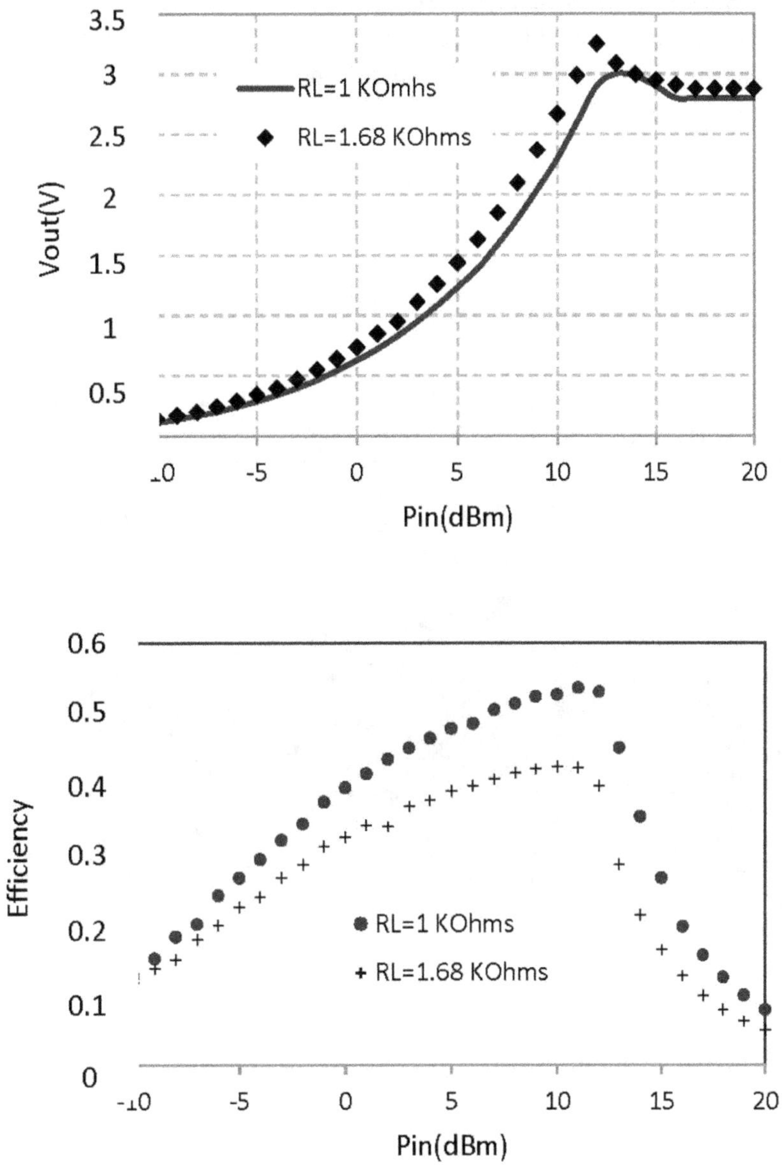

A single stage voltage doubler rectifier has been also designed and measured at 2.45GHz, Figure. 25. We adopted the same design guidelines of the series structure to optimize the performances. However, to minimize the rectifier size, we chose to suppress the input band-pass filter while keeping the same input matching topology.

The simulated DC output voltage of the voltage doubler is compared to the one of single series diode rectifier in Figure. 26(a) for a DC load of 1KΩ. The voltage doubler rectifier is dedicated to high input power application in comparison to the series topology. More than 3V DC voltage can be achieved for input power above 14dBm. This result is also validated by measurements as illustrated in Figure 26(b). In fact $V_{out}$ increases dramatically (more than 4V) for high input power (above 10dBm) in comparison with the series structure DC voltage curve.

As a conclusion, depending on the application, there are several guidelines that should be respected to ensure optimal performances in the specified power range and operation frequencies:

- Diode choice depending on its characteristics.
- Rectifying topology selection with respect to miniaturization trend (which technology?).
- The whole rectenna modelling in the same environment.

*Figure 25. Photograph of the voltage doubler rectifier*

*Figure 26. (a)Simulated $V_{out}$ for the two rectifiers, (b) measured $V_{out}$ of the voltage doubler rectifier*

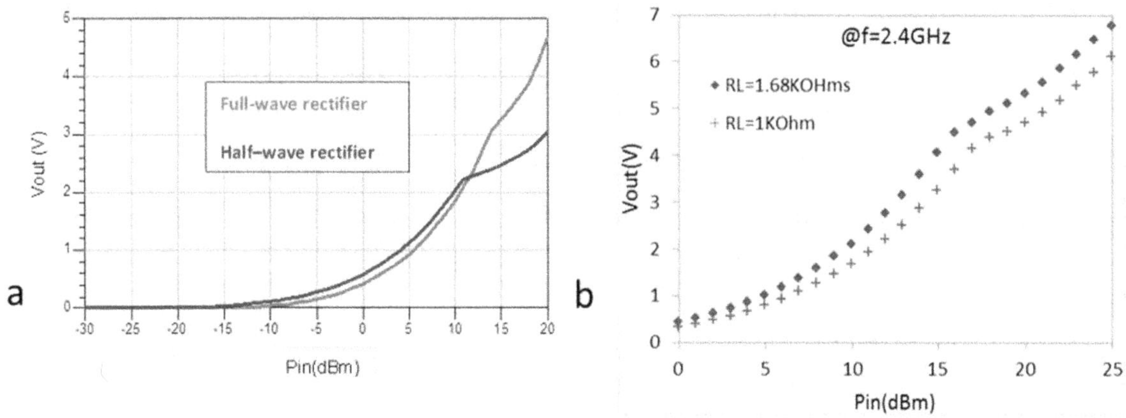


## CONCLUSION

This chapter presented an overview of wireless power transfer with focuses on the design of high efficiency low power rectifying circuits. A background survey of RF energy harvesting techniques has been firstly presented. Then, the performances of conventional rectifier topologies have been analyzed and discussed. The most effective rectenna designs, recently published in the state of the art, have been also reviewed. Design considerations for low power rectifier operations has been detailed and new high efficient low power rectifiers has been designed and measured in both the GSM and ISM bands.

## REFERENCES

Akkermans, J. A. G., van Beurden, M. C., Doodeman, G. J. N., & Visser, H. J. (2005). Analytical models for low-power rectenna design. *IEEE Antennas and Wireless Propagation Letters*, *4*(1), 187–190. doi:10.1109/LAWP.2005.850798

Al Agha, K., Bertin, M.-H., Dang, T., Guitton, A., Minet, P., Val, T., & Viollet, J.-B. (2009, October). Which wireless technology for industrial wireless sensor networks? The development of OCARI technology. *IEEE Transactions on Industrial Electronics*, *56*(10), 4266–4278. doi:10.1109/TIE.2009.2027253

Bouchouicha, D., Latrach, M., Dupont, F., & Ventura, L. (2010, Sept). An experimental evaluation of surrounding RF energy harvesting devices. *The 40th Microwave Conference*, 1381-1384.

Brown, W. C. (1984, September). The history of power transmission by radio waves. *IEEE Transactions on Microwave Theory and Techniques*, *32*(9), 1230–1242. doi:10.1109/TMTT.1984.1132833

Brown, W. C. (1991). An experimental low power density rectenna. *IEEE MTT-S International Microwave Symposium Digest*, 197-200.

*Diminution de l'expositionaux ondes électromgnétiques émises par les antennes relais de téléphonie mobile.* (2013) Technical report. Retrieved from http://www.developpement- durable.gouv.fr/IMG/pdf/rapport_COPIC_31_juillet_2013.pdf

Dolgov, A., Zane, R., & Papovic, Z. (2010). Power management system for online low power RF energy harvesting optimization. *Circuits and Systems I: Regular Papers, IEEE Transactions on, 57(7),* 1802-1811.

Douyère, A., Alicalapa, F., & Lan Sun Luk, J.-D. (2008). High efficiency microwave rectenna circuit: Modeling and design. *Electronics Letters*, *44*(24), 1409–1410. doi:10.1049/el:20081794

Edwards, C. (1947, Nov). Microwave converters. *Proc. IRE*, 35(11), 1181-1191.

Epp, L. W., Khan, A. R., Smith, H. K., & Smith, R. P. (2000, January). A compact dual polarized 8.51 GHz rectenna for high voltage (50V) actuator applications. *IEEE Transactions on Microwave Theory and Techniques*, *48*(1), 111–120. doi:10.1109/22.817479

Falkenstein, E. A. (2011). *Characterization and design of a low-power wireless power delivery system* (PhD dissertation). University of Colorado at Boulder, CO.

Falkenstein, E. A., Roberg, M., & Papovic, Z. (2012, July). Low-power wireless power delivery. *IEEE Transactions on Microwave Theory and Techniques*, *60*(7), 2277–2286. doi:10.1109/TMTT.2012.2193594

Fujino, Y., Ito, T., Fujita, M., Kaya, N., Matsumoto, H., Kawabata, K., & Onodera, T. et al. (1994, April). A driving test of a small DC motor with rectenna array. *IEICE Transactions on Communications*, *E77-B*(4), 526–528.

Georgiadis, A., Andia, G., & Collado, A. (2010). Rectenna deisgn and optimization using reciprocity theory and harmonic balance analysis for electromagnetic (EM) energy harvesting. *IEEE Antennas and Wireless Propagation Letters*, *9*, 444–446. doi:10.1109/LAWP.2010.2050131

Glaser, P. E. (1968). Power from the sun. *Science*, *162*(3856), 857–886. doi:10.1126/science.162.3856.857 PMID:17769070

Glenn, J. (1994). *The complete patents of Nikola Tesla*. New York: Barnes and Noble Books.

Gungor, V. C., & Hancke, G. P. (2009, Oct). Industrial wireless sensor networks: Challenges, design principles and technical approaches. *IEEE Transactions on Industrial Electronics*, *56*(10), 4258–4265. doi:10.1109/TIE.2009.2015754

Hagerty, J., Helmbrecht, F., McCalpin, W., Zane, R., & Papovic, Z. (2004, March). Recycling ambient microwave energy with broadband rectenna arrays. *IEEE Transactions on Microwave Theory and Techniques*, *52*(3), 1014–1024. doi:10.1109/TMTT.2004.823585

IEEE Standard for Safety Levels with Respect to Human Exposure to Radio Frequency Electromagnetic Fields, 3 kHz to 300 GHz. (1999). *IEEE Standard C95.1 edition*. Institute of Electrical and Electronics Engineers.

Imai, S., Tamaru, S., Fujimori, K., Sanagi, M., & Nogi, S. (2011, May). Efficiency and harmonics generation in microwave to dc conversion circuits of half-wave and full-wave rectifier types. In *IEEE MTT- S International Microwave Workshop Series on Innovative Wireless Power Transmission: Technologies, Systems, and Applications (IMWS)*, 15 –18. doi:10.1109/IMWS.2011.5877081

International Commission on Non-Ionizing Radiation Protection. (1998). Guidelines for limiting exposure to time-varying electric, magnetic, and electromagnetic fields (up to 300 GHz). *Health Physics*, *74*(4), 494–52. PMID:9525427

Karthaus, U., & Fisher, M. (2003, October). Fully integrated passive UHF RFID transpoder IC with 16.7-μW minimum RF input power. *IEEE Journal of Solid-State Circuits*, *38*(10), 1602–1608. doi:10.1109/JSSC.2003.817249

Kielkowski, R. M. (1994). *Inside SPICE - overcoming the obstacles of circuit simulation, chapter Understanding circuit simulation*. New York: McGraw-Hill.

Korolak, D., Taris, T., Deval, Y., Begueret, J. B., & Mariano, A. (2013). Bulk biased RF energy harvesting dedicated to 900MHz/2.4GHz ISM bands. *Workshop on Circuits and System design*, 176- 179.

Kotani, K., & Ito, T. (2007, Nov). High efficiency CMOS rectifier circuit with self-Vth-cancellation and power regulation functions for UHF RFIDs. *IEEE Asian Solid-State Circuits Conference*, (pp. 119-122). doi:10.1109/ASSCC.2007.4425746

Kotani, K., Sazaki, A., & Ito, T. (2009, November). High efficiency differential drive CMOS rectifier for UHF RFIDs. *IEEE Journal of Solid-State Circuits*, *44*(11), 3011–3018. doi:10.1109/JSSC.2009.2028955

Le, T., Mayaram, K., & Fiez, T. (2008, May). Efficient far field radio frequency energy harvesting for passively powered sensor networks. *IEEE Journal of Solid-State Circuits*, *43*(5), 1278–1302. doi:10.1109/JSSC.2008.920318

Lee, R.-G., Chen, K.-C., Chiang, S.-S., Lai, C.-C., Liu, H.-S., & Wei, M.-S. (2006, May). A backup routing with wireless sensor network for bridge monitoring system. *The 4th Annual Communication Networks and Services Research Conference (CNSR)*, (pp. 161-165).

Mabrouki, A., Latrach, M., & Lorrain, V. (2014, May). High efficiency low power design using zero bias schottky diodes. *13th Edition of IEEE Faible Tension Faibe Consommation*, (pp. 1-4). Monaco, France.

Mabrouki, A., Latrach, M., & Sayegh, Z. (2013, Sept). Design and experiment of RF rectifiers for wireless power transmission. In *13th Mediterranean Microwave Symposium*, (pp. 1-4). Saida, Lebanon. doi:10.1109/MMS.2013.6663079

Mainwaring, A., Culler, D., Polastre, J., Szewczyk, R., & Anderson, J. (2002). Wireless sensor networks for habitat monitoring. In *Proceedings of the 1st ACM International Workshop on Wireless sensor networks and applications, WSNA'02,* (pp. 88-97). New York, USA.

Marian, V., Menudier, C., Thevenot, M., Vollaire, C., Verdier, J., & Allard, B. (2011, June). Efficient design of rectifying antennas for low power detection. *MTT –S International Microwave Symposium Digest*, 1-4.

Marian, V. M., Vollaire, C., Verdier, J., & Allard, B. (2012). Low power rectenna systems for wireless energy transfer. In J. Agbinya (Ed.), *Wireless Power Transfer* (pp. 67–107). Melbourne, Australia: River Publishers.

McSpadden, J. O., & Chang, K. (1994). A dual polarized circular patch rectifying antenna at 2.45GHz for microwave power conversion and detection. In *IEEE MTT-S International Microwave Symposium Digest*, (pp. 1749-1752).

Merabet, B., Cirio, L., Takhedmit, H., Costa, F., Vollaire, C., Allard, B., & Picon, O. (2009). Low-cost converter for harvesting of microwave electromagnetic energy. In *Proceedings of the IEEE Energy Conversion Congress and Exposition*, (pp. 2592-2599). doi:10.1109/ECCE.2009.5316093

Mikeka, C., & Arai, H. (2011). Design Issues in Radio Frequency Energy Harvesting System. In T. Y. Kheng (Ed.), Sustainable energy harvesting technologies – Past, present and future, (pp. 235-256). Intech. doi:10.5772/25348

Nakamoto, H., Yamazaki, D., Yamamoto, T., Kurata, H., Yamada, S., Mukaida, K., & Gotoh, K. et al. (2007, January). A passice UHF RF identification CMOS tag IC using ferroelectric RAM in 0,35um technology. *IEEE Journal of Solid-State Circuits*, *42*(1), 101–110. doi:10.1109/JSSC.2006.886523

Paing, T., Morroni, J., Dolgov, A., Shin, J., Brannan, J., Zane, R., & Papovic, Z. (2007). Wirelessly powered wireless sensor platform. *European Microwave Conference*, (pp. 999-1002).

Papovic, Z. B. (2006, November). Wireless powering for low-power distributed sensors. *Serbian Journal of Electrical Engineering*, *3*(2), 149–162. doi:10.2298/SJEE0603149P

Ramanandraibe, E., Latrach, M., Abdouni, W., & Sharaiha, A. (2013, Sept). A half loop antenna associated with one SRR cell. *International Conference on Electromagnetics in Advanced applications*, (pp. 1442-1445). Torino, Italy. doi:10.1109/ICEAA.2013.6632488

Raytheon Company. (1977, Aug). *Electronic and mechanical improvement of the receiving terminal of a free-space microwave power transmission system* (Technical Report, PT – 4964, NASA Rep. CR-135194). Wayland MA: Raytheon Company.

Roberg, M., Reveyrand, T., Ramos, I., Falkenstein, E. A., & Papović, Z. (2012, December). High-Efficiency Harmonically Terminated Diode and Transistor Rectifiers. *IEEE Transactions on Microwave Theory and Techniques*, *60*(12), 4043–4052. doi:10.1109/TMTT.2012.2222919

Shinohara, N., & Matsumoto, H. (1998, March). Experimental study of large rectenna array for microwave energy transmisison. *IEEE Transactions on Microwave Theory and Techniques*, *46*(3), 261–267. doi:10.1109/22.661713

Takhedmit, H., Merabet, B., Cirio, L., Allard, B., Costa, F., Vollaire, C., & Picon, O. (2010). A 2.45-ghz low cost and efficient rectenna. *The Fourth European Conference Antennas and Propagation (EuCAP)*, (pp. 1-5). Barcelona, Spain.

Umeda, T., Toshida, H., Sekine, S., Fujita, Y., Suzuki, T., & Otaka, S. (2006, January). A 950-MHz rectifier circuit for sensor network tags with 10-m distance. *IEEE Journal of Solid-State Circuits*, *41*(1), 35–41. doi:10.1109/JSSC.2005.858620

Ungan, T., Le Polozec, X., Walker, W., & Reindl, L. (2009, Sept). RF energy harvesting design using high Q resonators. *IEEE MTT-S International Microwave Workshop on Wireless Sensing, Local Positioning, and RFID*, 1-4. doi:10.1109/IMWS2.2009.5307869

Walsh, C., Rondineau, S., Jankovic, M., Zhao, G., & Papovic, Z. (2005). A conformal 10ghz rectenna for wireless powering of piezoelectric sensor electronics. *IEEE MTT-S International Microwave Symposium Digest*, 4-12.

World Health Organization. (1999). *What are electromagnetic fields? Technical report*. WHO Regional Office for Europe.

Yoo, T., & Chang, K. (1992, June). Theoretical and experimental development of 10 and 35 GHz rectennas. *IEEE Transactions on Microwave Theory and Techniques*, *40*(6), 1259–1266. doi:10.1109/22.141359

Zhou, Y. (2013). *Contribution à la récupération d'énergie ambiante pour les objets communicants autonomes* (PhD dissertation). Université de nantes, France.

## KEY TERMS AND DEFINITIONS

**Antenna Effective Aperture:** The ability of antenna to extract energy from the electromagnetic wave. It is called also the effective area.

**Finite Difference Time Domain (FDTD):** A widely used simulation tool of electromagnetic phenomena characterized by the resolution of Maxwell's curl equations in the time domain.

**Miniaturization:** A trend to manufacture ever smaller mechanical, optical and electronic devices.

**Power Conversion Efficiency:** thE ratio of the DC power measured at the load to the received power injected at the input of the converter.

**Radio Frequency Identification (RFID):** A data collection technology that uses electronic tags for storing information. The tag is made of microchip combined with an antenna in a compact package.

**Rectenna:** The association of a receiving antenna that collects RF energy and a rectifying circuit that transforms it into DC voltage.

**Wireless Energy Transfer:** The process of sending electrical energy from one point to another without the use of a conducting environment.

# Section 2
# Piezoelectric Systems for Energy Harvesting

# Chapter 4
# Piezoelectric Energy Harvesting Skin and Its Application to Self-Powered Wireless Sensor Network

**Byeng Dong Youn**
*Seoul National University, South Korea*

**Heonjun Yoon**
*Seoul National University, South Korea*

**Hongjin Kim**
*Seoul National University, South Korea*

**Byung Chang Jung**
*Korea Institute of Machinery and Materials (KIMM), South Korea*

**Chulmin Cho**
*Mechatronics R&D Center, Samsung Electronics, South Korea*

**Yoon Young Kim**
*Seoul National University, South Korea*

## ABSTRACT

*Energy harvesting (EH) which scavenges electric power from ambient, otherwise wasted, energy sources has been explored to develop self-powered portable electronic devices. Vibration energy, a widely available ambient energy source, can be converted into electric power using a piezoelectric energy harvester that generates electric potential in response to applied mechanical strains. As a compact and durable design paradigm, a piezoelectric energy harvesting skin (PEH skin) which can be directly attached onto the surface of a vibrating engineered system has been proposed to scavenge electric power from vibration energy. The goal of this chapter is to describe the core technologies for the realization of the PEH skin from a system integration perspective as four parts: (a) modeling, (b) design, (c) manufacturing, and (d) demonstration. The readers will be able to learn the entire procedure of developing the PEH skin and applying it to self-powered wireless sensor network (WSN) through this chapter.*

DOI: 10.4018/978-1-4666-8254-2.ch004

## INTRODUCTION

The advances in wireless communications and low-power technology promote the use of wireless sensors. However, the limited life expectancy and high replacement cost of batteries make it difficult to use wireless sensors, although they have lots of benefits more than wired sensors (Anton & Sodano, 2007). Energy harvesting (EH) technology, which scavenges electric power from ambient, otherwise wasted, energy sources, has been explored to develop self-powered portable or wireless electronic devices and possibly eliminate battery replacement cost for wireless sensors (H. S. Kim, Kim, & Kim, 2011).

Vibration energy, one of widely available ambient energy sources, can be converted into electric power using piezoelectric, electromagnetic, electrostatic, and/or magnetostrictive transduction mechanisms. Various vibration sources with their maximum acceleration magnitude and frequency range were well listed by Roundy *et al.* (Roundy, Wright, & Rabaey, 2003). Among vibration-based energy harvesting mechanisms, piezoelectric energy harvesting has been most preferred due to its high energy density and no need for external equipment. A piezoelectric energy harvester can produce alternating current (AC) in response to applied mechanical strains.

Even though piezoelectric energy harvesters have been primarily designed as cantilever beams, they have some drawbacks from a practical point of view: (a) an additional space required for a proof mass and clamping device, (b) a great deal of vibration energy loss when clamping conditions become loosened after long time use, and (c) a fatigue failure expected due to excessive strains at a clamping part. As an alternative design paradigm of the cantilever beam, a piezoelectric energy harvesting skin (PEH skin) has been proposed which can be directly attached onto the surface of a vibrating engineered system and thus requires no need for clamping fixtures and proof mass (Lee & Youn, 2011b), as shown in Figure 1. This work has been featured on the website (PhysOrg.com, 2011) as an innovative design for a piezoelectric energy harvester.

It is of great importance to systematically design and fabricate the PEH skin to generate the sufficient amount of electric power for operating portable electronic devices in real time. For the purpose of designing the PEH skin and selecting best sites for installation, it is essential to preliminarily quantify harvestable electric power under a given vibration condition. So it is required to develop an electro-mechanical model of the PEH skin with the high predictive capability based on rigorous theories and

*Figure 1. Piezoelectric energy harvesting skin (PEH skin) for powering to operate portable and/or wireless electronic devices*

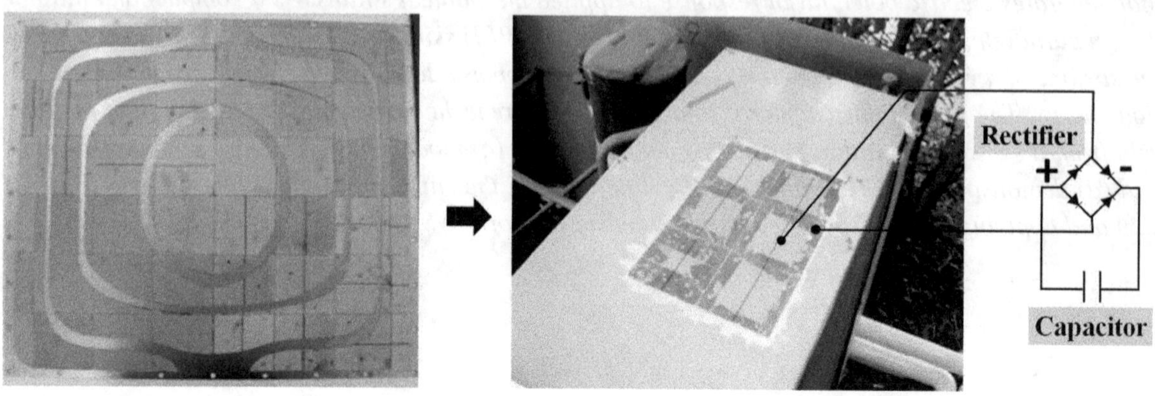

mechanics. In addition, because the PEH skin is fabricated by laminating thin piezoelectric layers onto a vibrating engineered system, the scientific and theoretical rationale for designing the PEH skin is needed to determine the optimal placement of the piezoelectric layers. Furthermore, the resonance frequency of the PEH skin and impedance matching should be considered to maximize harvestable electric power. On the other hand, the manufacturing process, such as a laser cutting, bonding, and curing of piezoelectric layers, should be standardized to fabricate reliable PEH skins while reducing production uncertainty for better performance. To successfully realize the PEH skin, therefore, it is necessary to thoroughly understand its core technologies, such as modeling, design, and manufacturing, and make a connection between them for establishing self-powered electronic devices in reality.

This chapter thus aims at comprehensively describing the sequentially executed steps for the realization of the PEH skin from a system integration perspective as four part: (a) modeling, (b) design, (c) manufacturing, and (d) demonstration. The rest of this chapter is composed of eight sections and organized as follows. A literature review on the fundamentals of piezoelectric energy harvesting is provided in the second section. The third section discusses how to analyze the electromechanical behavior of the PEH skin using an electromechanically-coupled analytical model and multiphysics finite element (FE) model. Furthermore, the fourth section describes the design methodology of the PEH skin to enhance the energy conversion efficiency. The fifth section explains the manufacturing process of the PEH skin, followed by the feasibility studies on real-time operating self-powered wireless sensors using the PEH skin in the sixth section. Finally, the future research directions and conclusion are given in the last two sections.

This chapter is designed to integrate our previous works (Jung et al., 2012; Lee & Youn, 2011a, 2011b; Lee, Youn, & Jung, 2009; Yoon & Youn, 2014; Yoon, Youn, & Kim, 2014).

## FUNDAMENTALS OF PIEZOELECTRIC ENERGY HARVESTING

Piezoelectric energy harvesting requires multidisciplinary research involving materials science, mechanical, and electrical engineering disciplines. As a result, extensive research efforts have been made to advance piezoelectric energy harvesting technology in a variety of fields as: (a) piezoelectric materials, (b) modeling and analysis, (c) mechanics-based design, (d) circuit configuration, and (e) system integration. To help the readers better understand core technologies for realizing the PEH skin, this section provides a literature review on the fundamentals of piezoelectric energy harvesting.

### Piezoelectric Materials

The prefix 'piezo' comes from the Greek 'piezein' which means to pressure or squeeze. Electricity is a physical phenomenon related to the flow of an electric charge. Therefore, piezoelectricity is "an interaction between electrical and mechanical behavior" (Ikeda, 1996).

As the direct piezoelectric effect, a piezoelectric material can produce electric polarization due to applied mechanical strain (see Figure 2). Piezoelectric energy harvesting and piezoelectric transducers correspond to this direct piezoelectric effect. The amount of the output voltage is proportional to the applied mechanical strain. Conversely, when electric polarization is applied, a piezoelectric material becomes strained. This is called the inverse piezoelectric effect.

In the IEEE Standard on Piezoelectricity (Standards Committee of the IEEE Ultrasonics, Ferroelectrics, & Frequency Control Society, 1988), the piezoelectric constitutive relations are given as:

*Figure 2. Electric polarization in response to applied mechanical strains (Courtesy of Prof. Soobum Lee)*

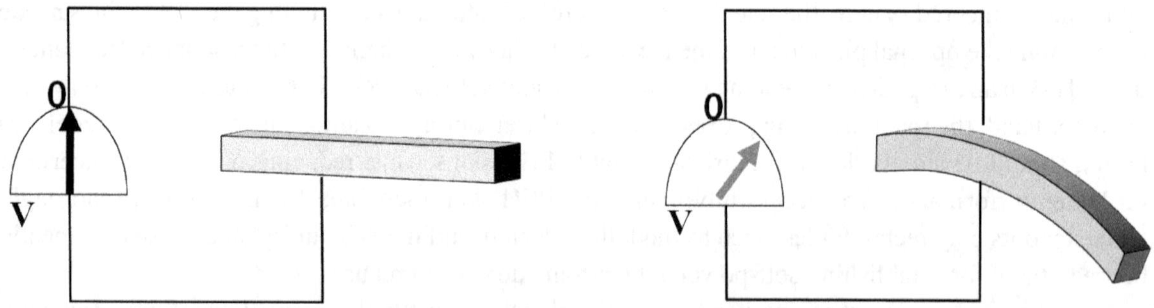


$$T_{ij} = c_{ijkl} S_{kl} - e_{kij} E_k$$

$$D_i = e_{ikl} S_{kl} + \varepsilon_{ij} E_k$$

These relations are also called the stress-charge form. In this case, the independent variables are strain $S_{kl}$ and electric field $E_k$. It should be noted that the elastic constant $c_{ijkl}$ and dielectric permittivity $\varepsilon_{ij}$ are coupled to each other (Holland & EerNisse, 1969). In equation, the dielectric permittivity is calculated at constant strain.

Piezoelectric materials include lead zirconate titanate (PZT), zinc oxide (ZnO), polyvinylidene difluoride (PVDF), lead magnesium niobate-lead titanate (PMN-PT), and polypropylene (PP) polymer. It is never enough to emphasize only the material issues in order to improve the mechanical and electrical properties of piezoelectric materials. For example, piezoelectric ceramics, such as PZT, have the high piezoelectric and dielectric constant, but are inherently brittle and less durable. Meanwhile, piezoelectric polymers, such as PVDF, have the high flexibility but low electromechanical coupling. For this reason, many material scientists have worked to fabricate flexible as well as electromechanically efficient piezoelectric materials based on the nanotechnology (X. Chen, Xu, Yao, & Shi, 2010; Y. Qi et al., 2010; Y. Qi et al., 2011).

## Modeling and Analysis

Many research efforts have been made to advance modeling and analysis of a piezoelectric energy harvester. One of the research interests is to provide analytical (or mathematical) models that better represent the physics of a piezoelectric energy harvester while assuming that the input vibration signal is a harmonic sinusoidal function. Round *et al.* developed a single degree of freedom lumped-parameter model for a cantilever beam; however, this model provides no dynamic strain distribution and modal analysis (Roundy & Wright, 2004). Sodano *et al.* formulated a single degree of freedom model and studied electric charge performance (Sodano, Park, & Inman, 2004). Because the lumped-parameter model might yield highly inaccurate results, Erturk *et al.* developed an amplitude correction factor which uses the ratio of tip-mass to beam mass to accurately predict the voltage response of the lumped-parameter model (Erturk & Inman, 2008). As an alternative modeling of a lumped-parameter model, DuToit *et al.* derived an approximate distributed-parameter model using the Rayleigh-Ritz discretization and a trial family of admissible functions (Dutoit & Wardle, 2007). Kim *et al.* developed a rigorous modeling for investigating the effect of a proof mass which serves to decrease the resonance frequency and to raise the applied strain (M. Kim, Hoegen, Dugundji, & Wardle, 2010). Based on the Euler-Bernoulli beam theory, Erturk *et al.* proposed an electromechanically-coupled distributed-parameter model and verified its accuracy with experimental observations (Erturk & Inman, 2009).

In practice, however, most realistic vibrations have physical uncertainty, such as the variation of an amplitude and driving frequency. Although the randomness considerably affects the variation in harvestable electric power, analytical models developed under the assumption of deterministic harmonic sinusoidal excitations cannot deal with the random nature in realistic vibration signals. This drives the research interest which is to analyze the amount of harvestable electric power of a piezoelectric energy harvester in a stochastic manner. Halvorsen formulated a stochastic description using a constant power spectral density for calculating electric power, proof mass displacement, and optimal load of resonant energy harvesters based on the Fokker-Plank equation (Halvorsen, 2008). Adhikari *et al.* presented a closed-form of a linear single degree of freedom model when the excitation is assumed to be white Gaussian noise (Adhikari, Friswell, & Inman, 2009). Zhao *et al.* predicted expected electric power and mean-square shunted displacement based on a constant power spectral density by employing the Fourier series representation or an Euler-Maruyama scheme to solve the stochastic differential equations (Zhao & Erturk, 2012). Yoon and Youn proposed stochastic quantification of the expected electric power using a statistical time-frequency analysis under non-stationary random vibrations (Yoon & Youn, 2014).

## Design Methodology

Various design methodologies of mechanical characteristics of a piezoelectric energy harvester have been proposed to enhance the amount of harvestable electric power. The most commonly used design of a piezoelectric energy harvester is a cantilever beam with single piezoelectric layer (unimorph) or two piezoelectric layers (bimorh) configurations. In general, the cantilever beam is utilized as 3-1 mode that the directions of the applied bending strain and output voltage are perpendicular to each other. Because the cantilever beam can achieve high mechanical strain and low natural frequency, they can produce relatively high electric power. Round *et al.* investigated the design considerations to generate high electric power with the dimensions of the tip mass, the length and thickness of the cantilever beam, and the optimal electrical resistance (Roundy et al., 2005). Goldschmidtboeing *et al.* found that a trapezoidal cantilever beam is more efficient than a rectangular cantilever beam due to the uniformly large strain at every point on the beam (Goldschmidtboeing & Woias, 2008). The importance of appropriate treatment of the flexible proof mass size has been addressed in micron-scale cantilever beam design due to effect on the electromechanical performance (M. Kim, Hong, Miller, Dugundji, & Wardle, 2011).

The output power can be maximized when the resonance frequency of a piezoelectric energy harvester matches the dominant excitation frequency of ambient vibrations. As mentioned in the last subsection, however, the excitation frequency of realistic vibrations may change as an operation condition of a engineered system. This fact means that vibration energy is spread over a wide spectrum. Therefore, this practical issue has been encouraged the exploration of methods to broaden the usable bandwidth of energy harvesters (Harne & Wang, 2013). Recently, nonlinear energy harvesters have been proposed to reliably scavenge electric power under broadband random vibrations. Cottone *et al.* demonstrated that a bistable oscillator can provide better performances compared to a linear oscillator in terms of the energy extracted from a generic wide spectrum vibration (Cottone, Vocca, & Gammaitoni, 2009).

Ferrari *et al.* proposed a nonlinear piezoelectric beam converter coupled to permanent magnets to create a bistable system which can exploit stochastic resonance with white-noise excitation (Ferrari et al., 2010). Stanton *et al.* validated the bistable inertial oscillator which consists of permanent magnets and a piezoelectric cantilever beam undergoing a homoclinic saddle-point bifurcation (Stanton, McGehee, & Mann, 2010). Arrieta *et al.* proposed to exploit the nonlinear behavior of a bistable composite plate with bonded piezoelectric layers (Arrieta, Hagedorn, Erturk, & Inman, 2010). Comprehensive reviews on nonlinear energy harvesting via bistable systems are available the article (Harne & Wang, 2013).

Besides nonlinear energy harvesting techniques, a variety of innovative designs have been proposed to overcome the practical limitation of the piezoelectric energy harvester with a single natural frequency under broadband multiple vibration modes. For example, a multiple cantilever beam structure was proposed which has many resonant frequencies due to the several different cantilevers (S. Qi, Shuttleworth, Olutunde Oyadiji, & Wright, 2010). In addition, it has been proven that a segmentation design can achieve high electric power by removing piezoelectric materials to avoid voltage cancellation at strain nodes (Lee et al., 2009).

On the other hand, computer simulation-based design can be used to optimize piezoelectric transduction performance within a design space under a given loading and boundary condition through the use of a finite element model. Rupp *et al.* developed a topology optimization method to design a multilayer piezoelectric energy harvester (Rupp, Evgrafov, Maute, & Dunn, 2009). In their work, the optimal electrical resistance was simultaneously calculated by treating this value as a design variable. Zheng *et al.* considered elastic materials as well as piezoelectric materials for the design considerations of energy harvesters to maximize the energy conversion factor under the topology optimization formulation (Zheng, Chang, & Gea, 2009). Chen *et al.* proposed a level set-based topology optimization approach to seek the optimal geometry for maximizing the energy conversion efficiency (S. K. Chen, Gonella, Chen, & Liu, 2010).

## Circuit Configuration

Electrical regulation should be optimized to maximize harvestable electric power, which is generally composed of three stages: (a) energy capture, (b) energy rectification, and (c) energy storage (Yoon & Youn, 2014), as shown in Figure 3.

One of the most important aspects in this electrical circuit configuration is the impedance matching between a piezoelectric energy harvester and electrical regulation. Moreover, because a piezoelectric energy harvester produces alternating current (AC) in accordance with the sign change of the curvatures of the dynamic strains, an AC/DC converter is needed use the output electric power to operate electronic devices.

Kasyap *et al.* developed the fly-back converter circuit, with which impedance could be modified to match that of a piezoelectric energy harvester (Kasyap et al., 2002). Ottman *et al.* studied the use of an adaptive step-down dc-dc converter to maximize the output electric power from a piezoelectric energy harvester (Ottman, Hofmann, Bhatt, & Lesieutre, 2002). Meninger *et al.* developed an energy harvesting system, with which the transduction process can be modified through the use of an additional capacitor to provide maximum energy transfer (Meninger, Mur-Miranda, Amirtharajah, Chandrakasan, & Lang, 2001). Guan *et al.* studied the efficiencies of the energy harvesting circuits considering the storage device voltages (Guan & Liao, 2007).

*Figure 3. Circuit configuration for piezoelectric energy harvesting (Reprinted from (Yoon, 2013), Copyright: Heonjun Yoon)*

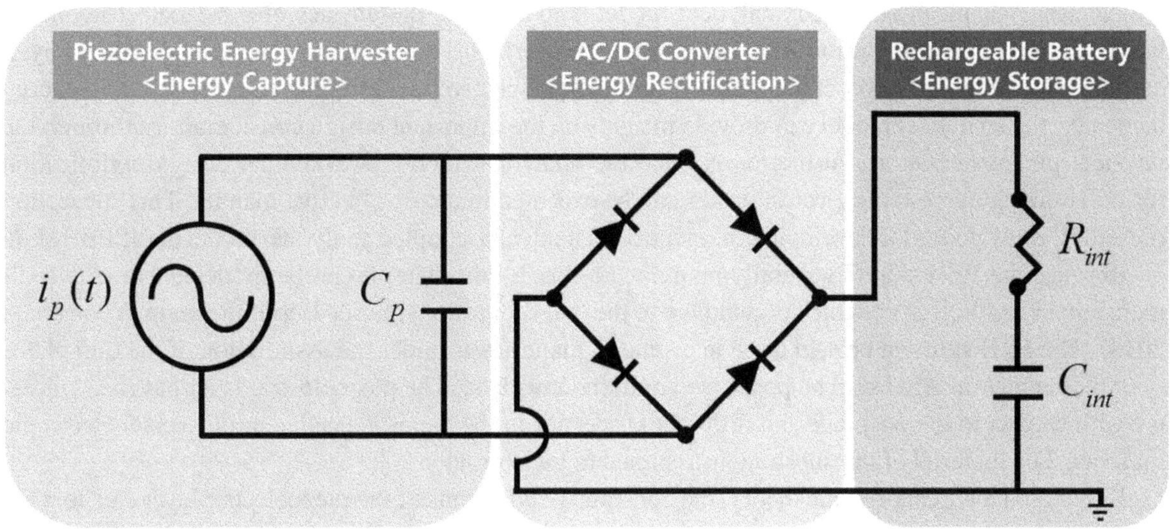


## System Integration

Many researchers have studied piezoelectric energy harvesters utilizing ambient energy sources from engineered systems and human movement. Elvin *et al.* proposed a self-powered damage detection system for a construction site using PVDF: a time varying load from a roller cart that can generate sufficient mechanical strains for energy harvesting (Elvin, Elvin, & Choi, 2003). Leland *et al.* mounted a piezoelectric energy harvester on a wooden staircase and generated electric power of around 30 $\mu$W (Leland, Lai, & Wright, 2004). Shoe-mounted energy harvester is to utilize human movement for generating electric power through pressure by heel strikes (Mateu & Moll, 2005; Shenck & Paradiso, 2001). Nuffer and Bein discussed a health monitoring system for vehicles using wireless sensors operated by a piezoelectric energy harvester (Nuffer & Bein, 2006). Other case studies of energy harvesting from vehicle engine and bridge vibrations can be found on the AmbioSystems website (Anton & Sodano, 2007). Granstrom *et al.* developed a piezoelectric polymer backpack strap that can generate electric power from the oscillating tension in the strap during walking (Granstrom, Feenstra, Sodano, & Farinholt, 2007). Lee *et al.* demonstrated that the PEH skin can scavenge electric power from vibration energy of a condensing unit in an outdoor air conditioner (Lee, Youn, & Giraud, 2010). According to experimental demonstration, the output power of 3.7 mW was obtained which is enough to operate wireless sensors.

# MODELING AND ANALYSIS OF PIEZOELECTRIC ENERGY HARVESTING SKIN

As mentioned in the last section, it is of importance to preliminarily quantify harvestable electric power under a given vibration condition. Therefore, this section discusses how to analyze the electromechanical behavior of the PEH skin with high-fidelity predictive capability and quantify harvestable electric power at an initial design.

## Electromechanically-Coupled Analytical Model

An electromechanically-coupled analytical model is to describe the physics of a piezoelectric energy harvester and represent relationships among the input variables and output responses (so called physical parameterization) (Yoon et al., 2014). From the perspective of designing the piezoelectric energy harvester, the analytical model can provide insights on the important design considerations through the physical parameterization. Furthermore, once the analytical model is available, design optimization for maximizing harvestable electric power can be executed in a cost-effective manner. This subsection presents the key points for developing the electromechanically-coupled analytical model of the PEH skin.

Because the PEH skin is generally manufactured as a thin plate, it can be modeled as the two-dimensional Kirchhoff plate which is analogue to the one-dimensional Euler-Bernoulli beam (Yoon et al., 2014). The PEH skin can be said to be in a state of plane stress under the assumption of the thin plate.

Assume that the PEH skin be positioned onto the $x$-$y$ plane. The piezoelectric layer has the symmetry with respect to the $x$-$y$ plane and different properties in the $z$-axis, so called a transversely isotropic material. The material of the substrate is assumed to be isotropic.

Based on the Kirchhoff plate theory, the constitutive equations of the piezoelectric layer can be written as a reduced matrix. The piezoelectric layer is considered to be polled along the $z$-axis. Because the only non-zero strains of the PEH skin occur in the $x$-$y$ plane and the poling direction of the piezoelectric layer is the $z$-axis, the directions of the applied mechanical strains and output voltage are perpendicular to each other, so called 3-1 mode.

The generalized Hamilton's principle (Meirovitch, 2001) is used to derive the electromechanically-coupled differential equation of motion for forced vibration of the PEH skin. The generalized Hamilton's principle states that the integral of the difference between the kinetic and potential energies should be minimum based on the calculus of variations. The kinetic and potential energies are expressed in terms of the transverse displacement of the PEH skin. Under the assumption of the base excitation, the transverse absolute displacement of the PEH skin can be described by the superposition of the base displacement and transverse relative displacement. It is worth noticing that the base excitation is considered as an external force.

To calculate electric current generated by the piezoelectric layer of the PEH skin in response to applied mechanical strain, Gauss's law can be used with respect to the electric displacement in an integral form (Ballas, 2007), which is the double definite integral over the area of the top surface of the piezoelectric layer. As shown in equation, the electric displacement $D_i$ is calculated with two variables which are the strain and electric field. The strain can be obtained from the strain-displacement relationships under the assumption of linear elastic behavior. On the other hand, when the piezoelectric layer is polled along the $z$-axis, the uniform electric field is applied through the thickness of the piezoelectric layer. As a result, electric current equation can be derived by substituting the strain and electric field into electric displacement in Gauss's law (Yoon et al., 2014).

In the strain-displacement relationship, the displacement means the transverse relative displacement of the PEH skin. Based on the expansion theorem, the transverse relative displacement of the PEH skin is approximated by a linear combination of the assumed modes. Due to the geometric non-homogeneity of the PEH skin, such as the discontinuity at the sides of the piezoelectric layer, it is tedious to find the exact solution for modal analysis of the PEH skin. Therefore, the Rayleigh-Ritz method, which is one of series discretization technique for approximating a distributed-parameter system, is implemented to obtain the assumed modes of the PEH skin (Yoon et al., 2014). That is why we use the term 'assumed mode', which is a simpler but closely related geometry against the PEH skin. The assumed modes of the PEH skin depend on the boundary condition and attachment of the piezoelectric layer.

Finally, the output voltage generated by the piezoelectric layer can be obtained by solving the differential equation of motion and electrical circuit equation simultaneously. Because the electromechanically-coupled analytical model of the PEH skin is developed based on the Kirchhoff plate theory, it can take account of the effect of the two-dimensional bending behaviors (e.g., the twisting mode, Poisson's ratio, and aspect ratio effect) on harvestable electric power. The detailed derivation procedure can be found in (Yoon et al., 2014).

## Finite Element Method

If the configuration of the PEH skin and/or the loading condition of a vibrating engineered system are complicated, it is better to perform multiphysics FE simulation rather than analytical approach. In this study, the FE model of the PEH skin was developed by using ANSYS® commercial software, as shown in Figure 4.

*Figure 4. Finite element model of piezoelectric energy harvesting skin (PEH skin): (a) geometry and (b) voltage coupling, (c) modal analysis for the fully clamped boundary condition, and (d) the output voltage prediction under open-circuit condition.*

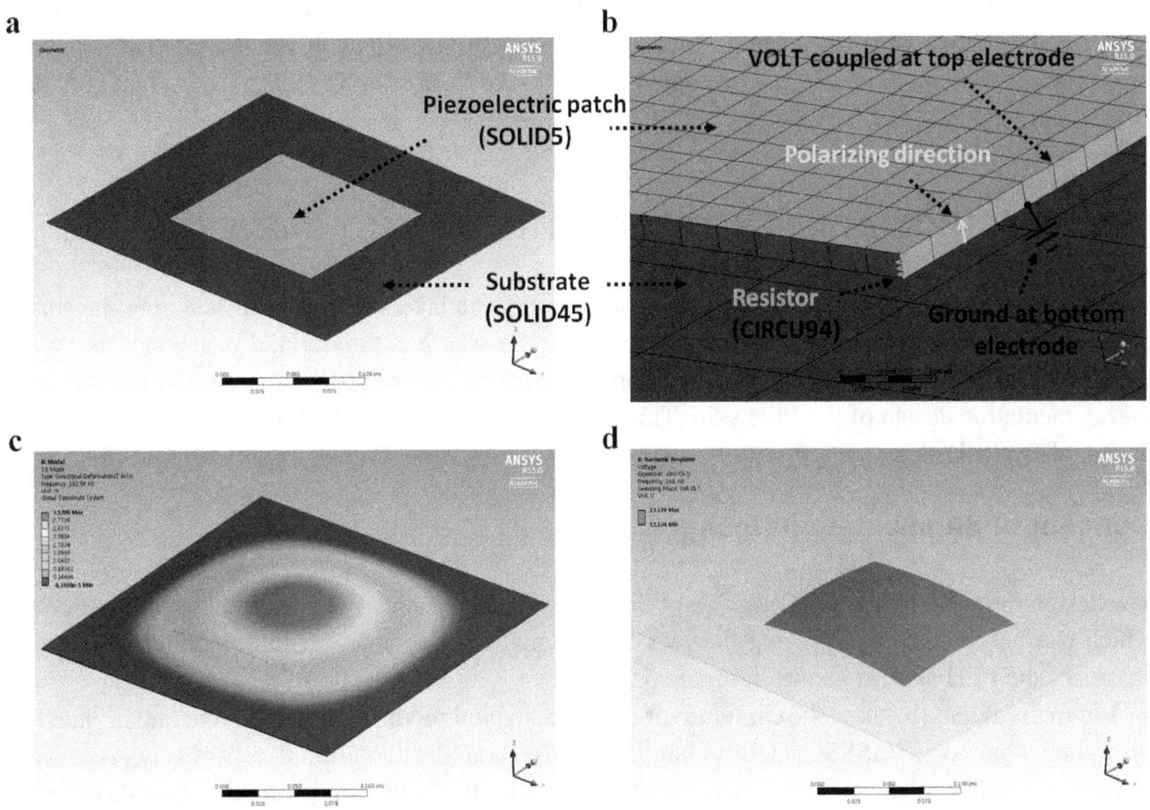

The electrode and adhesive material (e.g., conductive epoxy) layers are ignored in the FE model due to their extremely small thickness and a perfect bonding condition. As shown in Figure 4 (a) and (b), the piezoelectric layers (e.g., Piezo Systems, Inc. PSI-5H4E) can be modeled using SOLID5 element which has four coupled field degrees of freedom. Three degrees of freedom define the translational displacements (UX, UY, and UZ) based on the global coordinate system. The other degree of freedom indicates the electric potential difference (VOLT) between top and bottom electrodes. The substrate (e.g., STS304) can be modeled by SOLID45 element with three translation DOFs.

As shown in Figure 4 (b), the VOLT degree of freedom on the top surface of the piezoelectric layer is coupled to represent the top electrode, while that on the interface between the piezoelectric layer and the substrate is grounded (zero electric potential) to represent the bottom electrode. These two electrodes are connected with an external electrical load by CIRCU94 element.

Figure 4 (c) shows the second mode shape from modal analysis predicted by the FE model for the fully clamped boundary condition. In addition, Figure 4 (d) shows harmonic response for the output voltage (13.12 V) at open-circuit resonance frequency (146 Hz) when the external electrical load is 500 kΩ. The detail procedure for constructing the FE model of a piezoelectric energy harvester can be found in (De Marqui Junior, Erturk, & Inman, 2009; Lee & Youn, 2011b).

Once the FE model of the PEH skin is constructed, the model calibration is required to minimize the difference between the experimentally measured and predicted performances in the simulation by adjusting the material properties of the substrate and piezoelectric layer. The model calibration can be decomposed into two sequentially executed steps: (a) the first step for mechanical responses (e.g., the natural frequency) and (b) the second step for electrical responses (e.g., the output voltage and open-circuit resonance frequency). The resonance frequency is tuned under a mass constraint in the first step so that the Young's modulus and density are adjusted, while the voltage spectrum in a frequency domain is adjusted to match the experimental observation of the output voltage in the second step. The detail procedure of the calibration for the FE model can be found in (Lee et al., 2009; Youn, Jung, Xi, Kim, & Lee, 2011).

## DESIGN METHODOLOGY OF PIEZOELECTRIC ENERGY HARVESTING SKIN

As shown in Figure 1, the PEH skin consists of piezoelectric layers directly attached onto the surface of a vibrating engineered system as one embodiment. So where is the optimal placement that piezoelectric layers should be attached for generating higher electric power? This important question drives the segmentation design of the PEH skin. This section provides the methodology for the segmentation design of the PEH skin.

### Concept of Segmentation Design

The design rationale for the segmentation of the PEH skin can be clearly understood from the curvature which can be obtained by applying the two-dimensional Laplace's differential operator to the mode shapes of the PEH skin, as shown in Figure 5.

Figure 5 (a) and (b) show the curvature of the fundamental modes shape of a rectangular plate for a fully simply-supported (SSSS) and fully clamped (CCCC) boundary condition, respectively. As shown in Figure 5 (a) and (b), there is no inflection line in curvature for the fully simply-supported, while inflection

*Figure 5. Curvature of the fundamental mode shape for each boundary condition: (a) fully simply-supported and (b) fully clamped; Schematic top view of piezoelectric energy harvesting skin (PEH skin) configuration for each boundary condition: (c) fully simply-supported and (d) fully clamped.*

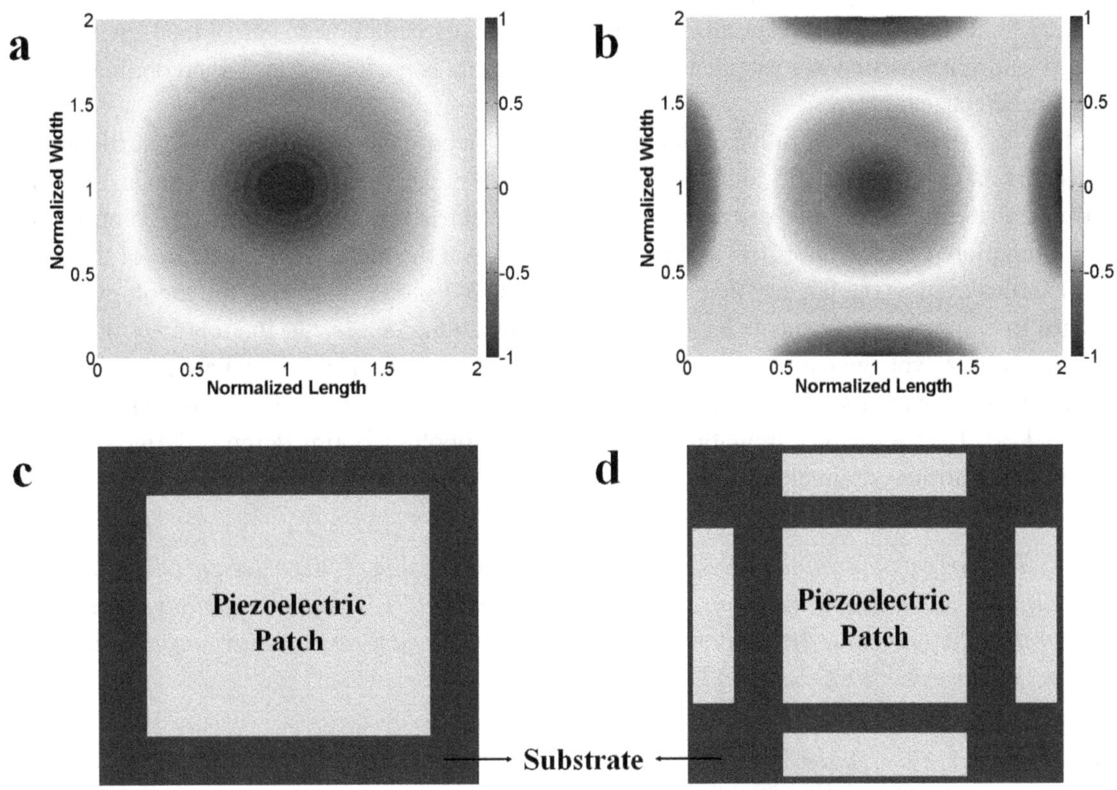


lines are developed in the fully clamped boundary condition. It is worth noticing that the fully clamped boundary condition has inflection lines in the fundamental mode unlike a fully simply-supported and cantilever (CFFF). Because the output voltage is proportional to the mechanical strain, when the curvature is positive, the output voltage is positive, and vice versa. This fact implies that the amount of the output voltage is dramatically reduced when the sign of the curvature changes. This electromechanical phenomenon is called voltage cancellation. In other words, voltage cancellation is caused by inconsistent voltage phase in the piezoelectric layer. For the purpose of avoiding voltage cancellation, therefore, the piezoelectric layer must be segmented along the inflection lines (Lee et al., 2009).

Figure 5 (c) and (d) show the segmentation design of the PEH skin suitable for each boundary condition. Figure 5 (c) shows no need of segmentation, while Figure 5 (d) shows the five-segmented piezoelectric layers for avoiding voltage cancellation. It can be concluded from this observation that the segmentation design of the piezoelectric layer depends on the curvature of the mode shapes of the PEH skin (Yoon et al., 2014).

## Design Optimization Process

The optimization process for the segmentation design of the PEH skin based on the FE model includes two main subtasks: (a) topology optimization and (b) shape optimization.

Figure 6 shows the topology optimization procedure for segmentation design of the PEH skin. In the topology optimization step, a conceptual design optimization is performed to find an optimal topology of the piezoelectric layer.

Initially the design space is fully covered with the piezoelectric layer. And then, the distribution of the piezoelectric layer is conceptually found by eliminating some piezoelectric materials. This design problem is formulated to find the optimal distribution of the piezoelectric layer subject to material constraint due to cost limitation (Lee & Youn, 2011a). Based on the fact that the higher absolute value of the in-plane normal strains ensures the larger voltage generation when 3-1 mode is utilized, the topology optimization tries to eliminate the elements having low in-plane strain. As a result, the topology optimization can also minimize voltage cancellation by eliminating the piezoelectric material around the inflection line where the amount of strains is locally smaller due to the sign change of the in-plane strains (Lee & Youn, 2011a). Therefore, the eliminating criterion for the topology optimization is defined as the sum of the in-plane normal strain along each direction (Lee & Youn, 2011a).

*Figure 6. Topology optimization procedure for segmentation design of piezoelectric energy harvesting skin (PEH skin): (a) target design space, (b) FE model for initial PEH skin, (c) topology optimization to remove inflection lines, and (d) the optimal external electrical load for power maximization (impedance matching) (Courtesy of Prof. Soobum Lee).*

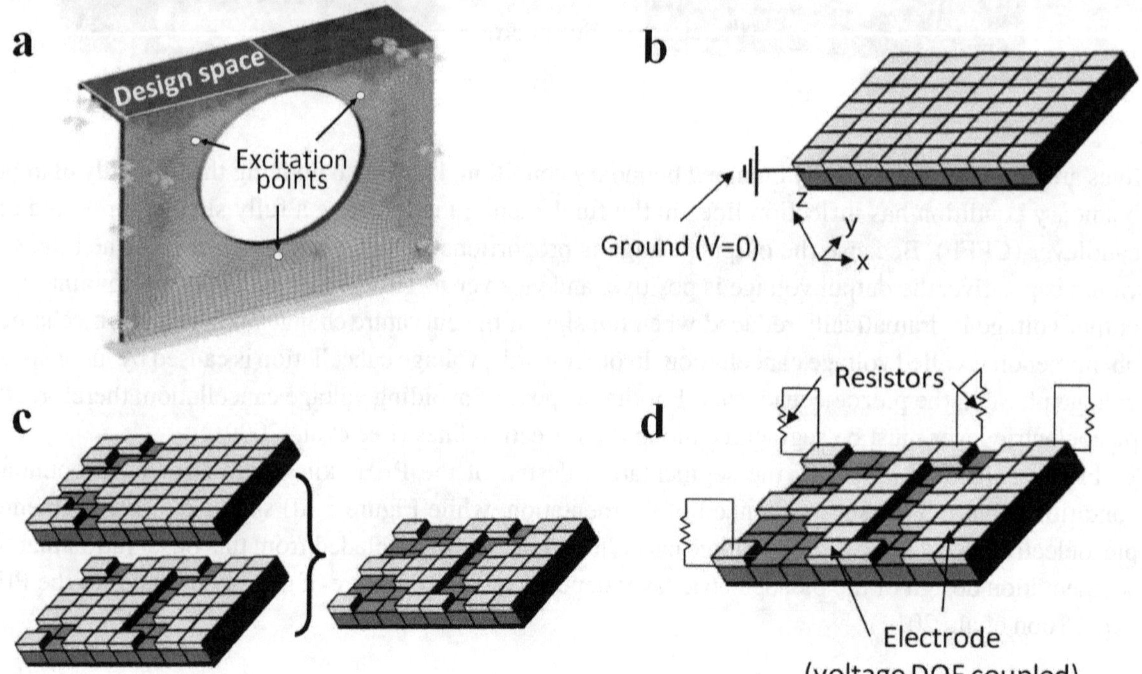

It is worth noticing that the inflection line can be also identified by the voltage phase angle at the top surface of the piezoelectric layer, because a significant change in the voltage phase angle exists across the inflection line (about 180°) (Lee & Youn, 2011b). After the piezoelectric layer design, the external electrical loads for each segment are calculated to maximize electric power.

The optimal shapes and placement of the piezoelectric layers found in the topology optimization, however, may be difficult to fabricate because manufacturability is not considered in this step. The shape optimization is thus performed to fill the lack of manufacturability in the topology optimization step. The detail procedure of the computational design methodology of the PEH skin is available in (Lee & Youn, 2011a).

## MANUFACTURING PROCESS OF PIEZOELECTRIC ENERGY HARVESTING SKIN

The manufacturing process of the PEH skin is composed of six steps: (a) preparation of a CAD file, (b) laser cutting of piezoelectric layers, (c) bonding, (d) curing, (e) wiring using a conductive tape, and (f) checking short circuitry as shown in Figure 7.

The first step is to prepare a two-dimensional CAD file of the PEH skin design in an actual size using a commercial software. In addition, the piezoelectric layers are assigned onto the substrate for laser cutting. In this study, a commercially available PZT sheet (Piezo Systems, Inc. PSI-5H4E and

*Figure 7. Manufacturing process of piezoelectric energy harvesting skin (PEH skin): (a) preparation of a CAD file, (b) laser cutting of piezoelectric layers, (c) bonding, (d) curing, (e) wiring using a conductive tape, and (f) checking short circuitry.*

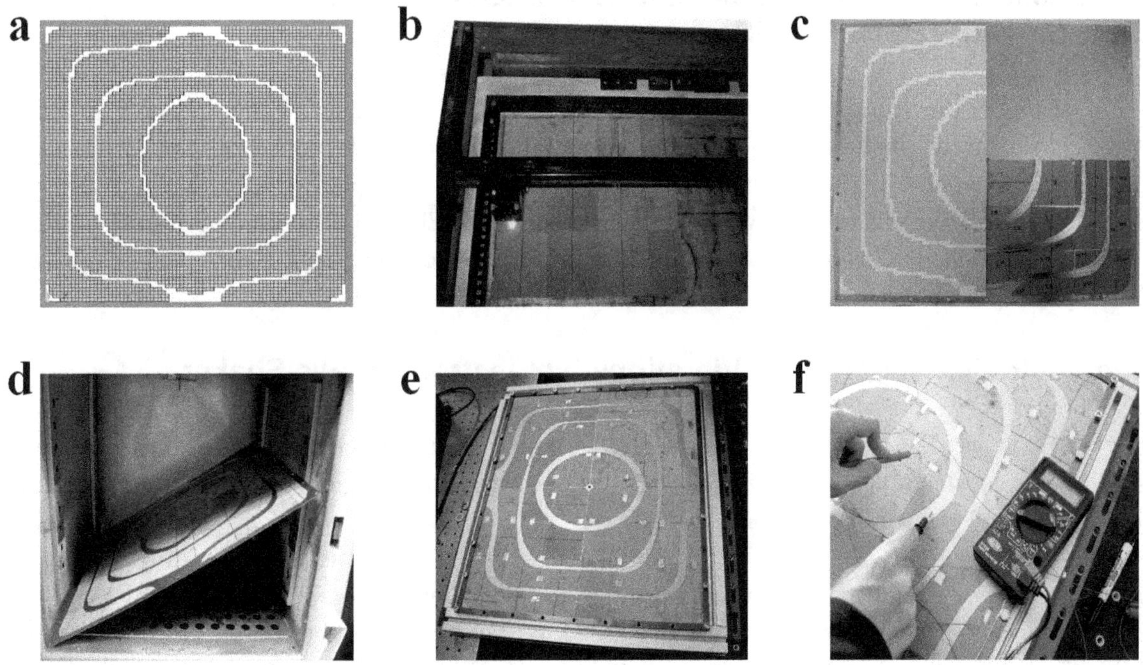

PSI-5A4E) was used as the piezoelectric layer. In the second step, PZT sheets are cut by a CNC laser cutter (Universal Laser Systems, Inc., M-300 Laser Platform). For the machining process, the power level and speed of laser should be carefully determined to prevent brittle fracture of PZT sheets. Make sure the poling direction of the PZT sheets to be aligned consistently. In the bonding process, a conductive epoxy (ITW Chemtronics, CW-2400) can be used to glue the segmented PZT sheets on the base structure. Make sure to prevent the overflow of the epoxy onto the top surface of PZT sheets, which causes an electric short. After the attachment of PZT sheets on the base structure, it should be cured for 15 minutes (70 °C) in a hot chamber. The process is repeatedly done for all area of the base structure. After wiring the laser-machined PZTs in one segment using a conductive tape, the short circuitry should be checked between the top and bottom electrode because the paste state epoxy may be squeezed out and reached at top electrode. The detail procedure of the manufacturing process of the PEH skin can be found in (Jung et al., 2012).

The sequentially executed procedures for manufacturing the PEH skin are summarized as follows:

**Step 1:** Prepare a two-dimensional CAD file of the skin design in an actual size using a drawing software.
**Step 2:** Load the CAD file to a CNC laser cutter.
**Step 3:** Cut PZT sheets using the CNC laser cutter.
**Step 4:** Sketch the PEH skin design on the base structure to locate PZT sheets.
**Step 5:** Attach the machined PZT sheets to the base structure using a conductive epoxy.
**Step 6:** Cure the epoxy for 15 minutes (70 °C) in a hot chamber with a uniformly distributed weight above PZT sheets.
**Step 7:** Wire the laser-machined PZTs in one segment using a conductive tape.
**Step 8:** Check if any PZT sheet is electrically short or not.

## EXPERIMENTAL DEMONSTRATION OF PIEZOELECTRIC ENERGY HARVESTING SKIN

The objective of this section is to validate whether the PEH skin can generate the sufficient amount of electric power from vibration energy to operate low-power portable electronics or wireless sensors in real-time. Therefore, two kinds of experimental demonstrations were performed with aim to check the feasibility of utilizing self-powered wireless sensors using the PEH skin: (a) harmonic vibration induced by an electromagnetic shaker (Jung et al., 2012) and (b) realistic vibration induced by an outdoor condensing unit (H. Kim, 2014).

### Demonstration 1: Harmonic Vibrations by Electromagnetic Shaker

The PEH skin was demonstrated to scavenge electric power from harmonic vibration (driving frequency is 50 Hz and acceleration level is 9.8 $m/s^2$) induced by an electromagnetic shaker. This study employed seven LEDs and one wireless sensor node, as shown in Figure 8.

As shown in Figure 8 (a), seven LEDs are light on when the base structure is vibrating. In addition, one wireless analog temperature sensor, transmitter, receiver, and monitoring system (USB connection for the receiver to a laptop computer) were also used for experimental demonstration, as shown in Figure 8 (b). The temperature sensor node requires 5.3 V for initiating and 2.7 V for transmitting the acquired

*Figure 8. Experimental demonstration of piezoelectric energy harvesting skin (PEH skin) under deterministic harmonic vibration: (a) seven LEDs and (b) one wireless sensor node.*

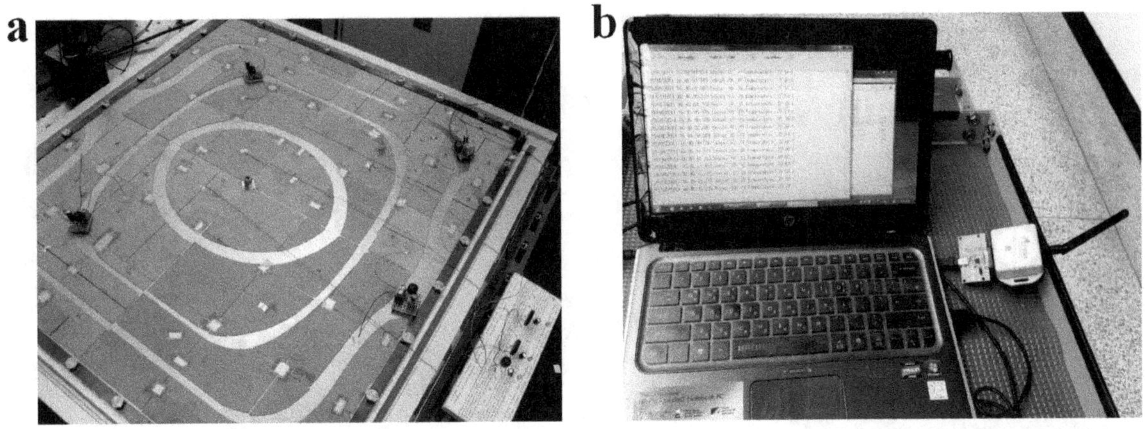


signals. Furthermore, it requires about 15 $\mu$W for continuous temperature monitoring. It was confirmed from experimental observations that the PEH skin can scavenge enough electric power to operate seven LEDs and one temperature sensor node. The detail procedure of the experimental demonstration in the case of the harmonic vibration induced by the electromagnetic shaker is available in (Jung et al., 2012).

## Demonstration 2: Realistic Vibrations by Outdoor Condensing Unit

Given realistic vibrations produced by an electromagnetic motor of the cooling fan inside (operating at 1170 ~ 1180 rpm), the top plate of the outdoor condensing unit was vibrating at its harmonics as shown in Figure 10 (a). Then, the operating deflection shapes (ODS) of the top plate was examined by using a scanning laser vibrometer (Polytec Gmbh., PSV-400). As shown in Figure 10 (b), three meaningful deflection shapes having large deformation were found at 33.8, 50.0, and 66.9 Hz and the corresponding peak acceleration levels were measured as 2.8, 2.4, and 0.9 g, respectively. It is noted that three frequencies measured are near the harmonics of the first peak frequency (16 ~ 17 Hz) by the cooling fan.

Based on the vibration characteristics of the top plate, the PEH skin was segmented into four regions by the superposition of strain inflection lines (where the sum of in-plane normal strains is zero) obtained from the first three harmonics while taking into account the manufacturability (H. Kim, 2014).

The experimental setup to check sensor powering capability of the fabricated PEH skin is shown in Figure 10. Five wireless sensors (four three-dimensional acceleration sensors and one analog temperature sensor from AmbioSystems, LLC.) were connected to wireless platforms (AmbioMote24 from AmbioSystems, LLC.) which were used to transmit sensor signals to a laptop computer. In addition, two receivers, each of which is connected to a different laptop, were used because one receiver could treat up to four sensor signals. So, the real-time signal from two acceleration sensors (ID: 19 and 41) were carried by a wireless platform (AmbioMote24-A), and two acceleration sensors and one temperature sensor (ID: 48, 52, and 40) were connected to the other platform (AmbioMote24-B). An oscilloscope (LT354M from LeCroy) was also used to measure the output voltage in real time.

*Figure 10. Experimental demonstration of a self-powered operation of a wireless sensor network (four accelerometers and one temperature sensor) by using PEH skin (Reprinted from (H. Kim, 2014), Copyright: Dr. Hongjin Kim).*

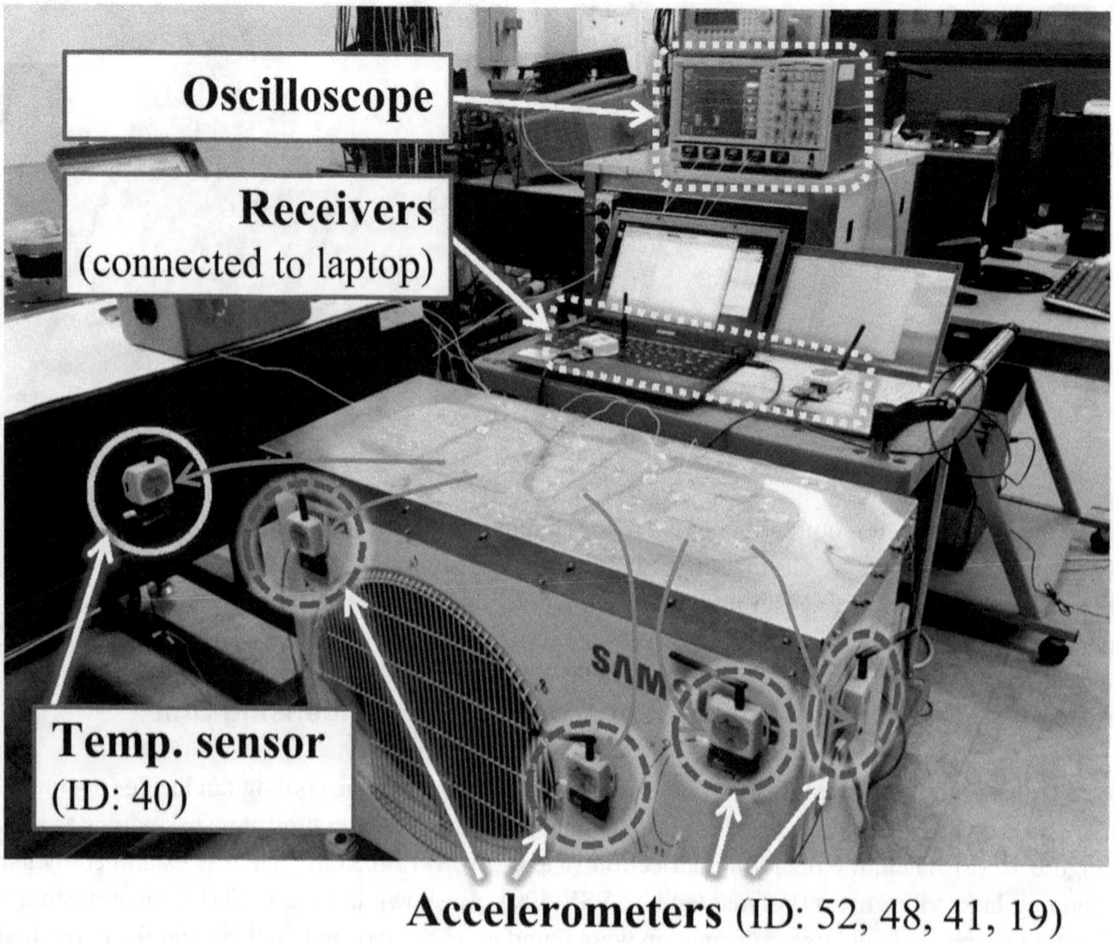


While the outdoor condensing unit is in operation, the output voltage from each PEH skin segment was measured as 4 ~ 5 $V_{pp}$. It was observed that all five wireless sensor signals were successfully transmitted to the laptop computers in real-time. Therefore, it is concluded that the fabricated PEH skin can harvest power from a realistic vibration, which is sufficient for real-time operation of a wireless sensor network.

## FUTURE RESEARCH DIRECTIONS

Although the technical advances described in this chapter successfully address critical challenges in the PEH skin, there are still several research topics where further investigations and developments are required to truly bring piezoelectric energy harvesting into an alternative solution for powering portable or wireless electronic devices. From a practical point of view, specific recommendations for the continuation of the research on piezoelectric energy harvesting are listed as follows.

*Figure 9. Identified vibration characteristics of the top plate of the outdoor condensing unit under operation: (a) a fast Fourier transformed result to indicate excitation frequencies and (b) three operating deflection shapes at frequencies of interest (Reprinted from (H. Kim, 2014), Copyright: Dr. Hongjin Kim).*

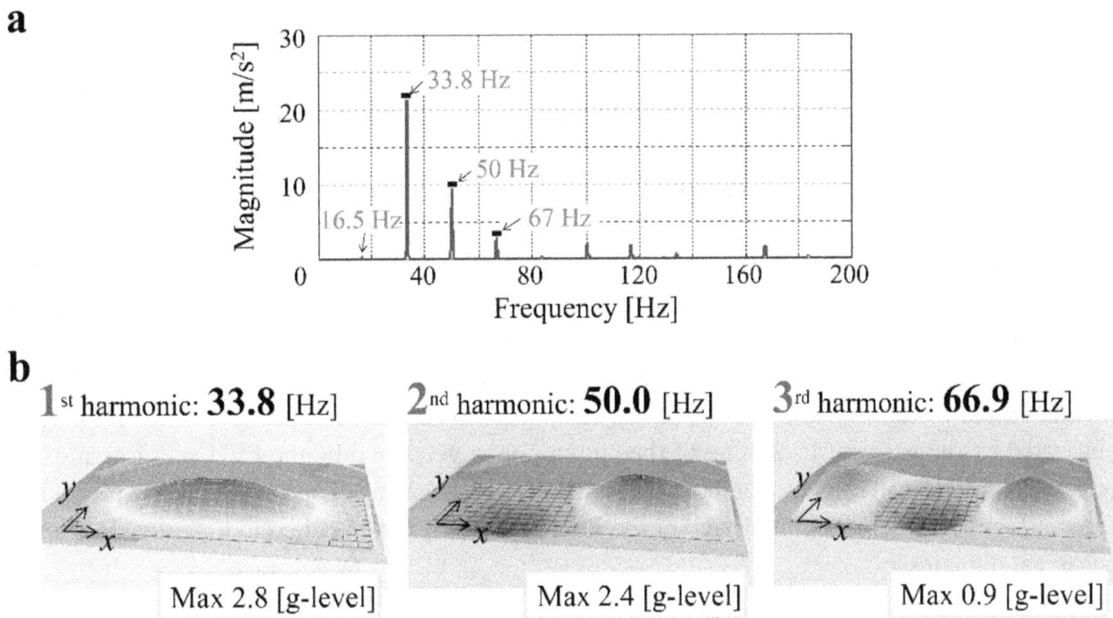


## Flexible Piezoelectric Energy Harvesting Skin

The PEH skin is required to be flexible for its application to the curved surface of a vibrating shell structure such as a pipeline. While materials scientists have made great progress in developing a flexible piezoelectric material based on the nanotechnology, however, its performance evaluation has been conducted without regard to the device configuration and/or vibration characteristics of the target practical application. For example, a piezoelectric energy harvester with higher piezoelectric constant does not necessarily yield increased electric voltage (M. Kim, Dugundji, & Wardle, 2013). Therefore, it is essential to design a flexible and efficient PEH skin at systems- as well as materials- levels.

Furthermore, future works should focus on combining multiscale simulation and electromechanical models to quantify harvestable electric power generated by flexible piezoelectric materials (e.g., nanocomposites, PZT nanofibers, PMN-PT thin films, and ZnO nanowire) of the PEH skin under a given vibration condition.

## Analysis and Design under Physical Uncertainty

The uncertainty propagation analysis is suggested to quantify the effect of the physical uncertainty sources (e.g., material properties and geometry) on the variation of the output performances (e.g., harvestable electric power and resonance frequency). To investigate the effect of physical uncertainty in the PEH skin, experimental observations for mechanical properties (e.g., elastic moduli, density, and manufacturing tolerance) and electrical properties (e.g., piezoelectric strain coefficient and dielectric permittivity) are required to be statistically characterized. In addition, statistical model calibration may be incorporated to adjust unknown random variables of the properties and update the electromechanical model.

Furthermore, because deterministic design optimization does not consider physical uncertainty in the material properties and manufacturing tolerance, it may result in unreliable solutions. In the design process, therefore, reliability-based design optimization (RBDO) is suggested to reliably scavenge electric power from vibration energy while accounting for physical uncertainty of the PEH skin.

## Advanced Segmentation Design

The segmentation design of the PEH skin may be sensitive to boundary conditions. For instance, a change in rivet tightness of the substrate may change the mechanical characteristics (e.g., acceleration amplitude, natural frequency, and inflection lines) and optimal placement of piezoelectric layers. Therefore, the sensitivity with respect to boundary conditions needs to be studied to design a robust and reliable PEH skin.

On the other hand, the vibration responses, such as the resonance frequency and deflection shape, are affected by the backward coupling of the output power generated by the PEH skin for an external electrical load. This fact implies that the inflection lines may be changed due to the inverse piezoelectric effect, and thus the segmentation design of the PEH skin needs to be studied in both the electrical and mechanical domains.

## Non-Stationary Random Vibrations

In practice, most realistic vibrations of practical engineered systems have the time-variant random nature in both amplitude and driving frequency, so called non-stationary random vibrations. Furthermore, unlike a cantilever beam having one edge, the PEH skin has four edges subjected to base excitation, and thus the vibration signal transferred to the PEH skin may be much more complicated.

This highlights two research needs. First, it is required for stochastic quantification of electric power generated by the PEH skin under non-stationary random vibrations. To characterize a non-stationary random vibration signal, a statistical time-frequency analysis is needed, which can represent how the energy of the vibration signal is distributed over both time and frequency domains (Yoon & Youn, 2014).

Second, it is required to develop nonlinear PEH skin for the purpose of achieving broadband energy harvesting by exploiting bistable mechanisms.

# CONCLUSION

As a new design concept for a piezoelectric energy harvester, the PEH skin has been recently proposed to scavenge electric power from vibrations induced by an engineered system. To make steps forward to sustainably operate portable or wireless electronic devices powered by the PEH skin without batteries, this chapter described the entire procedure of developing the PEH skin from a system integration perspective. With this purpose in mind, the core technologies (modeling, design, manufacturing, and demonstration) of the PEH skin were sequentially discussed. The plate theory-based electromechanically-coupled analytical model and multiphysics FE model were described for analysis of the PEH skin, followed by the design methodology for segmentation of the PEH skin to enhance the amount of the output power.

The topological design of the PEH skin can be obtained by removing the piezoelectric layers along the inflection lines from the strain mode shapes to minimize voltage cancellation. Shape optimization may be additionally followed for greater ease of manufacturing. Furthermore, the manufacturing process of the PEH skin was explained with the standard operating procedure (SOP). Finally, it can be concluded from experimental demonstrations that the amount of the output power generated by the PEH skin is sufficient to operate wireless sensors in real-time.

## REFERENCES


Adhikari, S., Friswell, M., & Inman, D. (2009). Piezoelectric energy harvesting from broadband random vibrations. *Smart Materials and Structures*, *18*(11), 115005. doi:10.1088/0964-1726/18/11/115005

Anton, S. R., & Sodano, H. A. (2007). A review of power harvesting using piezoelectric materials (2003–2006). *Smart Materials and Structures*, *16*(3), R1–R21. doi:10.1088/0964-1726/16/3/R01

Arrieta, A. F., Hagedorn, P., Erturk, A., & Inman, D. J. (2010). A piezoelectric bistable plate for non-linear broadband energy harvesting. *Applied Physics Letters*, *97*(10), 104102. doi:10.1063/1.3487780

Ballas, R. G. (2007). Piezoelectric multilayer beam bending actuators: Static and dynamic behavior and aspects of sensor integration. London: Springer London, Limited.

Chen, S. K., Gonella, S., Chen, W., & Liu, W. K. (2010). A level set approach for optimal design of smart energy harvesters. *Computer Methods in Applied Mechanics and Engineering*, *199*(37-40), 2532–2543. doi:10.1016/j.cma.2010.04.008

Chen, X., Xu, S., Yao, N., & Shi, Y. (2010). 1.6 V nanogenerator for mechanical energy harvesting using PZT nanofibers. *Nano Letters*, *10*(6), 2133–2137. doi:10.1021/nl100812k PMID:20499906

Cottone, F., Vocca, H., & Gammaitoni, L. (2009). Nonlinear energy harvesting. *Physical Review Letters*, *102*(8), 080601. doi:10.1103/PhysRevLett.102.080601 PMID:19257728

De Marqui Junior, C., Erturk, A., & Inman, D. J. (2009). An electromechanical finite element model for piezoelectric energy harvester plates. *Journal of Sound and Vibration*, *327*(1), 9–25. doi:10.1016/j.jsv.2009.05.015

Dutoit, N. E., & Wardle, B. L. (2007). Experimental verification of models for microfabricated piezoelectric vibration energy harvesters. *AIAA Journal*, *45*(5), 1126–1137. doi:10.2514/1.25047

Elvin, N., Elvin, A., & Choi, D. H. (2003). A self-powered damage detection sensor. *Journal of Strain Analysis for Engineering Design*, *38*(2), 115–124. doi:10.1243/030932403321163640

Erturk, A., & Inman, D. J. (2008). On mechanical modeling of cantilevered piezoelectric vibration energy harvesters. *Journal of Intelligent Material Systems and Structures*, *19*(11), 1311–1325. doi:10.1177/1045389X07085639

Erturk, A., & Inman, D. J. (2009). An experimentally validated bimorph cantilever model for piezoelectric energy harvesting from base excitations. *Smart Materials and Structures*, *18*(2), 025009. doi:10.1088/0964-1726/18/2/025009

Ferrari, M., Ferrari, V., Guizzetti, M., Andò, B., Baglio, S., & Trigona, C. (2010). Improved energy harvesting from wideband vibrations by nonlinear piezoelectric converters. *Sensors and Actuators. A, Physical*, *162*(2), 425–431. doi:10.1016/j.sna.2010.05.022

Goldschmidtboeing, F., & Woias, P. (2008). Characterization of different beam shapes for piezoelectric energy harvesting. *Journal of Micromechanics and Microengineering*, *18*(10), 104013. doi:10.1088/0960-1317/18/10/104013

Granstrom, J., Feenstra, J., Sodano, H. A., & Farinholt, K. (2007). Energy harvesting from a backpack instrumented with piezoelectric shoulder straps. *Smart Materials and Structures*, *16*(5), 1810–1820. doi:10.1088/0964-1726/16/5/036

Guan, M., & Liao, W. (2007). On the efficiencies of piezoelectric energy harvesting circuits towards storage device voltages. *Smart Materials and Structures*, *16*(2), 498–505. doi:10.1088/0964-1726/16/2/031

Halvorsen, E. (2008). Energy harvesters driven by broadband random vibrations. *Journal of Microelectromechanical Systems*, *17*(5), 1061–1071. doi:10.1109/JMEMS.2008.928709

Harne, R. L., & Wang, K. W. (2013). A review of the recent research on vibration energy harvesting via bistable systems. *Smart Materials and Structures*, *22*(2), 023001. doi:10.1088/0964-1726/22/2/023001

Holland, R., & EerNisse, E. P. (1969). *Design of resonant piezoelectric devices*. Cambridge, MA: MIT Press.

Ikeda, T. (1996). *Fundamentals of piezoelectricity*. Oxford University Press.

Jung, B., Cho, C., Yoon, H., Yoon, H., Youn, B. D., & Kim, Y. (2012). *Statistical Model Calibration for Energy Harvesting Skin Analysis and Design*. Paper presented at the 53rd AIAA/ASME/ASCE/AHS/ASC Structures, Structural Dynamics and Materials. doi:10.2514/6.2012-1833

Kasyap, A., Lim, J., Johnson, D., Horowitz, S., Nishida, T., Ngo, K., . . . Cattafesta, L. (2002). *Energy reclamation from a vibrating piezoceramic composite beam*. Paper presented at the International Congress on Sound and Vibration.

Kim, H. (2014). *Segmented piezoelectric skin energy harvesters* (Unpublished doctoral dissertation). Seoul National University, South Korea.

Kim, H. S., Kim, J.-H., & Kim, J. (2011). A review of piezoelectric energy harvesting based on vibration. *International Journal of Precision Engineering and Manufacturing*, *12*(6), 1129-1141.

Kim, M., Dugundji, J., & Wardle, B. L. (2013). Optimal design of piezoelectric materials and devices for energy harvesting. *Journal of the Korean Physical Society*, *62*(11), 1689–1695. doi:10.3938/jkps.62.1689

Kim, M., Hoegen, M., Dugundji, J., & Wardle, B. L. (2010). Modeling and experimental verification of proof mass effects on vibration energy harvester performance. *Smart Materials and Structures*, *19*(4), 045023. doi:10.1088/0964-1726/19/4/045023

Kim, M., Hong, S., Miller, D. J., Dugundji, J., & Wardle, B. L. (2011). Size effect of flexible proof mass on the mechanical behavior of micron-scale cantilevers for energy harvesting applications. *Applied Physics Letters*, *99*(24), 243506, 243506–243504. doi:10.1063/1.3663858

Lee, S., & Youn, B. D. (2011a). A design and experimental verification methodology for an energy harvester skin structure. *Smart Materials and Structures, 20*(5), 057001. doi:10.1088/0964-1726/20/5/057001

Lee, S., & Youn, B. D. (2011b). A new piezoelectric energy harvesting design concept: Multimodal energy harvesting skin. *IEEE Transactions on Ultrasonics, Ferroelectrics and Frequency Control, 58*(3), 629–645.

Lee, S., Youn, B. D., & Giraud, M. (2010). *Designing energy harvesting skin structure utilizing outdoor unit vibration.*

Lee, S., Youn, B. D., & Jung, B. C. (2009). Robust segment-type energy harvester and its application to a wireless sensor. *Smart Materials and Structures, 18*(9), 095021. doi:10.1088/0964-1726/18/9/095021

Leland, E. S., Lai, E. M., & Wright, P. K. (2004). *A self-powered wireless sensor for indoor environmental monitoring.* Paper presented at the Wireless Networking Symposium, University of Texas at Austin Department of Electrical & Computer Engineering.

Mateu, L., & Moll, F. (2005). Optimum piezoelectric bending beam structures for energy harvesting using shoe inserts. *Journal of Intelligent Material Systems and Structures, 16*(10), 835–845. doi:10.1177/1045389X05055280

Meirovitch, L. (2001). *Fundamentals of vibrations.* McGraw-Hill New York.

Meninger, S., Mur-Miranda, J. O., Amirtharajah, R., Chandrakasan, A. P., & Lang, J. H. (2001). Vibration-to-electric energy conversion. *IEEE Transactions on Very Large Scale Integration (VLSI) Systems, 9*(1), 64–76.

Nuffer, J., & Bein, T. (2006). *Applications of piezoelectric materials in transportation industry.* Paper presented at the Global Symposium on Innovative Solutions for the Advancement of the Transport Industry.

Ottman, G. K., Hofmann, H. F., Bhatt, A. C., & Lesieutre, G. A. (2002). Adaptive piezoelectric energy harvesting circuit for wireless remote power supply. *Power Electronics. IEEE Transactions on, 17*(5), 669–676.

PhysOrg.com. (2011). *Energy harvesting skin generates power from air conditioners.* Retrieved May 21, 2014, from http://phys.org/news/2011-04-energy-harvesting-skin-power-air.html

Qi, S., Shuttleworth, R., Olutunde Oyadiji, S., & Wright, J. (2010). Design of a multiresonant beam for broadband piezoelectric energy harvesting. *Smart Materials and Structures, 19*(9), 094009. doi:10.1088/0964-1726/19/9/094009

Qi, Y., Jafferis, N. T., Lyons, K. Jr, Lee, C. M., Ahmad, H., & McAlpine, M. C. (2010). Piezoelectric ribbons printed onto rubber for flexible energy conversion. *Nano Letters, 10*(2), 524–528. doi:10.1021/nl903377u PMID:20102189

Qi, Y., Kim, J., Nguyen, T. D., Lisko, B., Purohit, P. K., & McAlpine, M. C. (2011). Enhanced piezoelectricity and stretchability in energy harvesting devices fabricated from buckled PZT ribbons. *Nano Letters, 11*(3), 1331–1336. doi:10.1021/nl104412b PMID:21322604

Roundy, S., Leland, E. S., Baker, J., Carleton, E., Reilly, E., Lai, E., & Sundararajan, V. et al. (2005). Improving power output for vibration-based energy scavengers. *Pervasive Computing, IEEE, 4*(1), 28–36. doi:10.1109/MPRV.2005.14

Roundy, S., & Wright, P. K. (2004). A piezoelectric vibration based generator for wireless electronics. *Smart Materials and Structures, 13*(5), 1131–1142. doi:10.1088/0964-1726/13/5/018

Roundy, S., Wright, P. K., & Rabaey, J. M. (2003). *Energy scavenging for wireless sensor networks.*

Rupp, C. J., Evgrafov, A., Maute, K., & Dunn, M. L. (2009). Design of piezoelectric energy harvesting systems: A topology optimization approach based on multilayer plates and shells. *Journal of Intelligent Material Systems and Structures, 20*(16), 1923–1939. doi:10.1177/1045389X09341200

Shenck, N. S., & Paradiso, J. A. (2001). Energy scavenging with shoe-mounted piezoelectrics. *IEEE Micro, 21*(3), 30–42. doi:10.1109/40.928763

Sodano, H. A., Park, G., & Inman, D. (2004). Estimation of electric charge output for piezoelectric energy harvesting. *Strain, 40*(2), 49–58. doi:10.1111/j.1475-1305.2004.00120.x

Standards Committee of the IEEE Ultrasonics, Ferroelectrics, & Frequency Control Society. (1988). IEEE Standard on Piezoelectricity: An American National Standard. New York: Institute of Electrical and Electronics Engineers, Inc.

Stanton, S. C., McGehee, C. C., & Mann, B. P. (2010). Nonlinear dynamics for broadband energy harvesting: Investigation of a bistable piezoelectric inertial generator. *Physica D. Nonlinear Phenomena, 239*(10), 640–653. doi:10.1016/j.physd.2010.01.019

Yoon, H. (2013). *Analytical model for electric power prediction of piezoelectric energy harvesting* (Unpublished master's thesis). Seoul National University, South Korea.

Yoon, H., & Youn, B. D. (2014). Stochastic quantification of the electric power generated by a piezoelectric energy harvester using a time–frequency analysis under non-stationary random vibrations. *Smart Materials and Structures, 23*(4), 045035. doi:10.1088/0964-1726/23/4/045035

Yoon, H., Youn, B. D., & Kim, H. S. (2014). *Analysis of electromechanical performance of energy harvesting skin based on the Kirchhoff plate theory*. Paper presented at the ASME 2014 International Design Engineering Technical Conferences & Computers and Information in Engineering Conference. doi:10.1115/DETC2014-35433

Youn, B. D., Jung, B. C., Xi, Z., Kim, S. B., & Lee, W. (2011). A hierarchical framework for statistical model calibration in engineering product development. *Computer Methods in Applied Mechanics and Engineering, 200*(13), 1421–1431. doi:10.1016/j.cma.2010.12.012

Zhao, S., & Erturk, A. (2012). Electroelastic modeling and experimental validations of piezoelectric energy harvesting from broadband random vibrations of cantilevered bimorphs. *Smart Materials and Structures, 22*(1), 015002. doi:10.1088/0964-1726/22/1/015002

Zheng, B., Chang, C. J., & Gea, H. C. (2009). Topology optimization of energy harvesting devices using piezoelectric materials. *Structural and Multidisciplinary Optimization, 38*(1), 17–23. doi:10.1007/s00158-008-0265-0

## KEY TERMS AND DEFINITIONS

**Energy Harvesting:** Process by which electric power is converted from ambient, otherwise wasted, energy sources.

**Direct Piezoelectric Effect:** To produce electric potential by deformation of the electric dipoles in response to applied mechanical strains.

**Piezoelectric Energy Harvesting Skin (PEH Skin):** Piezoelectric energy harvester which can be directly attached onto a surface of an engineered system to scavenge electric power from vibration energy.

**Voltage Cancellation:** Electromechanical phenomenon that the amount of electric voltage is dramatically reduced when the sign of bending curvature of a piezoelectric material changes.

**Segmentation Design:** Design methodology which utilizes multiple modes by separating piezoelectric materials for the purpose of generating electric power efficiently.

**Electromechanically-Coupled Analytical Model:** To describe the physical behavior of a piezoelectric energy harvester and represent relationships among the input variables and output responses.

# Chapter 5
# Rainfall Energy Harvester

**Fabio Viola**
*University of Palermo, Italy*

**Pietro Romano**
*University of Palermo, Italy*

**Rosario Miceli**
*University of Palermo, Italy*

## ABSTRACT

*This chapter provides a detailed study on the harvest of the energy contained in raindrops by means of piezoelectric transducers. The energy harvester has the role of an electric source, able to recharge storage devices of small electronic components, such wireless sensors, by using the vibrational energy released by the drops hitting the transducer, reducing in such a way the chemical waste of conventional batteries. In technical literature, diverse studies agree on the level of suitable generated voltage on the electrodes of a piezoelectric transducer subjected to rainfall, but a complete characterization on the supplied power is still missing. This work, also to limit optimistic forecasts, takes into account the behavior of the transducers in different scenarios: subjected to real and artificial rainfall, standalone or in parallel configuration, in conventional geometries, due to the commercial format or in customized shape, free to move or with an imposed optimal deflection.*

## INTRODUCTION

In recent years an increasing attention in generating energy from the ambient vibrations, leds the study on the employment of smart materials. Low power consuming electronic devices, devoted to the wireless monitoring application, can be fed by the conversion of mechanical energy, otherwise wasted, into useful electric energy. The use of piezoelectric transduction mechanism over other electromagnetic and electrostatic alternatives seems to be one of the best strategy to ensure the requirements for the achievement of an amount of power and a reduced environmental impact. Piezoelectric materials have a lack of symmetry in the crystal lattice, which generates a dipole moment, function also of the stress acting on the structure. The lattice structure allows the fabrication of device both in macro and micro scale, thus

DOI: 10.4018/978-1-4666-8254-2.ch005

surpassing the limitations due to the number of windings to be used in the electromagnetic transducers. In this perspective at each point of interest in which a sensor must be placed, the kinetic energy of the ambient vibrations, energy normally wasted in environment without a useful purpose, can be coupled to the electromechanical transducer, which feeds electronic monitor apparatus. The vibration energy can be regarded as marginal energy, an energy of small entity, usable for utilities of reduced energy capacity, such as the transmission of information following the collection of the same by appropriate sensors.

A complete overview on the possibility of harvest energy and collect informations in different scenarios is difficoult to be provided, since novel approachs and innovative idea are continually being developed: Benasciutti and Moro (2010) studied the possibility of harvest energy from vibrating shoe-mounted piezoelectric cantilevers, such system can also be used to monitor weight distribution on the sole of the foot; Xiang, Wang, Shi and Zhang (2013) faced the harvesting of energy induced from the deformation of pavements due to moving vehicles; but also van den Ende, van de Wiel, Groen and van der Zwaag (2012) discussed the harvesting from automotive tires; and finally Hobeck and Inman (2012) proposed an innovative piezoelectric grass energy harvester.

Also the rainfall energy harvest has been faced by Jager, Guigon, Chaillout and Despesse (2008). The idea is to convert, by means of piezoelectric plates, the kinetic energy possessed by the drops of rainwater into electrical energy. A pioneering comparison of different piezoelectric materials, in order to investigate the possibility of energy generation water droplets energy sources for low power electronic devices, has been studied by Vatansever, Hadimani, Shah and Siores (2011) These studies confirm that the single drop of water hitting the piezoelectric plates generates voltages less than a dozen of volts (peak to peak), but no evaluation on power has been proposed. The drops of rain strike the piezoelectric material in a cantilever configuration, which may be subject to study to improve the energy produced as described by Benasciutti, Moro, Zelenika, and Brusa (2010) and again by Benasciutti, Moro, and Gallina (2013). Although the voltage peak to peak, produced by droplets, seems high enough to interact with electronic devices or rectifiers a more accurate characterization is required, in order to dispel excessive optimistic predictions. The concept of energy flow, presented by Liang and Liaoin (2011) and by Erturk and Inman (2011), clarifies the dissipation of energy during the harvesting process, in order to separete the electromechanical coupling coefficient of the system, natural frequencies, damping ratio and electric load. Useful schemes to model the electromechanical coupling have been proposed by Roundy and Wright (2004) and validated here for the raindrop harvester.

Objective of the chapter is to study the amount of power harvested from the rain, since such characterization is still missing. To better define the amount of power extracted from a single drop of rain, are compared different systems, consisting of commercial transducers, which are exposed to natural and artificial rain, and also are subject to suitable configurations, which differ in shape, for the fixing of the edges, for the deflection, in order to increase the extracted power.

## PIEZOELECTRIC TRANSDUCERS AND RAINDROP

Piezoelectricity is a form of coupling between electrical and mechanical behaviors present in many materials: the generation of an electric charge in certain non-conducting materials, such as quartz crystals and ceramics, when they are subjected to mechanical stress (such as pressure or vibration) is known as direct piezoelectric effect, whereas the generation of vibrations in such materials when they are subjected to an electric field is the coverse (inverse) effect.

The ability of piezoelectric materials to convert electrical energy into mechanical and vice versa depends on their crystalline structure. The necessary condition occurs because the piezoelectric effect is the absence of a center of symmetry in the crystal, which is responsible for charge separation between positive and negative ions and the formation of the Weiss domains, ie groups of dipoles with parallel orientation. Applying an electric field to a piezoelectric material, the Weiss domains are aligned in proportion to the field. Consequently, the size of the material change, by increasing or decreasing if the direction of the Weiss domains is the same as or opposite to the electric field. To describe this concept in simplistic terms, a stress (tensile or compressive) applied to a piezo crystal will alter the separation between the positive and negative charge sites in each elementary cell leading to a net polarization at the crystal surface. The effect is practically linear, i.e. the polarization varies directly with the applied stress, and at the same time direction dependent, so that compressive and tensile stresses will generate electric fields and hence voltages of opposite polarity. It is beyond the scope of this study provide an exhaustive description of the phenomenon and of the changes that have been developed to optimize the electrical performance and mechanical specifications; interesting discussion can be found in the work of Erturk and Inman (2012). The piezoelectric effect verifies the first law of the thermodynamics (principle of conservation of energy):

$$\dot{U} = T_{ij}\dot{S}_{ij} + E_i\dot{D}_i, \tag{1}$$

in which U is the stored energy density, $T_{ij}$ is the stress tensor, $S_{ij}$ is the strain tensor, $E_i$ is the electric field tensor, $D_i$ is the electric displacement tensor and the overdot represents the differentiation respect to time. By considering the material properties of the piezoelectric harvester, it is possible to define the following equations:

$$\begin{aligned} S &= sT + dE; \\ D &= dT + \varepsilon E\ , \end{aligned} \tag{2}$$

where the tensors depend on direction of the force and field, s compliance coefficient, d piezoelectric strain coefficient, ε permittivity. The transformations between elastic, electric and piezoelectric coefficient can be found in IEEE standards.

In this study, the energy harvester consists of a piezoelectric film on an epoxy cantilever sandwiched between electrodes that are used to collect the generated power. A water drop falling on the structure creates an impulsive force that brings the internal lattice structure of the piezoelectric element to deform, causing the loss of symmetry, and therefore the generation of small dipoles, which global effect is an impulsive voltage on electrodes. Mechanical vibrations follow the impact, stress is induced within the material, thus giving rise to an electrical source. A sheet of piezoelectric material has some limitations in the mechanical-electrical transduction for low-frequency signals, since the effects of the induced electric field, generated in the hitted region, are mitigated by the surrounding areas, and for large sheets effects are tenuous.

The behavior of generated pulses depends on the state of locking of the piezoelectric film: if it is bound by both ends, and only one.

Different studies in the literature show encouraging results with regard to the generation of electricity from water droplets (Jager et al., 2008). The piezoelectric transducers can reach tens of volts, but this result does not yet allow to attribute to them the character of power generators. The water drops continuity in the same place is very variable: there may be intervals of seconds (small rainfall) or fractions of seconds (downpour). Different performances due to variable drops dimensions (mass) and impact point make it hard to model the phenomenon. The voltage has a peak waveform, not a continuous voltage, so an equivalent average voltage has to be defined. For a power system the equivalent average current can be obtained by using a bridge rectifier and a smoothing capacity; for the theoretical model initially this approach has been not considered.

Different transducers configuration can be used. In figure 1 the two models locked by one (Viola et al., 2013) and two ends (Jager et al., 2008) are shown.

Larger collection surfaces can be used, made with piezoelectric material (Perera et al., 2014), or with an inert collecting diaphragm linked to piezoelectric cantilevers (Wong Ching-Hong et al. 2013), as shown in figure 2.

To evaluate the power output of a piezoelectric transducer it is necessary to define a range of possible stresses. The single drop of water can have a diameter that varies between 0.2 to 6 mm. Considering a

*Figure 1. Piezoelectric cantilevers, one edge bound on the left, two edges on the right*

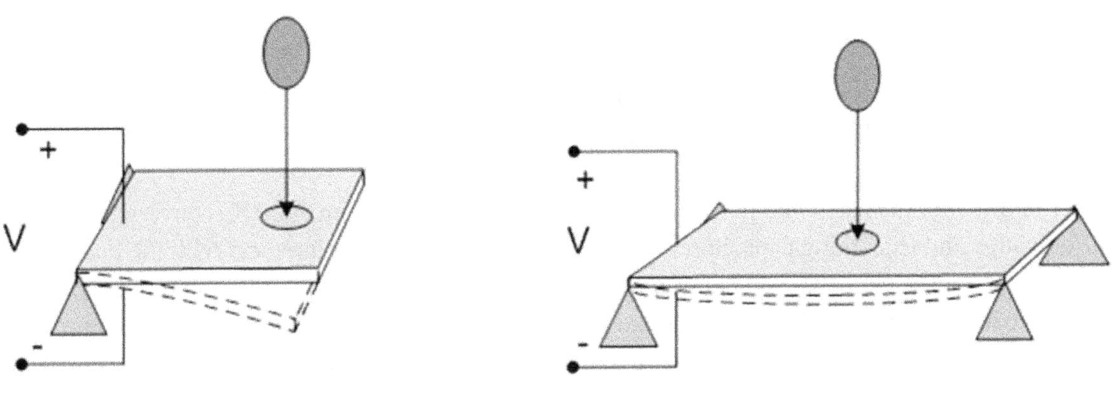


*Figure 2. Larger piezoelectric harvesting systems*

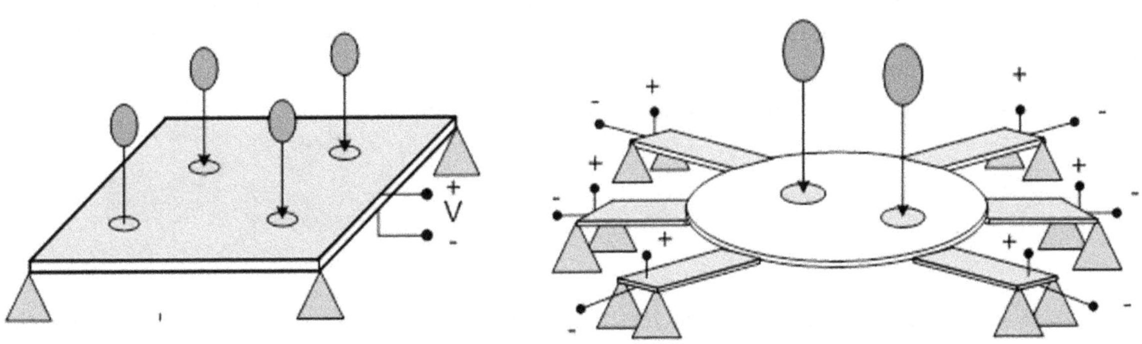

cruise speed on impact of approximately 2 m/s for the small drop and 9 m/s for the largest, it is possible to estimate the energy input: $E_{min} = 3.1\mu J$, $E_{max} = 0.063$ J. Also considering the interval of two seconds to have a successive drop, the power is: $P_{min} = 1.5\mu W$, $P_{max} = 0.031W$.

Being energy input little, no comparison can be made with a traditional photovoltaic system.

The harvestable power, however, is affected by several factors. The drop, while centering fully the piezoelectric film, is not able to transfer maximum energy as it is subject to the phenomenon of splashing: the collision is not complete since the main drop is separated itself into small drops leaving the impact surface. It must therefore associate an efficiency of a collision. In the same way we should introduce a performance of the electrical-mechanical system. The drop stresses the piezoelectric according to the 31 mode (electric field orthogonal to the strain of material) and not all the energy is converted into charges on the plates of the transducer. Finally, an electrical performance coefficient is to be introduced to take into account the losses of the rectifying bridge. The output power is given by:

$$P_{out} = \eta_{collision} \cdot \eta_{piezo} \cdot \eta_{rect} \cdot P_{max}. \tag{3}$$

The output power is certainly reduced, then the objective is to maximize it.

The transducers on which the experiments were conducted are Mide Volture V22B and the MEAS LDT1-028K. Volture sensors were mounted in a suitable support, caged to extremes, due to their fragility. The sensor Meas was mounted as a cantilever in a suitable support. Figure 3 and figure 4 show the piezoelectric transducers. The PVDF Meas is customable, the PZT Volture has a rigid structure and no modification can be introduced.

## CANTILEVER MODEL

In order to evaluate the harvested power, taken by the employment of an AC-DC converter, made by a rectifier bridge and smoothing capacitor, the shape of the waveform that interacts with the bridge has to be studied. In such a way, the equivalent model has been studied. In Figure 5 is reported the electro-mechanical scheme presented by Roundy and Wright (2004). The mechanical and electrical part are connected via a particular transformer, which as shown by Pinna et al. (2009), is far from usual electric transformer since current transformer ratio is not the reciprocal of voltage ratio. In the mechanical scheme the stress σ has the role of voltage; first derivative of strain, $\dot{S}$, that one of current. On the equivalent inductance $L_m$ the applied voltage is related to the second derivative of strain, on the equivalent resistance $R_m$ voltage is related to the first derivative and on the capacitance $C_m$ directly to the strain. The derivative is related to the longitudinal axis of the harvester, referred as axis1, not to time. Impact and following vibration are related to axis 3.

For a mode 31 equation 2 can be rewritten in the following relations:

$$\begin{aligned} S_1 &= s_{11}\sigma + d_{31}E_3; \\ D_3 &= d_{31}\sigma + \varepsilon E_3, \end{aligned} \tag{4}$$

with again S strain, s compliance coefficient, σ stress, d piezoelectric strain coefficient, E electric field, D displacement field, ε permittivity.

*Figure 3. Lead zirconate titanate (PZT) transducers: V22BL in the top and V22B in the bottom of the picture.Schematic of the cantilever under the piezoelectric*

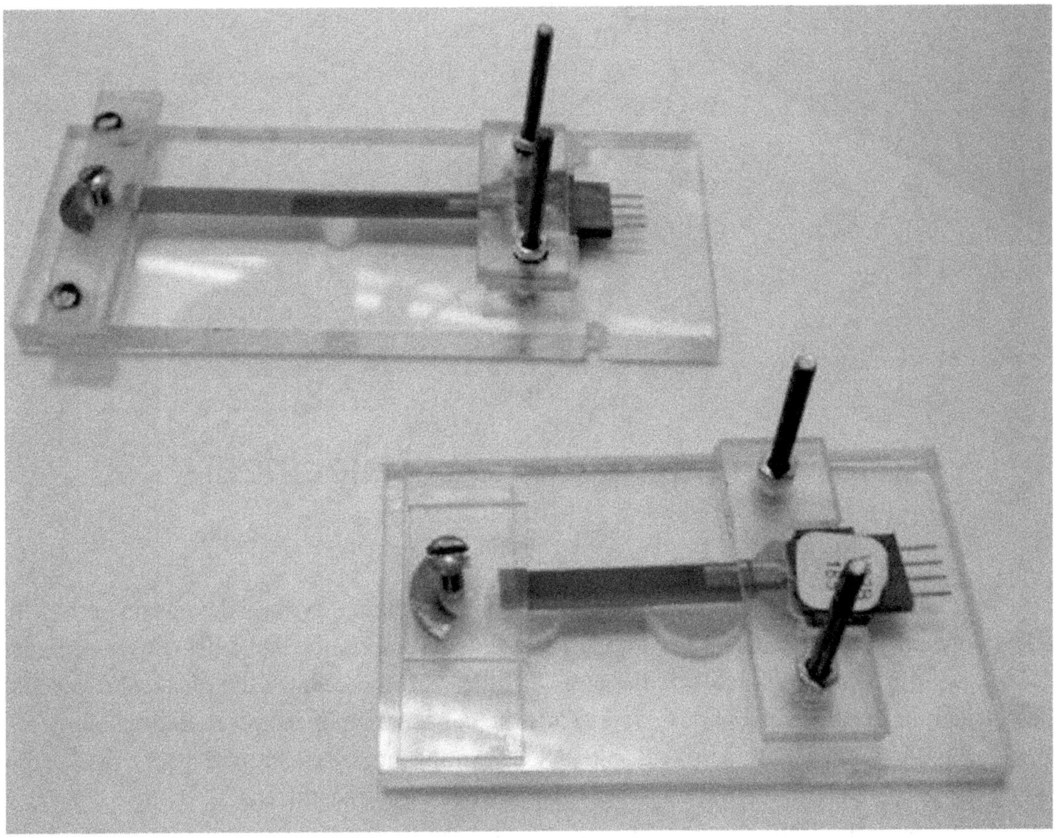

*Figure 4. Meas LDT1-028k polyvinylidene difluoride (PVDF) transducer*

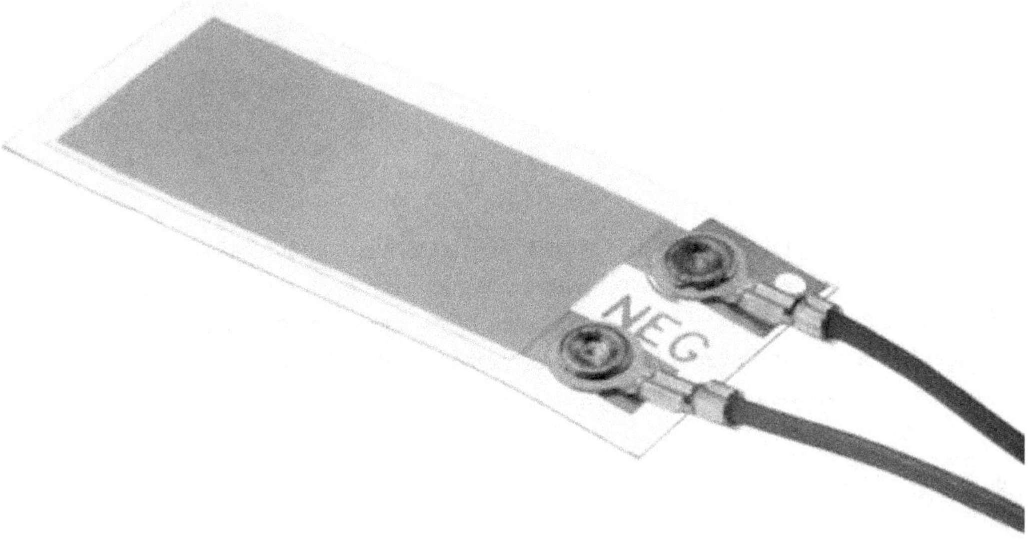

*Figure 5. Equivalent electro-mechanical scheme*

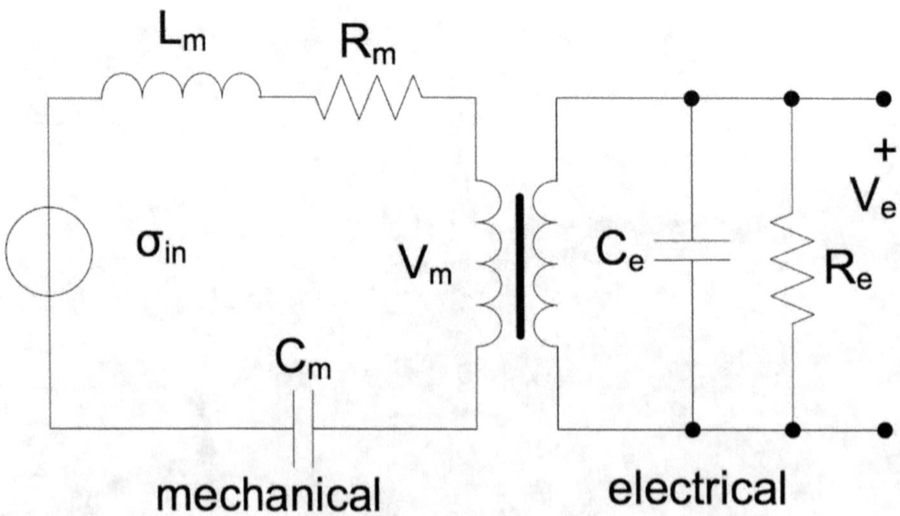


In the mechanical part, the inductor $L_m$ represents the equivalent mass and the inertia of the vibrating mass, $R_m$ represents the mechanical losses, $C_m$ represents the mechanical stiffness, stress generator $\sigma_{in}$ is due to external mechanical vibration, the equivalent transformer relates the physical quantities with those electrical. In the electrical part, $C_e$ represents the capacitance of the piezoelectric element, $R_e$ an external load and $V_e$ is the voltage across the piezoelectric transducer. Electric and mechanical parameters depend on the shape and the vibration mode of the piezoelectric transducer.

By applying the equivalent Kirchhoff's rule on first loop:

$$\sigma_{in} - L_m\ddot{S} - R_m\dot{S} - S/C_m = V_m, \tag{5}$$

the equivalent transformer constitutive equation are:

$$V_m = n_v V_e, \qquad I_m = \frac{1}{n_i} I_e. \tag{6}$$

The cantilever configuration is studied for the PVDF transducer, which has a free edge. Such harvester is made with only one piezoelectric layer, where complete bimorph models can be found in (Roundy at al., Erturk and Inman) The equivalent inductance $L_m$ is due to the geometrical configuration of the system, one edge of the cantilever is fixed, the free other is hitted by the mass of the drop of water, the inertia moment of the mass is evaluated by:

$$J = 2\left[\frac{wt_{mylar}^{3}}{12} + wt_{mylar}b^{2}\right] + \frac{wt_{piezo}^{3}}{12}, \tag{7}$$

with b distance between middle points of layers as shown in Figure 6.

The average stress in the beam is:

$$\sigma_{in} = \frac{1}{l_b}\int_{0}^{l_b}\frac{M(x)b}{J}dx, \tag{8}$$

where

$$M(x) = F_{in}\left(l_b + l_m/2 - x\right), \tag{9}$$

with lb length of cantilever, *F* force exerted by the drop.

$K_l$ is the geometrical coefficient which relates stress and external force:

$$\sigma_{in} = K_1 F_{in}. \tag{10}$$

During the vibrations $K_l$ relates also the stress due to the mass of water attached to the cantilever:

$$\sigma_L = K_1 F_m, \tag{11}$$

and in similar way it is related to the second derivative of the strain by:

*Figure 6. Layer of the PVDF cantilever*

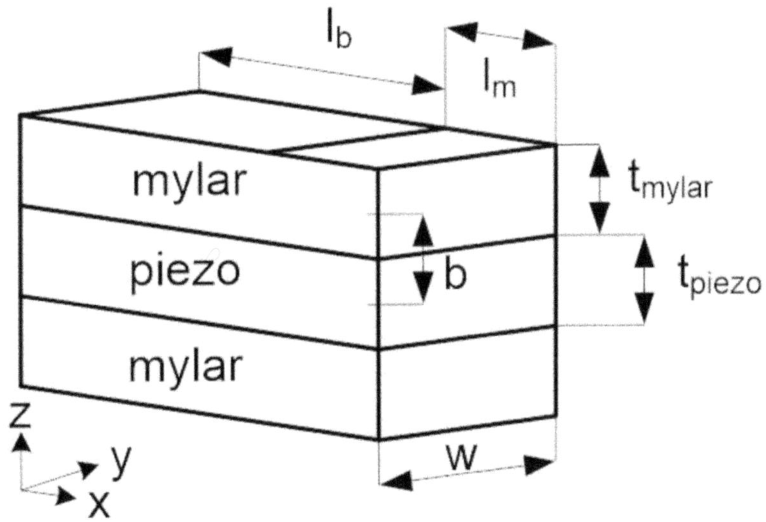

$$\sigma_L = K_1 K_2 m\ddot{S}, \tag{12}$$

with m mass of water and $K_2$ geometrical coefficient.

The equivalent inductance and resistance are:

$$L_m = K_1 K_2 m,\ R_m = K_1 K_2 b_m \tag{13}$$

with damping factor,

$$b_m = \sqrt{\frac{K}{m}} \Big/ \frac{m}{Q},\ K = 1/s_{11}$$

Young's module and Q =10.

The equivalent capacitance relates strain with stress by using the compliance constant $s_{11}$:

$$\sigma_C = \frac{S}{s_{11}} = \frac{1}{C_m} \int \dot{S} dx\,. \tag{14}$$

In order to evaluate the voltage ratio nv a strein zero configuration has been taken into account:

$$0 = s_{11}\sigma_T + d_{31}E_3, \tag{15}$$

Electric field is related to voltage:

$$\sigma_T = -\frac{d_{31}}{s_{11}} \int E_3 dz = -\frac{d_{31}}{s_{11}} \frac{V_e}{t_{piezo}}, \tag{16}$$

$$n_v = -\frac{d_{31}}{s_{11} t_{piezo}}. \tag{17}$$

The electric current is:

$$i = \frac{V_e}{R} + C_e \dot{V}, \tag{18}$$

with $C_e = \varepsilon \frac{w l_b}{t_{piezo}}$. (19)

Without considering an external electric field:

$$D = \frac{d_{31}}{s_{11}} S , \tag{20}$$

the charge on the electrodes is:

$$q = wl_b D = wl_b \frac{d_{31}}{s_{11}} S , \tag{21}$$

Current and current ratio are:

$$i = wl_b \frac{d_{31}}{s_{11}} \dot{S} = \frac{1}{n_i} \dot{S} . \tag{22}$$

The characterization has been made for a one-edge fixed cantilever but in a similar way the two-edge can be studied.

## TRANSFER FUNCTION

Mechanical and electrical equations are:

$$L_m \ddot{S} + R_m \dot{S} + S/C_m = \sigma_{in} + V_m , \tag{23}$$

$$wl_b \frac{d_{31}}{s_{11}} \dot{S} = \frac{V_e}{R} + C_e \dot{V} , \tag{24}$$

by operating the Laplace transform:

$$S\left( s^2 + \frac{b_m}{m} s + \frac{1}{mK_1K_2K} \right) = \frac{\sigma_{in}}{K_1K_2m} + \frac{d_{31}K}{t_{piezo}mK_1K_2} \tag{25}$$

$$S = \frac{\varepsilon}{t_{piezo}d_{31}Ks}\left( s + \frac{1}{R_eC_e} \right) V . \tag{26}$$

Voltage transfer function is:

$$V = \frac{A_1 s}{s^3 + B_2 s^2 + B_1 s + B_0} \tag{27}$$

with

$$A_1 = \frac{t_{piezo} d_{31} K \sigma_{in}}{\varepsilon m K_1 K_2} \tag{28}$$

$$B_2 = \frac{b_m}{m} + \frac{1}{R_e C_e}, \tag{29}$$

$$B_1 = \frac{1}{m K_1 K_2 K} + \frac{b_m}{m} - \frac{d_{31}^2 K^2}{\varepsilon m K_1 K_2}, \tag{30}$$

$$B_0 = \frac{1}{R_e C_e m K_1 K_2 K}. \tag{31}$$

In figure 7 the magnitude of Bode diagram is traced; the maximum is located for the $\omega_n = 260$ rad/s. Such behavior is principally due to mechanical properties of the cantilever, it is an average profile since it is been made taking into account the resitance only of the obsilloscope (1 MΩ).

## LABORATORY TEST

In order to perform the verification of the model and of the transfer function, an indoor experiment under simplified conditions has been conducted. The experiment reconstructs the falling of drops of water on the cantilever, then the output voltage is recorded and finally it is compared with the performance obtained through the model. To simulate the fall of rain, however, was used a pipette, which allows to create artificially a drop of calibrated water, with a default speed due to the height of fall, figure 8.

In the first place, it was necessary to fix the piezoelectric plate to a plane through a vise, in order to make stable the structure on one side and leave it completely free at the other, and this in order to keep intact the oscillatory properties of the material and therefore to reproduce the behavior of a cantilever beam. Furthermore, to make appropriate measures, the connectors of the lamina have been linked to the probe of an oscilloscope Lecroy model LT342L.

*Figure 7. Bode's diagram of the voltage transfer function*

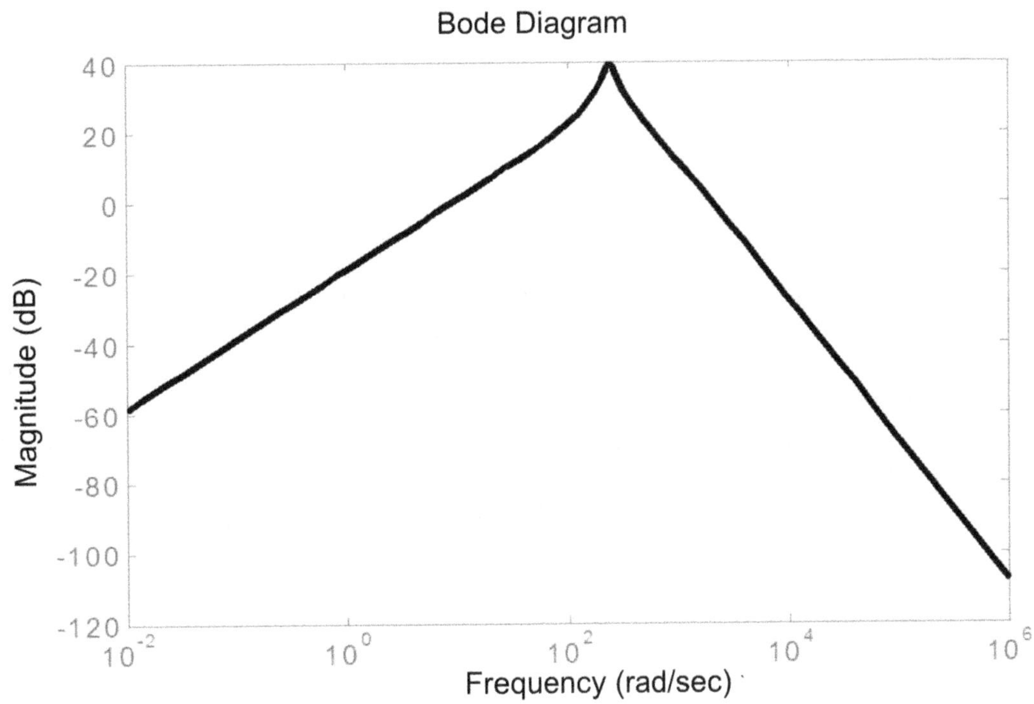


*Figure 8. Pipette, cantilever and vise.*

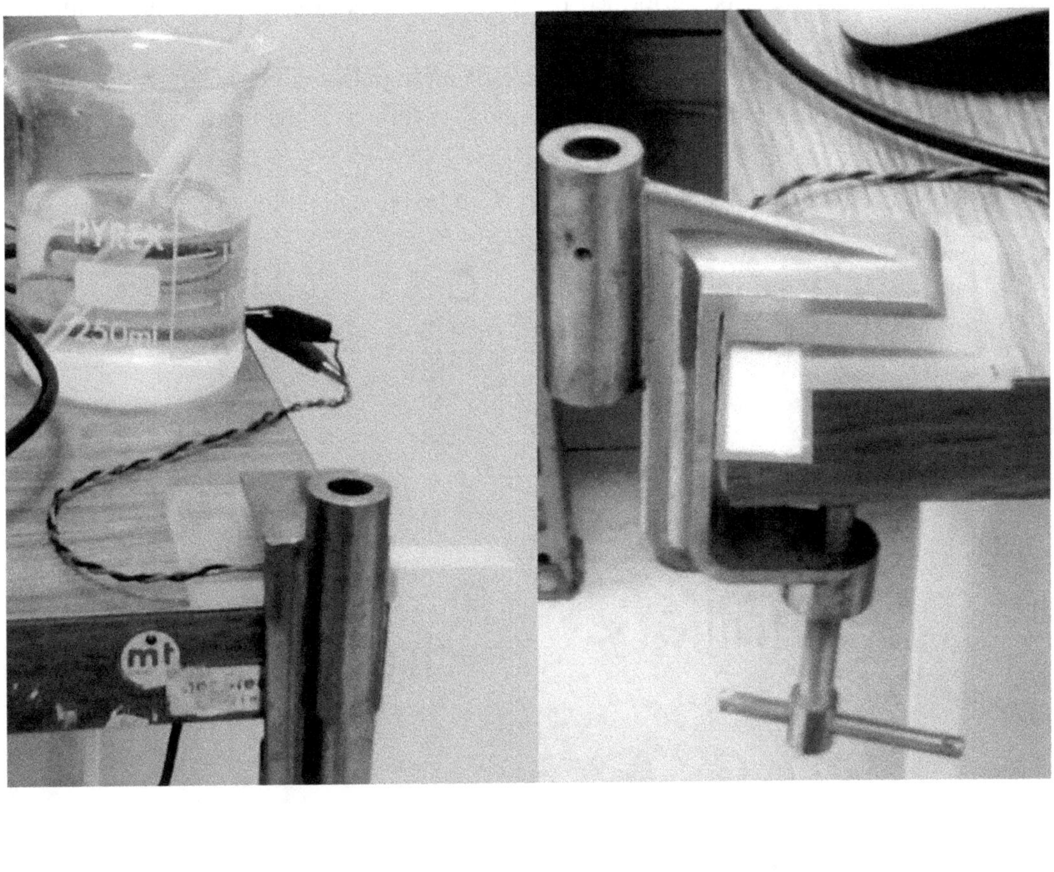

The pipette consists of a cylindrical capillary with thick walls, the end of which is cut along a section, Figure 9, where the drops are formed. In particular, there is a linking relationship between the radius of the droplet and that of the capillary:

$$R_{drop} = \left( \frac{3R_{capillary}\gamma}{2g\rho} \right)^{\frac{1}{3}} \tag{32}$$

where ρ is the surface tension of water, γ is the density of water and g is the gravity acceleration.

The use of the capillary for the creation of the droplet has a limit that concerns the movement of the liquid through the capillary. The process of ejection of the droplet must be relatively slow, in order to maintain a certain condition of static equilibrium, and to create drops similar to the ones of rain. A better performance, in the characterization of the water droplets, can be made by employing a disdrometer.

The first experiment was conducted at a height of 80 cm. The oscilloscope is set to 2V/division in order to obtain a more accurate measurement, the waveform of the generated voltage is shown in figure 10.

As a pulse stress the system, the system itself responds to the frequency that maximizes the output. A maximum of 6 V has been reached, this is due to the precision of fall in the edge area of the cantilever, to the use of a very large drop, to the absence of the phenomenon of splashing.

## MODEL VALIDATION

The response of the model can be performed by the SIMULINK program, in which a simple block diagram has been developed, figure 11.

The stress due to the water drop can be simulated by an impulsive force. A function similar to the delta of Dirac has been considered in the study. The pulse is defined as the limit for Δ that tends to zero

*Figure 9. Schematization of the release stages of the drop from the pipette*

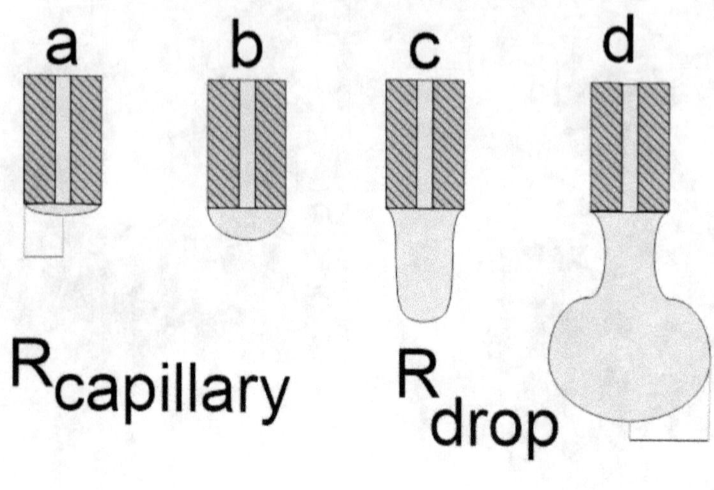

*Figure 10. Time profile of the recorded voltage*

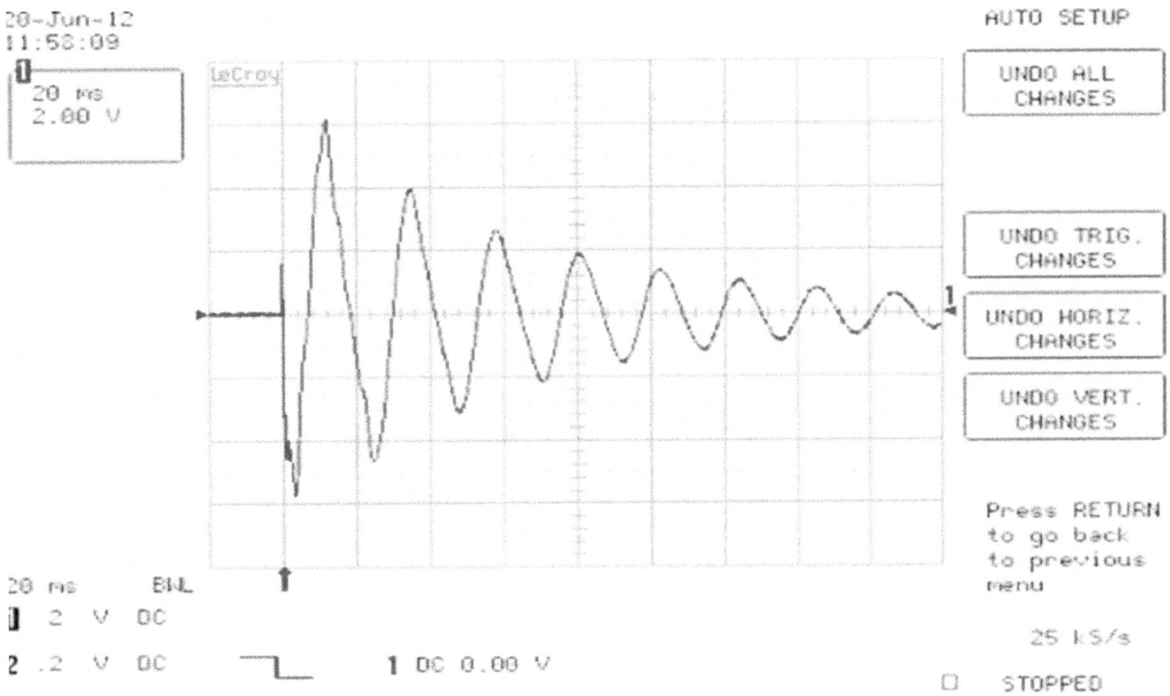


*Figure 11. Simulink model of the piezoelectric system*

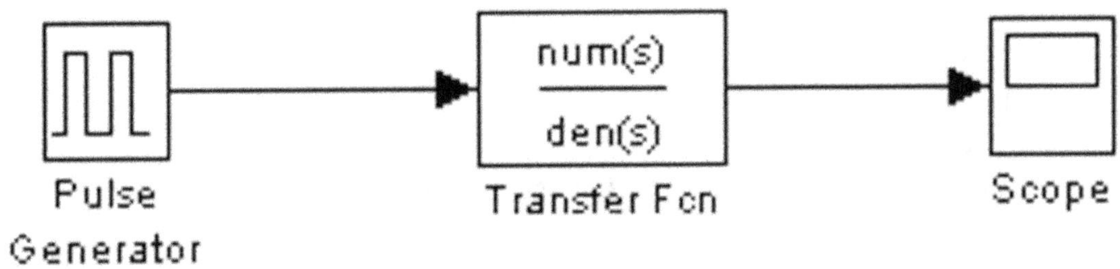


of the rectangular function with height 1/ Δ and amplitude Δ, such function can have an infinite value for t=t0 and outside the interval [t0, t0+Δ] the value is zero. Figure 12 shows the time profile of the pulse.

By considering an observation time equal to 10 s, the pulse generator has been considered as not ideal, and the impact of the drop on the piezoelectric has been set equal to 0.08% of the time interval. The mass is 0.12 g, final velocity is 1.24 m/s, force is 0.77 N.

Figure 13 shows the comparison between the recorded voltage and the simulated one. No splashing phenomenon has been observed.

Oscillations in the simulated system are very similar to those recorded. Furthermore, the voltage values obtained are similar.

Experiments in laboratory with different height of fall of the drop have been performed. These experiments have shown that by increasing height there is an increasing in the maximum value of the voltage produced by the drop, keeping constant its geometrical dimensions.

*Figure 12. Time-profile of the stress*

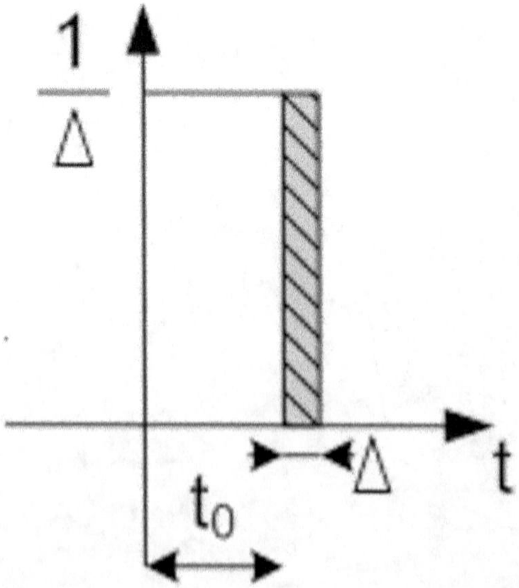


*Figure 13. Comparison between the time profiles of recorded and simulated voltage*

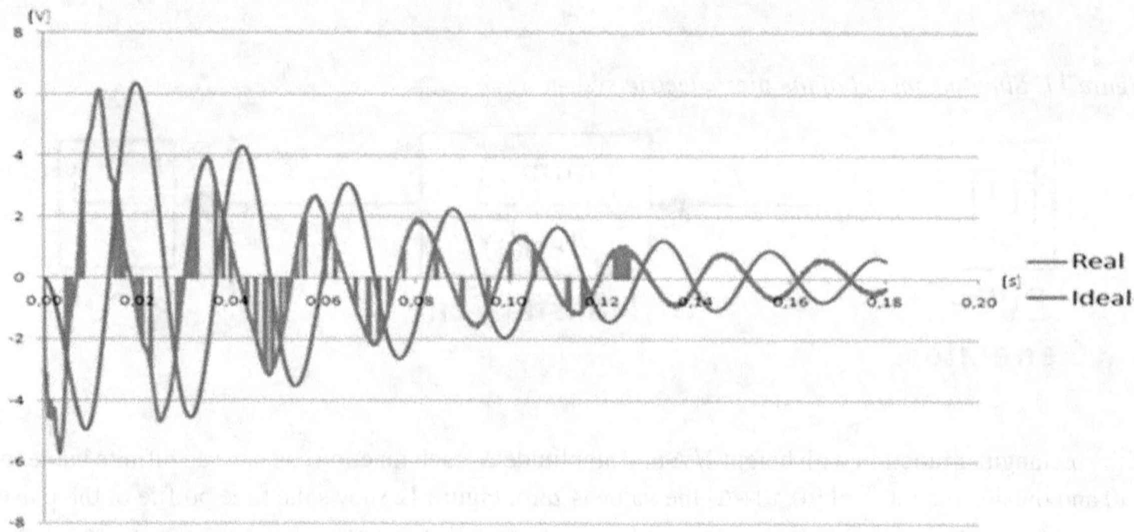


However, it has been found that as the speed of fall increases, the splash phenomenon occurs, and a loss of mechanical energy causes a divergence of the theoretical results compared to experimental ones.

## OPEN FIELD TEST

The piezoelectric transducers have been fixed on suitable support and were exposed to real and artificial rain. Different types of raindrop have been used since the real rain, in the period of the year in which it was used to extract power, had little diameter dimension, far from the dimensions of typical rain of other parts of the world. In this condition were measured the voltage values generated by the impact of the droplets. Roughly, some typical generated waveforms are definable: transducers bound to the both ends have generated waveforms more regular, in which a first pulse largest is followed by a second smaller and of opposite sign, figure 14.

Sometimes the second pulse has been followed by a third and sometimes it was not present. In some cases we have obtained single negative peak. This physical behavior has required a bit more attention since the physical stress did not change its direction, but it happened to the tension. It was found that the negative single peaks of tension occurred in correspondence with a state of piezoelectric plate already burdened with a water film, as a result of impact the compression status ranged, generating a negative peak of voltage.

The PVDF piezoelectric sensor has been used in cantilever configuration. An output voltage is represented in figure 15. The output voltage has an oscillating behavior, due to the presence of an underdamped system. The energy of the drop of water is absorbed and then released to the electrical system in a longer interval than that used by the Volture transducers.

In both cases, the performances were conditioned by the presence of an irregular water film on the transducers, which creates a mass that affects the damping behavior.

The voltage levels were maintained with maximum peaks of 4 volts during the different tests. Such voltage, though it seems high, refers to a peak to peak value, and it has been found for the artificial rain mainly. Another feature to be investigated is the frequency of the voltage waveform: cantilever transducers show an underdamped behaviour, transducers bound in both ends an overdumped one.

*Figure 14. Signals acquired for the PZT transducer: two present the typical behavior, a large positive pulse followed by a second and a third smaller; an only positive pulse. The waveform of the signals is often conditioned by the presence of the water film on the transducers.*

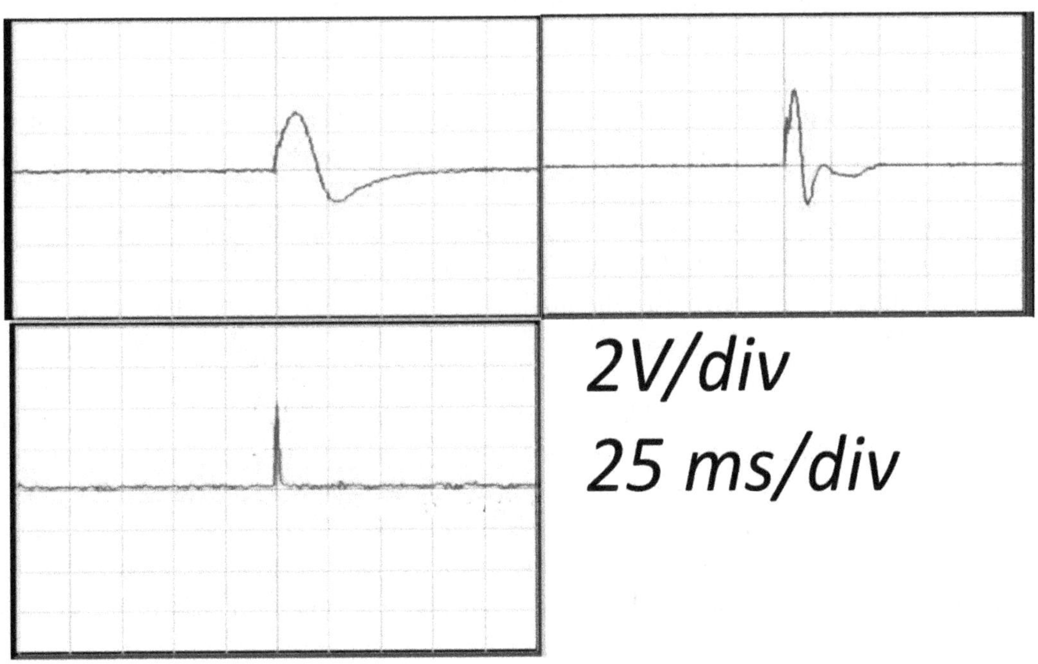

*Figure 15. Waveforms of the voltage given by a PVDF transducer in cantilever configuration. Oscillations are due to the particolar structure.*

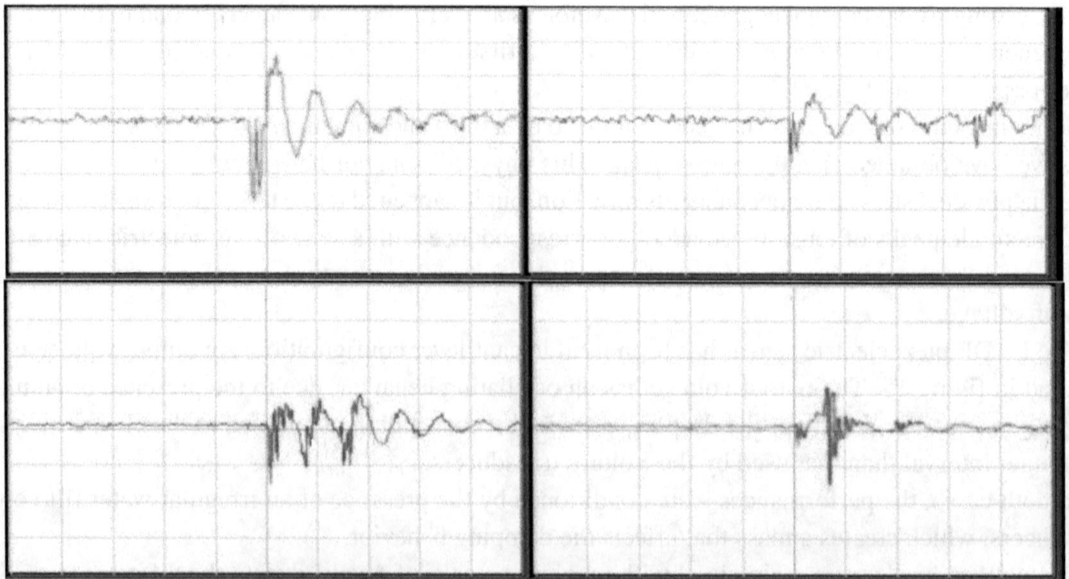


## POWER CHARACTERIZATION

To characterize the behavior of the transducers were carried out some measures by placing different resistive load connected to the electrodes. Ten measurements for each load were considered useful. The output parameter of the experiments is given by the power exchanged during the impact. In order to compare the behaviour of the different test the average power on the resistor has been taken into account and not the voltage time profile.

To compare the different behavior of the PZT and PVDF transducers, it was decided to refer both to the same conditions, thus using mostly artificial rain, which is more easily reproducible with the same characteristics.

In figure 16 the power extracted from the single drop of water, in the case of PZT transducers, for the loads of 10, 33, 47, 68, 82, 100, 180,270, 470 kΩ is represented.

In figure 17, the power extracted from the single drop of water, in the case of PVDF transducers, for the loads of 10, 22, 47, 100, 180, 470 kΩ, and for the same day test, is represented. Two cases have been studied a single PVDF transducer and two set in parallel, figure 18. Contrary to what could be expected maximum values of power are attributable to the case of single piezoelectric and not to the two parallel.

An explanation can be given to this behavior: the two drops of water do not affect the transducers simultaneously, so generating oscillating pulses in delay or with opposing phases, resulting in a reduction of voltage at the terminals. The experiences made with the same rain allow to make a comparative evaluation between similar solutions, but do not allow to derive absolute trends. The power behavior has then been studied by means of statistical analysis. More resistors have been added (10, 15, 22, 27, 33,

*Figure 16. Power extracted from single drop of water using the PZT transducers.*

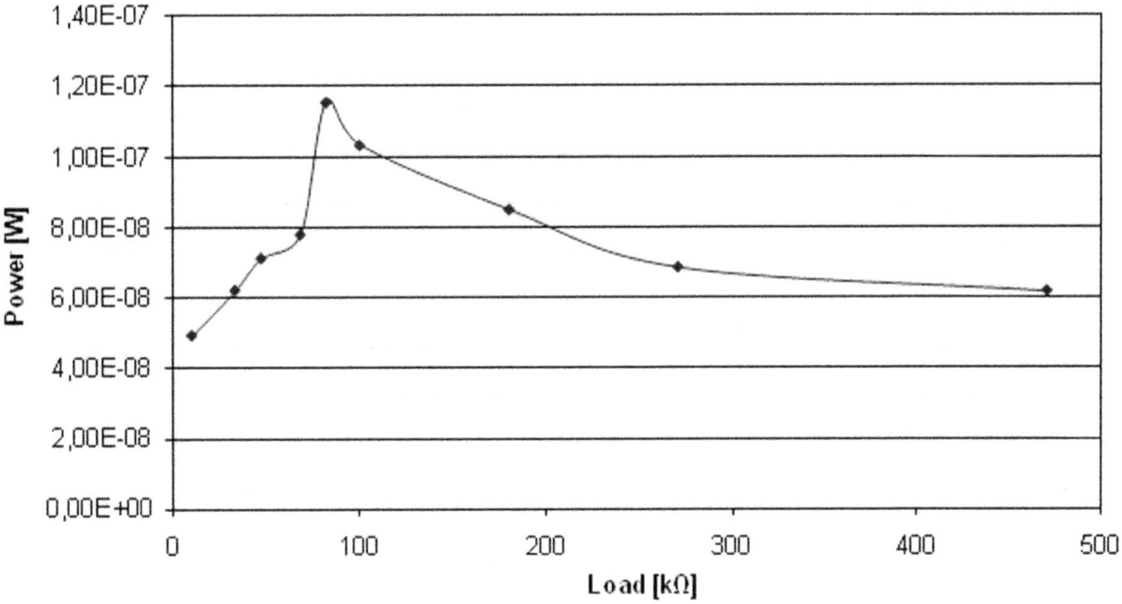


*Figure 17. Power extracted from single drop of water using a single PVDF transducer.*

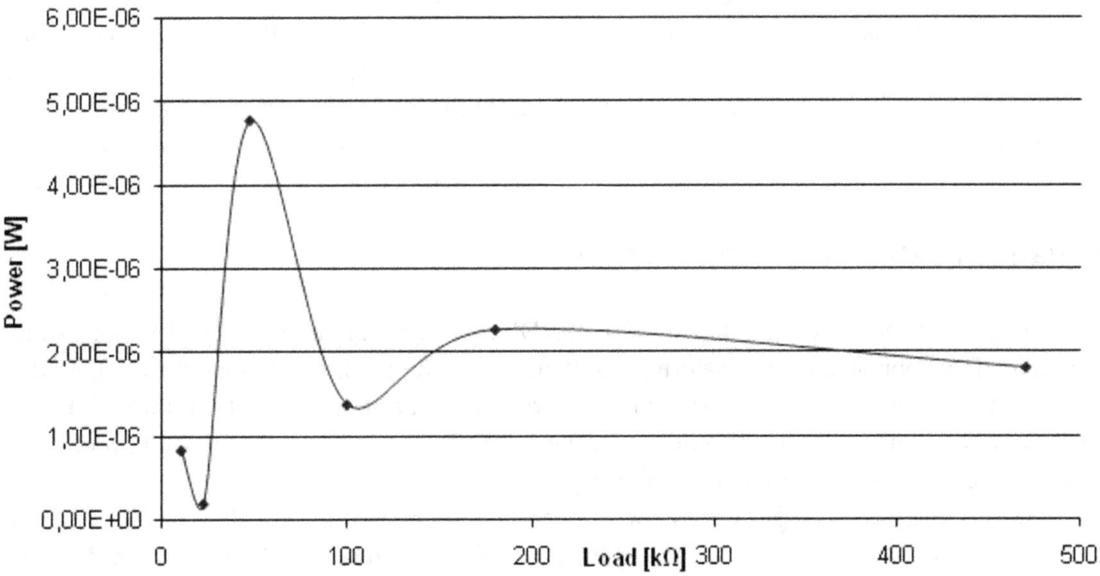

*Figure 18. Power extracted from single drop of water using two PVDF transducers in parallel configuration.*

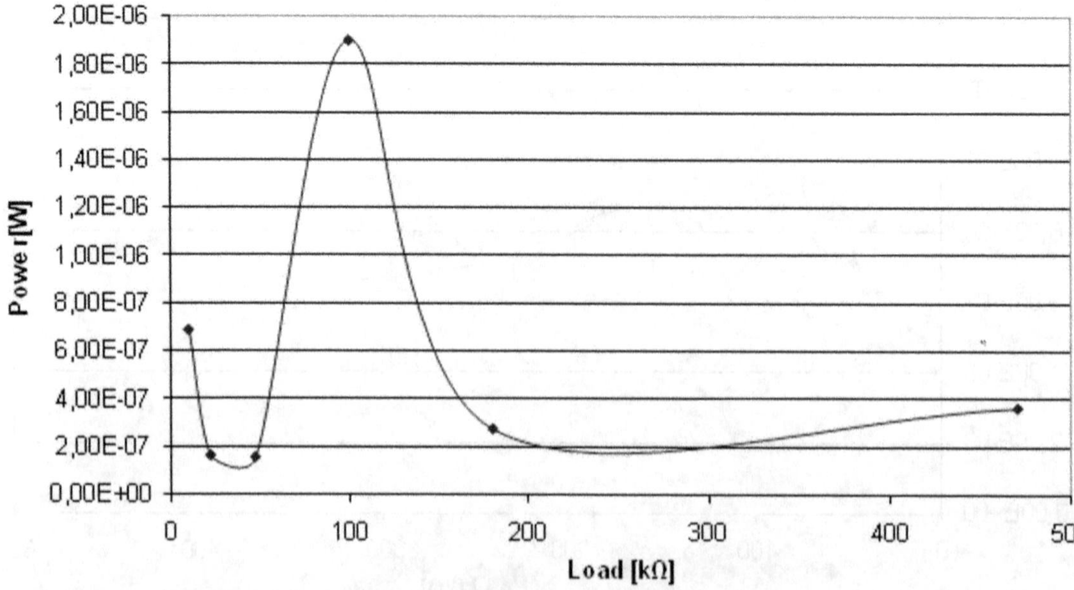


39, 47, 56, 68, 82, 100, 150, 180, 470 kΩ) and are considered as useful data those measures in which the power exceeds the parameter 2μW. It has been proved that loads between 47 kΩ and 100 kΩ have the highest possibility of extracting a power between 2 and 6 μW.

Once acquired the mean power behavior with a statistical analysis for the PVDF, the same evaluation has been performed also for the PZT transducer; it has shown a better performance around 100 kΩ, with a mean power of 0.13 μW. A comparison has been made on the levels of power extracted, assuming that the average of the performed measurements leds to have an impact with a equivalent drop and on similar surface, it can be noted that the power extracted differs by an order of magnitude, in favor of PVDF.

## Comparison between Different Stresses

Have been characterized also the behavior of the PVDF transducer in the presence of various stresses: natural rain, drops obtained from a watering can (height 4m), drops from raining channel (height 3m). The drops of water can reach the velocity of 9 m/s from a height of 4m, velocity of rainfall. Different water drops have been used in order to compare different raindrop, real rain is the one of the Sicily island, the other two can simulate the rain of northern regions.

In figure 19, the harvested power is shown. It has been found that the smaller voltage values were generated from drops of rain, those generated by drops of rain channel result about twice, and finally those of watering can are 5-6 times larger. The power can then be scaled by a factor of 4 for the drops of the gutter, and by a factor of almost 40 for the watering can. The explanation for this behaviour can be found in two particular aspects: first for the type of weather conditions, the natural rainfall had a reduced diameter compared to artificial rainfall and thus a lower kinetic energy; second, the natural rain, considering the same speed of fall, had a lateral component due to the wind that effected the exchange

*Figure 19. Power extracted from single drop of water in presence of various stresses: natural rain, drops from rain channel and drops from water can.*

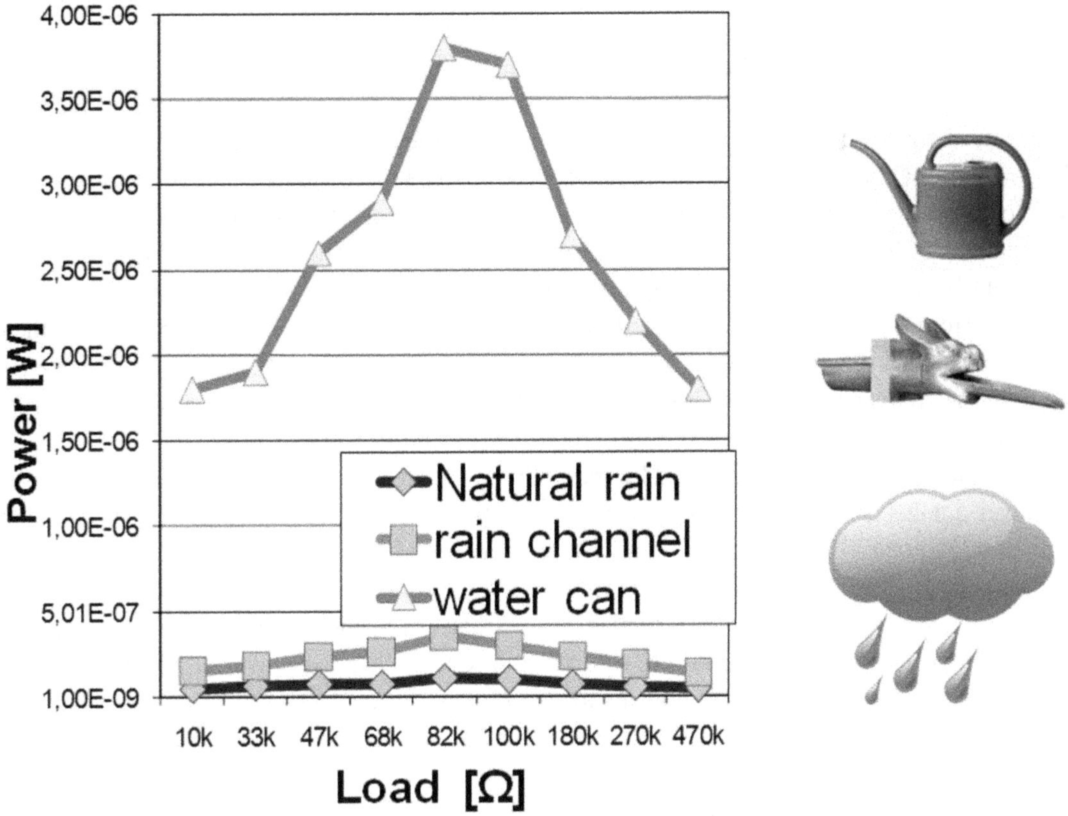


of kinetic energy during the impact. The artificial drops were always perpendicular to the plates thus maximizing efficiency. Thus, the experiments conducted in the laboratory may be dissimilar from real cases, one consideration has to be underlined: the Sicilian natural rain that was tested had a diameter less than 3 mm, for which the extracted power differs from the highest values obtained in the laboratory, with drops that simulate a precipitation of the center Europe.

The three examined cases confirm that the load that provides the best power transfer to the load is 82 kΩ.

## Maximizing the Power

In order to improve the performance of the harvester, some optimized solutions can be followed. The first implies the adjustment of impedance to compensate for the reactive part of the capacity. Additional solutions include the use of non-conventional geometries and the optimization of efficiency electromechanical imposing an optimized deflection.

By considering figure 5, it is possible to maximize the transfer of power by inserting an inductor in series with the resistor. A 82 kΩ resistor, which showed the best behavior in the transfer of power, has been chosen; three commercial inductors of 82μH, 820 μH and 82mH have been tested.

In figure 20 the harvested power, obtained for real rain, is shown. The mean harvested power decreases for 82 μH (-4%) and encrease for 820μH (+4%) and for 8.2mH (+8%). Considering that the frequency of the underdamped behavior does not exceed 150 Hertz and that the capacity is around few nF, the ideal inductive load is about a thousand H, but this load was not economically suitable.

An other way to improve the harvested power is to change the geometry of the cantilever. As suggested by Benasciutti, Moro, Zelenika and Brusa (2010) to maximize the energy conversion is not important to cover the largest area, but to subject the piezoelectric material close to its strength limits. The conventional rectangular harvester cantilever has a bending moment which decreases linearly from the fixed edge to the free one. A trapezoidal shape will allow to obtain a much regular moment, in such a way the PVDF harvester shape has been modified. In figure 21 the original and the custom shape of the meas PVDF harvester, shown in Figure 3, are schetked.

The custom shape has an area reduced of 20% respect the original one. During the tests the custom shape harvester has shown quite the same harvested power of the original one, the only differences were found on the frequencies of oscillation, which were predominantly higher. In such a way it is possible to deduce that the power can be maximized by using a smart geometry.

One other test has been conducted by mainteining the original rectangolar shape and pressing the cantilever against a suitably shaped profile, given by a cylindrical surface, as shown by Benasciutti, Moro and Gallina (2013). Figure 22 shows the cantilever and pressing surface.

Two surfaces with different curvature radius were considered, the first with $r_1$ = 500 mm and the second with $r_2$ = 800 mm. The first surface is one that has been constantly presented an increase in power, with a maximum at around 5%, while in the second case the power has remained unchanged. Such a way to maximize harvested power shows minor advantages.

*Figure 20. Power extracted from single drop of water in presence of different inductances: the harvested power decreases for 82 μH (-4%) and increase for 820 μH (+4%) and for 8.2mH (+8%).*

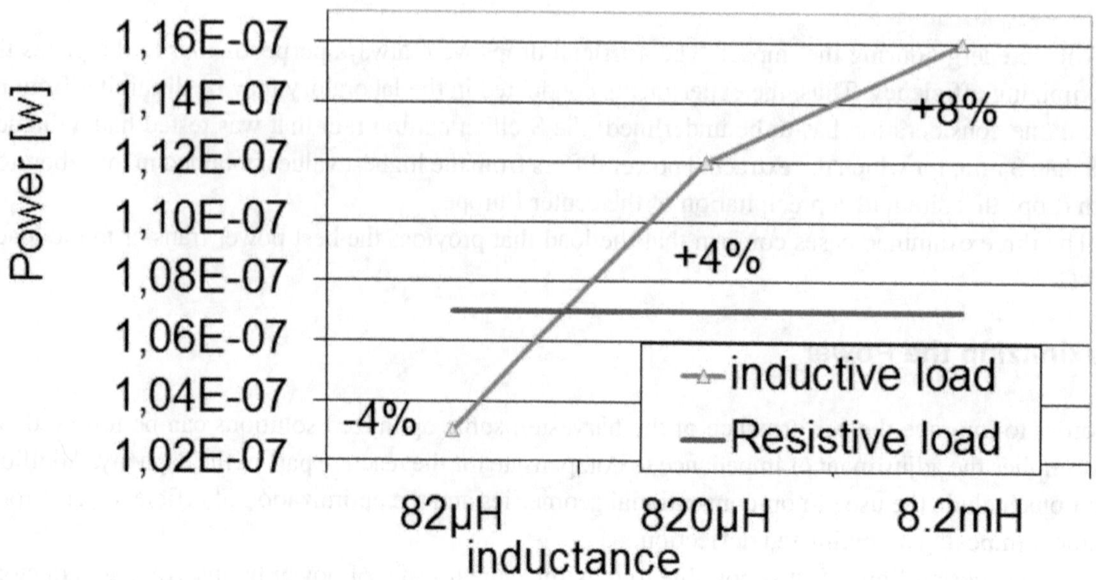

*Figure 21. Original PVDF cantilever: $L_e$ = 30 mm, $L_b$ = 34 mm, $L_{tot}$ = 4.2 mm, W =16 mm, $W_e$ = 12mm. Custom shape: $W_c$ = 8mm*

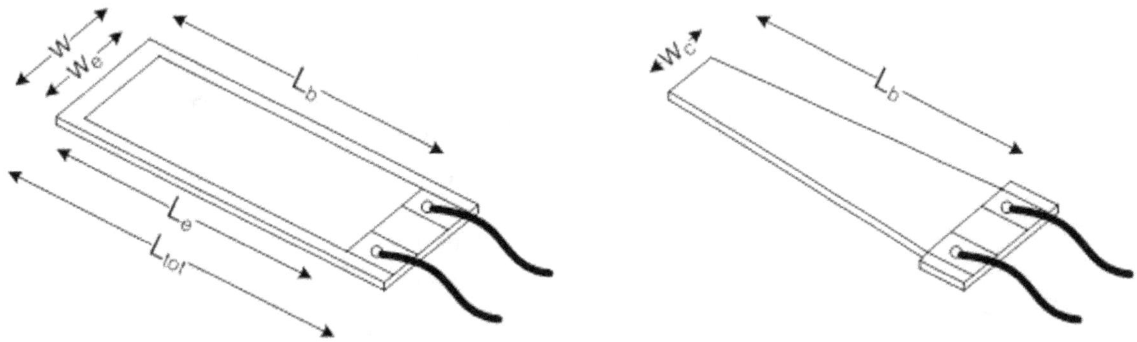


*Figure 22. PVDF cantilever and pressing surface*

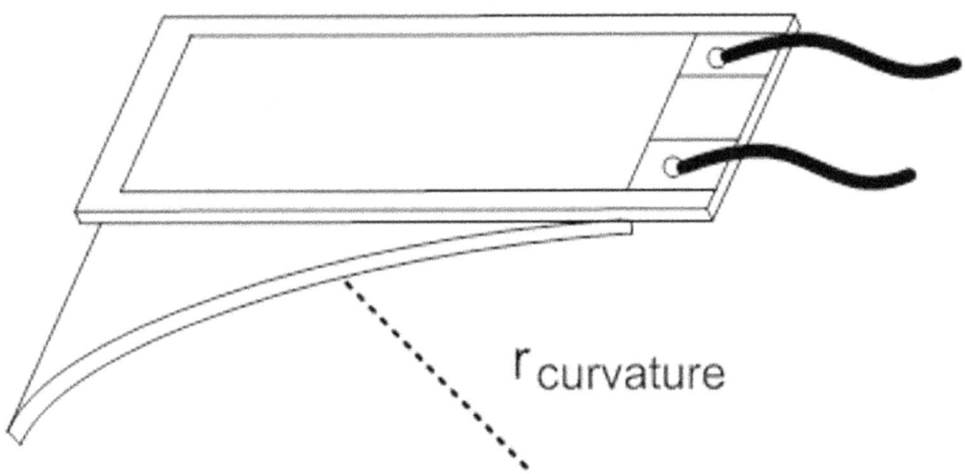


## APPLICATIONS

It is well known that the same piezoelectric devices may present different values of power extracted. A harvester, when attached to an industrial motor can draw power in order of mW, but, if attached to human body, power decreases to μW. It should of course read that in the first case the abundance of energy is mainly due to the presence of a system such as a motor in which electrical energy is already present, and there it is a recovery problem, unlike, the human body is devoid of a power supply, and here must then be generated. Similarly, a solar-powered sensor could enjoy power of tens of mW in outdoor locations, but less than tens of μW indoors. In this context sensor applications are born. The sensors devoted to the monitoring and control require a more careful design if they are to be connected in sites without energy supply.

Figure 23 shows a scheme of a raindrop energy harvester. The energy harvester is a piezoelectric beam in any cantilever configurations. The energy storage is made by a rectifier bridge and capacitors. Bridge is needed since the voltage waveform is alternate one. The bridge, in order to reduce the voltage

*Figure 23. Scheme of monitor system fed by raindrop energy harvester*

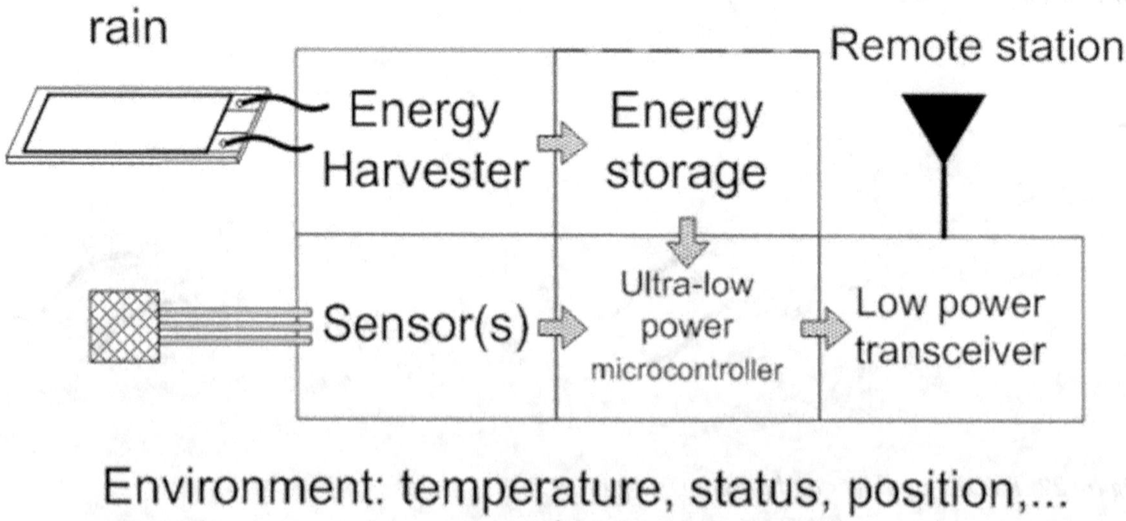


*Figure 24. Schematic and electronic circuit board of harvesting system employing the LT3588*

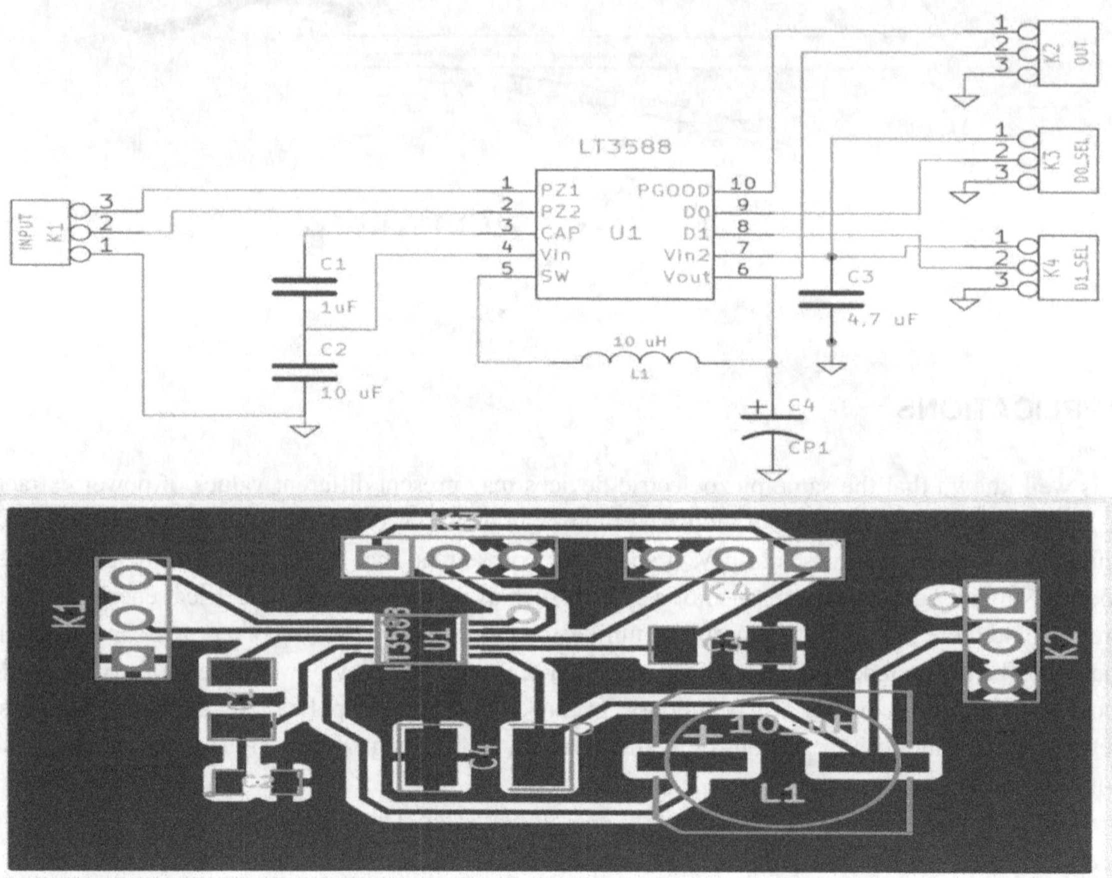

consumption, employs ultra-low threshold voltage metal-oxide semiconductor field-effect transistors (MOSFETS). In order to better store energy and gain a suitable voltage, micropower step-up low voltage booster modules have to be used. A capacitor receives the pulses generated by the rain and it charges progressively until reaching the voltage values suitable for supply to the sensors and the transmission system. Booster and energy management are regulated by the ultra-low power microcontroller. Integrated systems of rectifier bridge, storage and power management are commercially available at very low cost, such the pump charge LT3588 of Linear Tecnology.

In Figure 24 the schematics of the harvesting system employing LT3588 is shown.

In figure 25 the stepped output voltage, due to artificial rain, is shown.

## CONCLUSION

This chapter reports a detailed study on the piezoelectric energy harvesting of rainfall and its limits. Two types of commercial piezoelectric transducers were considered: Lead Zirconate Titanate (PZT) and polyvinylidene difluoride (PVDF) transducer. A comparison of harvested power has been made: PVDF

*Figure 25. Voltage output of the harvesting system subjected to artificial rain*

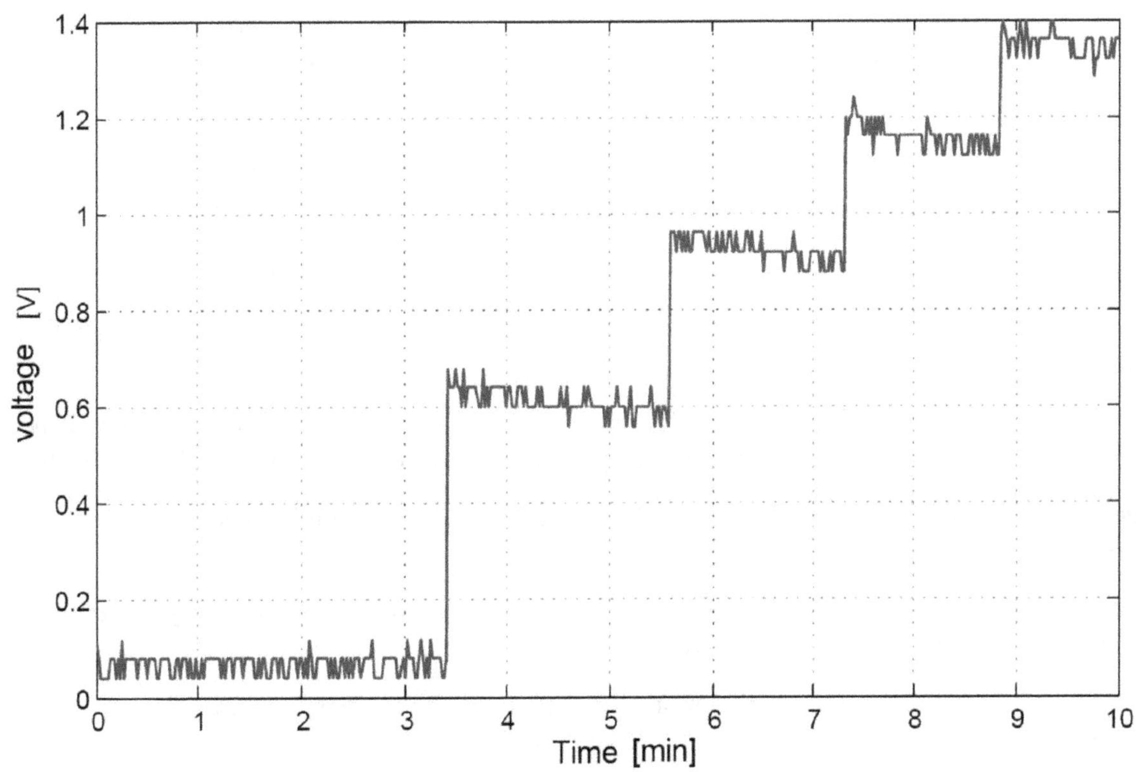

transducer generates a greater amount of power respect the PZT transducer. A circuital model of the PVDF cantilever configuration has been created and validated. For the PVDF harvester different stress conditions have been faced: from laboratory rain of a downpour to the low raindrop of Sicilian rain. In order to improve the harvested power technical solutions have been finally suggested.

## ACKNOWLEDGMENT

The authors acknowledge the financial support from University of Palermo and from PON i-Next.

## REFERENCES

Benasciutti, D., & Moro, L. (2010). Harvested power and sensitivity analysis of vibrating shoe-mounted piezoelectric cantilevers. *Smart Materials and Structures*, *19*(11), 115011. doi:10.1088/0964-1726/19/11/115011

Benasciutti, D., Moro, L., & Gallina, M. (2013). On the optimal bending deflection of piezoelectric scavengers. *Journal of Intelligent Material Systems and Structures*, *24*(5), 627–639. doi:10.1177/1045389X12469451

Benasciutti, D., Moro, L., Zelenika, S., & Brusa, E. (2010). Vibration energy scavenging via piezoelectric bimorphs of optimized shapes. *Microsystem Technologies*, *16*(5), 657–668. doi:10.1007/s00542-009-1000-5

*Datasheet of linear technology LT3588*. (n. d.). Retrieved from http://www.linear.com

*Datasheet of MEAS LDT1-028K*. (n. d.). Retrieved from http://www.meas-spec.com

*Datasheet of Mide Volture V22B*. (n. d.). Retrieved from http://www.mide.com/index.php

Erturk, A., & Inman, D. J. (2011). *Piezoelectric energy harvesting*. USA: John Wiley & Sons. doi:10.1002/9781119991151

Hobeck, J. D., & Inman, D. J. (2012). Artificial piezoelectric grass for energy harvesting from turbulence-induced vibration. *Smart Materials and Structures*, *21*(10), 105024. doi:10.1088/0964-1726/21/10/105024

Jager, T., Guigon, R., Chaillout, J. J., & Despesse, G. (2008). Harvesting raindrop energy: Theory. *Smart Materials and Structures*, *17*(1).

Jager, T., Guigon, R., Chaillout, J. J., & Despesse, G. (2008). Harvesting raindrop energy: Experimental study. *Smart Materials and Structures*, *17*(1).

Liang, J., & Liao, W. S. (2011). Energy flow in piezoelectric energy harvesting systems. *Smart Materials and Structures*, *20*(1), 015005. doi:10.1088/0964-1726/20/1/015005

Perera, K. C. R., Sampath, B.G., Dassanayake, V.P.C. & Hapuwatte, B. M. (2014). Harvesting of kinetic energy of the raindrops. *International Journal of Mathematical, Computational, Physical and Quantum Engineering*, *8*(2).

Pinna, L., Dahiya, R., & Valle, M. (2009). *SPICE model for piezoelectric bender generators.* Paper presented at 16th IEEE International Conference on Electronics, Circuits, and Systems, (ICECS). Hammamet, Tunisia.

Roundy, S., & Wright, P. K. (2004). A piezoelectric vibration based generator for wireless electronics. *Smart Materials and Structures, 22*(13), 1131–1142.

Standards Committee of the IEEE Ultrasonic, Ferroelectrics and Frequwncy Control Society. (1987). *IEEE Standard on Piezoelectricity.* IEEE New York.

Van den Ende, D. A., van de Wiel, H. J., Groen, W. A., & van der Zwaag, S. (2012). Direct strain energy harvesting in automobile tires using piezoelectric PZT–polymer composites. *Smart Materials and Structures, 21*(1), 015011. doi:10.1088/0964-1726/21/1/015011

Vatansever, D., Hadimani, R. L., Shah, T., & Siores, E. (2011). An investigation of energy harvesting from renewable sources with PVDF and PZT. *Smart Materials and Structures, 20*(5), 055019. doi:10.1088/0964-1726/20/5/055019

Viola, F., Romano, P., & Miceli, R., Acciari, G., & Spataro, C. (2014). *Piezoelectric model of rainfall energy harvester.* Paper presented at 2014 Ninth International Conference on Ecological Vehicles and Renewable Energies (EVER), Monaco.

Viola, F., Romano, P., Miceli, R., & Acciari, G. (2013). *Harvesting rainfall energy by means of piezoelectric transducer.* Paper presented at International Conference on CLEAN ELECTRICAL POWER Renewable Energy Resources Impact 2013, Alghero, Italy doi:10.1109/ICCEP.2013.6586952

Wong, C.-H., Dahari, Z., Abd Manaf, A., Sidek, O., Miskam, M.A., & Mohamed, J. J. (2013). Simulation of piezoelectric raindrop energy harvester. In TENCON Spring Conference, 2013 IEEE, (pp. 465 – 469). doi: 10.1109/TENCONSpring.2013.6584488

Xiang, H. J., Wang, J. J., Shi, Z. F., & Zhang, Z. W. (2013). Theoretical analysis of piezoelectric energy harvesting from traffic induced deformation of pavements. *Smart Materials and Structures, 22*(9), 095024. doi:10.1088/0964-1726/22/9/095024

## ADDITIONAL READING

Abdelaal, W. G. A. (2013). *Power enhancement for piezoelectric energy harvesting systems.* Saarbrücken, Germany: Lambert Academic Publishing.

Choi, S. B., & Han, Y. M. (2010). *Piezoelectric Actuators.* Boca Raton, FL: CRC Press.

## KEY TERMS AND DEFINITIONS

**Cantilever Harvester:** Flexible plate, subjected to constant vibrational stress.
**Electromechanical Model:** Model of the mechanical-electrical energy converter.

**Piezoelectric Effect:** Phenomenon whereby some crystalline materials are electrically polarized as a consequence of a mechanical deformation (direct piezoelectric effect), and vice versa deform elastically when subjected to the action of an electric field.

**PVDF:** Polyvinylidene fluoride piezoelectric polimer.

**PZT:** Lead zirconium titanate piezoelectric ceramic.

**Renewable Source:** Non-extinguishable source of energy, such wind, heat, sun and rain. Raindrop Energy: mechanical energy of the water drop.

**Storage System:** System composed of a rectifier of the waveform and a capacitor for the accumulation of electric charge.

# Chapter 6
# Energy Harvesting Aspects of Irregular Vibrating Forces Acting on Piezoelectric Devices

**Alessandro Massaro**
*Italian Institute of Technology (IIT), Italy*

## ABSTRACT

*After a brief introduction of piezoelectric materials, this chapter focuses on the characterization of vibrating freestanding piezoelectric AlN devices forced by different external forces acting simultaneously. The analyzed vibrating forces are applied mainly to piezoelectric freestanding structures stimulated by irregular vibration phenomena. Particular kinds of theoretical noise signals are commented. The goal of the chapter is to analyze the effect of the noise in order to model the chaotic vibrating system and to predict the output current signals. Moreover, the author also shows a possible alternative way to detect different vibrating force directions in the three dimensional space by means of curved piezoelectric layouts.*

## INTRODUCTION

Vibrating energy harvesting systems are commonly implemented by means of piezoelectric materials. Many piezoelectric materials have been studied by researchers during the past years. A piezoelectric material can convert energy between the mechanical and electrical fields through relationships between the stress **T**, the strain **S**, the electric **E**, and the electric displacement field **D** coupled with elasticity matrix, coupling matrix, and relative permittivity matrix which are characteristics of the piezoelectric material. Piezoelectric materials can be composed by different materials (composed/nano-composed materials) or can be constituted by thin films. Thin piezoelectric films can be etched by chemical wet methods and processed by photolithography in order to fabricate Micro Electro-Mechanical Systems (MEMS). The common piezoelectric thin films studied in literature are Lead Zirconate Titanate (PZT), Zinc Oxide (ZnO) and Aluminium Nitride (AlN). Considering $d_{31}$ PZT cantilevers at low frequencies, Glynne-Jones et al. (2001), have found a power density of $P = 0.017$ μW/mm$^3$ at a working frequency $f = 80$ Hz for an acceleration $A = 4g$, besides at $f = 70$ Hz for $A = 1g$ Isarakorn et al. (2008) have found

DOI: 10.4018/978-1-4666-8254-2.ch006

$P$ = 2.5 μW/mm$^3$. At higher frequencies, Fang et al. (2006) have estimated $P$ = 10.84 μW/mm$^3$ at $f$ = 608 Hz for $A$ = 1$g$, Renaud et al. (2007) have evaluated $P$ = 21.68 μW/mm$^3$ at $f$ = 1.8 KHz for $A$ = 1.9$g$, Shen et al. (2008) have found $P$ = 3.27 μW/mm$^3$ at $f$ = 462.5 Hz for $A$ = 2$g$, and, finally, Kim et al. (2008) have appraised $P$ = 0.3 μW/mm$^3$ at $f$ = 870 Hz for $A$ = 8$g$. Considering $d_{31}$ Aluminium Nitride cantilever beam, Marzencki et al. (2005) have found $P$ = 0.01 μW/mm$^3$ at $f$ = 204 Hz for $A$ = 8g, and Renaud et al. (2008) have measured $P$ = 0.22 μW/mm$^3$ at $f$ = 320 Hz for $A$ = 0.02$g$. At lower frequencies of $f$ = 64 Hz, Massaro et al. (2011a) have measured a higher power density of $P$ = 30.20 μW/mm$^3$ at a resonant frequency $f$ = 64 Hz and for $A$= 2$g$ by considering curved layouts in ring MEMS configuration able to resonate at low frequencies. Generally, in the real cases, the maximum density of power does not correspond to the vibrational force frequencies which is very low. Moreover, the vibrational sources are irregular and are totally different if compared with the controlled vibrational sources used during the experiments. Consecutively, the traditional reading electronic circuits are not so efficient for irregular signals. In this direction we propose to study in this chapter the behavior of freestanding piezoelectric structures vibrating by irregular forces in order to predict the output current signals. This prediction is useful for the design of correct reading electronic circuits. The structures proposed in this chapter takes into account Molybdenum/Aluminium Nitride/Molybdenum (Mo/AlN/Mo) structures, where the metallic Mo layers and the piezoelectric layers can be grown by means a radio frequency (RF) sputtering machine. The sputtering machine is composed by rotating holders containing the target materials to deposit. By means of a Mo target, the first metallic layer is grown on $SiO_2$ substrate which is centered (before to active the vacuum condition) in order to provide an uniform metallic deposition. By acting on the DC bias parameter it is possible also to change mechanical stress properties which are function of the used substrate material, and, provide surface adhesion properties. The Mo thickness is a function of sputtering deposition time. In order to deposit the AlN layer a pre-sputtering (pre-hating) process is necessary. After this stage it is important to control $N_2$ and Ar gases during the growth. The last metallic layer deposition is performed by furthermore rotating the sputtering holder again on Mo target, and by waiting before the sample cooling. The layer thicknesses will define the curvature of the final ring/helical freestanding structure as experimentally proved in Massaro et al. (2011b). After the deposition of the three layers a wet chemical process is required. Working in clean room, on the sample is applied a positive photoresist layer (coating) and, successively, the sample is processed by UV radiation and by chemical etching of the Mo (by $H_2O_2$), AlN (by $H_3PO_4$) and $SiO_2$ (by HF) layers as described in Massaro et al. (2011-b). A good approach is to check the surface color during the chemical etching process by means of a microscope: as described in Massaro et al. (2011b), the different observed colors will represent the effect of the etching process at each stage. Finally, an analysis by a profilometer is useful to check the layer thicknesses and an analysis by means of a Scanning Electron Microscope (SEM) is important in order to check the freestanding layout. The piezoelectric coefficient of AlN can be measured using piezoresponse force microscopy as discussed by Shin and Song (2010) or by means ($d_{33}$ coefficient) of acoustic wave propagation (see Ho et al. (2004)). The vibrating piezoelectric efficiency will be a function of the geometrical and physical properties of the used materials: the induced mechanical properties will define the vibration flexibility and the mechanical force resistance, moreover, the layouts will improve mainly the frequency resonance properties. However, as described in Massaro and Cingolani (2011c), the use of the proof mass allows to operate at lower frequencies if compared with the natural resonant frequencies of the freestanding cantilever beam. This aspect is significant especially when wind/ocean vibrating forces are considered. Usually it is no possible to find in natural sources regular vibrating forces coinciding with the natural resonant frequencies of the structure. For this reason it is very important to first study accurately the typology of source signals. The whole energy harvesting system can be stimulated by different vibrating source signals as discussed in Inmann (2000). The source vibrating signals (represented by carriers) can be modulated by discontinuous functions such as square/step waves representing the noise acting on the vibrating forces present in nature usually characterized by irregular noise signals due to a possible turbulence regime, and, by different low frequency waves (harmonics) acting simultaneously (we think for example to the irregular force of the wind, or of the ocean wave or of the human body motion). The total effect of external vibrating forces (including particular noises) applied to the piezoelectric beam, can be represented by a random noise function. The various directions of the vector forces can define the vibrating turbulence of the whole systems. Moreover, the magnitude of currents generated by the piezoelectric devices decreases in the time domain by means of an irregular damping effect which follows the signal modulation trend. In this chapter the theoretical effect of irregular vibration forces acting on AlN piezoelectric material is mainly discussed. Furthermore, by means of the elastic property together with the torsional degree of freedom, AlN curved layouts could exhibits very high power density at low frequency and for low intensities of vibration forces. The proposed study will be completed by discussing various experimental aspects of measurement.

## RANDOM VIBRATIONAL FORCES ACTING ON A PIEZOELECTRIC CANTILEVER BEAM

The modeling of a piezoelectric freestanding device is reported in Figure 1 (a), where is illustrated the sketch of the random external forces acting on the freestanding structure having a proof mass. The freestanding beam is characterized by a piezoelectric layer embedded between two metallic layers (see inset of Figure 1 (a)). External forces, which are characterized by different angular frequencies, act on the piezoelectric beam as different external vibrating sources. The total effect of an external vibrating force (including particular noises) applied to the piezoelectric beam can be represented by the following random noise function:

*Figure 1. (a) Model of random vibrating forces acting on a piezoelectric freestanding beam. (b) Block sketch of the whole model system.*

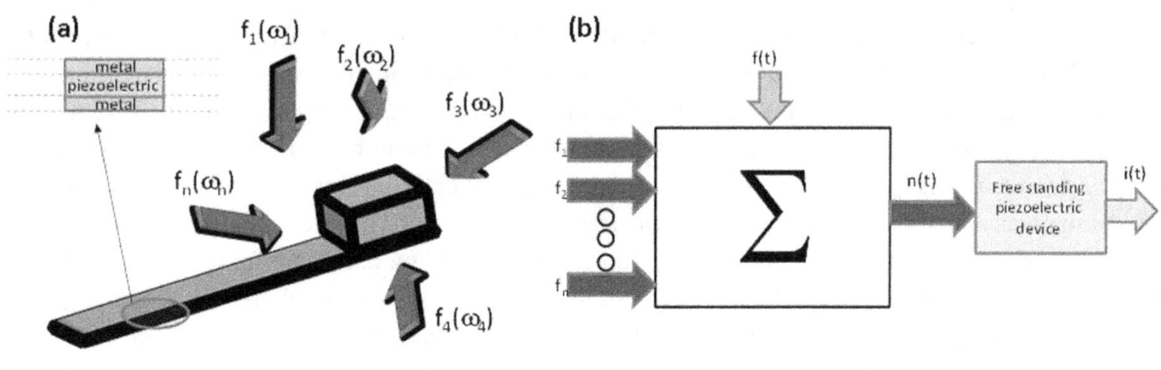

$$n(t)=\sum_{i=1}^{m} f_i=\sum_{i=1}^{m} f_{i0}\sin(\omega_i t-\vartheta_i) \tag{1}$$

where $f_{i0}$ are the vibrating wave amplitudes modulating the input signal, $\omega_i$ are the random angular frequencies (that usually do not coincide with the natural frequencies which are characteristic of the devices and provide maximum efficiency), $\theta_i$ are the phase shifts representing the signal noise with values between 0 and 2π, and *m* is the arbitrary number of iterations that create a random function. Each $f_{i0}$ can be associated to each vibrational external force: this is the case where different environmental forces act simultaneously on the piezoelectric system (for example when wind and ocean wave act in the same time). Otherwise $f_0$ can be associated to all the external forces as follows

$$n(t)=f_0\sum_{i=1}^{m}\sin(\omega_i t-\vartheta_i) \tag{2}$$

representing the case of an unique total environmental force amplitude related to different superimposed frequencies. The superimposed angular frequencies of the whole three dimensional space and the parameter $f_0$ can define the vibrating turbulence of the whole system. The complete sketch of the random vibrational forces acting on the cantilever beam is illustrated in Figure 1 (b): the total noise signal *n(t)* will represent the real input force of the freestanding devices which will generate an output current signal *i(t)* solution of the following equation system

$$\begin{aligned} i(t)&=\beta\dot{s}(t)-K\dot{i}(t)\\ Us(t)&=n(t)-\beta i(t)-m\dot{s}(t)+M\ddot{s}(t)\end{aligned} \tag{3}$$

where $\beta$ is a coefficient depending on the free standing surface, *K* is a model capacitance, *U* is the stiffness, *m* is the effective damping factor, and *M* is the effective mass of the system (see Massaro and Cingolani (2011c)). Equation 3 provides the generated current *i(t)* (the generated power is $Ri^2$ being *R* the external load) and the displacement *s(t)* of the cantilever beam for a generic vibration force effected by the noise function *n(t)*. Considering AlN as piezoelectric cantilever beam, we fix some geometrical and physical parameters obtaining the following coefficients of system Equation (3): $\beta$ = 0.000124, *K* = 9/25, *m* = 3.824680786*$10^{-7}$, *M* = 0.65 (cantilever beam having 1 mm of length, 50 μm of width, and thickness of 1000 nm). The equation system is solved by evaluating the constants of the differential equations assuming as initial conditions *i*(*t*=0)=0, and $\dot{s}$(*t*=0)=0 (solving the solutions at *t*=0). As first case we take into account a single (*m* = 1) harmonic (corresponding harmonic frequency: 50 Hz) as vibrating force having $f_0$ = 1, and, consecutively, *n*(*t*) = sin(100π**t*): in Figure 2 is illustrated the calculated output current versus the time steps with a clear damping effect which reduces the current intensity for high time steps. Assuming three vibrating forces (corresponding harmonic frequencies: 10 Hz, 20 Hz and 50 Hz), we calculate the electrical current trend of Figure 3: for simplicity we suppose $f_{10}$ = 1, $f_{20}$ = 1, $f_{30}$ = 1 and, consecutively, *n*(*t*) = sin(100π**t*-0.5*2π) + sin(40π**t*-0.3*2π) + sin(20π**t*-0.1*2π). In this figure is also reported the damping effect trend of the cantilever beam (see gray dashed line) at initial time steps. This trend is different if compared with the trend of Figure 2, and, is not observed for higher observation times as proved by Figure 4, where, for the particular considered time window, the

*Figure 2. Calculated current by considering one harmonic as vibrating force*

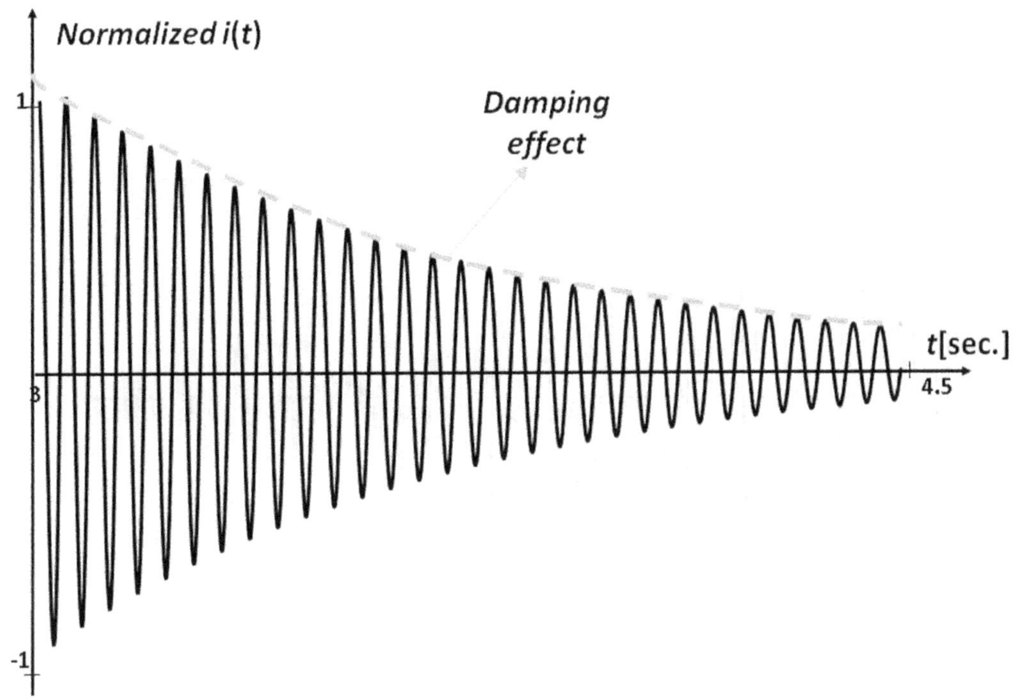


damping effect is negligible. The superposition of three harmonics ($m = 3$) with different phase shifts, introduces an evident signal noise which can be enhanced in a vibrating turbulence by means of complex external force amplitudes $f_{i0}$.

We examine now the effects of the wave amplitudes modulating the harmonics (carriers). In Figure 5 is illustrated an example of external force amplitude $f_{i0}$ defined as a step function indicating two different vertical directions of the external force (main vibrating directions of the cantilever beam). We observe that the various directions reported by the arrows of Figure 1 (a) of the vector forces add a furthermore turbulence to the whole systems if a three dimensional space is considered. In the case of a cantilever beam, a vibrating turbulence can be defined by an irregular amplitude $f_{i0}$ such as the function plotted in Figure 5. In this figure is indicated a period ($f_{i0} = 0$) where the system is not stimulated in the vibrational plane, and the directions of the external forces are defined on the top and on the bottom of the cantilever beam (see insets of the Figure 5); so, the function of $f_{i0}$ is to "switch" the direction of the external forces.

We apply for a first noise analysis the waveform $f_{i0}$ of Figure 6 (a) (acting on the bottom of the cantilever beam) obtaining a noise wave composed by two vibrating forces ($m = 2$) working at 20 Hz and 50 Hz. In this example we assume that only one harmonic is modulated by $f_{i0}$ of Figure 6 (a). In particular the total noise function is defined by $n(t)=1* \sin(100\pi*t-0.5*2\pi) + f_{i0}(t) * \sin(40\pi*t-0.3*2\pi)$, and it is plotted in Figure 6 (b), where the discontinuous trend corresponds with the time window where $f_{i0}$ is not null.

In Figure 7 is illustrated the normalized generated current: it is calculated for the AlN cantilever by considering the noise function of Figure 6 (b). In this figure it is possible to note that the plot obtained interpolating the peaks represents the ripple oscillating around the damping effect trend. This irregular ripple is mainly due to the noise signal; the ripple remains irregular also for high time steps proving that the electrical currents, solutions of Equation (3), are affected during all the time steps by the pulse $f_{i0}$

*Figure 3. Calculated current by superposition of three vibrating forces (three harmonics)*

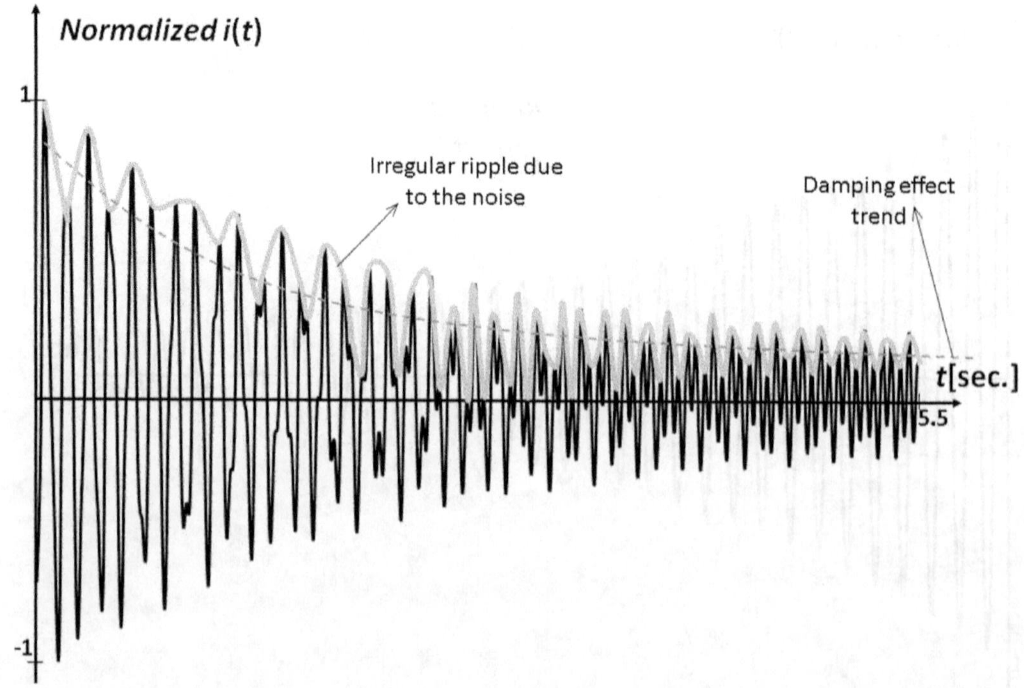


*Figure 4. Current signal calculated in a particular time window by superposition of three vibrating forces*

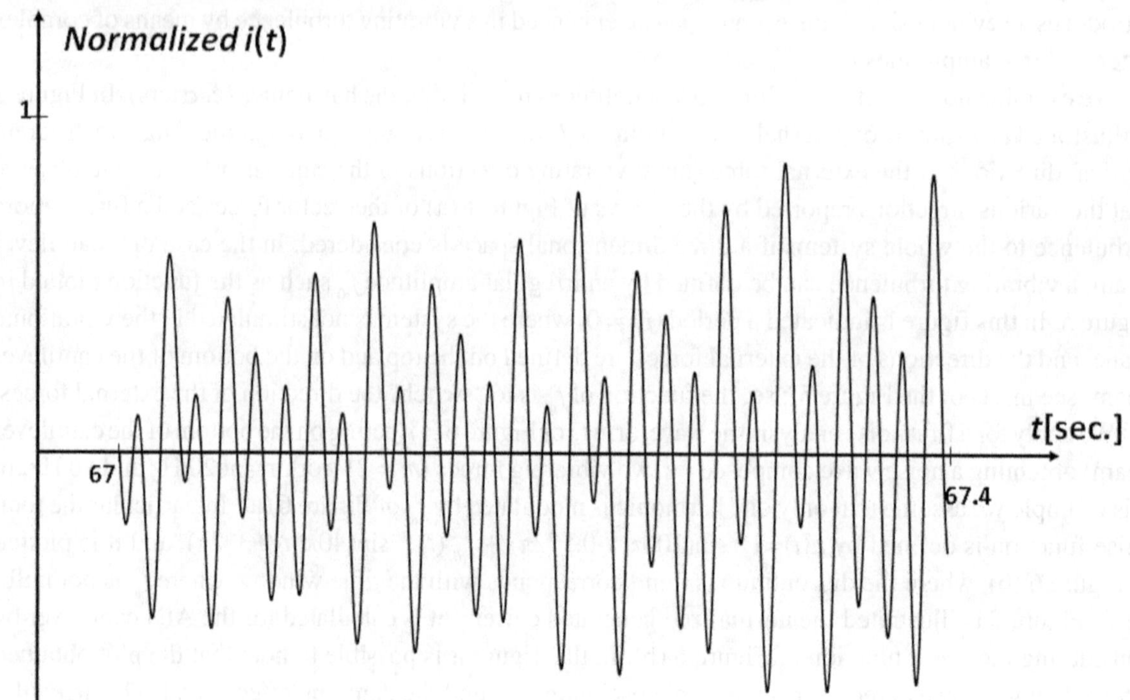

*Figure 5. Example of a force amplitude $f_{i0}(t)$ modulating a vibrating force. Insets: vibrating directions of applied forces. In the real cases t values are always positive.*

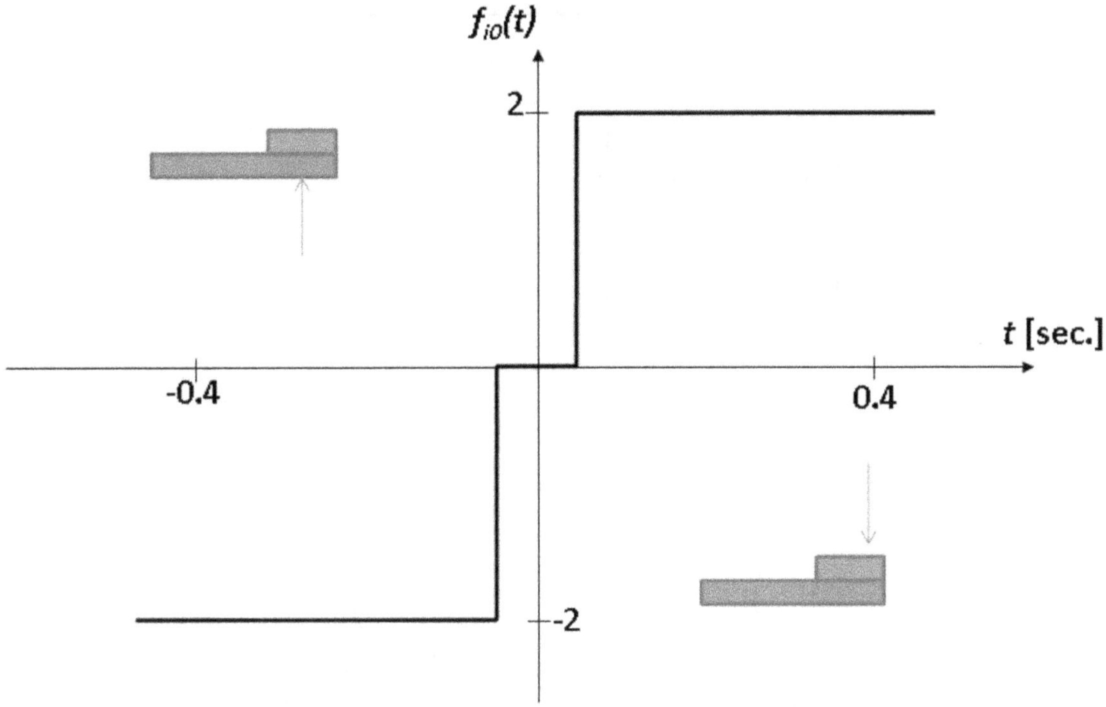


which acts only at the initial time steps. As also discussed in Massaro and Cingolani (2011c), without the application of $f_{i0}$, the current peaks will follow regularly the damping effect trend indicated by the dashed line and as also commented before in Figure2.

Concerning air exposed applications, a way to provide energy during the time is to minimize the air damping beam effect, the air damping proof mass effect, and the internal friction effect which is related to the interaction forces between the proof mass and the cantilever tip. This is possible by assuming for example the periodic function $f_{i0}(t)$ of Figure 8 (high external force intensity). By means of the use of a periodic pulse train, the signal can be amplified step by step in the time domain by the high level signals. In order to highlight the complex effect of the noise signal also for periodic signal modulation, the proposed waveform is applied initially for a modulation of a single harmonic defining the following noise function $n(t)=1*\sin(100\pi*t-0.5*2\pi)+f_{i0}(t)*\sin(40\pi*t-0.3*2\pi)$. The effect of $n(t)$ is shown in Figure 9 illustrating lines indicating periods where the current signal is absent. Applying $f_{i0}(t)$ to both harmonics the noise function becomes $n(t)=f_{i0}(t)*\sin(100\pi*t-0.5*2\pi)+f_{i0}(t)*\sin(40\pi*t-0.3*2\pi)$, and, the calculated electrical current follows the plot of Figure 10. In this case it is possible to note the same lines of Figure 9 and also larger periods where the current is absent. The presence of irregular periods proves that the noise generates during the time a very irregular output signal. We observe that this irregular behavior of the generated current could be stronger if more harmonics $h$ and more wave forms $f_i$ are considered ($n(t)=f_1(t)*h_1(t)+f_2(t)*h_2(t)+f_3(t)*h_3(t)+...+f_m(t)*h_m(t)$)

*Figure 6. (a) Example of $f_{i0}(t)$ modulating only one of the two considered harmonics (frequencies of $f_1$ = 50 Hz and $f_2$ = 20 Hz) having a phase shift of $\theta_1$=0.5*2π and $\theta_2$=0.3*2π, respectively, and (b) corresponding noise function n(t).*

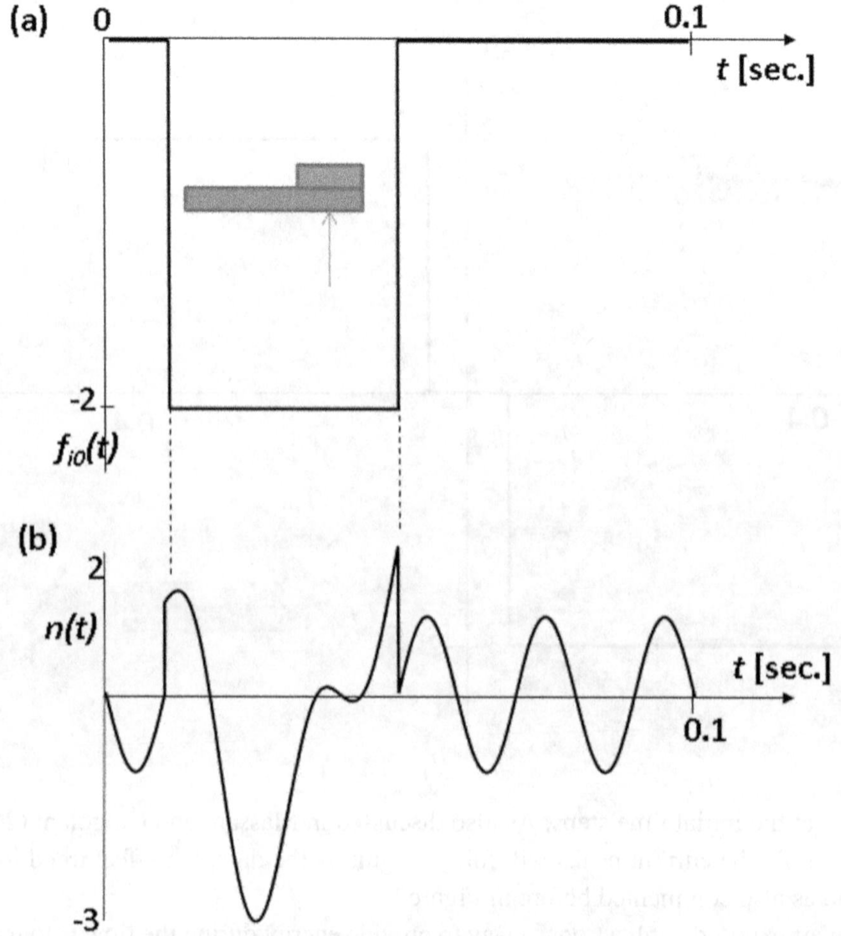


Concerning applications which involve chaotic/random vibration phenomena, different studies have been discussed. Impact of wind on piezoelectric MEMS at different velocities generates an output power function with critical speed velocities (He & Gao (2013)). Moreover, as commented by Ovejas et al. (2011) wind incidence on piezoelectric can be parallel or normal for different kinds of one side fixed or two side fixed setups (different oscillating behaviors). As discussed by Anvar and Li (2012), also ocean waves are characterized by an irregular amplitude trend. In ocean irregular thin flexible piezoelectric polymeric membranes (eels) can be used to extract energy using vortex shedding (Techet, Allen & Smits Techet (2002)). Finally, about rain as source energy, various impact situations (rain drop size and height) on piezoelectric flexible polymer have been studied by Guigon et al. (2008). All the listed studies can be modelled by a proper choice of the noise function parameters.

*Figure 7. Irregular normalized current signal calculated by the wave amplitude of Figure 6(a)*

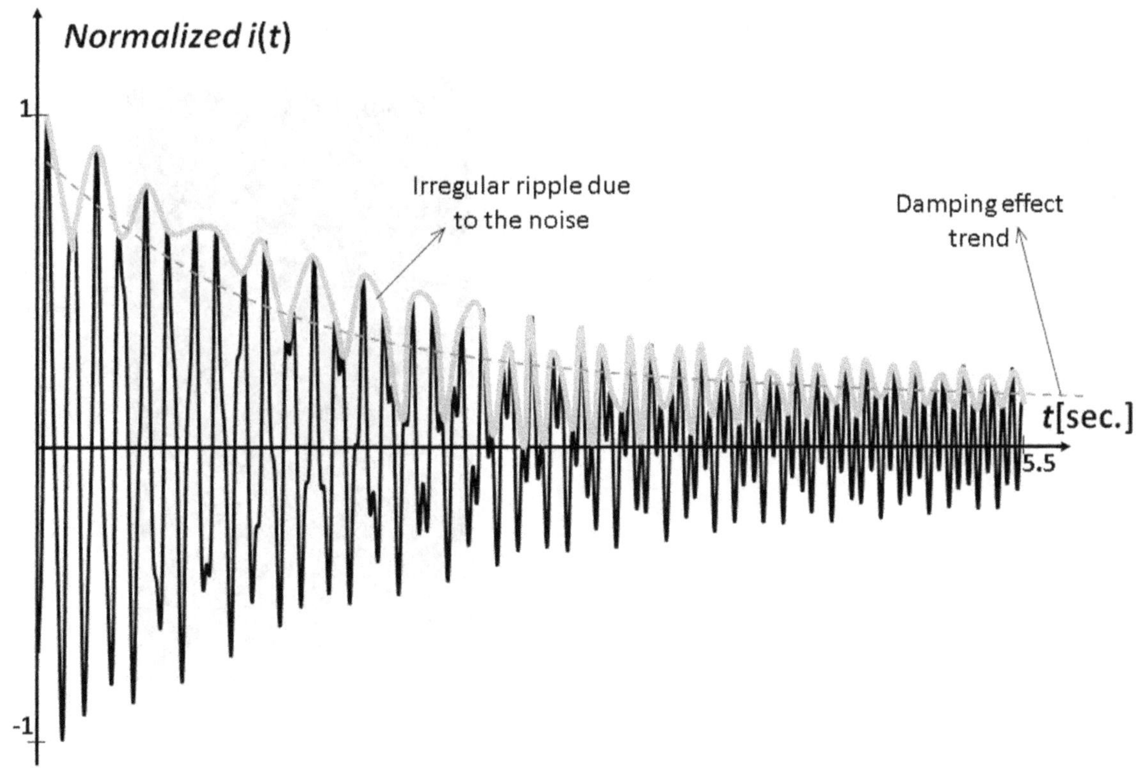


*Figure 8. Example of $f_{i0}(t)$ waveform as periodic pulse train*

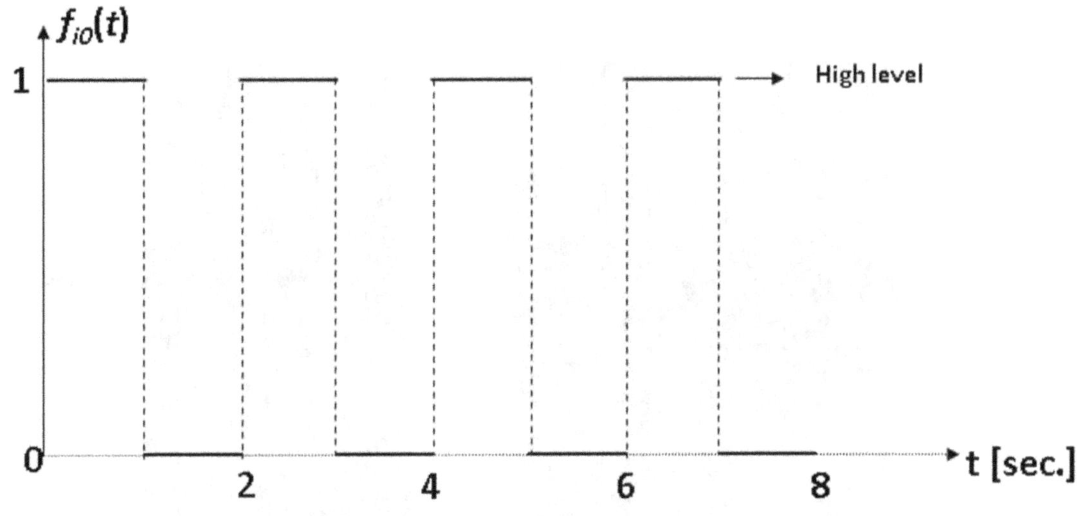

*Figure 9. Normalized current calculated for $f_{i0}(t)$ waveform of Figure 8 applied to a single harmonic*

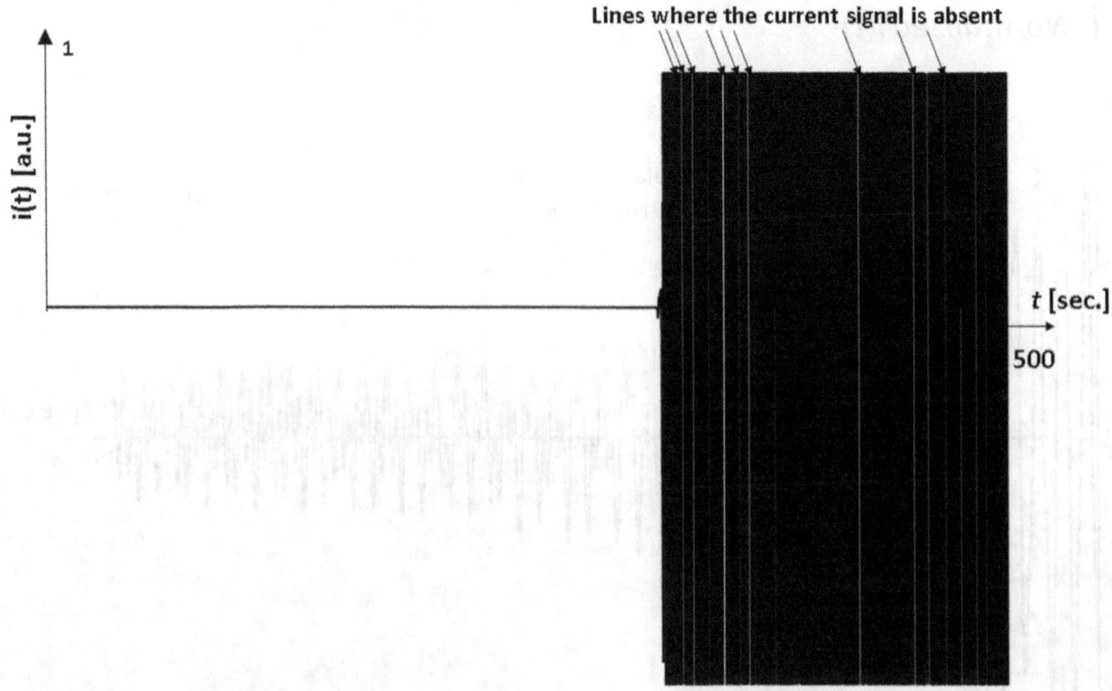


*Figure 10. Normalized current calculated for $f_{i0}(t)$ waveform of Figure 8 applied to both the harmonics*

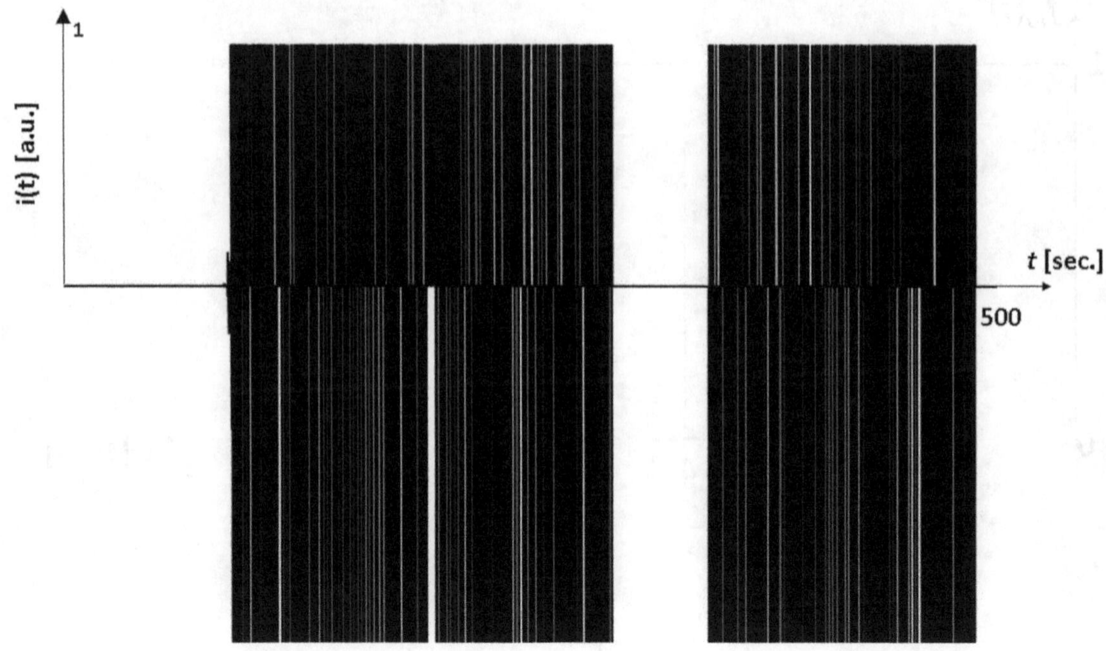

## CURVED LAYOUTS

The concept of torsional stresses induced by vibrations in an energy harvesting device has been discussed in Kulkarni et al. (2013) and primary in Massaro et al. (2011a). The torsional stress effect can be enhanced if curved or helical layouts are considered. This represents an important aspect when more vibrating forces operate simultaneously on the freestanding structure (see Figure 1 (a)) activating a chaotic vibration. AlN curved layouts represent an alternative way to collect vibrating energy power at low frequencies: the total external force $F(t)$ shakes the curved freestanding structures in the $(r,\theta)$ plane, the vibration, indicated by the function $s(t)$, will generate through the piezoelectric effect a voltage $V$ and a current $i$ signal at the output of the system. The curved layouts, by means of torsional modes, can collect the vibrational force coming from the three dimensional space: each noise function, discussed in the previous section for a single vibrational plane, can be defined in the $(x,y)$, $(x,z)$ and $(y,z)$ vibrational planes thus defining a very complex tip displacement trajectory. For this reason a preliminary study of torsional resonant mode is very important in order select the best vibration plane by adjusting the position of the device thus optimizing its efficiency. The equation system modeling the curved structure includes the strain-charge equation ($\mathbf{S}(r,\theta,z)=s_E\mathbf{T}(r,\theta,z)+d\mathbf{E}(r,\theta,z)$, $\mathbf{D}(r,\theta,z)=d^T\mathbf{T}(r,\theta,z)-\varepsilon_r^{\ T}\mathbf{E}(r,\theta,z)$ being $s_E$ the compliance matrix, $d^T$ the coupling matrix, and $\varepsilon_r^{\ T}$ the relative permittivity matrix) and the charge equation of the electrostatic defined in the $(r,\theta,z)$ domain. The system will be matched with an external load $Z = Xe^{j\phi}$ in order to transfer the maximum energy. Theoretically, for a ring layout, the oscillation of the fundamental mode happens in $(r,\theta)$ plane, and is found at low frequency (see the major theoretical peak 1 reported in the Figure 11). The fundamental mode is characterized by a displacement $s(t)$ in the $x$-$y$ plane (see inset of Figure 11). The other theoretical peaks (peaks 2, 3, 4) are related to the tip oscillations in all the $(r,\theta, z)$ space. We note that the main experimental peak could not correspond with the theoretical one. This is due to a possible shape different from the ring layout: an imperfect fabrication process could change locally the radius of curvature of the freestanding structure generating helical trend. This geometrical aspect produces a chaotic oscillation which allows to enhance better only some modes which could not correspond with the fundamental one. The calculated plot of Figure 11 is related to a ring curved structure with a radius of 85 μm. A three-dimensional (3D) Finite Element Method (FEM) approach implementing strain-charge equation and the charge equation of electrostatics in $(r,\theta, z)$ coordinate system was used for calculus. In the 3D FEM model the Mo layers are modelled by infinitely thin metallic planes as boundaries conditions. The analysis of the mode resonances is important in order to compare the vibrating source harmonics with the natural frequencies of the structure. In order to optimize the energy harvesting efficiency it is preferable to design different curved layouts where the torsional mode cover a broad frequency band. In this direction, an array composed by different curved layout could increase the total efficiency. The same principle of torsional mode efficiency for curved layouts can be also applied to improve well defined tip displacements in reconfigurable piezoelectric antennas operating in wireless systems: as discussed in Massaro et al. (2009), the antenna can be excited by the frequency corresponding a fundamental/ torsional mode in order to move the structure in a desired direction for electromagnetic radiation. Finally, we observe that ring layouts have been considered also for energy harvesting wearable applications addressed to knee-joint piezoelectric harvester (Pozzi & Zhu (2012)).

*Figure 11. FEM theoretical normalized voltage. Inset: sketch of the AlN curved layout and torsional moments of the tip.*

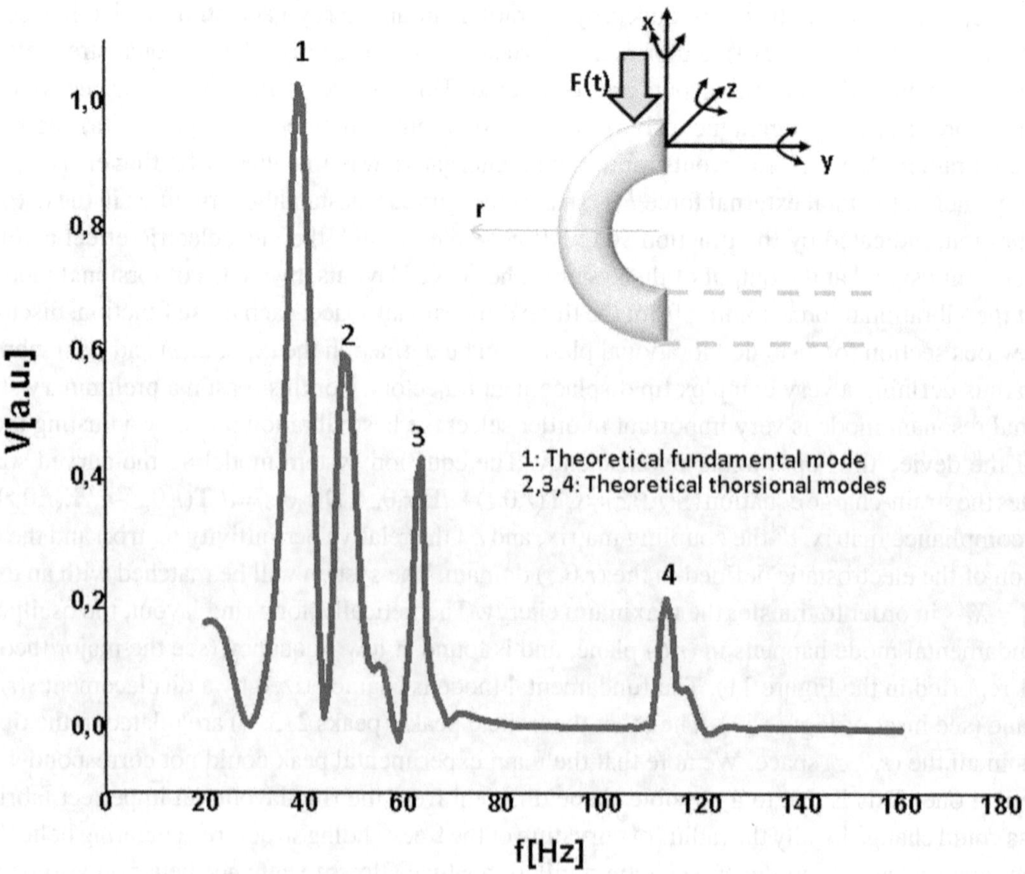


## SOME EXPERIMENTAL ASPECTS

After the fabrication of a freestanding beam, it is necessary to check the correct efficiency of the electric contacts. As discussed in Massaro et al. (2011b) the electrical contact area can be designed on a side of the sample in order to facilitate the measurements. By means of a micro-wires welded directly on the Mo layers, the piezoelectric measurements can be performed by the controlled vibration platform used in Massaro et al. (2011a), where the generated current signal is coupled to a lock-in amplifier instrument and by matching the output impedance by a proper electrical resistance able to maximize the piezoelectric output power. The piezoelectric resonance of a free standing device can be checked by means of a LCR meter and of a current/voltage probe station with tungsten tips able to measure the current/voltage characteristics, providing the equivalent electrical circuit (see comments in Massaro et al. (2011b)). Before the controlled measurements, it is possible to check the efficiency of the sample simply by observing the displacement of the tip (in the case of curved layouts), and analyzing the image detected by a microscope with a camera (Massaro, Cingolani & Passaseo (2011b)). We observe that in this check stage, we detect a voltage input signal, and we read an image of the total tip displacement. This preliminary test is necessary in order to check that there are not short-circuit effects due to the electrical contact of both Mo layers. An alternative way to check the tip displacement is reported in Figure 12 where a laser beam is focused on the micrometer width of the tip and is reflected on a screen: after the application of the voltage signal, the image of the laser beam could be shifted on a screen proving the cantilever displacement due to the piezoelectric effect. Considering irregular vibration forces, it is difficult to increase the energy harvesting efficiency: in this case the electrical resistance able to match the circuit impedance does not coincide with the resistance matched with the natural resonant frequency, the vibrating frequencies are different, and, the noise "covers" the useful vibrating source. For these reasons the use of a lock-in amplifier, and the choice of the output impedance is not simply. A statistical analysis of the random vibrating phenomena could help to integrate properly designed feedback circuits able to match the complex output impedance, and, arrays of different curved layouts could help to collect better the piezoelectric energy. Moreover, the study of the irregular damping effect discussed before, highlights that it is necessary to implement properly designed controlled/rectifier/amplifier circuits. The circuits should be designed according to the solutions found by Ramadass and Chandrakasan (2011) which have introduced new rectifier designs, by Hagedorn et al. (2013) which have provided criteria about impedance matching for broadband, by Yang and Tang (2009) which have studied accurate circuit modeling, by Lien and Shu (2012) which have analyzed equivalent impedance approach, and by Motter et al. (2012) which have used particular kind of rectifiers circuits. Of course the circuits will be defined univocally by the noise function initially studied. The circuit implementation can be performed experimentally by observing and simulating in laboratory the superposition of more vibrating forces: the simultaneous use of more vibrating source such as a personal computer fun controlled in frequency together with a controlled chemical shaker where will be placed the free standing piezoelectric device, will help to simulate in laboratory the effect of noise function introduced in Equation (2). The simulation of $n(t)$ signal switched by $f_0$ voltage generators will facilitate to check the correct feedback/stabilizer circuit: the accurate study of the irregular ripple trend of the current signal (see similar example of Figure 7), will define the minimum and the maximum oscillation thus defining

*Figure 12. Example of experimental check of a piezoelectric cantilever by laser beam reflection*

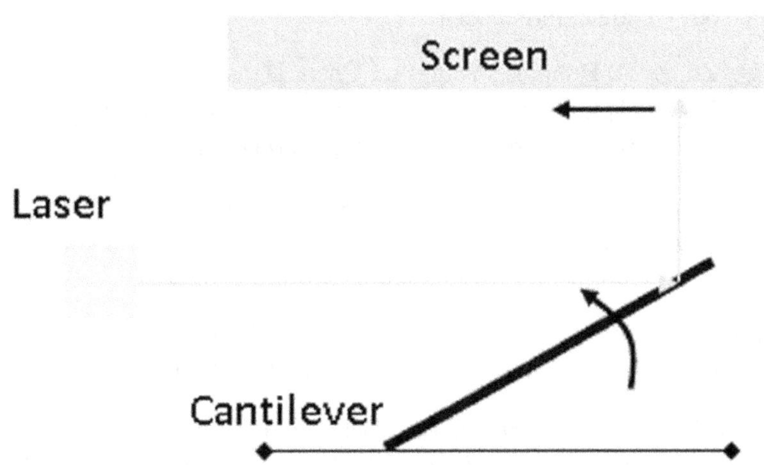

the reading error, and, supporting the choice of suitable electronic high sensitivity components. The exact reconstruction of external forces (statistic phase shift estimation, vibration frequencies estimation, wave amplitude reconstruction) represents the first important step to for an energy harvesting vibrating system addressed on high efficiency packaged components.

## CONCLUSION

The chapter describes the effects of noise signal acting on free standing AlN structures. After an introduction about AlN technology and characterization, some examples of output current trends have been discussed focusing the attention on one, two and three harmonics as vibrating sources. The effect of the wave amplitude modulating the vibration forces, has been commented by describing output current analytical results by means of the application of a single pulse wave amplitude and of a period pulse train. Very irregular generated currents have been observed, thus highlighting the problem to collect efficiently the produced energy in a vibrating turbulence condition. Moreover, due to the chaotic motion of free standing structures generated by complex vibrating forces, a solution by means of AlN curved layout has been commented, concluding that the use of curved layouts allows by means of torsional modes to collect better at low frequencies the vibration energy in the three dimensional space. Finally, some useful aspects for laboratory testing processes have been discussed by suggesting some preliminary experimental procedures.

## REFERENCES

Anvar, A., & Li, D. Y. (2012). Ocean wave kinetic energy harvesting system for automated sub sea sensors. *World Academy of Science. Engineering and Technology*, *6*, 9–13.

Balpande, S. S., Lande, S. B., & Rungta, S. (2009). Modeling of PZT based power harvester and power aware control approach for pervasive node BSN. In *Proceeding of International Advance Computing Conference* (IACC 2009). Patiala, India: IEEE.

Fang, H., Liu, J. Q., Zu, Z. Y., Dong, L., Wang, L., Chen, D., & Liu, Y. et al. (2006). Fabrication and performance of MEMS-based piezoelectric power generator for vibration energy harvesting. *Microelectronics Journal*, *37*(11), 1280–1284. doi:10.1016/j.mejo.2006.07.023

Glynne-Jones, P., Beeby, S. P., & White, N. M. (2001). Towards a piezoelectric vibration-powered microgenerator. *IEE Proceedings. Science Measurement and Technology*, *148*(2), 68–72. doi:10.1049/ip-smt:20010323

Guigon, R., Chaillout, J.-J., Jager, T., & Despesse, G. (2008). Harvesting raindrop energy: Experimental study. *Smart Materials and Structures*, *17*(1), 015039. doi:10.1088/0964-1726/17/01/015039

Hagedorn, F., Leicht, J., Sanchez, D., Hebn, T., & Manoli, Y. (2013). Impedance matching for broadband piezoelectric harvesting. *Journal of Physics: Conference Series*, *476*, 0120283. doi:10.1088/1742-6596/476/1/012083

He, X.-F., & Gao, J. (2013). Wind energy harvesting based on flow-induced-vibration and impact. *Microelectronic Engineering*, *111*, 82–86. doi:10.1016/j.mee.2013.02.009

Ho, C. J., Sing, T. K., & Li, P. C. (2004). Preferred orientation control and characterization of AlN thin films using reactive sputtering. *Tamkang Journal of Science and Engineering*, *7*, 1–4.

Inman, D. J. (2000). *Engineering vibration.* Prentice Hall.

Isarakorn, D., Briand, D., Gariglio, S., Sambri, A., Stucki, N., Triscone, J. M., & De Rooji, N. F. et al. (2008). A study on epitaxial piezoelectric thin films grown on silicon for energy scavenging applications. In *Proceedings of PowerMEMS*. Sendai, Japan: IOP.

Kim, H.-U., Bedekar, V., Islam, R. A., Lee, W., Leo, D., & Priya, S. (2008). Laser-machined piezoelectric cantilevers for mechanical energy harvesting. *Transaction on Ultrasonics, Ferroelectics, and Frequency Control*, *55*(9), 1900–1905. doi:10.1109/TUFFC.881 PMID:18986886

Kulkarni, V., Ben-Mrad, R., Eswar Prasad, S., & Nemana, S. (2013). A shear-mode energy harvesting device based on torsional stresses. *Transaction on Mechatronics*, *19*(3), 801–807. doi:10.1109/TMECH.2013.2259635

Lien, I. C., & Shu, Y. C. (2012). Array of piezoelectric energy harvesting by the equivalent Impedance approach. *Smart Materials and Structures*, *21*(8), 082001. doi:10.1088/0964-1726/21/8/082001

Marzencki, M., Basrour, S., Charlot, B., Spirkovich, S., & Colin, M. (2005). A MEMS piezoelectric vibration energy harvesting device. In Proceedings of PowerMEMS, (pp. 45-48). Tokyo, Japan: IOP.

Massaro, A., & Cingolani, R. (2011c). Parametric analysis of piezoelectric resonances of AlN layouts for vibrating energy harvesting. *Journal of Computational and Theoretical Nanoscience*, *8*(11), 1–8. doi:10.1166/jctn.2011.1955

Massaro, A., Cingolani, R., & Passaseo, A. (2011b). 3D FEM modelling and technology of piezoelectric ring MEMS antenna. *Progress in Electromagnetics Research*, *C*, 23, 123–135.

Massaro, A., De Guido, S., Ingrosso, I., Cingolani, R., De Vittorio, M., Cori, M., & Passaseo, A. et al. (2011a). Freestanding piezoelectric rings for high efficiency energy harvesting at low frequency. *Applied Physics Letters*, *98*(5), 053502. doi:10.1063/1.3551725

Massaro, A., Ingrosso, I., Giordano, C., Todaro, M. T., Cingolani, R., De Vittorio, M., & Passaseo, A. (2009). Micromachining Mo/ALN/Mo piezoelectric ring resonators. In *Proceedings of European Microwave Conference (EuMA)*. Rome, Italy: IEEE.

Motter, D., Lavarda, J. V., Dias, F. A., & Da Silva, S. (2012). Vibration energy harvesting using piezoelectric transducer and non-controlled rectifiers circuits. *Journal of the Brazilian Society of Mechanical Sciences and Engineering*, *34*(spe), 378–385. doi:10.1590/S1678-58782012000500006

Ovejas, V. J., & Cuadras, A. (2011). Multimodal piezoelectric wind energy harvesters. *Smart Materials and Structures*, *20*(8), 085030. doi:10.1088/0964-1726/20/8/085030

Pozzi, M., & Zhu, M. (2012). Characterization of a rotary piezoelectric energy harvester based on plucking excitation for knee-joint wearable applications. *Smart Materials and Structures*, *21*(5), 055004. doi:10.1088/0964-1726/21/5/055004

Ramadass, Y. K., & Chandrakasan, A. P. (2010). An efficient piezoelectric energy harvesting interface circuit using a bias-flip rectifier and shared inductor. *Journal of Solid-State Circuits*, *45*(1), 189–204. doi:10.1109/JSSC.2009.2034442

Renaud, M., Karakaya, K., Sterken, T., Fiorini, P., Hoof, C. V., & Puers, R. (2008). Fabrication, modelling and characterization of MEMS piezoelectric vibration harvesters. *Sensors and Actuators. A, Physical*, *145-146*, 380–386. doi:10.1016/j.sna.2007.11.005

Renaud, M., Sterken, T., Schmitz, A., Fiorini, P., Van Hoof, C., & Puers, R., R. (2007). Piezoelectric harvesters and MEMS technology: fabrication, modeling and measurements. In *Proceeding of International Conference of Solid State Sensors, Actuators and Microsystems*. New York: IEEE. doi:10.1109/SENSOR.2007.4300274

Shen, D., Park, J.-H., Ajitsaria, J., Choe, S.-Y., Wikle, H. C. III, & Kim, D.-J. (2008). The design, fabrication and evaluation of a MEMS PZT cantilever with an integrated Si proof mass for vibration energy harvesting. *Journal of Micromechanics and Microengineering*, *18*(5), 055017. doi:10.1088/0960-1317/18/5/055017

Shin, H., & Song, J. T. (2010). Piezoelectric coefficient measurement of AlN thin films at the nanometer scale using piezoresponse force microscopy. *Journal of the Korean Physical Society*, *56*(2), 580–585. doi:10.3938/jkps.56.580

Techet, A. H., Allen, J. J., & Smits, A. J. (2002). Piezoelectric eels for energy harvesting in the ocean. In *Proceedings of The Twelfth (2002) International Offshore and Polar Engineering Conference Kitakyushu*, Japan: ISOPE.

Yang, Y., & Tang, L. (2009). Equivalent circuit modeling of piezoelectric energy harvester. *Journal of Intelligent Material Systems and Structures*, *20*(18), 2223–2235. doi:10.1177/1045389X09351757

## KEY TERMS AND DEFINITIONS

**Cantilever Beam:** Micro-scale cantilever device.
**Chaotic Vibration System:** Chaotic vibrating elements due to the irregular vibration.
**Curved Layout:** Curved layout of free standing devices.
**Energy Harvesting MEMS:** Micro Electro-Mechanical Systems for energy harvesting applications.
**Piezoelectric Devices:** Devices made by piezoelectric materials.
**Signal Analysis:** Signal analysis and processing of electrical and voltage signals.
**Vibration Energy Harvesting Modeling:** Modelling of energy harvesting layouts.

# Chapter 7
# Modeling and Performance Enhancement of Low-Frequency Energy Harvesters

**Abdessattar Abdelkefi**
*New Mexico State University, USA*

## ABSTRACT

*There exist numerous low-frequency excitation sources, such as walking, breathing, and ocean waves, capable of providing viable amounts of mechanical energy to power many critical devices, including pacemakers, cell phones, MEMS devices, wireless sensors, and actuators. Harvesting significant energy levels from such sources can only be achieved through the design of devices capable of performing effective energy transfer mechanisms over low frequencies. In this chapter, two concepts of efficient low-frequency piezoelectric energy harvesters are presented, namely, variable-shaped piezoelectric energy harvesters and piezomagnetoelastic energy harvesters. Linear and nonlinear electromechanical models are developed and validated in this chapter. The results show that the quadratic shape can yield up to two times the energy harvested by a rectangular one. It is also demonstrated that depending on the available excitation frequency, an enhanced energy harvester can be tuned and optimized by changing the length of the piezoelectric material or by changing the distance between the two tip magnets.*

## INTRODUCTION

Harvesting energy from wasted mechanical energy has been the topic of several investigations in the past decade (Sodano, Park, & Inman, 2004; Anton & Sodano, 2007; Erturk, 2009; Karami, 2011; Abdelkefi, 2012). The ultimate goal of designing and fabricating these harvesters is to replace small batteries that have finite life span or would require expensive and difficult replacement. They are also used to operate self-powered devices including wireless and structural health monitoring sensors, cameras, pacemakers, data transmitters, and medical implants (Roundy & Wright, 2005; Clair, Bibo, Sennakesavababu, & Daqaq, 2010; Karami & Inman, 2012; Abdelkefi & Ghommem, 2013). These harvesters can be deployed in different locations including structure's surface, cell phones, pacemakers, buildings, bridges,

DOI: 10.4018/978-1-4666-8254-2.ch007

etc. Different transduction mechanisms have been used, such as electrostatic (Anton & Sodano, 2007), electromagnetic (Arnold, 2007; Karami, 2011), and piezoelectric (Sodano et al., 2004; Erturk, 2009). Because of its ease of application, non-reliance on external input voltage, and its suitability for designing small energy harvesters, the piezoelectric option has flourished in most of the recent investigations.

## Tuning and Broadband Mechanisms

Researchers have proposed different structural systems for energy harvesting from base or aeroelastic vibrations (Bryant & Garcia, 2009; Erturk, 2009; Abdelkefi, Nayfeh, & Hajj, 2012a; Abdelkefi & Nuhait, 2013; Zhao, Tang, & Yang, 2013; Dai, Abdelkefi, & Wang, 2014; Yan & Abdelkefi, 2014). Several investigations have been performed to design efficient piezoelectric energy harvesters from harmonic and random excitations. In order to exploit low-frequency excitation sources, such as walking, breathing, ocean waves, and the contraction and expansion of human muscles to power many critical devices inside and outside the human body, many concepts have been proposed to design efficient piezoelectric energy harvesters with a low natural frequency. One of the examples is to power modern pacemakers from heartbeat vibrations. In fact, these modern pacemakers require a one microwatt with a present dominant excitation frequency around 39 Hz (Karami & Inman, 2012). One particular focus was paid to design tunable energy harvesters to operate effectively at the available excitation frequency. To this end, different active and passive tuning mechanisms (Roundy & Wright, 2005; Shahruz, 2006; Challa, Prasad, Shi, & Fisher, 2008; Abdelkefi, Najar, Nayfeh, & Ben Ayed, 2011; Elvin & Elvin, 2011; Masana & Daqaq, 2011a; Lui, Lee, Kobayashi, Tay, J., & Quan, 2012; Daqaq, 2012) have been introduced to tune the natural frequency of the harvester to the available excitation frequency. For active tuning mechanisms, Roundy et al. (2005) showed that some of these mechanisms cannot be efficient in cases of fast-varying frequency or random excitations. In fact, they reported that the needed external power to operate the used actuator could be larger than the harvested power. For passive tuning mechanisms, Shahruz (2006) designed an energy harvester which was consisted of multiple cantilever beams with various lengths and tip masses attached to a common base. They showed that this device is capable of harvesting energy over a wide range of frequencies by properly selecting the length and tip mass of each beam.

Other research findings have proposed the use of monostable configurations for broadband transduction. In these systems, a nonlinear restoring force is introduced which can result in a significant change in the potential energy function of the harvester. This nonlinear restoring force can be obtained when considering the piezoelectric coupling in the harvester (Quinn, Triplett, Vakakis, & Bergman, 2011; Abdelkefi, Nayfeh, & Hajj, 2012b, 2012c) or when including an external magnetic force (Mann & Sims, 2008; Stanton, McGehee, & Mann, 2009; Barton, Burrow, & Clare, 2010; Erturk & Inman, 2011; Karami, 2011), or when applying an external axial loading force (Masana & Daqaq, 2011a). The presence of these stiffness nonlinearities results in the appearance of softening or hardening behaviors which lead to a wider range of resonant frequencies of the harvester. Considering the effects of the piezoelectric coupling, Abdelkefi et al. (2012b, 2012c) showed that broadband regions for the harvester's response can be obtained through hardening or softening responses due to the piezoelectric coupling, as shown, respectively, in Figures 1 and 2. An axially-loaded piezoelectric clamped-clamped beam was proposed by Masana and Daqaq (2011a) to tune the natural frequency of the harvester with the excitation frequency. They showed that, in the monostable region, an increase in the preload results in a decrease in the natural

*Figure 1. Wider range of resonant motions due to the softening effects for different forcing terms*

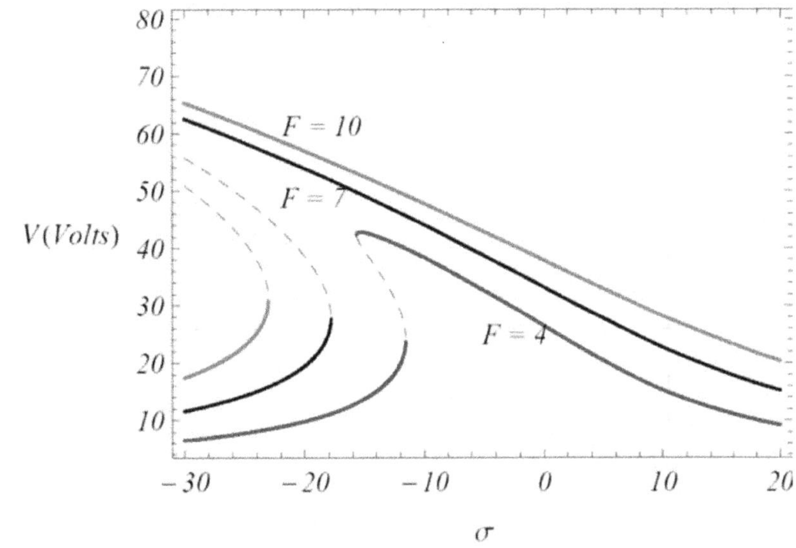


*Figure 2. Wider range of resonant motions due to the hardening effects for different forcing terms*

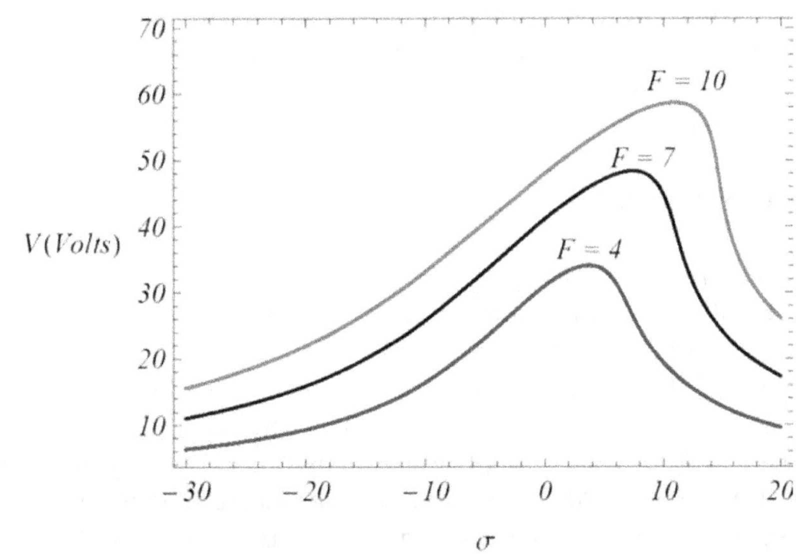


frequency of the harvester and an enhancement in the level of the harvested power. Mann and Sims (2008) proposed a magnetically-levitated inductive harvester which consists of two outer magnets to levitate an oscillating force. In this approach, the nonlinearity is introduced by the magnetic restoring force.

Bistable configurations have also been introduced in many research studies (Cottone, Vocca, & Gammaitoni, 2009; Stanton, McGehee, & Mann, 2010; Erturk & Inman, 2011; Masana & Daqaq, 2011b; Karami, 2011; Harne & Wang, 2013; Pellegrini, Tolou, Schenk, & Herder, 2013). In this case, a double-well potential energy function with two stable equilibria and one unstable saddle takes place (Cottone, et al., 2009), as shown in Figure 3. In the case of a magnetic restoring force, for a certain spacing distance

*Figure 3. Potential energy function of a bistable oscillator*

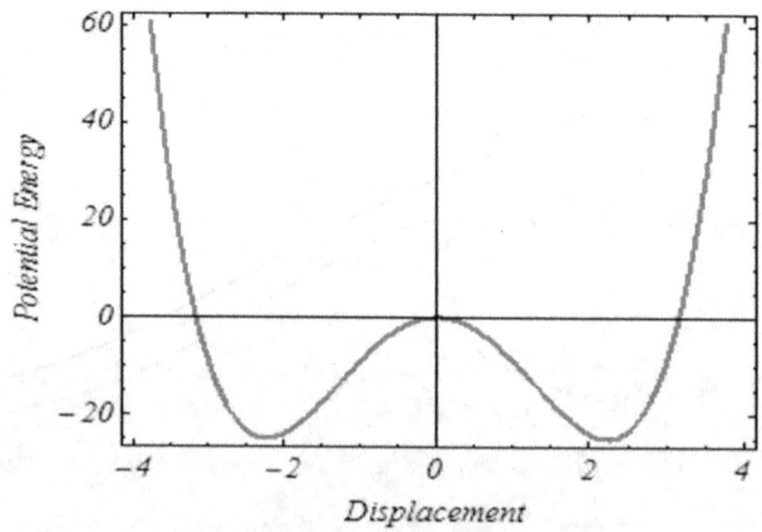


between the two magnets, the harvester's becomes bistable. In the case of an axial preload (Masana & Daqaq, 2011b), for a certain axial load, the harvester's becomes buckled and hence a bistable configuration takes place. As for these bistable harvesters, inter-well motions are present when enough mechanical energy is supplied to the harvester. Therefore, these types of harvesters can be designed when large-base excitation levels are present.

## Electromechanical Models

Generally, in the design of most of these energy harvesters, lumped spring-mass models or distributed-parameter models based on approximated mode shapes were used to represent the electromechanical modeling of these harvesters. In several investigations (DuToil, Wardle, & Kim, 2005; Roundy et al., 2005; Erturk & Inman, 2011; Daqaq, 2012), lumped-parameter models for linear and nonlinear energy harvesters were considered. These models are limited to the fundamental frequency of the harvester. In addition, they neglect the effects of the distributed beam-mass in the forcing amplitude as well as the moment associated to the tip mass. Also, they do not account for the effects of the dynamic mode shapes on the electrical response of the harvester. Erturk and Inman (2008) showed that the use of the traditional form of the lumped-parameter model leads to an underestimation in the level of the harvested power for cantilever beams under harmonic base excitations. This issue is particularly true in the cases where the proof mass is small. A correction factor depending on the ratio between the tip mass and the beam mass was introduced to improve the prediction capability of these lumped models.

As for the improved distributed-parameter models that are based on the Galerkin discretization, they were used in several studies to accurately predict the performance of the harvester (Erturk & Inman, 2009; Masana & Daqaq, 2011a; Abdelkefi, Barsallo, Tang, Yang, & Hajj, 2013; Ben Ayed, Abdelkefi, Najar, & Hajj, 2014). These models take into account the effects of the dynamic mode shapes, strain distribution, and higher vibration modes on the electromechanical response of the harvester. A common problem, when using these models, is the determination of the mode shapes and associated natural frequencies. In fact, some assumptions are taken to determine the mode shapes, such as neglecting the static position effects (Tang & Yang, 2012), assuming classical mode shapes of a cantilever beam with a tip mass for partially-covered piezoelectric cantilever beams (Song, Choi, Purekar, & Wereley, 2009, 2010; Masana & Daqaq, 2011a; Karami & Inman, 2011; Alamin Dow et al., 2012; Tang & Yang, 2012) or for truncated cantilever beams (Mateu & Moll, 2005; Goldschmidtboeing & Woias, 2008; Benasciutti, Moro, Zelenika, & Brusa, 2010). These approximated eigenfunctions can lead to erroneous predictions for the natural frequencies and the level of the harvester power.

## Motivations

For an efficient design of these energy harvesters, an accurate modeling of the internal and external forces is required. In fact, in the linear regime, energy harvesters are only able to efficiently harvest energy over a narrow range of frequencies. Therefore, an underestimation or overestimation of the system's parameters can result in a wrong prediction of this narrow range of frequencies which will result in a significant reduction in the level of the harvested power. Motivated by the need of developing accurate models for the design of low-frequency piezoelectric energy harvesters, static and dynamic analyses are performed for two different energy harvesters, namely, variable-shaped piezoelectric energy harvesters and piezomagnetoelastic energy harvesters. The first system is composed of metallic and piezoelectric layers (unimorph design) with a rigid mass attached to its free end. For this system, a linear distributed-parameter model is derived to investigate the effects of shape variations on the fundamental natural frequency of the harvester and its performance. As for the second harvester, it consists of a unimorph piezoelectric cantilever beam with a fixed magnet mass at the top of the proof magnet. A nonlinear electromechanical is developed and validated with experimental measurements. Then, a parametric study is performed to investigate the effects of the length of the piezoelectric material and spacing distance between the two magnets on the fundamental natural frequency of the harvester and its performance.

# VARIABLE-SHAPED PIEZOELECTRIC ENERGY HARVESTERS

## Mathematical Modeling

The energy harvester under investigation consists of a variable-shaped unimorph piezoelectric cantilever beam with a rigid mass attached to its free end, as shown in Figure 4. The beam is composed of a stainless steel substrate layer and piezoelectric layer and is subjected to a transverse harmonic displacement $Y(t)=Y_0 cos(\Omega t)$. The piezoelectric layer is bounded by two in-plane electrodes of negligible thicknesses connected to an electrical load resistance $R$. These electrodes are assumed to be perfectly conductive and cover the entire piezoelectric surface.

The beam has a variable symmetric width equal to $b(x)=2b_0\hat{b}(x)$, where $x$ represents the position along the beam's length, $2b_0$ denotes the width at the clamped side and $\hat{b}(x)$, is the nondimensional form of the half width function. In this chapter, linear and quadratic width shapes are considered which are, respectively, given by

*Figure 4. Schematic of a variable-shaped piezoelectric energy harvester which has a length L, trapezoidal width with a clamped side $2b_0$, total thickness $h_p+h_s$, and a tip mass M*

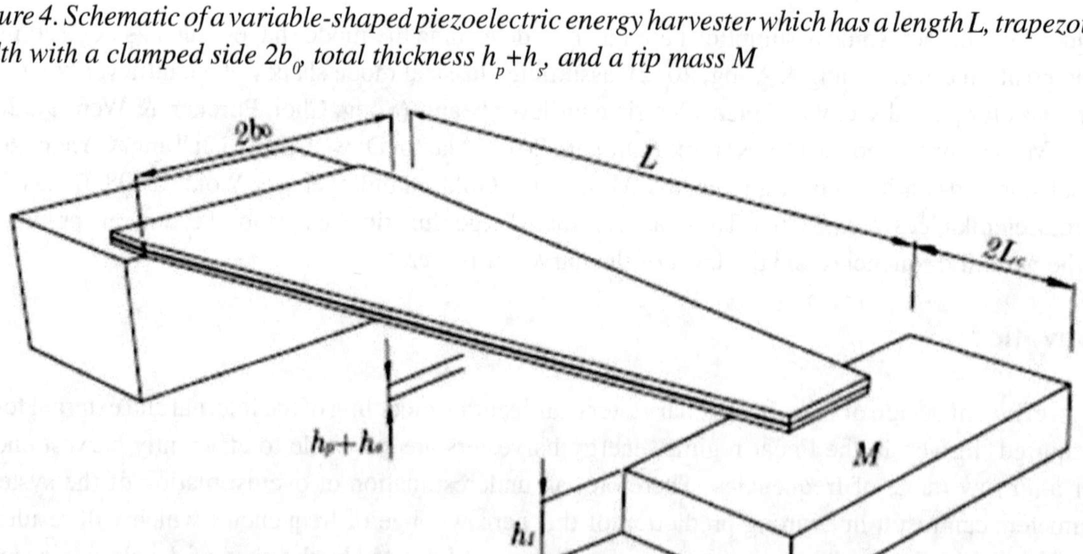


$$\hat{b}(x) = 1 + (\alpha_1 - 1)\frac{x}{L} \tag{1}$$

$$\hat{b}(x) = 1 + (-3 - \alpha_1 + 4\alpha_2)\frac{x}{L} + (2 + 2\alpha_1 - 4\alpha_2)(\frac{x}{L})^2 \tag{2}$$

where $\alpha_1 = \hat{b}$ (L) and $\alpha_2 = \hat{b}$ (L/2). The linear and quadratic representations of the width are plotted, respectively, in Figures 5(a) and (b).

*Figure 5. Geometric variables of the (a) linear and (b) quadratic width shape functions $\hat{b}(x)$*

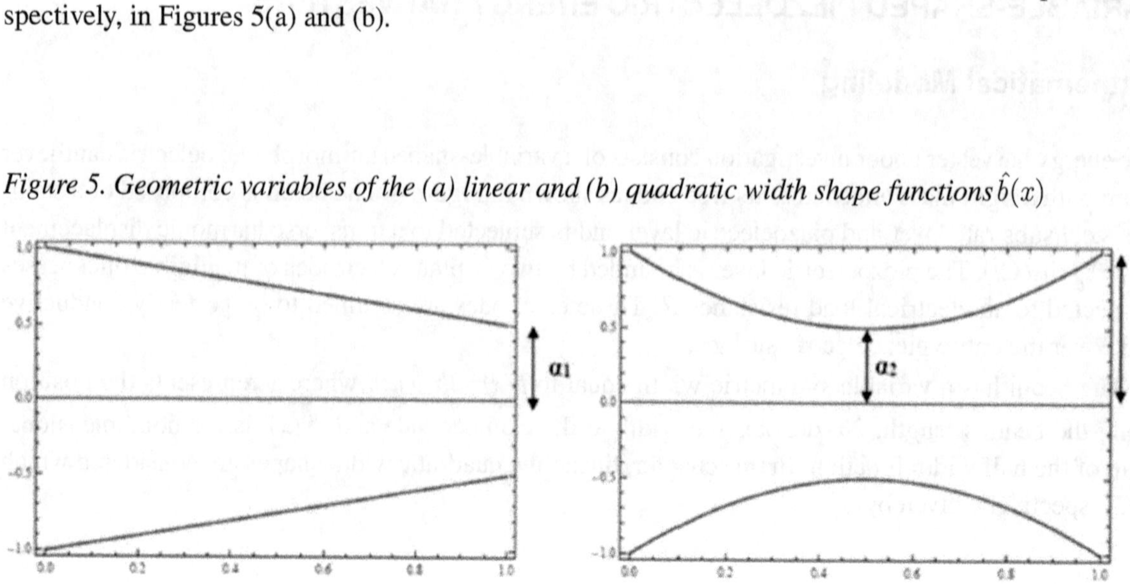

A linear distributed-parameter model is derived using Euler-Bernoulli beam theory *i.e.* the beam thickness is small compared to its length so that shear deformations and rotary inertia can be neglected. The strains are assumed to be continuous along the *y*-axis. However, the stresses are assumed to be varied with the thickness due to the changes in the material properties. The tip mass is modeled as a rigid body with mass *M* and mass moment of inertia *J*. Following Ben Ayed et al. (2014), after using the Hamilton's principle, the governing equation of motion and associated boundary conditions of the variable-shaped harvester are, respectively, given by

$$\begin{aligned} &Ey_3^*\left[b(x)v^{(4)} + 2b'(x)v''' + b''(x)v''\right] \\ &-e_{31}\left(\bar{y} - h_s - \frac{h_p}{2}\right)V(t)\left[b''(x)S(x) + 2b'(x)S'(x) + b(x)S''(x)\right] + m^* b(x)\ddot{v} = 0 \end{aligned} \tag{3}$$

and

$$\begin{aligned} &\delta v\Big|_{x=0} = 0 \\ &\delta v'\Big|_{x=0} = 0 \\ &\delta v\Big|_{x=L} \neq 0 \Rightarrow Ey_3^*\left(b'(L)v'' + b(L)v'''\right) - M(\ddot{v} + L_c\ddot{v}')\Big|_{x=L} = 0 \\ &\delta v'\Big|_{x=L} \neq 0 \Rightarrow -Ey_3^* v'' b(L) - ML_c(\ddot{v} + L_c\ddot{v}') - J\ddot{v}'\Big|_{x=L} = 0. \end{aligned} \tag{4}$$

where $v(x,t)$ is the transverse displacement,

$$m^* = \sum_{i=0}^{1}\int_{y_i}^{y_{i+1}} \rho(y)dy\,,\ Ey_3^* = \sum_{i=0}^{1}\int_{y_i}^{y_{i+1}} y^2 E(y)dy\,,$$

$E(y)$ and $\rho(y)$ denote, respectively, the effective Young's modulus and density and which depends on the system's material. $V(t)$ is the generated voltage, $S(x)=H(x)-H(x-L)$, $H(x)$ is the Heaviside function, and $L_c$ is the tip mass offset. The positions of the layers are defined with respect to the position of the neutral axis, as shown in Figure 6, and are noted by

$$y_0 = -\bar{y}\ ,\ y_1 = h_s - \bar{y} \text{ and } y_2 = (h_s + h_p) - \bar{y} \tag{5}$$

where

$$\bar{y} = \frac{(h_p + h_s)E_p h_p}{2(E_p h_p + E_s h_s)} + \frac{h_s}{2}$$

is the neutral axis position, $h_s$ and $h_p$ denote, respectively, the thicknesses of the steel and piezoelectric layers, the subscripts *s* and *p* refer, respectively, to the steel and piezoelectric layers.

*Figure 6. Neutral axis position*

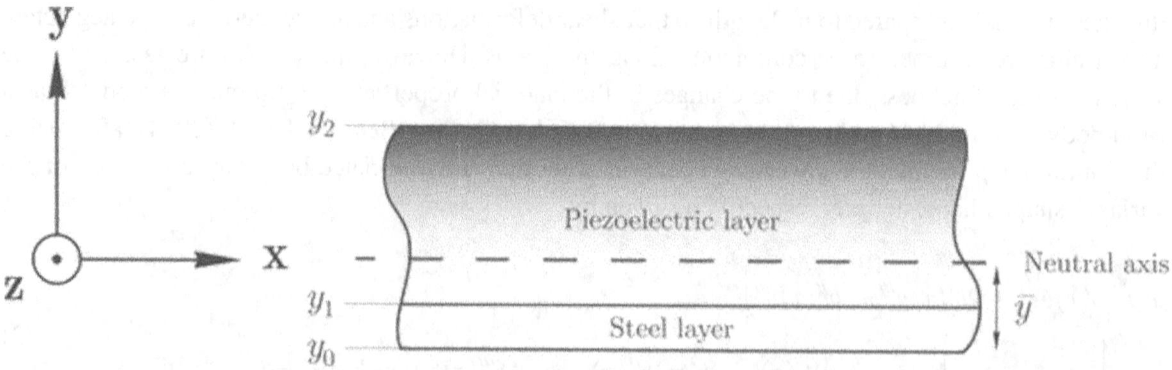


The following transformations are considered to write the nondimensional form of the equation of motion and associated boundary conditions

$$\hat{v} = \frac{v}{h},\ \hat{x} \to \frac{x}{L},\ \hat{t} \to \frac{t}{\tau},\ \hat{S}(\hat{x}) \to H(\hat{x}) - H(\hat{x}-1) \text{ and } \tau = L^2\sqrt{\frac{m^*}{Ey_3^*}}.$$

where $h$ is the total beam's thickness.

To simplify notation, hats are dropped and then the nondimensional equation of motion and associated boundary conditions of the harvester are given by

$$\begin{aligned} &v^{(4)} + c\dot{v} + 2\frac{b'(x)}{b(x)}v''' + \frac{b''(x)}{b(x)}v'' \\ &-\left(\bar{y} - h_s - \frac{h_p}{2}\right)\frac{e_{31}L^2}{Ey_3^* h}\left[\frac{b''(x)}{b(x)}S(x) + 2\frac{b'(x)}{b(x)}S'(x) + S''(x)\right]V(t)b''(x) + \ddot{v} = 0. \end{aligned} \tag{6}$$

and

$$\begin{aligned} &\delta v\Big|_{x=0} = 0 \\ &\delta v'\Big|_{x=0} = 0 \\ &2b_0 Ey_3^* \frac{h}{L^3}\left(b'(1)v'' + \alpha_1 v'''\right) - M\frac{h}{\tau^2}\left(\ddot{v} + \frac{L_c}{L}\ddot{v}'\right)\Big|_{x=1} = 0 \\ &2b_0 Ey_3^* v''\alpha_1 \frac{h}{L^2} + \frac{ML_c h}{\tau^2}\left(\ddot{v} + \frac{L_c}{L}\ddot{v}'\right) + \frac{h}{L\tau^2}J\ddot{v}'\Bigg|_{x=1} = 0. \end{aligned} \tag{7}$$

To relate the electrical and mechanical parts of the system, the Gauss law is used which expresses the charge conservation as follows

$$\frac{d}{dt}\int_A \mathbf{D.n}dA = \frac{V}{R} \tag{8}$$

where **n** is the vector normal to the plane (**x,z**) and **D** is the electric displacement vector. Following Ben Ayed et al. (2014), the nondimensional voltage-displacement equation based on the Gauss law is given by

$$2b_0 e_{31} h_p \frac{h}{L\tau}\int_0^1 b(x)\dot{v}''dx + 2b_0 \frac{L\varepsilon_{33}}{h_p \tau}\dot{V}\int_0^1 b(x)dx + \frac{V}{R} = 0 \tag{9}$$

where $e_{31}$ and $\varepsilon_{33}$ denote, respectively, the piezoelectric stress coefficient and the permittivity at constant strain of the piezoelectric material.

## Natural Frequencies and Mode Shapes

To develop a reduced-order model of this harvester, the exact structural natural frequencies and associated mode shapes are first determined. To this end, the damping and polarization terms are dropped and the transverse displacement is assumed as $v(x,t)=\phi(x)e^{i\omega t}$ where $\phi(x)$ is the mode shape and $\omega$ is its associated structural natural frequency. Therefore, equations (6) and (7) can be written as

$$\phi^{(4)} + 2\frac{b'(x)}{b(x)}\phi''' + \phi''\frac{b''(x)}{b(x)} - \omega^2\phi = 0 \tag{10}$$

and

$$\begin{aligned}
&\phi\Big|_{x=0} = 0\\
&\phi'\Big|_{x=0} = 0\\
&2b_0 E y_3^* \frac{1}{L^3}\left(b'(1)\phi'' + \alpha_1\phi'''\right) + \omega^2 M\frac{1}{\tau^2}\left(\phi + \frac{L_c}{L}\phi'\right)\Big|_{x=1} = 0\\
&2b_0 E y_3^* \alpha_1 \frac{1}{L^2}\phi'' - \frac{\omega^2 M L_c}{\tau^2}\left(\phi + \frac{L_c}{L}\phi'\right) - \frac{\omega^2}{L\tau^2}J\phi'\Bigg|_{x=1} = 0
\end{aligned} \tag{11}$$

To solve Equation (10), Differential Quadrature Method (DQM) is used (Tomasiello, 1998). This method is based on approximating the derivative of a function $w(x)$ with respect to a variable $x$ at a given sampling point as a weighted linear combination of the function values at all the sampling points in the spatial domain. Therefore, the differential equations are transformed into a set of algebraic equations. For a nondimensional variable that is comprised between 0 and 1, the $r^{th}$ derivative of $w$ at the sampling point is defined by

$$\left.\frac{d^r w}{dx^r}\right|_{\xi_i} = \sum_{j=1}^{n} A_{ij}^r w_j \tag{12}$$

The off-diagonal terms of the weighting coefficient matrix are defined using a recurrence relation, where the first matrix order is given by

$$A_{ij}^1 = \frac{\prod_{k=1,k\neq i}^{n}\left(\xi_i - \xi_k\right)}{\left(\xi_i - \xi_j\right)\prod_{k=1,k\neq j}^{n}\left(\xi_j - \xi_k\right)} i,j = 1,2,...,n \quad i \neq j \tag{13}$$

The higher-order weighting coefficient matrices up to *(n-1)* are defined as

$$A_{ij}^r = r\left[A_{ii}^{r-1}A_{ij}^1 - \frac{A_{ij}^{r-1}}{\left(\xi_i - \xi_j\right)}\right] i,j = 1,2,...,n \quad i \neq j \tag{14}$$

The diagonal terms turn out to be given by

$$A_{ii}^r = -\sum_{k=1,k\neq i}^{n} A_{ik}^r \tag{15}$$

The Chebyshev-Gauss-Lobatto grid points (Mittl & Jiwari, 2011) are used which have shown their efficiency for the convergence of the solution when compared to equally-spaced grid points. These grid points are given by

$$\xi_i = \frac{1}{2}\left[1 - \cos\left(\frac{i-1}{n-1}\pi\right)\right] i = 1,...,n \tag{16}$$

Applying the above discretization to Equations (10) and (11), one obtains

$$\sum_{j=1}^{n} A_{ij}^4\phi_j + 2\frac{b_i'}{b_i}\sum_{j=1}^{n} A_{ij}^3\phi_j + \frac{b_i''}{b_i}\sum_{j=1}^{n} A_{ij}^2\phi_j - \omega^2\phi_i = 0 \quad i = 3...n-2 \tag{17}$$

and

$$\begin{aligned}
&\phi_1 = 0 \\
&\sum_{j=1}^{n} A_{1j}^{1}\phi_j = 0 \\
&2b_0 E y_3^* \frac{1}{L^3}\left(b'(1)\sum_{j=1}^{n} A_{nj}^{2}\phi_j + \alpha_1 \sum_{j=1}^{n} A_{nj}^{3}\phi_j\right) + M\frac{\omega^2}{\tau^2}\left(\phi_n + \frac{L_c}{L}\sum_{j=1}^{n} A_{nj}^{1}\phi_j\right) = 0 \\
&2b_0 E y_3^* \alpha_1 \frac{1}{L^2}\sum_{j=1}^{n} A_{nj}^{2}\phi_j - \frac{\omega^2 M L_c}{\tau^2}\left(\phi_n + \frac{L_c}{L}\sum_{j=1}^{n} A_{nj}^{1}\phi_j\right) - \frac{\omega^2}{L\tau^2} J \sum_{j=1}^{n} A_{nj}^{1}\phi_j = 0
\end{aligned} \tag{18}$$

The system of Equations (17) and (18) is composed of *n-4* algebraic equations from the equation of motion and four equations from the boundary conditions for *n* unknowns. Writing these equations in a matrix form, one obtains

$$\left(F - \omega^2 [I]\right)[\phi] = [0] \tag{19}$$

where $[\phi]^T = [\phi_1, \phi_2, \ldots, \phi_n]$ . Clearly, the natural frequencies of the system are the square roots of the eigenvalues of the matrix $F$ and the mode shapes are its eigenvectors.

For a rectangular shape, the eigenvalue problem of the system (Equations (11) and (12)) can be easily analytically solved. In this case, Equations (11) and (12) are reduced to

$$\phi^{iv} - \omega^2 \phi = 0 \tag{20}$$

and

$$\begin{aligned}
&\phi(0) = 0 \\
&\phi'(0) = 0 \\
&\phi'''(1) + \frac{M L^3 \omega^2}{2b_0 E y_3^* \tau^2}\left(\phi(1) + \frac{L_c}{L}\phi'(1)\right) = 0 \\
&\phi''(1) - \frac{M L^2 \omega^2 L_c}{2b_0 E y_3^* \tau^2}\left(\phi(1) + \frac{L_c}{L}\phi'(1)\right) - \frac{J L \omega^2}{E y_3^* \tau^2}\phi'(1) = 0
\end{aligned} \tag{21}$$

The general solution of the Equation (20) has the following form

$$\phi(x) = A_1 \sin\left(\sqrt{\omega}x\right) + A_2 \cos\left(\sqrt{\omega}x\right) + A_3 \cosh\left(\sqrt{\omega}x\right) + A_4 \sinh\left(\sqrt{\omega}x\right) \tag{22}$$

where the $A_i$ are arbitrary constants. Substituting Equation (22) into Equation (21) and setting the determinant of the resulting coefficient matrix equal to zero, the characteristic algebraic equation governing the eigenvalues $\omega$ are easily obtained.

Considering the same geometric and physical properties as Ben Ayed et al. (2014), the first nondimensional natural frequency of the rectangular shape corresponds to 0.793521. Using DQM, Table 1 presents the variation of the first nondimensional natural frequency as a function of *n* for a rectangular shape case ($\alpha_1$=1). Clearly, there is a convergence at *n*=9. This nondimensional natural frequency based on the DQM method is exactly the same as the obtained one based on analytical solution.

Next, the effects of the truncated shape on the structural natural frequencies and associated mode shapes are investigated. To this end, the DQM method is applied to the linear and quadratic configurations of the cross-sectional width function. It should be mentioned that, for energy harvesting applications, researchers approximated their eigenfunctions of any beam's shape with the ones corresponding to a rectangular shape (Mateu & Moll, 2005; Goldschmidtboeing & Woias, 2008; Benasciutti et al., 2010). This statement becomes doubtful, especially when the truncated ratio is high.

The plotted curves in Figures 7(a) and (b) show the variations of the first natural frequency for the trapezoidal and quadratic configurations. Clearly, the parameters $\alpha_1$ and $\alpha_2$ significantly affect the structural natural frequency of the harvester. In fact, it follows from Figure 7(a) that a decrease in $\alpha_1$ (the ratio of width at the free and clamped end) in the trapezoidal configuration results in a decrease in the fundamental frequency which is beneficial for the low-frequency applications. As for the quadratic configuration, the 3D plot in Figure 7(b) shows the variation of the structural natural frequency when varying both $\alpha_1$ (the ratio of width at the free and clamped end) and $\alpha_2$ (the ratio of width at the half and clamped end). Inspecting this plot, a decrease in $\alpha_2$ is accompanied with a significant decrease in

*Table 1. Variation of the first nondiemnsional natural frequency as a function of n*

| | *n*=5 | *n*=7 | *n*=9 | *n*=11 |
|---|---|---|---|---|
| First nondimensional natural frequency | 0.803478 | 0.793259 | 0.793521 | 0.793521 |

*Figure 7. Variations of the first natural frequency as function of the shape parameters $\alpha_1$ and $\alpha_2$: (a) linear and (b) quadratic configurations*

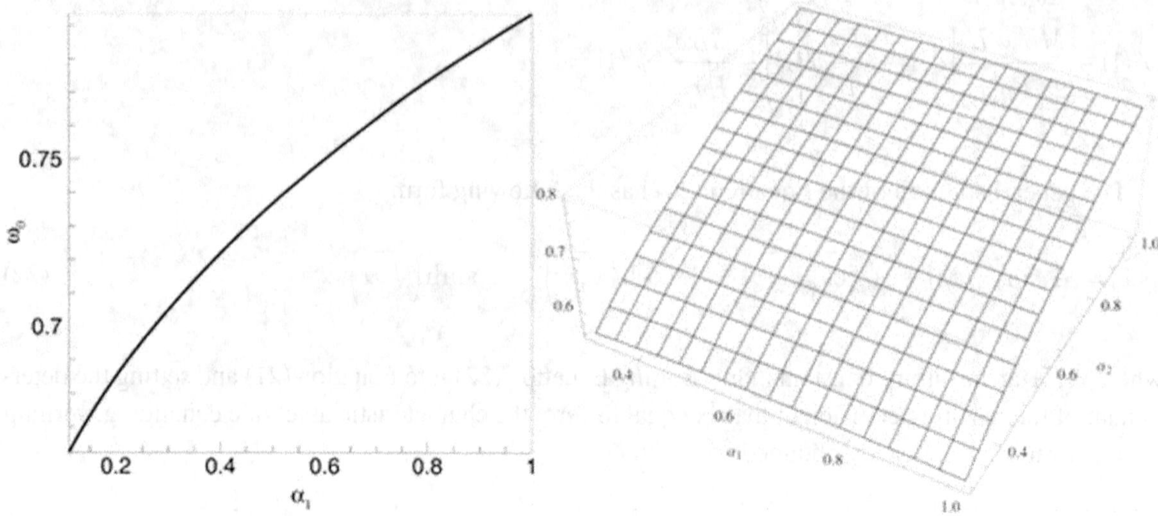

the value of the structural natural frequency for all values of $\alpha_1$. Unlike $\alpha_2$, the effect of $\alpha_1$ on the first natural frequency of the system is negligible for all values of $\alpha_2$. Consequently, monitoring the width of the beam gives the opportunity of having a tunable piezoelectric energy harvester whose fundamental frequency can be adjusted as a function of the available excitation frequency. Because the system behaves much better while excited at its resonance frequency, choosing specific shape parameters can enhance the performance of the harvester.

The influence of the shape parameters $\alpha_1$ and $\alpha_2$ on the mode shapes are presented in Figures 8(a) and (b). It is clear that the shape parameters strongly affect the first mode shape of the harvester. A considerable difference is obtained when the truncated ratio is high. The fact that significant variations in the natural frequencies and mode shapes can be obtained by varying the shape parameters $\alpha_1$ and $\alpha_2$ implies that rectangular beam counterparts should not be assumed.

## Reduced-Order Model

The Galerkin approach is used to derive the reduced-order model of the proposed system. To this end, the displacement $v(x,t)$ is expressed in the form $v(x,t)=\phi(x)q(t)$ where $\phi(x)$ is the fundamental mode shape of a truncated cantilever beam with a tip mass based on interpolation of the results of the Differential Quadrature Method and $q(t)$ is the modal mechanical response. Only one mode is considered in the Galerkin discretization because it is related to the choice of the excitation frequency which is generally low compared to the second and higher structural natural frequencies. To derive a reduced-order model and solve for the beam deflection and the output voltage, Ben Ayed et al. (2014) used the Euler-Lagrange equation and Gauss law (Equation (9)). Applying the Galerkin discretization, a linear distributed-parameter model is obtained as follows (Ben Ayed et al., 2014)

*Figure 8. Variations of the first mode shape as function of the shape parameters $\alpha_1$ and $\alpha_2$: (a) linear and (b) quadratic configurations*

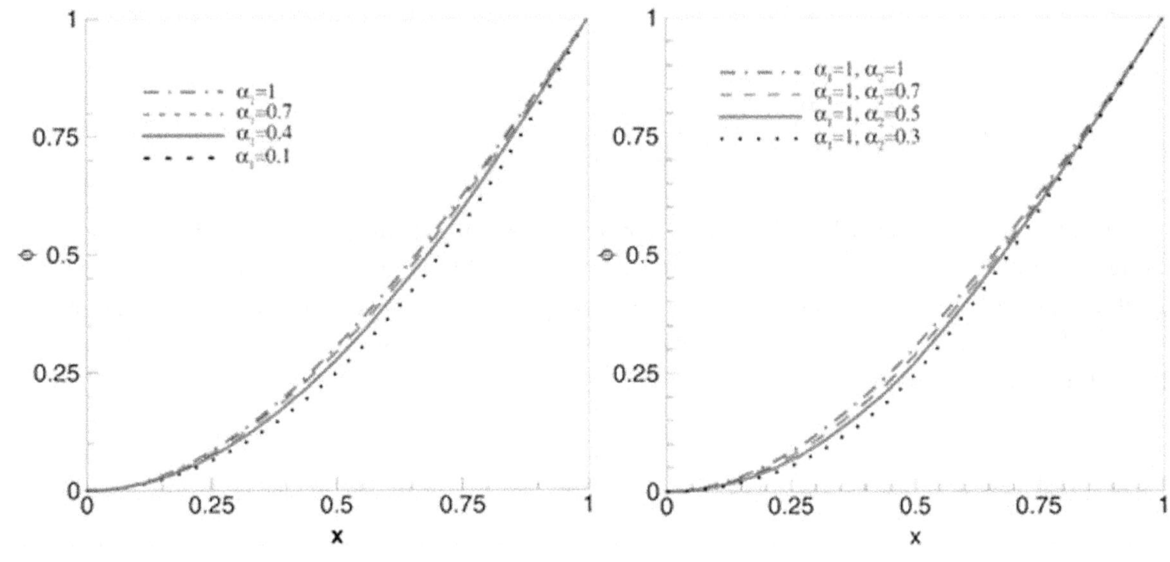

$$A_1\ddot{q} + c_1\dot{q} + B_1 q - K_1 V + F_1\ddot{Y} = 0 \tag{23}$$

$$2e_{31}b_0h_p\frac{h}{\tau L}\int_0^1 b(x)\phi''dx\dot{q} + 2b_0 \in_{33}^s \frac{\dot{V}L}{h_p\tau}\int_0^1 b(x)dx + \frac{V}{R} = 0 \tag{24}$$

where

$$\begin{aligned} A_1 &= \int_0^1 b(x)(\phi(x)^2)dx + \frac{M}{2b_0 mL}\left(\phi(1) + \frac{L_c}{L}\phi'(1)\right)^2 + \frac{J}{2b_0 mL^3}\phi'(1)^2 \\ B_1 &= \int_0^1 b(x)(\phi''(x))^2 dx \\ K_1 &= \frac{1}{2}e_{31}(y_1 + y_2)\frac{L^2}{hE_{y_3}}\int_0^1 b(x)\phi'' dx \\ F_1 &= \int_0^1 b(x)\phi(x)dx + \frac{M}{2b_0 mL}\left(\phi(1) + \frac{L_c}{L}\phi'(1)\right) \end{aligned} \tag{25}$$

For these types of linear forced systems, analytical solutions for the system's outputs can be obtained. Assuming that the base excitation can be expressed by $Y = Y_0\, e^{i\Omega t}$ where $\Omega$ is the excitation frequency, and that $q = Q\, e^{i\Omega t}$ and $V = V_0\, e^{i\Omega t}$. Equations (23) and (24) can then be written as

$$\begin{bmatrix} B_1 - A_1\Omega^2 + i\Omega c_1 & -K_1 \\ A_2 i\Omega & \frac{1}{R} + i\Omega B_2 \end{bmatrix}\begin{bmatrix} Q \\ V_0 \end{bmatrix} = \Omega^2 Y_0 \begin{bmatrix} F_1 \\ 0 \end{bmatrix} \tag{26}$$

where

$$A_2 = 2e_{31}b_0h_p\frac{h}{\tau L}\int_0^1 b(x)\phi''dx,\ B_2 = \frac{2b_0 \in_{33}^s L}{h_p\tau}\int_0^1 b(x)dx$$

Since the excitation term $Y = Y_0\, cos(\Omega t)$ represents the real part of the considered excitation, only the real part of the solutions of Equation (26) are considered to determine the expressions of the tip displacement $v(1,t) = \phi(1)q(t)$ and the generated voltage $V(t)$. The harvested power is then computed as $P=V^2/R$. Using the defined nondimensional parameters, the normalized harvested power $P_N$ is given by the following expression

$$P_N = \frac{V^2\tau^4}{Rh^2Y_0^2\Omega^4} \tag{27}$$

## Performance of the Variable-Shaped Harvester

The effects of the truncated shape parameters on the performance of the harvester are investigated. Frequency-response curves of the normalized harvested power for both linear and quadratic configurations are presented, respectively, in the plotted curves in Figures 9(a) and (b). It should be mentioned that the electrical load resistance is considered equal to $10^4$ Ω. Considering different values of $\alpha_1$ in the linear configuration, as shown in Figure 9(a), it is noted that a decrease in the value of $\alpha_1$ is accompanied by a left shift in the resonant peak of the harvester. This is expected because the structural natural frequency decreases when $\alpha_1$ becomes smaller, as shown in Figure 7(a). As for the quadratic configuration, the plotted curves in Figure 9(b) show that a decrease in $\alpha_2$ is accompanied by a significant decrease in the resonant frequency. This result is also expected because it was demonstrated in the eigenvalue analysis section that the smaller the value of $\alpha_2$, the smaller the value of the structural natural frequency, as shown in Figure 7(b). Not to mention that these frequency-response curves are presented for $\alpha_1$ =1 and

*Figure 9. Frequency-response curves for different beam geometries when $R=10^4$ Ω: (a) linear and (b) quadratic configurations when $\alpha_1=1$*

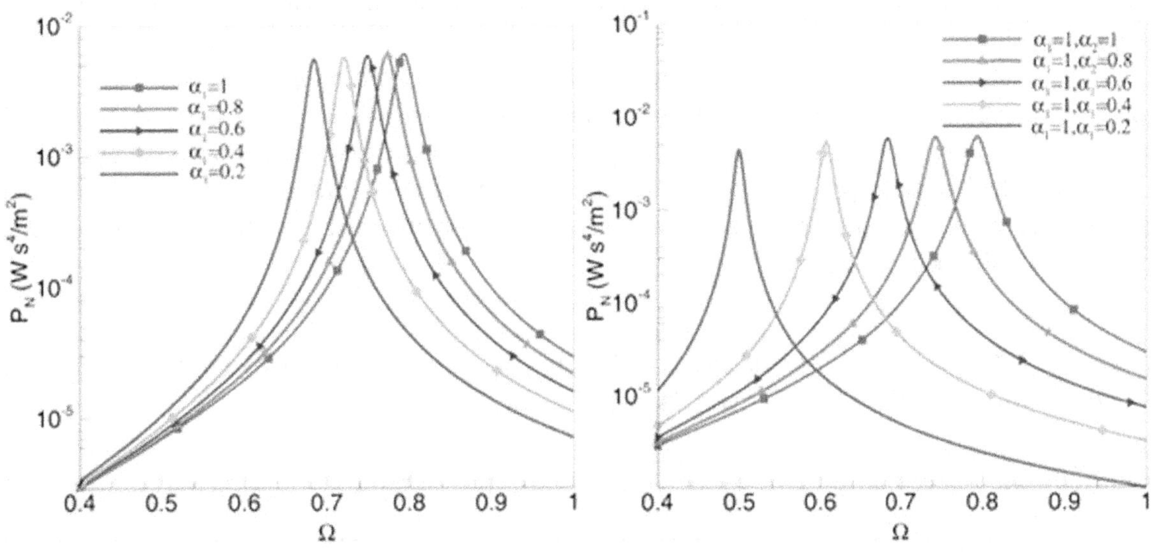


*Table 2. Resonant and structural frequencies for different values of $\alpha_2$ when $\alpha_1=1$ when $R=10^4$ Ω*

| | $\alpha_2$=1 | $\alpha_2$=0.6 | $\alpha_2$=0.2 |
|---|---|---|---|
| Nondimensional resonant frequency ($w_r$) | 0.793957 | 0.683125 | 0.499749 |
| Nondimensional structural frequency ($w_s$) | 0.793521 | 0.682980 | 0.499730 |

for different values of $\alpha_2$. Clearly, Table 2 shows that the resonant frequency is slightly different from the structural natural frequency of the harvester. This result is due to the fact that the electrical load resistance is set equal to $10^4\ \Omega$.

Mateu and Moll (2005) and Benasciutti et al. (2010) reported that the truncated shape is always better than the rectangular one in terms of levels of harvested power. In this research study, it is demonstrated that the load resistance is an important parameter to determine the best configuration in terms of harvester power. Clearly, the plotted curves in Figures 10(a) and (b) for the linear and quadratic configurations, respectively, show that the preference between the rectangular shape and the truncated one depends on the considered value of the load resistance. For instance, concerning the linear configuration presented in Figure 10(a), when the load resistance is between $10^2\ \Omega$ and $10^4\ \Omega$, the rectangular shape is better. On the other hand, the result is inverted for higher load resistance values ($R>10^4\ \Omega$). Clearly, there are two optimum values of the load resistance for which the highest harvested power levels are obtained. At these two optimum values, it is noted that the truncation ratio does not have a significant impact on the harvested power levels. It should be mentioned that all these plotted curves are considered under resonant excitation. As for the quadratic configuration, it follows from Figure 10(b) that similar tendency as the linear configuration is obtained. However, higher levels of the harvested power are obtained and significant differences between the considered shape parameters are observed. Clearly, the value of the shape parameter $\alpha_2$ strongly affects the maximum level of the harvested power as well as the optimum value of the load resistance.

In this study, it was demonstrated that variable-shaped piezoelectric energy harvesters are beneficial to design energy harvesters that can generate power at low-frequency excitations and with enhanced power density. Furthermore, the performed analysis shows that the load resistance has a strong effect on the level of the harvested power. It is also shown that truncated configurations can yield higher levels of harvested power than their rectangular counterparts. This is, however, only true for specific values of the load resistance.

*Figure 10. Effects of the shape parameters and load resistance on the performance of the harvester: (a) linear and (b) quadratic configurations when $\alpha_1=1$*

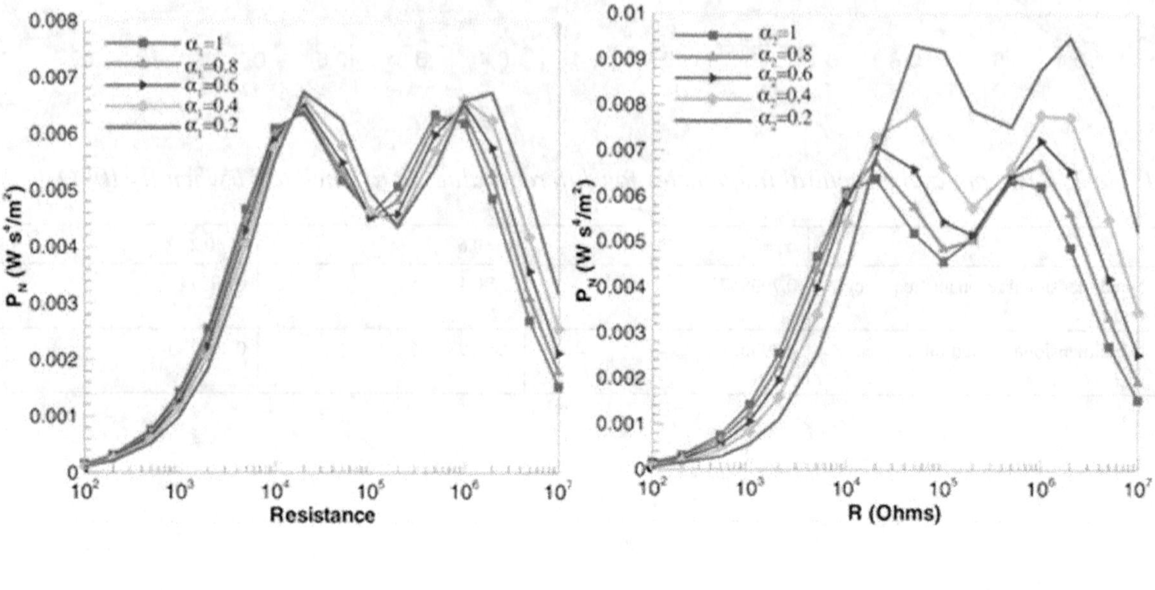

## PIEZOMAGNETOELASTIC ENERGY HARVESTERS

### Analytical Models Representations

The energy harvester under consideration consists of unimorph piezoelectric cantilever beam with a magnet tip mass. Another magnet mass is fixed at the enclosure of the device with a spacing distance $D_0$, as shown in Figure 11. The beam is composed of aluminum and piezoelectric layers and is subjected to a transverse harmonic displacement $Y(t)=Y_0 \cos(\Omega t)$. The piezoelectric layers are bounded by two-in-plane electrodes of negligible thicknesses connected to an electrical load resistance $R$. These electrodes are assumed to be perfectly conductive and cover the entire piezoelectric surface.

The response behavior of the harvester is modeled as an Euler-Bernoulli beam. The magnet tip mass is modeled as a rigid body with mass $M_t$ and mass moment of inertia $I_t$. The extended Hamilton principle is used to determine the equations of motion and associated boundary conditions of this system. This principle states that

$$\int_{t_1}^{t_2} (\delta T - \delta \Pi + \delta W_{nc}) dt = 0 \tag{28}$$

where $\delta$ is a variational operator, $t_2$-$t_1$ is any time interval, $T$ and $\Pi$ are, respectively, the kinetic and potential energies of the system, and $W_{nc}$ represents the work done by nonconservative forces.

The kinetic energy of the harvester is the sum of the kinetic energies of the beam and tip mass, which are expressed as

$$T = \frac{1}{2}\Big[\int_0^{L_1} m_1(\dot{v}+\dot{Y})^2 dx + \int_{L_1}^{L} m_2(\dot{v}+\dot{Y})^2)dx\Big] + \frac{1}{2}M_t(\dot{v}_L + \dot{Y} + L_c\dot{v}_L')^2 + \frac{1}{2}I_1\dot{v}_L'^2 \tag{29}$$

where $L$ is the length of the substrate beam, $L_1$ is the distance from the clamped side of the cantilever beam to the ending location of the piezoelectric layer, and $L_c$ is half the length of the tip mass. $m_1$ and $m_2$ are the masses of the beam per unit length which depend on the location of the piezoelectric layer:

*Figure 11. Schematic of the considered piezomagnetoelastic energy harvester under direct excitations $Y(t)$ and with a distance $D_0$ between the two tip magnets*

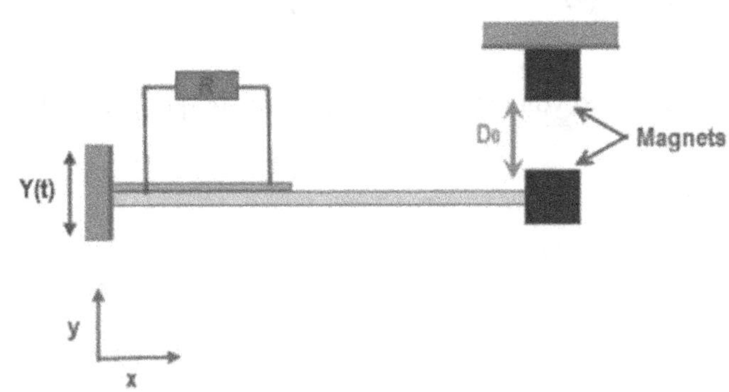

$$m_1 = b_1\rho_s h_s + b_2\rho_p h_p \text{ for } 0 \le x \le L_1 \text{ and } m_2 = b_1\rho_s h_s \text{ for } L_1 < x \le L$$

where $h_s$ and $h_p$ are, respectively, the thicknesses of the aluminum and piezoelectric layers. $\rho_s$ and $\rho_p$ are the densities of the aluminum and piezoelectric layers, respectively. $b_1$ and $b_2$ are, respectively, the width of the aluminum and piezoelectric layers.

The total potential energy of the harvester is given by

$$U = \frac{1}{2}\left(\int_0^L \int_{A_s} \sigma_{11}^s \varepsilon_{11} dA_s dx + \int_0^{L_1} \int_{A_p} \sigma_{11}^p \varepsilon_{11} dA_p dx - \int_0^{L_1} \int_{A_p} E_3 D_3 dA_p dx\right) \tag{30}$$

where $dA_s$ and $dA_p$ are, respectively, the cross-section areas of differential aluminum and piezoelectric beam elements. Using Hooke's law and linear constitutive equations of piezoelectricity, the strains and stresses in the aluminum and piezoelectric layers are, respectively, related by

$$\sigma_{11}^s = E_s \varepsilon_{11} \text{ and } \sigma_{11}^p = E_p \varepsilon_{11} - e_{31} E_3 \tag{31}$$

where $E_s$ and $E_p$ are Young's moduli at constant electric field and $E_3(t) = -V(t)/h_p$ is the electric field in the poling direction where $V(t)$ denotes the generated voltage.

The electric displacement $D_3$ is given by (Abdelkefi & Barsallo, 2014)

$$D_3 = e_{31}\varepsilon_{11} + \varepsilon_{33}^s E_3 \tag{32}$$

where $e_{31}$ is the piezoelectric stress coefficient and $\varepsilon^s_{33}$ is the permittivity at constant strain.

Using these expressions, the potential energy of the system can be written as

$$U = \frac{1}{2}\int_0^{L_1} \left[EI_1 v''^2\right] dx + \frac{1}{2}\int_{L_1}^{L} \left[EI_2 v''^2\right] dx - \int_0^{L_1} e_{31} b_2 \frac{(y_1 + y_2)}{2} v'' V(t) dx - \frac{1}{2}\frac{\varepsilon_{33} b_2 L_1}{h_p} V^2(t) \tag{33}$$

where

$$EI_1 = \frac{1}{3} E_s b_1 \left(y_1^3 - y_0^3\right) + \frac{1}{3} E_p b_2 \left(y_2^3 - y_1^3\right) \text{ and } EI_2 = \frac{1}{12} b_1 E_s h_s^3 .$$

Using Ohm's law, the generated voltage and the extracted charge $Q$ are related through

$$V(t) = R\frac{dQ}{dt} \tag{34}$$

The nonconservative work is divided into three parts, namely, the work due to the power delivered to the electrical load resistance $R$, the work done by the magnetic force, $F_{mag}$, and the work done by the viscous damping force. Therefore, the variation of the nonconservative work term is expressed as

$$\delta W_{nc} = -V\delta Q + F_{mag}(t)\delta v_L - \int_0^L c\dot{v}\delta v dx \tag{35}$$

where $c$ represents the viscous damping.

When the fixed magnet is on the top of the proof mass, the magnetic force can be expressed as (Tang and Yang, 2012)

$$F_{mag}(t) = -\frac{3\tau a_1 a_2}{2\pi(D_0 - v_L(t))^4} \tag{36}$$

where $\tau$ is the vacuum permeability, $a_1$ and $a_2$ are the moments of the magnetic dipoles, $D_0$ is the initial distance between the magnetic dipoles. When the magnets are repulsive, $a_1$ and $a_2$ are equal. On the other hand, if the magnets are attractive $a_1$=-$a_2$ . It should be mentioned that the expression of this magnetic force is obtained after assuming a magnetic dipole-dipole interaction between the tip magnet mass and the fixed magnet (Stanton et al., 2009) and assuming that the magnetic dipoles directions and the transverse deflection of the harvester are always vertically aligned.

Using the extended Hamilton principle, the governing equations of motion of the harvester are given by (Abdelkefi & Barsallo, 2014)

$$\begin{aligned} &EI\frac{\partial^4 v(x,t)}{\partial x^4} + c\frac{\partial v(x,t)}{\partial t} + m\frac{\partial^2 v(x,t)}{\partial t^2} + \left(\frac{d\delta(x)}{dx} - \frac{d\delta(x-L_1)}{dx}\right)\vartheta_p V(t) \\ &= -[m + M_t\delta(x-L)]\frac{\partial^2 Y(t)}{\partial t^2} + M_t L_c \frac{d\delta(x-L)}{dx}\frac{\partial^2 Y(t)}{\partial t^2} \end{aligned} \tag{37}$$

$$-e_{31}\frac{(y_1+y_2)}{2}b_2\int_0^{L_1}\frac{\partial^3 v(x,t)}{\partial t\partial x^2}dx - \frac{\epsilon_{33}^s b_2 L_1}{h_p}\frac{dV(t)}{dt} = \frac{V(t)}{R} \tag{38}$$

where

$$\vartheta_p = -e_{31}b_2\frac{(y_1+y_2)}{2}$$

and the stiffness $EI$ and mass of the beam per unit length $m$ are given by

$$EI = EI_1 \text{ and } m = m_1 \text{ for } 0 \le x < L_1 \text{ and } EI = EI_2 \text{ and } m = m_2 \text{ for } L_1 \le x < L.$$

The associated boundary conditions are given by

$$\begin{aligned} &v(0,t)=0 \\ &v'(0,t)=0 \\ &EI_2 v'''(L,t)=M_t\ddot{v}(L,t)+M_tL_c\ddot{v}'(L,t)-F_{mag}(t) \\ &EI_2 v''(L,t)=-M_tL_c\ddot{v}(L,t)-(I_t+M_tL_c^2)\ddot{v}'(L,t) \end{aligned} \tag{39}$$

## Derived Electromechanical Model

Static and eigenvalue problem analyses are performed to determine the exact mode shapes and associated natural frequencies of the harvester. The effects of the magnetic force and the piezoelectric sheet that does not cover the whole beam are considered.

In the presence of the magnetic force, the harvester undergoes a static position which depends on the distance between the two magnets and the moments of the magnetic dipoles (repulsive or attractive). To determine the static position, $v_s(x)$, of the harvester, we set the time derivatives and the piezoelectric effect in the equations of motion and associated boundary conditions equal to zero. Because the cantilever beam is partially covered by the piezoelectric layer, we divide the static position into two distinct regions

$$v_s = v_{s1} \text{ for } 0 \le x < L_1 \text{ and } v_s = v_{s2} \text{ for } L_1 \le x \le L \tag{40}$$

The resulting static problem for each region is given by

$$\begin{aligned} &v_{s1}^{iv}=0 \\ &v_{s2}^{iv}=0 \end{aligned} \tag{41}$$

with the following boundary conditions

$$\begin{aligned} &v_{s1}(0)=0\,,\ v_{s1}{}'(0)=0\,,\ v_{s1}(L_1)=v_{s2}(L_1),\ v_{s1}{}'(L_1)=v_{s2}{}'(L_1)\,,\ EI_1v_{s1}{}''(L_1)=EI_2v_{s2}{}''(L_1) \\ &EI_1v_{s1}{}'''(L_1)=EI_2v_{s2}{}'''(L_1),\ EI_2v_{s2}{}'''(L)=\frac{k}{(D_0-v_{s2}(L))^4},\ EI_2v_{s2}{}''(L)=0 \end{aligned} \tag{42}$$

where $k=\dfrac{3\tau a_1a_2}{2\pi}$.

The general solutions of $v_{s1}(x)$ and $v_{s2}(x)$ are given by

$$\begin{aligned} &v_{s1}(x)=A_1x^3+B_1x^2+C_1x+D_1 \\ &v_{s2}(x)=A_2x^3+B_2x^2+C_2x+D_2 \end{aligned} \tag{43}$$

Using the boundary conditions relations, the values of $A_1$, $A_2$, $B_1$, $B_2$, $C_1$, $C_2$, $D_1$, and $D_2$ are solved numerically.

After determining the static position of the harvester when subjected to a magnetic force, the exact mode shapes and associated natural frequencies are determined. To this end, the damping, forcing, and polarization terms are dropped from Equations (37) and (39). Furthermore, the harvester's deflection is decomposed into the sum of a static and a dynamic component as

$$v(x,t) = v_s(x) + u(x,t) \tag{44}$$

Substituting Equation (44) into Equations (37) and (39) and expanding the nonlinear magnetic force in a Taylor series expansion about $u$=0 as follows

$$F_{mag}(t) = -\frac{k}{(D_0 - v_s(L) - u(L,t))^4} = -\frac{k}{(D_0 - v_s(L))^4} - \frac{4ku(L,t)}{(D_0 - v_s(L))^5} \tag{45}$$

The governing equations of motion and associated boundary conditions of the dynamic component of the system are given by

$$EI\frac{\partial^4 u(x,t)}{\partial x^4} + m\frac{\partial^2 u(x,t)}{\partial t^2} = 0 \tag{46}$$

$$\begin{aligned} &u(0,t) = 0\,,\; u'(0,t) = 0 \\ &EI_2 u'''(L,t) = M_t \ddot{u}(L,t) + M_t L_c \ddot{u}'(L,t) + \frac{4ku(L,t)}{(D_0 - v_s(L))^5} \\ &EI_2 u''(L,t) = -M_t L_c \ddot{u}(L,t) - (I_t + M_t L_c^2)\ddot{u}'(L,t) \end{aligned} \tag{47}$$

Letting $v(x,t)=\phi(x)e^{i\omega t}$ where $\phi(x)$ is the mode shape and $\omega$ is its corresponding natural frequency. The mode shape is divided into two distinct regions because the piezoelectric material does not cover the whole beam as

$$\phi(x) = \phi_1(x) \text{ for } 0 \le x < L_1 \text{ and } \phi(x) = \phi_2(x) \text{ for } L_1 \le x \le L \tag{48}$$

The resulting eigenvalue problem for each region is then given by

$$\begin{aligned} &EI_1\phi_1^{iv} - m_1\omega^2\phi_1 = 0 \\ &EI_2\phi_2^{iv} - m_2\omega^2\phi_2 = 0 \end{aligned} \tag{49}$$

with the following boundary conditions

$$\phi_1(0)=0\,,\phi_1'(0)=0\,,\phi_1(L_1)=\phi_2(L_1)\,,\phi_1'(L_1)=\phi_2'(L_1)\,,\,EI_1\phi_1''(L_1)=EI_2\phi_2''(L_1)$$
$$EI_1\phi_1'''(L_1)=EI_2\phi_2'''(L_1),\,EI_2\phi_2'''(L)+\omega^2M_t\phi_2(L)+\omega^2M_tL_c\phi_2'(L)-\frac{4k\phi_2(L)}{(D_0-v_s(L))^5}=0 \tag{50}$$
$$EI_2\phi_2''(L)-\omega^2M_tL_c\phi_2(L)-\omega^2(I_t+M_tL_c^2)\phi_2'(L)=0$$

The general solutions of these mode shapes can be written as

$$\begin{aligned}\phi_1(x)&=A_{11}\sin\beta_1x+B_{11}\cos\beta_1x+C_{11}\sinh\beta_1x+D_{11}\cosh\beta_1x\\ \phi_2(x)&=A_{22}\sin\beta_2x+B_{22}\cos\beta_2x+C_{22}\sinh\beta_2x+D_{22}\cosh\beta_2x\end{aligned} \tag{51}$$

where the coefficients $\beta_1$ and $\beta_2$ are related by $\beta_1=\sqrt[4]{\frac{EI_2m_1}{EI_1m_2}}\beta_2$.

Normalizing the mode shapes using the following orthogonality conditions yields

$$\begin{aligned}&\int_0^{L_1}\phi_{1s}(x)m_1\phi_{1r}(x)dx+\int_{L_1}^{L}\phi_{2s}(x)m_2\phi_{2r}(x)dx+\phi_{2s}(L)M_t\phi_{2r}(L)+\phi_{2s'}(L)(I_t+M_tL_c^2)\phi_{2r'}(L)\\ &+\phi_{2s}(L)M_tL_c\phi_{2r'}(L)+\phi_{2s'}(L)M_tL_c\phi_{2r}(L)=\delta_{rs}\\ &\int_0^{L_1}\frac{d^2\phi_{1s}(x)}{dx^2}EI_1\frac{d^2\phi_{1r}(x)}{dx^2}dx+\int_{L_1}^{L}\frac{d^2\phi_{2s}(x)}{dx^2}EI_2\frac{d^2\phi_{2r}(x)}{dx^2}dx+\frac{4k\phi_{2s}(L)\phi_{2r}(L)}{(D_0-v_s(L))^5}=\delta_{rs}\omega_r^2\end{aligned} \tag{52}$$

where *s* and *r* are used to represent the vibration modes and $\delta_{rs}$ is the Kronecker delta, defined as unity when *s* is equal to *r* and zero otherwise. Then, the coefficients of the mode shapes are solved numerically.

After determining the exact mode shapes and associated natural frequencies, a reduced-order model is developed based on the Galerkin discretization. To this end, the deflection $v(x,t)$ is expressed as

$$v(x,t)=v_s(x)+\sum_{i=1}^{\infty}\phi_{ji}(x)q_i(t) \tag{53}$$

where $j$=1,2 depending on the value of $x$, $q_i(t)$ are the modal coordinates, and $\phi j_i(x)$ are the mode shapes.

Considering only one mode in the Galerkin discretization and applying the Euler-Lagrange equations which are given by

$$\frac{d}{dt}\frac{\partial\ell}{\partial\dot q}-\frac{\partial\ell}{\partial q}=\frac{\delta W_{nc}}{\delta q} \tag{54}$$

$$\frac{d}{dt}\frac{\partial\ell}{\partial\dot\lambda}-\frac{\partial\ell}{\partial\lambda}=\frac{\delta W_{nc}}{\delta\lambda}=-\frac{V}{R} \tag{55}$$

where $x= T\text{-}\Pi$ is the Lagrangian and $\dot{\lambda} = V$.

The following reduced-order model of the harvester is obtained

$$\ddot{q} + 2\xi\omega\dot{q} + \omega^2 q + \theta V + N_{mag} = f(t) \tag{56}$$

$$C_p\dot{V} + \frac{V}{R} - \theta\dot{q} = 0 \tag{57}$$

where $\xi$ is the mechanical damping coefficient, $f(t)$ denotes the first mode of the external force due to base excitation which is expressed as

$$f(t) = -[\int_0^{L_1} m_1\phi(x)dx + \int_{L_1}^{L} m_2\phi(x)dx + M_t(\phi(L) + L_c\phi'(L))]\frac{\partial^2 Y(t)}{\partial t^2} \tag{58}$$

and $N_{mag}$ represents the nonlinear contribution of the magnetic force which is equal to

$$k\phi(L)\left(\frac{1}{(D_0 - v_s(L) - \phi(L)q)^4} - \frac{1}{(D_0 - v_s(L))^4} - \frac{4\phi(L)q}{(D_0 - v_s(L))^5}\right)$$

The piezoelectric coupling term and capacitance of the harvester are, respectively, defined as

$$\theta = -e_{31}b_2\frac{(y_1 + y_2)}{2}\phi'(L_1) \text{ and } C_p = \frac{\varepsilon_{33}b_2L_1}{h_p}$$

## Approximated Electromechanical Model

An approximated nonlinear distributed-parameter model is considered in which the effects of the magnetic force and the partially-covered piezoelectric cantilever beam in the determination of the mode shapes are neglected. Hence, the classical mode shapes of a unimorph piezoelectric cantilever beam (fully-covered) are considered.

Neglecting the contribution of the magnetic force in Equation (50) and letting $v(x,t)=\psi(x)e^{i\gamma t}$ where $\psi(x)$ is the mode shape and $\gamma$ is its associated natural frequency, the approximated mode shapes are governed by

$$EI_0\psi^{iv} - m_0\gamma^2\psi = 0 \tag{59}$$

where

$$EI_0 = \frac{1}{3}E_s b_1\left(y_1^3 - y_0^3\right) + \frac{1}{3}E_p b_2\left(y_2^3 - y_1^3\right) \text{ and } m_0 = b_1\rho_s h_s + b_2\rho_p h_p.$$

The associated boundary conditions are given by

$$\begin{aligned} &\psi(0) = 0\,,\ \psi'(0) = 0 \\ &EI_0\psi'''(L) + \gamma^2 M_t\psi(L) + \gamma^2 M_t L_c\psi'(L) = 0 \\ &EI_0\psi''(L) - \gamma^2 M_t L_c\psi(L) - \gamma^2 (I_t + M_t L_c^2)\psi'(L) = 0 \end{aligned} \tag{60}$$

The approximated mode shape $\psi(x)$ can be expressed as

$$\psi(x) = A_0 \sin\beta_0 x + B_0 \cos\beta_0 x + C_0 \sinh\beta_0 x + D_0 \cosh\beta_0 x \tag{61}$$

These approximated mode shapes are normalized using the following orthogonality conditions

$$\begin{aligned} &\int_0^{L_1} \psi_s(x)m_1\psi_r(x)dx + \int_{L_1}^{L} \psi_s(x)m_2\psi_r(x)dx + \psi_s(L)M_t\psi_r(L) + \psi_{s'}(L)(I_t + M_t L_c^2)\psi_{r'}(L) + \\ &\psi_s(L)M_t L_c\psi_{r'}(L) + \psi_{s'}(L)M_t L_c\psi_r(L) = \delta_{rs} \\ &\int_0^{L_1} \frac{d^2\psi_s(x)}{dx^2} EI_1 \frac{d^2\psi_r(x)}{dx^2}dx + \int_{L_1}^{L} \frac{d^2\psi_s(x)}{dx^2} EI_2 \frac{d^2\psi_r(x)}{dx^2}dx + \frac{4k\psi_s(L)\psi_r(L)}{(D_0)^5} = \delta_{rs}\omega_{0_r}^2 \end{aligned} \tag{62}$$

where $\omega_0$ is the natural frequency of the harvester based on the approximated mode shapes.

After using only one mode in the Galerkin discretization ($v(x,t)=\psi(x)p(t)$) and applying the Euler-Lagrange equations as described in Equations (54) and (55), the reduced-order model of the harvester based on the approximated mode shapes are written as

$$\ddot{p} + 2\xi\omega_0\dot{p} + \omega_0^2 p + \theta_0 V + N_{0mag} = f_0(t) \tag{63}$$

$$C_p\dot{V} + \frac{V}{R} - \theta_0\dot{p} = 0 \tag{64}$$

where $\xi$ is the mechanical damping coefficient, the excitation forcing term is given by

$$f_0(t) = -[\int_0^{L_1} m_1\psi(x)dx + \int_{L_1}^{L} m_2\psi(x)dx + M_t(\psi(L) + L_c\psi'(L))]\frac{\partial^2 Y(t)}{\partial t^2}$$

and $N_0$ is given by

$$k\psi(L)(\frac{1}{(D_0-\psi(L)p)^4}-\frac{1}{D_0^4}-\frac{4\psi(L)p}{D_0^5})$$

## Lumped-Parameter Model

In addition to the derived and approximated nonlinear distributed-parameter models, a corrected lumped-parameter model of Tang and Yang (2012) is considered which is given by

$$M_{eq}\ddot{\eta}+C_{eq}\dot{\eta}+K_{eq}\gamma+\Theta V+\frac{k}{(D_0-\eta)^4}=-\mu M_{eq}\ddot{Y} \tag{65}$$

$$C^S\dot{V}+\frac{V}{R}-\Theta\dot{\eta}=0 \tag{66}$$

where $\eta(t)$ denotes the tip displacement of the harvester.

In this model, Tang and Yang (2012) used a correction factor $\mu$ in the forcing term (Erturk & Inman, 2008) and they identified the coefficients $M_{eq}$, $C_{eq}$, and $K_{eq}$ by using the classical mode shapes of non-covered cantilever beam with tip mass. It should be mentioned that the tip mass moment contribution as well as the rotary inertia term were neglected. Furthermore, the capacitance $C^S$ was considered from the manufacturer and the piezoelectric coupling $\Theta$ was experimentally identified through calculating the short-circuit current. It should also be noted that the differences between this presented lumped-parameter model and the one presented by Tang and Yang (2012) are the notations of some variables and parameters and the magnetic force which is considered here for a fixed magnet at the top of the moving harvester ($y$-axis orientation).

# RESULTS AND DISCUSSIONS

## Static and Eigenvalue Problems Comparisons

To show the importance of developing accurate models for the design of low-frequency piezoelectric energy harvesters, we first start comparing the static position, natural frequencies, and mode shapes when using different models. The parameters of the considered piezomagnetoelastic energy harvester which are based on the experiments of Tang and Yang (2012) where the spacing distance $D_0$ between the two magnets is set equal to 9.7 *mm*. In Figure 12(a), the harvester's static position along the beam axis is presented when using the derived model. For the chosen spacing distance $D_0$, it is clear that the magnet effects take place and a static position is present. This static position is always positive because the considered magnetic force is attractive ($a_1$=-$a_2$). It should be mentioned that for the other models,

*Figure 12. (a) Variation of the static position based on the derived model and (b) a comparison between the first mode shape obtained from derived and approximated models*

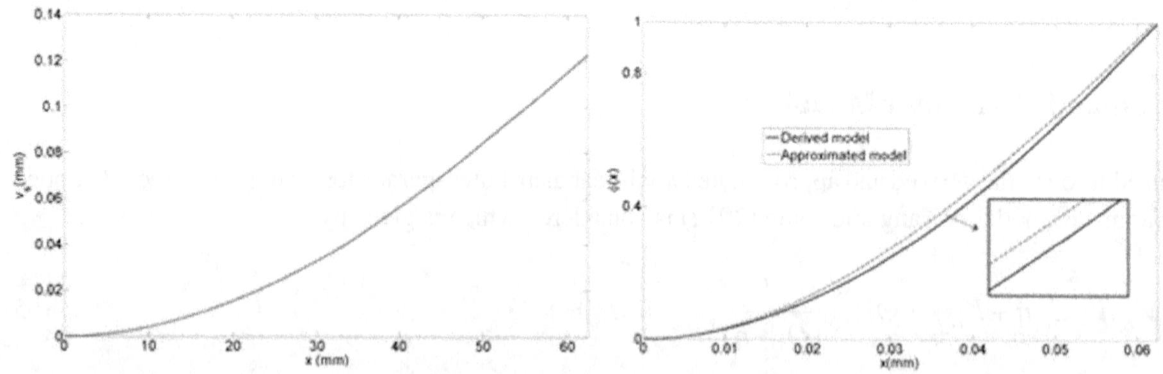


the associated static position is zero because the effects of the attractive magnetic force are neglected. In the presence of the static position, the eigenvalue problem analysis is modified and hence a more accurate prediction of the mode shapes and natural frequencies is obtained.

A comparison of the first mode shape obtained from the derived and approximated models is presented in Figure 12(b). It follows from this plot that there is a discrepancy in determining the first mode shape when neglecting the magnetic force contribution. Clearly, this discrepancy is more significant in the range between 25 *mm* and 55 *mm*. In fact, there is an overestimation in the values of the mode shape. This discrepancy will directly affect the fundamental natural frequency of the harvester as well as the piezoelectric coupling and hence the performance of the harvester.

Next, the effects of the load resistance on the coupled frequency and electromechanical damping for both approximated and derived models are determined. This investigation is performed through a linear analysis of the coupled electromechanical problem when neglecting the excitation force and the nonlinear contribution of the magnetic force. This linear analysis is performed based on the notations used in the derived model. The same procedure is used when performing the linear analysis based on the approximated model.

Introducing the following state variables:

$$\mathbf{Y} = \begin{bmatrix} Y_1 \\ Y_2 \\ Y_3 \end{bmatrix} = \begin{bmatrix} q \\ \dot{q} \\ V \end{bmatrix} \tag{67}$$

the linear reduced-order model can be expressed as

$$\begin{aligned} \dot{Y}_1 &= Y_2 \\ \dot{Y}_2 &= -2\xi\omega Y_2 - \omega^2 Y_1 - \theta Y_3 \\ \dot{Y}_3 &= -\frac{1}{RC_p} Y_3 + \frac{\theta}{C_p} Y_2 \end{aligned} \tag{68}$$

In a matrix form, these equations can be written as

$$\dot{\mathbf{Y}} = A(R)\mathbf{Y} \tag{69}$$

where

$$A(R) = \begin{bmatrix} 0 & 1 & 0 \\ -\omega^2 & -2\xi\omega & -\theta \\ 0 & \dfrac{\theta}{C_p} & -\dfrac{1}{RC_p} \end{bmatrix}$$

Clearly, the matrix *A(R)* has a set of three eigenvalues which are dependent on the value of the electrical load resistance. The first two eigenvalues are complex conjugate and are similar to those of a pure mechanical system in the absence of the piezoelectricity effect. The third eigenvalue is a result of the electromechanical coupling and is always real and negative. The real part of the first two eigenvalues represents the electromechanical damping coefficient and their positive imaginary part corresponds to the coupled frequency of the electromechanical system.

The dependence of the coupled frequency and the electromechanical damping on the value of the load resistance is presented in Figures 13(a) and (b), respectively. It follows from these plots that the variations of the coupled frequency and electromechanical damping have similar tendencies for both models. There is a significant increase that takes place when the load resistance is increased from $10^5\Omega$ to $10^6\Omega$ for both approximated and derived models. However, the approximated model overestimates the coupled frequency as well as the electromechanical damping of the harvester for all values of the load resistance. In fact, based on the approximated model, the coupled frequency is almost equal to 173.6 *rad/s* when the load resistance is set equal to $10^2\Omega$; we refer to this frequency as the linear short frequency. When the load resistance is near $10^8\Omega$, an increase in the coupled frequency is obtained to a value near 175.1 *rad/s*; we refer to this frequency as the linear open frequency. On the other hand, the short and open frequencies are, respectively, equal to 170.8 *rad/s* and 171.7 *rad/s* when considering the derived model. Therefore, an overestimation of more than 2% in the coupled frequency is obtained when using the approximated model instead of the derived model. This overestimation can result in a significant reduction in the level of the harvested power when using the approximated model to design the harvester for a specific range of excitation frequencies.

As for the electromechanical damping, the plotted curves in Figure 13(b) show that the electromechanical damping is maximized for specific values of the load resistance for both approximated and derived models. This region of electrical load resistance over which the electromechanical damping is relatively high coincides with the region over which a significant increase in the coupled frequency occurs. Away from this region, the electromechanical damping coefficient is relatively smaller for both models. However, the approximated model overestimates the electromechanical damping for all values of the electrical load resistance. This big discrepancy can attain an error of about 7% in the region of load resistance over which the electromechanical damping is maximum. Away from this region, the

*Figure 13. Variations of the (a) coupled frequency and (b) electromechanical damping as a function of the load resistance when $D_0$=9.7mm*

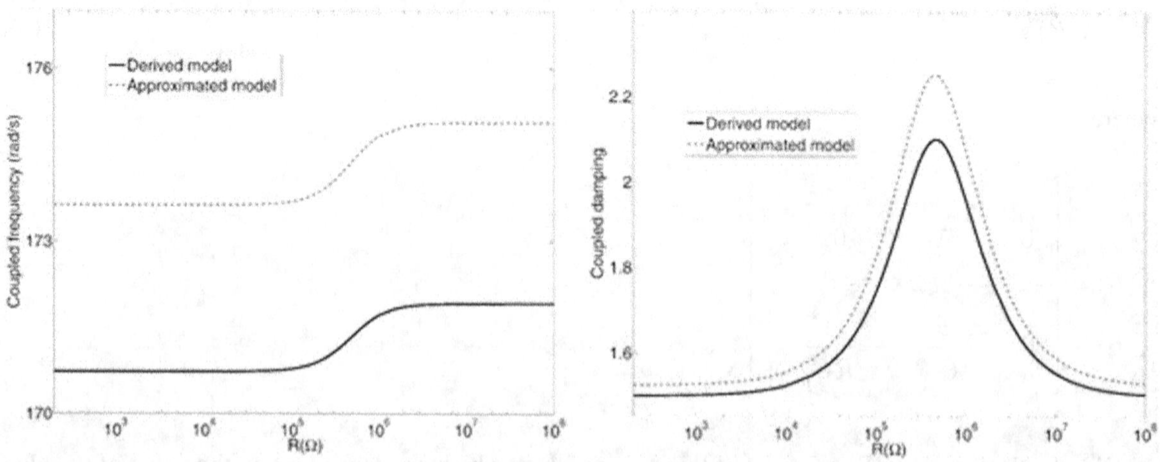


overestimation is lower. This discrepancy in the estimation of the electromechanical damping can lead to erroneous results in the prediction of the tip displacement and hence the associated level of the harvested power when using the approximated model.

## Model Validation

A comparison between the results of the approximated and derived models with the experimental measurements of Tang and Yang (2012) and their used corrected lumped-parameter model are presented. In Table 3, a comparison of the coupled frequency when the electrical load resistance is set equal to 4 $10^5\Omega$ and the spacing magnets distance is equal to 9.7 *mm* is shown. It follows from this table that there is a discrepancy in the coupled frequency when using the lumped-parameter or approximated distributed-parameter models and experimental measurements. In fact, the lumped model used by Tang and Yang (2012) underestimates the experimental value of the coupled frequency with an error of 1.2%. Furthermore, the approximated model overestimates this coupled frequency with an error equal to 2.1%. On the other hand, a very good agreement is obtained when using the derived model that is based on the exact mode shape (error less than 0.4%).

*Table 3. Coupled frequency comparisons between different models and experimental results when 4 $10^5\Omega$ and $D_0$=9.7 mm*

| | Frequency (*Hz*) | % Difference |
|---|---|---|
| Experiment | 27.15 | --- |
| Lumped model | 26.81 | 1.2 |
| Approximated model | 27.73 | 2.1 |
| Derived model | 27.25 | 0.36 |

In these types of harvesters, an under- or over-estimation of the coupled frequency with an error of 1 or 2% can result in wrong predictions of the narrow range of resonant frequencies as well as the levels of the harvested power. To investigate in details the possible erroneous predictions of the harvested power in the resonance region when using approximated models, we first plot, in Figures 14(a) and (b), the linear and logarithmic frequency-response curves of the average harvested power as obtained from different models and the experimental measurements of Tang and Yang (2012) when the electrical load resistance and the spacing magnets distance are, respectively, set equal to 4 $10^5\Omega$ and $D_0$=9.7 *mm* and when the root mean square of the base acceleration is set equal to 1 $m/s^2$. Inspecting these plots, it is noted that there is a left shift discrepancy in the coupled frequency of the harvester and an underestimation in the experimental value of the average harvested power at resonance when using the corrected lumped-parameter model. Table 4 shows error comparisons in the value of the average harvested power between different models and experimental measurements. Clearly, the use of the lumped-parameter model leads to wrong predictions in the average value of the harvested power in the resonance region. These erroneous predictions can result in an overestimation of 226.13% when the excitation frequency is set equal to 26.75 *Hz*. In addition, an underestimation of 58% is obtained when the excitation frequency is set equal to 27.7 *Hz*. Although the error in the coupled frequency is around 1%, a significant under- or over-estimation in the average value of the harvested power are obtained when using this lumped-parameter model. Consequently, a wrong design of the harvester is obtained if the corrected lumped-parameter model is used to represent the governing equations of the harvester.

When using the approximated model, inspecting Figures 14(a) and (b), there is an overestimation in both the coupled frequency and average value of harvested power at resonance. This approximated model leads to erroneous predictions when the excitation frequency is in the resonance region (between 26.75 *Hz* and 27.7 *Hz*), as shown in Table 4. An underestimation of 73.93% in the average value of the harvested power is obtained when the excitation frequency is set equal to 27 *Hz*. Furthermore, when the excitation frequency is set equal to 27.7 *Hz*, an overestimation of 119.3% is obtained in the prediction of the average value of the harvested power. Therefore, both lumped-parameter and approximated distributed-parameter models do not correctly predict the resonance region and the associated values of

*Figure 14. Comparisons of the frequency-response curves: (a) linear and (b) logarithmic between the used models and experimental measurements when $R=4\ 10^5\Omega$, $D_0$=9.7mm and $a_{rms}=1\ m/s^2$*

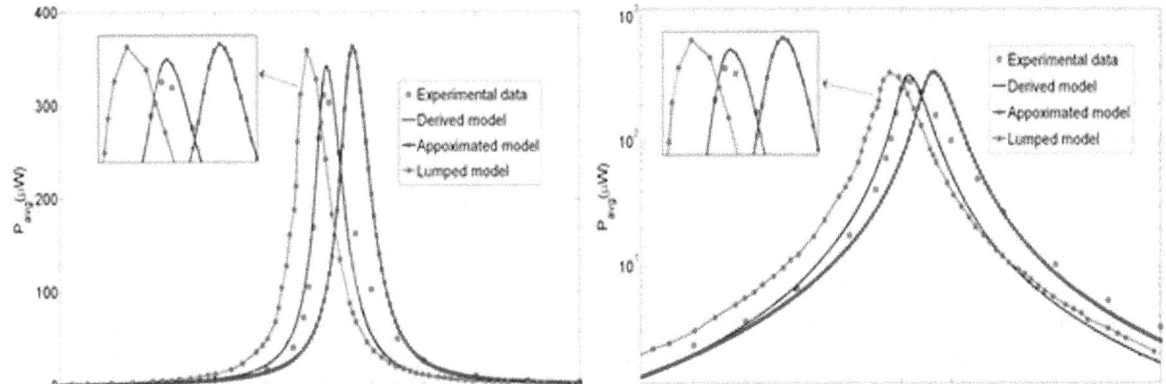

*Table 4. Average harvested power error comparisons between different models and experimental measurements when $R=4\ 10^5\Omega$, $D_0=9.7mm$ and $a_{rms}=1\ m/s^2$*

| | *26.75 Hz* | *27 Hz* | *27.25 Hz* | *27.7 Hz* |
|---|---|---|---|---|
| Lumped model | 226.13% | -9% | -40% | -58% |
| Approximated model | -57.5% | -73.93% | -40% | 119.3% |
| Derived model | ~0% | ~0% | ~~0% | -23% |

the average harvested power although the error in the coupled frequency is less than 2.1%. On contrary, a very good agreement is obtained in both the resonance range and the level of the harvested power when the derived model is used, as shown in Figures 14 (a) and (b) and Table 3.

Inspecting Figure 14(b), a clear softening behavior is obtained in the derived model as well as the experimental measurements. This behavior is well predicted by the derived model. On contrary, the lumped-parameter and approximated models have the same tendency as a linear system and the influence of the attractive magnetic force is not pronounced. The above comparisons between different analytical models and the experimental measurements show that the derived distributed-parameter model based on the exact mode shapes gives accurate predictions compared to the approximated distributed-parameter and lumped-parameter models. In the rest of this paper, only the derived distributed-parameter model is used in order to determine the effects of the length of the piezoelectric material and spacing distance on the natural frequency of the harvester and its performance.

## Effects of the Piezoelectric Layer Length on the Harvester's Response: Tunability

One of the most interesting parameters that can affect the response of the harvester is the length of the attached piezoelectric layer. Changing this length results in a variation in the capacitance of the harvester, the piezoelectric coupling, the natural frequency, the mode shape, and the forcing term. Therefore, the effects of varying the length of the piezoelectric material on the natural frequency and associated mode shape are investigated. This analysis is performed for two different system parameters. The first one has the same parameters as the experimental designed prototype with varying only the length of the piezoelectric sheet. This configuration is referred as first configuration. In the second system, the length of the aluminum beam, the thickness and width of the piezoelectric sheet, and the tip mass are, respectively, set equal to 70 *mm*, 0.356 *mm*, 1 *cm*, and 4.52 *g*. This configuration is referred as second configuration. It should be mentioned that this analysis is performed for very high values of the spacing distance $D_0$ and hence the magnetic force effect can be neglected.

Figure 15 shows the variations of the natural frequency of these two configurations with the length of the piezoelectric sheet. It follows from these plotted curves that a decrease in the length of the piezoelectric material results in a decrease in the value of the natural frequency for both configurations. This is beneficial in terms of obtaining low-frequency piezoelectric energy harvesters and increasing the power density of the associated harvester. It is also noted that there is an optimum value of the piezoelectric length in which the natural frequency is maximized. This value is near 53 *mm* for the first configuration and 60 *mm* for the second configuration. Furthermore, the rate of variation of the natural frequency near these optimum values is very small. However, for small length values of the piezoelectric sheet, the rate

*Figure 15. Variation of the structural natural frequency as a function of the piezoelectric length for the first and second configurations*

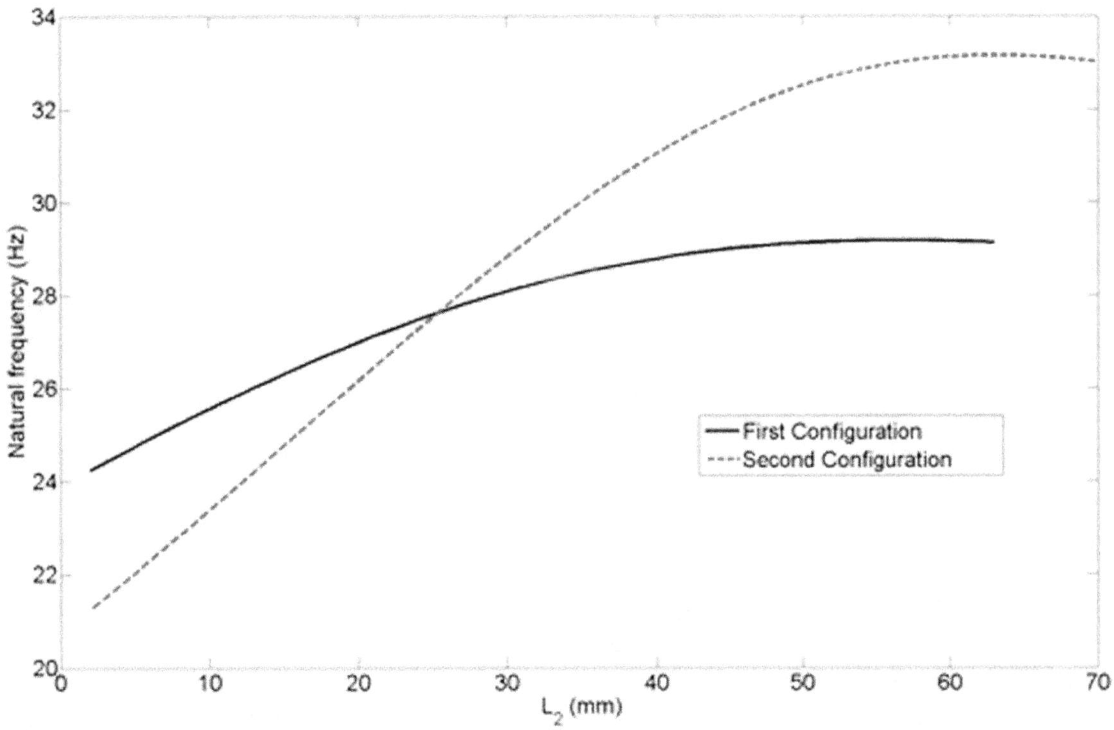


of variation of the natural frequency is important. Based on this analysis, depending on the available excitation frequency, the harvester can be passively tuned to match its natural frequency to this available excitation frequency.

The plotted curves in Figures 16(a) and (b) show the variations of the first mode shape with the length of the piezoelectric sheet for both configurations, respectively. Inspecting these plots, it is noted that there is a significant change in the mode shape when varying the length of the piezoelectric sheet and the difference between the mode shapes varies depending on the concerned region.

Frequency-response curves of the harvested power for different values of the piezoelectric lengths for the first and second configurations are presented in (Abdelkefi et al., 2013). They reported that there is a compromise between the length of the piezoelectric material, the electrical load resistance, and the available excitation frequency to design enhanced piezoelectric energy harvesters.

## Effects of the Magnets Spacing Distance on the Harvester's Response: Tunability and Broadband Response

Using the derived nonlinear electromechanical model, the effects of the spacing distance between the two attractive magnets on the static position of the harvester, its natural frequencies, and its performance are investigated. Using the system's parameters given in (Abdelkefi & Barsallo, 2014), the plotted curves in Figures 17(a) and (b) show the variations of the static deflection and the natural frequency of the harvester as a function of the spacing distance of the two attractive magnets. It follows from Figure

*Figure 16. Variations of the mode shapes as function of the length of the piezoelectric material: (a) first configuration and (b) second configuration*

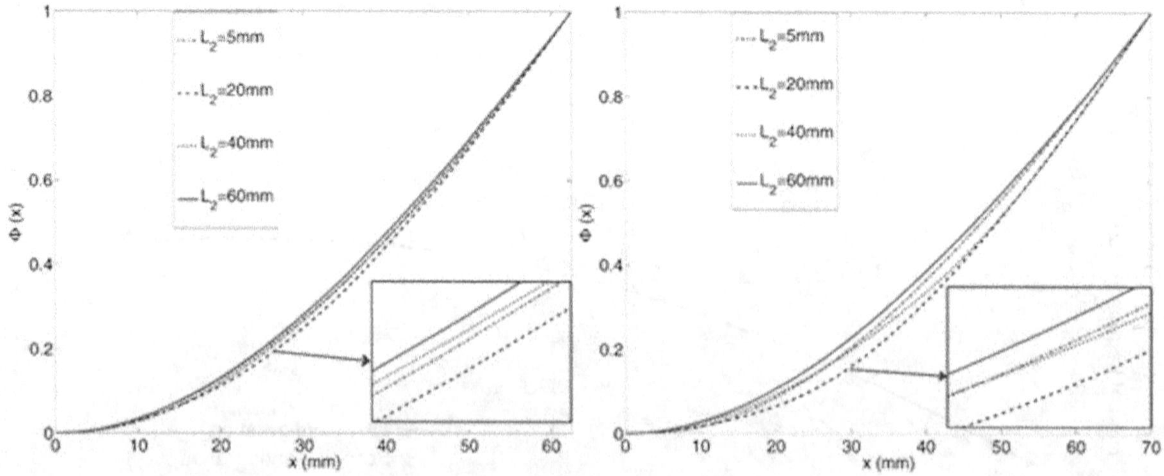


17(a) that an increase in the spacing distance $D_0$ is accompanied by a decrease in the static tip deflection value. This is due to the fact that the magnet force effect is reduced when the spacing distance between the two attractive magnets is increased. It should be mentioned that these results are shown for spacing distance values larger than 7 *mm*.

Considering the same range of spacing distance values and using the performed eigenvalue problem analysis (Equations (44)-(52)), the effects of the spacing distance between the two attractive magnets on the structural natural frequency of the harvester are presented in Figure 17(b). Clearly, three different regions are detected. For small values of the spacing distance ($D_0$ <12 *mm*), a sudden decrease in the natural frequency of the harvester is obtained which gives the opportunity to design low-frequency piezoelectric energy harvesters. In the second region (12 *mm*<$D_0$ <20 *mm*), the rate of change in the

*Figure 17. Variations of the (a) static position and (b) fundamental natural frequency of the harvester as a function of the spacing distance between the two attractive magnets*

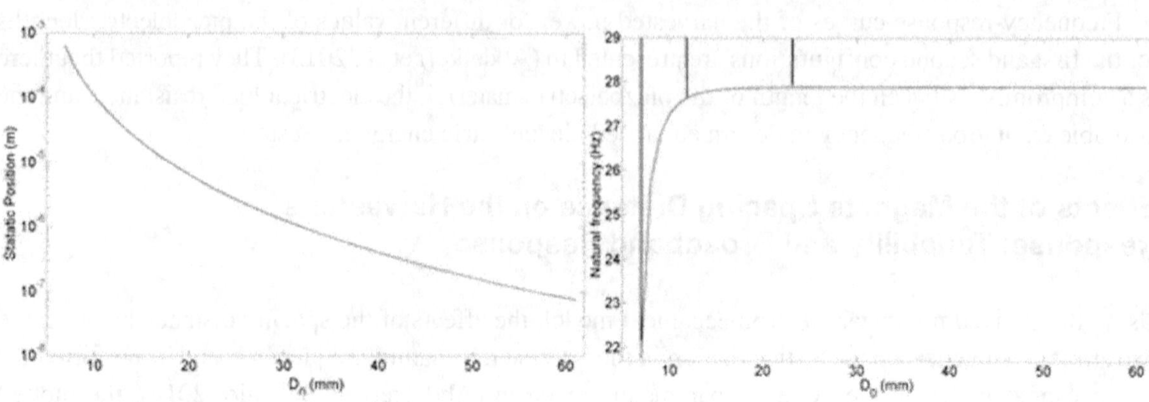

structural natural frequency is very low. For higher values of the spacing distance ($D_0$ >20 *mm*), the fundamental natural frequency of the harvester does not change, indicating the disappearance of the attractive magnetic force.

Next, the effects of the spacing distance $D_0$ on the performance of the harvester are investigated. To this end, the frequency-response curves of the average harvested power for different values of the spacing distance between the two attractive magnets are plotted in Figure 18 when the electrical load resistance and the root mean square value of base acceleration are set, respectively, equal to $10^6$ Ω and 1 $m/s^2$. It follows from this plot that a decrease in the spacing distance between the two attractive magnets is accompanied by a decrease in the resonant frequency. This is expected because it was demonstrated that the fundamental natural frequency decreases as the spacing distance between the two magnets is decreased. This result is very beneficial to tune the natural frequency of the harvester and match it to the available excitation frequency in the environment.

Furthermore, it is clear that a softening behavior takes place due to the presence of the nonlinear magnetic force. The presence of this behavior is especially important when the spacing distance $D_0$ is decreased. This is due to the fact that the nonlinear magnetic force becomes stronger when the spacing distance $D_0$ is decreased. To characterize this softening behavior, an increasing and decreasing sweep of the excitation frequency are performed. Clearly, wider hysteresis regions take place when the spacing distance becomes smaller. Therefore, this softening behavior gives the opportunity for a larger band-width of frequencies at which the system can generate high levels of the harvested power. To conclude, efficient energy harvesters can be designed to generate energy at broadband low-frequency excitations by choosing the best spacing distance between the two magnets.

*Figure 18. Frequency-response curves of the harvested power for various spacing distance values when using different initial conditions*

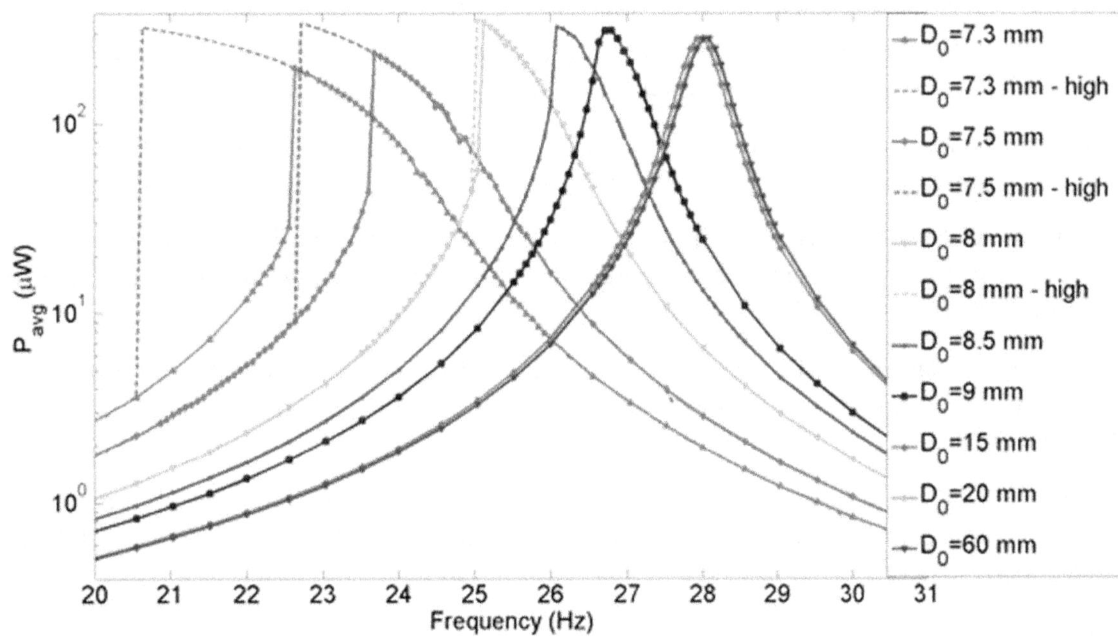

## CONCLUSION

Two concepts to design efficient low-frequency piezoelectric energy harvesters have been investigated, namely, variable-shaped piezoelectric energy harvesters and piezomagnetoelastic energy harvesters. Linear and nonlinear electromechanical models are developed and validated. For the variable-shaped energy harvester, it was shown that truncated configurations give the opportunity to design low-frequency systems with higher levels of harvested power than their rectangular counterparts. As for the piezomagnetoelastic energy harvester, it was demonstrated that there is a compromise between the spacing distance between the two magnets, the length of the piezoelectric material, the electrical load resistance, and the available excitation frequency to design enhanced low-frequency piezoelectric energy harvesters.

## REFERENCES

Abdelkefi, A. (2012). *Global nonlinear analysis of piezoelectric energy harvesting from ambient and aeroelastic vibrations* (PhD Dissertation). Virginia Tech, VA.

Abdelkefi, A., & Barsallo, N. (2014). Comparative modeling of low-frequency piezomagnetoelastic energy harvesters. *Journal of Intelligent Material Systems and Structures*. doi:10.1177/1045389X14523860

Abdelkefi, A., Barsallo, N., Tang, L., Yang, Y., & Hajj, M. R. (2013). Comparative modeling of low-frequency piezomagnetoelastic energy harvesters. *Journal of Intelligent Material Systems and Structures*. doi:10.1177/1045389X13507638

Abdelkefi, A., & Ghommem, M. (2013). Piezoelectric energy harvesting from morphing wing motions for micro air vehicles. *Theoretical & Applied Mechanics Letters*, *3*(5), 052001. doi:10.1063/2.1305204

Abdelkefi, A., Najar, F., Nayfeh, A. H., & Ben Ayed, S. (2011). An energy harvester using piezoelectric cantilever beams undergoing coupled bending–torsion vibrations. *Smart Materials and Structures*, *20*(11), 115007. doi:10.1088/0964-1726/20/11/115007

Abdelkefi, A., Nayfeh, A. H., & Hajj, M. R. (2012a). Modeling and analysis of piezoaeroelastic energy harvesters. *Nonlinear Dynamics*, *67*(2), 925–939. doi:10.1007/s11071-011-0035-1

Abdelkefi, A., Nayfeh, A. H., & Hajj, M. R. (2012b). Global nonlinear distributed-parameter model of parametrically excited piezoelectric energy harvesters. *Nonlinear Dynamics*, *67*(2), 1147–1160. doi:10.1007/s11071-011-0059-6

Abdelkefi, A., Nayfeh, A. H., & Hajj, M. R. (2012c). Effects of nonlinear piezoelectric coupling on energy harvesters under direct excitation. *Nonlinear Dynamics*, *67*(2), 1221–1232. doi:10.1007/s11071-011-0064-9

Abdelkefi, A., & Nuhait, A. O. (2013). Modeling and performance analysis of cambered wing-based piezoaeroelastic energy harvesters. *Smart Materials and Structures*, *22*(9), 095029. doi:10.1088/0964-1726/22/9/095029

Alamin Dow, A. B., Schneider, M., Koo, D., Rubaye, H. A., Bittner, A., Schmid, U., & Kherani, N. (2012). Modeling the performance of a micromachined piezoelectric energy harvester. *Microsystem Technologies*, *18*(7-8), 1035–1043. doi:10.1007/s00542-012-1436-x

Anton, S. R., & Sodano, H. A. (2007). A review of power harvesting using piezoelectric materials (2003-2006). *Smart Materials and Structures*, *16*(3), 1–21. doi:10.1088/0964-1726/16/3/R01

Arnold, D. (2007). Review of microscale magnetic power generation. *IEEE Transactions on Magnetics*, *43*(11), 3940–3951. doi:10.1109/TMAG.2007.906150

Barton, D., Burrow, S., & Clare, L. (2010). Energy harvesting from vibrations with a nonlinear oscillator. *Journal of Vibration and Acoustics*, *132*(2), 0210091. doi:10.1115/1.4000809

Ben Ayed, S., Abdelkefi, A., Najar, F., & Hajj, M. R. (2014). Design and performance of variable-shaped piezoelectric energy harvesters. *Journal of Intelligent Material Systems and Structures*, *25*(2), 174–186. doi:10.1177/1045389X13489365

Benasciutti, D., Moro, L., Zelenika, S., & Brusa, E. (2010). Vibration energy scavenging via piezoelectric bimorphs of optimized shapes. *Microsystem Technologies*, *16*(5), 657–668. doi:10.1007/s00542-009-1000-5

Bryant, M., & Garcia, E. (2009). Energy harvesting: A key to wireless sensor nodes. *Proceedings of the Society for Photo-Instrumentation Engineers*, *7493*, 74931W, 74931W-8. doi:10.1117/12.845784

Challa, V., Prasad, M., Shi, Y., & Fisher, F. (2008). A vibration energy harvesting device with bidirectional resonance frequency tunability. *Smart Materials and Structures*, *75*, 1–10.

Clair, D., Bibo, A., Sennakesavababu, V., & Daqaq, M. F. (2010). A scalable concept for micro-power generation using flow-induced self-excited oscillations. *Applied Physics Letters*, *96*, 144103. doi:10.1063/1.3385780

Cottone, F., Vocca, H., & Gammaitoni, L. (2009). Nonlinear energy harvesting. *Physical Review Letters*, *102*(8), 080601. doi:10.1103/PhysRevLett.102.080601 PMID:19257728

Dai, H. L., Abdelkefi, A., & Wang, L. (2014). Theoretical modeling and nonlinear analysis of piezoelectric energy harvesting from vortex-induced vibrations. *Journal of Intelligent Material Systems and Structures*, *25*(14), 1861–1874. doi:10.1177/1045389X14538329

Daqaq, M. F. (2012). On intentional introduction of stiffness nonlinearities for energy harvesting under white Gaussian excitations. *Nonlinear Dynamics*, *69*(3), 1063–1079. doi:10.1007/s11071-012-0327-0

DuToil, N. E., Wardle, B. L., & Kim, S. G. (2005). Design considerations for MEMS-scale piezoelectric mechanical vibration energy harvesters. *Integrated Ferroelectrics*, *71*(1), 121–160. doi:10.1080/10584580590964574

Elvin, N. G., & Elvin, A. A. (2011). Effects of axial forces on cantilever piezoelectric resonators for structural energy harvesting. *Strain*, *47*, 153–157. doi:10.1111/j.1475-1305.2010.00751.x

Erturk, A. (2009). *Electromechanical modeling of piezoelectric energy harvesters* (PhD Dissertation). Virginia Tech, VA.

Erturk, A., & Inman, D. J. (2008). On mechanical modeling of cantilevered piezoelectric vibration energy harvesters. *Journal of Intelligent Material Systems and Structures*, *19*(11), 1311–1325. doi:10.1177/1045389X07085639

Erturk, A., & Inman, D. J. (2011). Broadband piezoelectric power generation on high-energy orbits of the bistable Duffing oscillator with electromechanical coupling. *Journal of Sound and Vibration*, *330*(10), 2339–2353. doi:10.1016/j.jsv.2010.11.018

Goldschmidtboeing, F., & Woias, P. (2008). Characterization of different beam shapes for piezoelectric energy harvesting. *Journal of Micromechanics and Microengineering*, *18*(10), 104013. doi:10.1088/0960-1317/18/10/104013

Harne, R. L., & Wang, K. W. (2013). A review of the recent research on vibration energy harvesting via bistable systems. *Smart Materials and Structures*, *22*(2), 023001. doi:10.1088/0964-1726/22/2/023001

Karami, A. (2011). *Micro-scale and nonlinear vibrational energy harvesting* (PhD Dissertation). Virginia Tech, VA.

Karami, A., & Inman, D. J. (2011). Equivalent damping and frequency change for linear and nonlinear hybrid vibrational energy harvesting systems. *Journal of Sound and Vibration*, *330*(23), 5583–5597. doi:10.1016/j.jsv.2011.06.021

Karami, A., & Inman, D. J. (2012). Powering pacemakers from heartbeat vibrations using linear and nonlinear energy harvesters. *Applied Physics Letters*, *100*(4), 042901. doi:10.1063/1.3679102

Lui, H., Lee, C., Kobayashi, T., Tay, C. J., & Quan, C. (2012). Investigation of a MEMS piezoelectric energy harvester system with a frequency-widened-bandwidth mechanism introduced by mechanical stoppers. *Smart Materials and Structures*, *21*(3), 035005. doi:10.1088/0964-1726/21/3/035005

Mann, B. P., & Sims, N. D. (2009). Energy harvesting from the nonlinear oscillations of magnetic levitation. *Journal of Sound and Vibration*, *319*(1-2), 515–530. doi:10.1016/j.jsv.2008.06.011

Masana, R., & Daqaq, M. F. (2011a). Electromechanical modeling and nonlinear analysis of axially-loaded energy harvesters. *Journal of Vibration and Acoustics*, *133*(1), 011007. doi:10.1115/1.4002786

Masana, R., & Daqaq, M. F. (2011b). Relative performance of a vibratory energy harvester in mono- and bi-stable potentials. *Journal of Sound and Vibration*, *330*(24), 6036–6052. doi:10.1016/j.jsv.2011.07.031

Mateu, L., & Moll, F. (2005). Optimum piezoelectric bending beam structures for energy harvesting using shoe inserts. *Journal of Intelligent Material Systems and Structures*, *16*(10), 835–845. doi:10.1177/1045389X05055280

Pellegrini, S. P., Tolou, N., Schenk, M., & Herder, J. L. (2013). Bistable vibration energy harvesters: A review. *Journal of Intelligent Material Systems and Structures*, *24*(11), 1303–1312. doi:10.1177/1045389X12444940

Quinn, D., Triplett, L., Vakakis, D., & Bergman, L. (2011). Comparing linear and essentially nonlinear vibration-based energy harvesting. *Journal of Vibration and Acoustics*, *133*(1), 011001. doi:10.1115/1.4002782

Roundy, S., Leland, E. S., Baker, J., Carleton, E., Reilly, E., Lai, E., & Sundararajan, V. et al. (2005). Improving power output for vibration-based energy scavengers. *IEEE Pervasive Computing / IEEE Computer Society [and] IEEE Communications Society*, *4*(1), 28–36. doi:10.1109/MPRV.2005.14

Roundy, S., & Wright, P. K. (2005). A piezoelectric vibration based generator for wireless electronics. *Journal of Intelligent Material Systems and Structures*, *16*, 809–823. doi:10.1177/1045389X05054042

Shahruz, S. M. (2006). Design of mechanical band-pass filters for energy scavenging. *Journal of Sound and Vibration*, *292*(3-5), 987–998. doi:10.1016/j.jsv.2005.08.018

Sodano, H., Park, G., & Inman, D. J. (2004). A review of power harvesting from vibration using piezoelectric materials. *Journal of Shock Vibration Digest*, *36*(3), 197–205. doi:10.1177/0583102404043275

Song, H. J., Choi, Y.-T., Purekar, A. S., & Wereley, N. M. (2009). Performance evaluation of multi-tier energy harvesters using macro-fiber composite patches. *Journal of Intelligent Material Systems and Structures*, *20*(17), 2077–2088. doi:10.1177/1045389X09347017

Song, H. J., Choi, Y.-T., Wereley, N. M., & Purekar, A. S. (2010). Energy harvesting devices using macro-fiber composite materials. *Journal of Intelligent Material Systems and Structures*, *21*(6), 647–658. doi:10.1177/1045389X10361633

Stanton, S. C., McGehee, C. C., & Mann, B. P. (2009). Reversible hysteresis for broadband magneto-piezoelastic energy harvesting. *Applied Physics Letters*, *95*(17), 174103. doi:10.1063/1.3253710

Stanton, S. C., McGehee, C. C., & Mann, B. P. (2010). Nonlinear dynamics for broadband energy harvesting: Investigation of a bistable piezoelectric inertial generator. *Physica D. Nonlinear Phenomena*, *239*(10), 640–653. doi:10.1016/j.physd.2010.01.019

Tang, L. H., & Yang, Y. (2012). A nonlinear piezoelectric energy harvester with magnetic oscillator. *Applied Physics Letters*, *101*(9), 094102. doi:10.1063/1.4748794

Tomasiello, S. (1998). Differential quadrature method: Application to initial-boundary-value problems. *Journal of Sound and Vibration*, *218*(4), 573–585. doi:10.1006/jsvi.1998.1833

Yan, Z., & Abdelkefi, A. (2014). Nonlinear characterization of concurrent energy harvesting from galloping and base excitations. *Nonlinear Dynamics*, *77*(4), 1171–1189. doi:10.1007/s11071-014-1369-2

Zhao, L. Y., Tang, L. H., & Yang, Y. W. (2013). Comparison of modeling methods and parametric study for a piezoelectric wind energy harvester. *Smart Materials and Structures*, *22*(12), 125003. doi:10.1088/0964-1726/22/12/125003

## KEY TERMS AND DEFINITIONS

**Distributed-Parameter Model:** A mathematical representation for infinite-degree-of-freedom dynamical systems.

**Energy Harvesting:** The concept of generating energy from unused mechanical vibrations.

**Piezoelectricity:** The ability of certain materials to generate an AC voltage when subjected to mechanical stress.

**Low-Frequency Excitations:** Types of excitations with high period including walking, breathing, ocean waves, etc.

**Magnetic Coupling:** The interaction between two magnets or more that affects the behavior of a dynamical system.

**Nonlinear Dynamics:** The study of dynamical systems having nonlinear effects, such as magnetic force, geometric and inertia nonlinearities.

**Variable-Shaped Beams:** Cantilever beams with variable beam width.

# Section 3
# Thermal and Solar Systems for Energy Harvesting

# Chapter 8
# Energy Harvesting and Energy Conversion Devices Using Thermoelectric Materials

**Mihail O. Cernaianu**
*Horia Hulubei National Institute of Physics and Nuclear Engineering (IFIN-HH), Romania*

**Aurel Gontean**
*"Politehnica" University Timisoara, Romania*

## ABSTRACT

*The authors propose in this chapter an original, self-sustainable, power supply system for wireless monitoring applications that is powered from an energy harvesting device based on thermoelectric generators (TEGs). The energy harvesting system's purpose is to gather the waste heat from low temperature sources (<90°C), convert it to electrical energy and store it into rechargeable batteries. The energy harvesting system must be able to power a so-called condition monitoring system (CMS) that is used for the monitoring of heat dissipation equipment. The setup used for measurements (including mechanical details) and the experiments are described along with all the essential results of the research. The electronic system design is emphasized and various options are discussed.*

## INTRODUCTION

Energy harvesting and renewables energies is a major field of interest today for both academic and industrial companies. The chapter deals with a detailed analysis of TEMs (Thermoelectric Materials) and their applications. The internal parameters of a TEM were experimentally determined and based on the results an improved SPICE model was developed. The model is based on the classical governing thermodynamic equations where the parameters dependency of temperature and also the parasitic intrinsic elements were taken into account. A complete SPICE model was proposed and implemented, in which both the thermal and electrical components of a TEM and the complementary components

DOI: 10.4018/978-1-4666-8254-2.ch008

were modelled, along with the simulation of the convective heat transfer phenomena. The modelling and simulation parts are not presented in this chapter, but will be referred for a better understanding of the functioning of the system and its performance.

The chapter presents an in detail description of the mechanical subsystem designed and built by the authors. The subsystem purpose is to dissipate with the maximum efficiency the heat from the hot side of a thermoelectric module into the environment where the entire system is supposed to run. For this, a solution based on multiple cooling pipes (microchannel configuration) was chosen, that together with a closed water cooling circuit and a heat exchanger generates a thermo-syphon which requires no electrical energy to dissipate the heat. Another solution, based on heat pipes was also designed and tested for performance, yielding a good solution where a compact design is needed.

The electronics subsystem designed and built by the authors comprises in converters, storage stages for the harvested energy and an energy routing stage that can deliver the energy to different other subsystems (uC, sensors, Wi-Fi communication devices). The design followed an in depth process of simulations, experiments and optimizations in order to deliver the expected results.

## Renewable Energy Review and Trends

The need for renewable energy sources, other than fossil fuels, has become one of the most critical current issues that led to a significant research into alternative energy sources and energy conversion techniques.

The Renewable Energy Policy Network for the 21st century (REN21, 2011) convenes international multi-stakeholder leaders to enable a rapid global transition to renewable energy and is formed of a numerous steering committees from various countries around the world (US, Japan, Germany, Austria, UK, China, etc). Into their 2012 report, REN21 states that the total renewable power capacity, not including hydropower rose from 315 GW in 2010 to 390 GW in 2011, showing an important growth of 23% (REN21, 2012). This demonstrates the importance paid to this subject by countries all around the world and that efforts are being made to find alternative solutions to natural resources (e.g. petroleum and coal) that are getting depleted.

World primary energy consumption grew by 2.5% in 2011, as presented in Figure 1 by the BP Statistical review of world energy report 2012 (BP, 2012).

The prediction made by the "Science of sustainability" 2006 summary report (Figure 2), that the global energy consumption will reach 12 billion toe (tons of oil equivalent) proved to be correct (as represented in Figure 1) and is expected to increase by 60% until 2030, compared to the value from 1971, as pointed in (Summary Report, 2006) and represented in Figure 2. From this high amount of energy, a large portion is dissipated into ambient air as unused heat which leads to the necessity of techniques capable to recover the wasted heat.

The REN 21 status report states that thermoelectric generators (TEGs), devices that convert heat to electricity, are becoming cost effective and allow wood burning stoves to generate both heat and electricity enabling them to operate fans for improved combustion or provide electricity for other applications. In most applications (energy transportation, factories, thermal engines, electric generators, etc) enormous amounts of heat are wasted. Bitschi (2009) shows that this type of energy is difficult to use due to its low temperature although the amount is very large.

*Figure 1. World energy consumption in million tones oil equivalent, from '86 – 2011 (BP, 2012)*

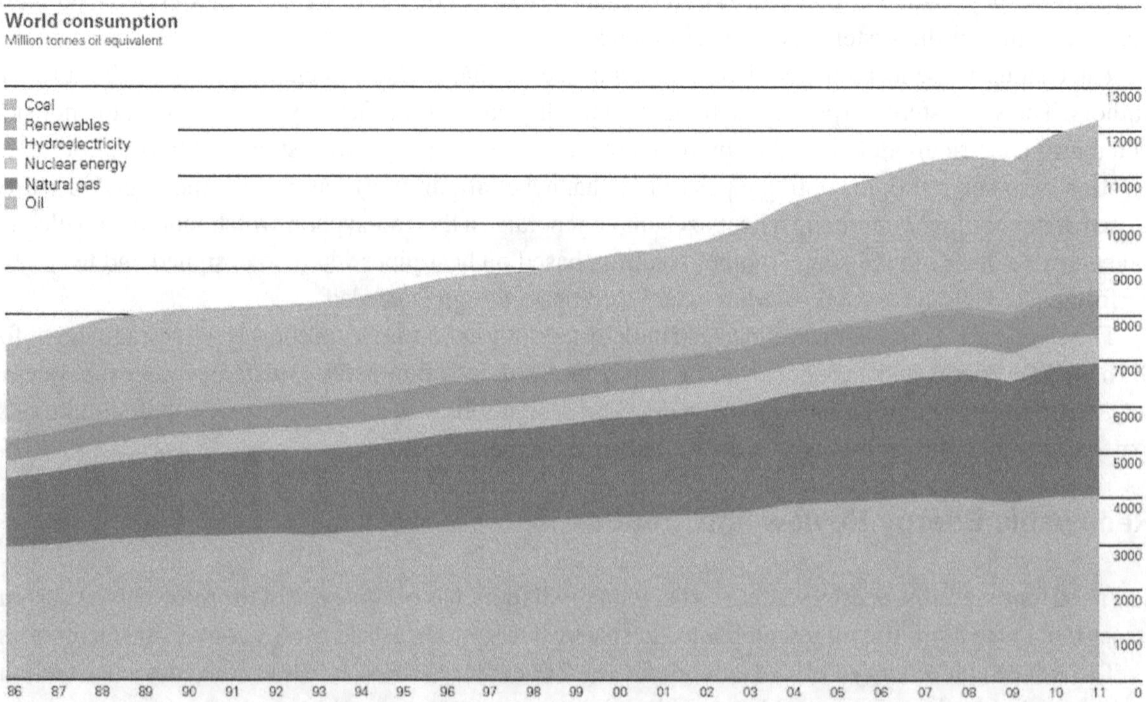


*Figure 2. Global primary energy consumption until 2030*

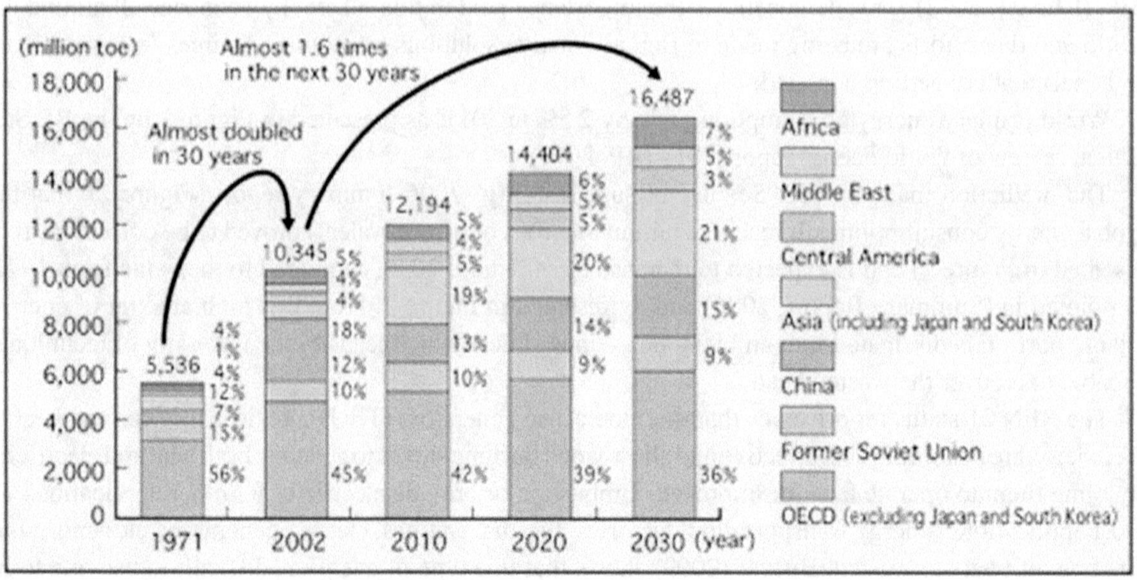

## TEG REVIEW

The TEGs employ direct thermal to electrical energy conversion, without the intermediate step of kinetic energy. With no moving parts, this technology offers an alternative of high potential, Bitschi (2009) claims that the improvements in material sciences and the progress of nanotechnology bring thermoelectric materials and therefore thermoelectric converters to renewed significance. The drawback of using TEG solutions in energy harvesting devices is the need for cooling systems to create the necessary temperature gradient that is proportional to the output power of the TEG. However, Yu and Zhao (2007) say that in practical situations a real thermoelectric generator combines thermoelectric modules with the specific heat exchangers in very large scale installations.

A TEG generates electricity due to the later called Seebeck effect that was first discovered in 1821 by the German physicist Thomas Johann Seebeck. Given a uniform temperature in a closed circuit that is formed by metals, the electromotive contact forces compensate each other and their algebraic sum is zero. Seebeck discovered that this compensation does not persist and that an electromotive force appears if one of the contact points between two metals has another temperature than the others, as Rowe (2006) explains. Conversely, the Peltier effect is the presence of heat at the contact region of two metals when they are electrified.

Conventional metallic thermocouples (Figure 3a) are made from metal or metal alloys. They generate small voltages, typically tens of micro volts per degree temperature difference when its junction is located in a hot or cold environment compared to the ambient, and generate small amounts of electrical power when connected to a resistive load. They are used extensively as sensors, in the measurement of temperature. The thermocouple also operates as a heat pump by the Peltier effect when a voltage

*Figure 3. Different thermoelectric devices: a) Conventional metal alloy thermocouple b) Bulk semiconductor c) Miniature thin film, Rowe, (2006).*

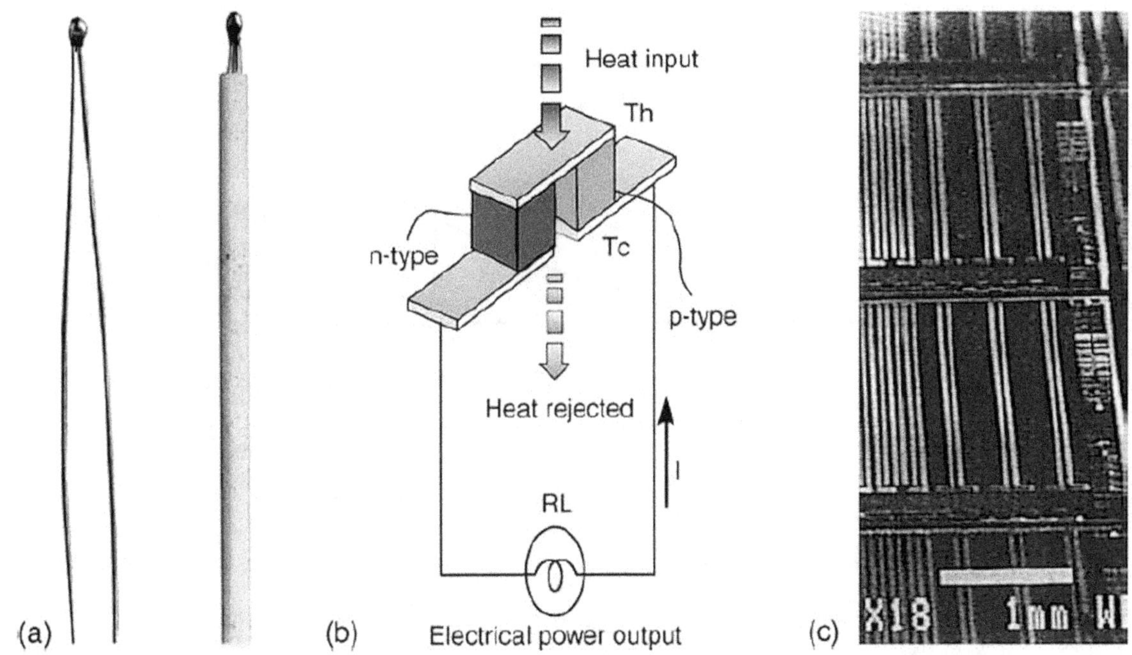

source is connected across the open ends so that an electric current is driven through the couple. Modern thermocouples benefit from the advances made in semiconductors materials. The material properties and geometry have been tailored specifically to meet the intended application requirements and possess Seebeck coefficients of hundreds of micro volts per degree, as seen in Rowe (2006).

In bulk devices for electricity generation or cooling applications, thermocouples have a typical geometry that consists of two ingot-shaped pellets (thermo-elements) of semiconducting material having dimensions in the order of millimeters connected at one end with an electrically conducting metal strap.

In the beginning, the thermoelectric materials were made of two joined metal components e.g. copper – constantan that has a Seebeck coefficient of approximately 41μV/K. Later on, the semiconductors became the choice for further development of thermoelectric materials. The reason can be explained by the help of Figure 4.

*Figure 4. Dependence of Seebeck coefficient, electrical conductivity and thermal conductivity on concentration of free carriers, Rowe (2006)*

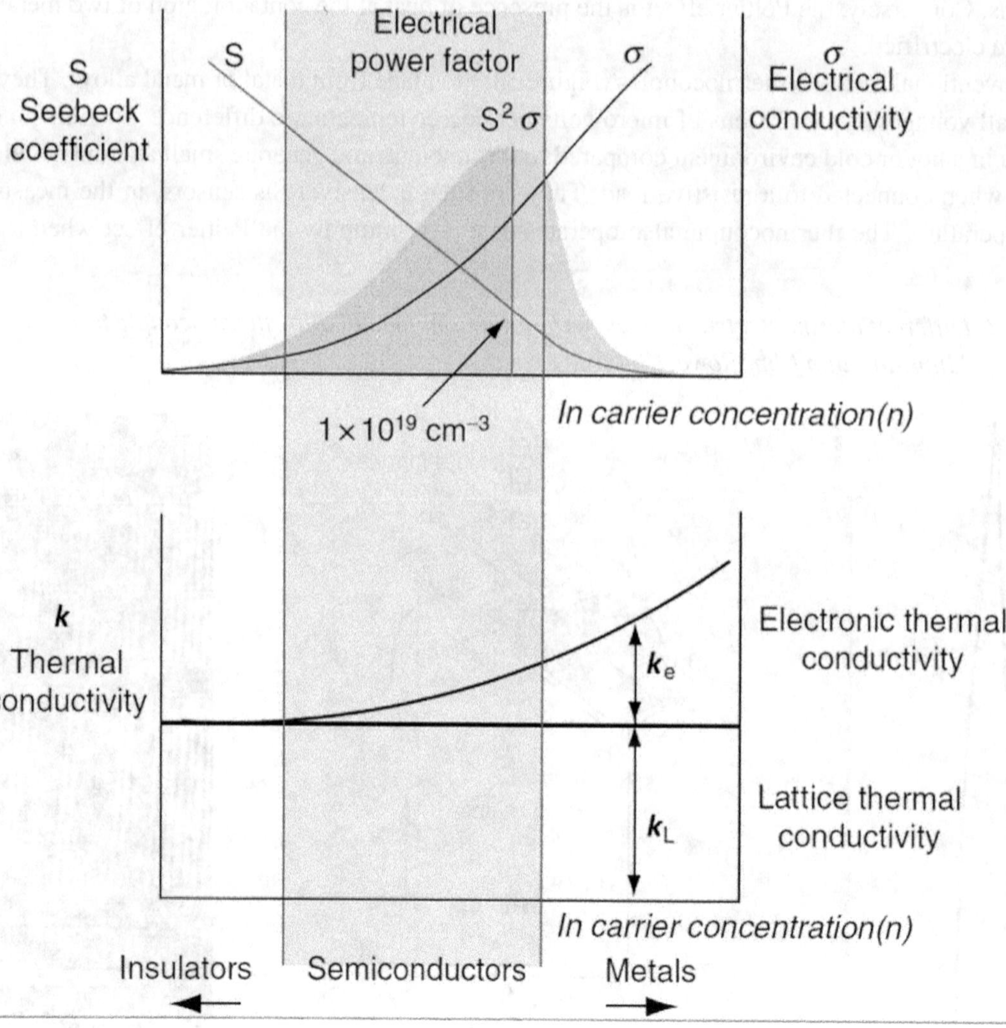

In order to obtain high ZT thermoelectric materials all the three parameters that comprise the ZT figure of merit must be improved in certain ways. The electrical conductivity σ is given as the product of the concentration and mobility of charge carriers. This parameter is high with metals, very low with insulators and has intermediate values in semiconductors. The $S^2\sigma$ factor, called "electrical power factor" has been introduced to facilitate the optimization of the thermoelectrical material. It can be seen that in semiconductors the Seebeck coefficient falls with carrier concentration while the electrical conductivity rises. The best values are indicated to be found when the carrier concentration reaches $10^{19}/cm^3$. Regarding the thermal conductivity, in semiconductors the electronic thermal conductivity is increasing with the carrier concentration while in insulators it has the lowest value. High efficient thermoelectric materials must have high electrical conductivity and low thermal conductivity. Therefore, semiconductor materials must be used. Rowe (2006) reports that the researcher's efforts are focused on reducing the lattice thermal conductivity by using crystal structures that conduct heat as glass and electricity like a crystal "phonon glass-electronic crystals".

In 2012, the company Hi-Z Technology presented a report showing the ZT figure of merit for the best known thermoelectric materials versus temperature (Figure 5). It can be observed that high temperature modules have been obtained that also exhibit high ZT.

In order to decrease the lattice thermal conductivity, other researchers are using nanocomposite structures and although high ZT materials have been achieved, Minnich (2008) claims they are not ready to be commercialized due to the high fabrication costs.

Thermoelectric technology has been used since 1930's as an alternative to batteries in devices such as gas operated radios and home generating stations, as explained by McCoy (2012). Later on, radio receivers were powered by TEGs that were heated from kerosene lamps. As technology evolved, different implementations were made where TEGs were used to recover the heat generated by the exhaust pipes of trucks and cars, as stated by Schock, et al. (2011) and Crane (2012).

A few watch companies (Citizen, Seiko) implemented in their products in the late 90's TEGs for harvesting the body heat and due to the difference between ambient air and body to generate the necessary energy to power the watch. The watches, however, were not appreciated very much, their usage being limited to colder regions of the world. Snyder (2008) presents the Seiko Thermic wristwatch design,

*Figure 5. Different alloys figure of merit as a function of temperature, McCoy (2012)*

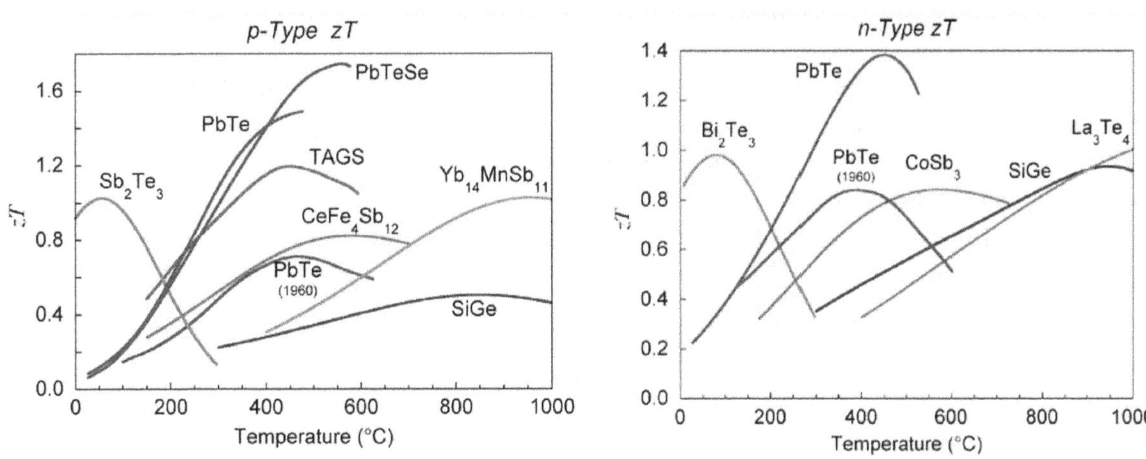

where microfabricated thermoelectric devices are used. The waste heat energy recovery from the human body is now again into attention, as presented in the 2011 Engineering and Physical Sciences Research Council 2011 report.

BMW received an award in 2008 for the development of a prototype TEG system (McCoy, 2012) that was able to recover 300W electric power from the engine exhaust system. With the advances made in material technology the power has been raised in 2011 to more than 600W as reported by Mazar (2012) Eder, Linde (2011). Amerigon, now Gentherm, a well-known thermoelectric manufacturer introduced the climate control seat technology based on TEMs that equipped the high end class of 2009 Jaguar XF mid-size sport cars. The company is also delivering working prototypes for BMW and Ford to improve the waste heat recovery. The concept of seat climate control based on TECs is also presented by Meisner (2011).

A commercial usage of TEMs can be found in portable refrigeration boxes where numerous manufacturers are using thermoelectric coolers in their designs.

Also, different research directions are presented by Jovanovic, Ghamaty and Bass (2006), Mateu, Codrea, Lucas, Pollak, and Spies (2006), and Russel (2011), where wireless TEG powered sensor nodes and TEC based cooling solutions for electronic packaging are investigated.

TEGs are also used in devices called radioisotope thermoelectric generators (RTEGs) that obtain their power by converting the heat released by the radioactive decay into electricity through the Seebeck effect. These devices have been used to power satellites and space probes over time (Apollo, Voyager I&II). One of the latest applications is powering the “Curiosity” Mars rover that was launched in 2011. The particular one used in this rover is a special case of RTEG developed by NASA and the DOE and is called a Multi-Mission Radioisotope Thermoelectric Generator (MMRTG). The heat from the natural radioactive decay of the inside plutonium 238 dioxide power source is applied to the hot thermocouple junction, while the cold junction is connected to a radiator that stands in the cold of outer space, as Taranovich (2012) explains.

## PARASITIC ELEMENTS IMPORTANCE IN TEMS

Due to the fact that in bulk devices the semiconductor pellets of a TEM are connected with electrically conducting metal straps and taking into consideration the difference between the *p* and *n* type semiconductor materials, parasitic components appear like in any semiconductor based device.

The overall equivalent circuit proposed by the author for a bulk thermoelectric module device is represented in Figure 6, where $R_m$, $L_{peq}$ and $C_{peq}$ represent the overall equivalent elements. Because of the Seebeck effect that appears only under a temperature gradient, the equivalent circuit also has an intrinsic power source, in series with the internal resistance and dependent on the Seebeck coefficient and the temperature difference.

Equation (1) offers the generated voltage through the Seebeck effect:

$$V_g = S \cdot (T_H - T_C) \tag{1}$$

*Figure 6. Schematic of the electrical network containing parasitic components*

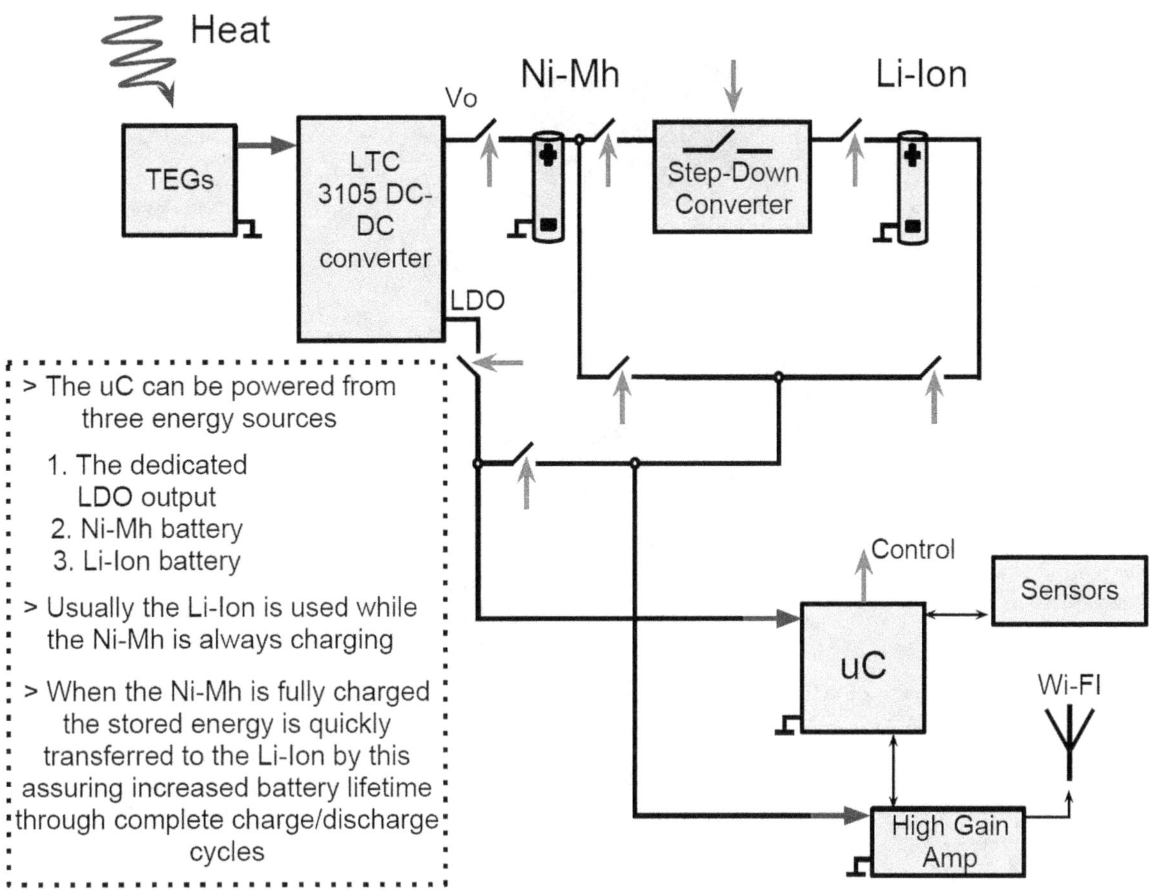


where: $S$ represents the Seebeck coefficient in [V/K], $T_H$ and $T_C$ the hot side and cold side temperatures respectively. However, if the two sides of the TEM are kept at the same temperature difference, the Seebeck generator will give no output voltage.

The circuit presented in Figure 7 refers only to the electrical part of a TEM, because only the electrical parasitic components are important for developing energy harvesting electronic circuits.

The TEMs intrinsic parasitic components are important when developing DC – DC converters and energy harvesting systems. These converters work with a switching frequency of 50 - 100 kHz (like the LTC3105 DC – DC converter) or even higher ($\geq$1 MHz) if the external L and C components need to have low values and dimensions. At such frequencies, the inductive components become important and the topology of the converter must be taken into account. In such cases, converters that have an inductance in series with the input source are preferable to those that employ a switch. Also, decoupling capacitors must be employed. However, for the boost converter topology, where an inductor is placed in series with the input power source, the possibility of using the TEM's internal inductor in order to reduce the number of additional components must be investigated.

*Figure 7. Equivalent electrical circuit of an internal TEM*

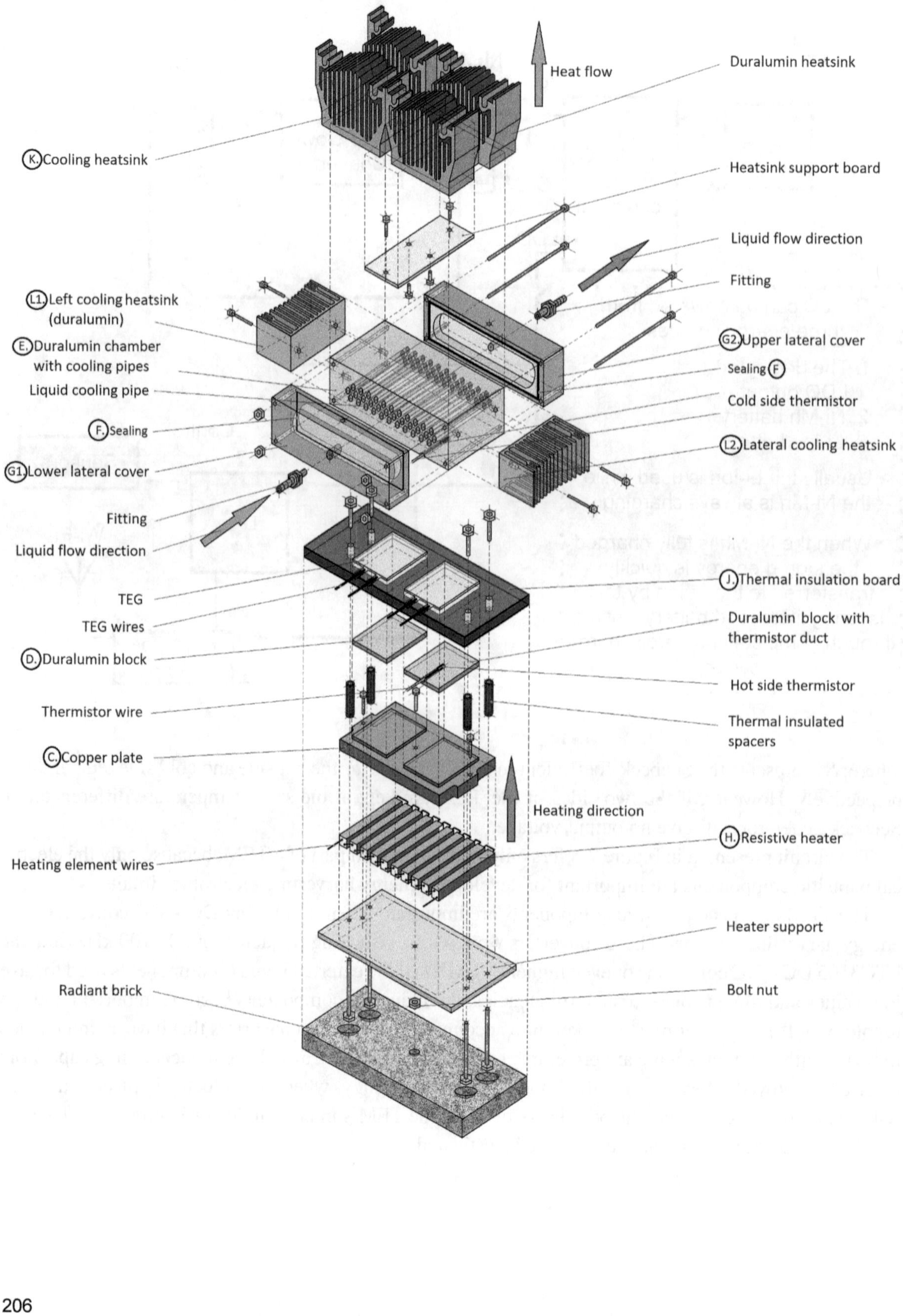

In the experiments conducted, it was found that the parasitic inductances have values in the range of μHs and the capacitances in the rage of tens of nFs, values that cannot be neglected in the design of DC-DC converters.

## NEW ENERGY HARVESTING SYSTEM PROPOSAL

The authors introduce an original, self-sustainable, power supply system for a wireless monitoring application that is powered from an energy harvesting device based on TEGs. The energy harvesting system purpose is to gather unused heat, convert it to electrical energy and store it into rechargeable batteries. The condition monitoring system (CMS) is to be used for the monitoring of heat dissipation equipment such as:

- Medium to high power electric generators.
- Wind turbine generators.
- Water turbines.
- Gas turbines.
- Boilers, exhaust pipes.

or any other equipment that needs to be monitored and can dissipate energy through heat.

The device can be used for monitoring various parameters such as vibrations, temperatures, pressures, pollution, gases etc. with the purpose of minimizing the repair costs and preventing equipment damage. The energy stored into rechargeable batteries is intended to be used for:

- Extending the autonomy of the wireless sensor network while no heat is present.
- Increasing the transmission power of the wireless transceiver.

The energy harvesting device is comprised of two major subsystems:

- The **main** subsystem is a mechanical device that considers also the TEGs and is intended to be placed where waste heat exists and is desired to be harvested.
- The **second** subsystem contains all the electronics necessary to harvest the maximum power available from the TEGs, store it into rechargeable batteries and use it accordingly.

The general system is presented in Figure 8. The output voltage generated by the LTC 3105 converter is used to charge a Ni-Mh rechargeable battery pack with low current in trickle charge regime (Linear Technology, 2014). Depending on the output current, the battery will be fully charged in a period starting from a couple of days up to two weeks. From the experiments, assuming a charging current of 5mA, that corresponds to a heat source with a temperature of 50°C, a 2.4V 1000mAh battery would charge in 9 days. The charging period of the Ni-Mh battery pack is lower than the discharging time of the Li-Ion battery. The Li-Ion energy will be mainly used to power the high gain wireless amplifier in order to increase the distance the system can transmit the measured data. The SST12CP11 high gain power amplifier is intended to be used. It provides 34dB gain with a maximum power consumption of 2.5W (5V and a maximum of 470mA). Assuming a case with a maximum transfer period of 10ms at maximum

*Figure 8. Proposed energy harvesting system, employing energy storage for high power Wi-Fi*

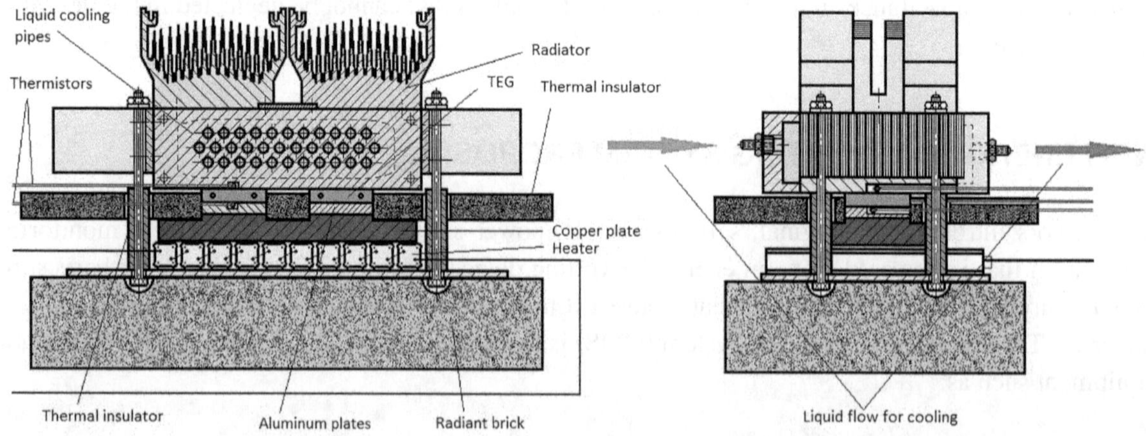


power – 2.5W, once every 1 min, we obtain a consumed energy of 1.5Ws. Taking into account that the Li-Ion battery will have a capacity of 1000mAh and provides a voltage of minimum 3.7V this functioning rate will deplete the battery in more than 300 days.

The energy consumption calculation proves that the proposed system will successfully achieve its purpose and the ratio between the available energy and consumption is excellent.

The microcontroller power management will be implemented in such way that the power consumption will be monitored by computation of the functioning and activity time. The design of the energy harvesting system is presented in detail below.

## THE ENERGY HARVESTING SYSTEM: DESCRIPTION

The next sections describe the detailed overview of a new developed energy harvesting system based on TEGs. Mechanical architecture, electronic design using several versions, implementation and results are discussed in the next sections.

### Mechanical Subsystem

For the experimental tryouts, the authors propose a modular test rig comprising of two thermoelectrical generators sandwiched between an aluminum radiator and a copper plate that is connected to an electrical heater. The test rig served also for parameters extraction so it was not optimized only for energy harvesting purposes and improvements can be further accomplished. The heater is made of serial connected power resistors that are controlled by a programmable power supply unit which provides a serial interface for communication. Two NTC thermistors are located each at the contact point of the TEG with the hot side and cold side, respectively (Figures 9 and 10).

The radiant brick along with the heater support are intended to force and direct the heat from the resistive heater through the copper plate (Figure 9). The copper plate is in direct contact with the two duralumin blocks that in turn are in contact with the two TEGs. One duralumin block contains a duct

*Figure 9. Proposed test rig exploded view*

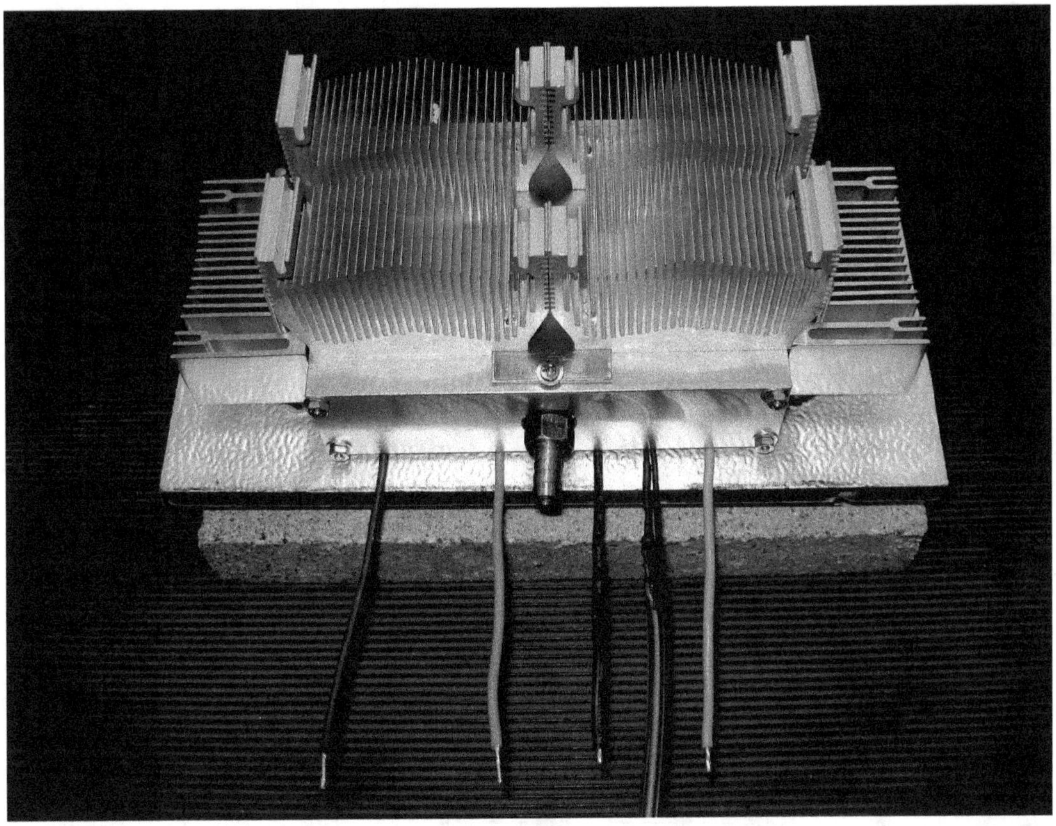

where a thermistor which provides the hot side's temperature of the TEGs is located. The cold sides of the two TEGs are in contact with a duralumin chamber that incorporates cooling pipes. The duralumin chamber also has a thermistor that will provide the cold side temperature of the TEGs. Four additional heatsinks are used to provide extra cooling. A third thermistor is located in the heating element (necessary when measuring the thermal conductance). Each thermistor employs a second order active low pass filter in order to remove the noise generated by the long wires and other interferences. The hot side is separated by the cold side through a 10mm thermal insulator board and an additional 3mm air layer (Figures 9 and 10). The thermal insulation board is manufactured by Selitherm (Selit, 2012) and has a thermal resistance of $0.25m^2K/W$. The insulation board is used to prevent the heat touching the cold side of the TEMs and it also permits the test rig to be used in parameters extraction or in the energy harvesting device as it will be detailed later.

Four spacers with low thermal conductivity, connected through bolts, are used to fasten all the elements and provide the necessary pressure to create a good physical contact between parts. Silver based, high thermal conductivity paste (higher than 4W/mK) is used between each contact point to ensure an increased thermal conductivity. The liquid pipes inside the duralumin chamber are used to cool the cold side of the TEMs when extracting the internal parameters or when the system is used as an energy harvesting device. The input and output fittings are connected to a heat exchanger through plastic pipes. When used as an energy harvesting system, the test rig is connected in thermosyphon setup, with distilled water as cooling agent.

*Figure 10. Proposed test rig section view*

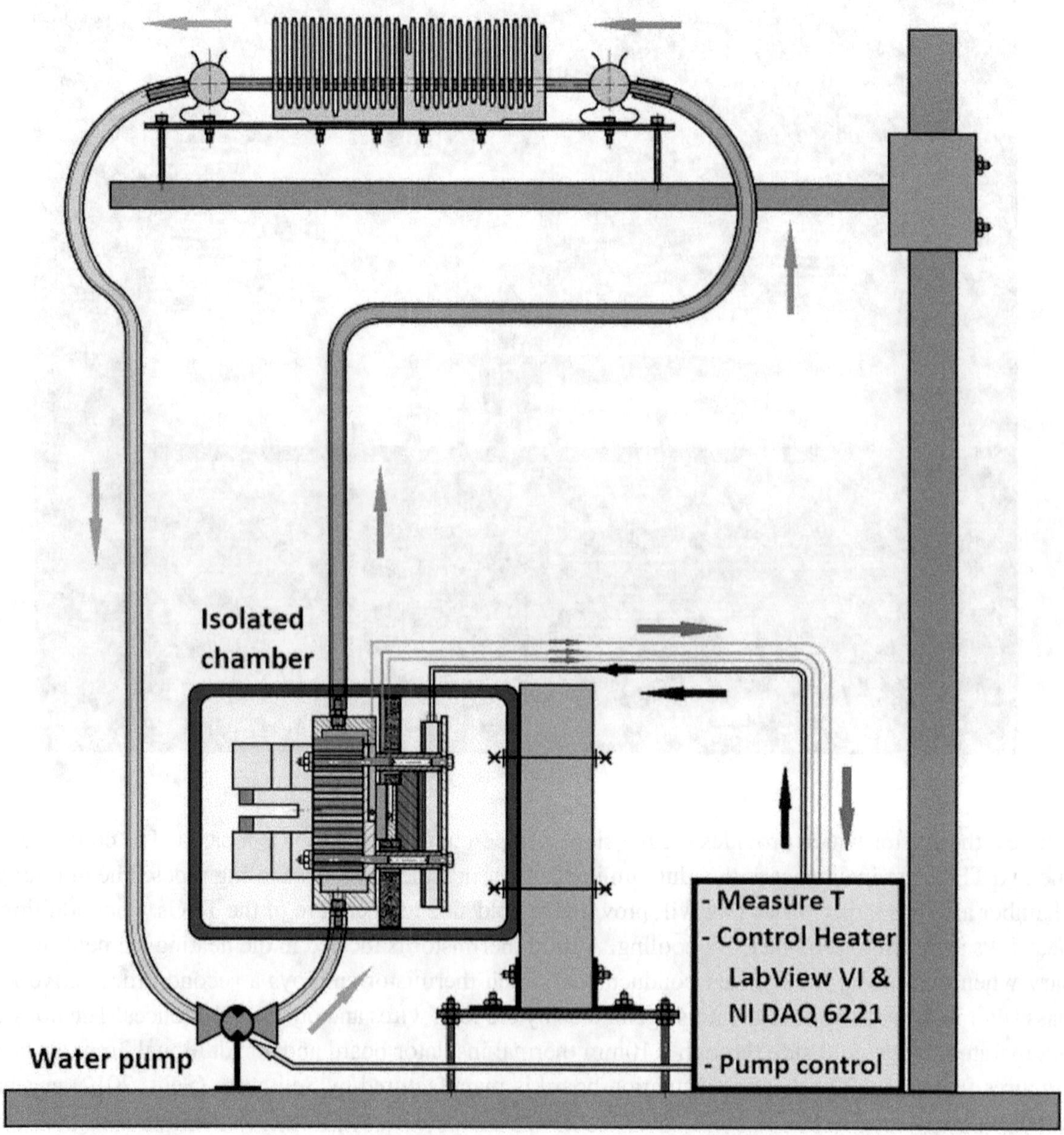


The experiments were conducted using two different TEM manufacturers. The first is Everredtronics with 127 thermocouples – TEG127-40A where the Seebeck coefficient value found on the manufacturers website is S = 0.04236 V/K. The Everredtronics module works as a TEG (Everredtronics, 2012).

The other manufacturer, Melcor, CP1.4-127-06L-RTV, is designed to be used as a heat pump or TEC (Melcor, 2012).

Two type of experimental tryouts were conducted, where the heat flow was generated by the power resistors and the thermistors placed on the heater and on the cold side of the TEMs recorded the temperature.

In one experiment, the cold side TEMs surface temperature was controlled with the help of the *E* chamber (Figure 11) that was connected through pipes to a cooled heat exchanger. A controllable liquid pump regulates the flow (in this case water) from the ***E*** chamber and the heat exchanger and consequently the temperature. The setup's general overview is presented in Figure 12. This experiment was useful to determine the real output parameter values for the chosen thermoelectric materials.

In the second experiment, the pump was not used and the system was tested in the real environment (without the isolated chamber), with the liquid cooling system functioning in the thermo-syphon mode, for energy harvesting purpose. The thermo-syphon cooling system was chosen to passively maintain the cold side temperature of the TEM. Further on, tests were also made with an in house designed heat-pipe cooling system that will be presented later.

## Electronics Subsystem

In the following the implementation of the electronics in the proposed energy harvesting system is described. Two novel solutions for a dynamic maximum power point tracking system are simulated and implemented in practice. Simulations were first done to fine tune the proposed energy harvest system and to assess its performance. The energy harvesting system performance is presented. An improved Cuk converter is also built that aims to harvest the energy from multiple serial coupled TEGs and store it into rechargeable batteries.

### The LTC3105 DC-DC Converter

The system purpose is to harvest unused heat, convert it to electrical energy and store it into rechargeable batteries. The energy stored is to be used in sensor measurements and long range data transmission.

*Figure 11. Proposed test rig main unit*

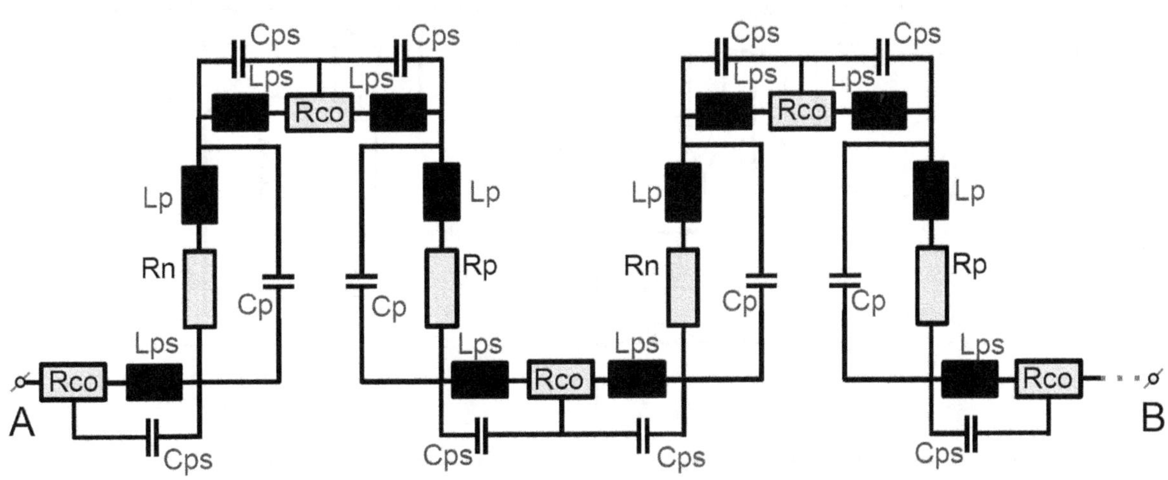

*Figure 12. Overview of the thermal conductivity measurement setup*

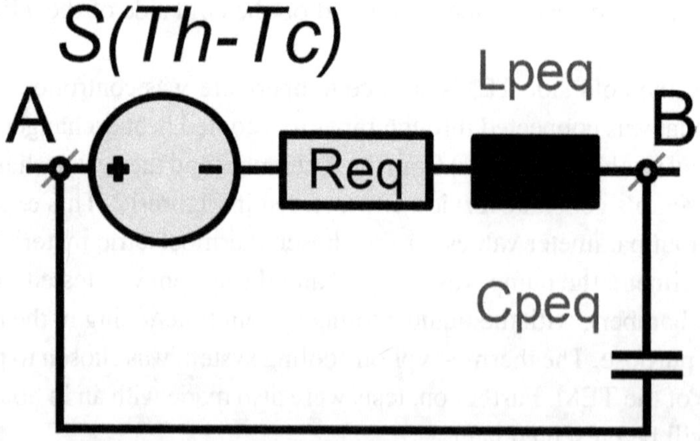


The energy harvesting system is comprised of two major units. The main unit was presented in Figure 9. The second unit contains the electronics necessary to convert the electrical power generated by the TEGs and store it into rechargeable batteries.

The electronics from the second unit is built around the LTC 3105 Step-Up DC-DC converter. This device allows a low start-up voltage of only 250mV with adjustable output voltage from 1.5 to 5.25V. Its internal block diagram is presented in Figure 13.

*Figure 13. Internal block diagram of the LTC 3105 DC-DC converter (Linear Technology, 2014)*

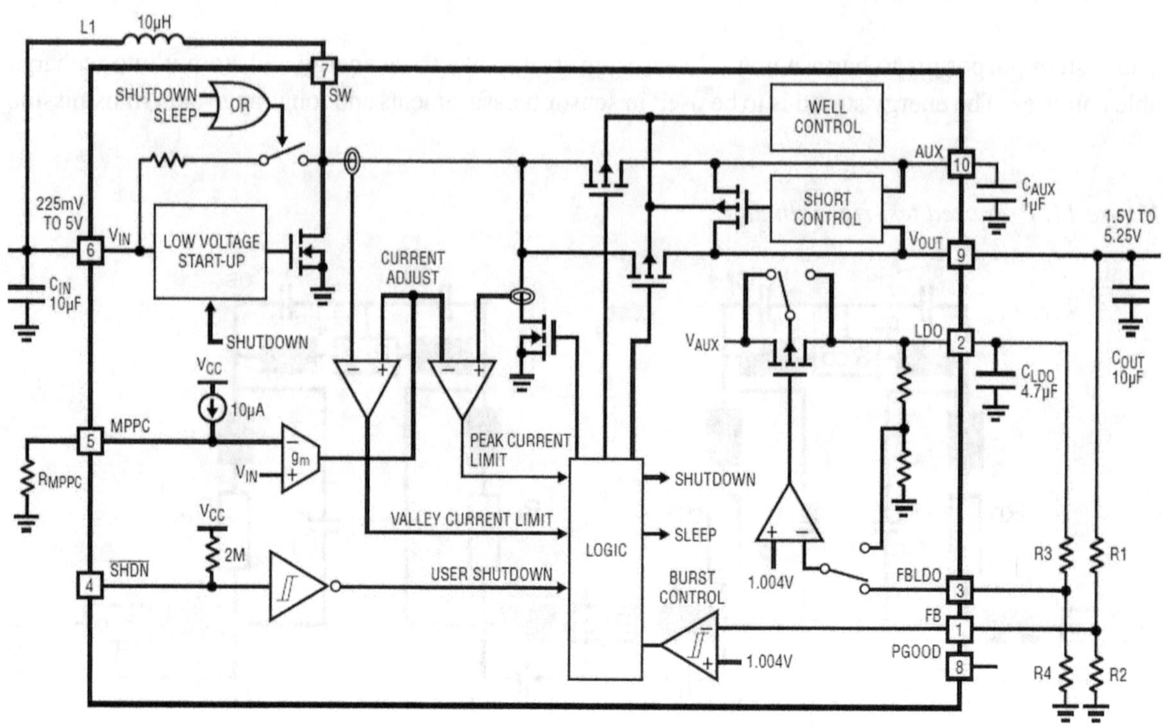

An important function of the LTC3105 is the built-in maximum power point tracking controller (MPPT) that enables the operation from low voltages and high impedance sources such as TEGs. This circuit allows the user to set the optimal input voltage operating point for a given power source, controlling the inductor current to maintain $V_{in}$ at the voltage set on the MPPT pin. The converter features an auxiliary LDO regulator that can output 6mA@2.2V and can be used as energy source for a low power microcontroller. The datasheet states that the maximum attainable current on the LDO pin is limited to 12mA.The LTC3105 has been optimized for use with high impedance power sources like thermoelectric generators (Linear Technology, 2014).

The energy harvesting system proposed in this work employs two serial connected TEGs (Figure 9). Although the LTC3105 can be used with multiple stacked power sources, the number is limited by the output voltage of 5.25V and the current limit of 400mA. If more than two TEGs are used as input power source, the maximum attainable temperature is to be taken into account as the input voltage needs to have a lower value than the output setpoint for the normal operation of the LTC3105. The proposed setup consists of two serial connected TEGs that provide a maximum of 2.9V at ΔT=23°C.

Taking into account that the internal parameters of the TEM (internal resistance, Seebeck coefficient, thermal conductance) vary with temperature, an autonomous way to track the maximum power point must be developed. The LTC3105 datasheet presents a couple of standard connections that take advantage of the internal current source of 10μA. The internal current source together with an externally connected resistor or diode generates a voltage on the MPPC pin that represents the minimum value the $V_{in}$ pin is maintained in order to extract the maximum power from the energy source. The converter datasheet presents a photovoltaic energy harvesting solution where the temperature tracking is made with the help of diodes that are thermally coupled to the photovoltaic panel (PV). In PV panels, the output voltage is inversely proportional to the temperature; moreover the output voltage of a TEG has a stronger variation with temperature than the PV does. This solution cannot be used on TEGs, or in the current proposed system where the output voltage varies from 0V to 3V.

In the present topic, two dynamic maximum power point tracking systems are described and implemented. The circuits are intended to be used together with the LTC3105 converter and TEG power sources. The devices purpose is to charge Ni-Mh or Li-Ion batteries although the generated output power can be used to charge a super-capacitor, like the LTC converter's datasheet specifies (Linear Technology, 2014).

## The Original Dynamic MPPT System that Uses the Existing Microcontroller

The dynamic maximum power point tracking system takes advantage of the microcontroller existing on the equipment (Figure 14). A thermistor that is thermally coupled with the TEGs is used to determine the actual hot side temperature.

A lookup table is implemented and stores the {output voltage, temperature value} pairs. The lookup table is used to compute the actual output voltage of the TEGs, based on the measured hot side temperature. Afterwards, the PWM output of the microcontroller together with an RC low pass filter is used as a DAC circuit to generate an output voltage. The operational amplifier (OA) connected in follower configuration is used to isolate the DAC output from the $R_{st}$ resistor. This configuration is needed when the microcontroller is not powered by the LTC3105 LDO regulator. If the existent heat source that powers the TEGs is not providing enough energy, the microcontroller will have no supply voltage. If the wasted heat energy increases and the TEGs output exceeds 0.3V, the internal 10μA current source becomes active and the LTC converter will try entering the normal operation mode. However, at this point,

*Figure 14. Proposed dynamic MPPT system that takes advantage of the microcontroller*

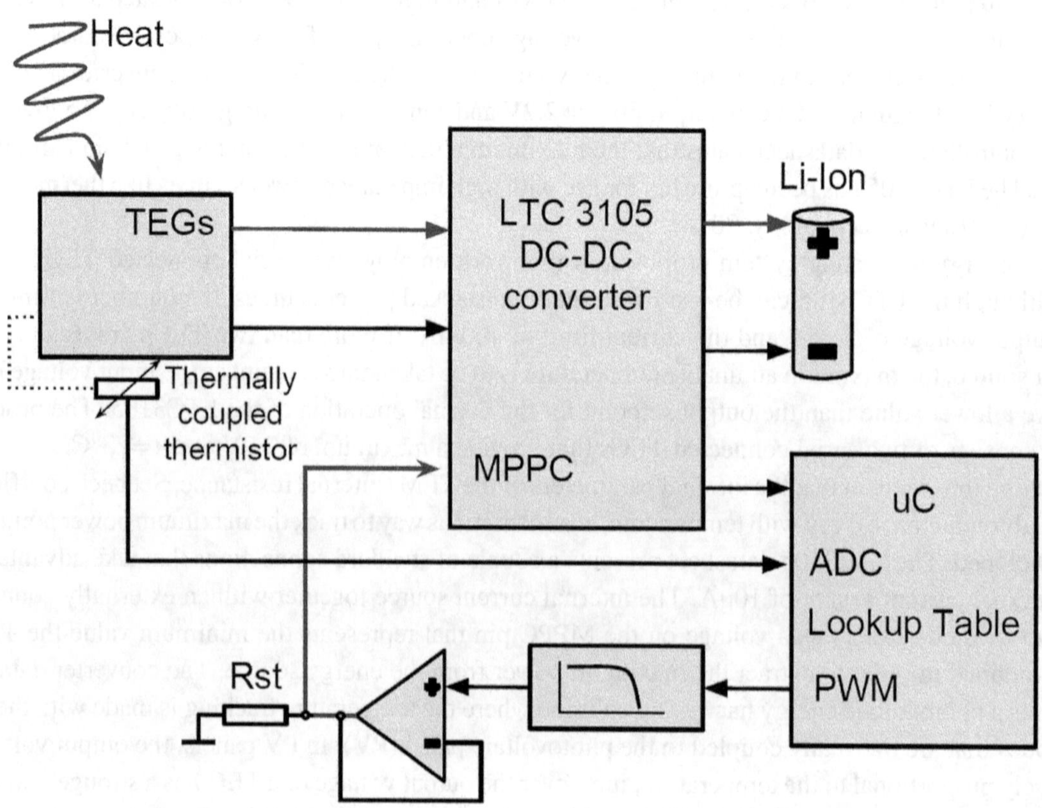


the $V_{MPPC}$ voltage is of great importance because if no resistance is connected at its output, the converter will not be able to enter the normal operation mode. The output voltage at the $V_{MPPC}$ pin is given by the 10μA internal current source and the resistor connected to the MPPC output ($V_{MPPC} = R_{st} \cdot 10\mu A$). This voltage represents the minimum value the input voltage is maintained in order to extract the maximum available power. As known, the maximum power extracted will occur when the load resistance equals the source resistance, or

$$V_{in} = \frac{V_{noLoad}}{2},$$

where $V_{in}$ is the LTC3105 input voltage and $V_{noLoad}$ is the output voltage of the TEG when no load is connected to it. In this way, when no power is available on the LDO output, the $R_{st}$ resistor will allow the converter to start operating from $V_{in} = 2 \cdot R_{st} \cdot 10\mu A$. After entering the normal operation mode when the microcontroller will be powered by the LDO output, the prescribed value for the $V_{MPPC}$ pin will be provided by the microcontroller and the DAC circuit, based on the temperature measurement. The OA follower has a second purpose in this case, that of sinking the 10μA the MPPC pin is sourcing.

## The Second Original Analog Dynamic MPPT System

In most cases, an analog solution, microcontroller free is desirable, due to several constraints that depend on the microcontroller's peripherals (e.g. only one PWM) or because specific designs employ no microcontroller. A completely autonomous analog solution for tracking the maximum power point is introduced in Figure 15.

The novelty of the circuit is to use a NTC thermistor and linearize its variation with $\boldsymbol{R_1}$ and $\boldsymbol{R_2}$ resistors. The necessary output offset adjustment is performed by modifying the gain of $\boldsymbol{OP_2}$ operational amplifier (OA). $OP_1$ is connected as a follower and has the purpose to isolate the $\boldsymbol{R_3}$ and $\boldsymbol{R_4}$ resistors from $\boldsymbol{R_1}$, $\boldsymbol{R_2}$ and **thermistor**. $\boldsymbol{R_{st}}$ is a startup resistor that permits the LTC3105 to start functioning even if $\boldsymbol{OP_1}$, $\boldsymbol{OP_2}$ and $\boldsymbol{OP_3}$ are not powered. This is necessary if the battery is depleted and the LDO output voltage setpoint was not reached by the converter to power the operational amplifiers. In case no voltage is applied to $\boldsymbol{OP_1}$, $\boldsymbol{OP_2}$ and $\boldsymbol{OP_3}$, the voltage on the $\boldsymbol{V_{MPPC}}$ input is $R_{st} \cdot 10\mu A = 240mV$ . $\boldsymbol{OP_3}$ is used as a second follower and has the purpose to isolate the rest of the resistor network from $\boldsymbol{R_{st}}$. It also ensures that the voltage on the MPPC pin is $R_{st} \cdot 10\mu A$ when no voltage is present on $\boldsymbol{V_o}$ output. After entering normal operation mode, when the output voltage reaches the setpoint value (2.2V in the actual case), $\boldsymbol{OP_1}$, $\boldsymbol{OP_2}$ and $\boldsymbol{OP_3}$ will be powered by the LDO output of the LTC converter. If the $\boldsymbol{V_o}$ output of the LTC

*Figure 15. Dynamic MPPT system*

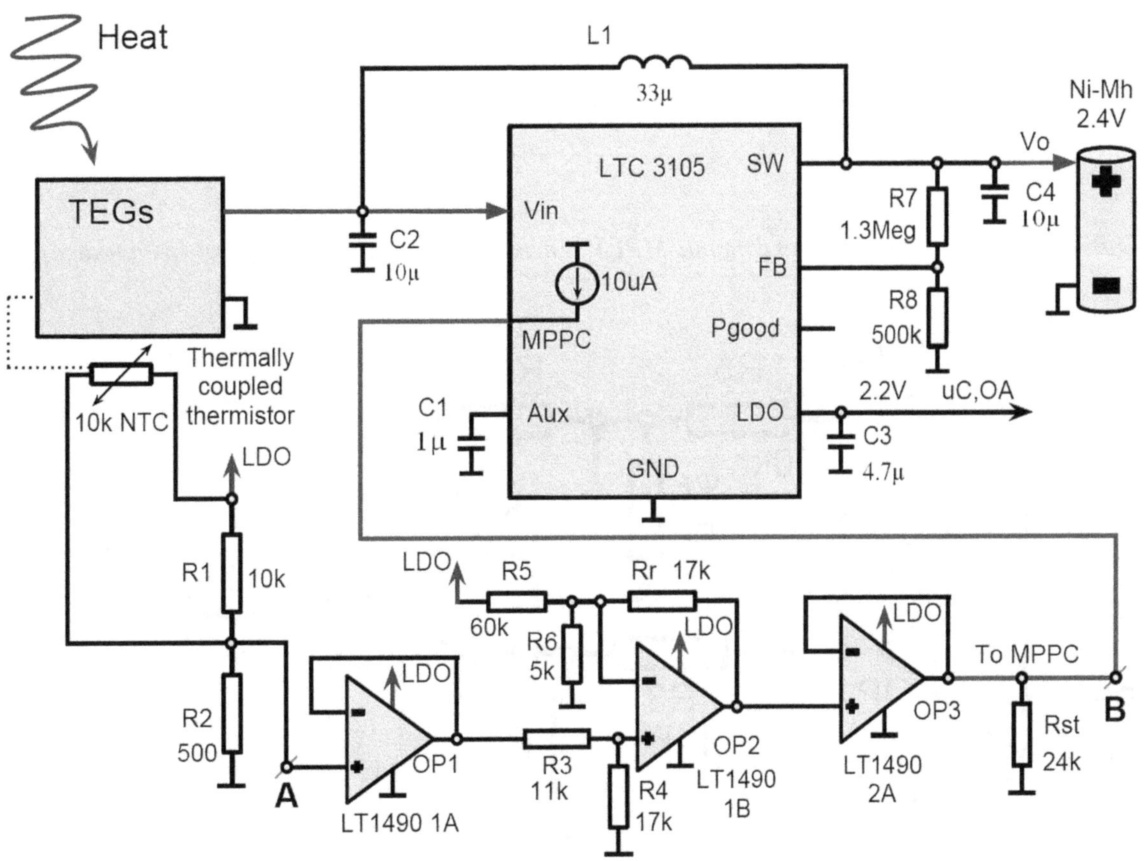

converter wouldn't be used to power a battery, this output could be used as a supply voltage for $OP_1$, $OP_2$ and $OP_3$. When a battery is charged by the LTC converter's output, the output voltage $V_o$ will be influenced by the variable battery voltage (e.g. 2.1V when the battery is discharged and more than 2.4V when it is fully charged, as measured in the current experiments). This variation will cause an erroneous $V_{MPPC}$ output, resulting in a lower efficiency in extracting the maximum available power.

In order to determine the correct slope for the output voltage of the $OP_2$ amplifier versus temperature, to linearize the NTC variation and determine the $R_{st}$ value, a NI LabView VI application (Figure 17) was developed to dynamically select the components values. This solution was adopted because it was considered a fast method to obtain a visual feedback of the output in response to the modified values. The upper left corner graph from Figure 17 presents the output voltage corresponding to the positive input pin of $OP_1$ follower (Figure 16). In the lower left corner the thermistors' characteristic versus temperature is presented. The graphic on the right side of Figure 17 shows the output voltage of $OP_2$ which is the prescribed voltage for the MPPC pin. The voltage output variation is represented as a function of TEGs hot side temperature. The offset and slope values can be adjusted from the $R_3$, $R_4$, $R_5$, $R_r$ resistors presented in Figure 17. The resistors values are represented as sliders in the VI application from Figure 17. The values are modified until the voltage - temperature variation corresponds to the experimentally determined characteristic.

Eventually, a simpler solution is possible (Figure 16). Manipulating the values of the resistors, the same functionality can be achieved using only a single OA IC. The operation (when no voltage is present on the $V_o$ output) is based on the equivalent resistance seen on the $V_{MPPC}$ pin. The voltage on $V_{MPPC}$ becomes:

$$(R_{st} + R_r + (R_5 + (R_{thermistor} \parallel R_1 + R_2)) \parallel R_6) \cdot 10\mu A \approx 230mV \quad (2)$$

Both solutions offered the same functionality in the tests performed.

*Figure 16. Alternative solution to dynamic MPPT system using a single OA IC (dual OA package)*

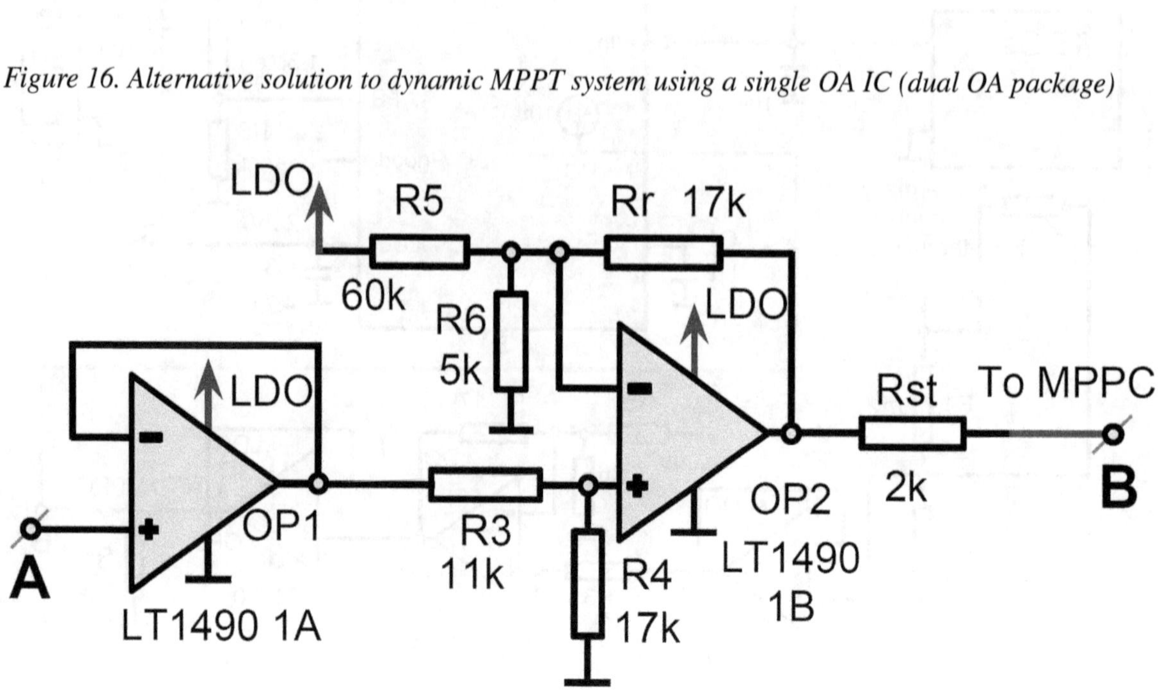

*Figure 17. LabView VI front panel describing the voltage-temperature characteristic at the output of $OP_2$ operational amplifier*

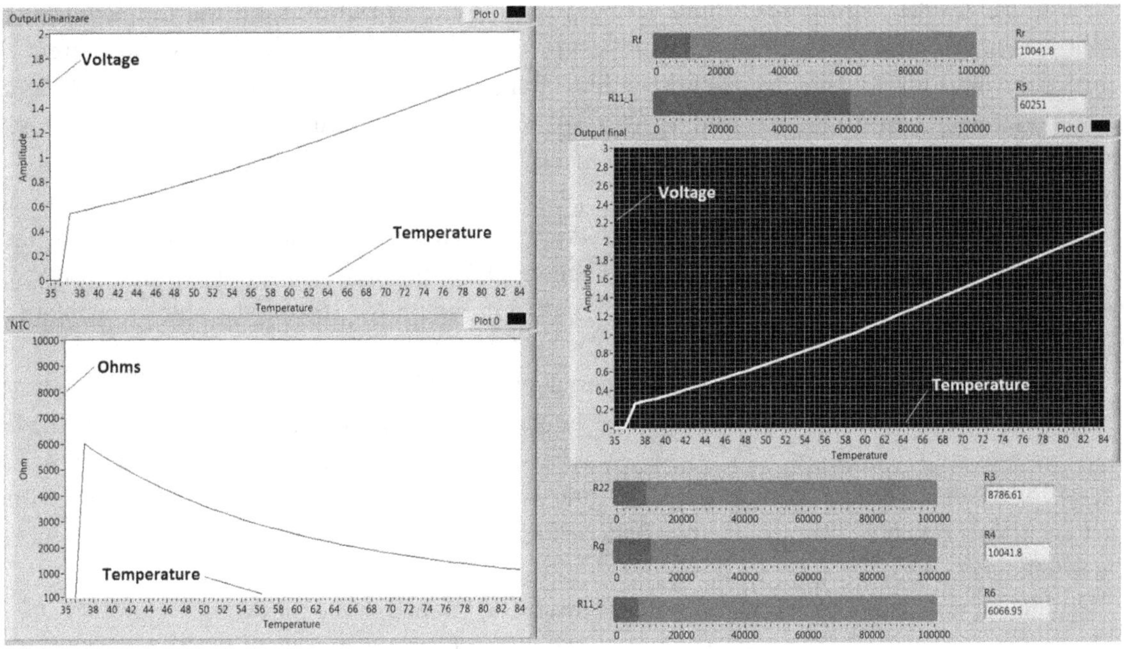


The circuits were both implemented and the TEM operating under heat flux was emulated through a programmable power source and a potentiometer (Figure 18). The control of the power source was implemented using LabVIEW and the measurements were made using the NI DAQ 6251.

The implemented TEG emulator disconnects the A point (Figure 18) from the rest of the circuit with the help of a relay. Afterwards, the $R_{load}$ resistor is connected and the dropout voltage is measured, in this was computing the $R_{int}$ resistor value. Using the lookup table, the corresponding output voltage is prescribed to the Agilent N6700 power source. The data from the lookup table was experimentally

*Figure 18. TEG emulator using a manual potentiometer and a programmable N6700B power supply*

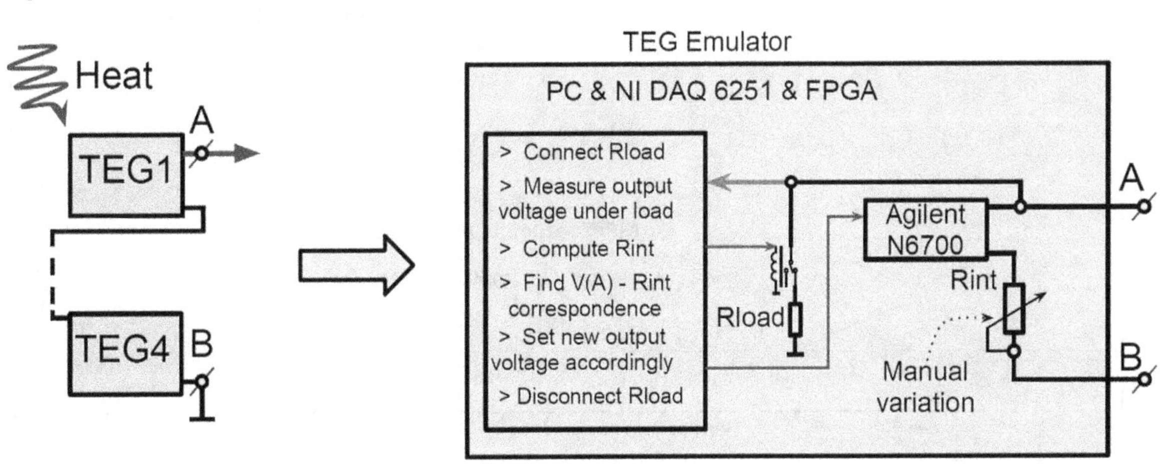

determined. By varying the internal resistance of the TEM, the behavior of the real device is emulated. According to the TEM's side's temperature difference, the output voltage and internal resistance changes.

For the steady-state behavior test, three voltages (0.4V, 1V, 2.2V) where fed to the LTC input and the output current was measured with a current probe on a variable resistive load. The maximum current was recorded up to the point where the output setpoint voltage remained at the prescribed value (3.6V). The output currents of 1.1mA, 10.5mA and 50mA respectively were accordingly to the simulation results with the same inputs and same output resistive loads. The simulation and experimental results for 1V input voltage are presented in Figure 19-21.

The dynamic tracking was also verified with the real system, the circuits being first tested without the output battery that would have powered the operational amplifiers. The dynamic MPPT proved to operate as expected with both configurations presented in Figure 15, 16 and was able to maintain the maximum output current.

The original energy harvesting system employing dynamic MPPT and a two stage energy storage system

Figure 22 presents the proposed system which employs a two stage energy storage solution. The first stage uses two battery packs of 2.4V Ni-Mh. A series of MOS switches can connect the battery packs in series or parallel. The idea is to charge each battery separately from the prescribed 3.6V LTC output. After each pack gets fully charged, the two batteries are connected in series and a 4.8V equivalent battery is obtained.

The idea of this setup has several advantages that are explained in the following.

The efficiency of LTC3105 converter versus the input voltage is presented in the datasheet in the two cases – 3V and 5V prescribed output. The two diagrams are presented in Figure 23. One can see that if the input voltage has low values (< 1.5V), a higher output voltage setpoint (5V) has a negative impact on the converter efficiency. Taking into account that the proposed energy harvesting device is designed to work with low temperature conditions (< 85°C) and that low temperatures are more prob-

*Figure 19. Simulation results for 1V input and 360Ω load*

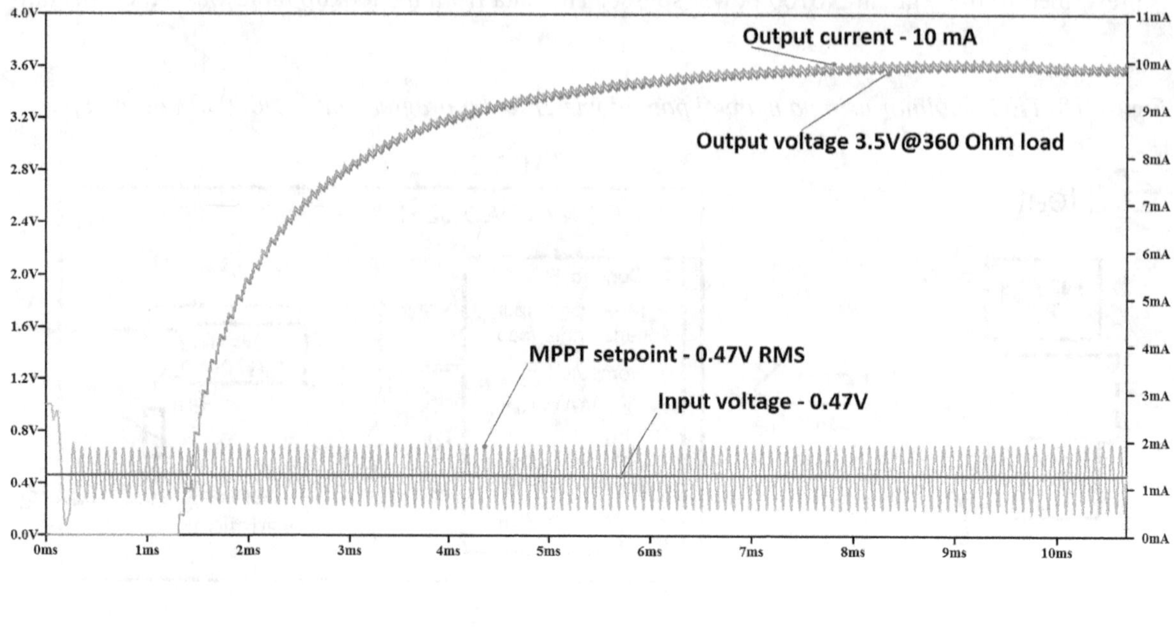

*Figure 20. Practical implementation results for 1V input and 360Ω load*

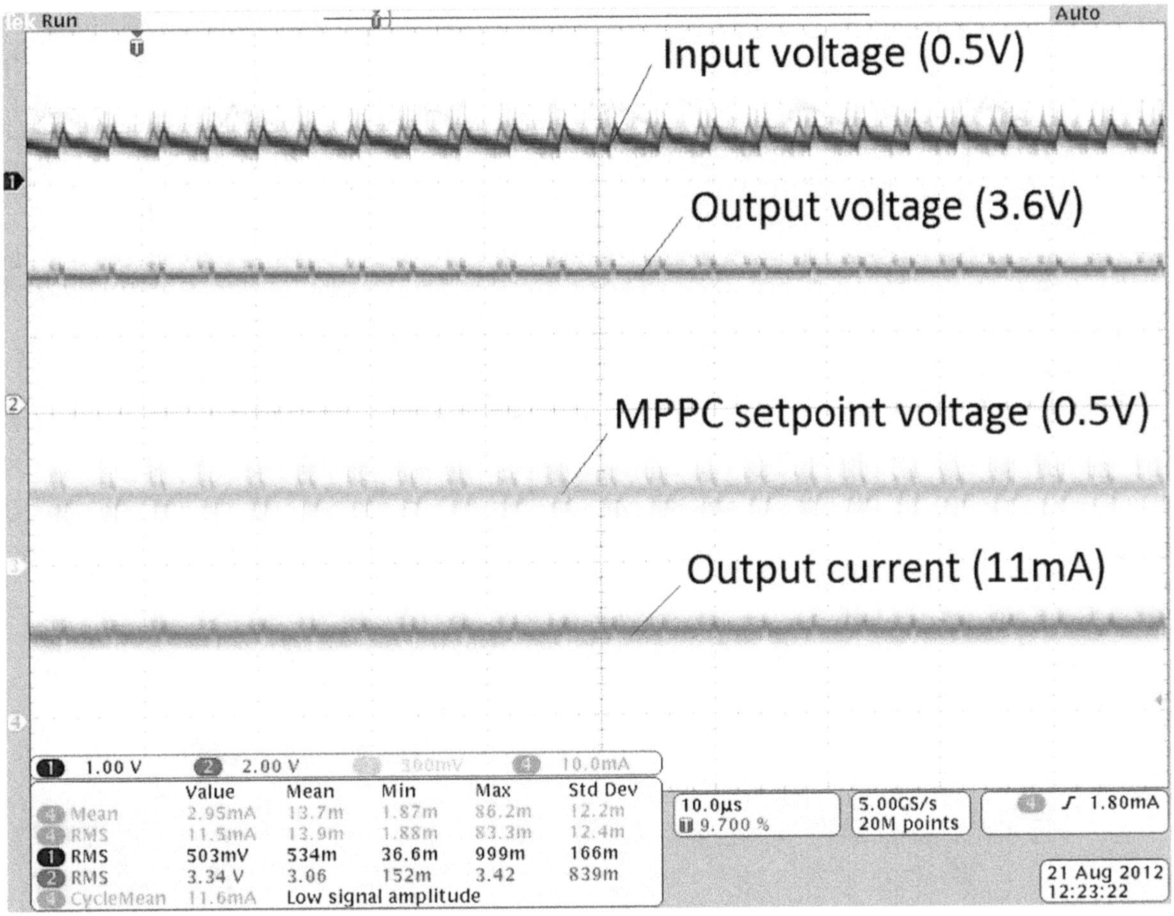


*Figure 21. Simulation results for 0.4V input and 2kΩ load*

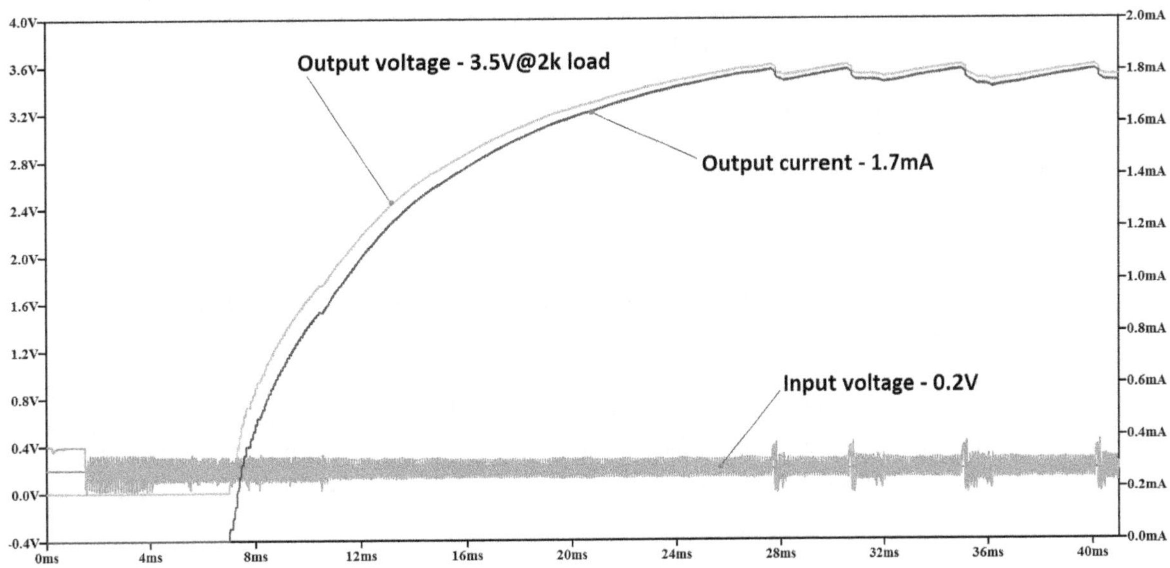

*Figure 22. The proposed energy harvesting system general overview*

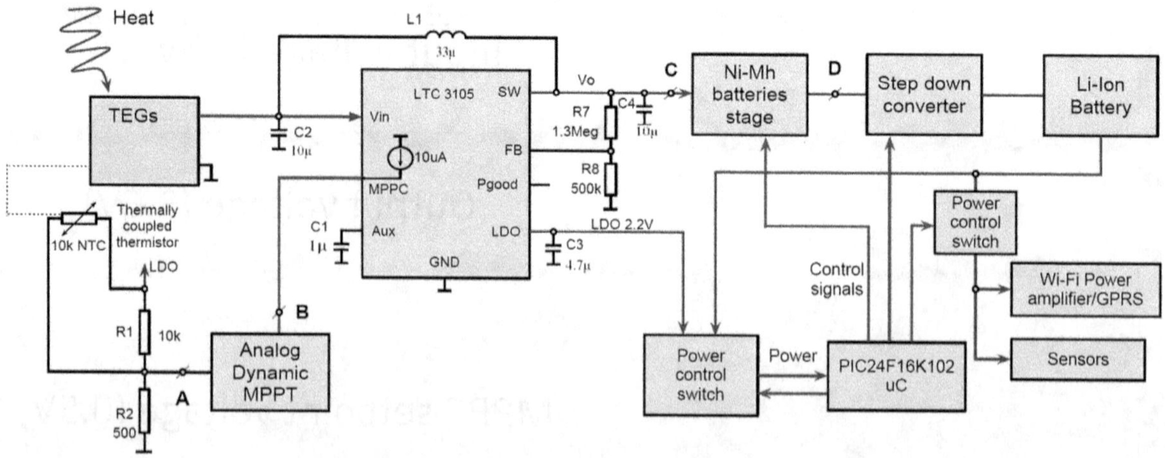


*Figure 23. LTC3105 datasheet detail representing the efficiency curve versus output current at different input voltages (Linear Technology 2014)*

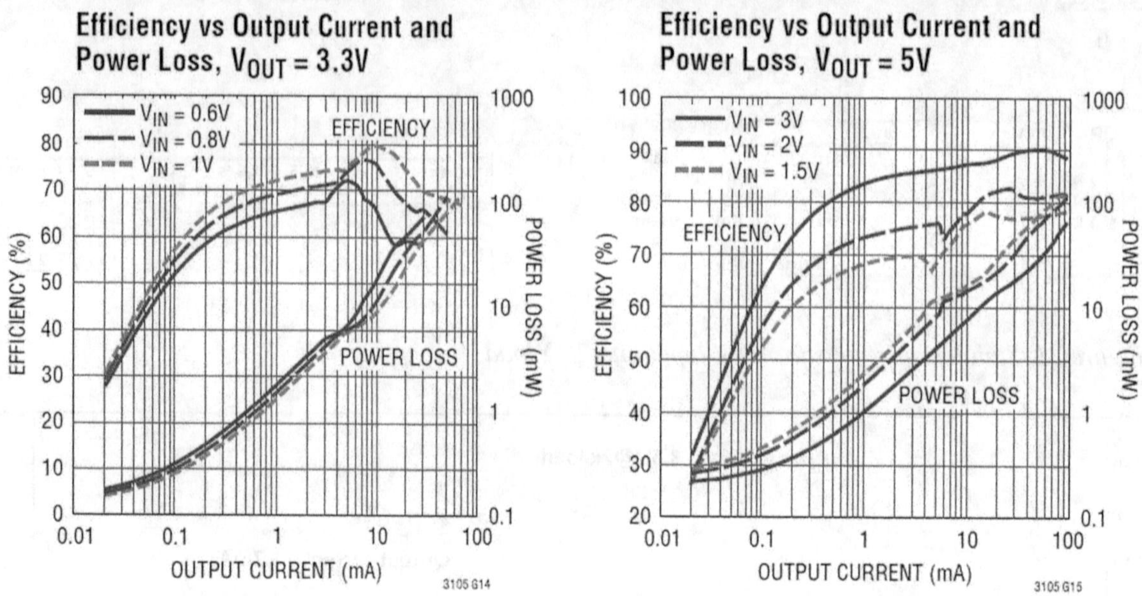


able than high ones, the output voltage of the TEGs in the proposed device will be centered on 1.5V. For this input voltage value, the efficiency is higher when a lower output voltage is prescribed to the LTC3105 converter output.

This fact was the first that led to the adoption of the 3.6V as the prescribed output voltage level. Also, this voltage is necessary to override the Ni-Mh battery pack voltage, even when it approaches full charge (>2.8V). With a higher output voltage, also a higher voltage battery pack can be used. In this case, a

3.6V Ni-Mh battery would be the maximum allowable value (a 4.8V would require the LTC3105 to provide more than 5.25V). Having a 3.6V Ni-Mh battery would require a step up –step down converter (to charge the second stage battery) that needs more components (with increased costs) than the currently employed step down topology.

Another reason for adopting the 3.6V output voltage and the Ni-Mh technology for the first storage stage is that the Ni-Mh can be charged from a low current source without damage (Energizer Battery Manufacturing Inc., 2008). In this way, by charging the battery with low current, trickle charge mode and combining this procedure with the full discharge process, the battery health can be maintained over a longer period. Also, the adoption of the required charging regime for the Li-Ion battery will also lead to the preservation of the batteries health.

Figure 24 presents the Ni-Mh batteries stage and the power paths controlled by the MOS switches. An n-MOS together with a P-Mos pair (e.g. $T_1$ and $T_2$ or $T_3$ and $T_4$) form a line switch. The diode present on the P-Mos drain to source terminals allows the battery to be decoupled from the LTC output when the switch is in the OFF state. $T_5$ and $T_8$ are used to put the batteries in series when they are completely charged and form a 4.8V battery.

Pull-down resistors were used to prevent the accidental turning on of the switches when no signal is available from the microcontroller. n-MOS power transistors (forward current >1A) with low $r_{dson}$ resistance were used to minimize the losses generated when charging the Li-Ion battery with currents higher than 1A.

*Figure 24. Detail of the Ni-Mh batteries stage design*

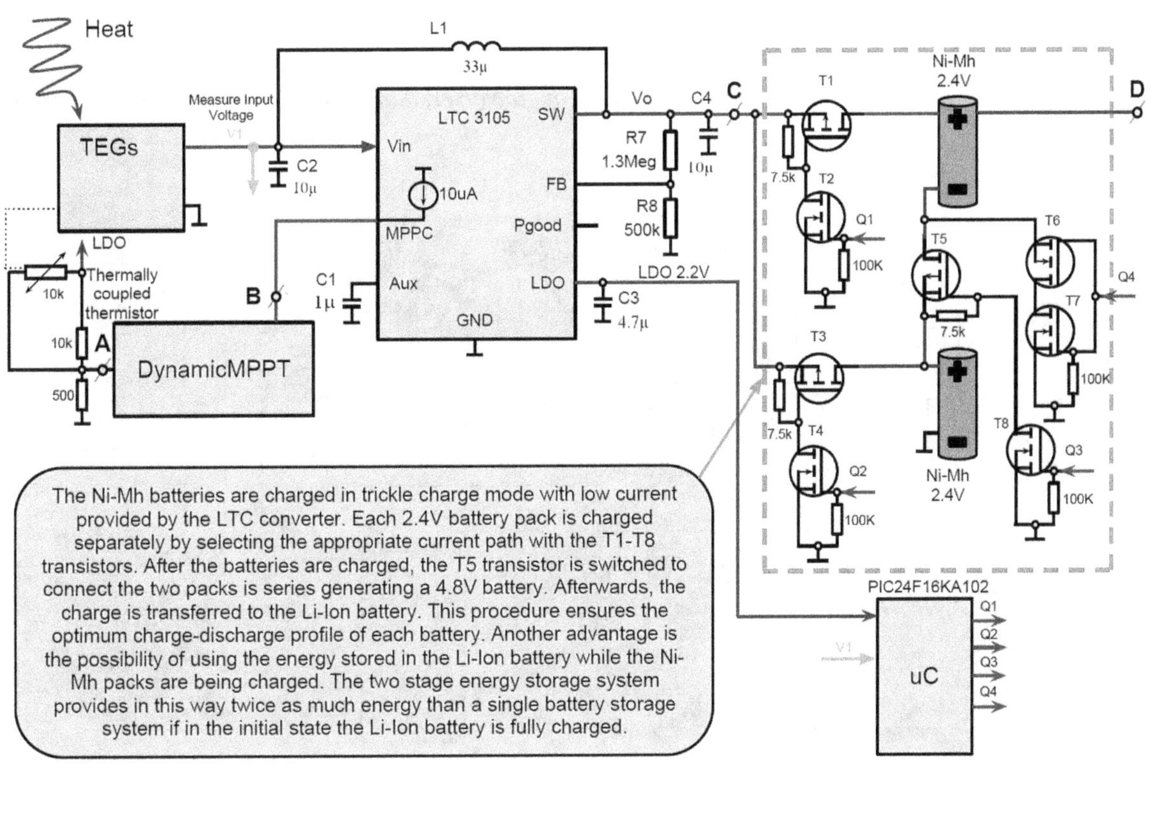

By measuring the input voltage value ($V_1$ in Figure 24), knowing the converter efficiency and integrating over time, the energy provided to the battery pack can be computed in order to determine the full charge moment.

Figure 25 presents the step down converter and the power switches array that are used to route the energy stored into the Li-Ion battery to different devices as Wi-Fi, sensors and microcontroller. The switches array that is used to control the batteries charging or to route the power from the Li-Ion battery will be referred as the Power Control Switch (PCS).

## The Buck Converter

The values for the converter components are calculated to meet the electrical requirements. The current programmed mode controller is chosen and designed to meet the stability criterion then simulated using analog circuits. The discrete form of the control loop is derived using the bilinear transform and implemented using NI Labview and the NI DAQ 6251 board. As required by the Li-Ion charging profile a constant voltage control loop is also implemented (Figure 26).

The Li-Ion battery was conceived and developed in Japan and was accepted immediately because of its high-energy density, good performance, and no memory effect as compared to nickel–cadmium (Ni–Cd) or nickel-metal hydride (Ni–MH) batteries. Li-Ion batteries have been used mainly for portable electronics and recently the application area has been extended to power tools and hybrid electric vehicles, as presented by Yoshio, Masaki, Brodd, Ralph J., Kozawa, Akiya (2009).

Li-Ion batteries are also widely used for energy storage devices despite the drawback that they require a complicated charging procedure if the lifetime of the battery is of concern. As stated in literature, a

*Figure 25. Detail of the Li-Ion battery stage, step down converter, additional devices and power control switch*

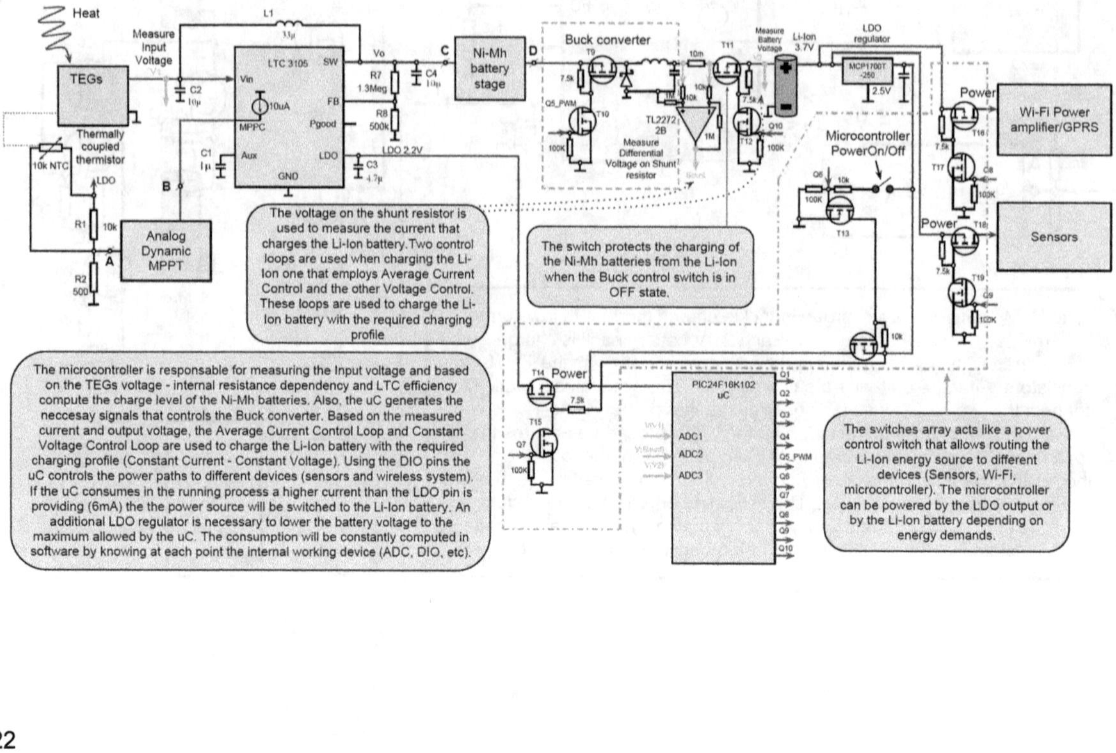

*Figure 26. Detail of the Buck converter design and control system*

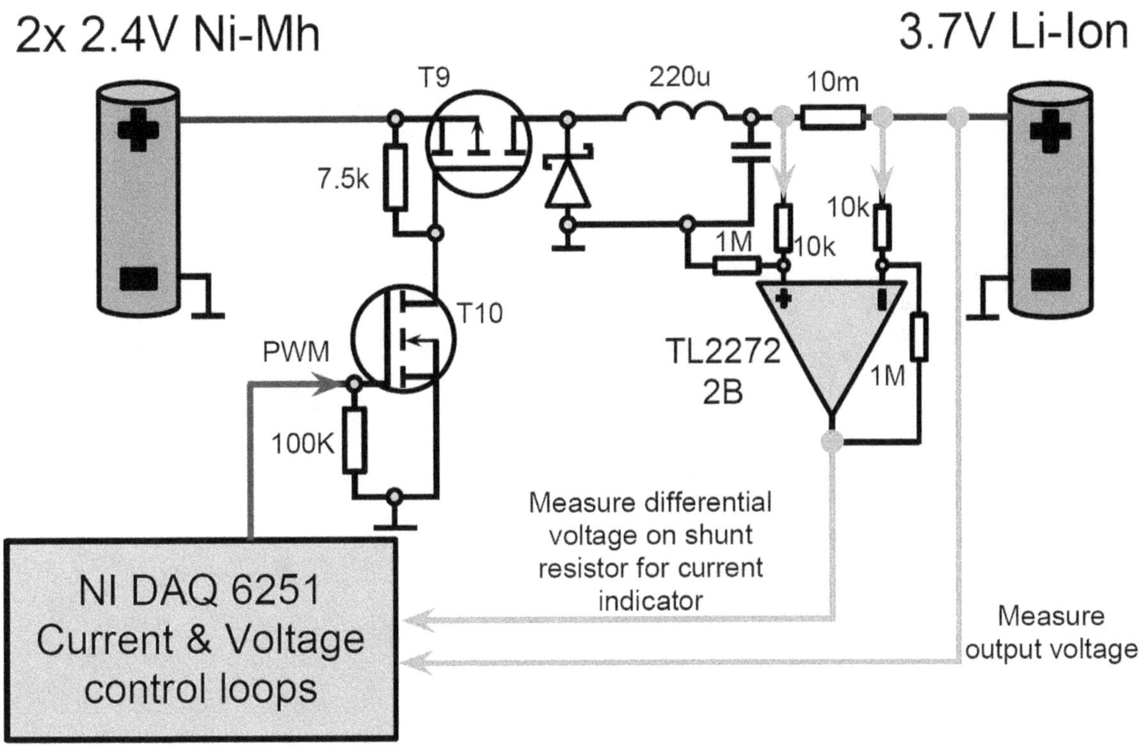


strict charging regime is necessary to safely charge Li-Ion batteries. Also, a protective circuit is used to prevent overcharge and over discharge. Charging is limited to a maximum voltage of 4.3 V/cell while the discharge voltage is limited between 2.7 – 3 V/cell.

The battery requires a controlled charge regime composed of two stage charge cycles – constant current and constant voltage (CC/CV). The first stage charges the battery at a constant current until the voltage reaches 4.1 to 4.2 V. When this value is reached, a constant voltage charge is initiated until the charge current reduces to 3% of the rated current. The charging regime is represented in Figure 27.

Another drawback of using Li-Ion batteries as energy storage devices is the need of a high charging current. This is hard to obtain in most cases, and sometimes the battery is replaced with a super capacitor or other batteries with different chemistry.

The maximum current that must be delivered to the rechargeable Li-Ion battery must be $1 \cdot C$ of its capacity which leads to a value of 1A for a 1000mAh battery capacity.

## Current Programmed Control

For the current application of Li-Ion battery charging, the average current control method (ACC) was chosen due to the fact that it has a large noise margin, no requirement for slope compensation, excellent voltage and current regulation and easy to obtain stability requirements. The goal is that the average value of the inductor current should follow a given reference value during a switching period. The principle of ACC is represented in Figure 28.

*Figure 27. Charging profile of a Li-Ion battery (Energizer Battery Manufacturing Inc., 2008)*

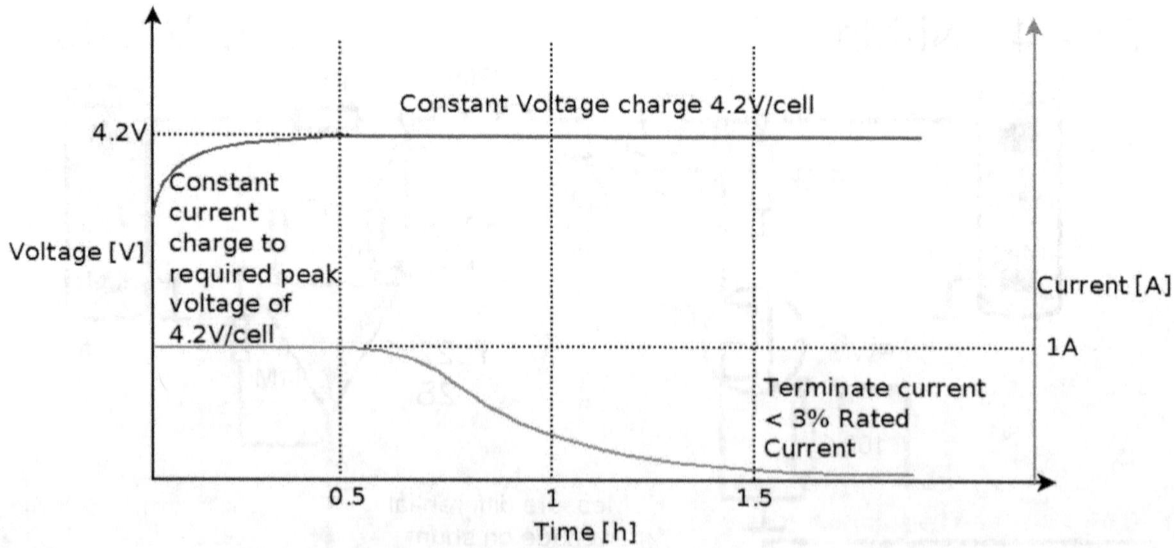


*Figure 28. Average current control loop topology amplifier structure*

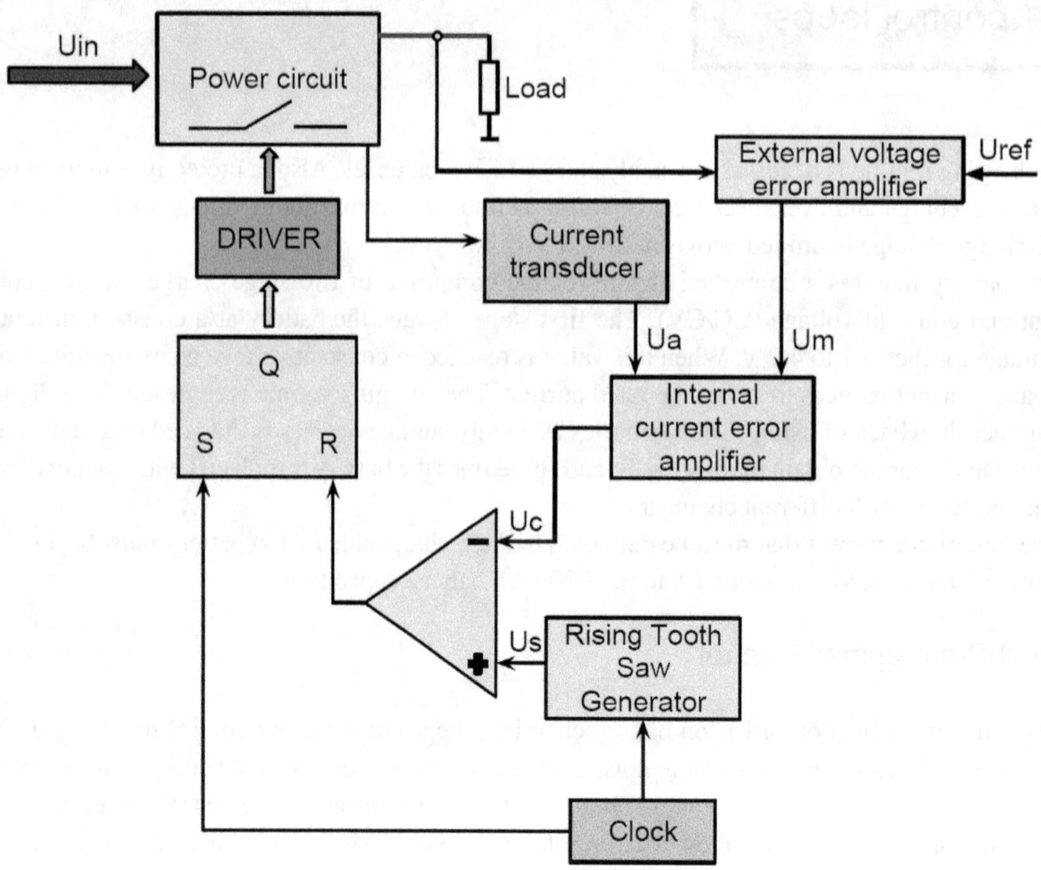

$$T(s) = H_{AE}(s) \cdot H_{PWM}(s) \cdot H_{CONV}(s) \cdot H_{CRTTRAD}(s) \quad (3)$$

The open loop transfer function for the system represented in Figure 28 while the error amplifier is presented in Figure 29.

A custom application was implemented for determining the analog component values to be used in the simulation.

For implementing the current programmed control loop in a digital counterpart the transfer function was converted to a discrete form. This approach is preferred because the converter can be driven by a microcontroller already existent on the energy storage device.

## Implementation and Experimental Results

The circuit was first simulated and then implemented. The flowchart diagram of the LabView implementation is represented in Figure 30. The ACC loop is implemented using shift registers and the constant voltage control loop using a built in PI regulator. The experimental setup can be seen in Figure 31.

Figure 32 shows the response (actual current and voltage) of the running charging process. In Figure 32, on the *left* side, it can be seen that the charging current is regulated at a constant value of 1 Amp. The constant current cycle ends when the voltage across the battery equals 4.2V, and then the constant voltage control loop comes into action, keeping the voltage across the battery at a constant value of 4.2V until the charge current reduces to 3% of the rated current – *right* side of the figure. The spike in the figures represents the switching from one control loop to another. The results show that the behavior of the experimental converter follows the simulated one (in ACC mode). The experimental results showed that the converter is able of maintaining a constant charge current and voltage even under perturbations.

*Figure 29. Error amplifier structure*

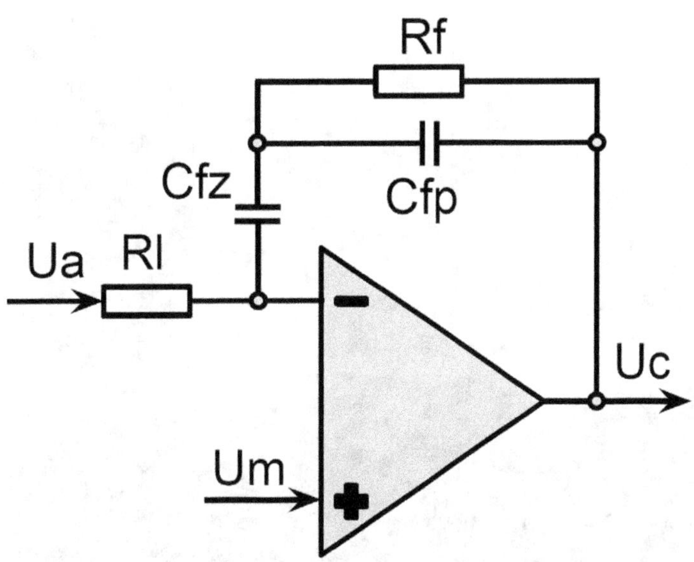

*Figure 30. Logic diagram of the digital control implemented with the NI DAQ 6221*

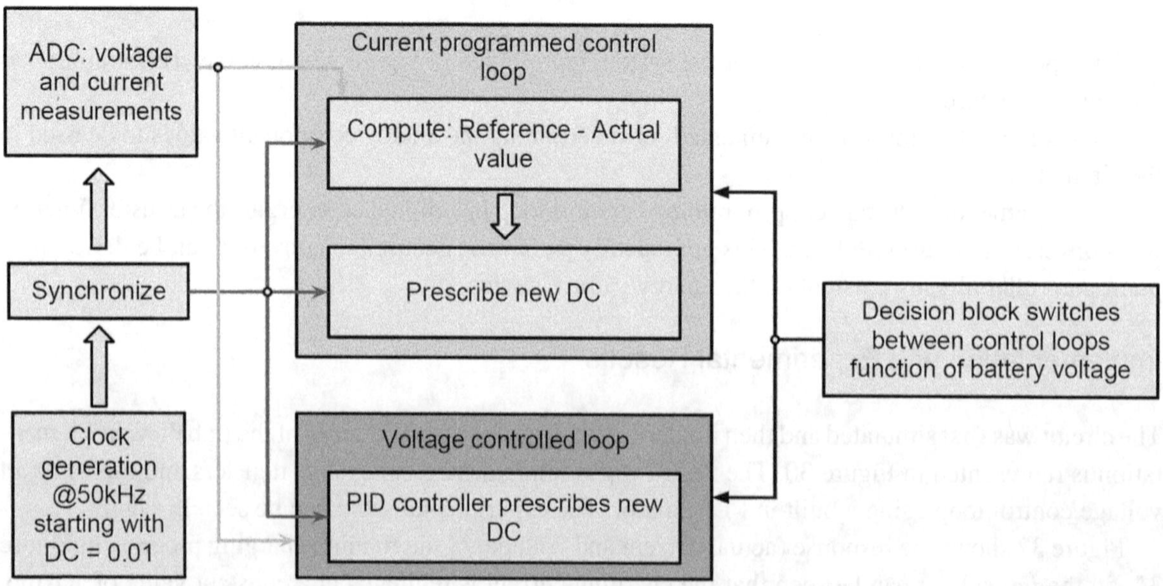


*Figure 31. Buck converter practical implementation*

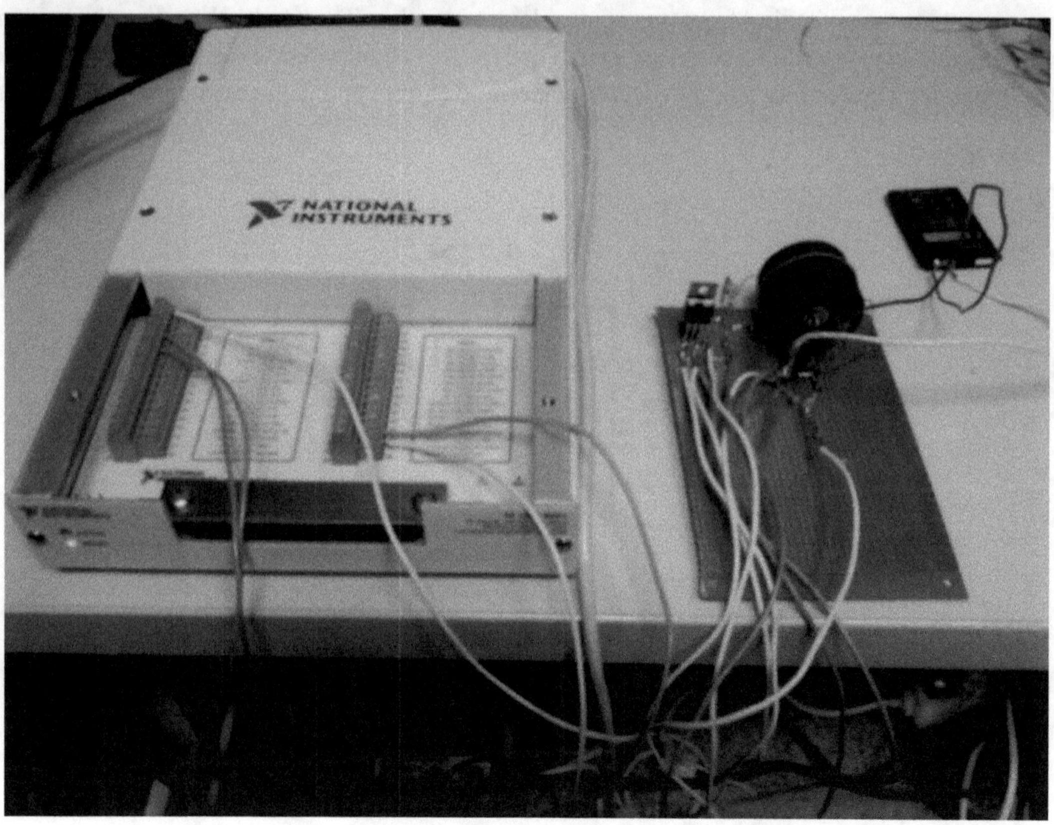

The chosen control methods: average current control and PI regulation of the constant voltage provide very good results for the proposed application of battery charging. The numerical implementation of these controls can be easily implemented in a microcontroller.

## Energy Harvesting System Simulation and Experimental Results

The proposed energy harvesting system presented in Figure24 was simulated in LTspice, where also the convective heat transfer was implemented. The system was tested with the thermosyphon cooling system described in Figure10. In this way, the natural environment conditions were also simulated. The simulation was necessary to assess the performance and compute the output parameters of the system under different conditions and in a fast manner compared to a real experiment. In the simulations, two TEG models were used – the one proposed by the authors (with varying parameters) and the constant parameters one. The first proved to generate the lowest errors. By correlating the experimental measured data to the model of the TEG, the electrical behavior of the device was accurately reproduced in simulation (Figure 33).

The comparisons between the experimental obtained values and the simulation ones are presented in the figures below. The experiments were conducted with the system operating in natural environment conditions.

In Figure 34, no load was connected to the TEGs output. One can observe that the simulated values follow the experimentally obtained ones.

The simulation model when the load is connected to the TEGs output is presented in Figure 35, where "TEG Positive Terminal" represents the output voltage from the TEGs model that is not displayed in the figure.

Figure 36 presents the temperature values from the simulation and experimental data, when the load (LTC 3105 converter with 360 Ω resistive output load) was connected to the TEGs output after 1000s since the recording procedure started. The output voltage of the LTC converter was set to 3.6V and the dynamic maximum power point tracking system was employed in both simulation and experiment.

*Figure 32. Front Panel of the running VI: The current loop is first active afterwards the voltage control loop is running.*

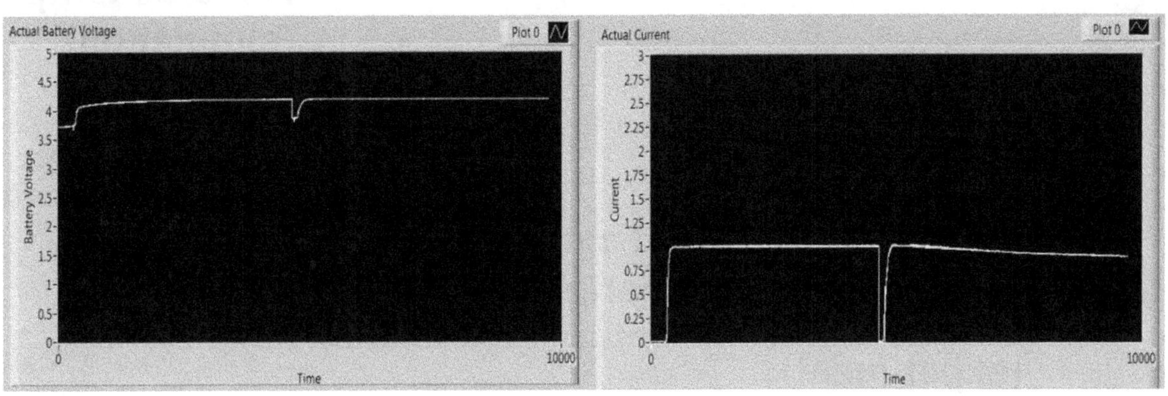

*Figure 33. Logic representation of the procedure that generates the input data for the heater temperatures in the SPICE simulator*

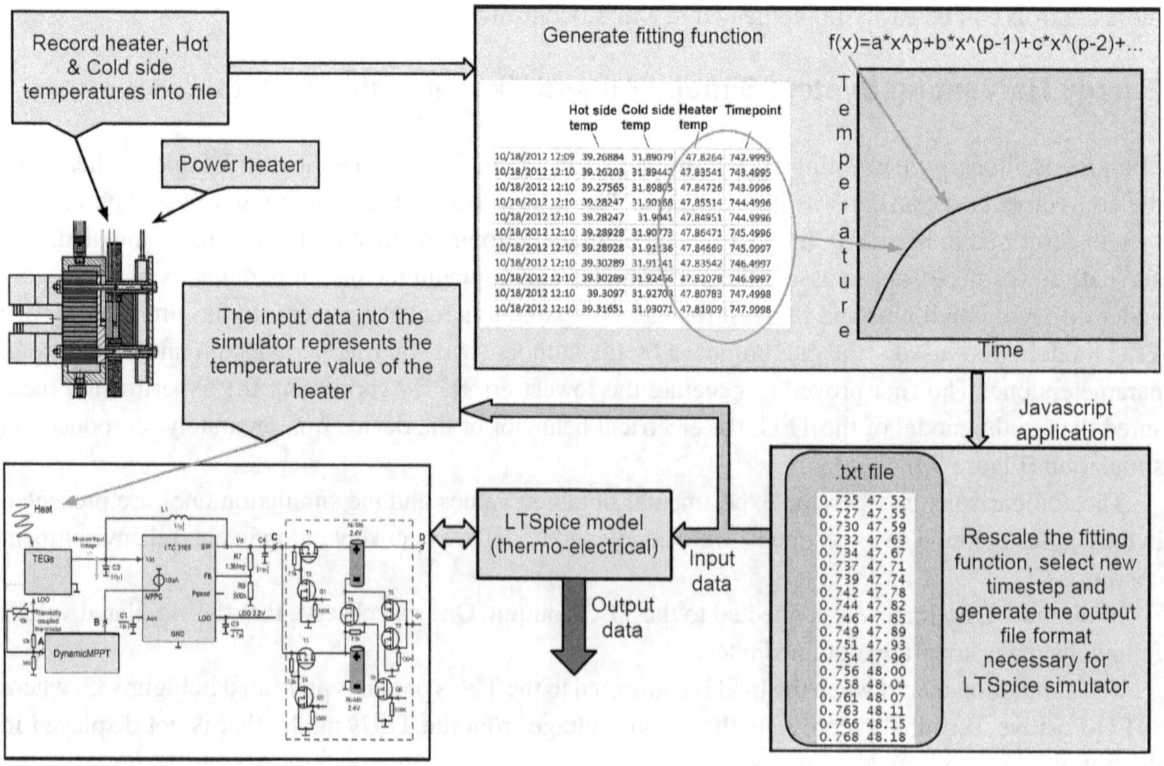


*Figure 34. Experimental and simulation results from the TEGs system without load, in natural environment*

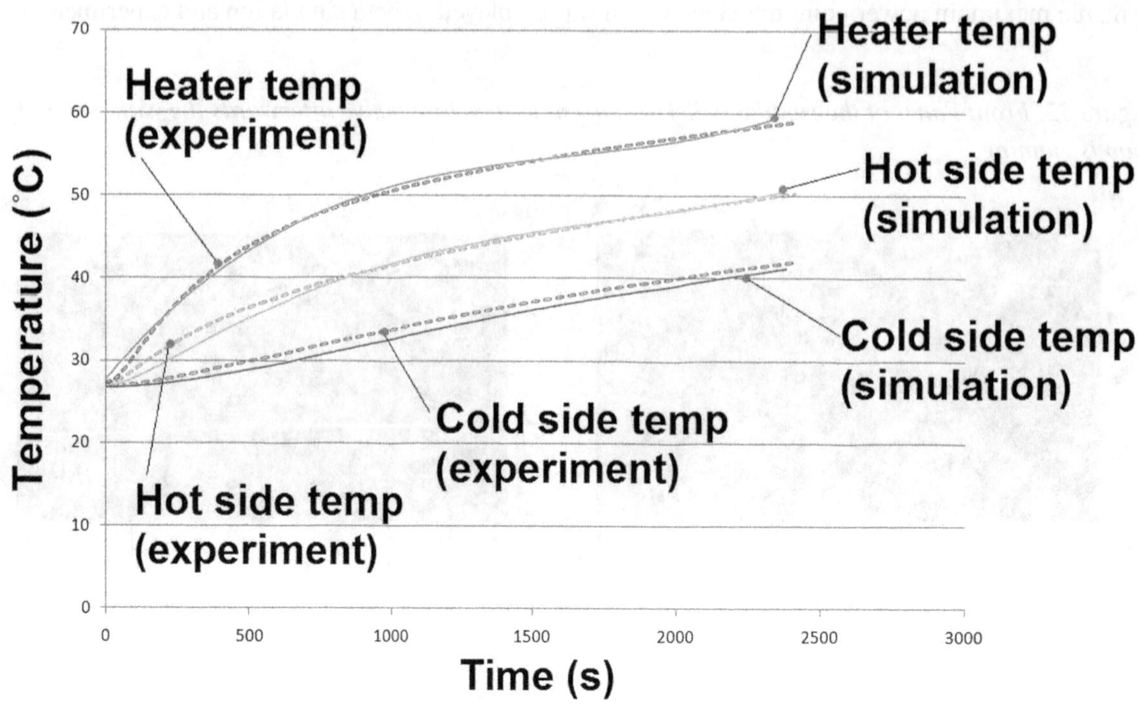

*Figure 35. Simulation model representing the DC-DC converter and load that are connected to the TEGs system's output*

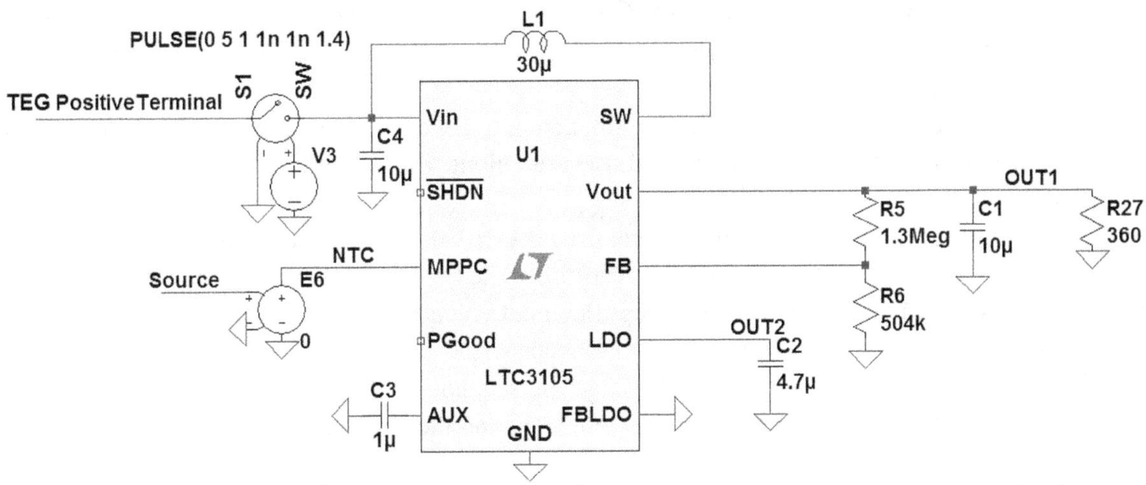


*Figure 36. Experimental and simulation results from the TEGs system with the load connected after 1000s, in natural environment*

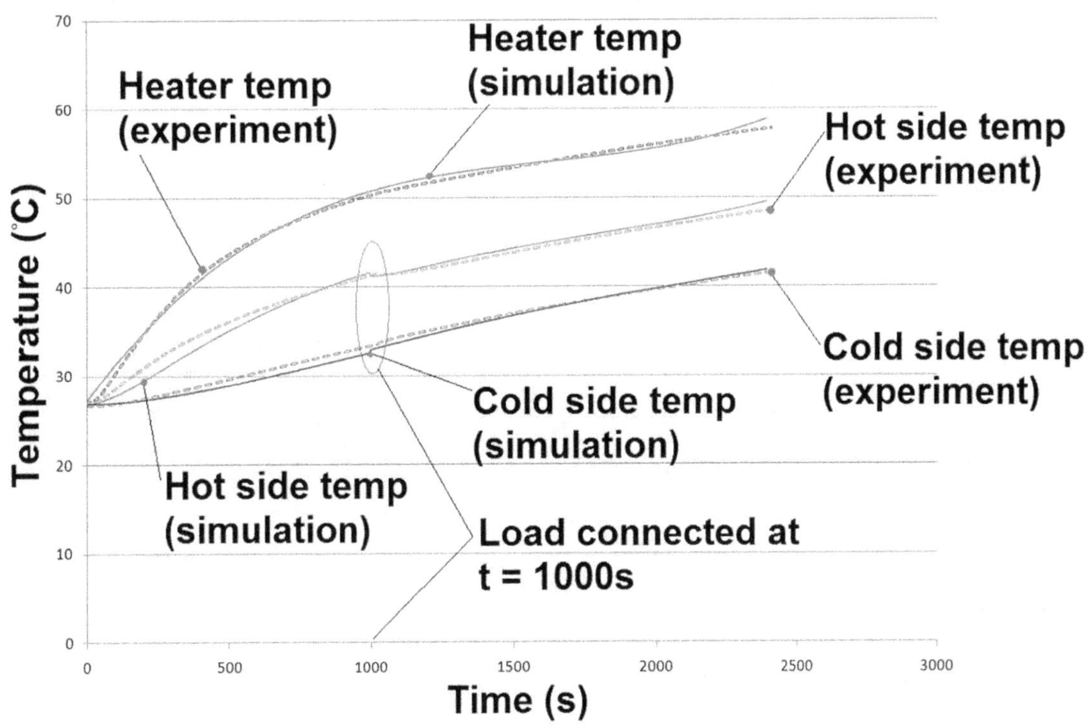

When the load is connected, one can see that the hot side of the TEGs temperature is getting lower while the cold side temperature increases. From the physical point of view, this phenomenon can be explained due to the diffusion process, as when the electrons start flowing (through the load connected to the TEG's output) from the hotter region of the TEG to the colder one, the first one losses heat while the latter absorbs heat.

Figure 37 presents the output voltage of the simulated model of the energy harvesting system as also the obtained experimental data, when the LTC converter along with the 360 Ω load is connected to the TEGs output.

Figures 38 – 41 present a series of experimental results with the TEGs system operating and charging the two Ni-Mh batteries. Figure 38 depicts the charging of the first Ni-Mh battery when the TEGs output voltage is 1V. The dynamic MPPT system keeps the input voltage to 0.5V in this case, maintaining the optimum power transfer to the output of the DC-DC converter.

In Figure 39, the PCS was controlled to charge the second Ni-Mh battery. Figure 40 presents the scenario when the battery charging process is completed and the two batteries are connected in series to generate a 4.8V Ni-Mh battery.

In Figure 41, the dynamic heating process of the TEGs system is presented. One can observe the increasing output voltage of the TEGs and in the same time the MPPT voltage. The DC-DC converter

*Figure 37. Simulation and experiment correlation of the TEGs system that is operating in natural environment with load*

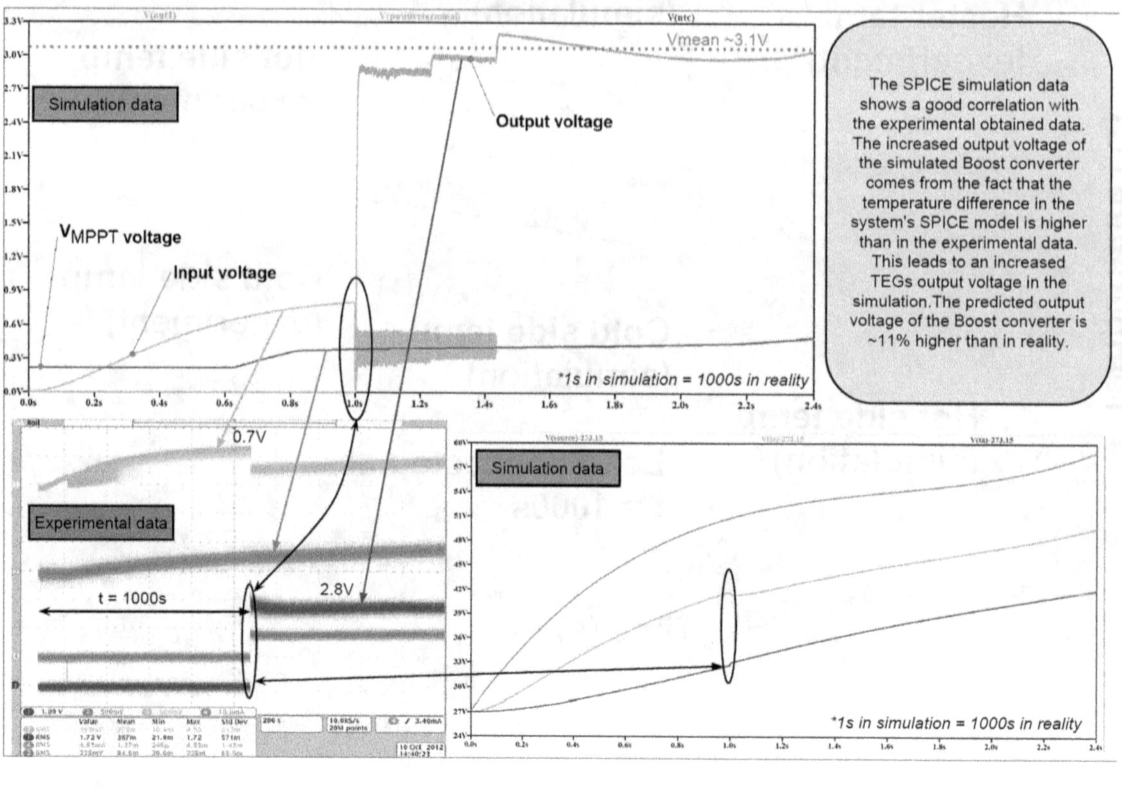

*Figure 38. Experimental results with 1V generated by the TEGs system and first battery charging*

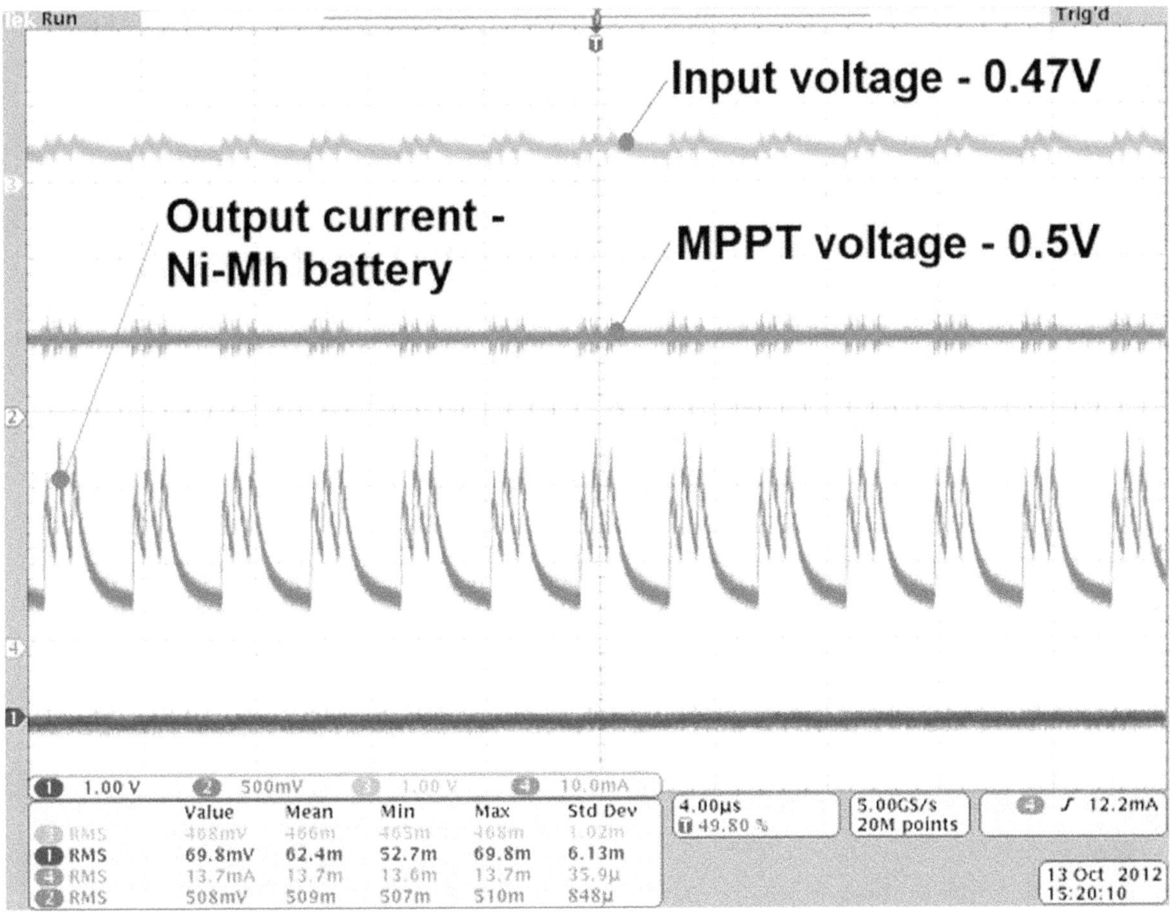


starts operating from ~ 0.3V but for this input voltage and with the Ni-Mh battery connected as load, the converter is not able to reach the output prescribed voltage of 3.6V. Only when the TEGs output voltage reaches ~ 0.5V, the LTC3105 output voltage attains 3.6V and the battery is starting to charge.

The proposed battery charging device operates as desired and the analog dynamic MPPT system improves the efficiency of the system by continuously adapting the voltage on the LTC3105 MPPT pin, in order to extract the maximum available power.

For the experiments conducted in this section, two NI DAQ boards were used – 6221 and 6251. The first, along with a specific built autonomous testing system controlled the P6100 power source that heated the power resistors and took care of recording the temperatures values and the input and output voltage values of the TEGs and the LTC converter. The second NI DAQ board was used to control the PCS system (Figure 42).

In Figure 42, the TEMs test rig can be observed on the left side of the picture, where the temperature and voltage values are monitored by one NI DAQ 6221 (not shown) while the electronics (dynamic MPPT, converter, batteries) are controlled by the second NI DAQ 6251 (lower right corner). The current and voltage values at the converter output are monitored by the oscilloscope (upper right corner).

*Figure 39. Experimental results with 1V generated by the TEGs system and second battery charging*

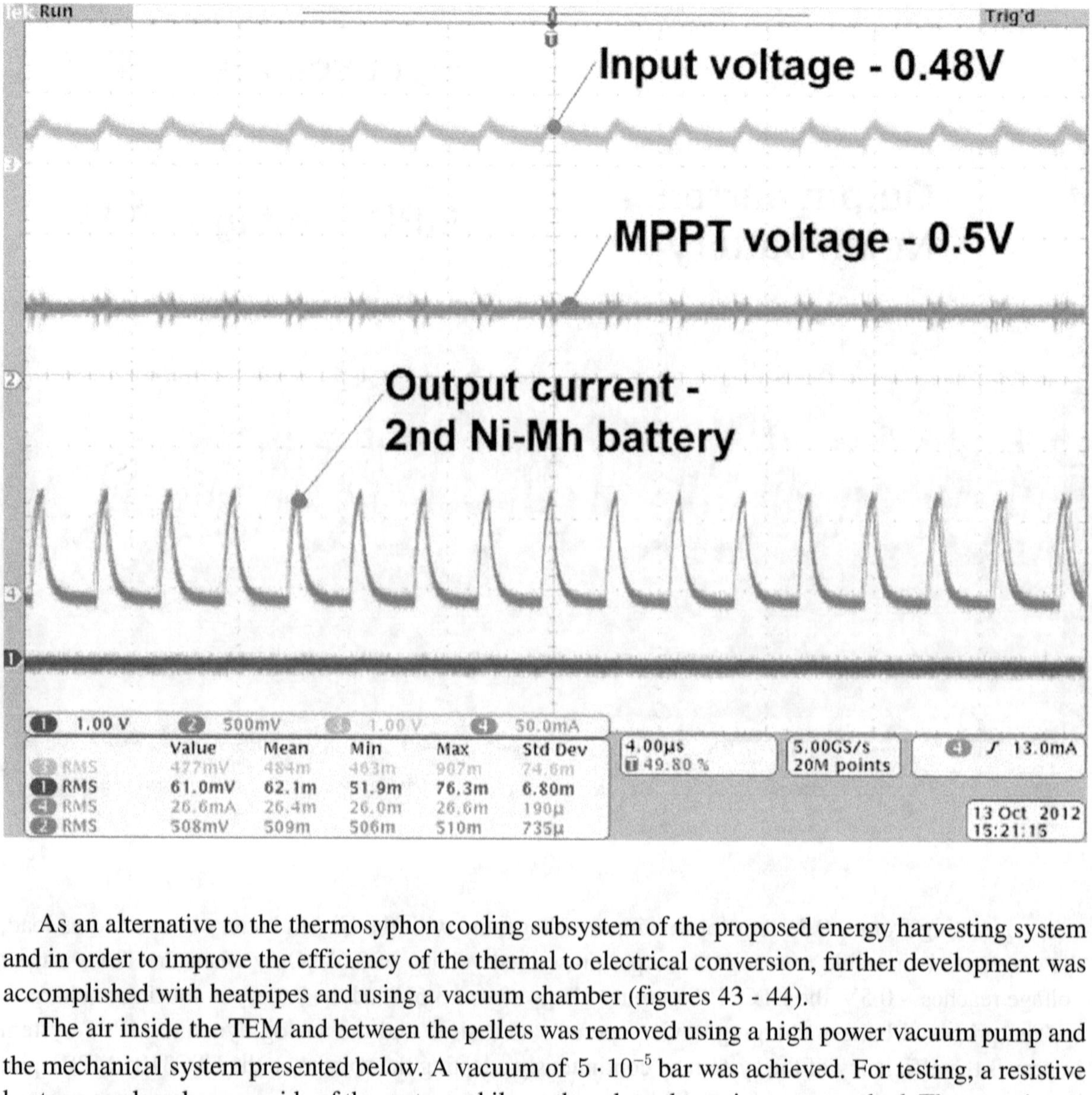


As an alternative to the thermosyphon cooling subsystem of the proposed energy harvesting system and in order to improve the efficiency of the thermal to electrical conversion, further development was accomplished with heatpipes and using a vacuum chamber (figures 43 - 44).

The air inside the TEM and between the pellets was removed using a high power vacuum pump and the mechanical system presented below. A vacuum of $5 \cdot 10^{-5}$ bar was achieved. For testing, a resistive heater was placed on one side of the system while on the other a heat pipe was attached. The experimental results show that the efficiency has been increased by up to 4% during the tests that were driven up to 80degC on the hot side of the system. The vacuumed TEG was able of providing a higher output voltage for the same temperature of the heat source. The system also protects the TEG from moisture while also generating a compact device.

## The Improved Ćuk Converter for Multiple TEGs

In order to extract the maximum available power from a TEG, efficient DC – DC converters are needed. Several devices of this kind already exist (e.g. LTC 3105, LTC 3108, MAXIM 17710, etc) but their us-

*Figure 40. Experimental results with 1V generated by the TEGs system and batteries connected in series with the help of the Power Control Switch*

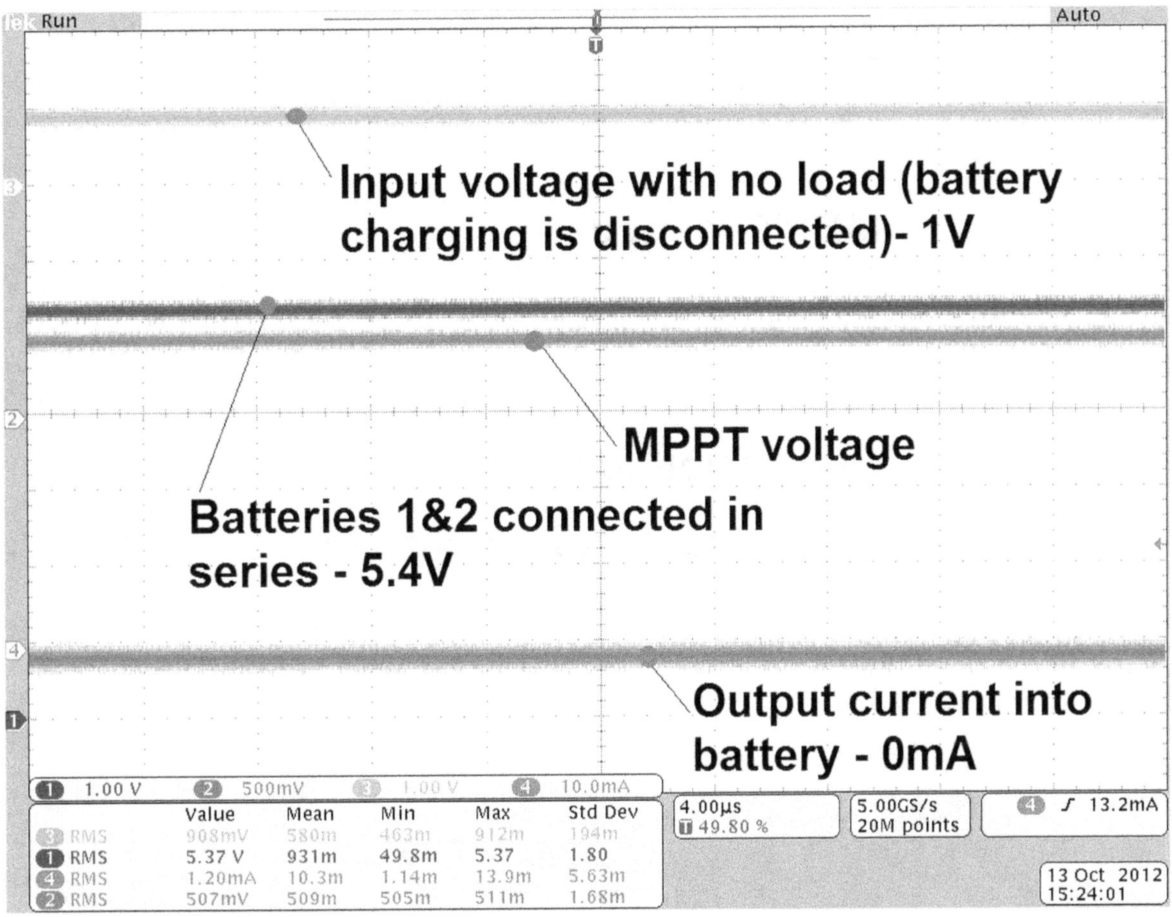


age is limited to one or two input energy sources – TEGs (Linear Technology, 2014), (Maxim, 2014). The LTC3105 for example, like presented in paragraph 5.1 has the output current limited to 400mA and the output voltage scalable from 1.5 to 5.25V. However, if the energy source has physically large dimensions (e.g. heat engines), several serial connected TEGs would provide more power than a single module would. For this kind of applications, custom DC – DC converters must be developed. In this way, a Ćuk converter that is supposed to operate from four (or a higher number) serial connected TEGs is proposed. The converter purpose is to charge a pack of either Li-Ion or Ni-MH batteries with maximum efficiency while adapting its impedance with that of the equivalent power source (TEGs). In this way, at any moment, the converter will work by transferring the maximum available power to the battery. In order to control this behavior, an indirect current control loop was developed. The loop is based on the continuous measurement of the TEGs output voltage and an a priori built lookup table that stores the output voltage – internal resistance correspondence of the TEGs. The correspondence has been experimentally measured and is necessary because a TEG's internal resistance and Seebeck coefficient varies with temperature.

*Figure 41. Experimental results with the dynamic heating of the TEGs system and the first battery connected for charging*

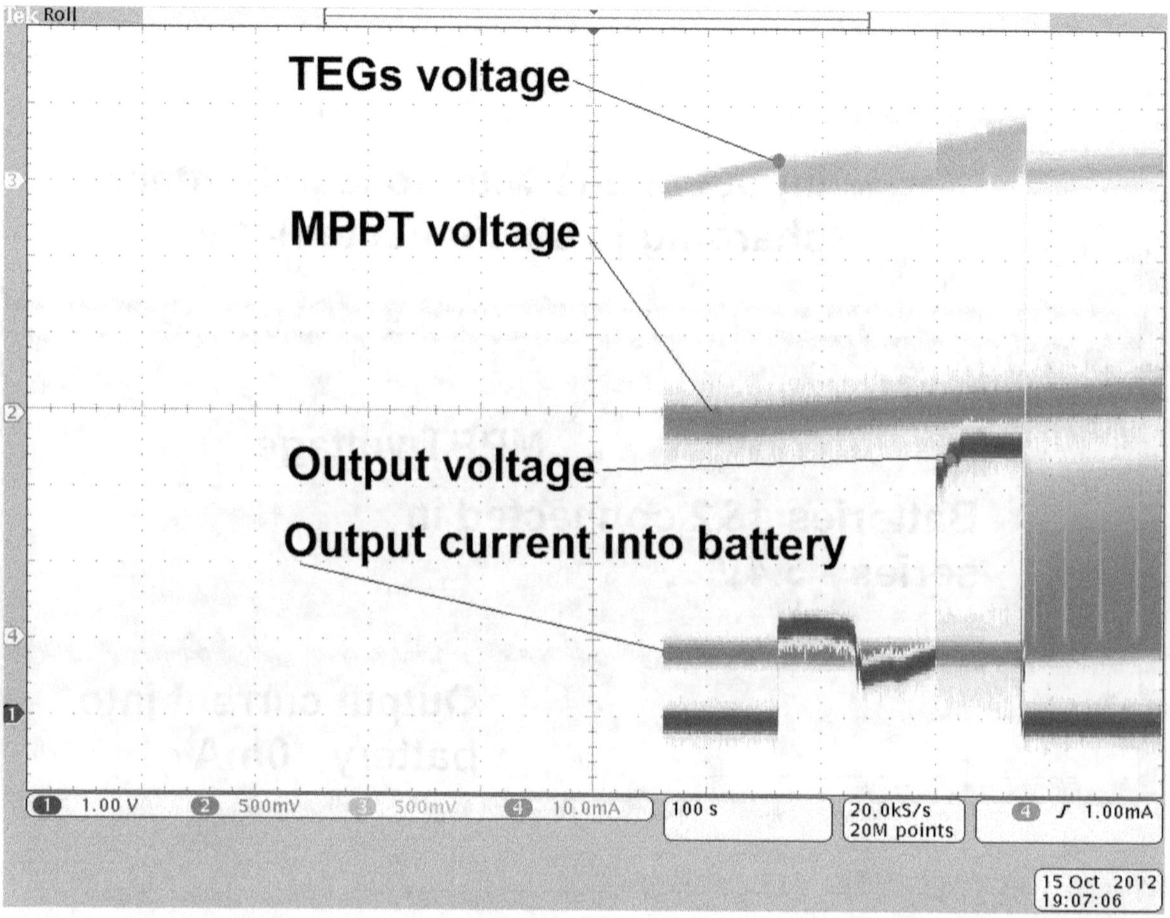


Figure 45 presents the proposed Ćuk design where, in order to increase efficiency, the original diode presented by Erickson (2001) has been replaced by an n-MOS switch. An isolation transformer is used to reverse the output voltage polarity and assure the operation of the second n-MOS. In standard configuration, Rashid (2007) shows that a Ćuk converter generates an output voltage with reversed polarity compared to the input. The control of the switches is performed by an 80 MHz Spartan 3E FPGA that is able to provide several nanosecond timing necessary when no cross conduction of the n-MOS switches is required. A small delay that depends on each of the transistor's rise and fall time needs to be implemented for dead time control. The need for such precise timing comes from the fact that the rise and fall time of any transistor differ, the falling time being usually higher than the rising one.

In order to measure the voltage necessary to control the duty cycle of the switches, a NI DAQ 6251 measurement board was used.

*Figure 42. Complete TEGs system experimental setup*

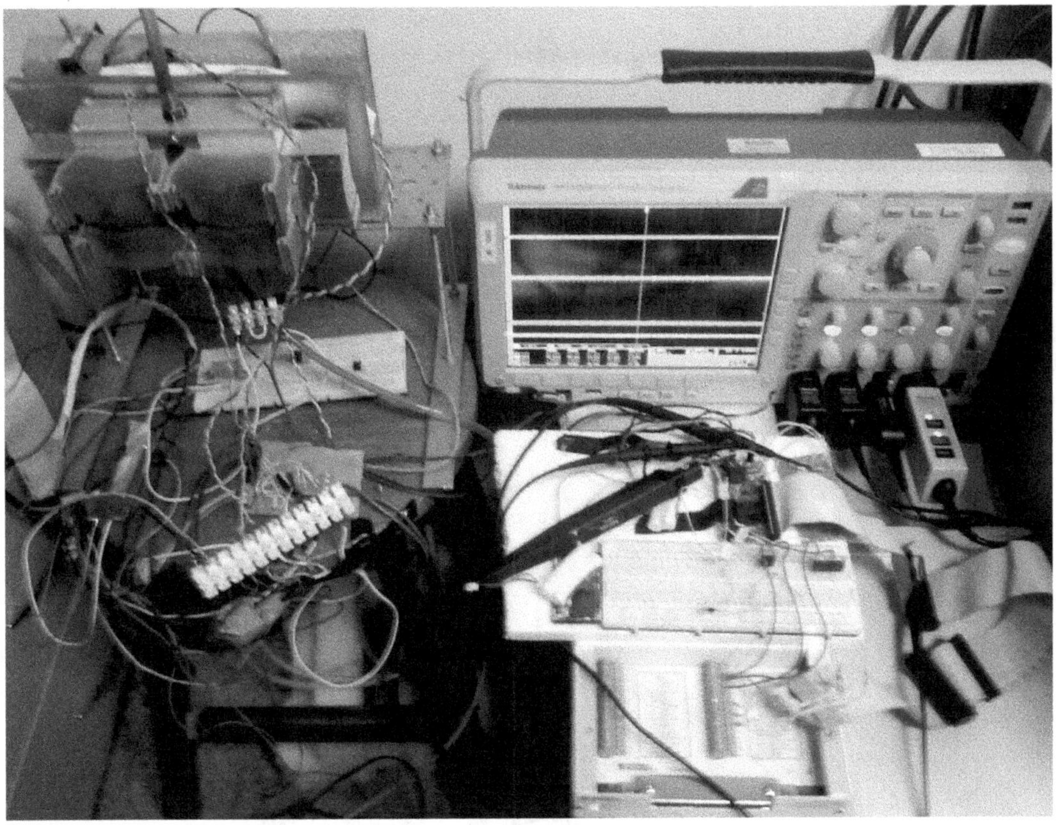

## The SPICE Simulation Model

At first, the simulation model was implemented under Linear Technology LTspice software (Figure 46). The component values and manufacturer were chosen from the simulator library and also represent the real components chosen in the real implementation. In the practical implementation, the transformer was specially designed and built for this specific application.

The simulated model employs both the Schottky diode and the n-MOS synchronous rectifier switch that replaces it. When the Schottky diode was used in simulation, the second's n-MOS switch control signal was connected to the $V_7$ power source (Figure 45) that outputs 0V signal. Two loads were alternatively used, the $R_2$ resistor and the $V_2$ battery.

A parametric step simulation was adopted to determine the correct duty cycle value that corresponds to the maximum transferred power between input and output.

The simulation results are presented in Figure 47, for 3V input voltage and the duty cycle corresponding to the maximum power transfer.

Figure 48 presents the simulation where the second n-MOS switch replaced the Schottky diode in order to improve the efficiency.

*Figure 43. Vacuumed TEG system cross sections*

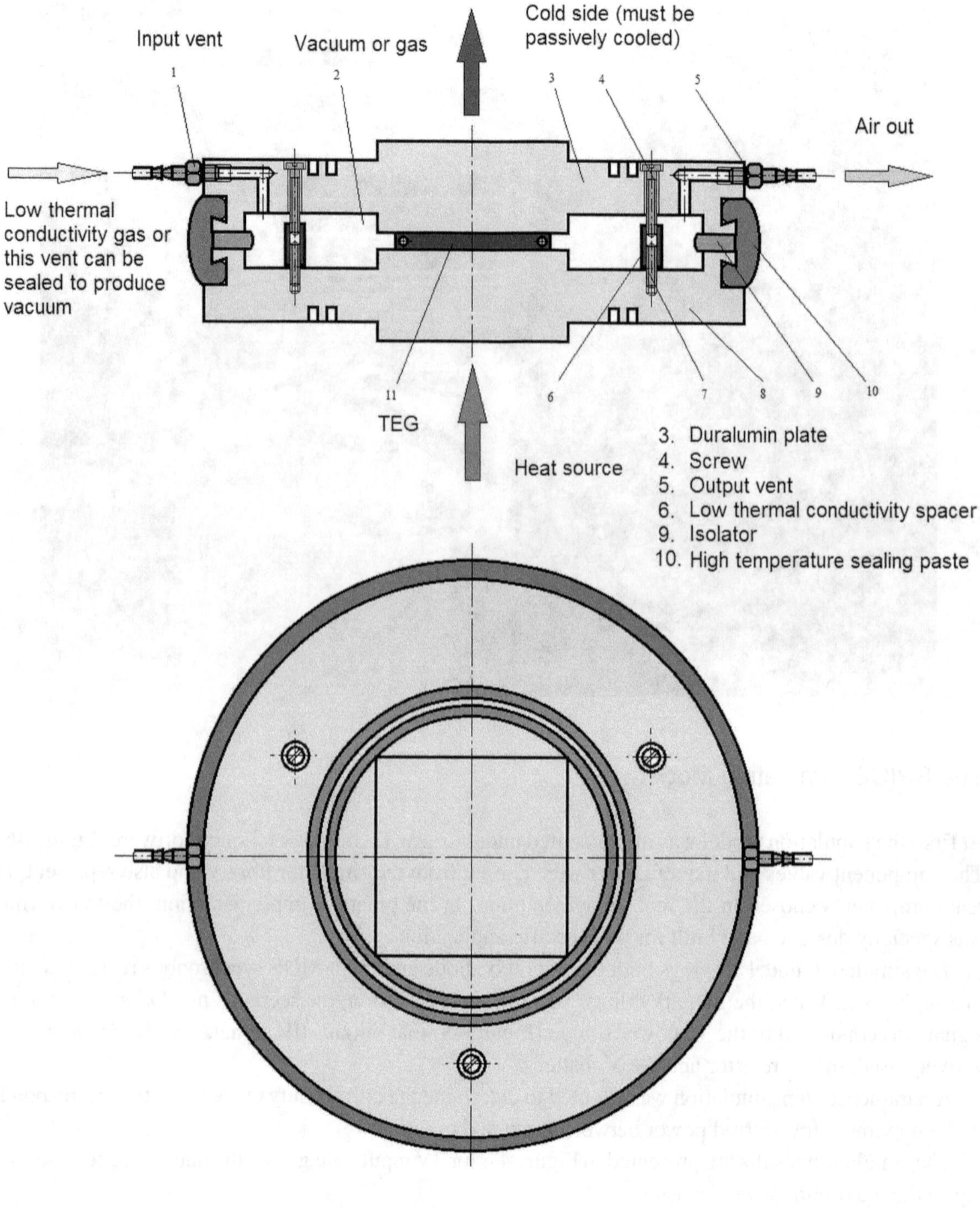

*Figure 44. Vacuumed TEG*

*Figure 45. Proposed Ćuk converter employing TEG emulator and FPGA dead time control*

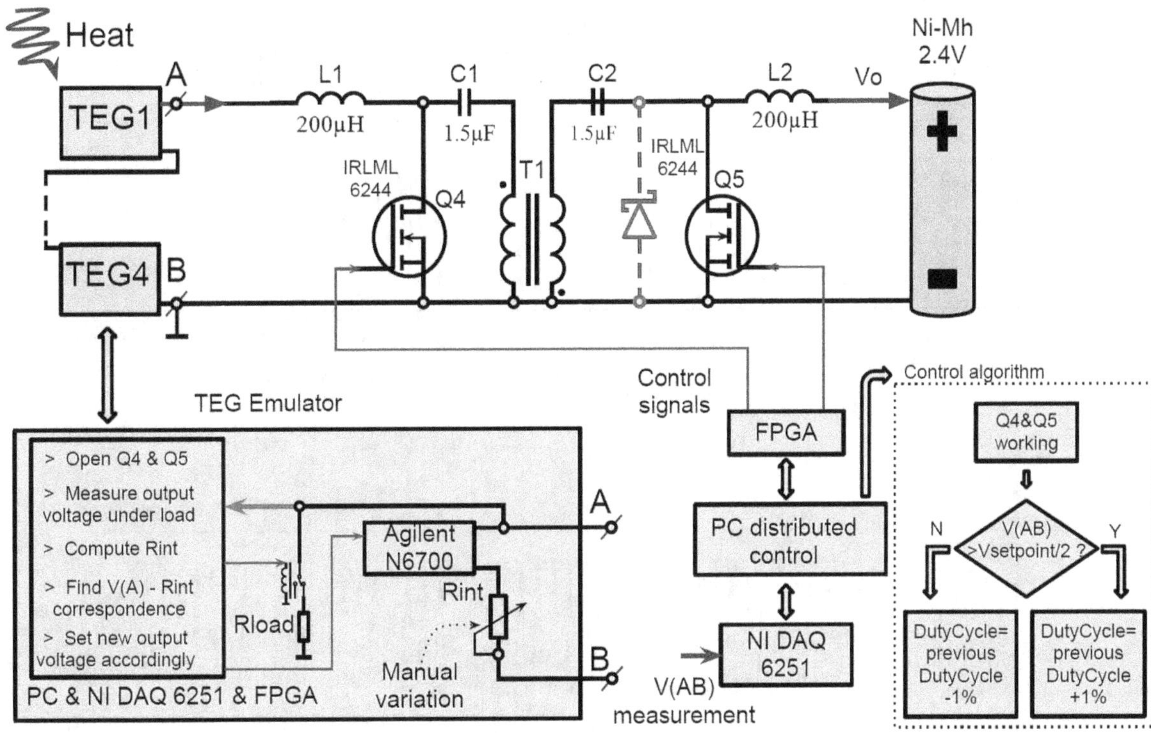

*Figure 46. Ćuk converter LTspice simulation model using real components values*

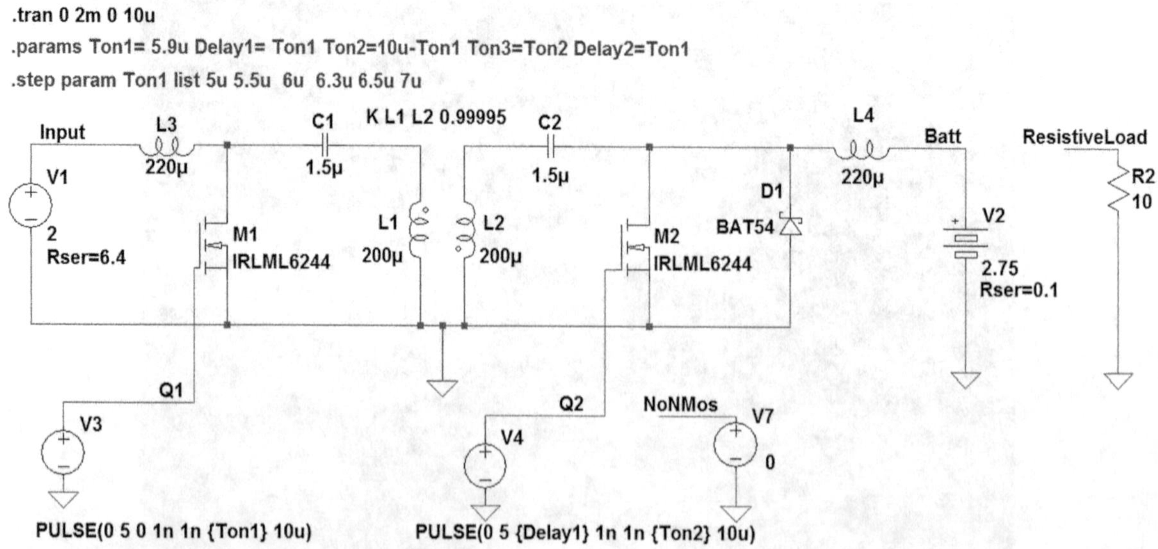


*Figure 47. Output voltage and current for 3V input and BAT54 Schottky diode*

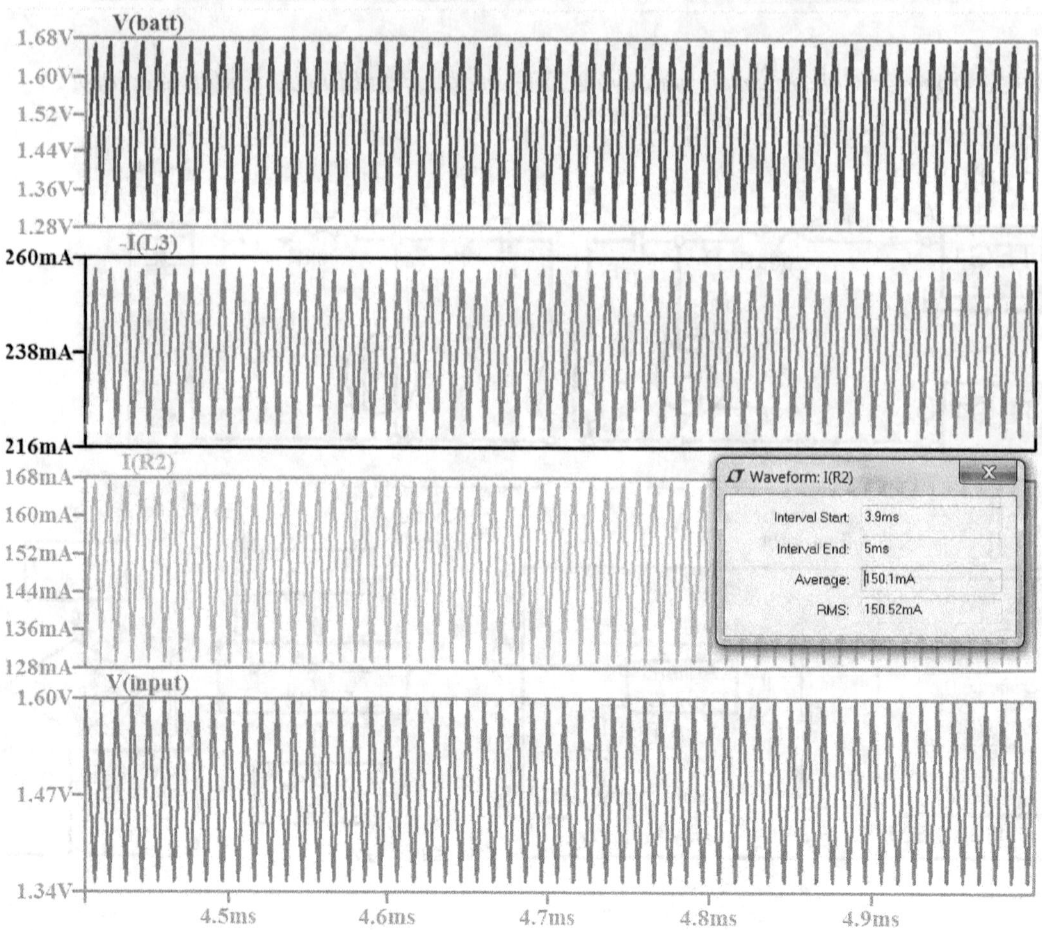

*Figure 48. Output voltage and current for 3V input and n-MOS switch that replaces the Schottky diode*

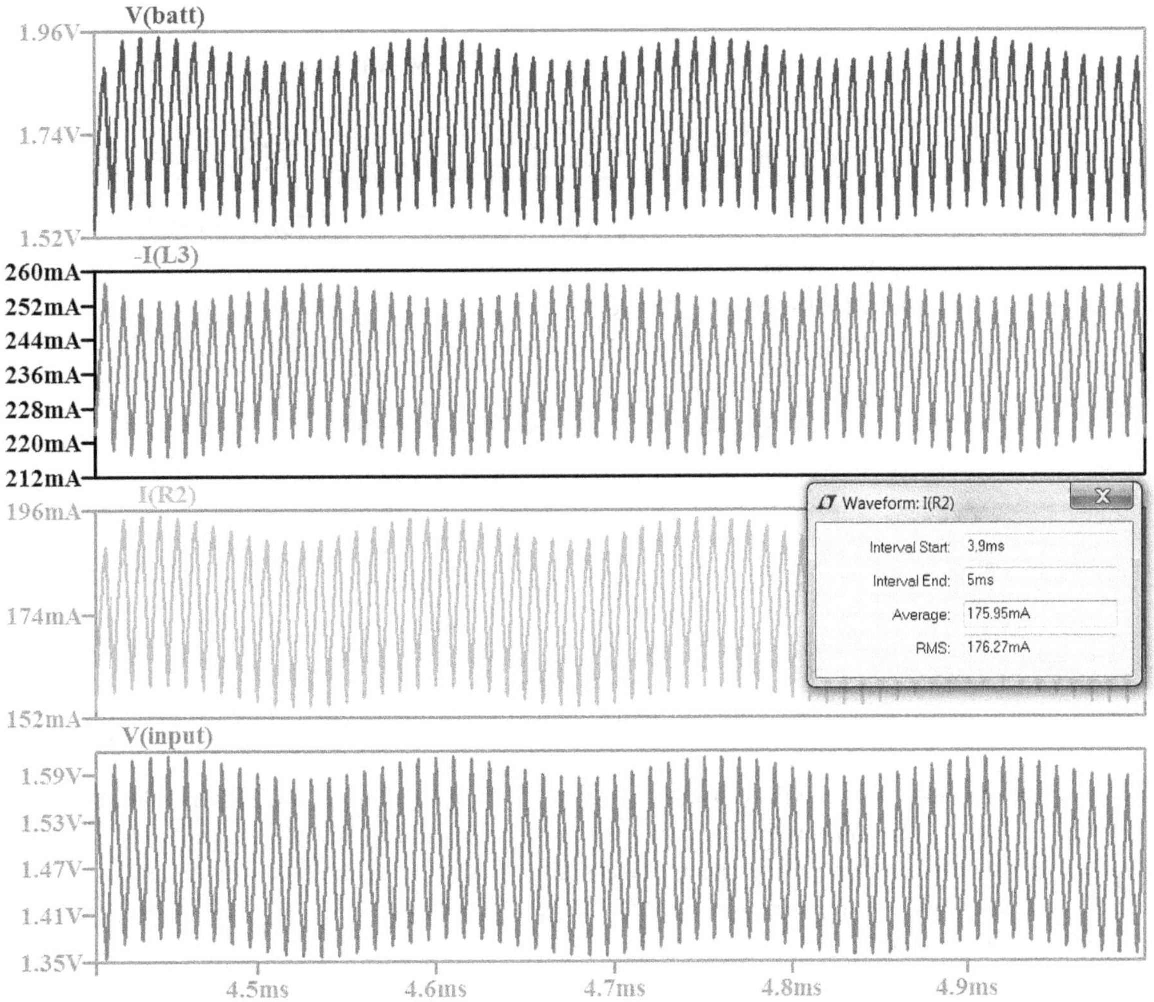


It can be seen that for the same input voltage, when the converter impedance is adapted for maximum power transfer, the current into the resistive load is 15% higher compared to the case when the Schottky diode was used.

The simulations were repeated with a 2.4V battery pack as load. The battery was simulated by a 2.75V power source in series with an internal resistance. The selected supply voltage value was chosen as experimentally determined with a 2.4V Ni-Mh battery pack.

Figure 48 present the converter efficiency with three different loads – the Ni-Mh battery, a 50Ω resistor and a 100Ω resistor. The efficiency is computed over an input range starting from 1.5V and up to 5V, when the duty cycle of the n-MOS switches was chosen to provide maximum power transfer between input and output.

One can see that the efficiency of the converter in the input range of 1.5-5V, when the Ni-Mh battery is used as load is around 90%.

*Figure 49. Efficiency vs. input voltage for three different loads – Ni-Mh battery, 50Ω load and 100Ω resistive load and when the Schottky diode is used*

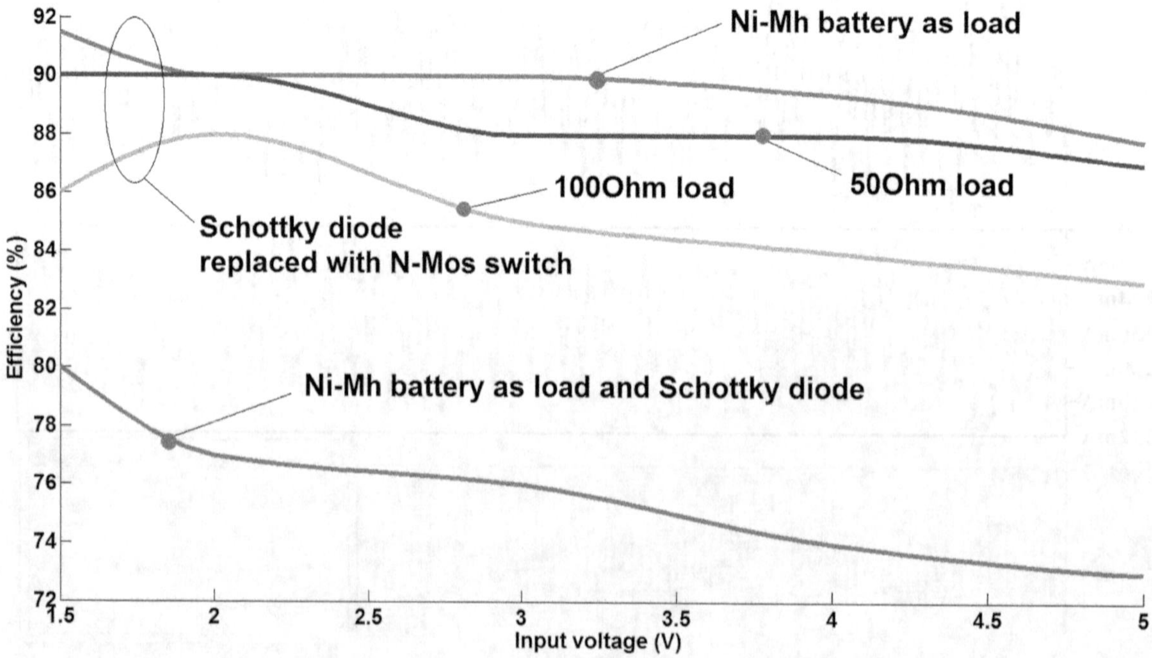


The efficiency when the Schottky diode is used instead of the second n-MOS switch is also presented in Figure 49. One can observe that in this case efficiency is up to 20% lower. While the input voltage increases, the decreasing slope of the efficiency is higher in the case of the Schottky diode. This is due to fact that as the input voltage is higher the duty cycle will get lower resulting in a longer conduction time for the diode and consequently a lower efficiency.

Simulation results prove that replacing the diode by the n-MOS switch leads to increased efficiency that reaches 90%. However, the simulation was performed considering ideal signal sources which in practical implementation will lead to a decrease in the efficiency.

Figure 50 displays the simulated efficiencies when the n-MOS switch replaced the diode and both switches were controlled by a 50kHz control signal. It can be observed that the efficiency for a 50Ω load is now lower, compared to the case presented in Figure 49. This can be explained by the commutation losses of the switches, that in the case of the 100kHz control signal are higher than the conduction losses.

## The FPGA Control

When replacing the diode from the original Ćuk topology with an n-MOS switch, the simultaneous conduction of the two transistors must be taken into account. Simultaneous conduction will appear because of the different rise and fall times of the n-MOS switches. The datasheet of the IRLML6244 transistor states that the typical total rise time is 12.4 ns while the fall time is 31ns. The datasheet values are generic ones and can vary between manufacturers and between different product samples which may lead to larger differences between the practical rise and fall time.

*Figure 50. Simulated efficiencies when a 50KHz control signal was used*

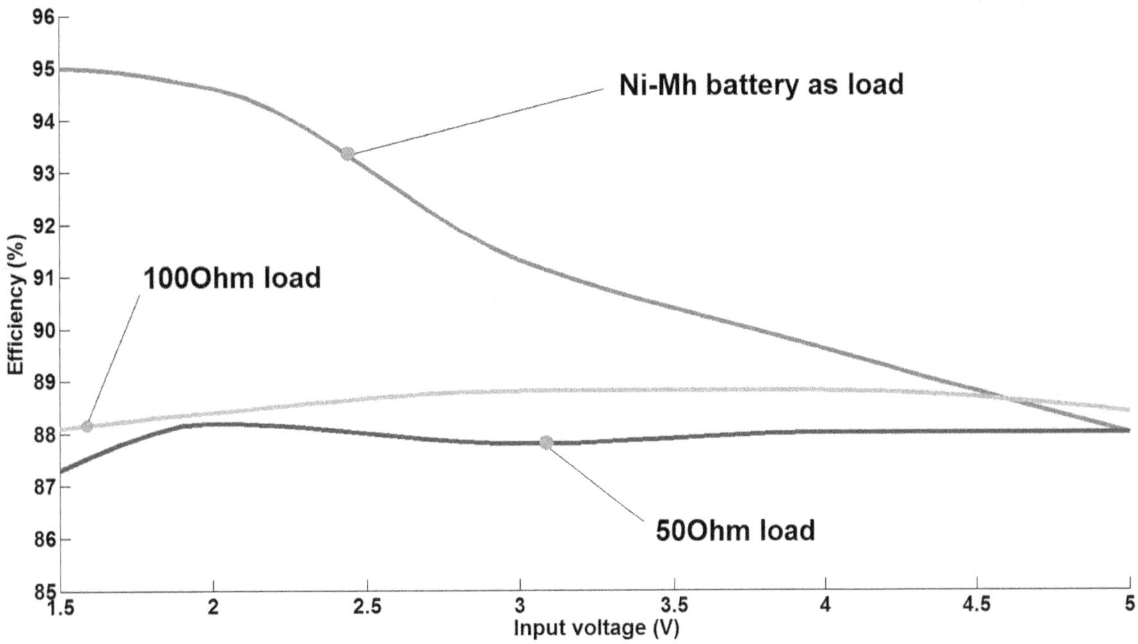


In order to prevent simultaneous conduction of the n-MOS switches nano-second resolution dead time control must be implemented. In the current work this is achieved with the help of an initially 50MHz SPARTAN 3E FPGA, on which an 80MHz oscillator was added. The increased oscillator frequency allows the FPGA to provide output signals with 12.5 ns resolution.

Figure 51 presents a 62.5ns delayed control signals that drive the n-MOS switches.

A LabVIEW FPGA VI (virtual instrument) was built to generate two control signals. For the n-MOS switch that replaces the diode, a user controlled delay that allows the adjustment of the control signal with an offset with respect to the first n-MOS switch, like in Figure 50, was built.

## The Control Algorithm and the Experimental Setup

In the experimental setup, the thermoelectrical generator input power source was emulated by a programmable Agilent N6700 power source in series with a variable resistor, the same as in the process explained in Figure 18. An autonomous application has been built in LabVIEW that, at specified time intervals, decouples the converter with the help of a relay (Figure 44), measures the output voltage of the emulated TEG (power source in series with the variable resistance) and modifies accordingly to the implemented lookup table the output voltage of the power source. In this way, a TEG that works under various temperature differences is emulated. Voltage measurements are performed with the help of NI DAQ 6251 board and LabVIEW. The NI DAQ is also responsible for driving the relay that connects or disconnects the load resistance in Figure 45, which in turn enables the computation of the actual internal resistance of the TEGs. In normal operation mode, when the converter charges the battery, the relay in Figure 45 is opened and the load resistance is disconnected. The NI DAQ board measures the input

*Figure 51. Oscilloscope view of the FPGA control signals output with 5 ticks delay (62.5ns)*

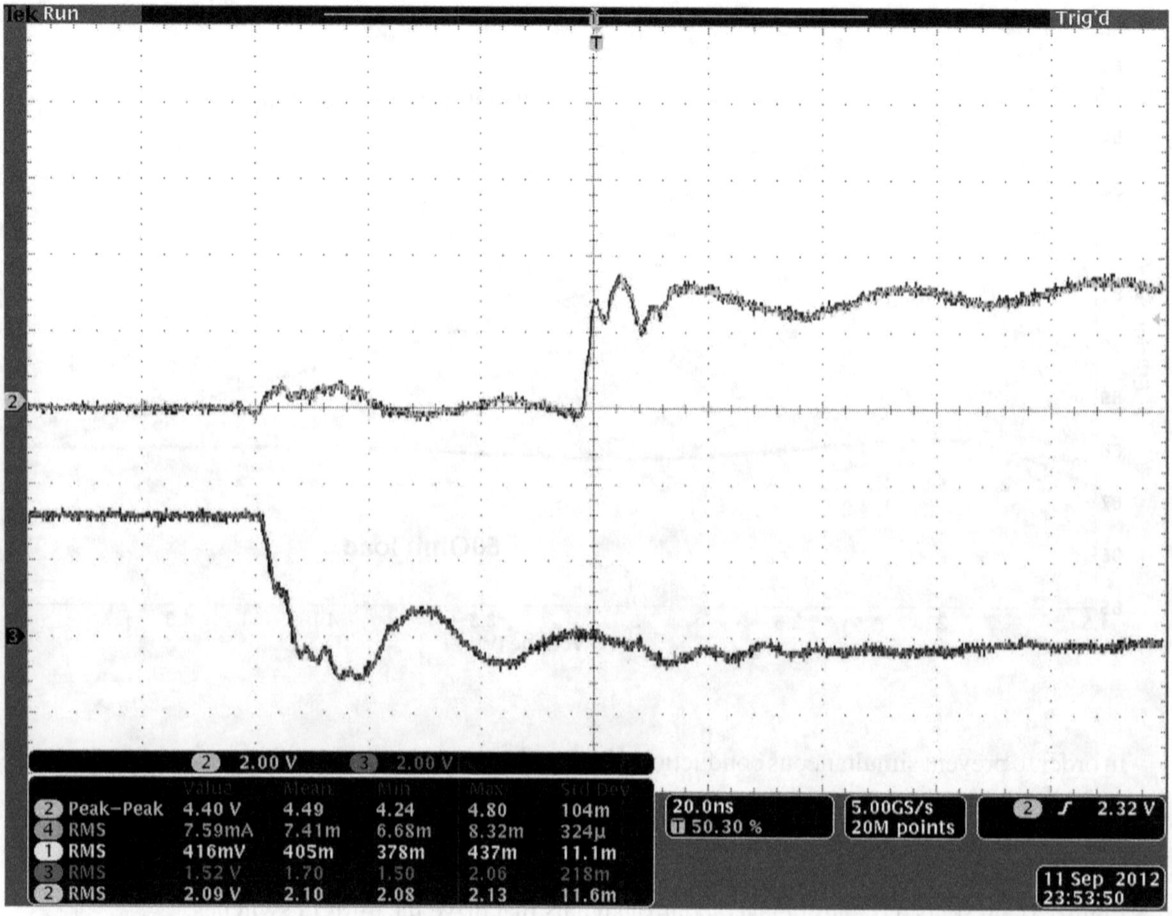


voltage of the converter that is the TEGs output voltage. The output voltage of the converter will also be measured by the NI DAQ board, which is helpful when observing the charging status of the battery.

A LabVIEW VI was built on the PC which accesses, through the NI driver, the SPARTAN 3E FPGA and also controls the NI DAQ 6251 board. The PC VI allows "on the fly" update of the FPGA control signals and allows the user to see the computed internal resistance $R_{int}$ of the emulated TEG and the input and output voltages of the converter.

The control algorithm that allows the proposed Ćuk converter to operate in the maximum power transfer regime was implemented in the LabVIEW PC VI. The maximum power transfer will occur when the resistance of the load equals the resistance of the source or

$$V_{in} = \frac{V_{noLoad}}{2},$$

where $\boldsymbol{V_{in}}$ is the Ćuk converter's input voltage and $V_{noLoad}$ is the output voltage of the TEG when no load is connected to it.

The control algorithm is implemented in such way that during normal operation, the measured input voltage of the converter is maintained to half the TEGs output voltage value, when no load is connected to it. The TEGs output voltage is measured at definite periods of time (10 seconds or more), when switch $Q_1$ is opened and the relay is connected, to check if the internal resistance has been modified.

The control algorithm increments or decrements with two ticks step the n-MOS switches duty cycle, until the desired setpoint has been reached. Taking into account that temperature variation is a slow process and no fast variations are expected, the proposed control algorithm proved effective in maintaining the prescribed setpoint, as will be presented in the following paragraph.

## Experimental Results

The converter has been implemented using the components specified in the LTspice simulation model. Using the emulated TEG device, different input voltages have been applied, as in the simulation. A 2.4V Ni-Mh battery has been used as load. The input and output voltages have been measured using the oscilloscope probes and the input and output currents by using a 1mA resolution Tektronix TCP0030 inductive current probe (Figure 52).

*Figure 52. Experimental setup: Cuk converter, SPARTAN FPGA, NI DAQ board*

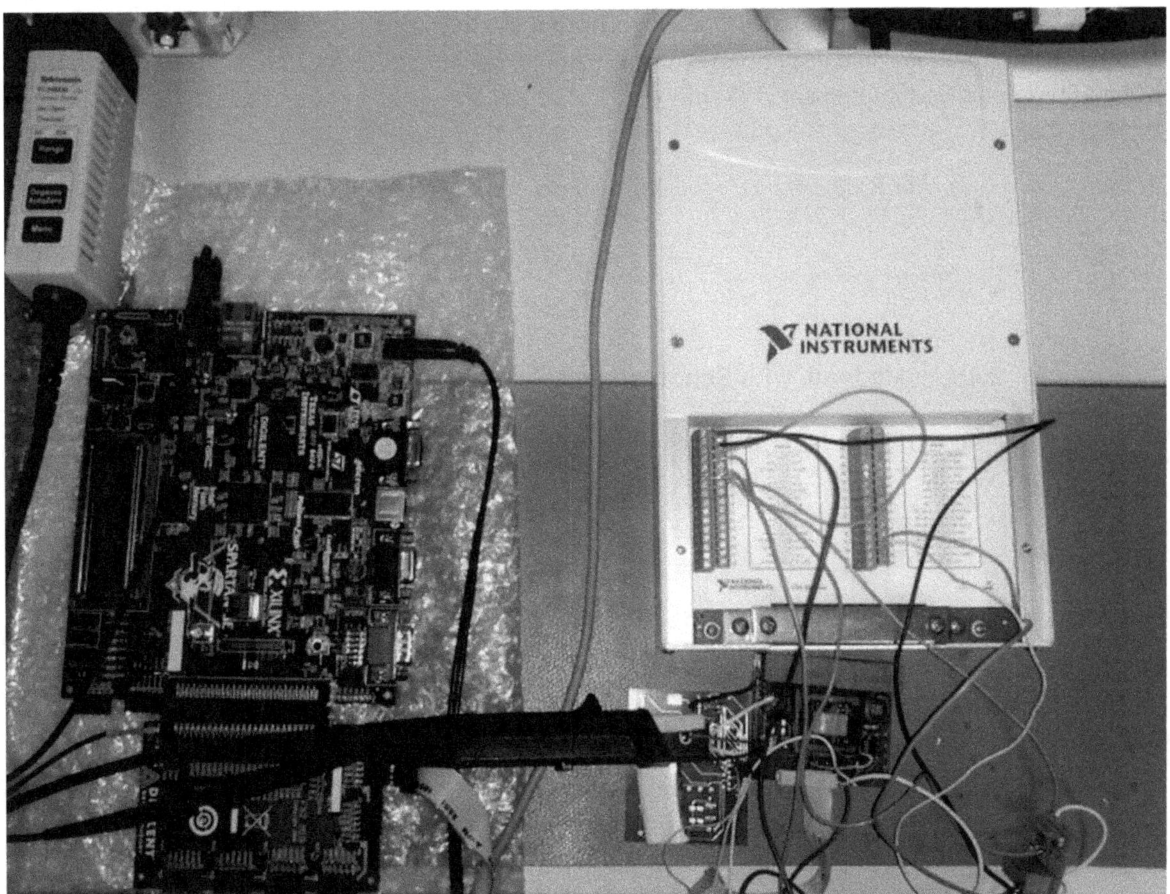

First, the experiments have been conducted using the original Schottky diode instead of the n-MOS switch. The results showed a lower efficiency by 9.5%, compared to the case when the n-MOS switch was used.

Figure 53 presents the experimentally computed efficiency versus input voltage, in two cases, with the n-MOS switch and the Schottky diode and with two different loads, the 2.4V Ni-Mh battery and a 100Ω resistor.

The experimentally determined efficiency is 7% lower than the simulated one (Figure 48) when the Ni-Mh battery was used as load. As mentioned, the simulation was conducted using ideal control signal sources with zero output impedance and infinite current sourcing. Due to the fact that the digital logic output of the FPGA has nonzero output impedance and a finite output current, a push-pull stage should be employed to improve the efficiency by rapidly charging the input capacitors of the $Q_4$ and $Q_5$ n-MOS switches.

Figure 54 presents the drain to source voltage of the two n-MOS switches with a 2.4V Ni-Mh battery pack load, without dead time control. Figure 55 shows how 100ns dead time control leads to no simultaneous conduction of the two switches.

Figures 56 and 57 presents the input voltage, output voltage and current through load for the two cases: with and without dead time control implementation.

One can observe that when dead time control is not implemented, large spikes appear in the input and output voltages. These are generated by the simultaneous conduction of the two n-MOS switches and are harmful to these switches as can exceed their rated drain to source voltage. Also, the output current is lower in the situation when dead time control is not implemented (82mA) with respect to the situation when 100ns delay was implemented (87mA).

*Figure 53. Experimentally determined efficiency versus input voltage of the proposed Ćuk converter*

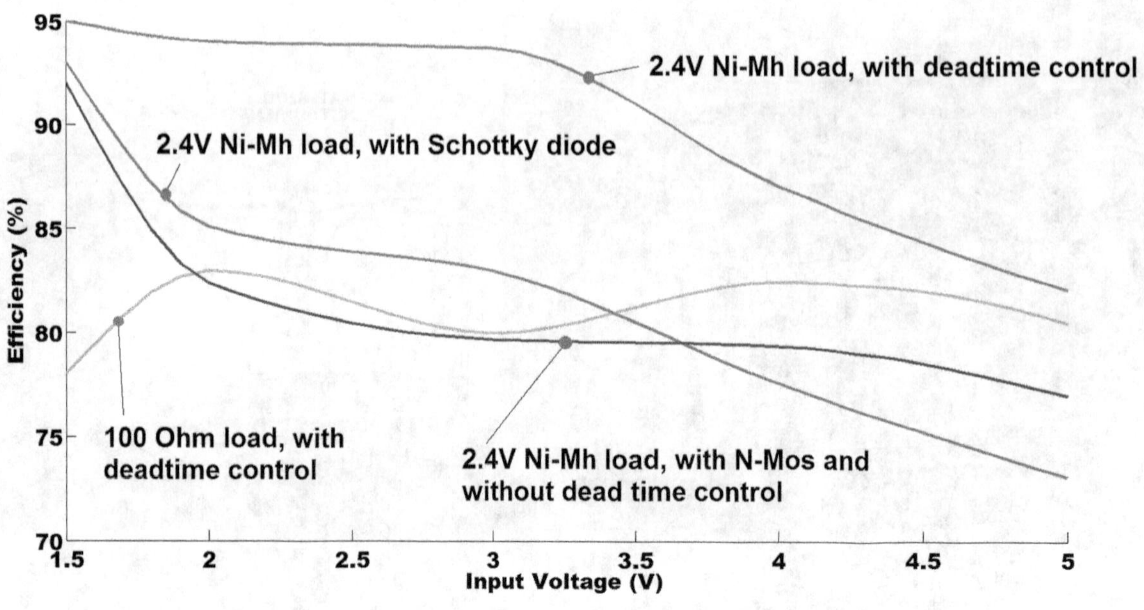

*Figure 54. $Q_4$ and $Q_5$ Drain to Source voltages and the simultaneous conduction when no dead time is implemented*

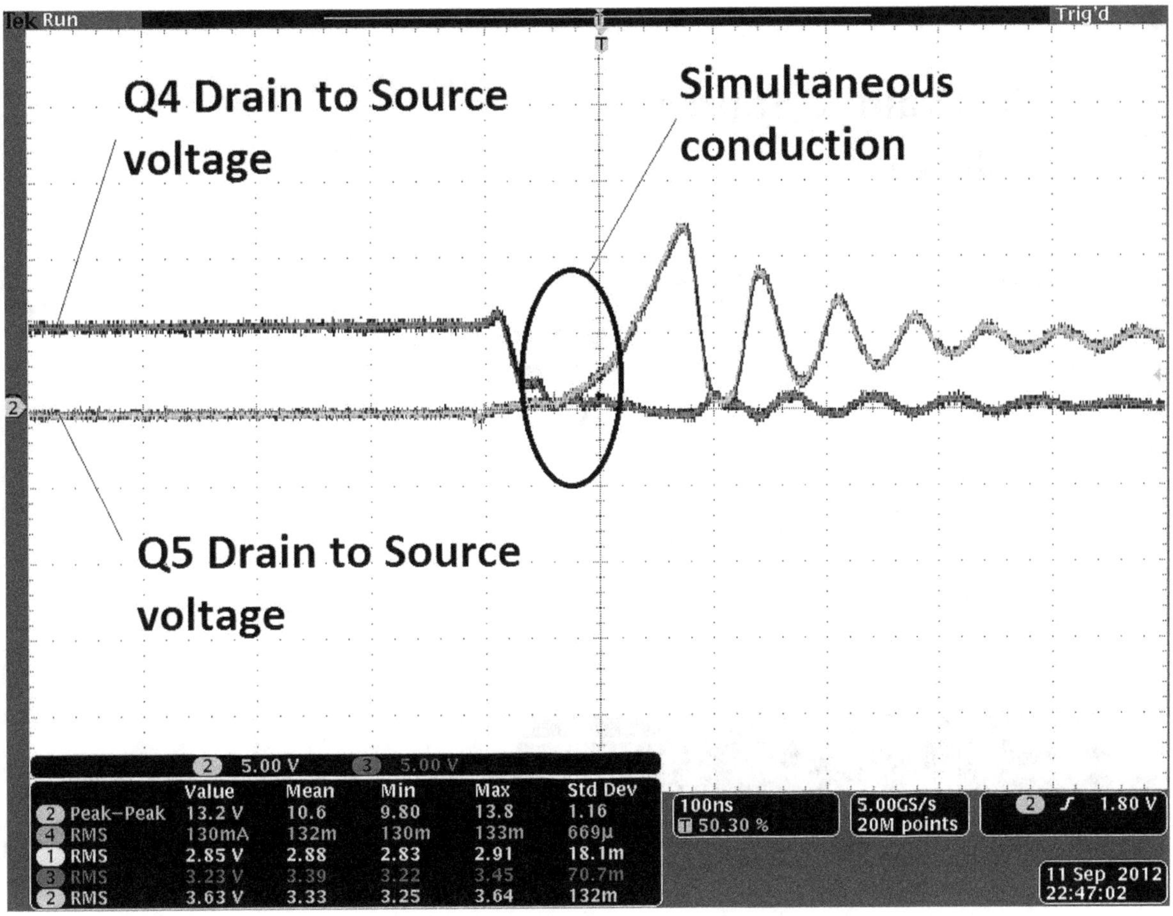


The experimental results show an increased efficiency with up to 12% when the dead time is present.

Simulation and experimental results show that when replacing the Schottky diode with an n-MOS switch with adequate dead time control, efficiency is increased with up to 9.5%. By employing a push-pull stage able to rapidly charge the input parasitic capacitors of the n-MOS switches, efficiency can be further increased.

The proposed autonomous control algorithm proved effective in maintaining the desired setpoint, while the TEG power source emulator was helpful in observing the dynamic behavior of the converter and in prescribing real values to the converter's input.

After the design phase, the precise dead time control can be implemented in an ASIC or advanced microcontroller device.

## Discussion

The design of the electronics part of the energy harvesting system started from the LTC 3105 DC-DC converter. The converter is designed to work with solar panels or TEGs and allows the adjustment of

*Figure 55. $Q_4$ and $Q_5$ Drain to Source voltages showing no simultaneous conduction with 100ns dead time*

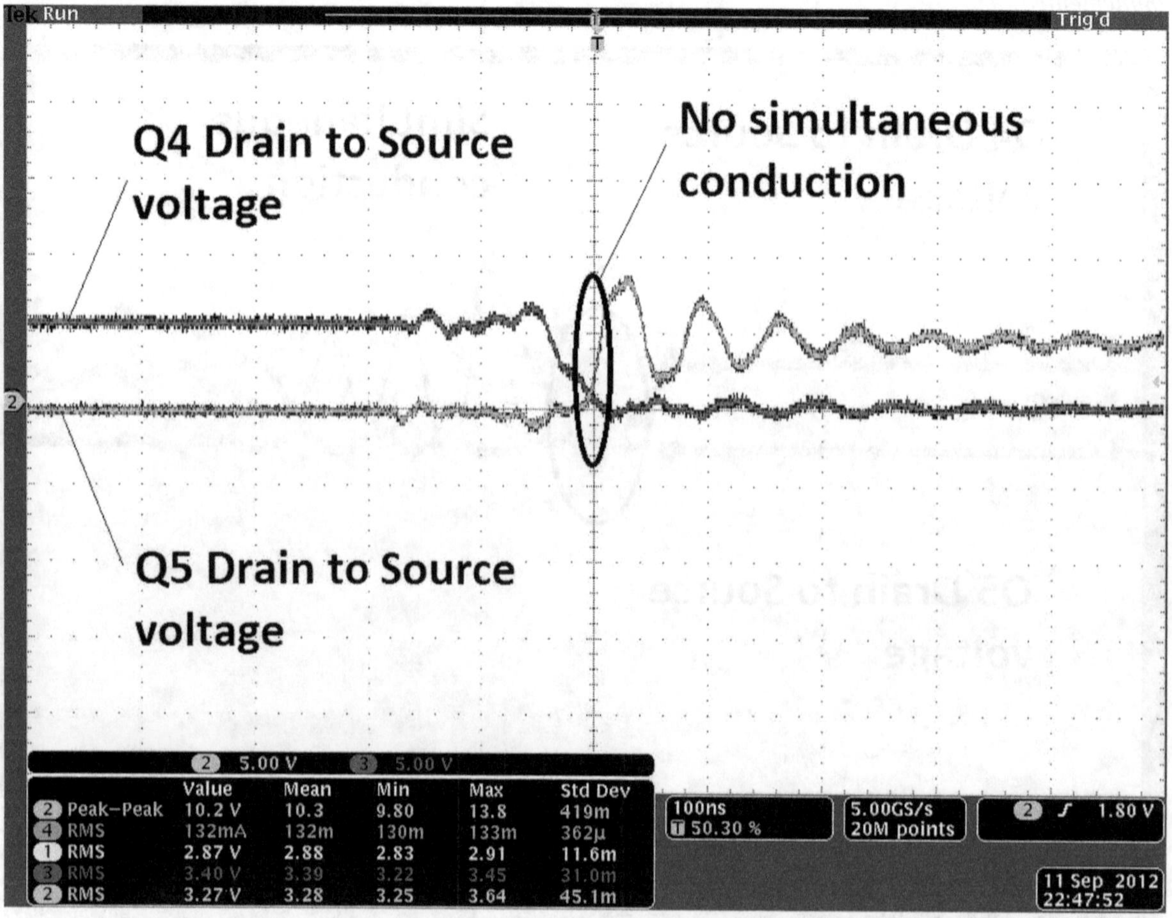


the output voltage value, offering high efficiency (~80%). The presence of an internal MPPT feature makes the LTC3105 converter an excellent device for harvesting low power energy source as it allows the maximum available power to be transferred to the output. The MPPT feature works by adjusting the equivalent impedance of the converter to match the one of the energy source. This is achieved by the MPPT voltage pin that prevents the input source from collapsing below the used programmable MPPT threshold. However, the datasheet application note of the device presents solutions that work only with solar panels. Taking into consideration that in the current application the TEGs internal resistance vary with temperature in an inverse way it does on solar panels (as the sunlight intensity increases the output voltage lowers), a specific circuitry must be designed. Two solutions, analog and digital were presented. The digital one takes advantage of the existing microcontroller that in the current application is existent on the system. However, in some cases a microcontroller is not available and an analog solution is preferred. The analog solution was first simulated in LTspice and it employs two or three OA devices depending on the user's choice. The solution is based on an NTC thermistor of which exponential temperature characteristic was linearized using two resistors. An ensemble of two OA realize the scaling and slope

*Figure 56. Experimental results for 5V input and with dead time control implementation*

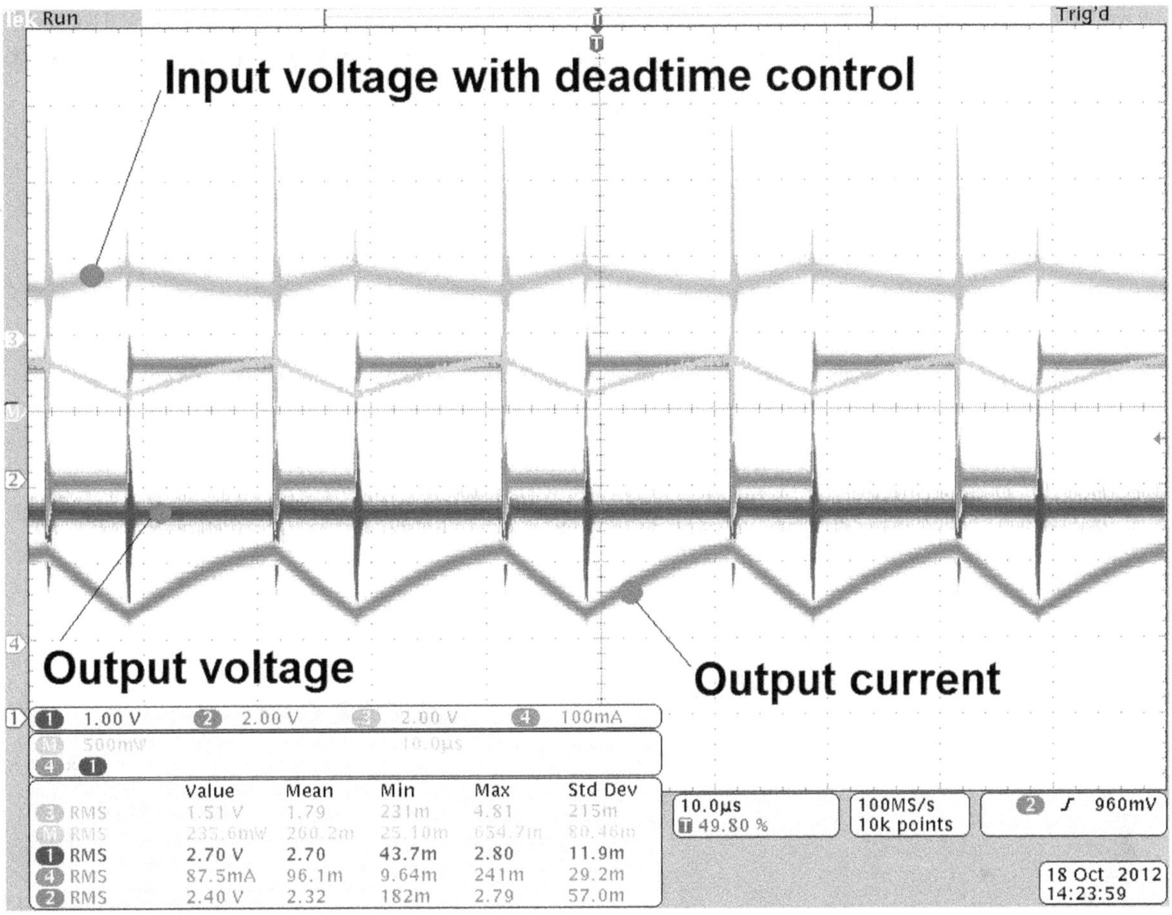


adjustment in such way that the output voltage of the proposed circuitry (input voltage to the MPPT pin) represents half the TEGs output voltage as it increases with temperature. In this way, a dynamic MPP system was developed. The powering of the OA was provided by the linear output voltage of the LTC converter. As this voltage output is only available when the input power source exceeds 0.25V, the dynamic MPP was designed such that when the circuit is not powered, the equivalent resistance seen by the MPPT pin to have the required value in order to enable the converter operation.

A dedicated LabVIEW application was implemented in order to help correlating the output voltage of the dynamic MPP to the experimentally determined characteristic of the TEGs voltage against temperature and find the necessary resistors values. The simulation results proved the correct functioning of the system. For the experimental circuit, very low powers OAs were chosen (~50µA $I_{cc}$/amplifier). In this way, the current consumption is below 0.5mA@2.2V. The circuit was implemented and proved to operate as in the simulation being able to maintain the maximum power transfer from TEGs to the LTC converter output. For the practical device testing, a TEG emulator device was designed and built. With the help of a programmable Agilent N6700 power source, a relay, a manual potentiometer and a

*Figure 57. Experimental results with 5V input voltage and without dead time control*

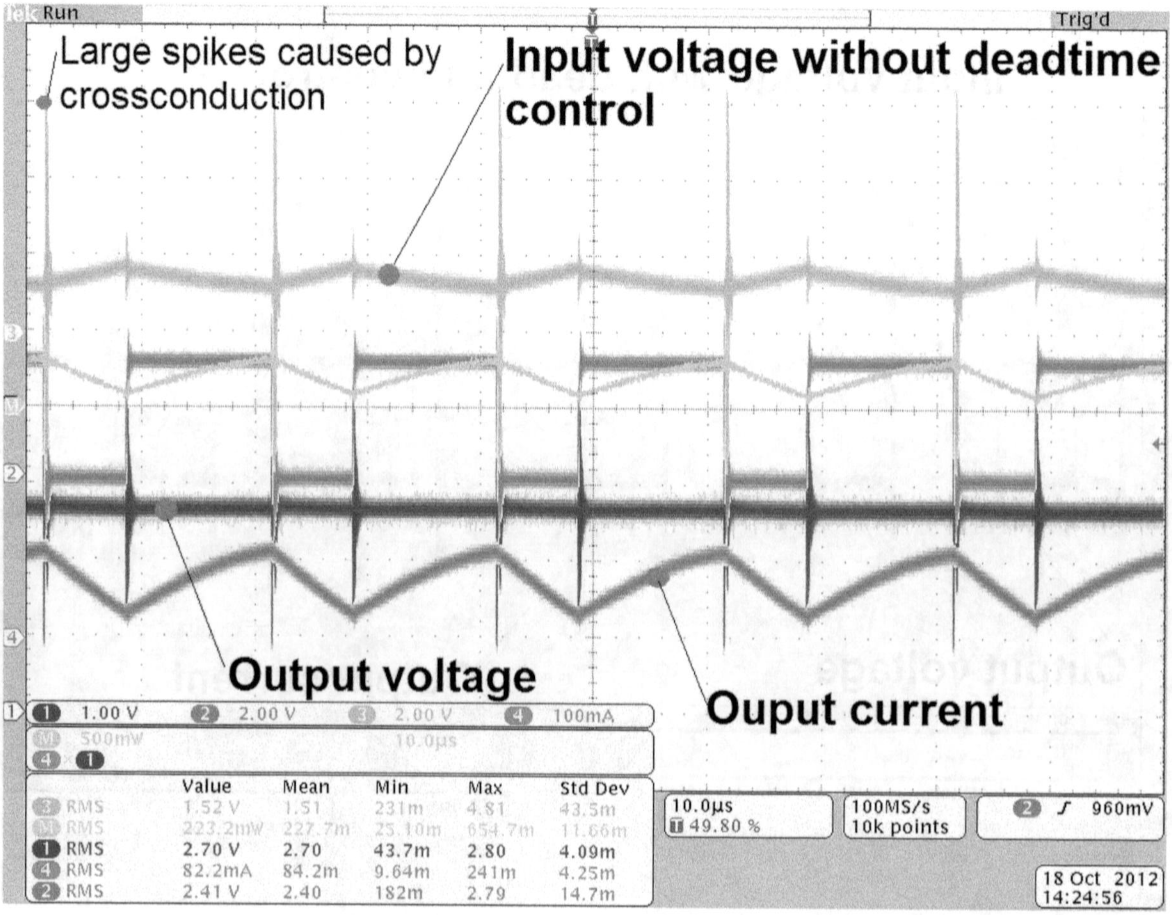


dedicated LabVIEW application an autonomous TEG emulator was implemented. The correspondence between the TEGs output voltage, internal resistance and temperature were stored in a lookup table and depending on the potentiometer value the N6700 power source generates the appropriate voltage. The potentiometer was manually varied and this action corresponds to a higher or lower temperature value on the TEGs faces.

In order to route the output of the LTC converter to different storage sources, a PCS (Power Control Switch) system was developed to sequentially charge two 2.4V Ni-Mh batteries. After their complete charge, the PCS decouples the input energy source and puts the two batteries in series generating a single 4.8V Ni-Mh battery. Also, the PCS is responsible for routing the energy from the second Li-Ion battery to different devices – GPRS/High power Wi-Fi, sensors, etc.

In order to generate a two stage charge storage, a step down converter was developed. This charges the second Li-Ion battery from the two 2.4V Ni-Mh batteries that are serially connected for this procedure. Two control algorithms were developed, one for constant current charge of the Li-Ion battery, the other for constant voltage. The constant current, constant voltage charge procedure represent the required charging regime of the Li-Ion battery in order to preserve its lifetime and capacity. For the current charge regime, the average current control mode (ACC) was employed. The control algorithm was first simulated using the standard analog devices after which the bilinear transform was used to implement the digital control in a microcontroller. The converter and the digital ACC mode were implemented in practice and the NI DAQ 6251 board was used to control the converter. The digital ACC mode was implemented in LabVIEW. The current was measured using a 10mΩ shunt resistor. For the constant voltage charge regime, a PI regulator was implemented in LabVIEW. The LabVIEW control algorithm also employed the charge mode switching (constant current to constant voltage) in an autonomous way. The practical results proved the correct operation of the device, the control algorithms being able to preserve the charging of the 3.7V Li-Ion battery even when perturbations are present. The efficiency of the Buck converter was not taken into account in this stage of the development. The efficiency can be improved if the Schottky diode is replaced by an n-MOS switch.

The experimental results validate the efficacy of the converter in transferring the charge from the first energy storage stage to the second, in this way also extending the lifetime of both batteries through complete charging/discharging cycles.

Further on, the TEGs system (mechanical parts and TEGs), the LTC converter, dynamic MPP system, PCS system, the batteries and power routes were tested in a real environment (with convective heat transfer and thermosyphon cooling). The PCS system was controlled in this stage by the NI DAQ 6251 board. The input and output voltage of the LTC3105 were measured by the NI DAQ 6221 board and also recorded with a TEKTRONIX DPO4104B oscilloscope along with the batteries charging current. The TEGs sides' temperatures were recorded and the values written into a file by the use of an autonomous testing system implemented in LabVIEW and with the NI DAQ 6221 as real world interface. The experimental obtained data were compared to the simulation proving a tight correspondence. All the implemented systems (dynamic MPP, Ni-Mh battery sequentially charge, PCS system, power routes) acted as desired, proving the correct operation of the entire system.

A comparison was made between the two TEG models – the one employing varying parameters and the constant parameters one. The first proved to generate the lowest errors. By correlating the experimental measured data to the model of the TEG, the electrical behavior of the device was accurately reproduced in simulation.

The largest error was observed in the prediction of the cold side temperature of the TEGs. This region is associated with the mechanical part of the TEGs system that has the most complex design. This leads to the conclusion that the thermal model of the mechanical setup can be further improved and that it is the one responsible for the erroneous results. If the system would be modelled in a dedicated FEM application, the errors could be reduced at the expense of a detailed and complex model of the mechanical system.

The simulation results of the proposed system are useful in the design of an intelligent and specific software power management system that takes into account the charging time of the batteries and how the energy should be used to increase the system efficiency. The detailed steps can further offer a design methodology to be followed when building similar systems.

Further on, an improved Ćuk converter is proposed. The converter aims to harvest the energy from multiple stacked TEGs and charge a battery by continuously adapting the converter impedance to match the one of the TEGs. This type of converter is needed when the waste heat energy spreads over a larger surface and/or the source wastes a high energy (has a high temperature). In this case, the dedicated low power devices are no longer of use as their output power is limited. For the proposed converter, the efficiency was of concern so several solutions were adopted in order to increase it. The Ćuk topology was chosen in order to obtain a step up – step down converter that is necessary when the input power source develops a high voltage and a specific voltage level is needed on the converter output. In order to improve efficiency, the diode existent in the original Ćuk topology was replaced by an n-MOS transistor. The transistors used for building the converter in practice were chosen to have very low $R_{dson}$ resistance (20mΩ), low voltage threshold (0.8V) and dead time control was implemented using an 80MHz FPGA to prevent the simultaneous cross conduction of the switches. The converter was first simulated and then implemented. The experimental results showed an average efficiency of 90% with a 2.4V Ni-Mh battery as load and input voltages from 1.5-5V. The dead time control necessity was analyzed and the experimental results were used to demonstrate the differences between the Schottky diode implementation and the n-MOS switch synchronous rectifier.

The converter impedance was matched to the TEGs internal resistance by imposing the voltage on the converter input to equal half the TEGs output voltage when no load was connected to it. In this way, no current sensor is needed to determine the maximum power transfer to the load.

Although the control was implemented in the design phase using a FPGA, it can be performed in an advanced microcontroller or ASIC after the required dead time delay between the control signals is correctly determined.

## CONCLUSION

In this work, a detailed description of the mechanical, thermal and electronic part of the proposed energy harvesting system is made. The authors implemented a novel analog dynamic MPPT system that is able of tracking the TEMs temperature and continuously adjusts the LTC3105 impedance in order to extract the maximum available power from the thermoelectrical generator. The circuit takes advantage of the internal MPPT from the LTC3105 converter. The implemented circuit extends the converter datasheet applications and can also be used in any other device that supports the same (MPPT) functionality.

The motivation for choosing a two stage energy storage system (first is using 2x2.4V Ni-Mh batteries while the second 1x3.7V Li-Ion) is explained through the increased efficiency of the battery charging procedure together with the increased battery lifetime and secondly by the available energy in a topology where one power source is used while the other is regenerating.

Further on, a complete self-sustainable wireless device for waste heat recovery was built, that aims to recover the waste heat produced by industrial machinery to power a conditional monitoring system which communicates with a base system via a high power Wi-Fi transceiver. The proposed device is completely autonomous and employs a two stage energy storage system. By using the simulation models in SPICE, the waste heat recovery device output power can be accurately predicted and various algorithms can be optimized for optimum power consumption.

The device was simulated and tested in practice, showing its effectiveness to charge batteries from temperatures as low as 35 °C.

Moreover, an improved efficiency Ćuk converter is implemented where the diode from the original topology is replaced by an n-MOS. The converter can be used to charge a battery from a waste heat energy source that has physically large dimensions typically when a number of TEMs can be serially connected. The necessary control logic for the converter is implemented in an FPGA along with 100ns dead time control. The experimental results show that when charging a 2.4V Ni-Mh battery, the efficiency of the proposed converter is increased to an average value of 90%, for the input voltage range of 1.5-5V. In the standard topology, with a Schottky diode, the efficiency is ~83% as determined in the experiments.

## FUTURE RESEARCH DIRECTIONS

The trend in the thermoelectric devices energy conversion is to develop new materials for better conversion efficiency. Multiple research are being made in this direction with promising results.

The present research is complementary to these directions, as it refers to the end user of the thermoelectric device that designs and optimizes energy harvesting systems. The thermo-electrical model developed in SPICE has an increased accuracy over the existent models as it takes into account the parameters variation vs temperature and also the parasitic components that exist in a common thermoelectrical device. An in-depth study of the parasitic inductance and capacitance components can be further made, using a FEM analysis of the materials used in the manufacturing of a device. Moreover, the opportunity of using the intrinsic parasitic elements of the thermoelectric module in connection with different high frequency DC-DC converter topologies can be further performed.

The thermoelectrical model developed in SPICE that also takes into account the free convection in natural environment, together with the electrical model of an energy storage solution can be used to implement and simulate intelligent algorithms for various parameters monitoring (e.g. vibrations, temperature, gas) devices. Therefore, self-sustainable wireless monitoring systems can be designed and fine-tuned, that dynamically adapt their modes of operation while maintaining their operation lifetime[1].

## REFERENCES

Bitschi, A. (2009). *Modelling of thermoelectric devices for electric power generation* (Ph.D. dissertation). Swiss Federal Institute of Technology Zurich, ETH Zurich.

BP. (June 2012). *Statistical Review of World Energy*. Retrieved November 2012 from http://www.bp.com/statisticalreview

Crane, D. T. (May 2012). *Thermoelectric waste heat recovery program for passenger vehicles,*. U.S. Department Energy Efficiency & Renewable Energy (EERE). Retrieved August 2014 from http://www1.eere.energy.gov

Eder, A., & Linde, M. (2011). *Efficient and Dynamic – The BMW Group Roadmap for the Application of Thermoelectric Generators*. Paper presented at 2nd Thermoelectrics Applications Workshop, San Diego. Retrieved August 2014 from http://www1.eere.energy.gov

Energizer Battery Manufacturing Inc. (2008). *Energizer charger handbook*. Retrieved September 2014 from http://data.energizer.com/PDFs/charger_appman.pdf

Erickson, R. W. (2001). *Fundamentals of Power Electronics* (2nd ed., pp. 440–482). New York: Kluwer Academic Publishers. doi:10.1007/b100747

Everredtronics. (2012). *TEG127-40A Technical Info*. Retrieved September 2014 from http://www.everredtronics.com/TEG.html

Jovanovic, V., Ghamaty, S., & Bass, J. (2006). Design, fabrication and testing of a novel energy harvesting thermoelectric power supply for wireless sensors. In *Proceedings of PWR2006 ASME Power, Atlanta, GA, USA* (pp. 643-651). doi:10.1115/POWER2006-88150

Mateu, L., Codrea, C., Lucas, N., Pollak, M., & Spies, P. (2006). Energy harvesting for wireless communication systems using thermogenerators. In *Proceedings of the XXI Conference on Design of Circuits and Integrated Systems (DCIS)*, Barcelona, Spain.

Maxim. (2014). *MAX17710 Energy-Harvesting Charger and Protector*. Retrieved September 2014 from http://www.maximintegrated.com/en/products/power/battery-management/MAX17710.html

Mazar, B. (2012). *State of the art prototype vehicle with a thermoelectric generator*. Paper presented at the TE Application Workshop, Baltimore. Retrieved August 2014 from http://www1.eere.energy.gov

McCoy, J. (Jan 2012). *Thermoelectric technology: Materials, processes, devices and systems*. ASM San Diego Chapter. Retrieved August 2014 from http://www.hi-z.com

Meisner, G. P. (2011). *Advanced thermoelectric materials and generator technology for automotive waste heat at GM*. Paper presented at Thermoelectrics Applications Workshop, San Diego.

Melcor. (2012). *CP1.0-127-05L-RTV Heat Pump*. Retrieved November 2012 from http://www.datasheetpro.com/part-info/CP1.0-127-05L-RTV/2981405

Minnich, A. (2008). *Modeling the Thermoelectric Properties of Bulk and Nanocomposite Thermoelectric Materials* (MSc dissertation). MIT, USA.

Rashid, M. H. (2007). *Power Electronics Handbook* (2nd ed.). Burlington, MA: Butterworth-Heinemann.

REN21. (2011). *Renewables 2011 Global Status Report*. Retrieved November 2012 from http://www.ren21.net

REN21. (2012). *Renewables 2012 Global Status Report*. Retrieved November 2012 from http://www.ren21.net

Report, S. (2006). *Science of Sustainability 2006 - A view from Japan*. Retrieved 1st November 2012 from http://www.sos2006.jp

Rowe, D. M. (2006). *Thermoelectrics Handbook, Macro to Nano*. Boca Raton, FL: CRC Press, Taylor & Francis Group.

Russel, K. (2011). *A Hybrid Thermoelectric Cooler Thermal Management System for Electronic Packaging* (MSc dissertation). McMaster University, Canada.

Schock, H., Caillet, T., Case, E., Fleurial, J. P., Hogan, T., Lyle, M., & Zhu, G. (Jan 2011). *Thermoelectric conversion of waste heat to electricity in an IC engine powered vehicle*. US Department of Energy, U.S. Department Energy Efficiency & Renewable Energy (EERE). Retrieved November 2012 from http://www.eere.energy.gov

Selit. (2012). *Selitherm Thermal Insulation Board*. Info Sheet. Retrieved November 2012 from http://www.selit.com

Snyder, J. (2008). Small thermoelectric generators. *The Electrochemical Society Interface, 17*(3), 54-56. Retrieved from http://www.electrochem.org/dl/interface/fal/fal08/fal08_p54-56.pdf

Taranovich, S. (2012). The Rover's Energy. *EDN*. Retrieved from http://www.edn.com/electronics-blogs/mission-to-mars--nasa-engineering-and-the-red-planet-/4392098/The-rover-s-energy

Technology, L. (2014) *LTC 3105 Step-Up DC/DC Converter*. Retrieved September 2014 from http://www.linear.com/product/LTC3105, 2011

Yoshio, M., Brodd, R. J., & Kozawa, A. (Eds.). (2009). Lithium-ion batteries science and technologies. New York: Springer Science + Business Media. doi:10.1007/978-0-387-34445-4

Yu, J., & Zhao, H. (2007). A numerical model for thermoelectric generator with the parallel-plate heat exchanger, Science Direct. *Journal of Power Sources*, *172*(1), 428–434. doi:10.1016/j.jpowsour.2007.07.045

## KEY TERMS AND DEFINITIONS

**Average Current Control (ACC):** A version of current mode control used in switching power supplies related to the introduction of a high gain current amplifier in the current loop, enabling less harmonic distortions and an excellent noise immunity, among other advantages.

**Heat Transfer Coefficient (HTC):** Proportionality coefficient between the heat flux and the temperature difference.

**Maximum Power Point Tracking (MPPT):** In general, the technique that is used to get the maximum available power from a device.

**Maximum Power Point Control (MPPC):** The control technique used to get the maximum available power from a device when it operates on a wide range of input values.

**Negative Temperature Coefficient (NTC):** In case of a temperature sensor, this refers to the lowering of the electrical resistivity with temperature increase.

**Thermoelectric Cooler (TEC):** A thermoelectric material to which power is supplied in order to extract the heat from one side and dissipate it on the other.

**Thermoelectric Generator (TEG):** A thermoelectric material that is placed under a temperature difference in order to harvest the heat dissipated and generate electricity.

## ENDNOTE

1 The present work is part of the thesis entitled "Analysis of themoelectrical modules for energy harvesting systems" by Dr. Mihail Cernaianu, accomplished under the supervision of Dr. Aurel Gontean at Politehnica University Timisoara, Romania, 2013.

# Chapter 9
# A Theory of Thermoelectric Energy Harvesting Systems

**Hal Edwards**
*Texas Instruments, USA*

**Toan Tran**
*Texas Instruments, USA*

**Jeff Debord**
*Texas Instruments, USA*

**Dave Freeman**
*Texas Instruments, USA*

**Kenneth Maggio**
*Texas Instrument, USA*

## ABSTRACT

*This chapter presents a study of thermoelectric energy harvesting with nano-sized thermopiles (nTE) in a planar 65 nm silicon CMOS process. These devices generated power from a 5C temperature difference at a density comparable to commercially available thermoelectric generators, following a metric used in the research literature (Hudak, 2008). By analyzing these devices as a thermoelectric harvesting system, the authors explore the impact of additional performance metrics such as heat source/sink thermal impedance, available heat flow density, and voltage stacking, providing a more comprehensive set of criteria for evaluating the suitability of a thermal harvesting technology. The authors use their thermoelectric system theory to consider the prospects for several thermoelectric energy harvesting applications.*

## INTRODUCTION

A vast quantity of waste heat is rejected by the heat engines used in industry, transportation, and buildings. The idea of converting some of this heat flow into electricity is compelling. Consider the internal combustion engine powering an automobile. Tens of kW of power is extracted from fuel at an efficiency of less than 40%, meaning that at least 60% of the fuel's chemical energy would be rejected to the environment as heat. If half of this waste heat could be coupled through a thermoelectric heat engine operating at half the Carnot efficiency, capturing half of the total temperature drop (say 100C), then electrical power equal to (1/2) x (60%) x (1/2) x (100C/500C) ~ 3% of the chemical energy of the fuel being consumed, potentially improving fuel economy by 10%.

DOI: 10.4018/978-1-4666-8254-2.ch009

This line of reasoning sparks interest in thermoelectric energy harvesting; but one quickly learns that the present state-of-the-art thermoelectric devices are significantly less efficient than assumed above, causing interest in large scale energy thermoelectric harvesting to fade.

However, there are some applications for which the cost of fuel is not the primary driver so efficiency is less important than the ability to achieve a certain power budget in a given usage environment. In situations such as wireless sensor nodes in the Internet of Things (IoT), there is a need for harvested power to avoid the cost of battery replacement. Wearable devices powered by body heat are also of interest, particularly if the harvested heat can extend usage lifetime or enable increased functionality by supplementing battery power.

In both of these applications, more power can be provided when thermoelectric efficiency is improved. So it is still an important topic of materials research to find thermoelectric materials and devices that support higher efficiency.

## Recent Developments in Thermoelectrics

In the early 1990s, it was proposed (Hicks, 1993) that by creating layered materials, thermoelectric efficiency could be improved. At the same time, new concepts in thermoelectric devices were introduced (Edwards, 1993; Nahum, 1994), some of which could in principle approach the Carnot limit (Edwards, 1995) under special circumstances.

More recently, nanostructuring has been found to improve silicon thermoelectric efficiency (Boukai, 2008; Hochbaum, 2008) by enhancing acoustic phonon scattering from interfaces. Based on this hypothesis and its compatibility with increased density (hence lower device cost), we developed nanoscale thermoelectric devices (nTE) in a 65 nm planar CMOS process in which silicon ridges as small as 80 nm are fabricated. Because these Si ridges are smaller than the room temperature acoustic phonon mean free path in silicon, it is expected that thermoelectric generating performance can be improved relative to bulk Si.

Material improvements are necessary, but unfortunately not sufficient. As we will see below, it also is important to understand how heat flows through a thermoelectric energy harvesting system and what constraints the heat flow places on the efficiency, size, and other aspects of thermoelectric devices and the thermal system built around them. So a primary objective of this article is to develop a theory of a thermoelectric energy harvesting system. Because nTE technology is an extreme example of device miniaturization, it is a good test case for a theory of a thermoelectric energy harvesting system.

## Thermoelectric Device Design

Figure 1 (a) shows the structure of an nTE device. N-type (n-TP) and p-type (p-TP) thermopiles are fabricated from the same silicon "active" regions used to form CMOS transistors and can be integrated on the same chip as CMOS circuitry without changing the manufacturing process or adding mask steps. Thermopiles connect electrically in series (1 → n-TP → p-TP → 2) and thermally in parallel. Junction isolation enables voltage stacking of the Seebeck voltage between neighboring cells (the deep DNWELL implant isolating the PWELL from the p-type silicon substrate is not shown). Each thermopile is 80 nm wide and 750 nm long, enabling a large number to be stacked electrically in series, although testing reported in this chapter used singly-stacked nTE's.

*Figure 1. (a). Cross section of a nTE device. (b) Schematic representation of the functional elements of a thermoelectric device.*

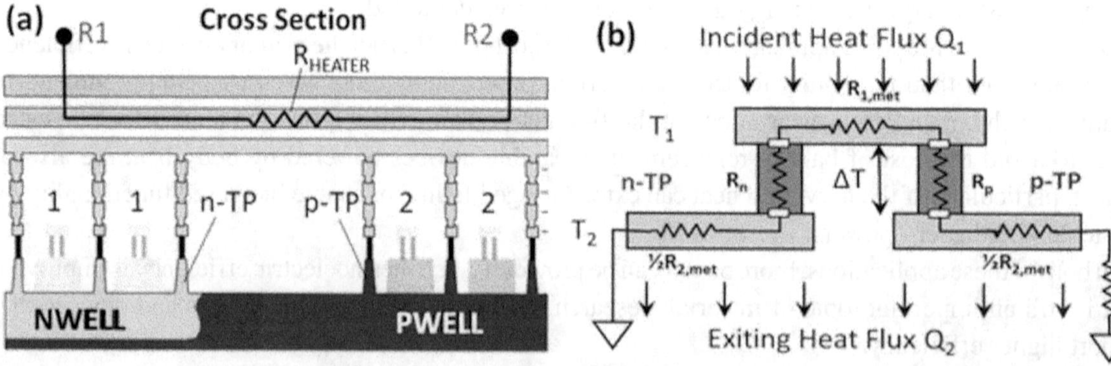


The nTE thermoelectric generators were tested using an integrated metal resistor constructed from the copper metal system. This resistor serves as both a heat source and a thermometer. In an actual harvester, the resistor shown in Figure 1(a) would be replaced by a heat exchanger that coupled heat from an external ambient heat source into the tops of the thermopiles. The heat sink as shown in Figure 1(a) is the wafer substrate. Heat source and heat sink could be reversed in a given application.

Figure 1 (b) shows parasitics such as metal and thermopile resistance that limit the current flow and hence the generated power, while the temperature drop $\Delta T$ that can be coupled across the thermopiles sets the voltage of the generator through the Seebeck effect. The phonon thermal conductivity $\kappa$ of the thermopiles allows heat to flow through the device without generating power (it is a heat leak) so one can begin to consider the device-level quantities (electrical resistance, thermal conductivity, and Seebeck coefficient of the thermopiles) which set the upper limit of thermoelectric generating efficiency.

In a real thermal harvesting application, there are additional parasitic effects to be considered, which is why we make the distinction in this chapter between a thermoelectric harvesting system and a thermoelectric harvesting device. Examination of Figure 1(a) shows the parallel heat leakage paths limiting the performance of the nTE device. For example, the dielectric encasing the device couples heat from the metal cage around the heater resistor to the Si substrate, bypassing the thermopiles. There also are series thermal bottlenecks, for example the thermal coupling between the heater resistor and the upper metal levels that are designed to couple heat down into the tops of the thermopiles. Other series thermal bottlenecks include the stacks of contact, via, and metal layers above the thermopiles, and the thermal spreading resistance from the bottom of the thermopiles into the silicon substrate. At a package or board level, thermal heat leaks around the device and series thermal impedance of heat spreaders or packaging must be considered.

## THEORY OF A THERMOELECTRIC SYSTEM

Thermoelectric research (Goldsmid, 1954; Edwards, 1999; Rowe, 2006) has focused on improving thermoelectric materials for refrigeration through the dimensionless figure of merit $zT_1 \equiv \alpha^2 T_1/\kappa\rho$, where $\alpha$ is the Seebeck coefficient, $\kappa$ is the thermopile thermal conductivity, and $\rho$ is its electrical resistivity. In the harvesting of waste heat, system-level considerations require attention as well.

## Theory of Thermoelectric Energy Harvesting Devices

Figure 2 shows the equivalent circuit for an ideal thermoelectric device. Ohmic losses and thermoelectric power are the coupling terms between thermal and electrical equivalent circuits. In the thermopiles, two parallel paths of heat flow are shown: conductive heat flow through K, and Peltier heat flow through $\pi_1$ and $\pi_2$. The Peltier heat flow, which is the heat carried by majority electrons and holes as they flow through the thermopiles, depends on the temperature. As it traverses to the cooler side, energy is transferred to the electrical equivalent circuit through the Seebeck effect. This is the energy that can be delivered to the load. Some electrical energy is dissipated through Ohmic heating, resulting in an energy flow back into the thermal equivalent circuit.

Heat from a heat source $Q_1$ at the hot end flows into the thermopiles where it is carried by the parallel mechanisms of conductance K and the Peltier effect

$\pi I = \alpha_{pn} T\, I$

*Figure 2. Equivalent circuits showing heat and current flow in an ideal thermoelectric system*

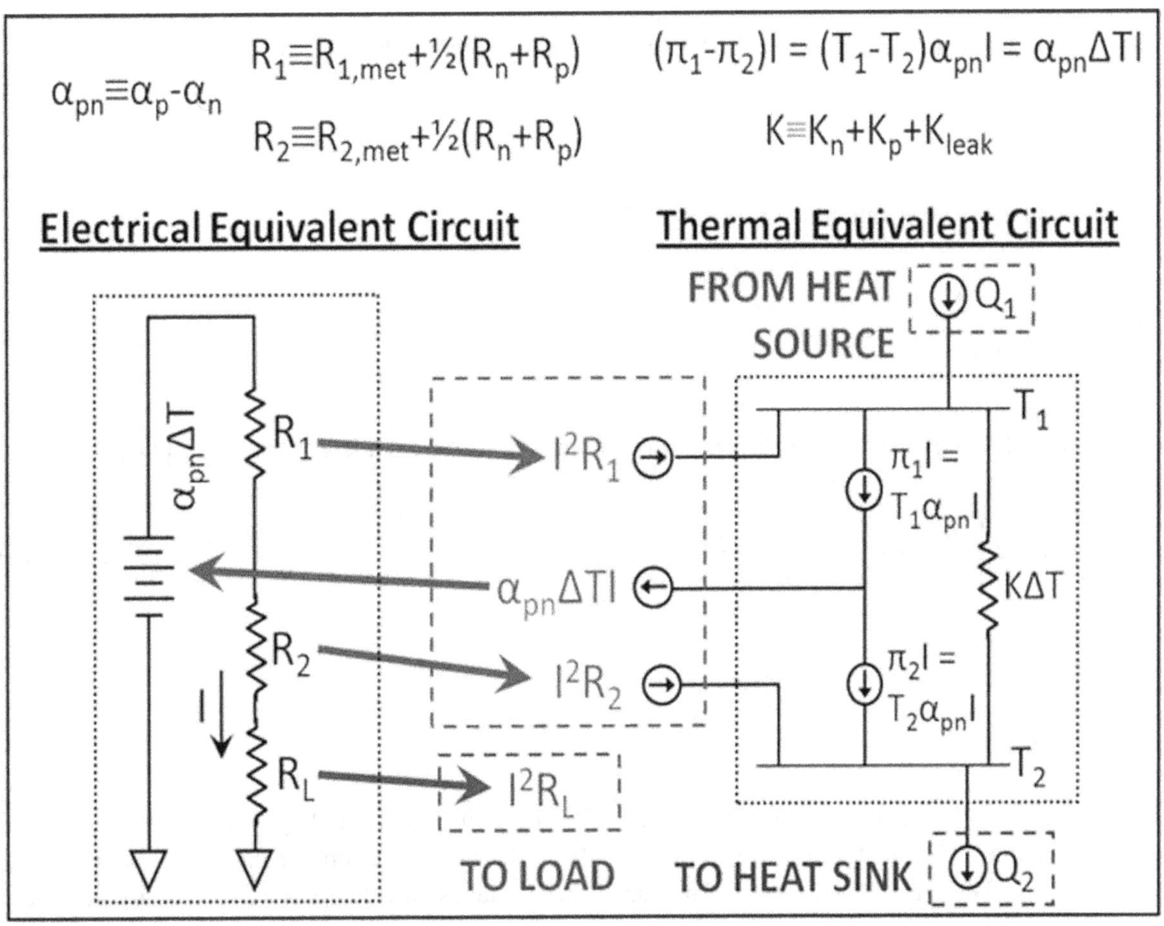

Some of the power is siphoned away to the electrical subcircuit. The remainder flows into the heat sink. The conservation of energy requires that the input heat $Q_1$ should equal the power $P_L$ delivered to the load plus the heat $Q_2$ rejected into the heat sink:

$$Q_1 = \mathrm{K}\Delta T + T_1\alpha_{pn}I - I^2R_1; \quad Q_2 = \mathrm{K}\Delta T + T_2\alpha_{pn}I + I^2R_2 \tag{1}$$

In the electrical subcircuit, the Seebeck voltage is applied to the lumped internal and load resistances:

$$\alpha_{pn}\Delta T = I\left(R_1 + R_2 + R_L\right) \tag{2}$$

Equation (2) may be solved for I, yielding the power delivered to the load:

$$P_L = I^2R_L = \frac{\left(\alpha_{pn}\Delta T\right)^2 R_L}{\left(R_1 + R_2 + R_L\right)^2} \tag{3}$$

The load resistance that maximizes $P_L$ can be found by solving $dP_L/dR_L=0$:

$$R_{L,MAX} = R_1 + R_2 \tag{4}$$

The load power generated under this condition is:

$$P_{L,MAX} = \frac{1}{4}\frac{\left(\alpha_{pn}\Delta T\right)^2}{R_1 + R_2} \tag{5}$$

The heat flow supplied by the heat source to maintain $\Delta T$ can be calculated from Eqns. (1), (2), and (4):

$$Q_{1,MAX} = \mathrm{K}\Delta T\left[1 + \frac{1}{4}ZT_1\left(2 + \frac{\Delta T}{T_1}\frac{R_1}{R_1 + R_2}\right)\right] \tag{6}$$

where

$ZT_1=\alpha_{pn}^2/(K(R_1+R_2))$

is the device-level dimensionless figure of merit (FOM). Z (in 1/K) is analogous to the material FOM $z=\alpha^2/\kappa\rho$ (in 1/K), with material parameters $\kappa$ (in W/cm-K) and $\rho$ (in $\Omega$-cm) replaced by the device parameters K (in W/K) and $R_1+R_2$ (in $\Omega$). The material FOM takes on values from $zT_1$~0.01 for thermally conductive semiconductors like Si to $zT_1$~1 for thermally insulating compound semiconductors such as

$Bi_2Te_3$. Geometrical factors area A and thickness t that appear in K=κA/t and R=ρt/A for a thermopile cancel in the product so the device FOM Z equals the material FOM z in the absence of parasitic effects.

The second term in square brackets of Eq. (6) indicates that more heat flows through the thermopiles than their thermal conductance K would indicate. This heat flow is carried by the Peltier effect as shown in Figure 2.

The thermal impedance of a thermoelectric device is:

$$\Theta_{TEG} \equiv \left(\frac{dQ_1}{dT_1}\right)^{-1} = \frac{1}{\mathrm{K}}\left(\frac{1}{1+\frac{1}{2}ZT_1}\right) \quad (7)$$

which is less than the thermopiles' phonon contribution 1/K due to the Peltier heat flow carried by electrons and holes. The generating efficiency η is:

$$\eta \equiv \frac{P_L}{Q_1} = \left(\frac{\Delta T}{T_1}\right)\frac{ZT_1}{4+ZT_1\left(2-\left(\Delta T/T_1\right)\left(R_1/\left(R_1+R_2\right)\right)\right)} \quad (8)$$

The leading factor in Eq. (8) is the Carnot efficiency of an ideal heat engine. The remaining factor is less than one, consistent with the second law of thermodynamics.

The preceding analysis uses ΔT as an independent variable, an appropriate assumption for applications with a fixed-$T_1$ heat source such as body heat. On the other hand, a heat source such as solar radiation may be more accurately represented as a heat flux $Q_1$, in which case $T_1$ will adjust to accommodate the thermoelectric device's ability to dissipate $Q_1$. To analyze this case, $\Theta_{TEG}$ from Eq. (7) can be used to calculate $\Delta T=\Theta_{TEG}Q_1$, then Eq. (5) determines $P_L$.

## Theory of Thermoelectric Energy Harvesting Systems

Generated power in Eq. (5) goes as the square of ΔT across the thermopiles, so it is desirable to reduce parasitic temperature drops in the path of heat flow. However, a real thermoelectric harvester will have a finite series thermal impedance $\Theta_S$ due to interfaces, case materials, or spreading resistance that share the total temperature difference available $\Delta T_A$ between the heat source and heat sink. Furthermore, there will be leakage paths $\Theta_{LK}$ due to mechanical supports, fill materials, or air convection that shunt heat around the thermopiles. Analysis of a thermoelectric system must consider these effects, which are summarized in Figure 3 (f). The temperature across the thermopiles becomes:

$$\Delta T = \Delta T_A \frac{1}{1+\left(\Theta_S/\Theta_{TEG}\right)+\left(\Theta_S/\Theta_{LK}\right)} \quad (9)$$

The thermopile thermal impedance $\Theta_{TEG}$ can be varied by changing the nTE array area. Eq. (9) shows that if a large thermopile array has a thermal impedance $\Theta_{TEG}$ that is much smaller than $\Theta_S$, then very little temperature drop will fall across the thermopiles and the Seebeck voltage will be small, thus

*Figure 3. Plots of Eq. (12) illustrate the power generation capability of a thermoelectric energy harvesting system. (a) and (b) are appropriate to the low temperature differences and low heat flows of body heat harvesting. (c) and (d) represent industrial use cases where higher temperatures and more intense heat flows are available. (e) summarizes energy harvesting for a wide range of applications assuming a high performance thermoelectric material with ZT=1. (f) shows the thermal equivalent circuit of a thermoelectric energy harvesting system.*

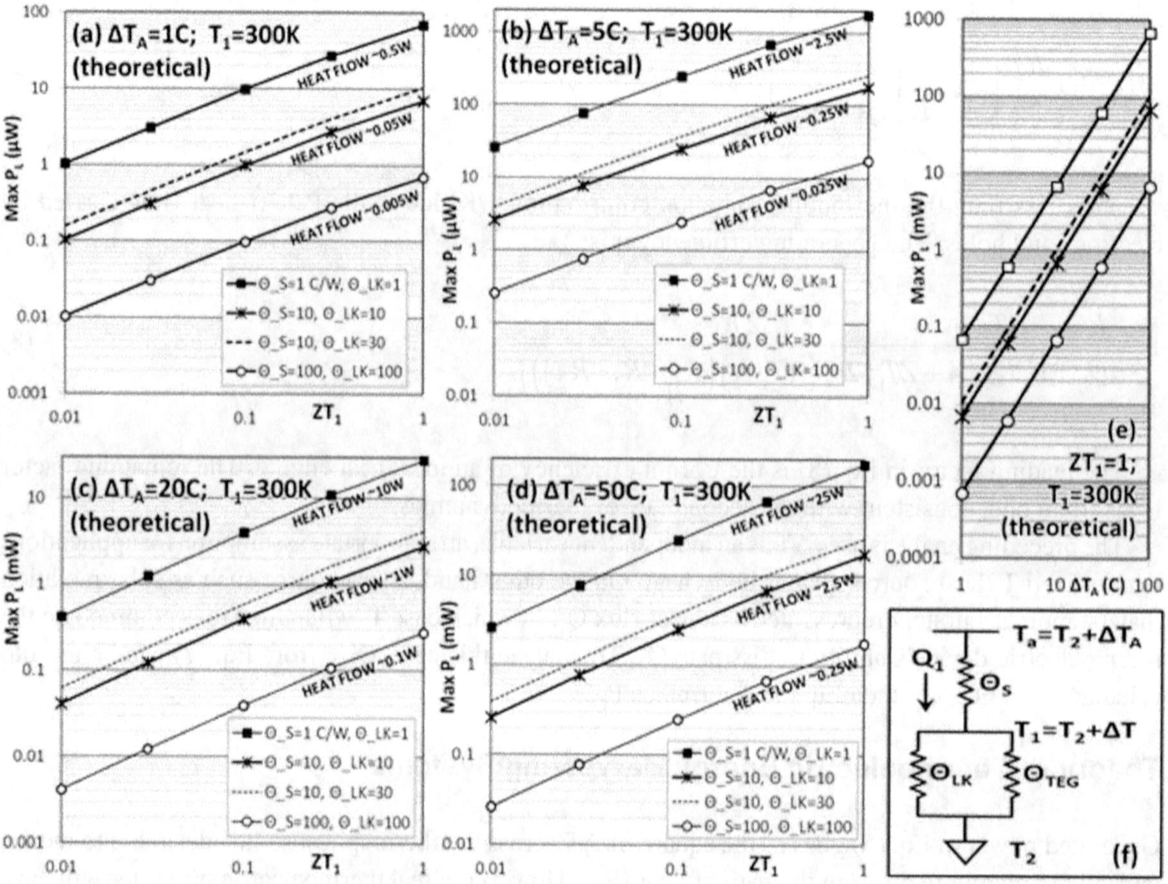


reducing the generated power which goes as Seebeck voltage squared per Eq. (3). Conversely, for a very small thermopile array, $\Theta_{TEG}$ can be made to exceed $\Theta_S$ so per Eq. (9) more temperature drop may be captured across the thermopiles. However, for a small thermopile array the electrical resistance will increase through R=ρt/A, so the generated current is reduced and again the generated power suffers.

The preceding arguments demonstrate that there is an optimum thermopile array size. Using Eq. (9) in Eq. (5) and including the area dependence of the nTE array resistance, we can find the optimum $\Theta_{TEG,MAX}$ at which the maximum generated power is obtained:

$$\Theta_{TEG,MAX} = \left(\Theta_S \,||\, \Theta_{LK}\right) \tag{10}$$

resulting in the temperature difference across the thermopiles at maximum generated power:

$$\Delta T_{MAX} = \frac{\Delta T_A}{2} \frac{\left(\Theta_{LK} \parallel \Theta_S\right)}{\Theta_S} \tag{11}$$

Eq. (11) illustrates that the series thermal impedance $\Theta_S$ will share the $\Delta T_A$ applied to the system by a heat source $Q_1$, leaving a reduced $\Delta T$ across the thermopiles.

The maximum power then becomes:

$$P_{L,MAX} = \frac{1}{16} \frac{T_1}{\Theta_S} \left( \frac{\Delta T_A}{T_1} \right)^2 \frac{1}{1+\left(\Theta_S / \Theta_{LK}\right)} \frac{ZT_1}{1+\frac{1}{2}ZT_1} \tag{12}$$

Note that even for no heat leak ($\Theta_{LK} = \infty$), the temperature drop $\Delta T$ across the thermopiles at maximum generated power is half of the temperature drop $\Delta T_A$ between the heat source and sink. This is analogous to the result in Eq. (4) that the thermopile array should be sized so as to achieve electrical match to the load. Since the generated power goes approximately as the square of $\Delta T_A \sim \Delta T/2$, there is an extra factor of (1/4) in Eq. (12) relative to the prefactor in Eq. (5). At each coupling event (thermal coupling to a heat flow; electrical coupling to a load), the maximum power that can be delivered to the load is reduced by a factor of four.

Equation (12) shows the best generating performance that can be expected for a thermoelectric harvesting system characterized by a thermoelectric device with $ZT_1$ that is connected between a heat source and sink differing in temperature by $\Delta T_A$, with series $\Theta_S$ and leak $\Theta_{LK}$ thermal impedances. Eq. (12) can be used to estimate harvesting capability in a general way, from an application perspective, without having to consider the internal design details of the thermoelectric device itself.

Equation (12) is plotted for several scenarios in Figure 3 (a)-(d). In Figure 3(a), an extremely small temperature difference of $\Delta T_A = 1C$ is used, showing that in order to obtain μW of power, a high z material and low thermal impedance would need to be used. The case of body heat energy harvesting is explored in Figure 3(b). The total temperature difference $\Delta T_A$ is estimated to be 5C, assuming that only part of the difference between the human body core temperature of 37C and a typical 21C room temperature can be captured. Higher temperature differences are explored in Figs. 3(c)-(d), where it may be possible to harvest mW of power from heat flows in industrial or automotive applications.

# RESULTS

## Extraction of Thermoelectric Parameters

The thermal impedance $\Theta_{TEG}$ of a nTE device can be measured using the hot strip method (Gustaffson, 1979). Consider a dielectric slab encasing a harvesting structure that includes a resistor R and a heat sink at temperature $T_2$. If a constant current I flows through R, the ohmic heat $I^2R$ will flow through the dielectric to the heat sink and the resistor's temperature $T_1$ will increase:

$$T_1 = T_2 + \Theta_R I^2 R \quad \text{or} \quad \Delta T = \Theta_R I^2 R \tag{13}$$

where $\Theta_R$ is the resistor's thermal impedance and $\Delta T = T_1 - T_2$. The resistance R increases with I through its temperature coefficient of resistance TCR:

$$\Delta T = \Theta_R I^2 R \left(1 + TCR \cdot \Delta T\right)$$

so:

$$\Delta T = \left(\Theta_R I^2 R\right) / \left(1 - \Theta_R I^2 R \cdot TCR\right) \tag{14}$$

The voltage that develops across the resistor is:

$$V = IR\left(1 + TCR \cdot \Delta T\right) = \frac{IR}{1 - \Theta_R I^2 R \cdot TCR} \tag{15}$$

Eq. (15) may be used to extract $\Theta_R$ from a V(I) curve. The resistor's temperature increase may be calculated using (14).

## Thermoelectric Device Characterization

Figure 4 shows typical operation data from nTE devices. Figure 4 (a) shows a load-line plot, in which a current source is applied to the thermopiles and voltage is monitored. A four-probe configuration is

*Figure 4. (a) Load line and (b) square law demonstration show nTE device characteristics.*

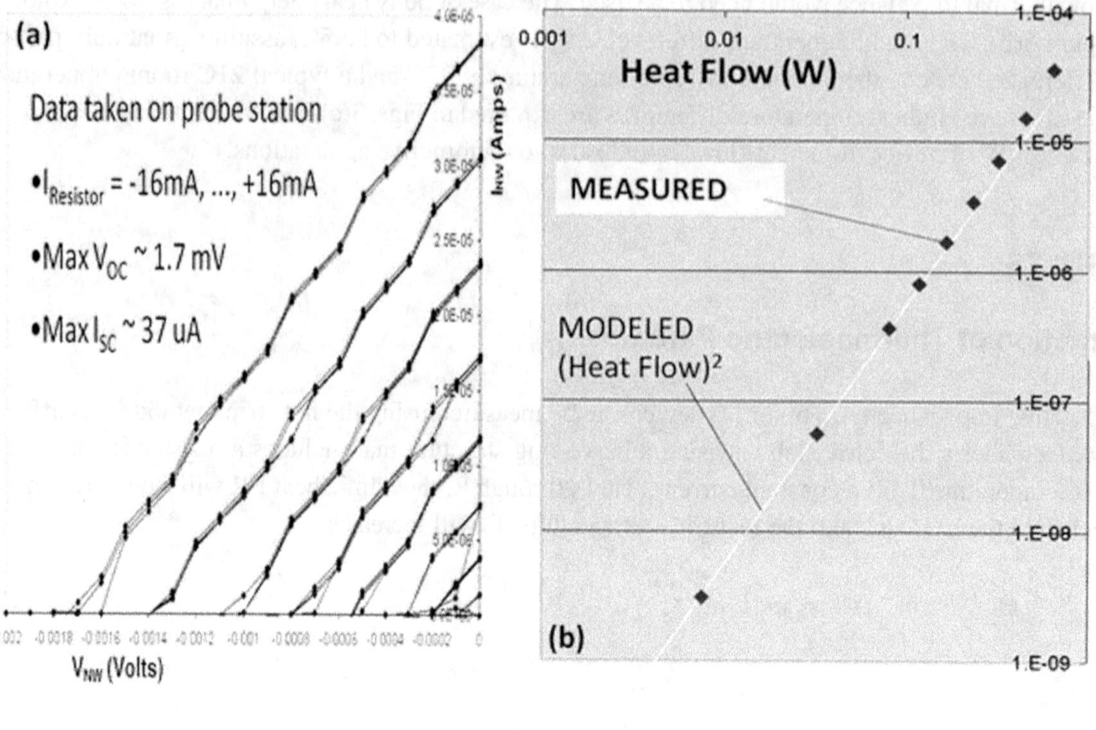

used in the nTE test structures to avoid contact resistance effects. For a resistor, a simple diagonal line through the origin would be expected. However, when heat flows from the heater resistor through the thermopiles (see Figure 1 (a)), a Seebeck voltage develops across the thermopiles which offsets the load line. The load lines in Figure 4 (a) were acquired with a series of heater current values. Note that negative and positive values of heater current cause the same shift, confirming that the shift is related to heat ($I^2R$) rather than some sort of leakage or other parasitic coupling between the heater resistor and the thermopiles.

There are several quantities of interest displayed in Figure 4(a). The point at which a load line intersects the current axis is the open-circuit voltage $V_{OC}$, which is indicative of the thermopiles' Seebeck coefficient (the battery in Figure 2) and the thermal coupling efficiency ($\Delta T/\Delta T_A$) which can be found from Eq. 9. The point where a load line intersects the voltage axis is the short-circuit current $I_{SC}$, which is indicative of the combined series resistances of the thermopiles and parasitics through Eq. 2.

The generated power (not shown in Figure 4(a)) is given by the product of current and voltage at each load line point (V,I). Each point in the load line also can be thought of as representing a load resistance $R_L$=V/I. It is easy to show that the power is a maximum for the point ($V_{OC}/2, I_{SC}/2$), which is a graphical way of illustrating the max power point load resistance in Eq. (4).

In Figure 4(b), we show an nTE array device in which the heat flow is increased over several orders of magnitude to test how well the generated power follows the theory developed above. The generated power shown in Figure 4(b) follows a square law relationship with heat flow over several orders of magnitude. This square law can be understood from Eqns. (6) and (12). Note that for this particular device, $ZT_1$ was low due to high parasitic leakage paths, so the second term of Eq. (6) is negligible, making the square law easy to observe.

## Thermoelectric Device Performance

To quantify the performance of the nTE devices, a series of device layouts was built on a 65nm CMOS testchip. Various geometrical parameters such as array dimensions (length, width, area $A_{ARRAY}$), thermopile dimensions (width and area $A_{TP}$), and thermopile packing fraction ($PF \equiv A_{TP} / A_{ARRAY}$) were varied. Several quantities were measured in these test structures. Thermopile array thermal impedance $\Theta_{TP}$ was measured in units of C/W by fitting four-probe I(V) curves of the heater resistor using Eq. (15). Four-probe resistance measurements quantified thermopile and heater resistance. Short-circuit current $I_{SC}$ and open-circuit voltage $V_{OC}$ were measured at several values of heat flow from the heater resistor. All of these parameters were stable and reproducible; variation across a 300mm wafer was typically on the order of 5%. Generated power is estimated using

$$P_{L,MAX}=V_{OC}^2/(4R_{TP}), \qquad (16)$$

which is equivalent to Eq. (5).

A method of normalizing thermopile array properties is needed in order to compare test structure results to application requirements. Conceptually, one can imagine building $N_{ARRAY}$ multiple instances of a given array size to multiply the generated power to $N_{ARRAY}$ $P_{L,MAX}$. The resistance would be reduced to $R_{TP} / N_{ARRAY}$. Heater resistor thermal impedance drops to $\Theta_{TP} / N_{ARRAY}$. Depending on the use of parallel or series electrical connection, the parallel array placement will multiply either $I_{SC}$ or $V_{OC}$ while leaving the other fixed. To facilitate analysis, $N_{ARRAY}$ was treated as a continuous variable rather than an integer, which due to the large number of thermopiles is a mild assumption. Furthermore, the specific series of test structures compared in Figure 5 all were built with nearly identical array areas, so the normalization error is negligible.

The system theory embodied in Eq. (12) shows the importance of the series thermal impedance $\Theta_S$ in setting the range of generated power since it restricts heat flow. The thermopile array thermal impedance that maximizes generated power was found to be given by Eq. (10) which, in the case of negligible heat leaks is $\Theta_S$. So to make a relevant comparison between our test results and an application of interest, we normalize our performance data in Figure 5 to a thermopile array size having $\Theta_{TEG}$=10C/W.

In a given nTE test structure during automated testing, a fixed input current I will produce a total heat in the heater resistor R that is equal to $I^2R$. Furthermore, each nTE test structure's heater resistor experiences a thermal impedance that depends on that particular device's geometry. Therefore, the raw harvesting data acquired from automated testing will have varying $\Delta T_{A,M}$. In order to compare nTE performance in the context of a given application's requirements, a way to scale the as-measured generated power back to a target $\Delta T_A$ is required. This is done by multiplying the generated power by the square of the ratio of temperature differences $(\Delta T_A/\Delta T_{A,M})^2$. The data points in Figure 5 are normalized to a body heat harvesting application of $\Delta T_A$=5C. Note that other studies often quote the temperature across the thermopile, which according to Eq. (11) would be less than half of $\Delta T_A$ (ie $\Delta T$<2.5C). We prefer to use the heat sink and source temperature difference as it is easier to measure and more closely linked to the application.

Figure 5 (a) illustrates the system theory derived above by examining a series of nTE device arrays with identical array dimensions and unit thermopile dimensions but with a varying number of thermopiles. It was constructed by deleting the tungsten contacts on the tops of thermopiles. This deactivates

*Figure 5. nTE data show system concepts. Data are normalized to a $\Theta_{TEG}$=10C/W array. Resistor temperature is normalized to $\Delta T$=5C using Eq. (5). (a) $P_L$ peaks with thermopile packing density. (b) A series of nTE layouts built with identical array area and identical total thermopile area $A_{TP}=W_{TP}L_{TP}$ but trading thermopile number $N_{TP}$ against width $W_{TP}$*

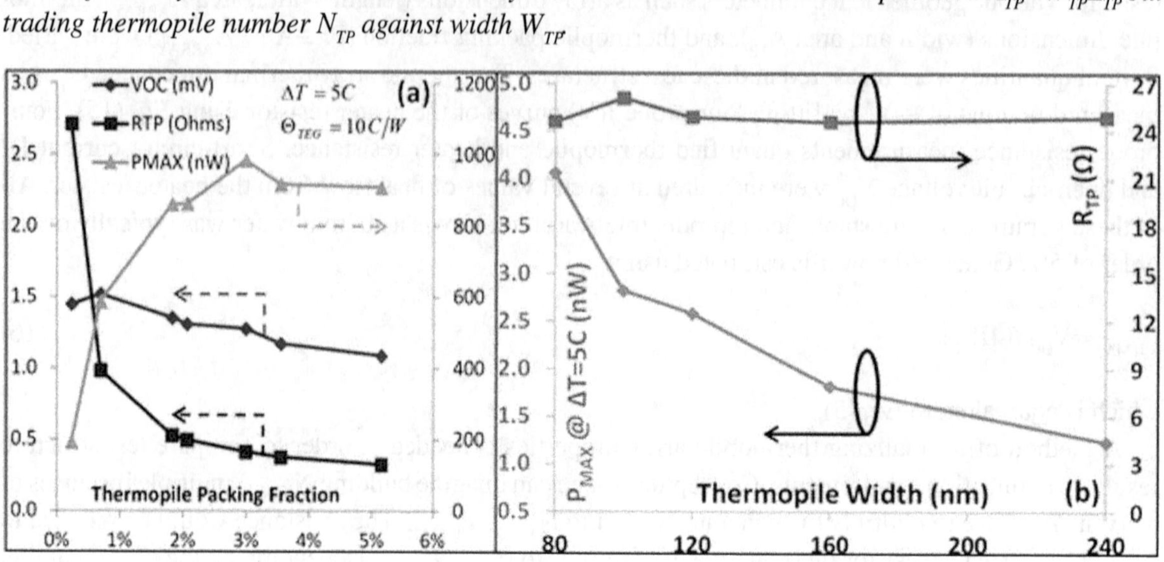

thermopiles selectively, so any variations among devices are due only to the packing fraction of thermopiles. The resistance of the thermopile arrays decreases with packing fraction because more thermopiles are added electrically in parallel as the packing fraction is increased.

The open-circuit voltage drops as packing fraction increases because adding thermopiles thermally in parallel reduces the thermal impedance of the thermopile array, reducing the fraction of the total temperature drop $\Delta T_A$ that falls across the thermopiles. In effect the thermal coupling efficiency is reduced by increasing the packing fraction, per Eq. (9). The generated power thus peaks as a function of packing fraction for an optimum array size as predicted by Eq. (10).

Figure 5 (b) reports the performance of a series of nTE test structures whose thermopile width varies in inverse proportion to the number of thermopiles. In other words, packing fraction and array dimensions are constant across this series of nTE test structures. The thermopile resistance, which goes as the total cross sectional area of thermopile perpendicular to the current path, is seen in Figure 5 (b) to be essentially constant across the data range. However, the generated power shown in Figure 5 (b) more than triples as the thermopile width is reduced from 240nm to 80nm despite having more perimeter, hence more heat leakage. This indicates that the thermopile itself is improved by the increased sidewall scattering, reducing its thermal conductivity as was studied in Si nanowires (Boukai, 2008).

This result supports our initial hypothesis that nanostructuring of silicon can improve thermoelectric performance by increasing phonon-sidewall scattering, reducing thermal conductivity. However, the silicon thermopiles are not yet small enough to test the largest improvement seen in Si nanowires (Boukai, 2008; Hochbaum, 2008). Nonetheless, to the best of the authors' knowledge, the present results are the first device-level verification of this concept (first reported in Edwards, 2013).

## DISCUSSION

### Assessment of nTE as a Body Heat Harvester

As discussed above, for body heat thermoelectric harvesting, we assume $\Delta T_A$=5C. The best nTE test structure layouts harvest 1-2 μW of power from a heat flow of ~500 W/cm$^2$ with nTE array sizes of 200-550μm. Due to the small size of these arrays of nTE devices, the density of generated power is very high, comparable to the best reports in the literature (Hudak, 2008). However, the heat flow density required to generate this power is 500 W/cm$^2$, so an intense heat source would be required to run this generator. Given the nTE array sizes mentioned above, the total heat flow is 200mW-1.5W. For comparison, consider a 10cm$^2$ wristwatch form factor. With Θ~100C/W at ΔT~5C, the heat flow is 50mW, insufficient to maintain 5C across even the most efficient nTE generators in the testchip.

Other applications may provide more intense and localized heat sources. For example, the use of a thermoelectrically powered thermal runaway detection/prevention circuit for power components may be more amenable to miniature thermopiles such as nTE.

### Miniaturization Improves Voltage Stacking

It is fairly straight-forward to design a startup circuit for a microchip if a voltage of 1V is available, so it is of interest to consider whether nTE could produce 1V to provde a heat-powered wakeup function. If all thermopiles can be stacked, then

$$V_{TOT}=(N_{n\text{-}TP}(-\alpha_n)+N_{p\text{-}TP}\alpha_p)\Delta T \quad (17)$$

For the nTE test structures having the highest $P_L$ values, the thermal impedance of each thermopile was $\Theta_{TP}=2{\bullet}10^6$C/W±20%, so it is expected that $N_{n+p}$=100,000 n-TP/p-TP thermocouple pairs would be needed to create a $\Theta_{TEG}$=10C/W array. The measured open-circuit voltage of these devices with $\Delta T_A$=5C was $\alpha_{pn}{\bullet}$5C=1.1mV±10%, so only ~900 must be in series to supply $V_{TOT}$=1V. In fact, even for $\Theta_{TEG}$=100C/W ($N_{n+p}$=10,000) and a 1C differential ($\alpha_{pn}{\bullet}$1C=0.22mV), it should be possible to attain more than 1V using nTE.

It is worth considering how much power could be generated in this situation. The open circuit voltage for the 10,000 n+p thermopiles with 1C described above would be 2.2V. Each n+p pair of thermopiles had about 2kΩ of resistance, so the series combination would have 20MΩ of internal series resistance. Plugging in open-circuit voltage for αΔT in Eq. (5), we get 240 nW of power delivered to a load, which may be sufficient to pre-charge a capacitor for a wake up function.

Conventional thermoelectric devices cannot stack as many thermopiles in series because each thermopile is large and has a low thermal impedance, meaning that it takes fewer thermopiles to match the application's series thermal impedance. The difference is due to lateral patterning through the geometrical factor A/t in K=κA/t, R=ρt/A. nTE thermopiles are 80nm by 750nm and 400nm thick, so nTE's A/t is 70 times smaller than those in the Seiko body heat powered watch (Hudak, 2008).

## Prospects for Thermoelectric Energy Harvesting

The generated power $P_L$ is limited to a fraction of the available heat flow $Q_1$ which in turn is controlled by the total thermal impedance seen by the heat source as indicated by the leading factor ($1/\Theta_S$) in Eq. (12). To put this into perspective, a ~10cm² microprocessor heat sink with thermal grease typically experiences a thermal impedance of less than 1 C/W. A ~1cm³ chip cooling fin might spec 10-20 C/W at working temperatures (hot enough to drive convection), but a 1cm² air interface near room temperature has a weaker coupling of 100-1000C/W, depending on air flow, humidity, and other factors.

Consider powering a ~10cm² device with a wristwatch form factor from body heat ($\Delta T_A$=5C) having a thermoelectric generator matched to $\Theta_S$~100C/W. From Figure 3 (b), this means that $ZT_1$ would need to exceed 0.1 to obtain $P_L$>1μW, assuming low leakage ($\Theta_{LK}$>>100C/W). This constrains the support materials that can be used to separate the "hot side" from the "cold side". For example, if 1mm³ ceramic standoffs with κ≈1W/m-K were used to separate hot and cold sides by 1mm, it would only take 10 standoffs to create a 100C/W heat leak. A thermoelectric powered wristwatch by Seiko (Hudak, 2008) used an array of machined $Bi_2Te_3$ thermopiles 80μm by 80μm and 600μm tall, consistent with the $ZT_1$>0.1 and high $\Theta_{LK}$ requirements above. The prospects for body heat powered systems would seem to be in the μW range. The use of a high $zT_1$ material and a large surface area, say a ~100cm² area covering a large fraction of a wide wristband, might enable 10's of μW to be generated. Clearly for such a large area thermoelectric device targeting a consumer application, the use of low-cost manufacturing methods would be important.

For industrial IoT applications, more temperature difference and heat flow may be available. For example, heat sinks a few inches across with $\Theta_S$ ~1 C/W are commercially available. If a material with $zT_1$~0.3 were used and the heat leak were moderate ($\Theta_{LK}$~$\Theta_S$), then according to Figure 3 (c)-(d) a temperature difference of 20C could drive ~10mW of power generation which may be enough to drive a wireless sensor node. Note that the efficiency is still quite low even for these higher temperatures since a heat flow of 10W is required. The heat source would have to be a fairly significant piece of equipment such as a pump. Care would need to be taken that the harvester not cause overheating problems in the equipment it is harvesting from, since it would add thermal impedance to the path of heat rejection to the ambient.

Larger scale thermoelectric harvesting projects have been explored (Rowe, 2006). As alluded to in the introduction, there are many sources of waste heat such as auto tailpipe exhaust or industrial process waste heat. These reject heat at 100s of C above ambient temperatures, leading to the hope that some power could be recovered, improving system efficiency or cost. However, there are some issues to be addressed. First, high temperature systems place strong demands on the materials used, although the radioisotope thermopile generators (RTG's) used in deep-space probes demonstrate that reliable high temperature thermoelectric devices can be made. Second, the efficiency of thermoelectric harvesting predicted by Eq. (12) and plotted in Figure 3 (a)-(d) is fairly low so the cost may outstrip the benefit. For example, solar thermal generation using semiconductor thermopiles makes no sense even if the thermopiles materials could sustain elevated temperatures because solar cells exist that can harvest power at relatively much higher efficiency (10's of percent).

A more subtle requirement for large-scale thermoelectric energy harvesting is the intimate mechanical coupling from heat source to thermopile to heat sink. This means that as the entire system undergoes thermal stress at high temperature, this stress must be transferred across the thermopiles, because they must be coupled with low thermal impedance between the heat source and the heat sink. Any strain relief mechanism that were used would have to have an extremely low thermal impedance to avoid restricting heat flow. Support mechanisms between the heat source and sink would have to have a high thermal impedance to avoid shunting heat flow around the thermopiles. As a result of these challenging constraints, studies of high temperature industrial applications of thermoelectric harvesting have reported relatively rapid wear and delamination failures after a period of weeks or months (Rowe, 2006). These problems, put together, represent significant hurdles to large-scale thermoelectric energy harvesting.

## FUTURE RESEARCH DIRECTIONS

Improving the $zT_1$ of thermoelectric materials is an ongoing area of research, with particularly high activity levels in nanostructured materials (Rowe, 2006; Hudak, 2008). Semiconductor nanowires seem to be particularly interesting in this regard because the thermal conductivity can be reduced substantially without sacrificing electrical resistivity or Seebeck coefficient. For example, we have shown in Figure 6 that an array of 80nm wide silicon thermopiles generate about three times as much power as a matched 240nm wide thermopiles, supporting the concept of thermal conductivity reduction by phonon-sidewall scattering. If devices could be built that incorporate the much smaller geometry nanowires studied in (Boukai, 2008; Hochbaum, 2008), up to two orders of magnitude improvement could be achieved relative to bulk silicon properties, which in principle could make silicon competitive with the performance of the compound semiconductors such as $Bi_2Te_3$ conventionally used for thermoelectric conversion. The obvious advantage of using silicon is the large installed base of silicon microelectronic manufacturing capability, leading to the possibility of low cost manufacturing methods that might meet the price requirements of commercial applications such as wearable devices.

The system theory derived above highlights the challenges in thermal system design which would have to be overcome. For example, as the thermopiles are improved they will become less thermally conductive, so other parallel thermal heat leakage paths become more important to the overall thermoelectric system efficiency, as indicated in Eq. (12). Because nanostructuring works by increasing thermopile perimeter, it also will be necessary to suppress any heat leaks that scale with thermopile perimeter such as cladding or support material needed to contain the nanowires in a device.

As noted for the case of diffuse heat flows such as body heat, it is also important to significantly reduce the fabrication cost of high-$zT_1$ thermoelectric materials and devices. Efficient and low-cost heat couplers that can couple a diffuse heat flow into a microfabricated thermoelectric device without excessive losses through heat leaks may be a way to enable system cost reduction through miniaturization.

The identification of applications with a heat source that has an innately high heat density should fit with the cost advantages of miniaturized thermoelectric devices as well. Thermal runaway protection for high power components may be such an application. It also may be possible to find sensing applications for miniaturized thermoelectric devices. The use of a miniaturized thermoelectric energy harvester as a heat flow sensor may also be of interest if appropriate applications can be identified. For example, thermoelectrically powered pilot light sensors for residential water heater control are available commercially. This is an application that senses heat flow while being powered by a highly concentrated heat source.

## CONCLUSION

We have provided the first device-level confirmation of the benefits of silicon nanostructuring in thermoelectric devices. We also have developed device-level and system-level theories to judge the suitability of a given thermoelectric technology to an application. We have shown using the example of nTE technology that a simple metric such as high density of generated power does not always imply suitability for a given thermal energy harvesting application.

When evaluating a potential energy harvesting technology, it is necessary to consider a variety of application requirements:

- Material ($zT_1$) and device ($ZT_1$) figures of merit.
- Series thermal impedance of the heat source and sink.
- The voltage stacking capability of a thermally-matched thermopile array size must meet the application voltage requirement.
- Once sufficient thermopiles are stacked in series to achieve the application voltage requirement, the internal series resistance of the stacked thermopile array must be low enough to supply sufficient electrical current to support the application's power requirement.
- Size mismatch between the heat source/sink and a thermopile array sized to achieve thermal match (Eq. 14).
- Series thermal impedance introduced by spreading structures used to alleviate size mismatch.
- Heat leakage paths caused by support structures, fill material, spreading structures, or air convection.

## ACKNOWLEDGMENT

We would like to acknowledge our colleagues: Ryan Hultquist, Nancy Little, and Brad Sucher for process support; Chuck Roth and Ivan Coronado for testing and electrical measurements; Dave Truty (STX Cadware) for testchip layout; Don Plumpton, Doug Prinslow and Cissy St John for testchip execution; and the leadership of Mark Denissen, Chris Bowles, Kyle Flessner, Thierry Bosson, and Taylor Efland.

## REFERENCES

Boukai, A., Bunimovich, Y., Tahir-Kheli, J., Yu, J., Goddard, W. III, & Heath, J. (2008). Si nanowires as efficient thermoelectric materials. *Nature*, *451*(7175), 168–171. doi:10.1038/nature06458 PMID:18185583

Edwards, H., Debord, J., Tran, T., Freeman, D., & Maggio, K. (2013), Performance metrics for thermoelectric energy harvesting studied using a novel planar 65 nm silicon CMOS-based thermopile. *Proceedings of the 2013 IEEE Sensors Conference*, Baltimore, MD, Nov 3-6, 2013. doi:10.1109/ICSENS.2013.6688611

Edwards, H., Niu, Q., & de Lozanne, A. (1993). A quantum dot refrigerator. *Applied Physics Letters*, *63*(12), 1815–1817. doi:10.1063/1.110672

Edwards, H., Niu, Q., & de Lozanne, A. (1999). Thermoelectric conversion. In *Wiley encyclopedia of electrical and electronics engineering* (Vol. 22, pp. 80–90). New York: John Wiley and Sons. doi:10.1002/047134608X.W3044

Edwards, H., Niu, Q., Georgakis, G., & de Lozanne, A. (1995). Cryogenic cooling using tunneling structures with sharp energy features. *Physical Review B: Condensed Matter and Materials Physics*, *52*(8), 5714–5736. doi:10.1103/PhysRevB.52.5714 PMID:9981758

Goldsmid, H., & Douglas, R. (1954). The use of semiconductors in thermoelectric refrigeration. *British Journal of Applied Physics*, *5*(11), 386–390. doi:10.1088/0508-3443/5/11/303

Gustafsson, S., Karawacki, E., & Khan, M. (1979). Transient hot-strip method for simultaneously measuring thermal conductivity and thermal diffusivity of solids and fluids. *Journal of Physics. D, Applied Physics*, *12*(9), 1411–1421. doi:10.1088/0022-3727/12/9/003

Hicks, L., & Dresselhaus, M. (1993). Effect of quantum-well structures on the thermoelectric figure of merit. *Physical Review B: Condensed Matter and Materials Physics*, *47*(19), 12727–12731. doi:10.1103/PhysRevB.47.12727 PMID:10005469

Hochbaum, A., Chen, R., Delgado, R., Liang, W., Garnett, E., Najarian, M., & Yang, P. et al. (2008). Enhanced thermoelectric performance of rough silicon nanowires. *Nature*, *451*(7175), 163–167. doi:10.1038/nature06381 PMID:18185582

Hudak, N., & Amatucci, G. (2008). Small-scale energy harvesting through thermoelectric, vibration, and radiofrequency power conversion. *Journal of Applied Physics*, *103*(10), 101301–101324. doi:10.1063/1.2918987

Nahum, M., Eiles, T., & Martinis, J. (1994). Electronic microrefrigerator based on a normal-insulator-superconductor tunnel junction. *Applied Physics Letters*, *65*(24), 3123–3125. doi:10.1063/1.112456

Rowe, D. (Ed.). (2010). *Thermoelectrics handbook: macro to nano*. New York: CRC Press, Taylor & Francis.

## KEY TERMS AND DEFINITIONS

**Efficiency:** In a generation process, the efficiency is the fraction of fuel energy which may be delivered to a load. In a *thermoelectric energy harvesting system*, the "fuel energy" is the available heat flow Q across a temperature difference ΔT. The energy delivered to the load is the product of the voltage applied to the load $V_L$ and the current through the load $I_L$.

**Intensive or Extensive Parameter:** An *intensive* material parameter is a material property that is independent of the size, shape or volume of the material. Hence intensive material parameters such as thermal conductivity or electrical resistivity or density are often tabulated in reference books. An *extensive* parameter does depend on the size, shape, or volume of the material. Examples of extensive parameters include electrical resistance of a circuit element, the thermal conductance of a device or junction, and the weight of an object.

**Seebeck Effect:** When a pair of dis-similar materials p and n is joined together and a temperature difference ΔT exists between them, a voltage $V=\alpha_{pn}\Delta T$ will appear across their junction due to the Seebeck effect.

**Thermal Conductance (K):** The *extensive* parameter of a device or interface that determines the heat Q that flows through the device or interface in response to an applied temperature difference ΔT. If the device or interface is of a rectangular shape of area A perpendicular to the direction of heat flow and thickness t, then Fourier's Law becomes $Q=K\Delta T$, where the thermal conductance may be written in terms of the thermal conductivity κ as $K=\kappa A/t$, which is analogous (but reciprocal) to the relationship between electrical resistance (R in Ω) and electrical resistivity (ρ in Ω-cm).

**Thermal Conductivity (κ):** Thermal conductivity is the *intensive* material property that governs heat conduction in materials, in particular setting the required temperature gradient $\nabla T$ needed to drive a given heat flow q through the material. This relationship is embodied in Fourier's Law $q=\kappa\nabla T$. The typical units of thermal conductivity are W/cm-K or W/m-K.

**Thermoelectric Energy Harvesting:** The use of the Seebeck effect to extract electrical power from heat flowing through a junction of dis-similar materials.

**Thermopower ($\alpha_{pn}$):** Also known as the Seebeck coefficient, thermopower is the property of a junction of dis-similar materials p and n that gives rise to their Seebeck effect. The typical units of thermopower are μV/K.

# Chapter 10
# Silver Coatings with Protective Transparent Films for Solar Concentrators

**Monserrat Gutiérrez Muñoz**
*CIDETEQ, S.C., Mexico*

**Ernesto González De León**
*Instituto Tecnológico de Tepic, Mexico*

**Jose de Jesus Perez Bueno**
*CIDETEQ, S.C., Mexico*

**Yunny Meas**
*CIDETEQ, S.C., Mexico*

**Guy Stremsdoerfer**
*Ecole Centrale de Lyon, France*

## ABSTRACT

*The use of solar energy as a renewable source is one of the most promising resources to generate electricity. The viability of concentrating solar power (CSP) systems depends on the development of highly reflective materials that are inexpensive and maintain their optical properties for extended lifetimes under outdoor environments. In this chapter, the implementation of flexible polymer substrates plated with silver by the Dynamic Chemical Plating technique (DCP) is proposed because of its low cost, and easy and rapid deposition, in addition to the high speed at which the deposit is made. However, the deposits made under this technique have certain nanoscale imperfections, which begin to exist certain permeability of substances that can stain the silver over time, so a study of this feature is performed, to help assess their durability.*

## INTRODUCTION

The Sun is the source of life and origin of almost all other forms of energy that man has used since the beginning of history. It would be irrational not to try to take advantage by all possible techniques, this free, clean and inexhaustible energy source, which can liberate us from dependence on oil or other unsafe, polluting or simply exhaustible alternatives.

It is vital to continue with the development and improvement of technology to collect, accumulate and distribute solar energy, for the very conditions that make it definitely competitive on a global scale.

DOI: 10.4018/978-1-4666-8254-2.ch010

Solar radiation can be converted to electricity in two ways: either through solar panels that generate electricity directly, or converting solar radiation into heat energy and then convert it into electricity. Regarding the second, is having a great commercial prosperity in a number of countries with high insolation levels. These energy recovery systems are called Concentrating Solar Power (CSP).

CSP systems use mirrors or lenses (solar concentrators) with tracking systems to focus a large area of sunlight onto a small area. The concentrated light is then used as heat or as a heat source for a conventional power plant (solar thermoelectricity). Solar concentrators represent about 60% of the cost of a CSP, which represent an important motivation to study them and decrease their cost (Wang, 2010).

Structure concentrators consist of a support structure, a substrate, a reflective agent (steel, aluminum or silver) and a transparent protective layer which enables preservation of the properties of the substrate. Its features should be: low production cost, long life in environment condition, high transmittance, among other (Barlev, Vidu, & Stroeve, 2011).

In this chapter, different configurations are considered, using aluminum as the substrate, or, the silver plating on polymer by Dynamic Chemical Plating technique (DCP) with a protective coating. Further, several coatings were prepared by sol-gel technique and their optical properties were studied as well as feasibility for use in solar technology.

## BACKGROUND

### Historic

212 BC, Archimedes used mirrors for the first time to concentrate the Sun's rays. In 1615, a small solar powered motor was developed by Salomon De Caux, this motor consisted of glass lenses and an air-tight metal vessel. Lavoisier and Joseph Priestley developed the theory of combustion by concentrating solar radiation on a test tube for gas collection. In 1878, Paris, a small solar power plant was exhibited; the plant was made of parabolic dish concentrators. The first CSP was installed in 1913, in Egypt as a parabolic trough solar field for pumping water. Because of the oil crisis in 1970s lot of pilot CSP plants were built. Then, the first commercial CSP was operated in California, USA (1984-1991) It was not until 2006 that interest was once again rekindled for the development of large scale CSP plants. (Behar, Khellaf, & Mohammedi, 2013).

The central solar thermal electricity produced concentration similarly to conventional power plants: using steam to drive a turbine. The difference between them lies in the origin of energy instead of fossil fuel using solar radiation. Solar concentrators with tracking systems focus a large area of sunlight onto a small area. Thus, four main elements are required: Solar concentrator, receiver, some form of transportation or storage, and power conversion.

Techniques use solar radiation to electrical energy conversion using an intermediate conversion into thermal energy are classified into two groups: those that concentrate solar radiation along a line and concentrate solar radiation at a point. For linear concentration, two technologies that are used: parabolic trough and linear Fresnel reflector. Pointed concentration, with much higher concentration ratios and the ability to track two axes, there are two applicable technologies: the central tower and parabolic dish using Stirling engines (Fend et al., 2003).

## Parabolic Trough

For this technology, curved as a parabola reflectors are used (Figure 1a). Normally, channels are designed to follow the sun in one axis, predominantly north-south. The receiver comprises the absorber tube (usually metal) inside an evacuated glass envelope. The absorber tube is generally a coated stainless steel tube, with a spectrally selective coating that absorbs the solar (short wave) irradiation well, but emits very little infrared (long wave) radiation. This helps to reduce heat loss. Through these tubes the heat transfer fluid is circulating, as synthetic heat transfer oil. The fluid is heated to about 400° C by the concentrated solar and pumped through a series of heat exchangers to produce superheated steam, which is converted to electrical energy in a conventional steam generator turbine (Zhang, Baeyens, Degrève, & Cacères, 2013).

## Fresnel Mirrors

Linear Fresnel collectors (LFCs) are similar to parabolic trough collectors, but use a series of long flat, or slightly curved, mirrors placed at different angles to concentrate the sunlight on either side of a fixed receiver (located several metres above the primary mirror field) (Hoffschmidt et al., 2012). A set of nearly flat reflectors concentrate the solar radiation invested high linear receivers (Figure 1b). The water flowing through the steam becomes receptors. This linear concentration system is similar to a parabolic trough, with the advantage of low cost reflectors and structural support, fixed fluid seals, a separate reflector system receiver, and longer focal lengths that enable the use of plane mirrors. This technology is seen as a potential alternative lower cost to the parabolic trough technology for the production of solar process heat.

## Central Tower

A joint circular heliostats (large mirrors with solar tracking systems) concentrates sunlight onto a central receiver mounted on a tower (Figure 2a). A heat transfer medium in this central receiver absorbs the highly concentrated radiation reflected by the heliostats and converted to thermal energy that is used to generate

*Figure 1. Linear concentration a) parabolic trough, b) Fresnel*

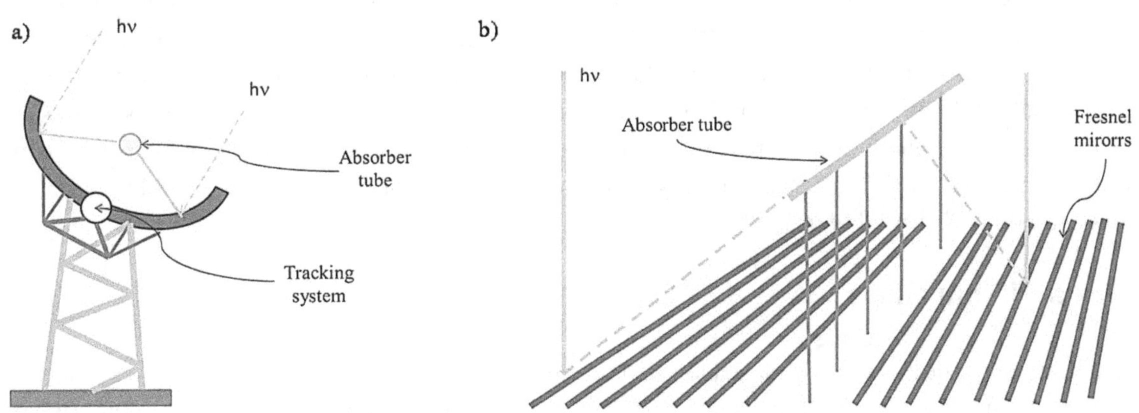

*Figure 2. Pointed concentration a) central tower, b) dish stirling*

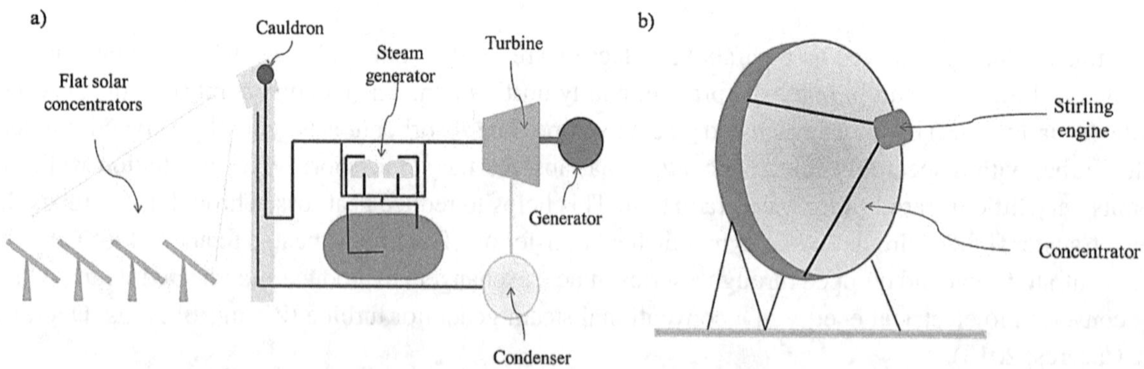


superheated steam for the turbine. Nowadays, heat transfer fluids have been developed include water / steam, molten salt and air. Air or gas is used to pressure at very high temperatures of about 1000° C or more as a heat transfer medium, it can also be used to directly replace natural gas in a gas turbine, which takes full advantage of the magnificent cycle (60% and more) of modern combined cycle gas and steam.

The main advantages of this technology are: The higher temperatures can potentially allow greater efficiency of the steam cycle and reduce water consumption for cooling the condenser; the higher temperature also makes the use of thermal energy storage more attractive in order to achieve schedulable power generation; and Higher temperatures will also allow greater temperature differentials in the storage system, reducing costs or allowing greater storage for the same cost (Hoffschmidt et al., 2012).

## Dish Stirling

A parabolic reflector disc shaped concentrates sunlight onto a receiver located at the focal point of a disk (Figure 2b). The concentrated rays radiation absorbed in the receiver to heat a fluid or gas (air) to about 750 ° C. The fluid or gas is used to generate electricity in a small piston or Stirling engine or a micro turbine, connected to the receiver.

## The Solar Field

Key components to reduce the solar field cost are support structures, including foundations, mirrors and receivers. The viability of solar concentration technology requires the development of reflector materials that are low cost and maintain high specular reflectance for extended lifetimes under severe outdoor environments including ultraviolet irradiation, air, rain, dirt, sand storms, low and high temperatures and depending on the technology must take different forms: curved or flat. Which represent a real challenge for the researchers. At present, mainly four types of solar concentrators are used, which will be described below (Kennedy & Terwilliger, 2005):

## Thick and Thin Glass

Thick glass concentrators are constructed as a conventional glass mirror (Figure 3). The silver layer is extremely vulnerable to air pollutants and particles of moisture or salt present in coastal environments.

*Figure 3. Glass mirror structure*

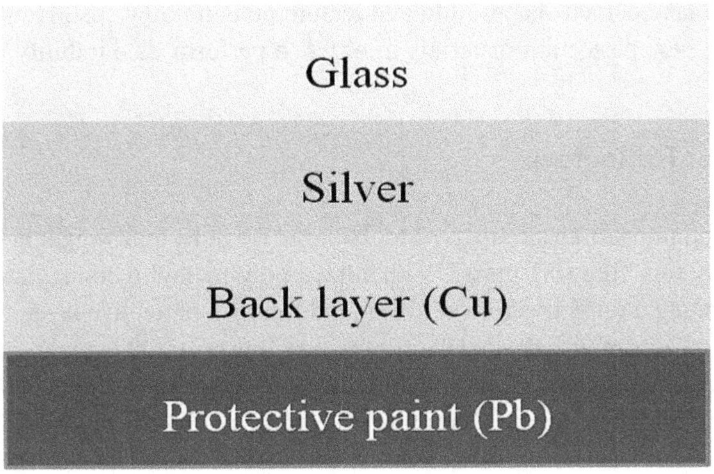


To protect silver, later copper layer and a layer of protective paint is applied. Copper has a significant protective role because it slows the tarnishing of silver and provides an area where the paint adheres. The protective coating protects the copper from abrasion and corrosion, thereby extending the life of the Ag-Cu mirrors. CSP companies have developed the thin glass mirrors; glass thickness is about 1mm and is bonded to metal substrates. Thin glass concentrators have the same structure as the thick glass mirrors, therefore, are protected by a layer of copper and a subsequent layer of protective paint.

Although, glass concentrators present lot of advantages as excellent durability in terms of reflective layer corrosion, are heavy and brittle which increase the cost. Further, curved shapes are difficult to make and require slumped glass, which is expensive.

## Aluminized Reflectors

Utilize a polished aluminum substrate, a layer of reflective aluminum improved and especially the formation of a protective oxide (alumina). Because of this last layer, another coating initially was not used, but the low durability of the reflectivity forced to add an acrylic polymer coating (Figure 4).

*Figure 4. Aluminized reflectors structure*

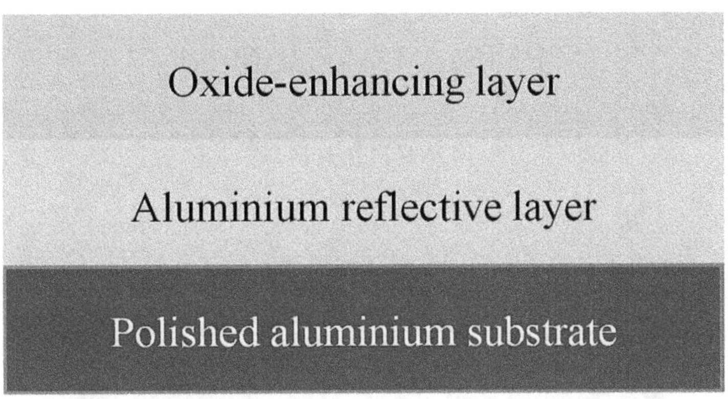

Aluminized reflectors do not have the disadvantages of glass concentrators as they are lightweight, flexible so, they can take curved shapes, thin and feature potential low cost. However, is necessary to continue developing new protection materials in order to perform its durability (Fend, Jorgensen, & Küster, 2000).

## Silvered Polymer Reflectors

Polymer base concentrators are an alternative for obtaining flexible, lightweight hubs. These concentrators consist of a very thin film (0.1 mm) UV-stabilized poly-methyl-methacrylate (PMMA) which is deposited by evaporating a reflective film of silver, and finally a protective layer of copper (Figure 5).

Polymer reflectors, accomplish desired characteristics: low cost, lightweight, resistant, greater flexibility. However, as aluminum mirrors, their viability depends on the development of protection coating.

For mirrors, cost reductions may be accomplished by moving from heavy silver-backed glass mirror reflectors to lightweight front-surface advanced reflectors. The advantages of thin-film reflectors are that they are potentially less expensive, will be lighter in weight and have a higher reflectance. They can also be used as part of the support structure. However, their long-term performance needs to be proven. Ensuring that the surface is resistant to repeated washing will require attention. In addition to these new reflectors, there is also work underway to produce thinner, lighter glass mirrors. The CSP technologies are capital intensive, about half of the capital cost of a solar thermal plant is invested in solar concentrators. This causes the reduction of the cost of the concentrators is fundamental to achieving the objectives of energy costs. The cost of the reflector for the four CSP technologies represents approximately 30% of the arena hub. However, the structural costs are closely related to the cost of the reflector representing 50% of the cost of a dish or heliostat and 75% of the cost of a parabolic trough. Therefore, the improvement of solar reflectors, play an important role in achieving cost reduction required in solar collectors. Besides the CSP systems must operate reliably for decades in extremely harsh environmental conditions including UV and low radiation, wind, rain, sandstorms, dirt and high temperatures. This is a formidable challenge in the manufacture of inexpensive and long lasting elements (Debije & Verbunt, 2012) .

Despite the considerable efforts made in the development of solar collectors, there remains a need for a viable solar collector cost of competing with other technologies that is durable and has good optical properties, this challenge remains on the development of a resistant and transparent coating (Behar et al., 2013) .

*Figure 5. Silvered polymer reflectors structure*

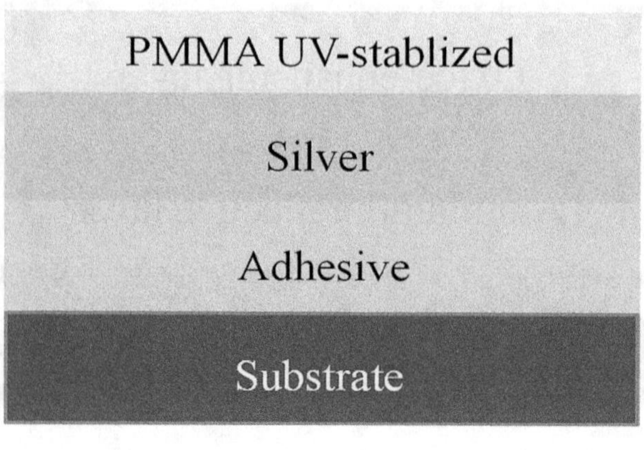

For this reason, the main objective of this chapter is to develop a resistant coating with the optical and mechanical properties required, optimizing polymer and aluminum reflectors (Schissel, Kennedy, & Goggin, 1995). Two different structures are proposed, first consist in using a polymer as subtract which is silver metalized by Dynamic Chemical Plating, and finally a protecting coating is deposited by dip coating (Figure 6). Second structure, is using aluminum as a substrate, and a developed coating.

## DEVELOPMENT OF SOLAR CONCENTRATORS

### Polymer Substrate, Silver Deposit

Dynamic Chemical Plating technology (DCP) is the controlled reduction of metal cation electrons. Two solutions were prepared, the first containing the metal salts of silver and the second electron reserves as a reductant. With the help of a device, aqueous solutions are projected in the form of droplets (Figure 7). The droplets coalescence leads to the formation of a liquid film coating, which wets the surface exposed to the projection. This new technology, is fundamental in order to cheapen solar concentrators, furthermore the ease to apply over large areas, is a friendly environmental technique.

Silver deposit, as can be seen in Figure 8, the optical profilometry shows a uniform silver film, the image, was processed with the software Gwyddion, obtaining an average roughness of 8.69 nm.

However, as a new technique, the deposit presents some troubles over the time, some "failures" are formed on the surface of the silver, which has been growing and leads to a complete detachment of the silver. This "failures" have been analized by Laser Scanning Confocal Microscopy (LSCMS), images were obtained in two and three dimensions (Figures 9a and 2b, respectively). Analyzing, red areas correspond to silver; it is proposed that all the black areas correspond to some other compound, in this case a silver compound; and the areas which present green color correspond to the fluorescent emission of the substrate (PMMA), the two peaks are located in 471 nm and 519 nm. Around the green peaks, a greater number of black areas are presented. For these reasons, is presented as a hypothesis the reaction of silver with some other element that can be presented when de deposition is made, or can be on the atmosphere, and after a certain reaction time, leads to the detachment of silver.

*Figure 6. Structure proposed for polymer concentrators*

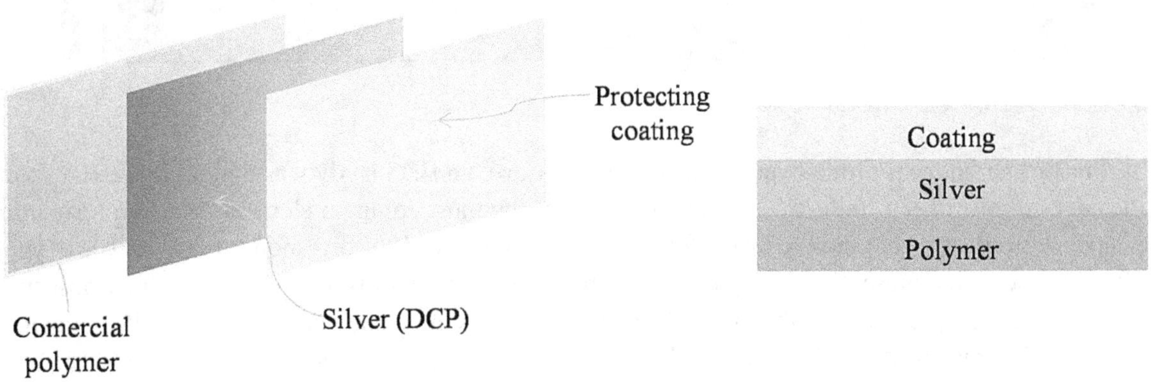

*Figure 7. Dynamic chemical plating*

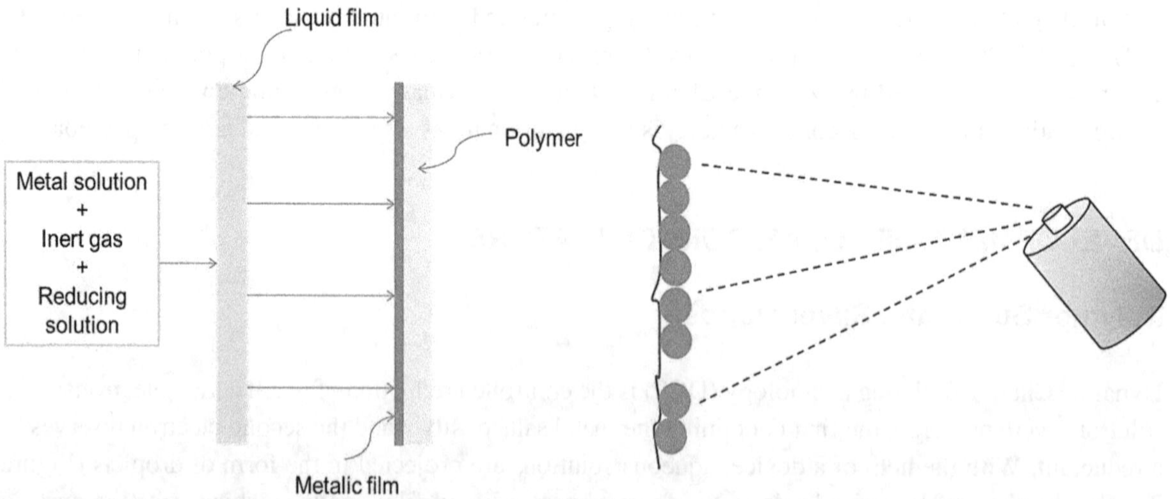


*Figure 8. Optical profilometry of silver deposition by DCP*

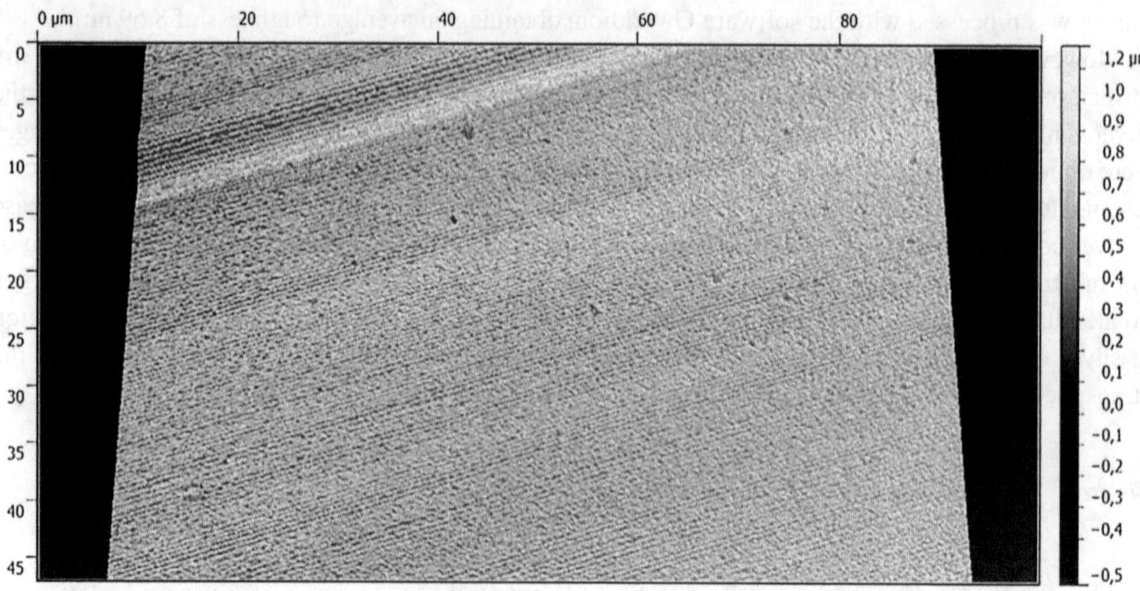

For further analysis of the hypotheses discussed above, an EDS study was done, Figure 10a shows the edges of one of the "failures", and the presence of the most common elements that can form compounds with silver where showed on the following images, where 10b corresponds to silver, 10c oxygen, 10d chlorine, 10e represents to phosphor, 10f carbon, 10g sulfur and 10h nitrogen. Carefully observing these figures, the presence of silver is evident on the whole sample, and compound of sulfur, nitrogen, and phosphor may be discarded, on the other hand the most significant presence is given by a chlorine compound and in the second place some oxide is taking place mainly on the "failure".

*Figure 9. Confocal microscopy image of "failure" in the silver deposit a) two dimensional, b) three-dimensional.*

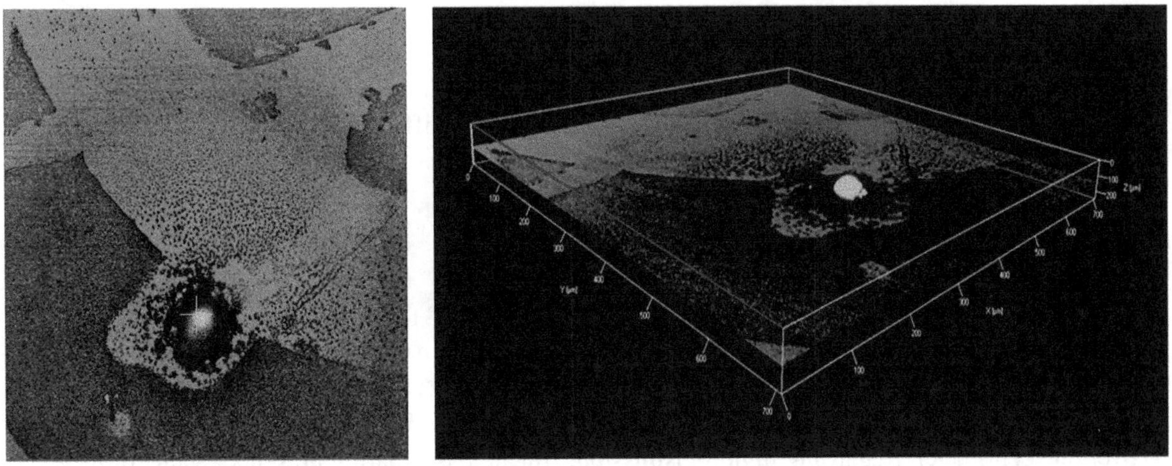

*Figure 10. a) SEM of "failure" EDS b) silver, c) oxygen, d) chlorine, e) phosphor, f) carbon, g) sulfur h) nitrogen*

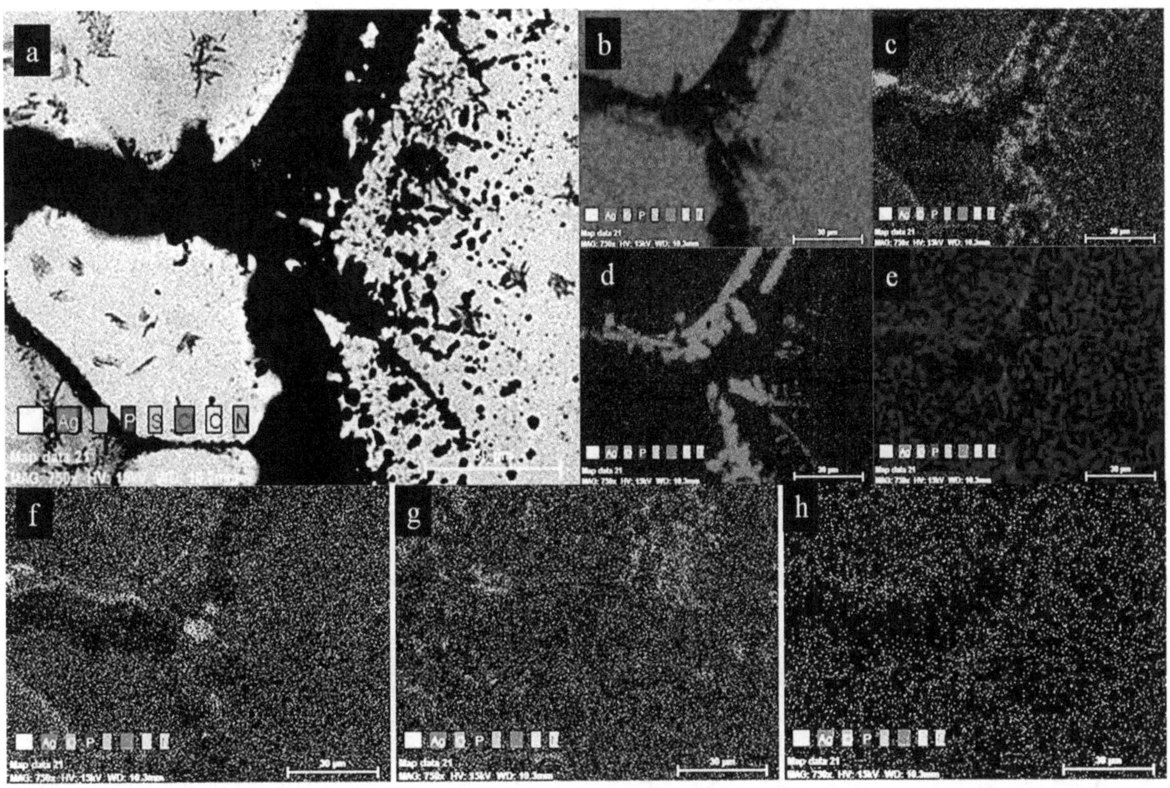

As can be seen, silver deposit made by DCP could causes a reactive silver, for this reason it is necessary to apply another treatment like flame in order to reduce this behavior.

## Coatings for, Glass and Silver Substrates

In order to analyze optical properties as light transmition, four different coatings were developed, and applied on conventional glass slides, and DCP silver.

### Silica Sol-Gel Coatings, Silicate Base

In order to elaborate the transparent coating the following volume proportions were mixed: 65% of colloidal silica, 17.5% of sodium metasilicate (Na2SiO3) and 17.5% of sodium hydroxide (NaOH).

Figures 11a and 11b, show the coating on glass and thickness, respectively. The thickness of this coating was about 30 microns. The thickness of such coatings can be varied from less than one micron to over one centimeter. Regarding light transmission, silicate base coating presents a value of 87%.

Figure 12 shows results obtained on silver, an uniform transparent coating, although it is a very cheap coating a thermal post treatment mots be done in order to prevent the coating from being solved in water.

### Silica Sol-Gel Coating, TEOS Base

TEOS sol-gel was prepared with the following volume proportions: 42% tetraethoxysilane (TEOS), 42% EtOH, 13% water and 3% nitric acid at 5%.

In Figure 13a a silica coating is observed even without fracture. Figure 9b shows the roughness and thickness of the coating. In this case, the coating thickness was 4.5 microns. However, this type of coating is unlikely to have a thickness greater than 1 micrometer, since the surface tension due to solvent loss processes end up fracturing.

*Figure 11. a) Surface image silica sol-gel coating from silicate 70X, glass as substrate. b) Profilometry graph showing the thickness of the coating, the substrate being the horizontal line to the left, at 0 μm.*

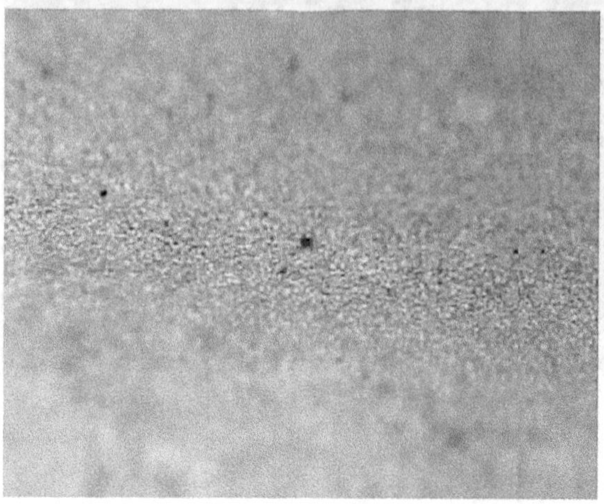

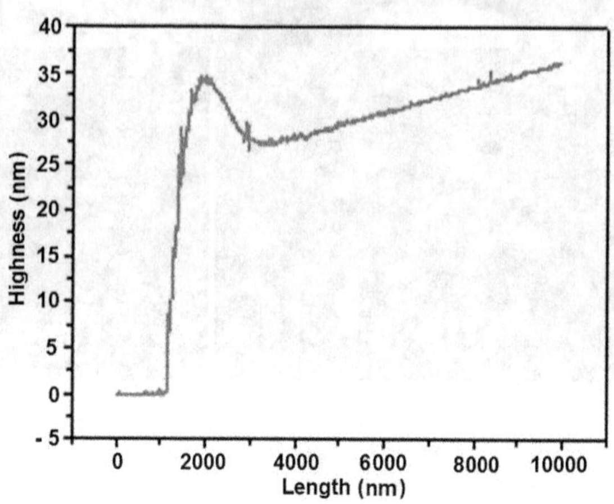

*Figure 12. a) Surface image silica sol-gel coating, silver as substrate. b) Surface image silica sol-gel coating from silicate 70X, silver as subtrate*

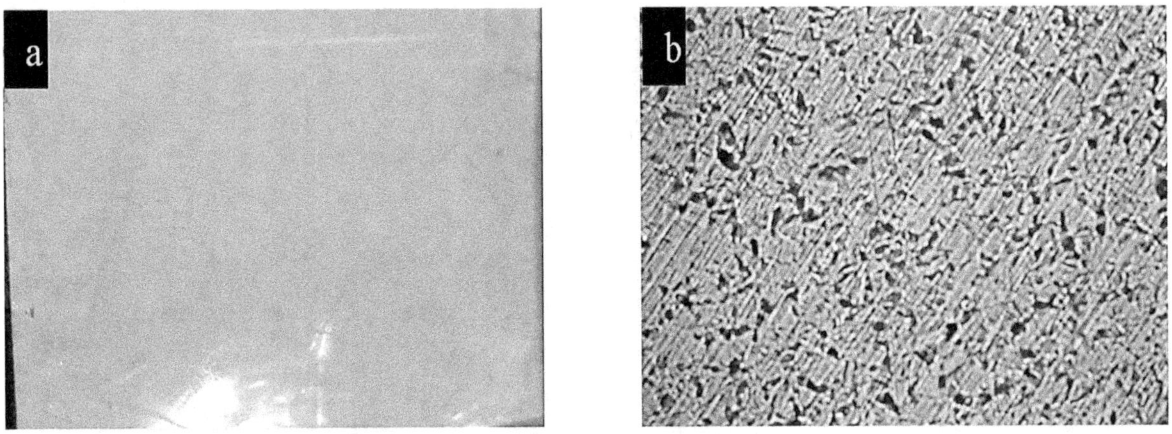


*Figure 13. a)Image surface of a sol-gel silica coating from TEOS, 70X glass as substrate b) Profilometry graph showing the thickness of the coating, the substrate being the horizontal line to the left at -4 μm.*

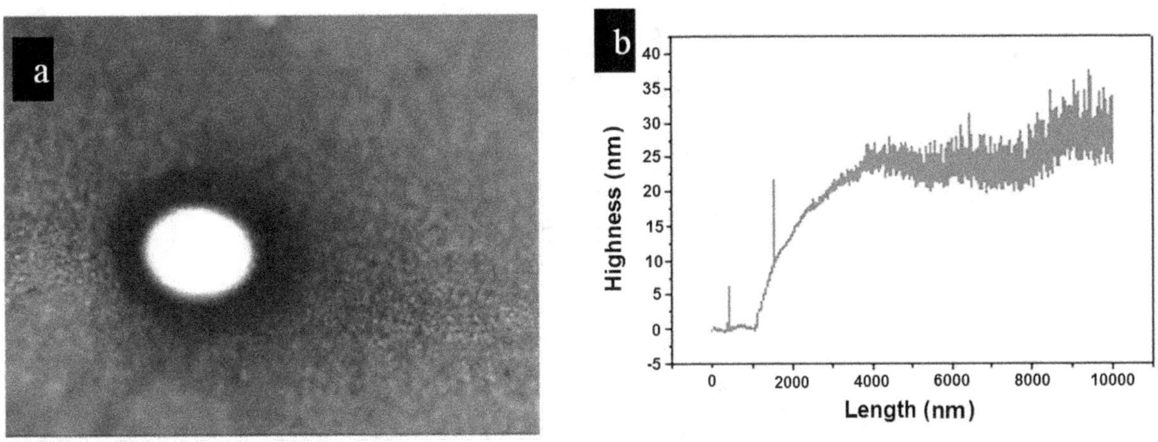


The fractures mentioned above, can be observed in a more properly form in Figure 14b, therefore it is really important to control the thickness to avoid this phenomenon.

## Hybrid Coating: PMMA-Silica, Sol-Gel TEOS_

Another way to prevent fractures obtained with inorganic coatings, and increase its elasticity, is the formation of hybrid coatings. In this case a coating was used with the following proportions by volume: 50% PMMA on diethyl oxalate and 50% TEOS sol-gel

Figure 15 a, a transparent hybrid coating PMMA-SiO2. White circle corresponds to light scattering by an air bubble, which has been caused by air entrapment in a viscous solution. Figure 15b shows the graph of profilometry. The coating has been unusually rough surface. Or ruffling roughness effects, may appear in the coating of this type.

*Figure 14. a) Surface image TEOS sol-gel coating, silver as substrate. b) Surface image TEOS sol-gel coating from silicate 70X, silver as subtrate*

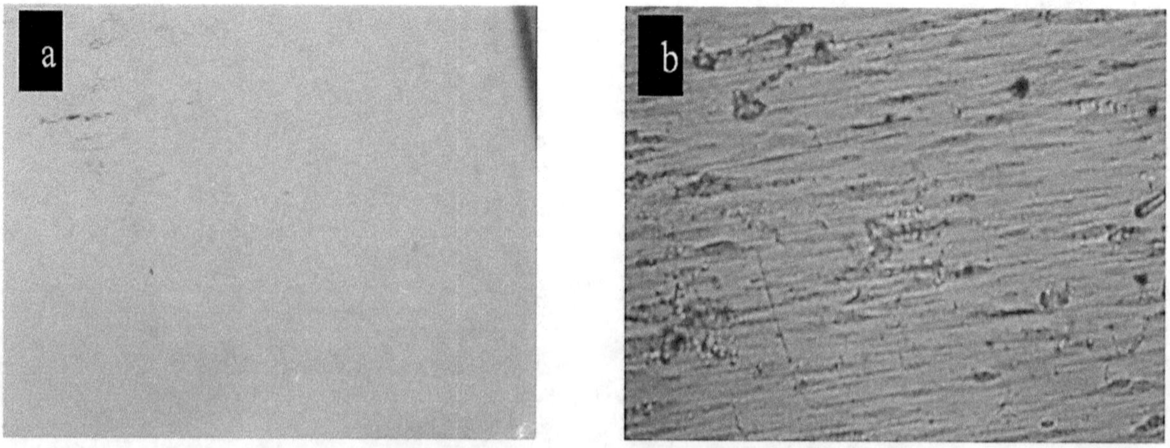


*Figure 15. a) Surface image of a hybrid coating 70X, glass as substrate b) Profilometry graph showing the thickness of the coating, the substrate being the horizontal line to the left, at 0 μm.*

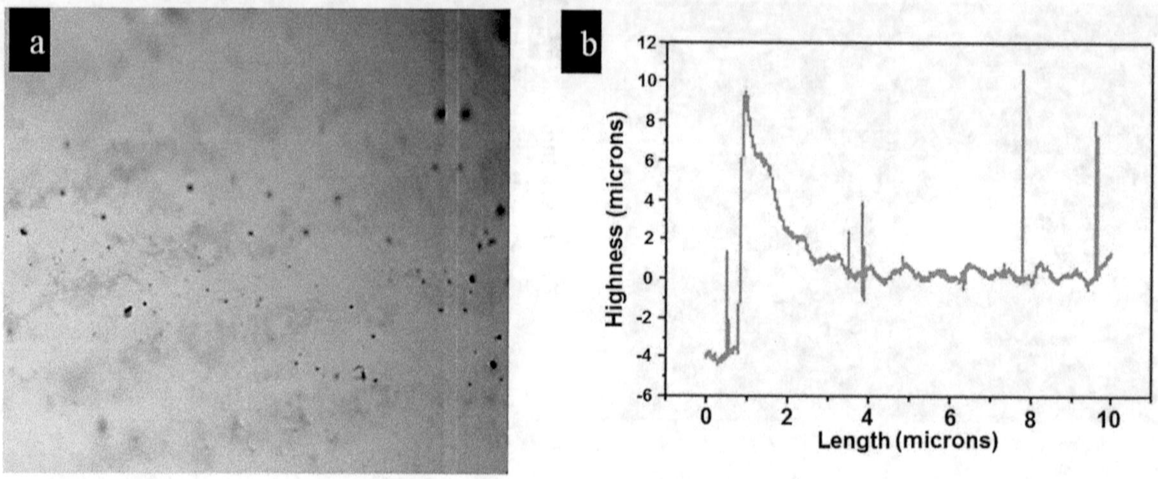


As in case 1, this type of coating can be varied from less than one micron to over one centimeter. In the case of Figure 15b, the coating thickness was about 25 μm.

As had been mentioned, DCP silver is very reactive, for this reason, when a solvent is applied on it the "failures" discussed above appear more prominently, as can be observed in Figure 16. This phenomenon can be avoided by a pretreatment or by using a less aggressive solvent.

## Aluminum Complex Coatings

The objective of the experimentation reported in this section is to obtain a transparent coating of an aluminum precursor, optionally may have a high mechanical resistance to peel. Should comply with it

*Figure 16. a) Surface image hybrid coating, silver as substrate. b) Surface image hybrid coating from silicate 70X, silver as subtrate*

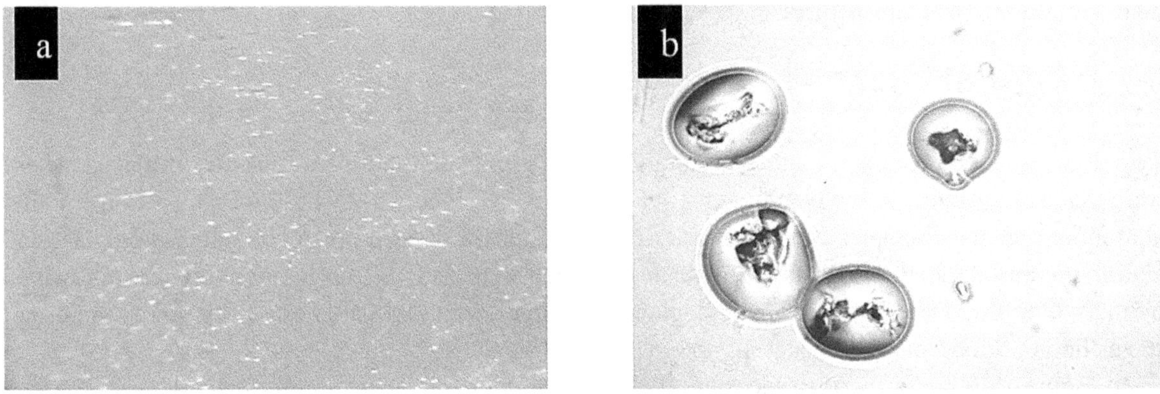


could be feasible for use as protective top layer for solar concentrators. Obtaining a fourth type of protective coating against abrasion was tested based on a precursor of aluminum (nitrate, acetate, aluminum isopropoxide or trisecbutoxide.

Figure 17 shows the materials obtained with aluminum complexes on the aluminum substrate (row 1) and glass (line 2). Columns: a) aluminum nitrate, b) aluminum isopropoxide, c) aluminum trisecbutóxido d) aluminum acetate. The behavior of the solutions prepared with the four aluminum precursors is observed. The first three precursors (a, b) shows white appearance, while the fourth was aluminum acetate which showed clear.

*Figure 17. Coatings on aluminum substrate (file 1) and glass (file 2) of a) aluminum nitrate, b) aluminum isopropoxide, c) aluminum trisecbutóxido, d) aluminum acetate*

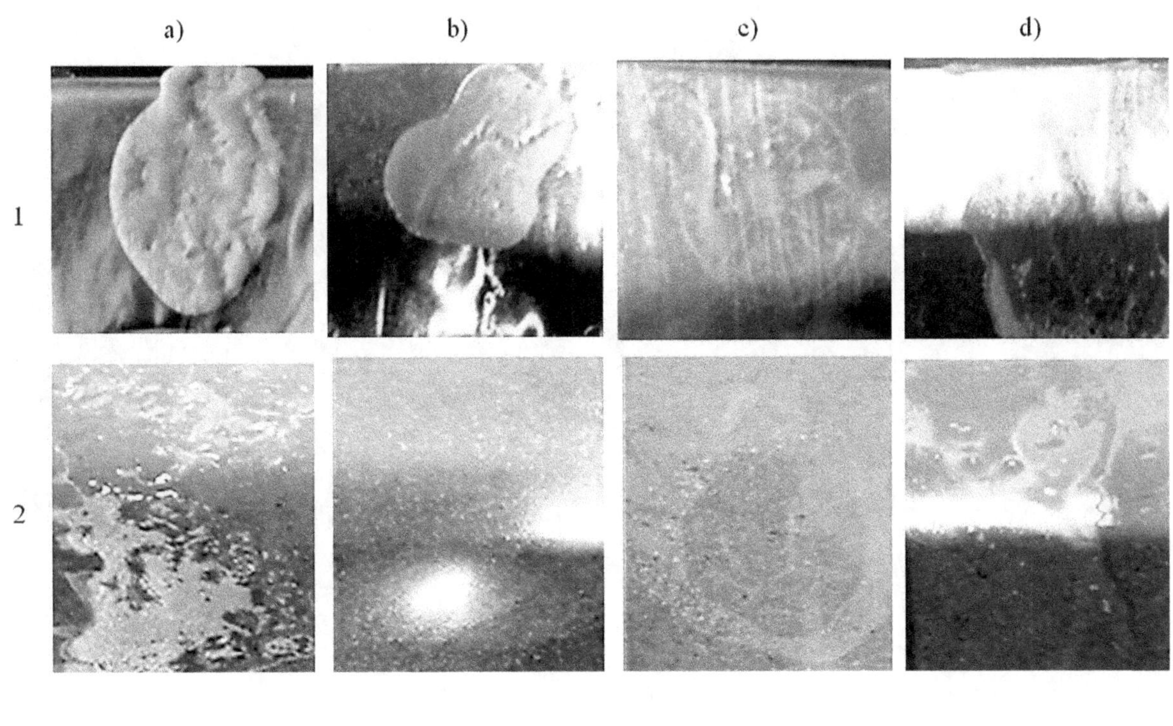

Although the only transparent coating is produced with aluminum acetate as a precursor, presents the same disadvantage that the coating prepared with TEOS precursor, the thickness should not be greater than 1 micron to prevent fractures.

## Vacuum Drying

Two drops of each precursor solution of aluminum on glass, aluminum and plastic are placed. These were treated in vacuum in the lyophilizer (LABCONCO FreeZone 2.5) to 1510 mPa for 2 hours. Other aluminum precursor samples were prepared only now 2 drops of each precursor solution on glass and aluminum. In order to thermally treat it with an electric grill to 288 ° C, this temperature was reaffirmed by thermography. This in order to assess changes during drying, which is one of the most important stages in the formation of the material.

In Figure 18 three types of surfaces are illustrated: a) glass, b) plastics and c) aluminum foil under vacuum. In which trisecbutoxido isopropoxide and aluminum nitrate were detached from the surface when the vacuum generation. Aluminum acetate Figure 17. Provided favorable characteristics such as transparency, for further study.

*Figure 18. Aluminum precursor materials on glass and aluminum substrates, vacuum dried emplenado lyophilization equipment. I) Trisecbutóxido aluminum, II) isopropróxido aluminum III) aluminum itrato. IV) aluminum acetate.*

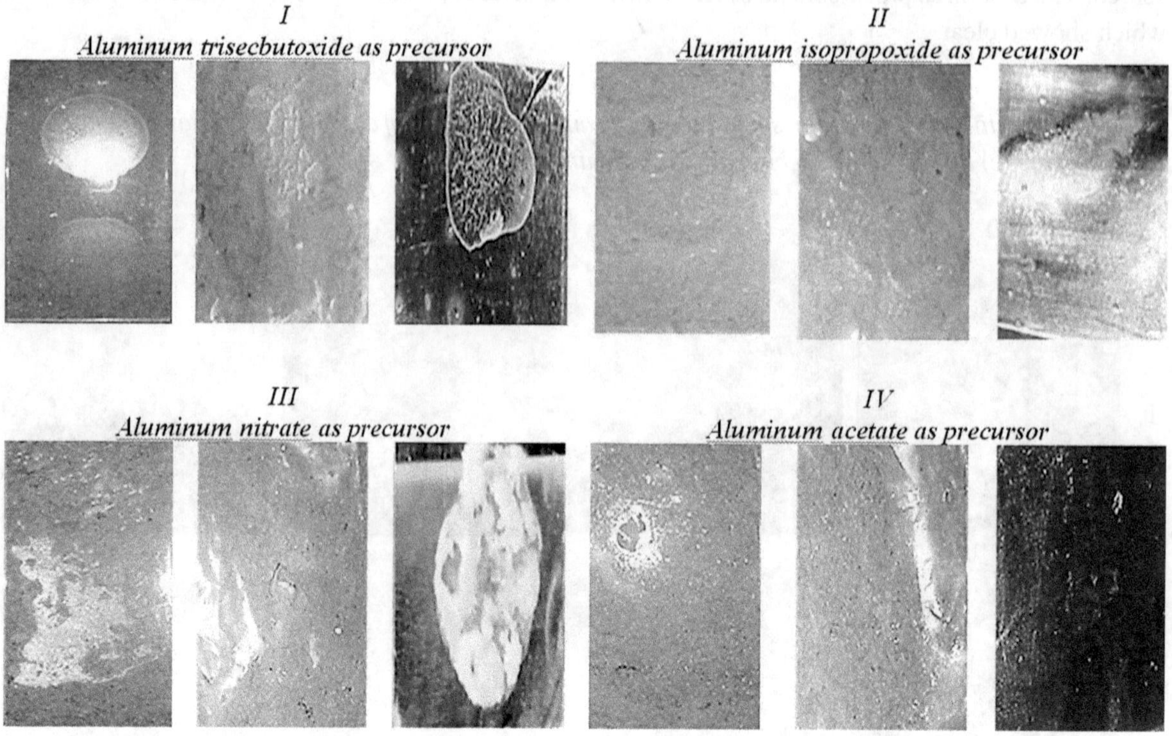

We conclude that the vacuum drying technique is not a viable option for processing aluminum solutions, when looking for high transparency and coatings without porosity or defects. The vapor pressure of the solvent is high enough for drying fast and without control. Still, in the case of aluminum acetate were again transparent coatings with few surface defects.

## SOLUTIONS AND RECOMMENDATIONS

1. To avoid "failures" in both silver and coatings that use aggressive solvents it is necessary to apply another treatment as flame which can reduce silver completely.
2. If inorganic coatings are used the thickness must be controlled by a method that allows obtain thicknesses less than one micron.

## FUTURE RESEARCH DIRECTIONS

1. Further analysis should be done to identify the chemical composition into the silver film failures, such as: x ray fluorescence, XPS, and nuclear magnetical resonance-solids
2. The sol-gel technique should be scaled to larger areas keeping uniformity and thickness for the coatings
3. Other polymers are being investigated looking for optical and performance properties. Specially, further improving their weathering and wearing characteristics
4. The PMMA possesses optical characteristics for this kind of application but there are adhesion concerns when exposed to water. The use of silane coupling agents may improve this condition.
5. The aluminum acetate based coatings shown could be improved in order to obtain transparent, homogenous, crack-free coatings.

## CONCLUSION

Silver deposit by Dynamic Chemical Plating technology is a very promising alternative for solar concentrators, using a very thin film. It is important to protect silver with a coating which can fulfill the desired characteristics. In work to explore alternatives purely organic coatings for use in protecting metal surfaces highly reflective of solar concentrators. For this, coatings were tested both silica from silicates as TEOS and hybrid coatings, specifically the case of PMMA-SiO2. The latter because of the high transparency of the said polymer and low UV resistance to prolonged exposure to sunlight may, as previous studies showing that surrounds the silica increased polymer resistance to weathering.

- Transparent coatings were obtained.
- The development of the hybrid film requires humidity control.
- The aluminum acetate precursor has showed the possibility of being useful in the preparation of alumina as a clear high transmittance coat.

## REFERENCES

Barlev, D., Vidu, R., & Stroeve, P. (2011). Innovation in concentrated solar power. *Solar Energy Materials and Solar Cells*, *95*(10), 2703–2725. doi:10.1016/j.solmat.2011.05.020

Behar, O., Khellaf, A., & Mohammedi, K. (2013). A review of studies on central receiver solar thermal power plants. *Renewable & Sustainable Energy Reviews*, *23*, 12–39. doi:10.1016/j.rser.2013.02.017

Chan, C. M., Cao, G. Z., Fong, H., & Sarikaya, M. (2000). *Coatings on polyester*, 148–154.

Debije, M. G., & Verbunt, P. P. C. (2012). *Thirty years of luminescent solar concentrator research: Solar energy for the built environment*. Advanced Energy Materials; doi:10.1002/aenm.201100554

Fend, T., Hoffschmidt, B., Jorgensen, G., Küster, H., Krüger, D., Pitz-Paal, R., & Riffelmann, K. J. et al. (2003). Comparative assessment of solar concentrator materials. *Solar Energy*, *74*(2), 149–155. doi:10.1016/S0038-092X(03)00116-6

Fend, T., Jorgensen, G., & Küster, H. (2000). Applicability of highly reflective aluminium coil for solar concentrators. *Solar Energy*, *68*(4), 361–370. doi:10.1016/S0038-092X(00)00027-X

Hoffschmidt, B., Alexopoulos, S., Rau, C., Sattler, J., Anthrakidis, A., Boura, C., & Hilger, P. et al. (2012). Concentrating solar power. *Comprehensive Renewable Energy*, *3*(2), 595–636. doi:10.1016/B978-0-08-087872-0.00319-X

Kennedy, C. E., & Terwilliger, K. (2005). Optical durability of candidate solar reflectors. *Journal of Solar Energy Engineering*, *127*(May), 262. doi:10.1115/1.1861926

Schissel, P., Kennedy, C., & Goggin, R. (1995). Role of inorganic oxide interlayers in improving the adhesion of sputtered silver film on PMMA. *Journal of Adhesion Science and Technology*, *9*(4), 413–424. doi:10.1163/156856195X00356

Wang, Z. (2010). Prospectives for China's solar thermal power technology development. *Energy*, *35*(11), 4417–4420. doi:10.1016/j.energy.2009.04.004

Zhang, H. L., Baeyens, J., Degrève, J., & Cacères, G. (2013). Concentrated solar power plants: Review and design methodology. *Renewable & Sustainable Energy Reviews*, *22*, 466–481. doi:10.1016/j.rser.2013.01.032

## ADDITIONAL READING

Baek, S. H., Kim, S. B., Shin, J. K., & Hyun Kim, J. (2012). Preparation of hybrid silicon wire and planar solar cells having ZnO antireflection coating by all-solution processes. *Solar Energy Materials and Solar Cells*, *96*, 251–256. doi:10.1016/j.solmat.2011.10.007

Belleville, P. (2010). Functional coatings : The sol-gel approach. *Comptes Rendus. Chimie*, *13*(1-2), 97–105. doi:10.1016/j.crci.2009.12.005

Butel, G. P., Coughenour, B. M., Macleod, H. A., Kennedy, C. E., Olbert, B. H., & Angel, J. R. P. (2011). Second-surface silvered glass solar mirrors of very high reflectance. *Proceedings of the Society for Photo-Instrumentation Engineers*, 81080L, 81080L–9. doi:10.1117/12.894373

Chanfreau, S., Moreau, N., Ni, L., Chemtob, A., & Croutxe, C. (n.d.). *EIS Characterization of UV-Cured Hybrid Sol-Gel Coatings for Corrosion Protection of Aluminum Alloy 2024-T3*, 5–7.

Cui, B. H., Zayat, M., Parejo, P. G., & Levy, D. (2008). *Highly efficient inorganic transparent UV-protective thin-film coating by low temperature sol-gel procedure for application on heat-sensitive substrates*, *65–68*. doi:10.1002/adma.200701960

Eren, B., Marot, L., Litnovsky, A., Matveeva, M., Steiner, R., Emberger, V., & Meyer, E. et al. (2011). Reflective metallic coatings for first mirrors on ITER. *Fusion Engineering and Design*, *86*(9-11), 2593–2596. doi:10.1016/j.fusengdes.2010.12.038

Glöß, D., Frach, P., Gottfried, C., Klinkenberg, S., & Liebig, J. (2008). *Multifunctional high-reflective and antireflective layer systems with easy-to-clean properties*, *516*, 4487–4489. doi:10.1016/j.tsf.2007.05.097

Hebrink, T. J. (2012). *8 durable polymeric films for increasing the performance of concentrators*. Third Generation Photovoltaics.

Huang, M., Hameiri, Z., Gong, H., Wong, W.-C., Aberle, A. G., & Mueller, T. (2014). Hybrid silver nanoparticle and transparent conductive oxide structure for silicon solar cell applications. *Physica Status Solidi (RRL) - Rapid Research Letters*, *5*. doi:10.1002/pssr.201400005

Ketonen, M. (2011). *UV curing of hybrid coatings: The sol-gel process*.

Krebs, F. C. (2009). Fabrication and processing of polymer solar cells: A review of printing and coating techniques. *Solar Energy Materials and Solar Cells*. doi:10.1016/j.solmat.2008.10.004

Liu, J.-J., Ho, W.-J., Lee, Y.-Y., & Chang, C.-M. (2014). Simulation and fabrication of SiO2/graded-index TiO2 antireflection coating for triple-junction GaAs solar cells by using the hybrid deposition process. *Thin Solid Films*, 3–8. doi:10.1016/j.tsf.2014.05.024

Maschinenbau, V. D. F., & Energietechnik, V. (1977). *Sol-Gel preparation and characterization of corundum based ceramic oxidation protection coatings*.

Mohelnikova, J. (2009). Materials for reflective coatings of window glass applications. *Construction & Building Materials*, *23*(5), 1993–1998. doi:10.1016/j.conbuildmat.2008.08.033

Nielsen, K. H., Orzol, D. K., Koynov, S., Carney, S., Hultstein, E., & Wondraczek, L. (2014). Large area, low cost anti-reflective coating for solar glasses. *Solar Energy Materials and Solar Cells*, *128*, 283–288. doi:10.1016/j.solmat.2014.05.034

Pahud, C., Biron, R., Ballif, C., Haug, F., Escarre, J., & Söderström, K. (2011). Highly reflective nano-textured sputtered silver back reflector for flexible high-efficiency n – i – p thin-film silicon solar cells. *Solar Energy Materials & Solar Cells*, *95*, 3585–3591. doi:10.1016/j.solmat.2011.09.016

Purcar, V., Stamatin, I., Cinteza, O., Petcu, C., Raditoiu, V., Ghiurea, M., … Andronie, A. (2012). Fabrication of hydrophobic and antireflective coatings based on hybrid silica films by sol – gel process. *Surface & Coatings Technology, 206*, 4449–4454. doi:10.1016/j.surfcoat.2012.04.094

Schmauder, T., & Küper, S. (n.d.). *Applications of Highly Reflective Silver Coatings for Car Headlamp Reflectors.*

Schottner, G. (2001). *Hybrid Sol - Gel-Derived Polymers: Applications of Multifunctional Materials*, 3422–3435.

Thomas, N., & Wolfe, J. (2000). *UV-Shifted Durable Silver Coating 1 for Astronomical Mirrors.*

Wojcik, A. B., Schuster, K., Kobelke, J., Chojetzki, C., Michels, C., Rose, K., & Matthewson, M. J. (n.d.). *Novel Hybrid Glass Protective Coatings for High Temperature Applications*, 368–372.

Wright, M., & Uddin, A. (2012). Organic-inorganic hybrid solar cells: A comparative review. *Solar Energy Materials and Solar Cells, 107*, 87–111. doi:10.1016/j.solmat.2012.07.006

Xue, C., Jia, S., Chen, H., & Wang, M. (2008). *Superhydrophobic cotton fabrics prepared by sol – gel coating of TiO 2 and surface hydrophobization*, 035001. doi:10.1088/1468-6996/9/3/035001

Yang, M., Zhou, C., & Jiang, D. (2008). Infrared Physics & Technology Comparison of different strategies to realize highly reflective thin film coatings at 1064 nm. *Infrared Physics & Technology, 51*(6), 572–575. doi:10.1016/j.infrared.2008.09.002

Ye, L., Zhang, Y., Zhang, X., Hu, T., Ji, R., Ding, B., & Jiang, B. (2013). Sol-gel preparation of SiO2/TiO2/SiO 2-TiO2 broadband antireflective coating for solar cell cover glass. *Solar Energy Materials and Solar Cells, 111*, 160–164. doi:10.1016/j.solmat.2012.12.037

Yu, Y.-Y., & Ciou, C.-Y. (2013). Hybrid solar cells based on colloidal nanocrystals and conjugated polymers. *Thin Solid Films, 544*, 318–323. doi:10.1016/j.tsf.2013.03.084

## KEY TERMS AND DEFINITIONS

**Aluminum Complex Coatings:** Coatings created by sol-gel process, using different precursors.

**Hybrid Coating:** Coating created by two or more compounds of different nature as organic and inorganic.

**Microscopia Confocal:** Technique to obtain high-resolution images based in fluorescent markers.

**Optical Profilometry:** Technique for measuring roughness and reliefs on surface.

**Precursor:** Indispensable compound to produce another compound by a chemical reaction.

**Sol-Gel:** Creation of a network of oxides by progressive condensation reaction of molecular precursors in a liquid medium.

**Substrate:** Object of the action of other reagents.

Section 4

# Non-Conventional Systems and Materials for Energy Harvesting

# Chapter 11
# Biomechanical Energy Harvesting:
## Design, Testing, and Future Trends in Healthcare and Human–Machines Interfacing

**Giorgio De Pasquale**
*Politecnico di Torino, Italy*

## ABSTRACT

*Portable electronic systems and wearable sensor networks are offering increasing opportunities in fields like healthcare, medicine, sport, human-machine interfacing and data sharing. The technological research is looking for innovative design solutions able to improve performances and portability of wearable systems. The power supply strategy is crucial to improve lifetime, reduce maintenance, preserve the environment and reduce costs of smart distributed electronic systems applied to the body. The conversion of biomechanical energy of limbs and joints to electricity has the potential to solve much of the actual limitations. The design and building of wearable energy harvesters for wearable applications require different approaches respect to traditional vibratory energy harvesters. This chapter focuses on transduction materials, modeling strategies, experimental setups, and data analysis for the design of biomechanical energy harvesters; a case study based on system integration and miniaturization is also described for applications in the field of human-machines interfacing.*

## INTRODUCTION

Many industrial and academic researches are currently addressed to design and to optimize the technologies related to portable and wireless electronics. Some of the fields mainly attracted by this category of devices are health care, medicine, bioengineering, human-machine interface (HMI), insurance, military, sport and entertainment. In medical applications, for instance, distributed remote sensors for body monitoring with self-powering can reduce sensitively the cost of medical services. The repetitive and

DOI: 10.4018/978-1-4666-8254-2.ch011

scheduled procedures of patients health monitoring normally performed in hospitals could be replaced with automatic detections of physiological parameters by smart self-powered sensors. Remote sensing techniques applied to biological parameters of people are now feasible by using diagnosis devices for patients with specific pathologies that do not require permanence inside hospitals. In those cases, the remote monitoring of specific physiological parameters, through wireless sensors powered with rechargeable batteries, allows to monitor the evolution of clinical processes staying at home with increased comfort, wellness and cheapness. The "cloud data management" of the remotely sensed information could be used for normal patients, civilians, soldiers, etc. and save much of the doctors' time and healthcare costs. In US, for instance, more than 3 million nurses are spending at least 5 minutes a day each on one patient to check basic vital parameters (Clanaman, 2013). This corresponds to millions of hours that can be potentially saved by adopting wearable sensors networks in hospitals, for instance in the fields of mental health (Barak et al., 2008), respiratory conditions (Jaana et al., 2009), smoking cessation (Myung et al., 2009), cardiovascular diseases (Neubeck, 2009), diabetes (Tran et al. 2008), etc.

The design constraints characterizing wearable devices are generally the small size, the light weight and the low power consumption. This last specification, in particular, suggests using energy harvesting from human body motion to generate the required power to supply the worn electronic devices. The design and building of sensing nodes characterized by high integration and reliability, self-powering, low cost, and wearability have the potential to extensively detect biological parameters of people for the mentioned fields. Furthermore, they will allow size and weight reduction of devices, prevention of batteries recharge, replacement and disposal. In portable sensing nodes, the battery duration is generally short, due to the size and weight constraints. The biomechanical energy harvester (EH) makes the nodes energetically autonomous in normal conditions, and slows down the battery discharge in critical situations of limited mobility.

The main research efforts are currently focused on efficiency improvement, electronics miniaturization, improvement of biocompatibility and portability (small invasiveness on body motions) and data communication and management according to clinical requirements.

About the working principles of biomechanical energy harvesting, the general approach is to convert the kinetic energy associated to the motion of limbs and joints to electricity by appropriate transducers. These electro-mechanical transducers are based on piezoelectric materials and on electro-magnetic inductors. Within this frame, many different design solutions can be conceived, depending to the requirements of the measurement and to the application. Some examples are the extended monitoring of body temperature, the calculation of mobility indexes, the monitoring of breathing, metabolic functions, etc. In most clinical practices, the wireless detection is already used, for instance with telemetry strategies; the physiological parameters of interest are measured in remote by wireless sensors and then stored. In this case, present limitations are the reduced portability of devices, which are uncomfortable to wear, large and heavy because of the batteries. The batteries need periodic recharge or replacement, and in case of discharge, the data acquisition is stopped. However, the main limitation is the reduced portability of data: the measured values are transmitted to a receiver that stores them in a local memory, from where they are downloaded by the medical staff. Instead, the internet connection of the wearable sensors, for instance by using portable devices like smartphones, will make the data accessible everywhere in real time. This improvement could eliminate completely the restrictions to the mobility of those patients who need permanent medical supervision or periodic monitoring of physiological data. Of course, the described technology makes sense only with local power generation to guarantee the system supply and comfortable portability, and the most attractive power source available is the biomechanical power of the patient.

The mentioned design constraints of dimensions, weight and power consumptions have been focused several times in the literature for wearable devices (Teng et al., 2008; Yoo et al., 2010; Ramadan et al., 2014; De Pasquale et al., 2011, 2012). The main issue is the limited amount of energy that can be harvested with these imposed portability constraints. Periodical replacements of batteries are unpractical and expensive, and batteries disposal impacts on the environment. As stated before, the alternative is represented by self-powering strategies with biomechanical energy harvesting (Mitcheson et al. 2008). Kymissis et al. (1998) introduced an integrated flexible piezoelectric bimorph into the sole of a shoe, Granstrom et al. (2007) conceived and built an EH embedded into a backpack (whose size and weight are not completely satisfactory). A joint structure mounted on the knee was used by Pozzi and Zhu (2011). By analyzing these devices, their common limitations are the size and the invasive effects on the normal motion. Preliminary investigations and characterizations of innovative transducers and high flexible materials have been carried out by the author in previous activities (De Pasquale & Somà, 2013), where composites made with polymeric matrix and embedded piezoelectric fibers were studied. These materials promise suitability for energy harvesting from large bending motions with good reliability (Priya, 2007; Lee et al., 2012), and represent a valid alternative to traditional piezoceramics (Klimiec et al. 2008; Shu et al. 2006). The simulation and modeling of piezoelectric-based devices have been investigated in literature (Mossi et al., 2006; Elvin et al., 2006). The review paper of Mitcheson et al. (Mitcheson et al., 2007, 2008) summarizes some designs and models of EH, as well as the properties of some laboratory prototypes. The author describes the theory of electro-mechanical transduction for kinetic energy harvesting in (De Pasquale, 2013).

In order to define some crucial points of the design of biomechanical energy harvesting, the most important specifications can be summarized as follows: a) the needing to transform movements into electric energy through appropriate electromechanical transducers (piezoelectric or magnetic-inductive); b) the needing to provide temporal continuity to the electrical supply, despite irregular and random motions; c) the needing to satisfy the energetic requirements of the user; d) the needing to respond to the requirements imposed by the application (sampling frequency, transmission frequency, accuracy, reliability, etc).

This chapter describes the design approach of biomechanical EHs. The main focus is given to the system-oriented design, which provides correct strategies to define the harvester specifications in relation to the applicative environments (operative conditions) and to the required performances of sampling and transmission frequency. The same harvester may work properly in some context and may not in other contexts. The duty cycle of the system is the crucial parameter to verify the balance of generated/consumed power for each specific activity of the biomechanical system, including all the components before and after the EH. The description of experiments, test benches design and building, and prototypes fabrication is the next part of the chapter, where the main focus is given to the practical issues to be solved in the verification process of performances and reliability. Test benches and test methodologies are described, as well as the main requirements of testing repeatability and experimental confidence with real operative conditions. Testing on the human body is then described by presenting some results from biomechanical EHs applied to the human body. The state of the art will be critically analyzed, by means of applicative-oriented approach and harvesting systems feasibility. The main problems of portability, comfort and integration are analyzed with particular focus on innovative materials (flexible circuits, conductors embedded in fabrics, conductive fabrics, etc).

## STATE OF ART AND GROUND-BREAKING PERSPECTIVES

The thermoelectric energy source has been used in the past for harvesting energy from human body (Böttner et al., 2004; Pop et al., 2008; Stark, 2006; Su et al., 2010; Xie et al., 2010; Leonov, 2011), but its main drawback is the small power density of final devices. They require large surfaces to produce the energy needed for reasonable duty cycles, and the portability of the devices is strongly reduced.

The available solutions for wearable sensor nodes are based on the power storage in batteries, which can be replaced or eventually recharged by wire. The main limitation of this approach is the small amount of energy imposed by portability constraints. The alternative to periodical batteries replacement/recharge of batteries is represented by self-powering strategies with biomechanical energy harvesting (Mitcheson et al., 2008). The advances in electronic technology provide the progressive reduction of power consumption of components and suggest considering the biomechanical energy harvesting as new power source.

The kinetic energy associated to the walking activity has been converted into electricity by wearable harvesting devices. The prototype developed by Bionic Power (Bionic Power), is 750 g heavy and produces about 12 W; it has been developed for military applications and serves to supply GPS navigators. The particular application field does not require any particular optimization of the wearability and comfort of use. The kinetic motion of arms is the energy source of the generator introduced in (Xie et al., 2012), where a permanent magnet is rotating alternatively and induces current in an array of coils; the authors mention an output power of dozen of milliwatts.

The development of piezoelectric fabrics with integrated active fibers has the potential to integrate the energy harvester in clothes without interfere with normal movements of the body (Guillot, 2007). Similar results with slightly lower integration level can be obtained with composite materials with piezoelectric fibers (Sakamoto et al., 2011; Jadidian at al., 1996; Malik et al., 2003).

The simulation and modeling of piezoelectric-based devices have been investigated in literature (Leland et al., 2006; Ajitasaria et al., 2007).

The review proposed by Mitcheson et al. (2007) reports the main design and modeling approaches for energy harvesting and the list of some laboratory prototypes. The author illustrates the theories of electro-mechanical transduction for kinetic energy harvesting in (De Pasquale et al., 2013). Among the others, some prototypes were specifically designed for the human body (Renaud et al., 2005; Lu et al., 2011; Kulah et al., 2008; McKnight et al., 2002) or for portable devices as bicycles (Goh et al., 2011), mobile electronics (Somà et al., 2011) or sky equipment (Somà et al., 2011).

The medical practices based on remote techniques are called "telemedicine"; a large literature is available about this topic, for instance in the fields of mental health (Barak et al., 2008), respiratory conditions (Jaana et al. 2009), smoking cessation (Myung et al., 2009), cardiovascular diseases (Neubeck et al., 2009), diabetes (Tran et al., 2009), etc. In the veterinary science there are some experiences of telemetry applied to physiology monitoring: detection of biomarkers in beagle dogs (Baird et al., 2014), tracking of sea turtles (Thums et al., 2013), cardiovascular detection of monkeys (Chaves et al., 2006), etc. The biomechanical energy harvesting from animals has some target applications in sensors used for breeding monitoring. These devices, currently supplied with batteries, are mainly used to sense body temperature and motion for detecting metabolic disorders, general diseases, important events alert (parturition), rumination, etc. (Milkline). The size and weight of actual devices confine their application to big animals (bovine, swine, ovine), while solutions for small sized animals as poultry are still missing and can be achieved with energy harvesting.

The biomechanical energy harvesting has high impact ground-breaking perspective for the future. Just one of the possible examples is represented by HMI. Within this context, glove-based systems are the most immediate tool to convert gestures to logic commands for machines. Among classical applications there are manufacturing, robotics, wearable devices control, design and 3D modeling, data management, medicine and health care, and sign language understanding. Glove HMI can be defined as sets of sensors and electronics integrated in a support worn on the user's hand, which are able to send the measured parameters to a receiver for data processing purposes.

Within human-machines communications strategies, passive interfaces are based on the detection of signals and markers positioned on the body and have limited performances. On the other hand, active interfaces are based on sensors or actuators that are able to send different commands to the receiver depending on the user inputs. This last typology of HMI has improved potentialities but, on the other hand, requires electric supply. The simpler approach to provide energy to the wearable device is by storing this energy into batteries, which can be rechargeable or non-rechargeable. Unfortunately, this solution is affected by strong drawbacks: periodic maintenance of the interfacing device (batteries recharge or replacement) are required, mass and weight increasing, reduced portability and comfort of use, disposal issues. A valid alternative strategy to provide the required power to the active interface is by generating energy directly on place by converting the biomechanical energy of the human body. The next step is the design of integrated energy harvesters able to convert the biomechanical kinetic energy into electricity. This will open to the next generation of HMIs, were the user and the machine will merge together in correspondence of their interface.

The first prototypes of glove interface were proposed in the 1970s and 1980s, as the MIT-LED glove and the Sayre Glove (Sturman et al., 1994). In the manufacturing applications, glove systems are generally used to interact with virtual environments generated by softwares through computer monitors. Here, hands are used instead of mouse and keyboard to organize and move virtual objects. Daimler-Benz is one of the first companies that adopted glove HMI: the operators selected the furniture of cars interiors by using them in combination with virtual reality programs. Maintenance operations simulations on aircrafts and for personnel training were conducted by Boeing with virtual reality and glove interface (Steffan et al., 1998). The performances of astronauts and pilots were also tested with similar softwares to take confidence with the cockpit (Doerr et al., 2007).

About the design and 3D modeling, HMI have been proposed in combination with CAD (computer aided design) softwares, which are largely used by engineers and architects to support their design activities. The motion of hands can be directly converted in logic commands to create shapes and 3D objects, without using the computer peripherals (Keefe et al., 2011). The devices introduced in the literature have some limitations in terms of data transmission by wire, power supply by wire and uncomfortable use due to low integration of components. The introduction of wearable EHs demonstrated high potential in reducing the weight of the device (battery size is reduced) and eliminating physical connections with the machine.

The control of articulated systems and robots can becomes potentially easier and more natural by implementing HMI; these beneficial effects are much more evident in robots with multiple joints where many degrees of freedom are involved. For instance, the configuration of fingers and the motion of hands has been used to impose control commands to a mobile robot through the CyberGlove interface (Iba et al., 1999). Similarly, the system designed by NASA-DARPA (Diftler et al., 2003) can send remote commands to an anthropomorphic robot used in the space for extravehicular operations; it uses a helmet equipped with a stereo screen, a microphone and two CyberGlove interfaces linked to the robot.

In the medicine and health care fields, electronic HMI gloves are used for the rehabilitation of functional assessment and motion of hands: some indicative data are acquired (pinch and grip strength, temperature sensitivity, joints angles ranges, etc.) and processed to evaluate the pathology evolution (Burdea et al., 2007). The wearable device, in these cases, has the main function of sensing unit instead of interface. Similar approach is used to record the motion of hands to verify the respect of ergonomic prescriptions; here the kinematics of upper arms are detected and stored and then used to define biomechanical models for optimizing tasks and environments (Allen et al., 1995).

The contribution of energy harvesting in the HMIs applied to portable electronics has high potential impact, due to the requirements of miniaturization. Normally, the size of components cannot be reduced beyond a certain limit imposed by the handling requirements. For this reason, the future trend is to push the technology to replace traditional peripherals with wearable ones for text entry and pointing (Edminson et al., 2002; Rosenberg et al., 1999).

In the sign language understanding, HMI systems were used to send alphanumerical symbols to the machine accordingly to the gestural code used by the deaf community (Sturman et al., 1994). Different strategies were implemented, in static and dynamic postures of the hand, from few tens to several thousand of detected signs. Specific variants were provided for the Japanese alphabet and national languages (Takahashi et al., 1991).

In applications related to data management, virtual images and contents can be represented with increased immediateness by means of HMI systems; furthermore, the interaction of the user with the data is sensitively improved. For instance, with the Data Glove has been possible to visualize and interact with the simulated airflow around an aircraft at NASA laboratories: the user could grab and move each streamline of the flow and activate real time simulations (Bryson et al., 1992). Similar operations were conducted with fluid dynamics in indoor environments (Malkawi et al., 2004) and with medical data (Tani et al., 2007).

## BIOMECHANICAL ENERGY HARVESTING

### Design Specifications and System Architecture

In wearable devices for sensing and data transmission, the main limitations imposed by current technologies are associated to the power supply, which is normally provided by storage batteries. The design strategy based on the presence of batteries is affected by reduced portability of the device because of its increased size and weight. This drawback reduces the application to the everyday life of people without invasive effects. Furthermore, additional appreciated peculiarities of wearable devices and sensors are the high integration level of components, the miniaturization, the portability, the reliability, the possibility to reuse, and the low cost. The biomechanical energy harvesting design strategy leads to the development and integration of power generators embedded in the wearable device. The energetic autonomy provides high value to the system functionality and positively impacts on most of the mentioned requirements.

By considering a wearable sensing node with self-powering, the specifications needed are achievable with an integrated system including some basic functional parts: sensing, power generation and transmission. The scheme reported in Fig. 1 identifies the functional blocks of each node of wearable sensing networks. The sensing and transmission parts require electronic design oriented to the implementation of low power architectures with specific low consumption components. The power generation is extremely challenging and requires the main effort in the research.

*Figure 1. Self-powered sensing node schematics*

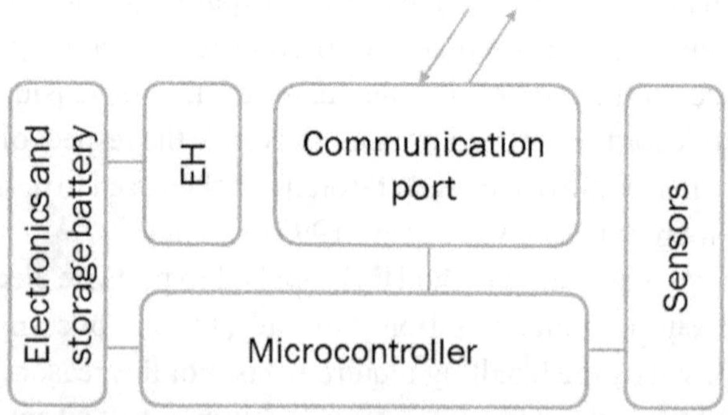


The conversion of the biomechanical motion of human body (or, for some applications, of animals' body) to electric energy need to solve some design problems and is based on some constraints caused by the application itself. The biomechanical EH provides electric output by converting the kinetic energy of human activities. The randomly shaped voltage coming from the harvester must be processed by dedicated electronics that provides rectification and leveling up to (or down to) the battery charge voltage. Many strategies can be used at this purpose, mainly based on diodes bridges and leveling capacitors, or by using more complicated operational controls in step-up/step-down circuits. The energy stored in the battery is used to supply the utilizers when the generation level is too small; the battery capacitance also provides increased stability to the supplied voltage. Microcontrollers are used to manage the activities of sensors (to define sampling parameters) and of communication ports (to define communication parameters).

## Transduction Principles

The transduction principle is the first design variable and, for wearable EHs, the options are limited to magnetic-inductive and piezoelectric strategies. Magnetic-inductive EHs are based on the Faraday's law applied to the traditional configuration of permanent magnet and coil in relative motion and its application is generally limited to the activities inducing alternate motions or vibrations such as walking or running. In this case, the motion of the body can be used as excitation for spring-mass-damper systems, as inductive harvesters. Different design variants have been presented in the literature, including fixed coil and oscillating magnet or vice-versa. The most important parameters are the size and the material of the magnet, responsible of the magnetic field intensity, and the coil geometry (number of turns, length, coil area, wire diameter). Smaller is the wire diameter, higher is the number of coils in the available space; however, on the contrary, the wire diameter is proportional to the resistance and limits the output current. Rare earth permanent magnets (as neodymium iron boron, NdFeB) provide the best performances in terms of power density of the harvester.

In case of inductive harvesters, the design issues are related to the lowering of resonance frequency (in the range 3-7 Hz) of the generator, which must be tuned to the vibration frequency of the excitation. Size and weight are also crucial because it is hard to miniaturize this typology of generators due to the components availability limited to the millimeter scale. The main advantage of inductive generators is the high ratio between output current over output voltage, which facilitates the charging of storage batteries in short times with rough signal conditioning.

Instead of inductive generators, piezoelectric EHs can be used. This typology of transducer is normally applied to vibrating systems (cantilever beams with proof mass on the tip) to create an alternative to the other generators. However, in case of biomechanical energy harvesting, the properties of some piezoelectric materials open to the possibility of building non-vibratory generators. This solution allows generating power from the motion of the single human body joint, regardless to the type of activity and to the dynamic properties of the generator. The high flexibility of some innovative transducers leads to distributed energy harvesting systems on the whole body with improved efficiency of biomechanical energy conversion for wider ranges of body motions. The low stiffness and light weight are then suitable for preserving the comfort and wearability of the harvesting systems without impacting on normal activities. Differently from inductive EHs, wearable piezoelectric generators require dedicated design procedures, which are different from those applied in vibrating piezo EHs. For this reason, the main focus of this section is oriented to wearable piezoelectric generators, which are the only considered in the following; the reader is addressed to traditional kinematic EHs design literature (De Pasquale, 2013) for more accurate descriptions of inductive harvesters.

## Design Process

The preliminary design activity is addressed to investigate the biomechanics and to study the body kinematics. The analysis of the motion of human limbs and joints in combination to some activities provides the energetic information about the best candidate locations for the harvester on the body. The joints excited by typing on keyboards are definitely different than those excited by running; however, the same joints are subjected to different motions (in amplitude, frequency and angular velocity) in different activities (e.g. squeezing a tennis ball or playing the piano).

The traditional tools for the modeling of human limbs are derived by the inverse kinematic approach used in robots design. The body is modeled with a constrained system with articulated rigid bodies, in order to define a kinematic chain; the kinematics and the dynamics of the chain is described by nonlinear equations which can be solved with analytic formulations or with dedicated software tools (multibody modeling). Other relevant kinematics information about human joints can be derived by physiology and anatomy disciplines, which codify the ranges of motion depending on age, gender and geographic provenience.

After identifying the location of the harvester depending on the energy associated to the joints in the specific activity, the next step is to define the excitation mode of the transducer. Normally, three actuation modes are suitable for biomechanical energy harvesting: axial loading, bending, and their combination. By considering the ratio between the electric output and the imposed displacement, the first solution provides the best performance. Instead, the pure bending requires large angular motions to get small strains of the piezoelectric transducer. However, pure axial load causes high resistive forces that oppose to the normal motion of limbs and require additional muscle forces to move the joint. Instead, flexion of low-stiffness piezoelectric transducers applied to the joints does not require increasing of muscular force.

The electric output conversion and power management is normally conducted with dedicated electronic circuits that are designed and dimensioned according to the amount of power to be converted. The random motion of joints produces strongly irregular electric power output from the piezoelectric EH. The generated voltage must be rectified and increased/decreased to the desired level in order to charge the storage battery. This operation is carried out by diodes bridges, leveling capacitors and step-up/step-down circuits. The function of the storage battery is to provide a reservoir of energy in case the EH is not working for some time, and also to provide leveled voltage supply to the utilizers. Compared to normal batteries, the storage batteries of energy harvesting systems are smaller and lighter and do not require maintenance and active charging by the user.

The duty cycle (ratio between power generated and consumed by the self-powered system) is the more relevant design parameter that indicates the level of autonomy of the system for a given activity. It also indicates the time needed to produce the amount of energy required to supply the utilizers for 1 s. Generally, the global energetic efficiency, which impacts on the duty cycle, is mainly affected by the design of the EH (e.g. large transducers with non-uniform strain of the material increase the parasitic capacitance of the system), by electric connections, and by circuits and electric components losses.

## MODELING OF THE BODY KINEMATICS

### Human Body Biomechanics

The reference system applied to human body is usually defined by the three Cartesian planes called "*transversal*" (or horizontal), "*frontal*" and "*sagittal*". The transversal plane separates the upper and lower parts of the body, the frontal plane divides the front from the back sides of the body and the sagittal plane separates right left parts in two symmetrical portions. The motion of limbs and other body parts is named accordingly to the joint involved and to the reference system planes. Normally, the motions acting along the transversal plane around a vertical (or longitudinal) axis are defined *internal/external rotations* (or *pronation/supination*); the motions along the frontal plane with respect to a sagittal axis are named *abductions/adductions*; motion along the sagittal plane around a frontal axis are named *flexions/extensions*. Some joints having multiple degrees of freedom are interested to the motion of *circumduction*.

The human joints are classified in *synarthrosis*, *diarthrosis* and *amphiarthrosis*. Synarthroses are links between adjacent bones made of connective tissues and, depending to the case considered, they permit very limited relative motion or do not permit any motion (e.g. union between vertebral bones, cranial sutures, etc). Diarthroses are the joints with wider mobility and they are characterized by bones interfaces separated by cavity (e.g. elbow joint, knee joint, etc); depending to the shape, this joint may have different number of degrees of freedom. Amphiarthroses are continuous joints characterized by very limited mobility, connected by ligaments or elastic cartilage (e.g. sacroiliac joint, tibiofibular joint, etc).

For every joint, the number of degrees of freedom corresponds to the number of planes along which the motion is allowed. Alternatively, the number of degrees of freedom also corresponds to the number of axes around which the motion is allowed.

The most suitable human joints for hosting wearable EHs are diarthroses. They are divided, depending to the shape, in:

1. *Enarthrosis*: joint between a ball-shaped end of a bone and a spheroid cavity of the second bone (e.g. hip joint interesting the head of femur and the acetabulum). This joint allows rotation degree of freedom, then the movable bone can be oriented along infinite directions around the same nominal point (center of rotation), the movements allowed are flexion/extension, abduction/adduction, internal/external rotation and circumduction. The joint allows motions along three planes, then the number of degrees of freedom is three (rotation).

2. *Condylarthrosis*: similar to enarthrosis but the convex and concave parts of the joint have elliptical shape instead of spherical (e.g. radiocarpal joint); the two surfaces have different radius of curvature and allows motion along two orthogonal planes. This joint allows flexion/extension, abduction/adduction and circumduction motions. The joint allows motions along two planes (sagittal and frontal), then the number of degrees of freedom is two (rotation).
3. *Sellar* (saddle) joint: it connects two surfaces with different curvatures, concave and convex and arranged specularly (e.g. carpometacarpal thumb joint). This joint allows motions in the sagittal and frontal planes (flexion/extension, abduction/adduction and circumduction) while the axial rotation is constrained (biaxial joint). The joint allows motions along two planes (sagittal and frontal), then the number of degrees of freedom is two (rotation).
4. *Trochoid* (pivot) joint: joint between a cylinder-shaped end of a bone and a cylindrical concave surface, where the convex articular surface is parallel with the longitudinal axis of the bone (e.g. radioulnar joint). The joint allows internal/external rotations (pronation/supination) around the axis of the bone. The joint allows motions around one axis (longitudinal or vertical) and on the transversal plane, then the number of degrees of freedom is one (rotation).
5. *Hinge joint* (ginglymus): joint between a cylinder-shaped end of a bone and a cylindrical concave surface, where the convex articular surface is orthogonal with the longitudinal axis of the bone (e.g. interphalangeal joints, humeroulnar joint). This joint allows flexural/extension motions. The joint allows motions around one axis (horizontal) and on the sagittal plane, then the number of degrees of freedom is one (rotation).
6. *Arthrodial* (plane) joint: this joint connects two plane surfaces with light convexities or concavities and allows small sliding motions; there are not reference axes of the joint (e.g. between second and fifth sternocostal joints). The joint allows sliding and gliding motions between the articular surfaces. The joint allows motions along three planes, then the number of degrees of freedom is three (translation).

The joint is identified by the number of its degrees of freedom, which can vary between one and three. This leads to the alternative definition of *uniaxial*, *biaxial* and *triaxial joint*. The amplitude of motion associated to the joint is called "range of motion": its value is reported to a reference position and varies depending to the age and gender. By considering only one rotation of the joint, the average rotation axis is given by the geometrical center of the curvature. Also, by considering one single motion of the joint, the rotation axis is oriented orthogonally to the plane along which the motion occurs. Theoretically, this axis is situated at the geometrical center of curvature of the convex part of the joint. Actually, the surfaces irregularities of the joint cause the migration of the rotation axis during the joint motion; for this reasons it is better to define the joint rotation axis as "instant rotation axis" and the rotation point as "instant rotation center".

For energy harvesting purposes, the following joints are the most attractive in case of piezoelectric transduction:

- Hand joints, in particular inter-phalangeal joints and metacarpo-phalangeal joints;
- Elbow joint, in particular the humero-ulnar joint;
- Knee joint.

The main reasons making these joints particularly suitable for harvesting energy are the large range of motion (elbow and knee), the high joints density in small volume (hand), and the joint activation in typical human activities as walking/running/cycling/climbing stairs (elbow and knee), typing/gripping/manipulating (hand).

The modeling requires some input data about biometrical dimensions of the human body. The database of anthropometric measurements is available in the International Standard normative. For instance, Tabs. 1 and 2 reports the measurements of upper limbs averaged over a large population. Also, the anthropometric dimensions are divided per gender and geographic provenience.

## Hand Joints

Human hand is the most sophisticated part of the entire body motion system. In the hand there is the higher number of joints, muscles and haptic receptors in the smaller volume. The candidate locations for hosting wearable EHs are situated on the fingers, where up to three joints are present. As reported in Fig. 2, fingers from 2 (index finger) to 5 (little finger) have three joints, namely the *metacarpo-phalangeal* (MCP), the *proximal-interphalangeal* (PIP) and the *distal-interphalangeal* (DIP) *joints*. The finger 1 (thumb) has two joints instead. In the following, some experimental results about the output of EH applied to the PIP joints are reported. The PIP joint in fact has the wider bending angle, it is the most used in usual operations and its location is the most convenient for hosting the piezo transducers by the system design viewpoint.

The calculation of the energy associated to fingers bending motion and the measurement of the energy generated by the harvester allows defining the energetic efficiency of the biomechanical conversion device. The evaluation of the energy produced by the motion of fingers must be conducted in relation the specific activity.

*Table 1. Distance between shoulder and elbow (International ISO)*

| Country | Gender | Samples | Mean (mm) | Standard Deviation (mm) |
|---|---|---|---|---|
| Italy | MALE | 1950 | 322 | 22 |
| | FEMALE | 1922 | 303 | 22 |
| Japan | MALE | 2871 | 338.4 | 15.3 |
| | FEMALE | 2460 | 311.8 | 14.4 |
| Kenya | MALE | 131 | 362.8 | 28.3 |
| | FEMALE | 74 | 338.2 | 23.9 |
| Korea | MALE | 2612 | 337.8 | 15.6 |
| | FEMALE | 2613 | 307.3 | 15.2 |
| The Netherlands | MALE | 555 | 376.6 | 23.0 |
| | FEMALE | 683 | 342.5 | 20.0 |
| Thailand | MALE | 1246 | 348 | 17 |
| | FEMALE | - | - | - |

*Table 2. Distance between elbow and wrist (International ISO)*

| Country | Gender | Samples | Mean (mm) | Standard Deviation (mm) |
|---|---|---|---|---|
| Italy | MALE | 1899 | 247 | 22 |
| | FEMALE | 1575 | 224 | 17 |
| Japan | MALE | 2868 | 261.3 | 14.7 |
| | FEMALE | 2455 | 235.6 | 13.2 |
| Kenya | MALE | 130 | 303.1 | 27.8 |
| | FEMALE | 74 | 282.9 | 19.7 |
| Korea | MALE | 2612 | 267.2 | 14.1 |
| | FEMALE | 2611 | 242.8 | 13.0 |
| The Netherlands | MALE | - | - | - |
| | FEMALE | - | - | - |
| Thailand | MALE | 1246 | 278 | 17 |
| | FEMALE | 1170 | 240 | 12 |

*Figure 2. Joints of human fingers*

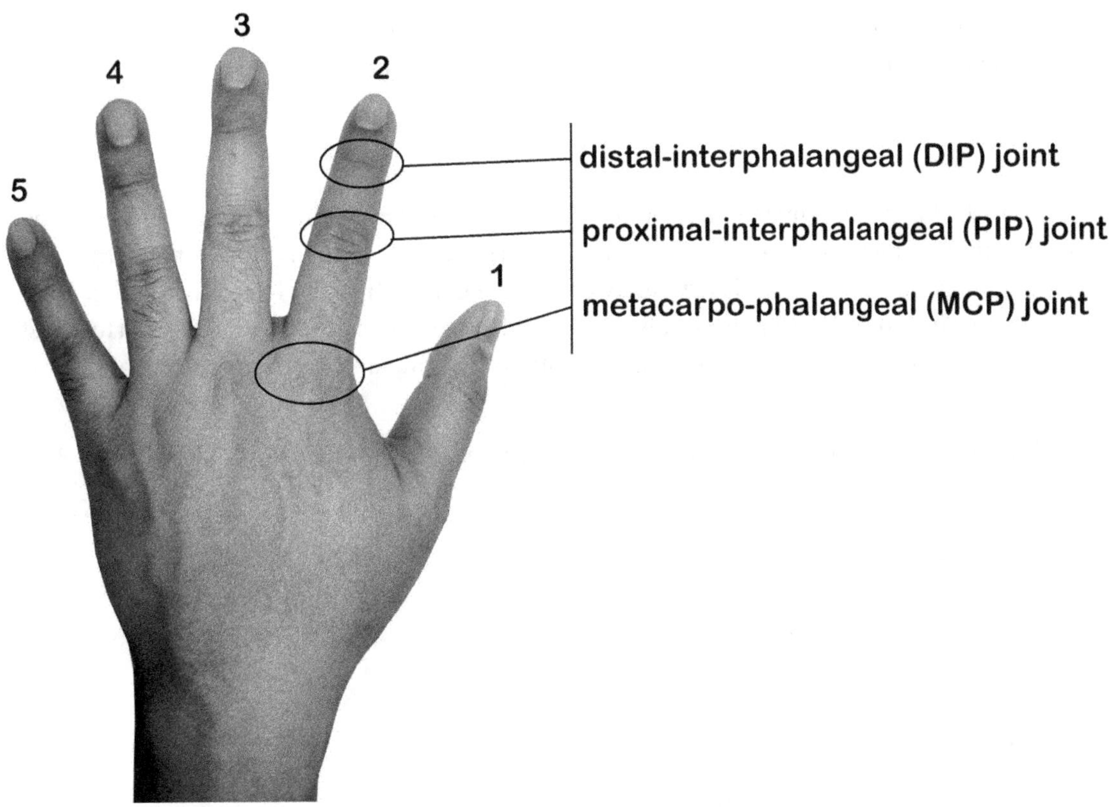

## Elbow Joint

The elbow is composed by two joints:

- **Humero-Ulnar Joint:** Is a hinge joint (or ginglymus) that allows flexural/extension motion of the forearm along the sagittal plane and identifies one degree of freedom;
- **Radio-Ulnar Joint:** Is a trochoid (or pivot) joint that allows internal/external rotations (pronation/supination) of the forearm and identifies one degree of freedom.

In particular, the humero-ulnar joint is suitable for harvesting energy by piezoelectric transducers due to the large motion range covered (Fig. 3). The reference position for this joint corresponds to the limb relaxed along the body. During the flexion, the humero-ulnar joint produces the motion of the front side of the forearm towards the front side of the arm. In case of active motion (i.e. motion produced by the work of muscles), the flexion motion range of this joint is quantified in 145° (for passive motion, i.e. motion induced by external work, the range is 160°) as shown in Fig. 4. The motion is limited by the contact between soft tissues of the limb and by muscles contraction (Pirola, 2003).

The opposite travel (the extension), consists in the motion of the forearm away from the arm; similarly to flexion, also extension occurs on the sagittal plane around a horizontal axis. Generally, the reference position of the limb corresponds to the extended position. This configuration of the joint is called "*absolute extension*". Sometimes, especially women and children can overcome the reference position by an additional extension travel from 5° to 15°, which takes the name of '*hyperextension*' (Fig. 3).

## Knee Joint

This joint is the largest of human body and it is formed by the conjunction of tibia, femur and patella, arranged within ligament structures and close to the menisci, which is interposed between femur and tibia. The kinematics of this joint is extremely complicated as well as the loads distribution among its components during normal activities as walking. Several models were developed in the past to investigate the load distribution on bones, the force-length relationship of ligaments, the dynamics of the articular

*Figure 3. Reference position, flexion and hyperextension positions and angles of elbow joint*

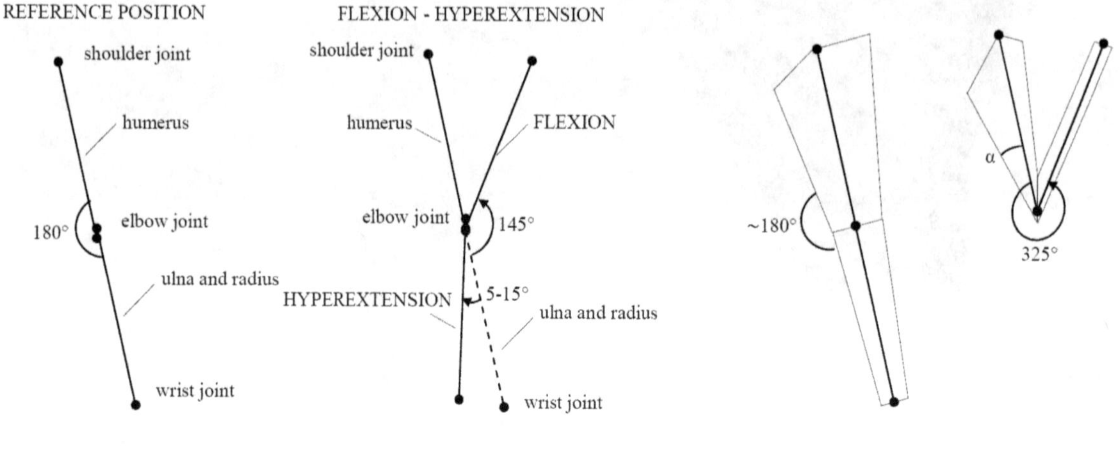

*Figure 4. Elbow joint biomechanics (Pirola, 2003)*

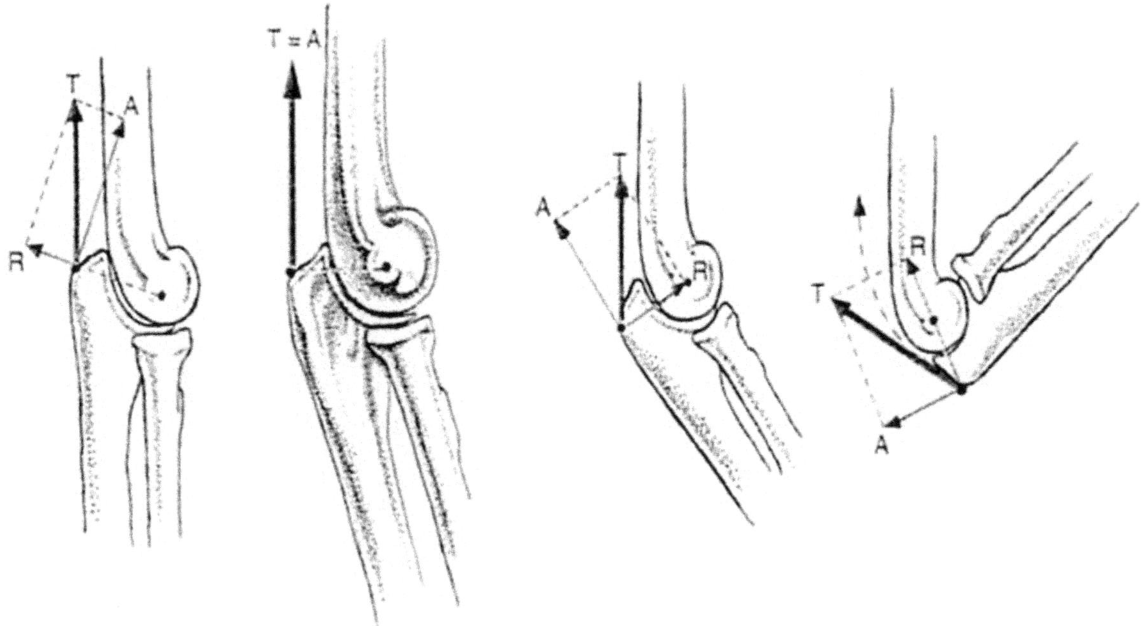


cartilage and the geometry of the surfaces. The two-dimensional four-bar-linkage model (Huson, 1974) represents the ligaments between tibia and femur with rigid bars connected with hinges. This simple approach considers the ligaments length as constant during the joint motion. The most complicated models describe the knee joint in three-dimensions and consider tibia and femur as rigid bodies connected by deformable links (the ligaments) and include the contact between the bones surfaces. The results of mathematical models normally require experimental validation.

The actual motion of the knee joint is a combination of its passive motion characteristics and the external loads; then, the knee-joint motion must be studied during specific tasks (walking, running, etc). The level of detail of the modeling must be adjusted to the final target of energy harvesting. At this purpose, the description of the motion of single bones and ligaments by equations in the time domain is probably useless. The interesting specifications for designing energy generators are limited to flexion and rotation angles, motion frequency and angular velocities. The bending radius of the joint is also important.

Due to the internal structure of the joint, the flexion of the knee is accompanied by a rotation of the tibia around its axis. The experimental values of angles involved are reported in literature (Blankevoort et al., 1991): for flexion angles of the knee ranging from 0° to 95°, the internal tibial rotation varies within 8° to 30° and the external tibial rotation within 6° and 22°. These values are referred to four specimens subjected to ± 3 Nm torque (passive motion).

*Figure 5. Migration of contact point in articular surface and center of rotation in knee joint (left) and motion of tibia during flexion (right) (Forni et al., 1974).*

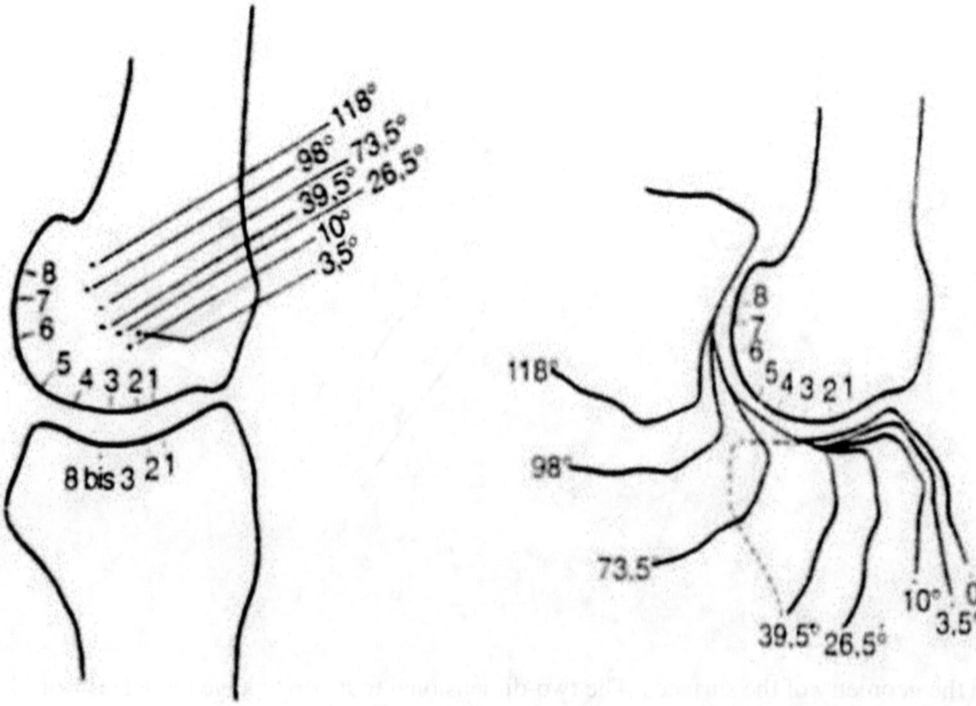


Figure 5 (left) reports the variation of the contact point between articular surfaces and the migration of the center of rotation (flexion). The same figure (right) indicates eight positions of tibia during the flexion motion of the knee from 0° to 118°. Figure 6 reports eight positions of the femur during the flexion motion from 0° to 118° from the left side view.

## MATERIALS AND TRANSDUCERS

### Piezoelectric Plates

In wearable energy harvesting applications, system requirements impose severe restrictions to the mass and volume of the generator; furthermore, the low invasiveness of transducers on the normal motion of hands is crucial. The most diffused piezoelectric materials have fragile behavior and are accompanied by high stiffness values; the first mentioned property is not compatible with applications where large bending motions are imposed and probably causes cracks initiation and propagation and reliability issues. The second property leads to high forces for actuating the transducers, which cause uncomfortable usage and increasing muscular forces needed.

*Figure 6. Femur positions during flexion of knee joint (Forni et al., 1974).*

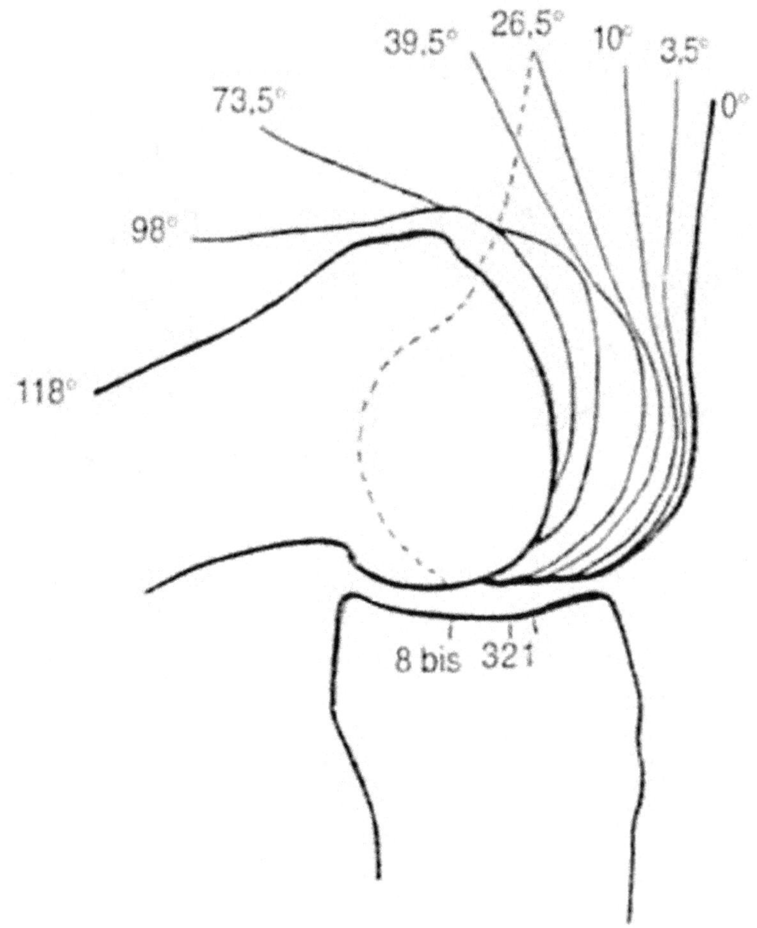


Innovative piezoelectric generators with large bending ratios, small thickness, low stiffness and high reliability are available in the form of foils coated with polymer. For instance, the transducer reported in Fig. 7a (Physik Instrumente, PIC252) is based on piezo foil with two metalized surfaces connected to the electrodes and inserted in a polymeric package. The material of the package and the small thickness of the foil allow large bending of the transducer with very small applied force, as demonstrated by Fig. 7b. The transducer has the properties reported in Tab. 3. Laboratory test results conducted on a dedicated bench revealed good performances of the transducer in terms of conversion efficiency and reliability under repetitive cycles loading. The high flexibility of the transducer assures the preservation of joints mobility without increasing the muscular force to complete the natural motion.

## Piezoelectric Fiber Composite Materials

Traditional electro-mechanical transducers based on piezoelectric materials are usually rigid, heavy and brittle. These characteristics do not fit with the requirements of energy harvesting from human body motion where angular travels of joints are very large. Alternative materials with high conversion efficiency, good reliability and high flexibility are needed. This suggests moving from bulk piezo materials

*Figure 7. Piezoelectric transducer PIC252 (Physik Instrumente) composed by internal active foil and polymeric package: internal structure and components (a), and verification of bending flexibility (b)*

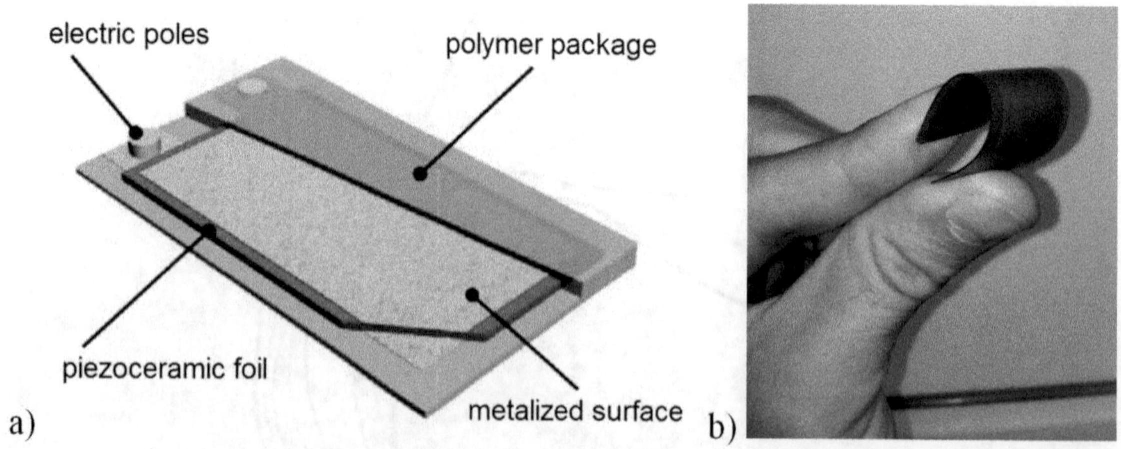


*Table 3. Geometrical and electro-mechanical properties of the piezoelectric transducer PIC252 (Physik Instrumente)*

| Property | Value | Unit |
|---|---|---|
| Length | 61 | mm |
| Width | 35 | mm |
| Thickness | 0.4 | mm |
| Bending radius | 12 | mm |
| Piezo layer thickness | 100 | μm |
| Young's module | 16.4 | GPa |
| Electrical capacitance | 150 | nF |
| Temperature range | -20 / +180 | °C |
| Mass | 2.1 | g |

and piezo sheets to composite materials with polymeric matrix and piezoelectric fibers embedded. The technology called "viscose suspension spinning process" (VSSP) allows producing fibers with diameters ranging from 10 to 250 μm having all the desirable properties of ceramic piezo materials (electrical, thermal, chemical), but without their typical detrimental characteristics (brittleness, weight, low reliability). An example of this typology of transducers is commercialized by Advanced Cerametrics Inc. (Fig. 8). They have substrate in kapton and PZT-5A type fibers. The author provided some experimental characterization results of these transducers in (De Pasquale et al., 2013), where 0.3 μW were measured for 45° bending angle at 1 rad/s. The electro-mechanical properties are summarized in Tab. 4.

*Figure 8. Piezoelectric fiber transducer (Advanced Cerametrics Inc).*

*Table 4. Geometrical and electro-mechanical properties of the piezoelectric fiber composite transducer (Advanced Cerametrics Inc)*

| Property | Value | Unit |
|---|---|---|
| dimensions | 132x14x1.1 | $mm^3$ |
| piezoelectric charge coefficient (d33) | 550 | pC/N |
| electromechanical coupling factor (k33) | 0.67 | - |
| young's module (Y33) | $2.44 \cdot 10^{10}$ | $N/m^2$ |
| yield strength | 157 | MPa |
| blocking force at 1kV | 1 | N |

## TEST BENCHES DESIGN AND BUILDING

### Dynamic Test Bench Reproducing Walking/Running Conditions

The dynamic test bench is normally used to reproduce the excitation provided by walking or running activities and its main application is in validating the performance of vibrating EHs. This is the case, for instance, of inductive generators based on permanent magnets oscillating inside a wire coil. The experimental characterization requirements in this case are the generation of cyclic excitations to run the test in repeatable conditions. The scheme reported in Fig. 9 describes the test bench: a wave function generator is used to produce the control signal, which is sent to the electro-mechanical shaker after power amplification. The EH connected to the shaker produces electric output when excited; the output values are measured by the acquisition system and stored in a PC. At the same time, the acceleration of the shaker and the harvester position are measured to calculate the frequency response function of the system. The same configuration can be used also for piezoelectric EHs. In this case, it is often required to lower the resonance frequency of the generator to fit the walking/running frequency (from 3 to 12 Hz); at this purpose a proof mass applied on the tip of piezoelectric cantilevers works as resonance tuning system.

For more accurate results, the real acceleration profile of a walking person is measured and stored; then, the same signal is used to generate a command profile voltage to actuate the electro-mechanical shaker. The setup of the experimental bench for the dynamic excitation of EH is reported in Fig. 10.

*Figure 9. Functional scheme of the test bench reproducing the dynamic excitation of walking/running*

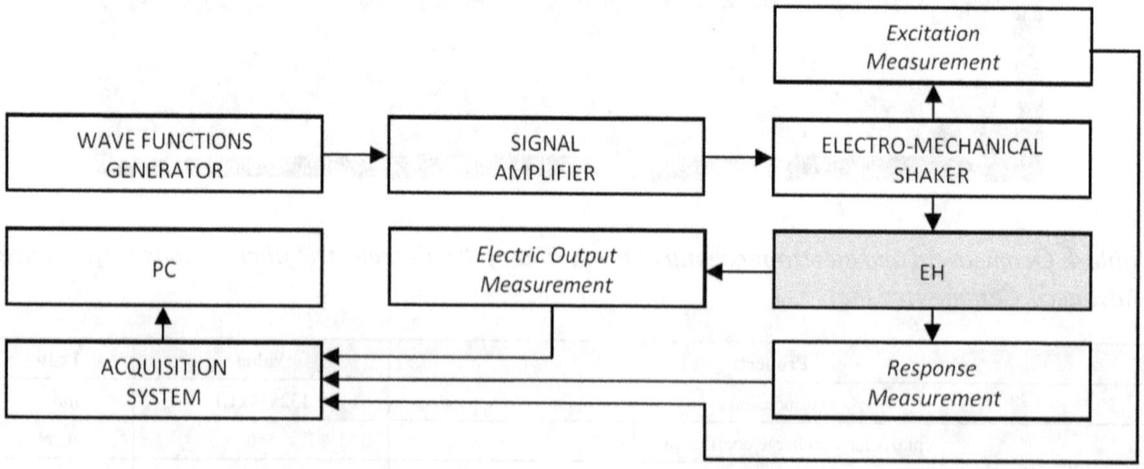


*Figure 10. Dynamic test bench setup*

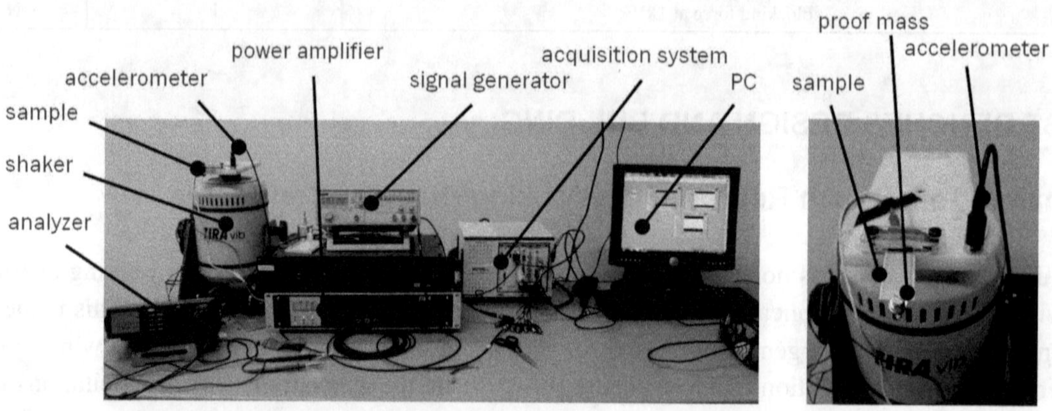


## Simulator of Human Joints Kinematics

This simple test bench has the goal of reproducing the kinematics of human joints; it is particularly useful to characterize piezoelectric EHs subjected to large deflection angles. The angular motions of elbow and knee joints, properly simplified according to the assumptions reported before, can be reproduced by means of this testing machine. The structure of the bench is represented in Fig. 11 and it is composed by two beams connected with a hinge providing the rotational degree of freedom. The piezoelectric transducer is mounted on two metallic planes connected to the beams; by using a polymeric sliding support, the transducer can bend without applied axial forces. This actuation method is able to reproduce faithfully the real loading condition, corresponding to the case of transducer applied to garments or directly to the skin. Figure 12 shows some phases of the experimental tests.

The joint simulator is moved manually by the operator during the experiments; the motion is controlled by a variable capacitance accelerometer (Endevco 7596A), which is used as inclinometer by

*Figure 11. Sketch of the artificial limb with inclination detection; the accelerometer is situated at the distance r from the center of rotation of the limb*

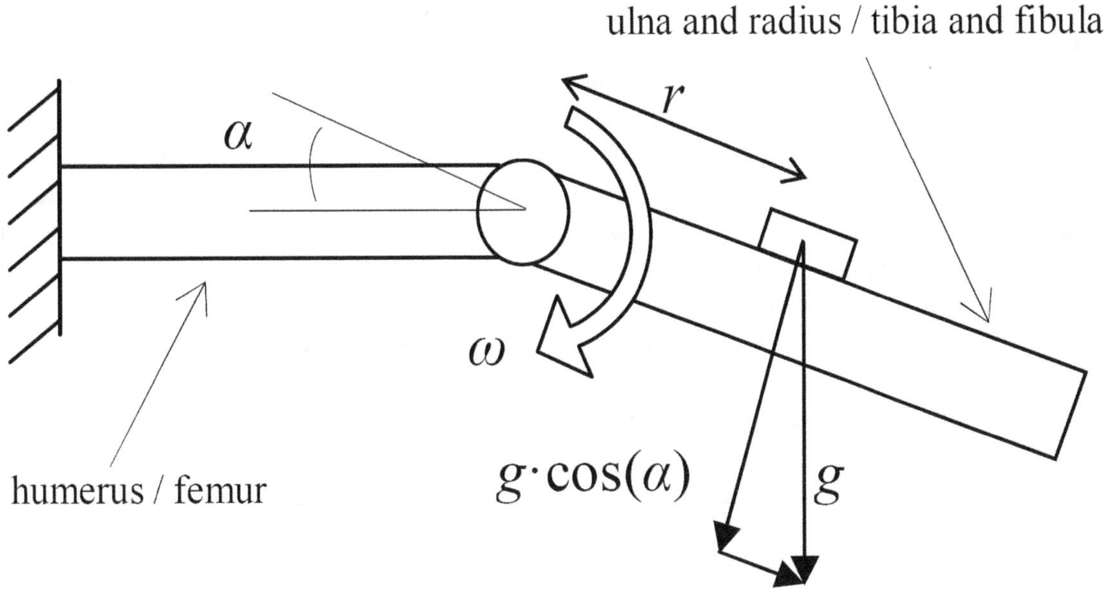


*Figure 12. Experimental phases of piezo fiber transducers characterization on the test bench reproducing the artificial human limb with joint*

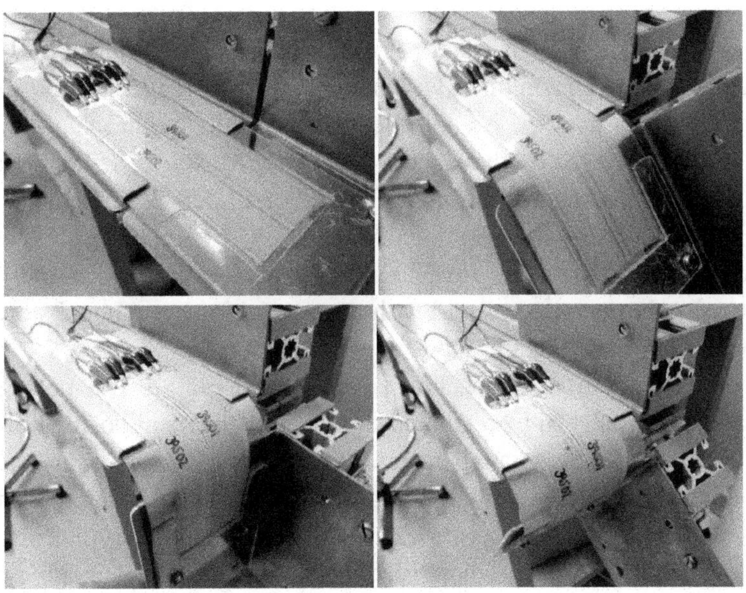

detecting the instant angular position of the movable half of the limb through the actual component of the acceleration of gravity. Then, by calculating the derivative of the angle, the angular velocity can be obtained for flexion and extension motions. Two piezoelectric transducers are bended simultaneously:

the first one is used to measure and store the output voltage in open circuit configuration; instead, the voltage drop across a known resistive load connected in parallel is measured on the second transducer to estimate the output current. In piezoelectric transducers, the function describing the output power with reference to the applied load has an absolute maximum; for this reason, the electric resistance has been previously characterized to find the optimum load value, which is used for measuring the output power. Finally, the combination of measured voltage and current provides the instant output power during the bending motion of the artificial joint. The output signals (voltage in open circuit configuration, voltage drop across the optimum resistance and angular position) are measured simultaneously with the National Instruments PXIe-4497 (24 bit, 204.8kS/s) analog-to-digital acquisition system in order to calculate the instant output power and the relative instant angular velocity.

The instant value of acceleration measured by the sensor on the limb is the combination of the component of $g$ orthogonal to the limb (that is used to calculate the instant value of the angle), as shown in Fig. 11, and of the tangential acceleration of the circular motion, which has the same direction. In the simplified case of uniform circular motion, the amount of this last component is $a_t = r\dot{\omega}$ ; by considering that the maximum angular velocity detected during the tests is about 4 rad/s and that the circular motion radius is $r = 0.05$ m, it results $a_t = 0.105$ m/s$^2$. Because of the value of $a_t$ is about 1% of the acceleration of gravity ($g$), this last value is neglected during experimental data processing. The dynamic response of the transducer in terms of resonance frequency and quality factor is preliminary characterized with electromechanical shaker. Then, in resonance condition, the optimum electric load is determined for maximizing the output power.

## Motorized Limb with Electronic Control

This test bench, designed and built by the author specifically for biomechanical EHs characterization, is able to reproduce the motion of human limbs with imposed angular range, angular velocity and bending radius of the sample. The test machine is controlled electronically by means of a microcontroller and electronic circuit and it is actuated by the electric DC motor Maxon 118778. The scheme of the test bench is reported in Fig. 13. The microcontroller ATmega328 (32KB flash memory, 2KB SRAM, 1KB EEPROM, 16 MHz clock speed) is programmed with the PC by imposing the desired experimental settings. The electric motor actuates the artificial limb and excites the EH sample. The electric output of the EH (current and voltage) are measured with the acquisition system and stored; also the instant position of the limb is measured with an encoder.

The design and building of the testing machine are reported in Fig. 14. The motorized plate reproducing the movable part of the limb can be regulated to change the vertical position of the sample, and at the same time, its distance from the rotation center. This allows varying the curvature of the sample in different testing conditions. Figure 15 shows the strategy adopted to hold the sample to the machine: the piezo transducer is supported by a polymeric flexible layer that reproduces the curvature of the human joint at increasing rotation angles. The sample is connected to this substrate only at one end; the other end is attached to an elastic element (made with elastomeric material) that works as axial load modulator. The opposite end of this axial spring is held to the machine structure and can be moved: depending to the position of the spring constraint, the axial load transferred by this element to the sample varies from 0 to the maximum value.

*Figure 13. Diagram of motorized limb for reproducing large bending angles of human joints.*

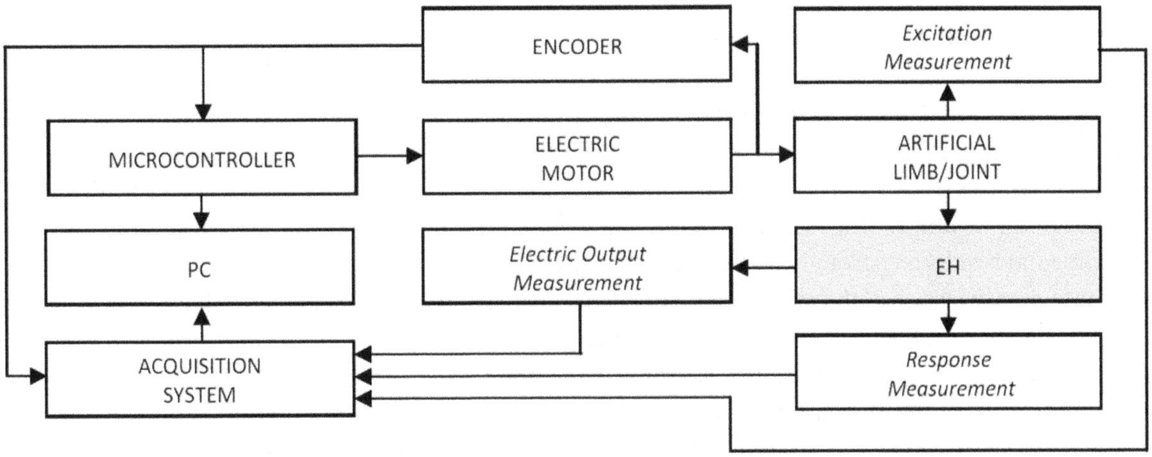


*Figure 14. Motorized test bench of reproducing the motion of human limbs: design and building.*

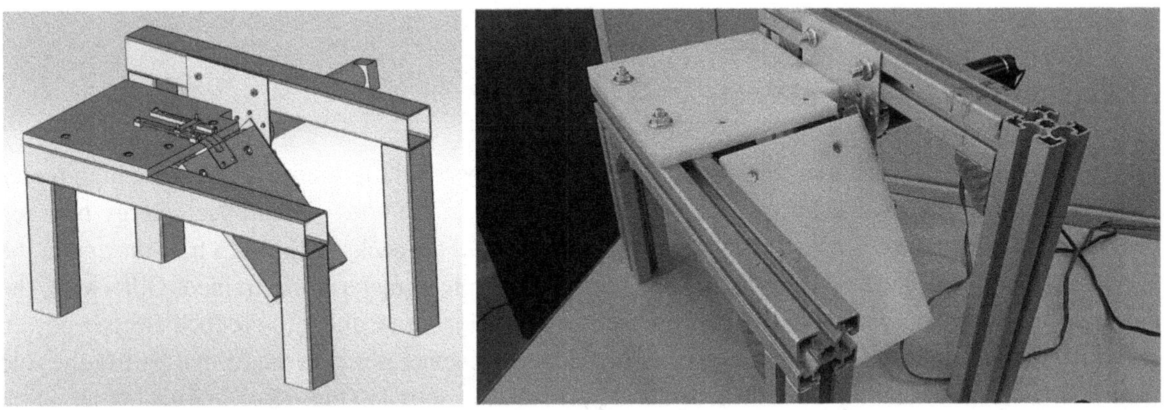

*Figure 15. Sample holding system*

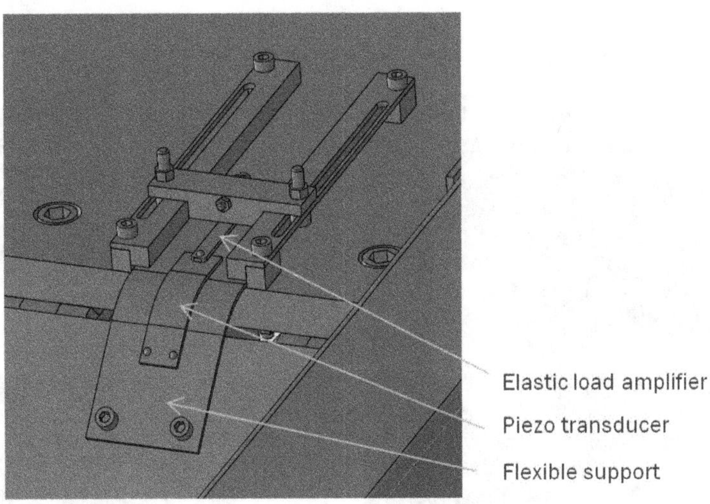

# TEST ON HUMAN BODY

## Biomechanical Energy of Hands

In experimental analysis of biomechanical EHs performances, the alternative to test benches are measurements conducted on the human body. The advantage of this approach is the high faithfulness of testing conditions (excitation, constraints, kinematics, etc.); however the results are not perfectly repeatable because it is almost impossible to reproduce several times the same motion of any joint. Then, the experimental results are strongly related to the body characteristics of the operator conducing the tests.

The characterization of the piezoelectric transducer PIC252 (Physik Instrumente) has been performed with tests on human body. The preliminary characterization of the piezo transducer on shaker allowed defining the optimum electric load to maximize the output power. This characterization has been conducted by lowering the resonance of the piezo in cantilever configuration to the value of 4 Hz in order to simulate the quasi-static excitation conditions on the hands joints. The properties of the transducer are the same reported in Tab. 3. The piezo transducer is applied on a glove in correspondence of MCP and PIP joints of fingers from 2 to 5; then the hand of the operator is moved alternatively in opened-closed position. Finally, the output power is measured. Figure 16 shows two phases of the test of the piezo transducer on the human hand.

## Biomechanical Energy of Feet

The application of energy harvesting systems to sneakers is particularly promising because high forces are transferred to the ground in correspondence to this part of the body. The impact energy between heel and ground can be converted to electricity by using small air packs with piezo transducers on the surface: when the air pack is compressed and released, the piezo material is strained. Otherwise, the deformation of the sole can be transferred to an embedded piezo transducer to produce electric power. By following this approach, the best candidate location of the generator is in the frontal part of the sole (Fig. 17), where the bending is maximized during the walking activity (Fig. 18).

*Figure 16. Experimental characterization of the piezoelectric transducer PIC252 (Physik Instrumente) on the human hand*

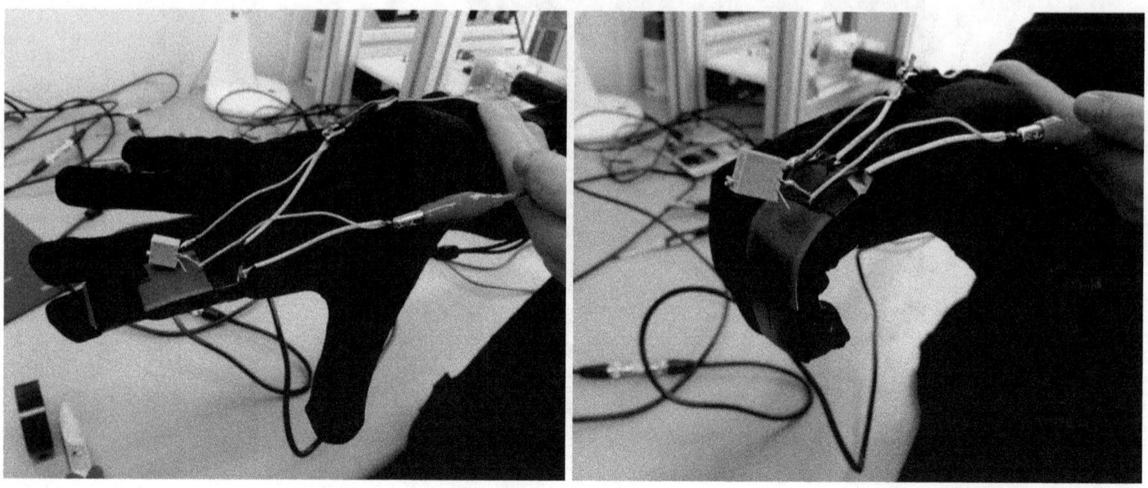

*Figure 17. Candidate position of the piezo generator for converting the sole bending in electricity*

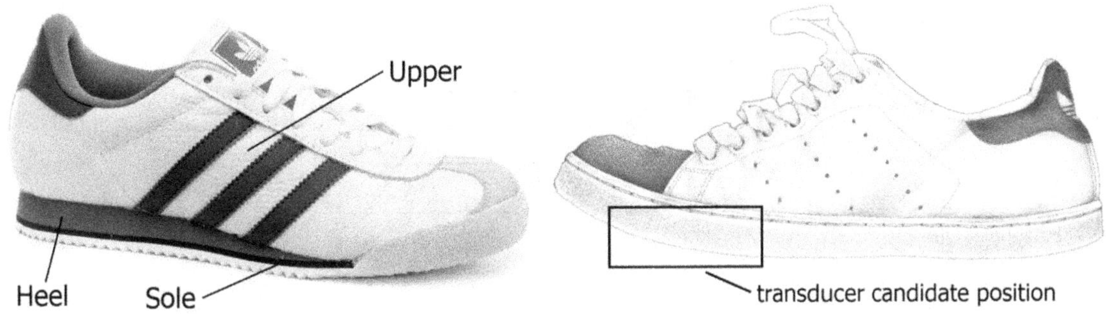


*Figure 18. Kinematics of walking (Pirola, 2003)*

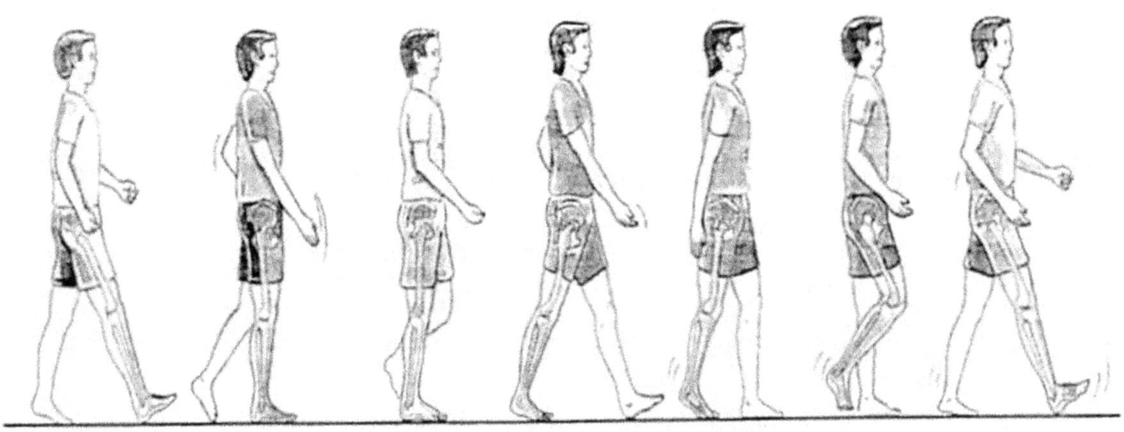

Three main design solutions can be considered in connecting the piezoelectric transducer to the sneaker sole: a) connection with relative sliding, b) connection without relative sliding (glued), and c) elastic axial connection. In the case of connection with relative sliding, the piezo transducer is constrained to the sole only at one end and its bending is induced by the deformation of the sole. The constructive solution is, for example, a cavity in the sole where the transducer is inserted (Fig. 19a).

The stress and strain distributions into the piezo material are the same of a cantilever (Fig. 19b); the strain is given by the relation

$$\varepsilon^{\max}(x) = \frac{\sigma(x)}{E} = \pm\frac{6Fx}{Ebh^2} \tag{1}$$

with the symbols explained in Tab. 5.

The variation of the maximum stress and strain along the transducer for different applied forces follows the trends reported in Fig. 20, which suggest to use a bimorph piezo generator for this type of sole connection. The properties of the most diffused sneaker soles, reported in Tab. 5, are considered.

*Figure 19. Connection with relative sliding (a) and stress/strain distribution in the piezo transducer cross section (b).*

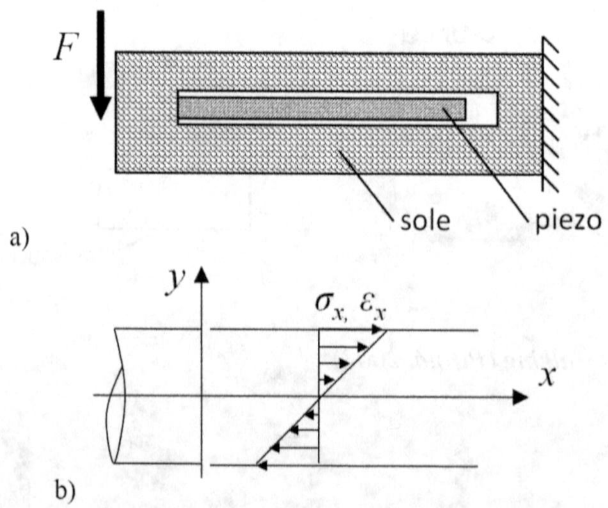


*Table 5. Geometrical and material properties of piezo transducer and sole*

| Property | Symbol | Value | Unit |
|---|---|---|---|
| sole width | $B_s$ | 100 | mm |
| sole thickness | $H_s$ | 13 | mm |
| sole Young's module (rubber) | $E_s$ | 20 | MPa |
| piezo transducer dimensions | $l \times b \times h$ | 60x35x0.5 | mm |
| piezo transducer Young's module | $E$ | 23300 | MPa |

*Figure 20. Maximum stress (a) and strain (b) along the transducer for different applied forces in case of connection with relative sliding with the sole*

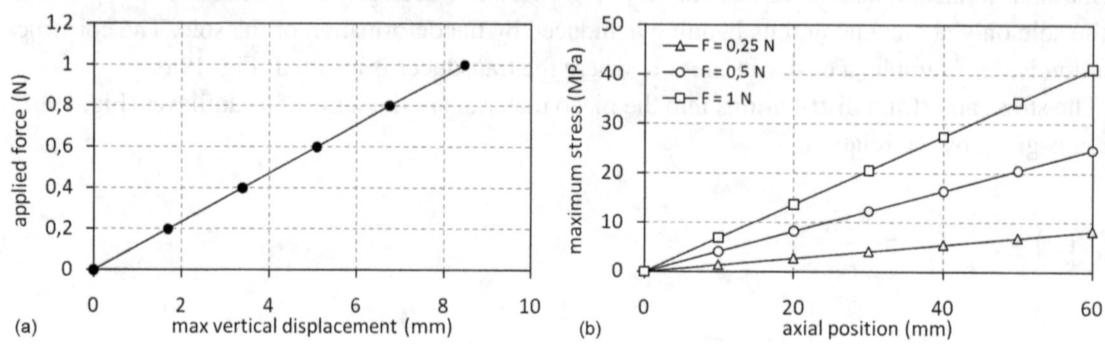


The limitation of this type of connection is that the piezo material is not uniformly loaded and the charge generation at its electrodes is not maximized.

To excite the piezo material in the axial direction and provide more uniform loading, the connection without relative sliding can be used. Here the piezo transducer is glued on the surface of the sole and its deformed shape follows the bending imposed by the sole at its external side. From the constructive viewpoint, this configuration is simply obtainable by gluing the piezo under the sole (Fig. 21a).

The stress and strain distributions into the piezo material are approximately constant, by considering that the thickness of the piezo is much smaller than the thickness of the sole (Fig. 21b); the strain is given by the relation

$$\sigma(x) = E\varepsilon_s(x) = \frac{6Fx}{bh^2} \tag{2}$$

The effects of glued surfaces must be considered carefully to preserve the reliability of the system. In fact, the large difference between the mechanical strength of the piezoelectric transducer and the sole may cause large stress gradients at their interface. The maximum values of the stress difference in piezo material and sole material at the interface is reported in Fig. 22 for increasing applied forces (calculated with the properties listed in Tab. 5). This stress level is applied in cyclical mode and the involved materials may experience fatigue behavior causing possible reduction of the efficiency of the generator. The stress distribution in the material suggests using a unimorph piezo generator for this type of sole connection.

The configuration with elastic axial connection has the goal of removing the stress gradients at the piezo-sole interface and, at the same time, of applying axial loading to the structure to increase the EH efficiency. In this case, the transducer is constrained at one end and connected to an elastic element on the other end. By adjusting the stiffness of the elastic element, the axial load on the piezo material can be modulated. The constructive version of this design is reported in Fig. 23a.

The stress and strain distributions into the piezo material are the sum of the flexural component and of the axial component provided by the spring; they are reported in Fig. 23b for increasing values of the applied force. The stress distribution in the material suggests using a unimorph piezo generator for this type of sole connection. The stress and strain are given by the relations

*Figure 21. Connection without relative sliding (a) and stress/strain distribution in the piezo transducer cross section (b)*

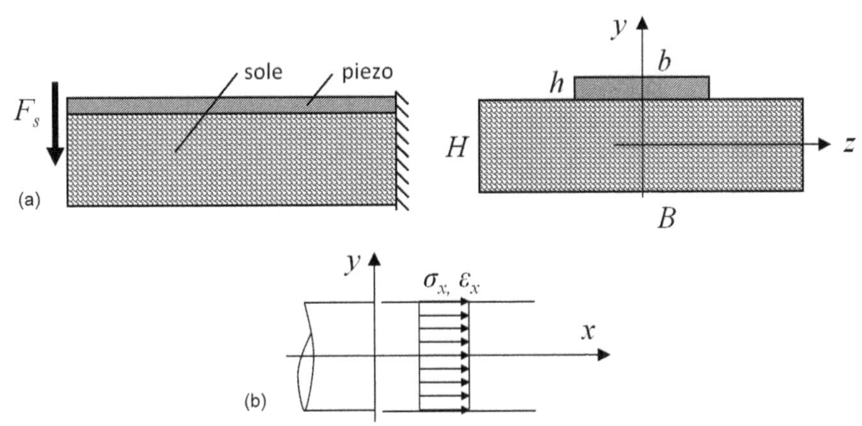

*Figure 22. Maximum stress difference between piezoelectric and sole materials at the interface for variable applied forces*

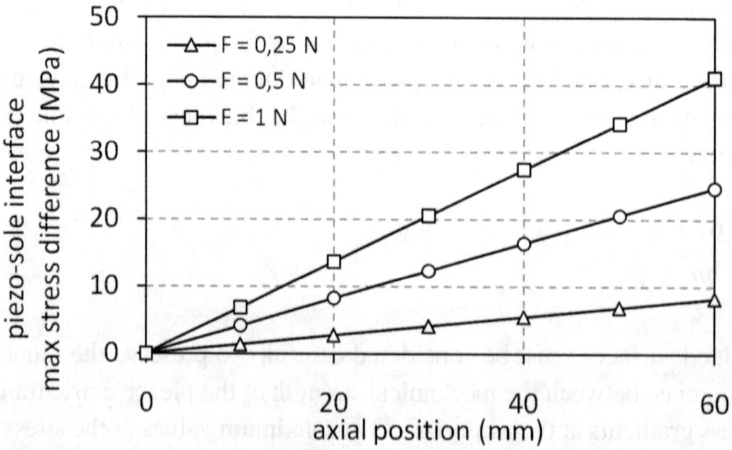


*Figure 23. Configuration with elastic axial connection (a) and stress/strain distribution in the piezo transducer cross section (b)*

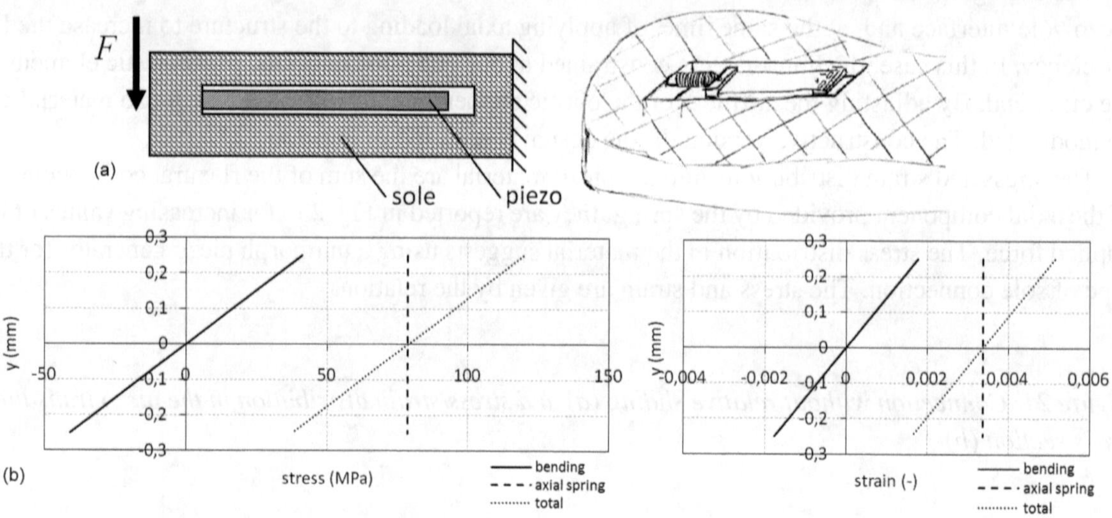


$$\sigma(x) = E\varepsilon_s(x) = \frac{6Fx}{bh^2} \tag{3}$$

$$\sigma^{\max}(x) = \frac{3F_s l^2 k}{E_s bhBH^2} + \frac{6Fx}{bh^2} \tag{4}$$

$$\varepsilon^{\max}(x) = \frac{3F_s l^2 k}{EE_s bhBH^2} + \frac{6Fx}{Ebh^2} \tag{5}$$

# EXPERIMENTAL RESULTS

## Measurements on Test Benches

In experimental tests conducted with the simulator of human joints, the piezoelectric transducer is subjected to controlled flexion with imposed angular velocity and angular travel length.

Before the test, the dynamic response of the transducer in terms of resonance frequency and quality factor is preliminary characterized with electromechanical shaker. The experimental value of the flexural resonance frequency is 6.3 Hz when the free vibration length of the transducer is 114 mm (Fig. 24a). Then, in resonance conditions, the optimum electric load is determined for maximizing the output power (Fig. 24b).

The optimum resistive load at resonance is 915.6 kΩ. The reported experimental values are referred to a piezoelectric fiber composite material with the properties listed in Tab. 4. The resonance frequency has been detected on the cantilever configuration of the transducer with 114 mm free length and 0.35 g acceleration. The optimum resistance is characterized at 0.5 g acceleration and the same free length.

The output power sensitivity to the input acceleration has been tested on the shaker in resonance conditions (6.3 Hz), with the optimum resistive load (915.6 kΩ) and 114 mm free length of the cantilever: the results are reported in Fig. 25.

With the simulator of human limbs, the instant value of the acceleration is measured on the rotating part of the machine by a capacitive acceleration sensor, which is able to detect the static (gravity, *g*) and dynamic components. The resulting value of the acceleration is a combination of the component of *g* orthogonal to the limb (that is used to calculate the instant value of the angle), as shown in Fig. 11, and of the tangential acceleration of the circular motion, which has the same direction. In the simplified circumstance of uniform circular motion, the amount of this last component is $a_t = r\dot{\omega}$; by considering that the maximum angular velocity detected during the tests is about 4 rad/s and that the circular motion radius is $r = 0.05$ m, it results $a_t = 0.105$ m/s$^2$. Because of the value of $a_t$ is about 1% of the acceleration of gravity, this last value is neglected in experimental data processing.

The results of experimental tests on the artificial limb are reported in Figs. 26-28; the output values of open circuit voltage, current and power are reported together with the instant angular velocity for variable bending angles of the joint (45°, 90° and 135° respectively) and by considering both flexion and extension motions. Figure 29 collects the previous results by showing how the half-power value of the electric output varies according to the angular velocity applied to the joint, to the angular amplitude and to the motion direction (flexion or extension).

## Tests on Human Body

Also for the tests on human body, some of the properties of the piezoelectric transducer are firstly characterized by using the electromechanical shaker. The transducer is actuated in the cantilever configuration with the shaker and the electric output is measured. The goal of this preliminary characterization

*Figure 24. Dynamic tests on electro-mechanical shaker: detection of resonance frequency of the piezo-electric fiber composite transducer (a) and characterization of optimum resistive load (b)*

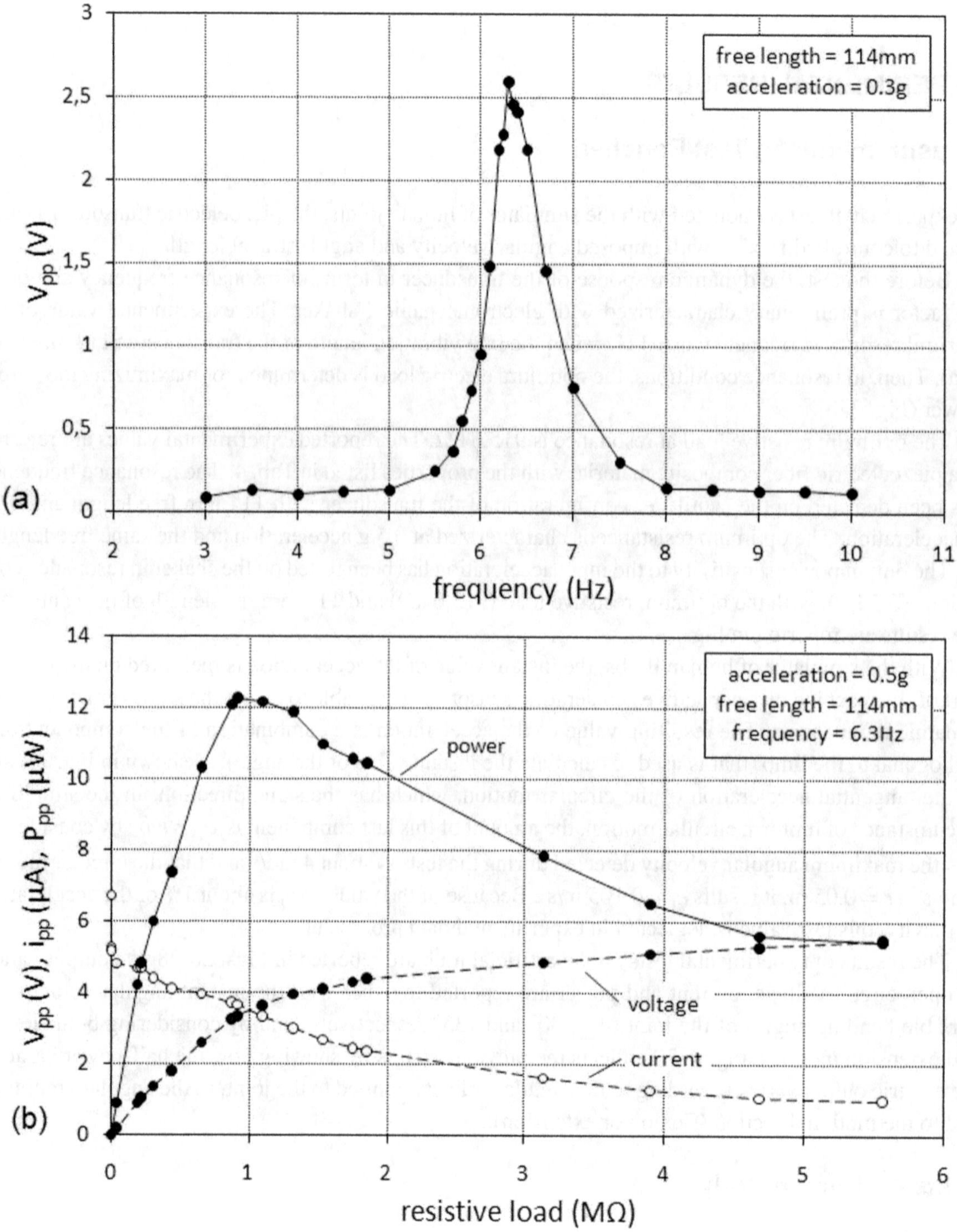

*Figure 25. Dynamic tests on electro-mechanical shaker: electric output at variable accelerations*

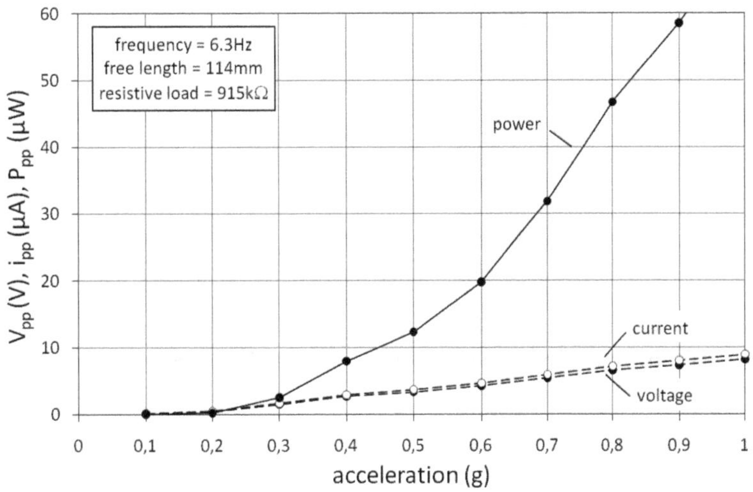


is to identify the optimum resistive load that maximizes the generated power. The application of piezo transducer to fingers, for instance, suggests that the EH is excited at very low operative frequencies in the final configuration (in the order of few Hz). Then, during the preliminary dynamic characterization, a proof mass is added on the tip of the cantilever to lower the resonance of the beam and to simulate quasi-static excitation conditions. The piezoelectric transducer considered is the foil described by Fig. 7 and having the properties listed in Tab. 3. By adding the mass $m = 30$ g, the measured resonance frequency is 10 Hz, as reported in Fig. 30a, with 0.3 g acceleration. The resulting optimum resistive load for the piezo transducer in quasi-static excitation conditions is $R_{opt} = 216$ kΩ, as reported in Fig. 30b.

Other tests are conducted with the piezo transducers applied to fingers in their final configuration, which is in correspondence to the PIP joints of fingers 2 to 4. Four piezoelectric transducers are attached to the fingers and their electric output is measured separately for the opening-closing motion of the hand. The test consists in repeating the cyclic movement of the fingers for 10 s; in the meanwhile, the output parameters of open circuit voltage ($V_0$) and of voltage difference ($V$) across $R_{opt}$ are measured and stored. These values of voltage and current are reported in Fig. 31.

For each finger, the generated current is calculated as

$$i = \frac{V}{R_{opt}} \tag{6}$$

and the instant power is calculated as

$$P = V_0 i \tag{7}$$

The measured values of current and power are reported in Fig. 32.

The cumulative power is obtained as the integral of the instant power over time as

*Figure 26. Flexion-extension test results on artificial joint at 45° amplitude*

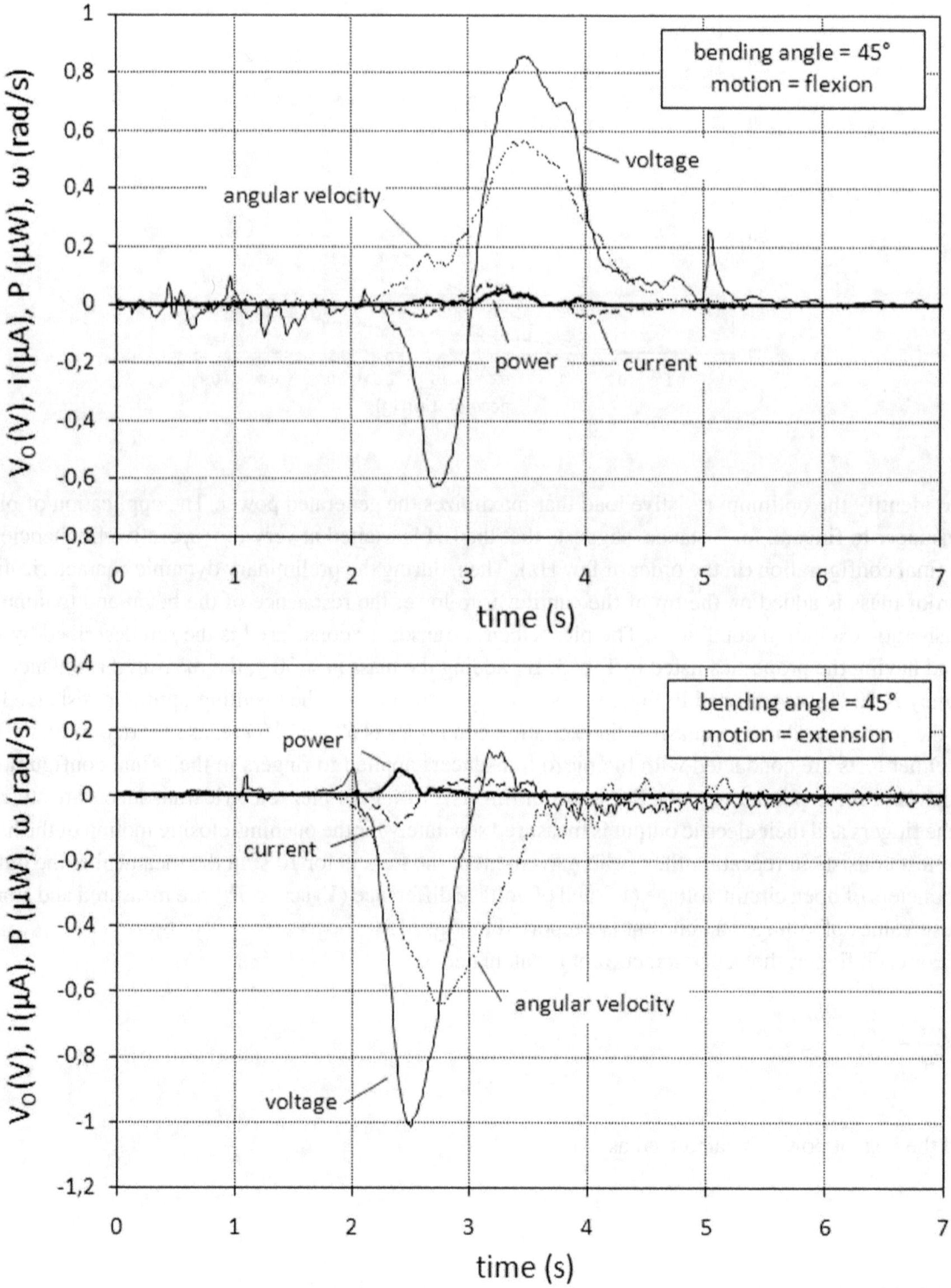

*Figure 27. Flexion-extension test results on artificial joint at 90° amplitude*

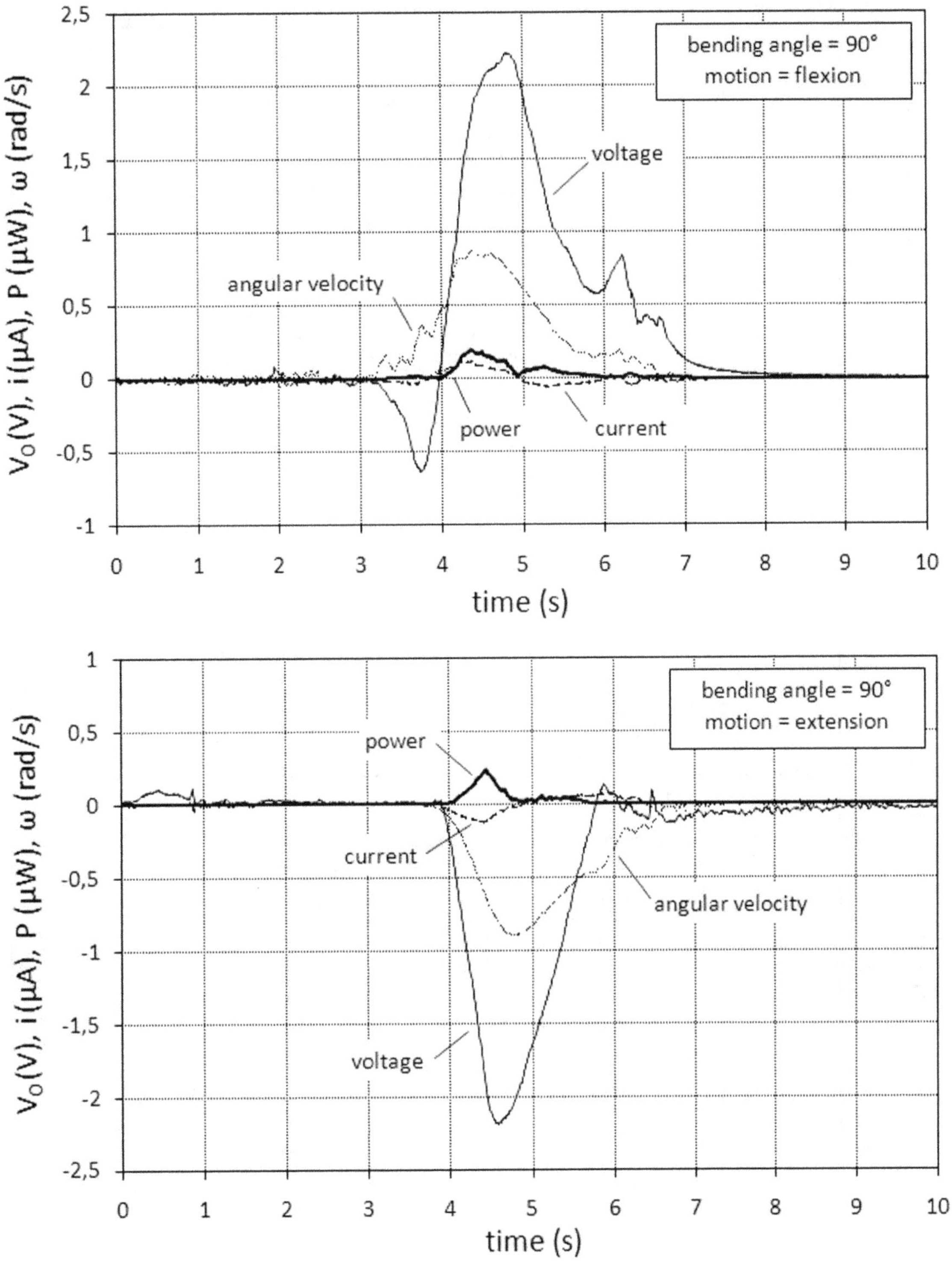

*Figure 28. Flexion-extension test results on artificial joint at 135° amplitude*

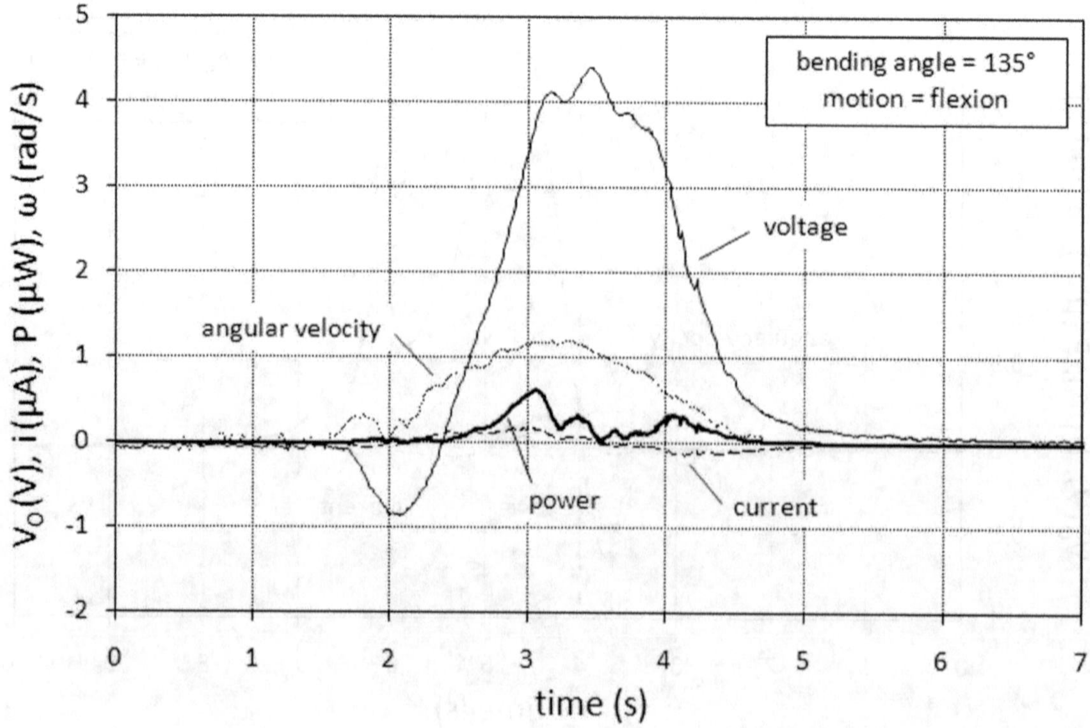


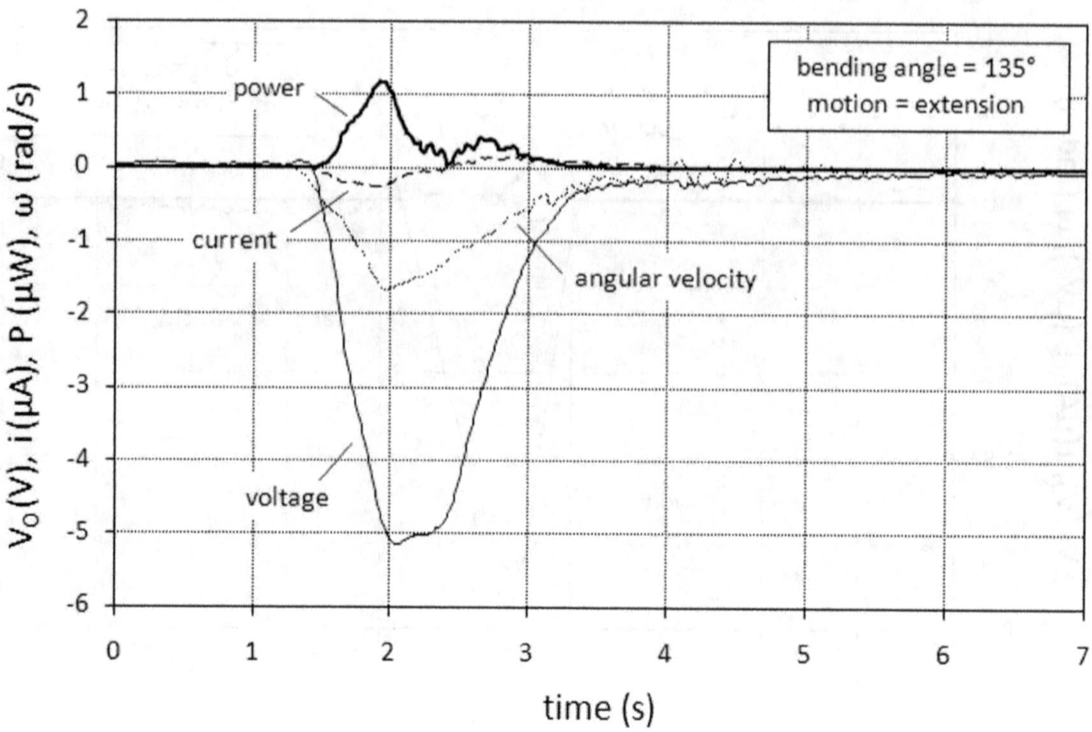

*Figure 29. Half-power value of electric output at variable angular velocities and bending angles of the artificial joint*

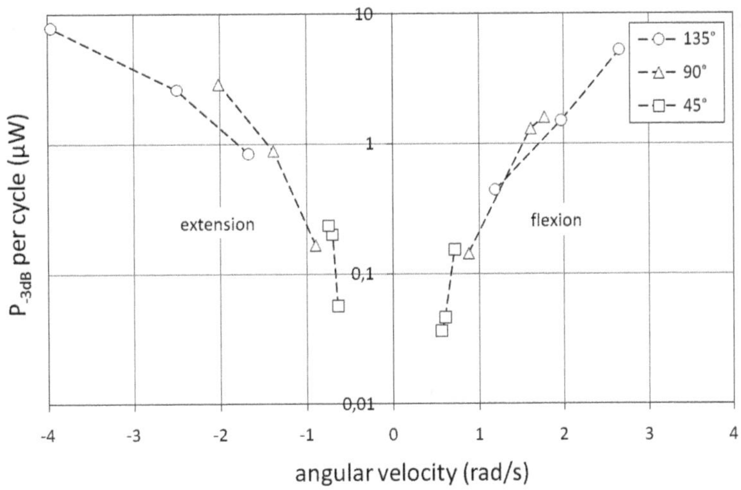


$$P_{\text{cumul}}(t) = \sum_{T=0}^{t} P(T) \cdot \Delta t \tag{8}$$

where $t_j$ are the sampling times and

$$\Delta t = t_{j+1} - t_j \tag{9}$$

The average instant power is defined as

$$\overline{P}_{\text{inst}}(t) = \frac{P_{\text{cumul}}(t)}{t} = \frac{\sum_{T=0}^{t} P(T) \cdot \Delta t}{t} \tag{10}$$

The measured values of the cumulative power and of the average instant power are reported in Fig. 33.

Additional experimental results have been obtained by characterizing piezoelectric transducers applied to the sole of sneakers during walking activity. In this case, the energy associated to the bending of the rubber sole and caused by cyclic contact between feet and the ground is converted to electricity. The piezo transducer is applied to the lower surface of the sole in a small cavity that prevents the contact between the EH and the ground, to avoid the damage of the active material. The connection configuration of the piezo to the sole allows the sliding between their surfaces (that corresponds to the first case previously described). Before the test, by using the electromechanical shaker, the optimum load resistance has been characterized with the same procedure already described before; the value $R_{opt}$ = 100 kΩ has been found. Then, output voltage and current of the transducer have been measured, with the EH applied under the sole in the location indicated by Fig. 17. The resulting experimental values are reported in Fig. 34. Then, the output power and cumulative output power have been calculated and plotted in Figs. 35.

*Figure 30. Characterization of piezoelectric transducer with 30 g proof mass at 0.3 g acceleration on the shaker with sinus actuation: resonance frequency (a) and optimum resistive load characterization (b)*

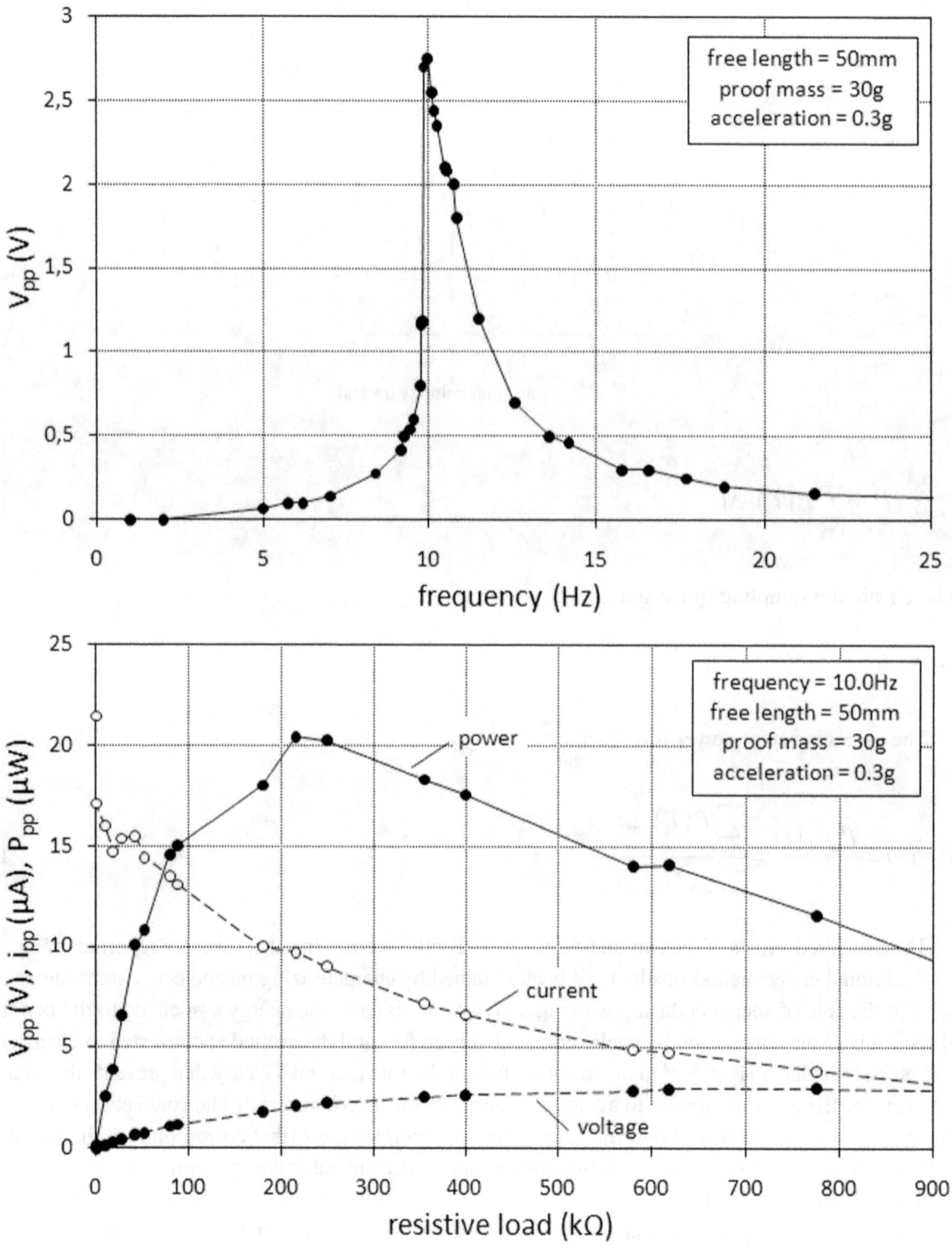

*Figure 31. Measured values of the output voltage for each finger: open circuit voltage (a) and voltage across $R_{opt}$ (b)*

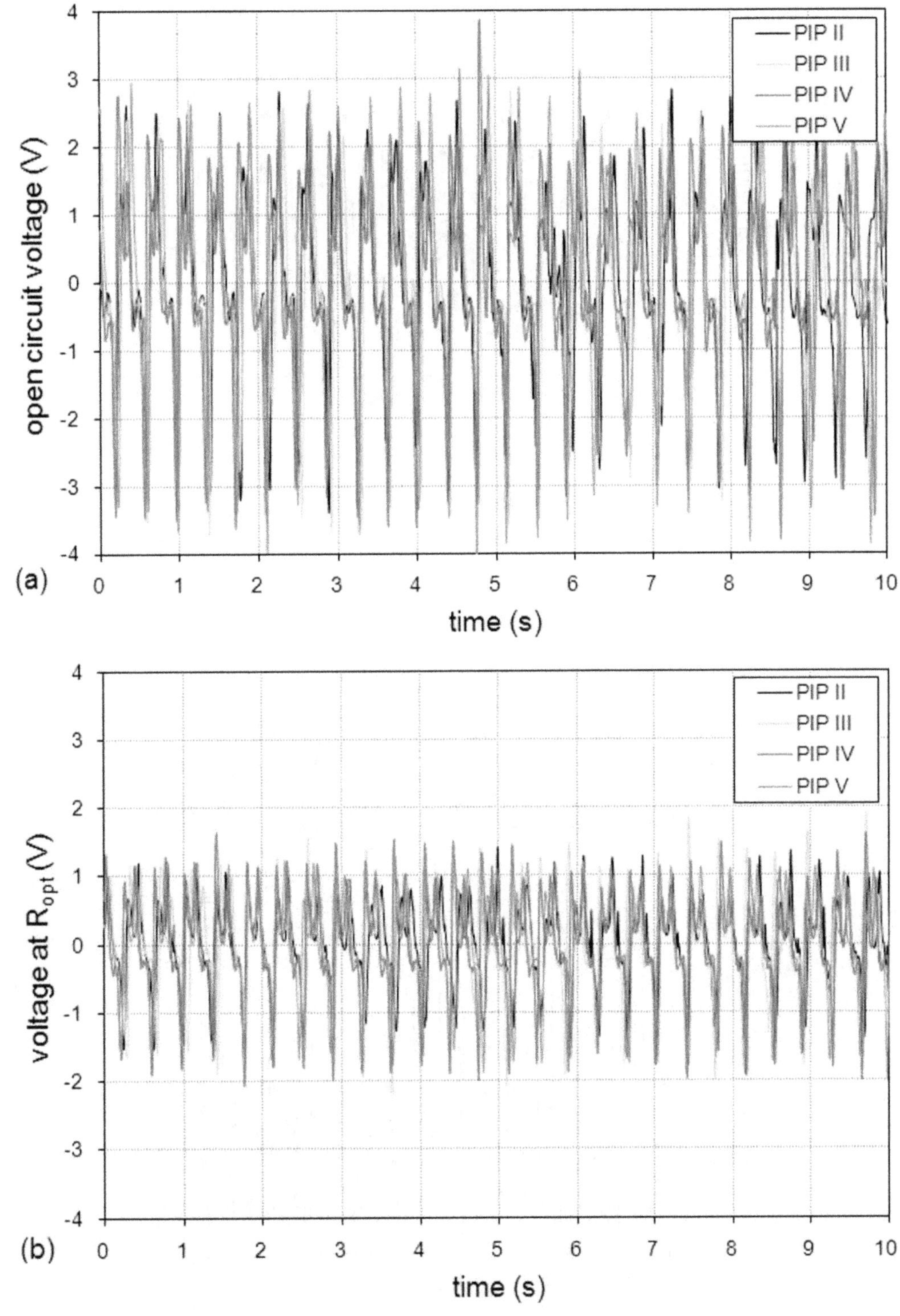

*Figure 32. Measured values of output current (a) and power (b) generated by the harvesters for each finger, and total output power generated by all the harvesters (c)*

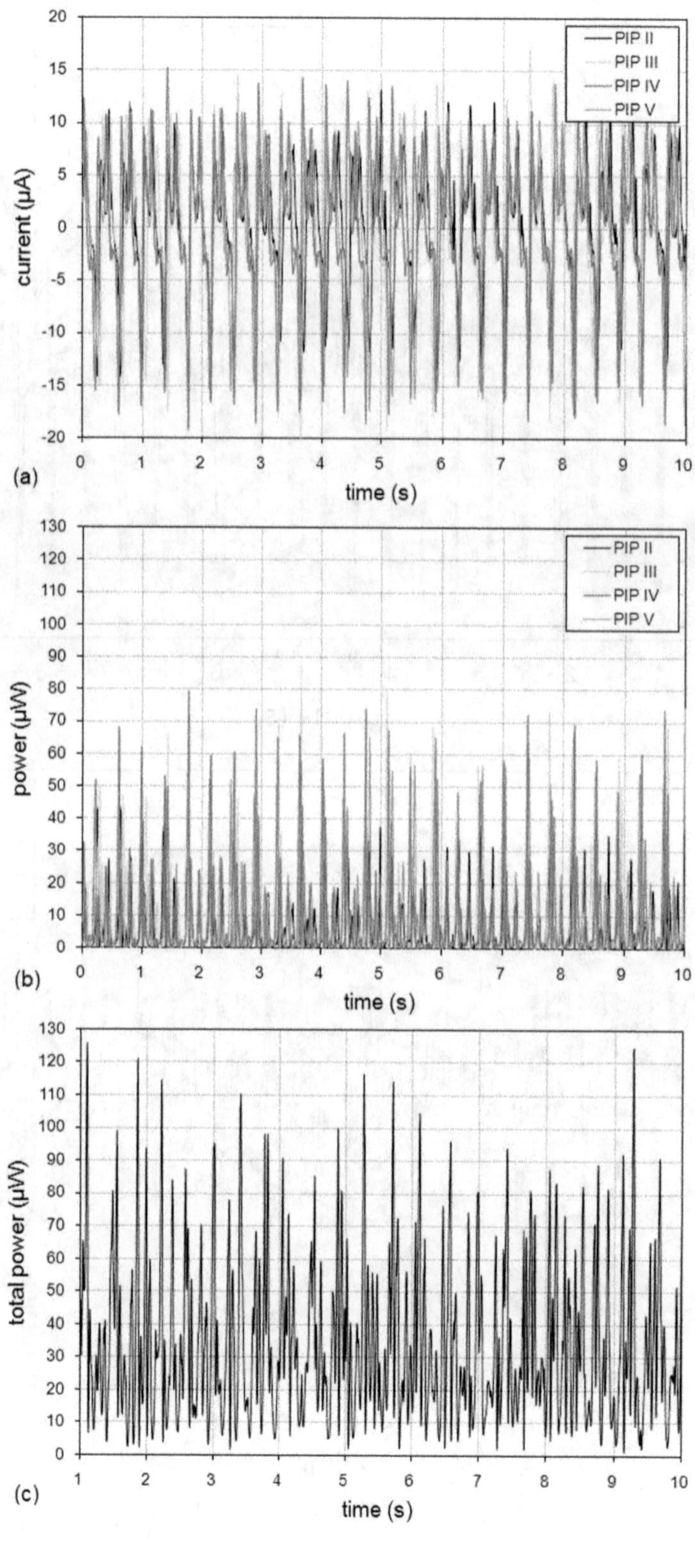

*Figure 33. Cumulative power (a) and average instant power (b) for each finger and for all the harvesters applied to the hand fingers*

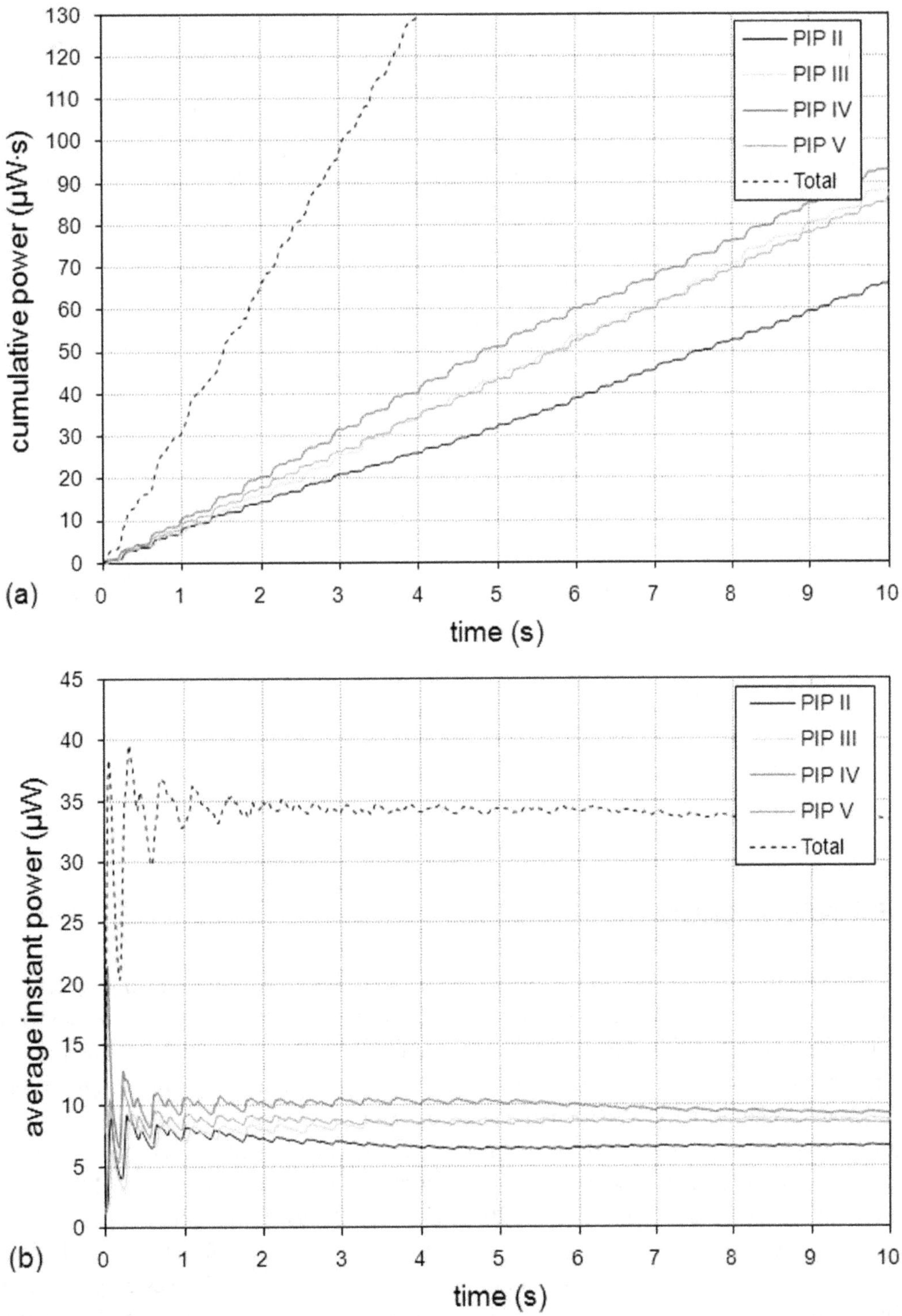

*Figure 34. Output voltage (a) and output current across the optimum resistance $R_{opt}$ (b) of piezoelectric EH applied under the sole in walking activity*

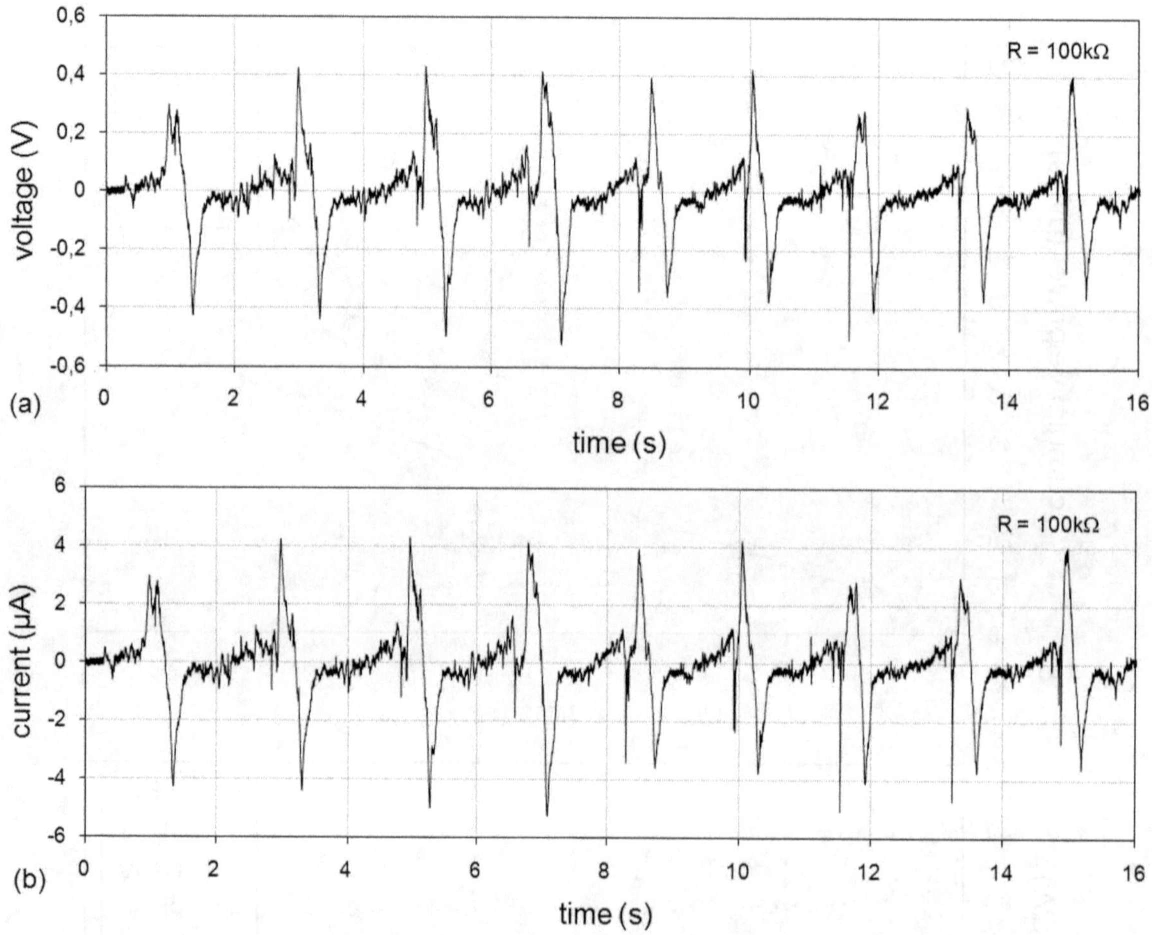


## SYSTEM INTEGRATION

The piezoelectric transducers previously tested are integrated in a wearable sensing system to verify the energetic suitability of the integrated device and to estimate its duty cycle. The piezoelectric fibers composite materials are used in the wearable integrated system; they are coupled with simple MEMS (electro-mechanical systems) sensors for human motion detection applied to garments. The experimental results previously reported have been used to reference the electric output of the generators. By considering the motion of joints associated to the walking activity at 1.3 m/s velocity, every cycle (corresponding to one complete step) includes elbows rotation (about 40° at 4 rad/s (Riemer et al., 2011)) and knees rotation (about 55° at 5 rad/s (Riemer et al., 2011; Li et al., 2009)) in both flexion-extension modes. From the experimental measurement, it results about 0.22 and 0.33 μW for flexion and extension at the maximum velocity (< 1 rad/s) and 45° amplitude. By including in the harvested power estimation all the limbs, by applying three combined transducers per joint (allowed by the high flexibility) and by correcting the experimental results accordingly to the actual bending velocity, it results about 13.20 μW output power per cycle of walking.

*Figure 35. Instant power (a) and cumulative power (b) in presence of the optimum resistance $R_{opt}$ of piezoelectric EH applied under the sole in walking activity*

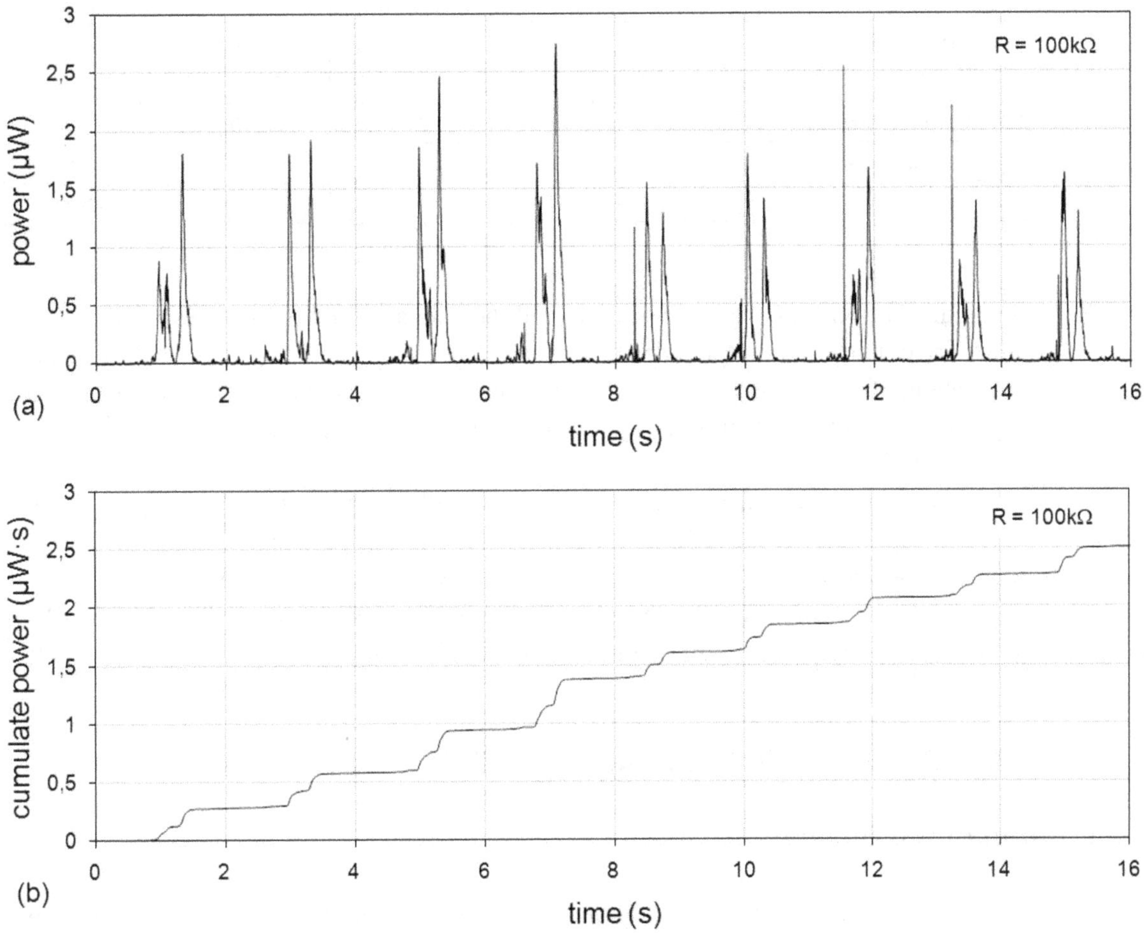


The simple sensing system considered is composed by the Atmel ATmega328 chip working as micro-controller (embedded in the Lilypad 328 Arduino wearable device), a three-axis MEMS accelerometer (±2 to ±16 g programmable full-scale range), a temperature sensor (0.5 °C accuracy) and a RF module (30 m range, 250.000 bps rate, 2.54 GHz). The power consumed by single components is indicated in Tab. 6. The energy generated by the harvester is temporarily accumulated in a storage battery, which works as reservoir and levels the power supply to the load.

The following operation modes are considered: stand-by (only the microcontroller is supplied, in the power saving mode), sensing (the microcontroller is active, accelerometer and temperature sensor are supplied) and transmission (microcontroller and RF module are active). In the stand-by mode, the self-powered system is able to store in the battery the difference between the harvested power and the power consumed by the microcontroller: 12.83 μW/cycle. One cycle of walking includes flexion and extension motion of all the limbs and three transducers are supposed to be connected to the elbows and knees. The generation rate described allows supplying the operations of acceleration and temperature sensing for 1 s every 235 cycles, which corresponds to about 300 m of walking distance (with 0.8 m

*Table 6. Electrical specifications of the wearable system components*

| Device | Status | Voltage | Current (mA) | Power (mW) |
|---|---|---|---|---|
| Microcontroller | Active | 3.65 | 0.20 | 0.73 |
| | Stand-by | 3.65 | $1 \cdot 10^{-4}$ | $3.65 \cdot 10^{-4}$ |
| 3-axis MEMS accelerometer | Active | 2.91 | 0.50 | 1.46 |
| Temperature sensor | Active | 4.10 | 0.20 | 0.82 |
| RF module | Active | 3.10 | 45.00 | 139.50 |

average step length). The RF module is able to transmit the data acquired by the sensors; the receiver can be installed, for instance, on portable device (smartphones, etc.). The period needed to support a transmission of about 100 ms, the time required for connection, transmission and disconnection, is about 1092 cycles. The corresponding walking distance needed to harvest the power required to data transmission is about 1350 m.

The duty cycle estimation provided is influenced by the walking speed: faster is the gait, higher is the harvested power and more frequent the sensing and transmission operations allowed. On the other hand, additional sensors and devices connected to the system will increase the power consumption and the recharging time. For instance, the same microcontroller is able to support interfacing with pH sensors, blood pressure sensors, electrocardiography (ECG) and electromyography (EMG) sensors, etc.

## CASE STUDY OF WEARABLE HUMAN-MACHINE INTERFACE

### Description of the Device

Human hands are one of the most efficient and effective mechanical interfaces in the animal realm with external environment. The hands are used by everybody for sending messages, emotions, and commands, for perceiving the world around us and for interacting with objects and other people. These activities are done almost continuously and every day. Then, it sounds natural to believe that the simplest and easiest way to interact also with machines is through the direct use of hands without intermediate peripherals (buttons, joysticks, keyboards or mice). These peripherals were born together with the first personal computers, when the only possibility to communicate with electronic devices was through monitors or displays. Unfortunately, in many operative cases (especially in the industrial field) the situation remained almost unaltered even at present times and people and machines continue to communicate by means of obsolete channels. Often, the opportunity to increase the quality and the quantity of the information shared is not exploited at all. These considerations are valid for many machines, systems, implants, etc. that usually contribute to our everyday life. In many practical cases of interfacing, the consequence is that the dialogue with the machine implies the modification of user's natural behavior, of his spontaneous posture and of his lifestyle in general. The machine behaves as a stranger body that enters in the human life, and the user has to follow its requirements of communication in order to use its potential.

With the goal to improve the situation, in the last 30 years, researchers introduced a number of new typologies of interfaces between the humans and the machines and the HMI discipline is at now one of the most active in finding solutions to overcome these technological frontiers. At the beginning, the most diffused concept was to wear the human body, and in particular the hands, with electro-mechanical devices able to sense the movements and the positions of the limb, to convert this data with a coded conversion protocol and finally to control the machine. At this purpose, many electronic glove systems were introduced, starting from the 1970s, which were designed in particular to measure the hand configuration, the bending of fingers joints and their rotation angles. The accompanying software provided the required interpretation of gestures and each configuration of the limbs were translated to commands for the machine. In these electronic wearable devices, the sensors used were preliminary piezoresistive or based on light emission and detection (optical fibers). Since those years the crucial design specification of power supply started to be focused. The very first sensing wearable systems were really covered by wires, connectors and cables; the last versions instead tried to integrate electric conductors with storage batteries. It is evident that the next frontier of the HMI research is to integrate power generators in the devices, for instance by converting the kinetic energy of human body to electricity by means of biomechanical EHs. In glove-based systems, the kinetic energy of fingers could be converted and stored in temporary batteries and then used to supply the smart interface.

In the last years, the PC was the favorite machine to experiment new ways of communication, under the pushing of software engineering for virtual reality, computer graphics, computer animation, etc. Instead, the impact of wearable communication interfaces to mechanical systems, extensive machines, implants and industrial equipments is still weak and almost unexplored. The concept of wearable interface for mechanical, medical and industrial systems is the natural evolution of the remote controllers which are generally used presently. The miniaturization of the remote electric interface and its integration in cloths or gloves provides some immediate benefits: the comfort of use, the presence of the communication system directly on the body, the continuous interaction between the user and the machine by simple defined gestures and without using external devices as controllers, the wireless interfacing. The embedded power generation promises to increase the potential and performances of wearable interfaces thanks to the autonomy from external supply sources and from periodic recharge of batteries (with related disposal issues and weight problems) and thanks to the possibility to supply compact feedback systems. These advanced, highly integrated systems have the potential to provide compact, energetically autonomous, wearable, low cost and long life solutions for the next generation of interfaces with the machines.

The applicative research reported in this section has potential links with some fields close to the HMI. The power generation from the motion of human body, for instance, has tremendous impacts on the devices for physiological parameters monitoring (self-powered sensors networks) for medicine and telemedicine applications. Similarly, the biomechanical energy harvesting has potential interest in the monitoring of biological parameters in breeding animals for the control of the food chain, and in the geo-referencing of people. Regardless the application, common goals of the research are material properties optimization to build effective biomechanical transducers, the design and development of integrated systems for the energy conversion, the increase of bio-harvesters efficiency and reliability.

This applicative case study demonstrates (De Pasquale et al.) the feasibility of the above mentioned highly integrated, self-powered HMI supported by the latest technologies defines some application fields in the industrial and medical sectors and provides additional contributions to the development of HMI systems for future machines and technology platforms. A prototype of glove-based HMI is designed and manufactured by the author by using new generation materials coming from the latest technologies: flexible and washable electric wires weaved into the fabric, conductive and washable fabrics, and high-flexible piezoelectric plates. The HMI introduced, thanks to the optical communication between a light emitter integrated in the wearable interface and the machine, is able to convert the user's commands to machine logic inputs by means of dedicated software. The glove interface is self-powered through four piezoelectric transducers integrated in the fabric, which are able to convert the kinetic energy associated to the bending of finger joints into electric energy. The design and dimensioning of the integrated EH is based on the experimental tests described in the previous section of this chapter. The optical strategy used for the interfacing requires low power consumption and provides favorable duty cycles, despite other communications strategies could be used (e.g. radio frequency), with variable duty cycles.

The proposed HMI prototype is conceived for those applications where gloves are used continuously: agriculture, vehicles control, robots and manipulator driving, plants management, processes control and many other industrial fields. The HMI can be used to activate a press, to control a garbage truck, to activate the tools of farm machines, to move a crane, to open a motorized window, etc.

The concept of wearable interface can be intended as the migration of a machine portion to the human body to benefit the communication and its immediateness. The interface role is the same of remote controllers that is to send commands to a receiver. Different strategies and components are suitable to this purpose, including radio frequency transmitters, inertial systems, optical signals, etc. The optical communication strategy has been selected for building the HMI prototype due to some peculiarities: the power consumed by light emitters is lower than other common communication solutions; the optical channel works in analog mode avoiding multiple sampling, data management and packaging prior to the transmission; the optical communication is well suited for indoor working environments and for operations in small space, which are typical conditions of industrial environments; the optical communication is not influenced by electromagnetic interferences (differently from RF devices).

Figure 36 shows the functional components of the glove-based HMI, which include the piezoelectric transducers applied to the back of the fingers at the PIP (proximal inter-phalangeal) joints, the electric wires embedded in the fabric connecting the transducers, the aluminum package containing the electronic power management unit, the switch made by conductive fabric for operating the communication, and the optical communication port.

## Towards Powerless Active Motion Capture

Passive optical interfaces are consolidated since long time in HMI glove applications. They give the possibility to conduct simple motion capture operations and gesture interpretation, which are coded to

*Figure 36. Description of the functional components of the glove HMI*

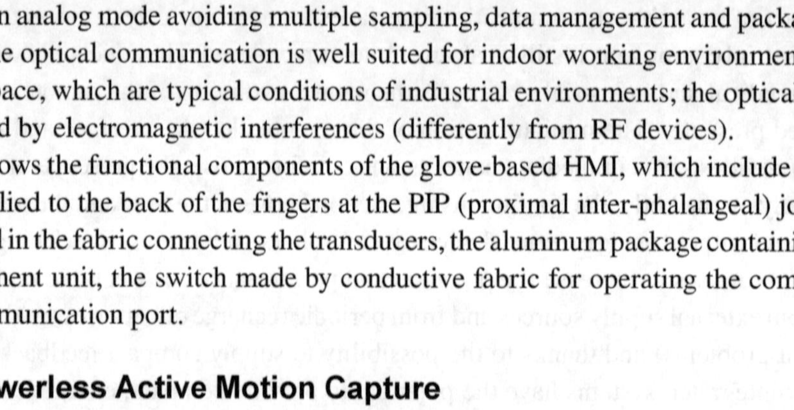

identify specific commands. More precisely, the passive motion capture can be obtained with a number of solutions already used. For instance, in computer animation, biometrics and biodynamic studies, typical passive motion capturing strategies based on optical markers, 3D trackers, colored adhesive pointers, etc.

Instead, the optical communication of the glove HMI proposed in this case study is active, which means that it is able to send coded sequences of inputs to the machine by changing the status of the port (in this case, on/off). This solution reduces drastically the number of gestures to be memorized by the user to control the interface. Also, passive ports are affected by the common problems of wrong gestures interpretation due to sensors accuracy and multi-channels data management; this kind of problem is overcome with the active port of this prototype.

Active interfaces need integrated electronics able to change the status of the communication channel depending to the user's input. Examples of this strategy are the sequential lighting of LEDs or the combination of lightings and movements. The same concept is valid for RF transmitters or inertial systems when used to send information instead of optical ports.

The HMI prototype presented in this paper includes the energy harvesting, the energy management circuit and smart controller of optical port, everything designed to respect the main requirements of wearability, compactness (integration of components) and comfort of use. Also, dedicated software is created to send the optical commands to the machine. The optical input imposed by the user through the HMI glove is detected by a receiver (for instance, a webcam). Then, the software recognizes the optical input and uses the commands for setting in real time the parameters of a microcontroller that controls the machine. The software is virtually invisible for the user, who directly controls the machine without intermediate interfaces like monitors or displays. The movements (or the operations) of the machine are used by the user as direct feedbacks of its original commands, according to the desired immediateness and naturalness of the dialogue.

## Materials, Technologies, and Components

The integration of transducers and electronic components to wearable supports as fabric causes some technological problems. Flexibility of electronic circuits and conductors is one critical requirement, due to the needing of preserving the comfort of the system when it is worn. Other constraints are resistance and compatibility with hostile environments, which require suitable protections (against dust, impacts, etc.), contact with water or other liquids (both in working operations and washing), small mass and good reliability. The integration of the EH to the HMI glove must be carefully considered; in case of piezoelectric transduction, the stiffness of the harvester should not obstacle the normal joints rotation. The mechanical strength of the transducer and its reliability under multiple load cycles must be considered in the design as well.

In the design of the glove HMI, special effort is provided to the integration of piezoelectric transducers, electric conductors and electric controls, as well as to the miniaturization of electronic components that include the power management circuit and the storage batteries. Electric parts are included into a protective aluminum package able to provide protection against mechanical dust, mechanical wear and impacts. The package is extremely compact and small sized (55x46x10 $mm^3$ volume), and the aluminum material provides strength and lightness. It is situated on the back side of the hand, which is an area almost unused during normal manual operations.

The basic material of the glove is technical fabric, which provides impermeability, transpiration, wear resistance and high elasticity. The electrical connections among the four piezoelectric transducers situated on the fingers are fabricated with the technique of the fabric-embedded wires: electrically conductive filaments are inserted in the texture so that they are mechanically and esthetically integrated in the fabric. The flexural deformability of the conductive wire (Fabrickit, electrical resistance = 0.3 Ω/m) is the same of the fabric meaning that the stiffness of the fabric is not increased after the wires application. The electric switch command that activates the light emitter is realized with conductive fabric instead of discrete electric components, which improves the integration of the system. The conductive metalized nylon fabric is applied to the inner part of thumb finger and is used as electric bridge to close the circuit that provides current to the LED optical port. This conductive fabric (Statex, surface resistance < 0.02 Ω, abrasion resistance = 500.000 cycles, temperature range = –30 / +90 °C, thickness = 0.1 mm, specific mass = 77 g/m$^2$) is normally used in radio frequency shields and is made of nylon support covered by three Sn/Cu/Ag layers. The light emitter used as optical interface is the high efficiency LED L-383IDT (Kingbright, size = 9.6x5x2.5 mm$^3$, forward voltage = 1.7 V, forward current = 2.5 mA, wavelength = 627 nm, capacitance = 15 pF).

## CONCLUSION

The design approach for biomechanical EHs is described, by considering some relevant aspects of the biomechanics of human body through simplified modeling and the properties of transduction materials. The applicative context of this technology is associated with uncertainties in excitation conditions, repeatability of operational specifications, and actual electric output values. At this purpose, experimental validation of the results predicted by models is almost always needed during the design process. Then, dedicated test benches are built to reproduce the human limbs motion with repeatable conditions and selectable parameters. Other typologies of experiments are conducted directly on the human body.

The feasibility of energy harvesting from the motion of human body has been demonstrated. The integration of wearable generators to sensors and systems for the human body is proposed in the glove HMI prototype. In the future, specific research efforts are needed, especially focused on the single EH device coupled to its wearable electronics, in order to increase the conversion efficiency and to optimize the duty cycle of the integrated system. In general, the activities described in this chapter demonstrate that it is very important, in the designing and prototyping phases, to consider the complete wearable system. In fact, the right dimensioning can be obtained only by including all the information about human body activities involved (excitation source), transducer typology and material, transducer loading mode, conditioning electronics (energy dissipation), battery storage capacity (energy reservoir), utilizers typology, and system requirements (energy consumption). The integration and miniaturization of components also need additional contributions, as well as the research on materials for increasing the conversion efficiency of transducers. The performances optimization of biomechanical EHs will probably increase the number of application fields and improve data portability and the opportunities to interface people together and with machines in the near future.

## REFERENCES

Ajitasaria, J., Choe, S. Y., Shen, D., & Kim, D. J. (2007). Modeling and analysis of a bimorph piezoelectric cantilever beam for voltage generation. *Smart Materials and Structures*, *16*(2), 447–454. doi:10.1088/0964-1726/16/2/024

Allen, C., Karam, K. Z., Cam, P., Hill, M., & Tindle, T. (1995). Application of virtual reality devices to the quantitative assessment of manual assembly forces in a factory environment. In *Proceedings of IEEE International Conference on Industrial Electronics, Control and Instrumentation*, *2*, 1048-1053. doi:10.1109/IECON.1995.483940

Baird, T. J., Aulbach, A., O'Donohue, K. P., Dalton, J. A., & Gauvin, D. V. (2014). Background variability in standard clinical pathology biomarkers in beagle dogs instrumented with chronic indwelling telemetry devices. *Journal of Pharmacological and Toxicological Methods*, *69*(3), 223–228. doi:10.1016/j.vascn.2014.01.005 PMID:24508734

Barak, A., Hen, L., Boniel-Nissim, M., & Shapira, N. (2008). A comprehensive review and a meta-analysis of the effectiveness of internet-based psychotherapeutic interventions. *Journal of Technology in Human Services*, *26*(2-4), 109–114. doi:10.1080/15228830802094429

Bionic-Power. (n. d.). *Website*. Retrieved from www.bionic-power.com

Blankevoort, L., Huiskes, R., & de Lange, A. (1991). The envelope of passive knee-joint motion. In L. Blankevoort, Passive motion characteristics of the human knee joint: experiments and computer simulations. Katholieke Universiteit te Nijmegen.

Böttner, H., Nurnus, J., Gavrikov, A., Schubert, A., & Schlereth, K. H. (2004). New thermoelectric components using microsystems technologies. *Journal of Microelectromechanical Systems*, *13*(3), 414–420. doi:10.1109/JMEMS.2004.828740

Bryson, S., & Levit, C. (1992). The virtual wind tunnel. *IEEE Computer Graphics and Applications*, *12*(4), 25–34. doi:10.1109/38.144824

Burdea, G. C., Weiss, P. L. T., & Thalmann, D. (2007). Guest editorial special theme on virtual rehabilitation. *IEEE Transactions on Neural Systems and Rehabilitation Engineering*, *15*(1), 1. doi:10.1109/TNSRE.2007.891367

Chaves, A. A., Keller, W. J., O'Sullivan, S., Mixson, L., & Briscoe, R. J. (2006). Cardiovascular monkey telemetry: Sensitivity to detect QT interval prolongation. *Journal of Pharmacological and Toxicological Methods*, *54*(2), 150–158. doi:10.1016/j.vascn.2006.03.004 PMID:16679034

Clanaman, S. (2013). MEMS medical gadgets for Obamacare impress CES visitors. *MEMS Journal*. Retrieved from www.memsjournal.com

De Pasquale, G. (2013). Energy harvesters for powering wireless systems. In D. Uttamchandani (Ed.), *Handbook of MEMS for wireless and mobile applications* (pp. 345–400). Cambridge, UK: Woodhead Publishing. doi:10.1533/9780857098610.2.345

De Pasquale, G., Kim, S.G., & De Pasquale, D. (Under review). GoldFinger: wireless wearable human-machine interface powered by biomechanical energy harvesting with dedicated software. *International Journal of Mechanical Sciences.*

De Pasquale, G., & Somà, A. (2011). *Device for diagnosing railway bogies by applying an energy-autonomous measuring and transmitting bolt, and corresponding control method.* PCT Patent Number WO2011/117718.

De Pasquale, G., & Somà, A. (2013). Energy harvesting from human motion with piezo fibers for the body monitoring by MEMS sensors. In *Proceedings of Symposium on Design, Test, Integration and Packaging of MEMS/MOEMS* (pp. 79-84). Barcelona, Spain.

De Pasquale, G., Somà, A., & Fraccarollo, F. (2012). Piezoelectric energy harvesting for autonomous sensors network on safety-improved railway vehicles. *Proceedings of the Institution of Mechanical Engineers. Part C, Journal of Mechanical Engineering Science, 226*(4), 1107–1117. doi:10.1177/0954406211418158

De Pasquale, G., Somà, A., & Fraccarollo, F. (2013). Comparison between piezoelectric and magnetic strategies for wearable energy harvesting. *Journal of Physics: Conference Series, 476*, 012097. doi:10.1088/1742-6596/476/1/012097

De Pasquale, G., Somà, A., & Zampieri, N. (2012). Design, simulation, and testing of energy harvesters with magnetic suspensions for the generation of electricity from freight trains vibrations. *Journal of Computational and Nonlinear Dynamics, 7*(4), 041011. doi:10.1115/1.4006920

Diftler, M. A., Culbert, C. J., Platt, R., & Bluethmann, W. J. (2003). Evolution of the NASA/DARPA robonaut control system. In *Proceedings of IEEE International Conference on Robotics and Automation, 2*, 2543-2548. doi:10.1109/ROBOT.2003.1241975

Doerr, K. U., Rademacher, H., Huesgen, S., & Kubbat, W. (2007). Evaluation of a low-cost 3D sound system for immersive virtual reality training system. *IEEE Transactions on Visualization and Computer Graphics, 13*(2), 204–212. doi:10.1109/TVCG.2007.37 PMID:17218739

Edmison, J., Jones, M., Nakad, Z., & Martin, T. (2002). Using piezoelectric materials for wearable electronic textiles. In *Proceedings of IEEE International Symposium on Wearable Computers*, (41-48). doi:10.1109/ISWC.2002.1167217

Elvin, N. G., Lajnef, N., & Elvin, A. A. (2006). Feasibility of structural monitoring with vibration powered sensors. *Smart Materials and Structures, 15*(4), 977–986. doi:10.1088/0964-1726/15/4/011

Forni, I., & Cappellini, O. (1974). *Compendio di meccanica articolare.* Urbino, Italy: Agraria Editore.

Goh, J., & Shoji, M. (2011). *Electric power generator for a bicycle.* US Patent Number US 8053912.

Granstrom, J., Feenstra, J., Sodano, H. A., & Farinholt, K. (2007). Energy harvesting from a backpack instrumented with piezoelectric shoulder straps. *Smart Materials and Structures, 16*(5), 1810–1820. doi:10.1088/0964-1726/16/5/036

Guillot, F. M., Beckham, H. W., & Leisen, J. (2007). *Piezoelectric fabrics for energy harvesting.* National Textile Center.

Huson, A. (1974). Biomechanische problem des kniegelenks. *Ortopädie*, *3*, 119–126.

Iba, S., Weghe, J. M. V., Paredis, C. J. J., & Khosla, P. K. (1999). An architecture for gesture-based control of mobile robots. In *Proceedings International Conference On Intelligent Robots and Systems*, (pp. 851-857). doi:10.1109/IROS.1999.812786

International Standard ISO 7250. (n. d.). *Basic human body measurements for technological design. Part 1 (Body measurement definitions and landmarks) and Part 2 (Statistical summaries of body measurements from individual ISO populations).*

Jaana, M., Paré, G., & Sicotte, C. (2009). Home telemonitoring for respiratory conditions: A systematic review. *The American Journal of Managed Care*, *15*, 313–320. PMID:19435399

Jadidian, B., Janas, V., Safari, A., French, J., Weitz, G., Luke, J., & Cass, R. (1996). Development of fine scale piezoelectric ceramic/polymer composites via incorporation of fine PZT fibers. In *IEEE International Symposium on Application of Ferroelectrics*, 1, 31-34. doi:10.1109/ISAF.1996.602705

Keefe, D. F., Feliz, D. A., Moscovich, T., Laidlaw, D. H., & LaViola, J. J. (2001). Cave painting: A fully immersive 3D artistic medium and interactive experience. In *Proceedings of Symposium on Interactive 3D Graphics and Games* (pp. 85-93).

Klimiec, E., Zaraska, W., Zaraska, K., Gąsiorski, K. P., Sadowski, T., & Pajda, M. (2008). Piezoelectric polymer films as power converters for human powered electronics. *Microelectronics and Reliability*, *48*(6), 897–901. doi:10.1016/j.microrel.2008.04.001

Kulah, H., & Najafi, K. (2008). Energy scavenging from low-frequency vibrations by using frequency up-conversion for wireless sensor applications. *IEEE Sensors Journal*, *8*(3), 261–268. doi:10.1109/JSEN.2008.917125

Kymissis, J., Kendall, C., Paradiso, J., & Gershenfeld, N. (1998). Parasitic power harvesting in shoes. In *Proceedings of IEEE International Symposium on Wearable Computers* (pp. 132-139).

Lee, M., Chen, C. Y., Wang, S., Cha, S. N., Park, Y. J., Kim, J. M., & Wang, Z. L. et al. (2012). A hybrid piezoelectric structure for wearable nanogenerators. *Advanced Materials*, *24*(13), 1759–1764. doi:10.1002/adma.201200150 PMID:22396355

Leland, E. S., & Wright, P. K. (2006). Resonance tuning of piezoelectric vibration energy scavenging generators using compressive axial preload. *Smart Materials and Structures*, *15*(5), 1413–1420. doi:10.1088/0964-1726/15/5/030

Leonov, V. (2011). Energy harvesting for self-powered wearable devices. In A. Bonfiglio & D. De Rossi, (Eds.), Wearable monitoring systems. New York: Springer. doi:10.1007/978-1-4419-7384-9_2

Li, Q., Naing, V., & Donelan, J. M. (2009). Development of a biomechanical energy harvester. *Journal of Neuroengineering and Rehabilitation*, *6*. PMID:19549313

Lu, C., & Tsui, C. Y. Ki, & W.H (2011). Vibration energy scavenging system with maximum power tracking for micropower applications. In Proceedings of Very Large Scale Integration Systems (pp. 2109-2119).

Malik, N., & Ray, M. C. (2003). Effective coefficients of piezoelectric fiber-reinforced composites. *AIAA Journal*, *41*(4), 704–710. doi:10.2514/2.2001

Malkawi, A., & Srinivasan, R. (2004). Multimodal human-computer interaction for immersive visualization: Integrating speech-gesture recognitions and augmented reality for indoor environments. In *Proceedings International Conference on Computer Graphics* (pp. 171-176).

McKnight, W. H. (2002). *Energy-harvesting device using electrostrictive polymers.* US Patent Number US 6433465B1. Milkline. Retrieved from http://www.milkline.com

Mitcheson, P. D., Green, T. C., & Yeatman, E. M. (2007). Power processing circuits for electromagnetic, electrostatic and piezoelectric inertial energy scavengers. *Microsystem Technologies-Micro-and Nanosystems-Information Storage and Processing Systems*, *13*, 1629–1635.

Mitcheson, P. D., Yeatman, E. Y., Rao, G. K., Holmes, A. S., & Green, T. C. (2008). Energy harvesting from human and machine motion from wireless electronic devices. In Proceedings of IEEE (pp. 1457-1486).

Mossi, K., Mouhli, M., Smith, B. F., Mane, P. P., & Bryant, R. G. (2006). Shape modeling and validation of stress-biased piezoelectric actuators. *Smart Materials and Structures*, *15*(6), 1785–1793. doi:10.1088/0964-1726/15/6/033

Myung, S. K., McDonnell, D. D., Kazinets, G., Seo, H. G., & Moskowitz, J. M. (2009). Concerns about a meta-analysis of computer smoking cessation programs – Reply. *Archives of Internal Medicine*, *169*, 929–937. doi:10.1001/archinternmed.2009.109 PMID:19468084

Neubeck, L., Redfern, J., Fernandez, R., Briffa, T., Bauman, A., & Freedman, S. B. (2009). Telehealth interventions for the secondary prevention of coronary heart disease: A systematic review. *European Journal of Cardiovascular Prevention and Rehabilitation*, *16*(3), 281–289. doi:10.1097/HJR.0b013e32832a4e7a PMID:19407659

Pirola, V. (2003). *Cinesiologia*. Milano, Italy: Edi.Ermes Editors.

Pop, V., van de Molengraft, J., Schnitzler, F., Penders, J., van Schaijk, R., & Vullers, R. (2008). Power optimization for wireless autonomous transducer solutions. In Proceedings of PowerMEMS (pp. 141-144).

Pozzi, M., & Zhu, M. (2011). *Pizzicato excitation for wearable energy harvesters. Solar & Alternative Energy*. SPIE Newsroom.

Priya, S. (2007). Advances in energy harvesting using low profile piezoelectric transducers. *Journal of Electroceramics*, *19*(1), 167–182. doi:10.1007/s10832-007-9043-4

Ramadan, K. S., Sameoto, D., & Evoy, S. (2014). A review of piezoelectric polymers as functional materials for electromechanical transducers. *Smart Materials and Structures*, *23*(3), 033001. doi:10.1088/0964-1726/23/3/033001

Renaud, M., Sterken, T., Fiorini, P., Puers, R., Baert, K., & van Hoof, C. (2005). Scavenging energy form human body, design of a piezoelectric transducer. In *Proceedings of 13th International Conference on Solid-State Sensors, Actuators and Microsystems* (pp. 784-787). Seoul- Korea.

Riemer, R., & Shapiro, A. (2011). Biomechanical energy harvesting from human motion: Theory, state of the art, design guidelines, and future directions. *Journal of Neuroengineering and Rehabilitation, 8*. PMID:21521509

Rosenberg, R., & Slater, M. (1999). The chording glove: A glove-based text input device. *IEEE Transactions on Systems, Man, and Cybernetics - Part C, 29*(2), 186–191. doi:10.1109/5326.760563

Sakamoto, W. K., Souza, E., & Das-Gupta, D. K. (2011). Electroactive properties of flexible piezoelectric composites. *Materials Research, 4*(3), 201–2014. doi:10.1590/S1516-14392001000300010

Shu, Y. C., & Lien, I. C. (2006). Analysis of power output for piezoelectric energy harvesting systems. *Smart Materials and Structures, 15*(6), 1499–1512. doi:10.1088/0964-1726/15/6/001

Somà, A., De Pasquale, G., & Fraccarollo, F. (2011). *Ski- or trekking-racket equipped with magnetic-inductive harvester device for electric generation.* PCT Patent Number PCT/IT1012/000280.

Stark, I. (2006). Thermal energy harvesting with Thermo Life. In *Proceedings of the International Workshop on Wearable and Implantable Body Sensor Networks.* doi:10.1109/BSN.2006.37

Steffan, R., Schull, U., & Kuhlen, T. (1998). Integration of virtual reality based assembly simulation into CAD/CAM environments. In *Proceedings of Annual Conference of the IEEE Industrial Electronics Society* (pp. 2535-2537). doi:10.1109/IECON.1998.724128

Sturman, D. J., & Zeltzer, D. (1994). A survey of glove-based input. *IEEE Computer Graphics and Applications, 14*(1), 30–39. doi:10.1109/38.250916

Su, J., Vullers, R. J. M., Goedbloed, M., van Andel, Y., Leonov, V., & Wang, Z. (2010). Thermoelectric energy harvester fabricated by Stepper. *Microelectronic Engineering, 87*(5-8), 1242–1244. doi:10.1016/j.mee.2009.11.135

Takahashi, T., & Kishino, F. (1991). Hand gesture coding based on experiments using a hand gesture interface device. *SIGCHI Bullettin, 23*(2), 67–74. doi:10.1145/122488.122499

Tani, B., Maia, R. F., & von Wangenheim, A. (2007). A gesture interface for radiological workstation. In *Proceedings of International Symposium on Computer-Based Medical Systems* (pp. 27-32).

Teng, X. F., Zheng, Y. T., Poon, C. C., & Bonato, P. (2008). Wearable medical systems for p-health. *IEEE Reviews in Biomedical Engineering, 1*, 62–74. doi:10.1109/RBME.2008.2008248 PMID:22274900

Thums, M., Whiting, S. D., Reisser, J. W., McMahon, C. R., & Meekan, M. G. (2013). Tracking sea turtle hatchlings - A pilot study using acoustic telemetry. *Journal of Experimental Marine Biology and Ecology, 440*, 156–163. doi:10.1016/j.jembe.2012.12.006

Tran, K., Polisena, J., Coyle, D., Coyle, K., Kluge, E., Cimon, K., . . . Scott, R. (2008). Home telehealth for chronic disease management (Technology report n. 113). Ottawa, Canada: Canadian Agency for Drugs and Technologies in Health.

Xie, J., Lee, C., & Feng, H. (2010). Design, fabrication, and characterization of CMOS MEMS-Based thermoelectric power generators. *Journal of Microelectromechanical Systems, 19*(2), 317–324. doi:10.1109/JMEMS.2010.2041035

Xie, L., & Du, R. (2012). Harvest human kinetic energy to power portable electronics. *Journal of Mechanical Science and Technology*, *26*(7), 2005–2008. doi:10.1007/s12206-012-0503-7

Yoo, J., Yan, L., Lee, S., Kim, Y., & Yoo, H. J. (2010). A 5.2 mW self-configured wearable body sensor network controller and a 12 μW wirelessly powered sensor for a continuous health monitoring system. *IEEE Journal of Solid-State Circuits*, *45*(1), 178–188. doi:10.1109/JSSC.2009.2034440

## KEY TERMS AND DEFINITIONS

**Artificial Limb:** Machine or test bench reproducing the kinematics and forces of the limbs of human body.

**Biomechanical Energy:** The energy associated to the motion of biological systems, and in particular to the human body.

**Biomechanics:** The study of the mechanics of a living body, especially the forces exerted by muscles on the skeletal structure.

**Human-Machines Interface:** Systems and softwares used to improve the mutual interaction between humans and machines.

**Piezoelectric Fiber Composite Materials:** Materials composed by inactive matrix and active piezoelectric fibers used to convert mechanical energy to electric energy.

**Piezoelectric Transducers:** Materials with piezoelectric properties used to convert mechanical energy to electric energy.

**Remote Monitoring:** The measurement, analysis and evaluation of biological parameters by health care professionals from remote locations.

**Telemedicine:** Discipline for the evaluation, diagnosis and treatment of patients in remote locations by using telecommunication technology.

**Wearable Sensors:** Devices applied to the body which are used to measure physical parameters thanks to their dedicated design.

# Chapter 12
# Nonlinear Dynamics and Design of Aeroelastic Energy Harvesters

**Abdessattar Abdelkefi**
*New Mexico State University, USA*

## ABSTRACT

*The concept of harvesting energy from flow-induced vibrations has received a great deal of attention in the last few years. This technology would help in the replacement of small batteries that require expensive and time consuming maintenance and development of self-powered electronic devices, such as health monitoring sensors, medical implants, data transmitters, wireless sensors, and cameras. In this chapter, a particular focus is paid to the concept of harvesting energy from aeroelastic instabilities, such as flutter in airfoil sections, vortex-induced vibrations in circular cylinders, and galloping in prismatic structures. Nonlinear electroaeroelastic models for these energy harvesters are derived and validated with experimental measurements. It is shown how linear and nonlinear analyses can be used to breach traditional barriers in the design and performance enhancement of these aeroelastic energy harvesters, characterization of their behaviors, and identification of the contribution of different types of nonlinearities.*

## INTRODUCTION

Energy harvesting through conversion of wasted mechanical energy to electrical power has received growing attention in the last few years (Sodano, Park, & Inman, 2004; Beeby, Tudor, & White, 2006; Anton & Sodano, 2007; Cook-Chennault, Thambi, & Sastry, 2008; Tang & Yang, 2012; Abdelkefi, 2012). The ultimate goal in this research field is to power sensors or electronic devices by converting ambient energy. This technology is useful in wireless applications in which external battery charging requires expensive and time consuming maintenance. This technology could also potentially extend the missions of such aircraft systems, such as unmanned air vehicles (UAVs) and micro air vehicles (MAVs) (Magoteaux, Sanders, & Sodano, 2008; Abdelkefi & Ghommem, 2013). These harvesters can be deployed in many

DOI: 10.4018/978-1-4666-8254-2.ch012

locations including urban areas, high wind areas, ventilation outlets, rivers, ducts of buildings, lifting components in aircraft structures, etc. Although different transduction mechanisms can be used to convert wasted mechanical vibrations to electrical energy, piezoelectric transduction has been deemed the most suitable option for small devices because it can be placed in small volumes and used over a wide range of frequencies. In addition, the piezoelectric transduction does not need an external input voltage.

To harvest energy from wasted mechanical energy available in the environment, several concepts have been proposed. The most common concept consists of a unimorph or bimorph piezoelectric cantilever beam subjected to external excitations applied at its base. Resonant motions are obtained and then converted to electrical energy when the excitation frequency matches the natural frequency of the harvester. In their review paper, Anton & Sodano (2007) presented and discussed different approaches to enhance the performance of these harvesters, such as the use of efficient piezoelectric materials, use of distinct mode coupling, and optimization of the power conditioning circuitry. For the same purpose, different other strategies have been introduced including multi-modal systems (Abdelkefi, Najar, Nayfeh, & Ben Ayed, 2011), beam geometry (Goldschmidtboeing & Woias, 2008; Ben Ayed, Abdelkefi, Najar, & Hajj, 2014; Rosa & De Marqui, 2014), magnetic coupling (Karami & Inman, 2011; Tang & Yang, 2012), and bistable configurations (Stanton, McGehee, & Mann, 2009; Karami & Inman, 2011; Masana & Daqaq, 2011).

The above studies focused on harvesting energy from base excitations. However, there have also been interest in the concept of harvesting energy from flow-induced vibrations including flutter of airfoil sections (Bryant & Garcia, 2009; Erturk, Vieira, De Marqui, & Inman, 2010; De Marqui, Erturk, & Inman, 2011; Sousa, de M Anicezio, De Marqui, & Erturk, 2011; Abdelkefi & Nuhait, 2013), vortex-induced vibrations (VIVs) of circular cylinders (Akaydin, Elvin, & Andreopoulos, 2010, 2012; Abdelkefi, Hajj, & Nayfeh, 2012a; Mehmood, Abdelkefi, Hajj, Nayfeh, Akhtar, & Nuhait, 2013; Mackowski & Williamson, 2013; Dai, Abdelkefi, & Wang, 2014), galloping of prismatic structures (Sirohi & Mahadik, 2011; Abdelkefi, Hajj, & Nayfeh, 2013a; Zhao, Tang, & Yang, 2013; Yang, Zhao, & Tang, 2013; Bibo & Daqaq, 2014), and wake galloping of parallel cylinders (Jung & Lee, 2011; Abdelkefi, Scanlon, McDowell, & Hajj, 2013b). In this concept, the piezoelectric energy harvester is placed in a flow field and excited to undergo large limit-cycle oscillation amplitudes that can be transformed to electrical energy. Depending on the operating flow speed, these piezoaeroelastic energy harvesters can be designed and deployed in different locations.

In the case of airfoil- or wing-based piezoaeroelastic energy harvesting, in general, an airfoil section is attached to the end of a piezoelectric cantilever beam. As the flow speed increases, there may be a critical speed at which self-excited motions take place because the structural damping is insufficient to damp out motions due to coupled aerodynamic effects. Several studies have focused on the conversion of aeroelastic vibrations to electrical power. Bryant and Garcia (2009) demonstrated theoretically and experimentally the possibility of harvesting energy from a typical airfoil section. Erturk et al. (2010) focused on the effects of piezoelectric power generation on the linear flutter speed for a wing section with piezoceramic layers attached onto plunge stiffness beams. De Marqui et al. (2011) developed finite element models for piezoelectric cantilevered plates under airflow excitation based on the doublet-lattice methods. These research studies have stressed the linear characteristics of the wing-based energy harvester with a particular focus on the flutter boundary. The obtained oscillations at the linear flutter speed are non-realistic scenario because of the presence of inherent structural and/or aerodynamic nonlinearities in these aeroelastic systems. These nonlinearities can lead to various behaviors (Dowell & Tang, 2002; Gilliat, Strganac, & Kurdila, 2003; Vasconcellos & Abdelkefi, 2014), such as limit cycle oscillations, multiple equilibria, bifurcations, chaos, and various types of resonances (internal and super/subharmonic). These nonlinearities can lead to the presence of supercritical or subcritical Hopf bifurcations, as shown in Figure 1. For the purpose of energy harvesting, it is important to generate energy at low speeds and, as such, to decrease the flutter speed. Furthermore, it is better to attain limit-cycle oscillations at low flow speeds while avoiding subcritical instabilities, which can lead to fatigue failures. Abdelkefi et al. (2012b) considered nonlinear aspects of wing-based energy harvesters and focused on generating energy at low freestream velocities through limit-cycle oscillations and eliminating subcritical Hopf bifurcations. Sousa et al. (2011) presented modeling and experiments of wing-based energy harvesters having combined nonlinearities (freeplay and cubic) in the nonlinear torsional spring. They showed that the existence of the freeplay nonlinearity reduces the cut-in speed by 2 m/s and produces LCOs below the linear flutter speed. They also reported that the response amplitude increases with increasing the airflow speed and decreases with increasing the nonlinear-linear stiffness ratio. Abdelkefi and Hajj (2013) investigated experimentally the effects of controlled freeplay nonlinearity on harvesting energy from a wing-based energy harvesting system. They demonstrated that increasing the freeplay nonlinearity gap results in a decrease in the cut-in speed through a subcritical instability and gives the possibility to harvest energy at low wind speeds.

As for energy harvesting from vortex-induced vibrations, a circular cylinder is attached to the end of a piezoelectric cantilever beam and when the fluid flows past this cylinder, the flow exhibits the so-called Karman vortex street. This street consists of vortices that shed from both sides of the bluff body. The alternate shedding of these vortices causes unsteady aerodynamic forces on the body. When these vortices are shed at a frequency near the natural frequency of the harvester, lock-in or synchronization takes place and resonant transverse vibrations occur (see Figure 2). These vibrations are referred to as vortex-induced vibrations (VIVs). In the past few years, this phenomenon has been considered by several researchers in order to convert aeroelastic vibrations to usable form of electrical energy. Akaydin et al. (2010) placed a flexible unimorph piezoelectric material (polyvinylidene fluoride (PVDF)) in the wake of a cylinder, which was subjected to cross flows. In such way, the PVDF was induced to vibrate by the

*Figure 1. Schematic of limit-cycle oscillations response of a wing-based energy harvester*

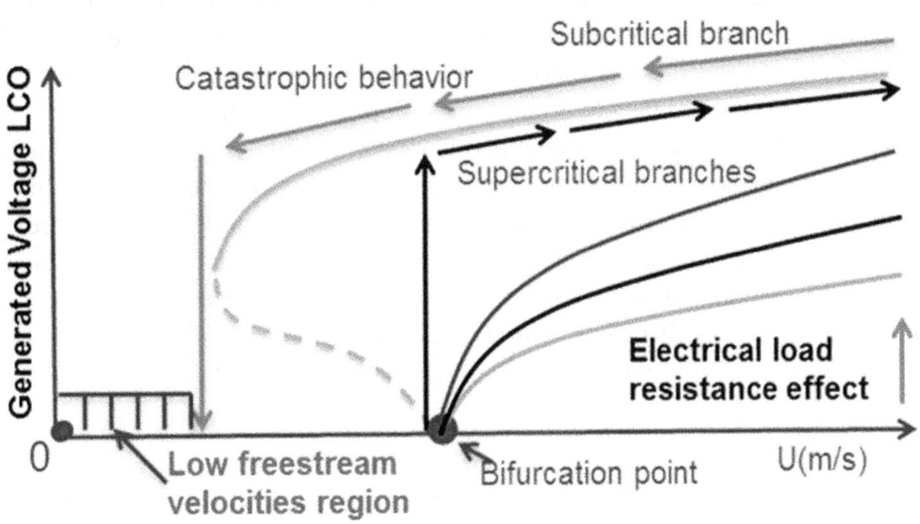

*Figure 2. Schematic of VIV phenomenon in circular cylinder-based energy harvester*

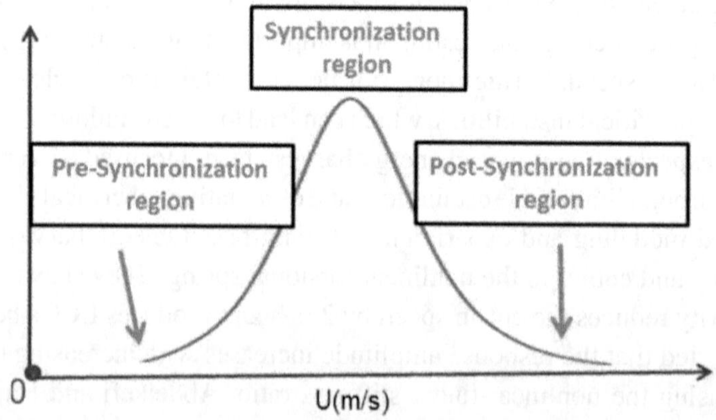


vortex street to harvest energy. Abdelkefi et al. (2012a) and Mehmood et al. (2013) investigated the possibility of harvesting energy from VIV oscillations of elastically-mounted circular cylinders. They used mathematical models that account for the coupled lift force, cylinder's motion, and generated voltage. They investigated the effects of the load resistance and Reynolds number on the synchronization region and the performance of the harvester. Grouthier et al. (2012) discussed the possibility of harvesting energy from VIV of long tensioned cables. Akaydin et al. (2012) performed an experimental research of a self-excited piezoelectric energy harvester subjected to a uniform flow. They reported that 0.1 *mW* can be harvested from VIV oscillations at a wind speed of 1.192 *m/s*.

In the case of harvesting energy from galloping oscillations, a prismatic structure is attached to the free end of a piezoelectric cantilever beam. This approach is known as transverse galloping and it is known by the presence of large-amplitude oscillations. This phenomenon is similar to the flutter of airfoil sections. The difference is that the transverse galloping approach is a one degree-of-freedom system and the flutter of airfoil sections is a two- or three-degree-of-freedom system. This approach has been recently proposed in many studies. Sirohi and Mahadik (2011) proposed harvesting energy from transverse galloping of a structure that has an equilateral triangle section and generates more than 50 *mW* at a wind speed of 11.6 *mph*. The effects of the cross-section geometry and ambient temperature on the onset speed of galloping and the performance of galloping energy harvesters were investigated by Abdelkefi et al. (2013a, 2013c). Yang et al. (2013) and Bibo and Daqaq (2014) experimentally investigated the effects of the cross-section geometry on the performance of galloping energy harvesters.

This chapter presents nonlinear electroaeroelastic modeling and design of these aeroelastic energy harvesters. To determine the accuracy of the developed mathematical models, the obtained numerical results are compared with experimental measurements. Linear and nonlinear analyses are performed to investigate the effects of the system's parameters on the onset of instability and the performance of these harvesters. The nonlinear normal form is derived and used to characterize the type of instability and to determine the importance of different structural or/and aerodynamic nonlinearities in the harvester's response. Finally, qualitative and quantitative comparisons are presented and discussed.

## FLUTTER-BASED AEROELASTIC ENERGY HARVESTER

### Mathematical Modeling

The aeroelastic energy harvester under consideration consists of a rigid airfoil constrained to rotate about the elastic axis (pitch motion) and translate vertically (plunge motion), as shown in Figure 3. The harvester is supported by linear and nonlinear torsional and flexural springs. Due to the presence of a piezoelectric coupling in the plunge degree of freedom and a resistive load in the electrical domain, the well-known governing equations of a two-degree-of-freedom typical section airfoil are modified. Including these modifications, the governing equations of the piezoaeroelastic system are expressed as (Erturk et al. 2010; Abdelkefi, Nayfeh, & Hajj, 2012c)

$$m_T\ddot{h} + m_W x_\alpha b\ddot{\alpha} + c_h\dot{h} + k_h(h)h - \frac{\theta}{l}V = -L \tag{1}$$

$$m_W x_\alpha b\ddot{h} + I_\alpha\ddot{\alpha} + c_\alpha\dot{\alpha} + k_\alpha(\alpha)\alpha = M \tag{2}$$

$$C_p\dot{V} + \frac{V}{R} + \theta\dot{h} = 0 \tag{3}$$

where $h$ is the plunge deflection, $\alpha$ is the pitch angle, $m_T$ is the total mass of the wing with its support structure, $m_W$ is the wing mass alone, $I_\alpha$ is the mass moment of inertia about the elastic axis, $b$ is the half chord length, $l$ is the wing span, $x_\alpha$ is the dimensionless distance between the center of mass and the elastic axis, $c_h$ and $c_\alpha$ are, respectively, the plunge and pitch structural damping coefficients, $L$ and $M$ are the aerodynamic lift and moment about the elastic axis, $R$ is the electrical load resistance, $V$ is the voltage across this load resistance, $C_p$ is the capacitance of the piezoelectric layer, $\theta$ is the electromechanical coupling term. We should mention that the capacitance and electromechanical coupling depend on the

*Figure 3. Schematic of flutter-based energy harvester: (a) three-dimensional and (b) two-dimensional*

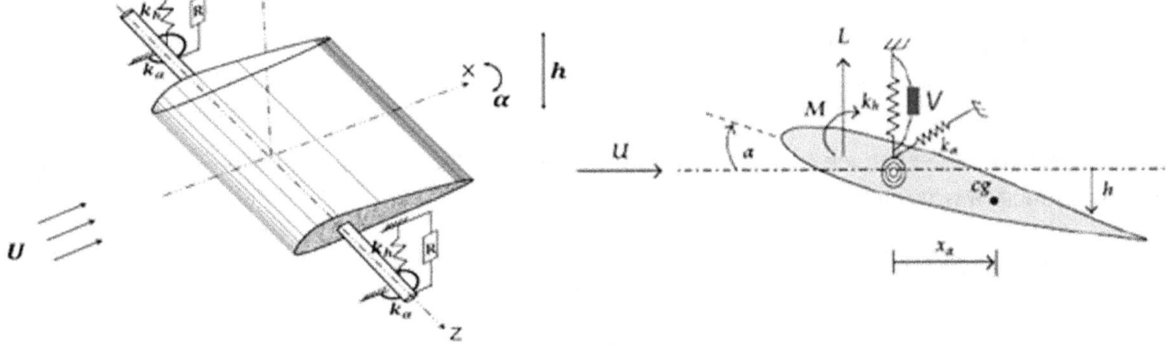

number of piezoelectric layers and their connection (in parallel or in series). $k_h$ and $k_\alpha$ are the structural stiffnesses for the plunge and pitch motions, respectively. These stiffnesses can be experimentally realized by using nonlinear springs (Abdelkefi, Vasconcellos, Nayfeh, & Hajj, 2013d), freeplays (Sousa et al. 2011; Vasconcellos, Abdelkefi, Marques, & Hajj, 2012), or their combination. For simplicity, the stiffnesses of the plunge and pitch motions are, respectively, represented in polynomial forms as

$$k_\alpha(\alpha) = k_{\alpha 0} + k_{\alpha 2}\alpha^2 \tag{4}$$

$$k_h(h) = k_{h0} + k_{h2}h^2 \tag{5}$$

## Aerodynamic Loads Representation

As for the aerodynamic loads $L$ and $M$, different representations can be used which depend on the unsteadiness of the flow, stall effects, and three-dimensional effects. When the aspect ratio, which is defined as the span length over the chord, is high, a two-dimensional approximation of the aerodynamic loads can be assumed. For low-aspect ratio wings, a more accurate representation of the three-dimensional effects and unsteady wake must be used. Abdelkefi et al. (2014a) investigated the concept of harvesting energy from low-aspect ratio wings. They used the three-dimensional unsteady vortex-lattice method (3D UVLM) to model the unsteady aerodynamic loads. Furthermore, they used a strong coupling scheme that is based on the Hamming's fourth-order predictor-corrector method to treat the aerodynamic loads, structure's response and piezoelectric transduction as a single dynamical system.

In this chapter, a two-dimensional airfoil section is considered. In this case, the unsteady aerodynamic loads can be modeled using the Theodorsen's approach (Theodorsen, 1935). Hence, the lift and moment on a wing section are given by

$$L = \pi\rho b^2\left[\ddot{h} + U\dot{\alpha} - ba\ddot{\alpha}\right] + 2\pi\rho UbQC \tag{6}$$

$$M_\alpha = \pi\rho b^2\left[ba\ddot{h} - Ub\left(\frac{1}{2} - a\right)\dot{\alpha} - b^2\left(\frac{1}{8} + a^2\right)\ddot{\alpha}\right] + 2\pi\rho b^2 U(a + \frac{1}{2})QC \tag{7}$$

where $U$ is the freesteram velocity, $C$ is the Theodorsen function, and

$$Q = U\alpha + \dot{h} + \dot{\alpha}b\left(\frac{1}{2} - a\right) \tag{8}$$

is the effective angle of attack multiplied by the freestream velocity. In the frequency domain, the Theodorsen function is written as

$$C(k) = F(k) + iG(k) \tag{9}$$

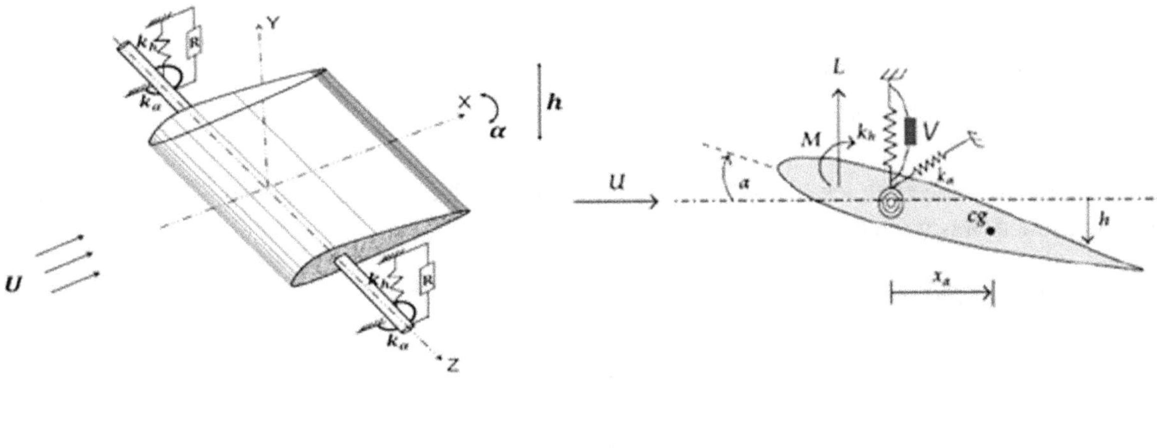

where $k=\omega b/U$ is the reduced frequency of harmonic oscillations, $\omega$ is the oscillation frequency, and $F$ and $G$ are functions of the Hankel and modified Bessel functions. Two different representations of the aerodynamic loads were used, namely, the quasi-steady approximation and the unsteady representation based on the Duhamel formulation. All details about the derivation of the aerodynamic loads based on the unsteady formulation are presented in (Trickey, 2000; Abdelkefi et al., 2013d). In this chapter, the quasi-steady approximation is used to model the lift and moment. In this representation, the Theodorsen function $C$ is set equal to unity. This assumption neglects any lag between the unsteady oscillations and its effect on the aerodynamic loads, thereby limiting the modeling to small reduced frequency values. Furthermore, the non-circulatory effects are assumed negligible compared to the circulatory effects. In addition, because high amplitudes in the angle of attack can be obtained, a quasi-steady representation with a stall model is considered. Consequently, the aerodynamic loads can be expressed as

$$L = 2\pi\rho U^2 b(\alpha_{eff} - c_s\alpha_{eff}^3) \tag{10}$$

$$M = 2\pi\left(a + \frac{1}{2}\right)\rho U^2 b^2(\alpha_{eff} - c_s\alpha_{eff}^3) \tag{11}$$

where $c_s$ is a nonlinear parameter associated with stall and $\alpha_{eff}=Q/U$.

## Normalized Problem

The following state variables are introduced

$$\mathbf{X} = \begin{bmatrix} X_1 \\ X_2 \\ X_3 \\ X_4 \\ X_5 \end{bmatrix} = \begin{bmatrix} h \\ \dot{h} \\ \alpha \\ \dot{\alpha} \\ V \end{bmatrix} \tag{12}$$

and the equations of motion can be rewritten as

$$\begin{aligned}
&\dot{X}_1 = X_2 \\
&\dot{X}_2 = -\frac{I_\alpha k_{h0}}{d}X_1 - (c_1 + d_1U)X_2 - (k_1U^2 - \frac{m_W x_\alpha b k_{\alpha 0}}{d})X_3 - (c_2 + d_2U)X_4 \\
&-\theta_1 X_5 - N_{1\alpha}(X) - N_{1h}(X) + N_{NL1} \\
&\dot{X}_3 = X_4 \\
&\dot{X}_4 = \frac{m_W x_\alpha b k_{h0}}{d}X_1 - (c_3 + d_3U)X_2 - (k_2U^2 + \frac{k_{\alpha 0}m_T}{d})X_3 - (c_4 + d_4U)X_4 \\
&-\theta_2 X_5 - N_{2\alpha}(X) - N_{2h}(X) + N_{NL2} \\
&\dot{X}_5 = -\frac{\theta}{C_p}X_2 - \frac{1}{RC_p}X_5
\end{aligned} \tag{13}$$

where

$$
\begin{aligned}
d &= m_T I_\alpha - (m_w x_\alpha b)^2 \\
c_1 &= [I_\alpha c_h] / d \\
d_1 &= [I_\alpha \rho b c_{l\alpha} + \rho b^3 m_w x_\alpha c_{m\alpha}] / d \\
c_2 &= [-m_w x_\alpha b c_\alpha] / d \\
d_2 &= [I_\alpha \rho b^2 c_{l\alpha} (\frac{1}{2} - a) + m_w x_\alpha b^4 \rho c_{m\alpha} (\frac{1}{2} - a)] / d \\
c_3 &= [-m_w x_\alpha b c_h] / d \\
d_3 &= [-m_w x_\alpha b \rho b c_{l\alpha} - m_T c_{m\alpha} \rho b^2] / d \\
c_4 &= [m_T c_\alpha] / d \\
d_4 &= [-m_T b^3 \rho c_{m\alpha} (\frac{1}{2} - a) - m_w x_\alpha b^3 \rho c_{l\alpha} (\frac{1}{2} - a)] / d \\
k_1 &= [I_\alpha \rho b c_{l\alpha} + m_w x_\alpha b^3 \rho c_{m\alpha}] / d \\
k_2 &= -[\rho b^2 c_{l\alpha} m_w x_\alpha + m_T \rho b^2 c_{m\alpha}] / d \\
N_{1h} &= I_\alpha [k_{h2} X_1^3] / d \\
N_{1\alpha} &= -m_w x_\alpha b [k_{\alpha 2} X_3^3] / d \\
N_{2h} &= -m_w x_\alpha b [k_{h2} X_1^3] / d \\
N_{2\alpha} &= m_T [k_{\alpha 2} X_3^3] / d \\
N_{NL1} &= [\rho U^2 c_s b (c_{l\alpha} I_\alpha + m_w x_\alpha b^2 c_{m\alpha}) \alpha_{eff}(X)] / d \\
N_{NL2} &= -[c_s \rho U^2 b^2 (c_{l\alpha} m_w x_\alpha + m_T c_{m\alpha}) \alpha_{eff}(X)] / d
\end{aligned} \tag{14}
$$

In vector form, these equations of motion can be expressed as

$$\dot{\mathbf{X}} = B(U)\mathbf{X} + \mathbf{C}(\mathbf{X}, \mathbf{X}, \mathbf{X})$$

where **C(X,X,X)** is a cubic vector function of the state variables and

$$
B(U) = \begin{bmatrix}
0 & 1 & 0 & 0 & 0 \\
-\frac{I_\alpha k_{h0}}{d} & -(c_1 + d_1 U) & -(k_1 U^2 - \frac{m_W x_\alpha b k_{\alpha 0}}{d}) & -(c_2 + d_2 U) & -\theta_1 \\
0 & 0 & 0 & 1 & 0 \\
\frac{m_W x_\alpha b k_{h0}}{d} & -(c_3 + d_3 U) & -(k_2 U^2 + \frac{k_{\alpha 0} m_T}{d}) & -(c_4 + d_4 U) & -\theta_2 \\
0 & -\frac{\theta}{C_p} & 0 & 0 & -\frac{1}{R C_p}
\end{bmatrix}
$$

## Linear Analysis: Flutter Speed vs System's Parameters

The effects of the system's parameters on the linear flutter speed are determined through a linear analysis in which the variations of the real and imaginary parts of the eigenvalues of the matrix *B(U)* as function of the system's parameters are investigated. *B(U)* has a set of five eigenvalues. These eigenvalues are arranged so that the first four are complex conjugates. The fifth eigenvalue which is a result of the electromechanical coupling is always real and negative. The real parts of these eigenvalues represent the damping coefficients and the positive imaginary parts correspond to the coupled frequencies of the aeroelastic energy harvester. The speed, $U_f$, for which one or more eigenvalues have zero real parts, corresponds to the onset of the instability and is termed the linear flutter speed.

The design of these flutter-based energy harvesters is the most important step in order to harvest energy at different operating speeds. The plotted curves in Figures 4, 5, and 6 show the variations of the linear flutter speeds as functions of the linear torsional spring, eccentricity, and load resistance. Clearly, the choice of the linear spring coefficients as well as the eccentricity plays an important role in determining the flutter speed. In fact, a decrease in the linear flexural spring is accompanied by a decrease in the flutter speed. Furthermore, there is an optimum value for the linear torsional spring at which the flutter speed is a minimum, as shown in Figure 4.

It follows from Figure 5 that the eccentricity can be optimized to minimize the flutter speed. Furthermore, when the eccentricity approaches zero, the flutter speed tends to infinity and the system is stable for any wind speed. Unlike the significant effects of the linear spring coefficients and the eccentricity on the flutter speed, the load resistance has a negligible on the flutter speed, as shown in Figure 6.

*Figure 4. Variations of the linear flutter speed as a function of the linear torsional spring for different values of the linear flexural spring*

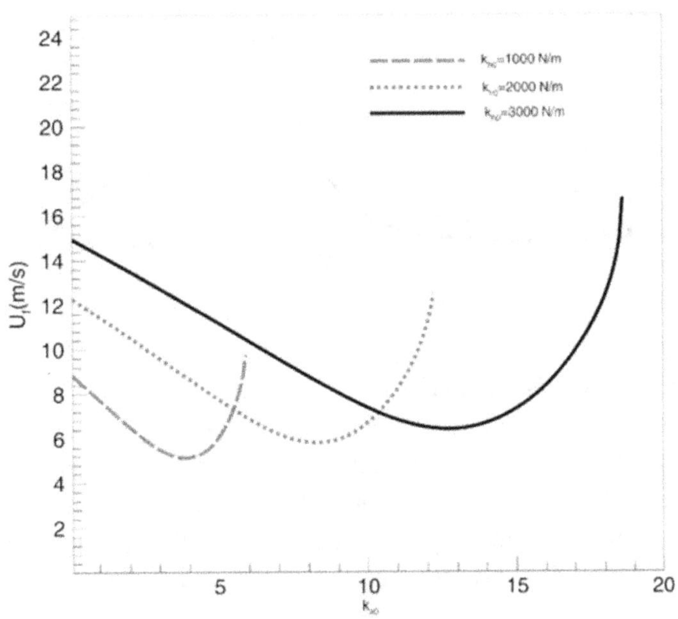

*Figure 5. Variation of the linear flutter speed as a function of the eccentricity*

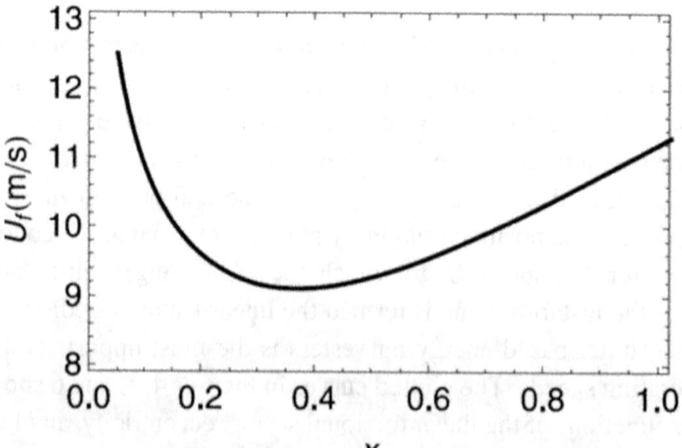


*Figure 6. Variation of the linear flutter speed as a function of the electrical load resistance*

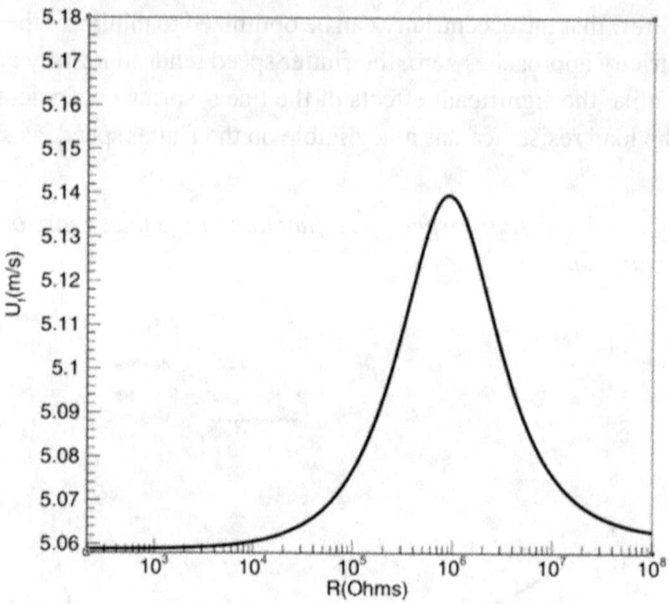


## Normal Form of the Hopf Bifurcation

To characterize the type of instability and determine the importance of different present nonlinearities on the level of the harvested power, the normal form of the Hopf bifurcation near the flutter speed $U_f$ is derived. To this end, a perturbation term $\sigma_U U_f$ is added to the flutter speed and hence express the wind speed as $U= U_f + \sigma_U U_f$. Then, the matrix $B(U)$ can be written as $B(U)=B(U_f)+ \sigma_U B_1(U_f)$ where

$$B_1(U_f) = \begin{bmatrix} 0 & 0 & 0 & 0 & 0 \\ 0 & -d_1U_f & -2k_1U_f^2 & -d_2U_f & 0 \\ 0 & 0 & 0 & 0 & 0 \\ 0 & -d_3U_f & -2k_2U_f^2 & -d_4U_f & 0 \\ 0 & 0 & 0 & 0 & 0 \end{bmatrix}$$

Near bifurcation, the equations of motion can be rewritten as

$$\dot{\mathbf{X}} = B(U_f)\mathbf{X} + \sigma_U B_1(U_f)\mathbf{X} + \mathbf{C}(\mathbf{X},\mathbf{X},\mathbf{X}) \tag{15}$$

where

$$\mathbf{C}^T = [0, C_2, 0, C_4, 0]$$

Letting $G$ be the matrix whose columns are the eigenvectors of the matrix $B(U_f)$ corresponding to the eigenvalues $j\omega_1$-$\mu_1$, - $j\omega_1$-$\mu_1$, $j\omega_2$, - $j\omega_2$, and $-\mu_5$. Then, defining a new vector $\mathbf{Y}$ such that $\mathbf{X} = G\,\mathbf{Y}$, equation (15) can be rewritten as

$$G\dot{\mathbf{Y}} = B(U_f)G\mathbf{Y} + \sigma_U B_1(U_f)G\mathbf{Y} + \mathbf{C}(G\mathbf{Y}, G\mathbf{Y}, G\mathbf{Y}) \tag{16}$$

Multiplying (16) from the left by the inverse of $G^{-1}$of $G$ yields

$$\dot{\mathbf{Y}} = J\mathbf{Y} + \sigma_U K\mathbf{Y} + G^{-1}\mathbf{C}(G\mathbf{Y}, G\mathbf{Y}, G\mathbf{Y}) \tag{17}$$

where $J= G^{-1}B(U_f)\, G$ is a diagonal matrix whose elements are the eigenvalues $j\omega_1$-$\mu_1$, - $j\omega_1$-$\mu_1$, $j\omega_2$, - $j\omega_2$, and $-\mu_5$ and $K= G^{-1}B_1(U_f)\, G$. It is noted that $Y_2$ and $Y_4$ are the complex conjugates of $Y_1$ and $Y_3$, respectively. Consequently, equation (17) can be written in a component form as

$$\dot{Y}_1 = j\omega_1 Y_1 - \mu_1 Y_1 + \sigma_U \sum_1^5 K_{1i} Y_i + N_1(\mathbf{Y}) \tag{18}$$

$$\dot{Y}_3 = j\omega_2 Y_3 + \sigma_U \sum_1^5 K_{3i} Y_i + N_3(\mathbf{Y}) \tag{19}$$

$$\dot{Y}_5 = -\mu_5 Y_5 + \sigma_U \sum_1^5 K_{5i} Y_i + N_5(\mathbf{Y}) \tag{20}$$

where the $N_i(\mathbf{Y})$ are trilinear functions of the components of **Y.**

According to the Center-manifold theorem, there exists a center manifold (Nayfeh, 2011)

$$Y_1 = H_1(Y_3,\bar{Y}_3),\ \bar{Y}_1 = \bar{H}_1(Y_3,\bar{Y}_3),\ Y_5 = H_5(Y_3,\bar{Y}_3)$$

Moreover, the dynamics of the system are similar to those of the center manifold; that is,

$$\dot{Y}_3 = j\omega_2 Y_3 + \sigma_U \left(K_{31}H_1 + K_{32}\bar{H}_1 + K_{33}Y_3 + K_{34}\bar{Y}_3 + K_{35}H_5\right) + N_3(H_1,\bar{H}_1,Y_3,\bar{Y}_3,H_5) \tag{21}$$

Because $\sigma_U$ is small and $N_1$ and $N_5$ are cubic functions of the components of **Y**, $H_1$ and $H_5$ are zero to the third approximation. Therefore, keeping only the resonance terms in equation (21), the complex-valued nonlinear normal form is then obtained as

$$\dot{Y}_3 = j\omega_2 Y_3 + \sigma_U K_{33} Y_3 + \alpha_e Y_3^2 \bar{Y}_3 \tag{22}$$

where $\alpha_e$ is the effective nonlinearity which depends on the cubic nonlinear spring coefficients and the stall parameter. Table 1 presents the real and imaginary part of the effective nonlinearity for two different configurations when the stall parameter and electrical load resistance are set equal to 10.833 and $10^6\Omega$, respectively.

Next, $Y_3$ is expressed in the polar form

$$Y_3 = \frac{1}{2} a e^{j(\omega_2 t + \gamma)} \tag{23}$$

where $a$ is the amplitude of oscillations and $\gamma$ is its phase. Substituting equation (23) into equation (22) and separating the real and imaginary parts, one obtains the following real-valued normal form:

$$\dot{a} = \beta_r a + \frac{1}{4}\alpha_{er} a^3 \tag{24}$$

*Table 1. Parameters of the two considered configurations when the stall parameter and electrical load resistance are set equal to 10.833 and $10^6\Omega$, respectively*

| | First Configuration | Second Configuration |
|---|---|---|
| $k_{\alpha 0}$ | 4 | 6.833 |
| $k_{h0}$ | 1000 | 2844.4 |
| $U_f$ | 5.13 | 9.18 |
| $\beta_r$ | 0.79 $\sigma_U$ | 3.93 $\sigma_U$ |
| $\beta_i$ | 0.83 $\sigma_U$ | 3.02 $\sigma_U$ |
| $\alpha_{er}$ | $-2.11\ 10^{-5}-1.17\ 10^{-13}\ k_{h2}-1.82\ 10^{-7}\ k_{\alpha 2}$ | $-4.61\ 10^{-5}-3.52\ 10^{-11}\ k_{h2}+9.04\ 10^{-7}\ k_{\alpha 2}$ |
| $\alpha_{ei}$ | $-3\ 10^{-5}+4.06\ 10^{-11}\ k_{h2}+5.30\ 10^{-6}\ k_{\alpha 2}$ | $-4\ 10^{-5}+1.25\ 10^{-11}\ k_{h2}+1.55\ 10^{-6}\ k_{\alpha 2}$ |

$$\dot{\gamma} = \beta_i + \frac{1}{4}\alpha_{ei}a^2 \tag{25}$$

where $\beta = \sigma_U K_{33}$ and the subscripts $r$ and $i$ denote, respectively, the real and imaginary parts.

The equilibrium solutions of equation (24) are:

$$a = 0 \text{ and } a = \pm\sqrt{\frac{-4\beta_r}{\alpha_{er}}} \tag{26}$$

where $a$=0 is the trivial solution. The other solutions are nontrivial. The origin is asymptotically stable for $\beta_r < 0$ or $\beta_r = 0$ and $\alpha_{er}$ <0, unstable for $\beta_r > 0$ or $\beta_r = 0$ and $\alpha_{er}$ >0. The nontrivial solutions exist when $\beta_r\alpha_{er}$ <0. They are stable (supercritical Hopf bifurcation) for $\beta_r > 0$ and $\alpha_{er}$ <0 and unstable (subcritical Hopf bifurcation) for $\beta_r < 0$ and $\alpha_{er}$ >0.

To check the accuracy of the analytical solution derived by the nonlinear normal form analysis, the nonlinear normal form amplitudes of the plunge motion are compared with their counterparts from numerical integration of the equations of motion when considering the values of the second configuration as presented in Table 1 when $k_{\alpha 2}$=0 *N.m* and $k_{h2}$=67.685 *N/m*. Clearly, the real part of the effective nonlinearity is negative in this case. Hence, the system's has a supercritical instability. It follows from Figure 7 that there is a very good agreement between the amplitudes of LCO near the bifurcation. On the other hand, it is noted that there is a relatively large disagreement of the outputs as the wind speed is increased further beyond the flutter speed. This is expected because the normal form of the Hopf bifurcation is applicable only near the onset of the bifurcation.

*Figure 7. Comparison between of the plunge amplitudes when considering analytical prediction using the nonlinear normal form and numerical solution*

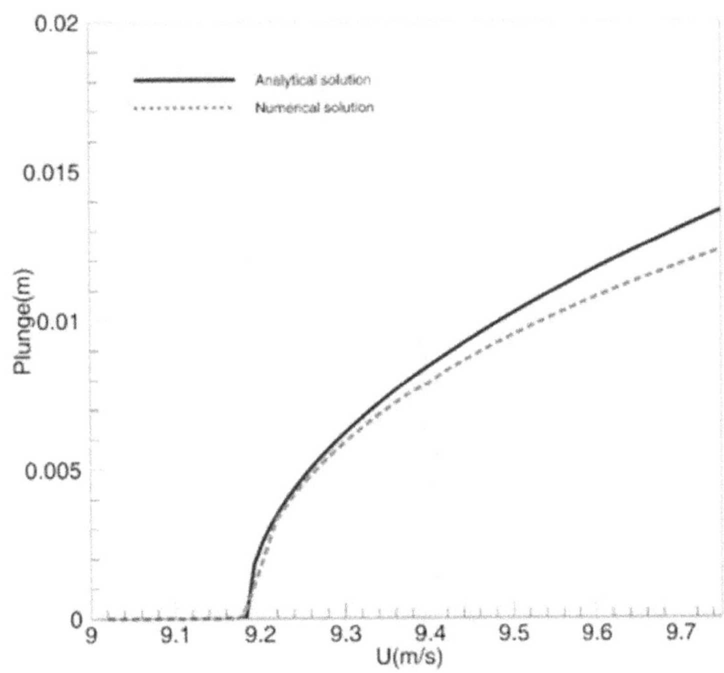

## Nonlinear Analysis: System's Performance

Inspecting the expressions of the real part of the effective nonlinearity in Table 1, it is noted that it depends on the nonlinear torsional and flexural spring coefficients. As can be deduced from Table 1, the sensitivity of the system's outputs to variations in the nonlinear torsional spring coefficient is more important than its counterpart in the nonlinear flexural spring coefficient. Clearly, the nonlinear normal form is very beneficial in designing the nonlinear spring coefficients in order to obtain a supercritical Hopf bifurcation and to maximize the level of the harvested power. In fact, these coefficients can be designed in such as way the real part of the effective nonlinearity is always negative and its absolute value is minimized and hence the generated voltage is maximized. From Table 1, it is clear that the harvester of the first configuration has always a supercritical instability when considering hard spring values ($k_{\alpha 2}>0$ and $k_{h2}>0$). However, there is a critical value of the nonlinear torsional spring for which the harvester of the second configuration changes instability from supercritical to subcritical one.

As for the second configuration, the plotted curves in Figure 8 show the increasing-decreasing bifurcation diagrams of the plunge motion when $k_{\alpha 2}=60$ *N.m* and $k_{\alpha 2}=100$ *N.m.* Clearly, a hysteresis region takes place for these two cases. This is a characteristic of the subcritical instability. This result is predicted by the nonlinear normal form. In fact, when considering these two nonlinear torsional spring values, the real part of the effective nonlinearity is positive, indicating the subcritical Hopf bifurcation.

Figures 9(a) and (b) show the variations of the harvested power and pitch angle as function of the eccentricity for two different values of the nonlinear torsional spring and when the freestream velocity is set equal to *5m/s*. It should be mentioned that the used parameters are given in (Abdelkefi, Nayfeh, & Hajj, 2012d). It follows from these plots that an increase in the eccentricity is followed by a decrease in

*Figure 8. Increasing-decreasing plunge amplitude bifurcation diagrams of the second configurationwhen* $R=10^6\Omega$

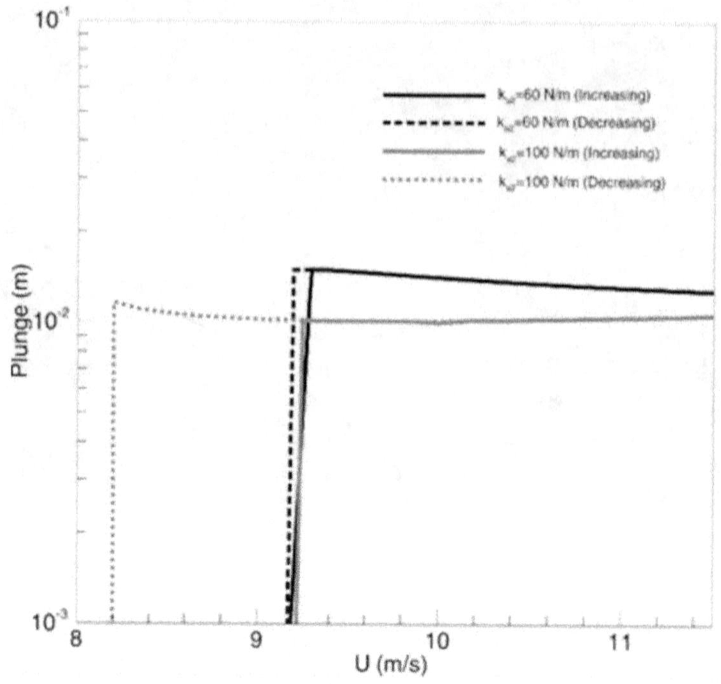

*Figure 9. Variations of the (a) harvested power and (b) pitch amplitude with the eccentricity for different values of $k_{\alpha 2}$ when U =5 m/s and R = $10^6\Omega$*

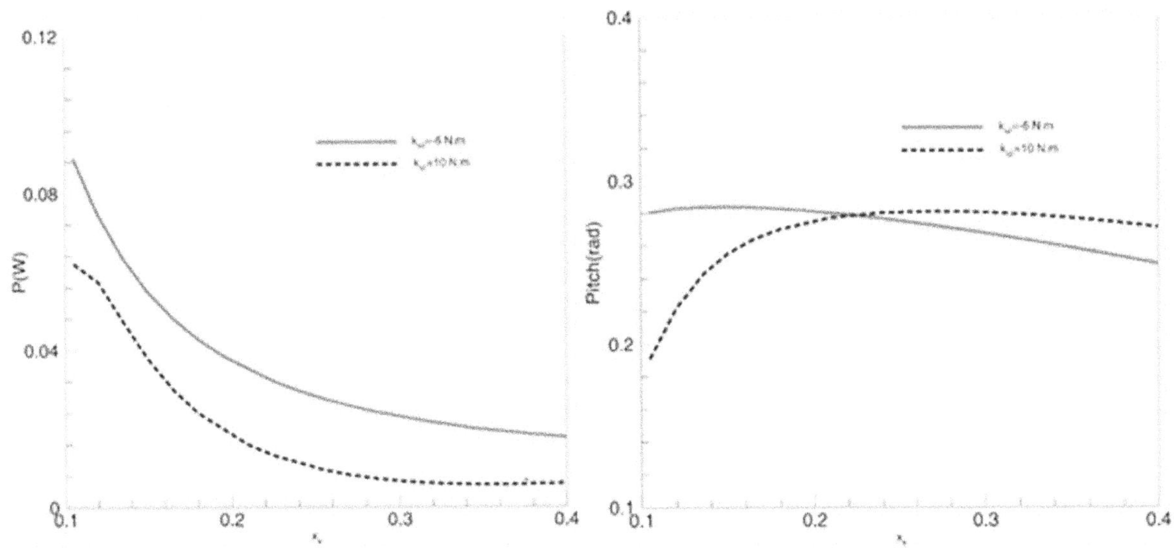


the harvested power which is directly related to the plunge amplitude (Abdelkefi et al., 2012d). The results also show that the harvested power is the highest for small eccentricity for which the pitch amplitude is small. Clearly, a proper choice of the eccentricity can result in a significant increase in the level of the harvested power and a significant decrease in the pitch amplitude.

Figure 10 shows the variations of the harvested power as a function of the electrical load resistance for different values of the eccentricity when *U=5m/s*. It is noted that, for every eccentricity, there is an optimum value for the load resistance that maximizes the harvested power.

In conclusion, linear and nonlinear analyses have been performed to investigate the effects of the system's parameters of a flutter-based energy harvester on its linear flutter speed and performance. It was demonstrated that the linear spring coefficients as well as the eccentricity can be optimized to harvest energy at low wind speeds. It was also shown that the normal form of the Hopf bifurcation is beneficial to design enhanced flutter-based energy harvesters. Moreover, the results showed that a decrease in the eccentricity is accompanied by a significant increase in the level of the harvested power. However, it impacts the speed at which energy can be harvested. It is also concluded that the harvested power can be increased by properly choosing the electrical load resistance.

## VIV-BASED AEROELASTIC ENERGY HARVESTER

### Nonlinear Electromechanical Modeling

The problem of harvesting energy from vortex-induced vibrations of a circular cylinder is considered. As shown in Figure 11, the harvester is composed of a bimorph piezoelectric cantilever beam with a circular cylinder attached to its free end. The red wireframe indicates the placement of the piezoelectric sheets on both sides of the cantilever beam. The piezoelectric sheets are bounded by two-in plane electrodes

*Figure 10. Variation of the harvested power as a function of the load resistance for different values of $x_\alpha$ when $U$ =5 m/s and $k_{\alpha 2}$=-6 N.m*

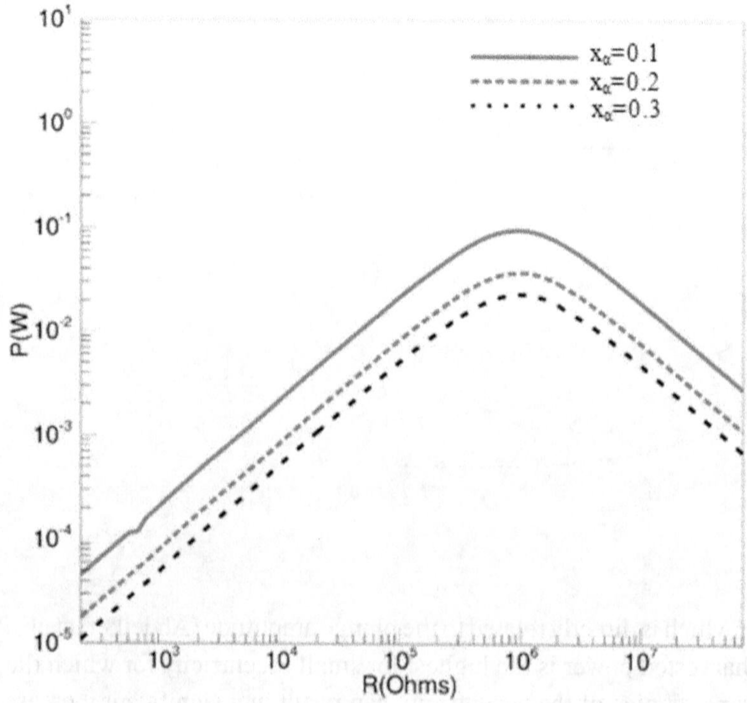


*Figure 11. Schematic of the VIV-based energy harvester*

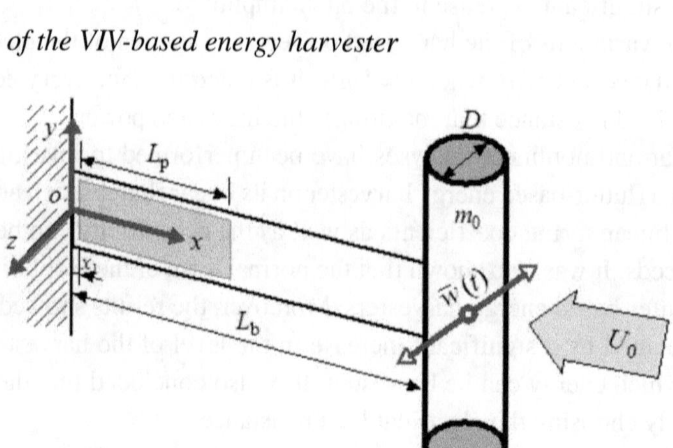


of negligible thickness connected in series to an electrical load resistance. Different electromechanical models can be used to represent the system's response. In this chapter, a nonlinear distributed-parameter model is developed based on the Euler-Lagrange principle to determine the electromechanical governing equations of the VIV-based energy harvester. Our objective is to validate our numerical results with the experimental measurements of Akaydin et al. (2012).

A reduced–order model of the VIV-based energy harvester is developed by applying the Galerkin procedure to the Lagrangian and virtual work. Following Dai et al. (2014), the equations of motion of the harvester can be written as follows

$$\ddot{r}_i + 2\varsigma_i\omega_i\dot{r}_i + \left(\frac{C_D\rho_0 DU_0L_0}{2}\left[\varphi_i\left(L_b\right) + D/2\,\varphi_i'\left(L_b\right)\right]\left[\sum_{j=1}^{n}\left(\varphi_j\left(L_b\right) + D/2\,\varphi_j'\left(L_b\right)\right)\dot{r}_j\right]\right) + \omega_i^2 r_i - \theta_i V = \alpha_i q$$

$i=1,2,3\ldots n$ (27)

$$C_p\dot{V} + \frac{V}{R} + \sum_{i=1}^{n}\theta_i\dot{r}_i = 0 \tag{28}$$

where

$$\alpha_i = C_{L0}\rho_0 DU_0^2L_0\left[\varphi_i\left(L_b\right) + D/2\,\varphi_i'\left(L_b\right)\right]/4 \tag{29}$$

$$C_p = \frac{\varepsilon_{33}w_pL_p}{2t_p} \tag{30}$$

$$\theta_i = \frac{e_{31}w_p\left(\mathrm{t_p} + \mathrm{t_b}\right)}{2}\left(\varphi_i'(x_1 + L_p) - \varphi_i'(x_1)\right) \tag{31}$$

where $n$ denotes the number of modes used in the Galerkin discretization, $r_i(t)$ are the modal coordinates of the displacement, and $\varphi_i(x)$ are the mode shapes of a cantilever beam with a tip mass, which can be obtained by satisfying the boundary conditions. $C_{L0}$ and $C_D$ are the steady lift and mean sectional drag coefficients which can be, respectively, considered equal to 0.3 and 1.2 in the region of well-developed wakes (Facchinetti, de Langre, & Biolley, 2004). $\rho_0$ is the density of fluid and $L_0$ is the length of the cylinder. $e_{31}$ is the piezoelectric stress coefficient, $R$ is the electrical load resistance, and $C_p$ is the capacitance of the harvester.

## Fluctuating Lift Force Representation

One of the challenges in these types of problems is the determination of the fluctuating lift force representation. Several researchers (Schulz & Kalilinderis, 1998; Yang, Preidikman, & Balaras, 2008; Akhtar, 2009; Mehmood et al., 2012, 2013) used direct numerical simulations (DNS) to solve Navier-Stokes and determine the lift force. In these numerical simulations, one cannot solve for the motion of the cylinder and the generated voltage without determining the fluid loads. Yet, one cannot determine the fluid loads without the knowledge of the motion of the cylinder and the effects of energy harvesting on this motion. This complication necessitates that the fluid force equation which govern the dynamics of the flow field, the cylinder's response equation and the voltage equation, which govern the dynamics of the cylinder and generated voltage, be solved in a coupled manner. To overcome this difficulty, the Hamming fourth-order predictor–corrector technique (Carnahan Luther, & Wilkes, 1969; Nuhait & Mook, 2010; Ghommem, Abdelkefi, Nuhait, & Hajj, 2012; Mehmood et al., 2012) was used. In this technique, the fluid loads (output of the CFD code) are coupled to ODEs governing the cylinder's motion and voltage output. The predicted state of the cylinder, which depends on the load resistance, is then used in the CFD code to compute the new fluid loads. These loads are then used to compute the new states of the cylinder and voltage using the corrector scheme. More details about this methodology can be found in Mehmood et al. (2013).

In this chapter, a phenomenological representation is used to model the fluctuating lift force. The widely employed approach or representation is the so-called wake oscillator model, which was first put forward by Birkhoff and Zarantanello (1957). The ensuing researches of using a nonlinear oscillator equation to model the lift are Hartlen and Currie (1970) and Skop and Griffin (1975). They used, respectively, Rayleigh and van der Pol oscillators to model the lift force and coupled this oscillator with the cylinder's motion by a velocity term. Subsequently, the excellent insights into the workings of wake oscillator model were contributed by Skop and Balasubramanian (1997) and Facchinetti et al. (2004). Skop and Balasubramanian (1997) recommended a stall term in the description of the cross-flow force whereas Facchinetti et al. (2004) systematically investigated the pattern of the force induced by the cylinder's motion on the wake oscillator. Facchinetti et al. (2004) found that the acceleration coupling representation is better to model the lock-in region, which was confirmed by Violette et al. (2007) through comparison with numerical simulations and experiments.

It should be stated that the wake oscillator employed in the present study is based on the model of Facchinetti et al. (2004), that is

$$\ddot{q} + \lambda\omega_s\left(q^2 - 1\right)\dot{q} + \omega_s^2 q = \frac{A}{D}\frac{\mathrm{d}^2\overline{w}(t)}{\mathrm{d}t^2} = \frac{A}{D}\left[\left.\frac{\partial^2 w(x,t)}{\partial t^2}\right|_{x=L_b} + \frac{D}{2}\left.\frac{\partial^3 w(x,t)}{\partial x\partial t^2}\right|_{x=L_b}\right] \tag{32}$$

where $q(t)$ describes the behavior of the near wake and stands for the lift acting on the cylinder. $\overline{w}(t)$ is the displacement of the cylinder and $w(x,t)$ is the displacement of the cantilever beam. $D$ is the diameter of the cylinder. The values of $\lambda$ and $A$ are constants and are identified from the experimental measurements. $\omega_s$ is the vortex-shedding frequency, which is defined as $\omega_s=2\pi S_t U_0/D$, $S_t$ being the Strouhal number which is determined by the Reynolds numbers range (Chen, 1987) and $U_0$ represents the freestream velocity.

Using the Galerkin approach to discretize the cylinder's displacement, the expression of the wake oscillator can be rewritten as (Dai et al., 2014)

$$\ddot{q}+\lambda\omega_s\left(q^2-1\right)\dot{q}+\omega_s^2 q=\frac{A}{D}\sum_{i=1}^{n}\left(\left[\varphi_i\left(L_b\right)+\frac{D}{2}\varphi_i'\left(L_b\right)\right]\ddot{r}_i\right) \tag{33}$$

## Model Validation

In order to get convergent solutions, the required number of modes in the Galerkin discretization is first determined. To this end, variations of the harvested power as a function of the freestream velocity are plotted when considering different number of modes in the Galerkin approach and when the electrical load resistance is set equal to 2.46 $10^6 \Omega$, as shown in Figure 12. It follows from this plot that the harvested power values converge as the number of modes is increased. It is clear that considering only one mode, two modes, or three modes in the Galerkin discretization is not sufficient to get a convergence. In fact, there is an underestimation in the power output. Clearly, considering four or five modes gives almost the same values for the harvested power. Hence, misleading and erroneous results in the predictions of the voltage and power outputs can be obtained when lumping all system parameters in one mode approximation in the Galerkin approach. In the rest of this chapter, we consider in the numerical predictions five modes in the Galerkin discretization.

To determine the accuracy of the developed reduced-order model for the VIV-based energy harvester, our numerical predictions which are based on five modes in the Galerkin approach are compared with the experimental measurements of Akaydin et al. (2012). It should be mentioned that the coefficients $\lambda$ and $A$ are identified to be equal to 0.24 and 12, respectively. These values are in good agreement with the suggested values by Facchinetti et al. (2004). Setting the load resistance equals to 2.46 $10^6$ $\Omega$, the variations of the average harvested power as a function of the freestream velocity as obtained from our derived model and the experimental measurements of Akaydin et al. (2012) are plotted in Figure 13.

*Figure 12. Variation of the average harvested power as a function of the freestream velocity for different modes in the Galerkin approach when R=2.46 $10^6$ $\Omega$*

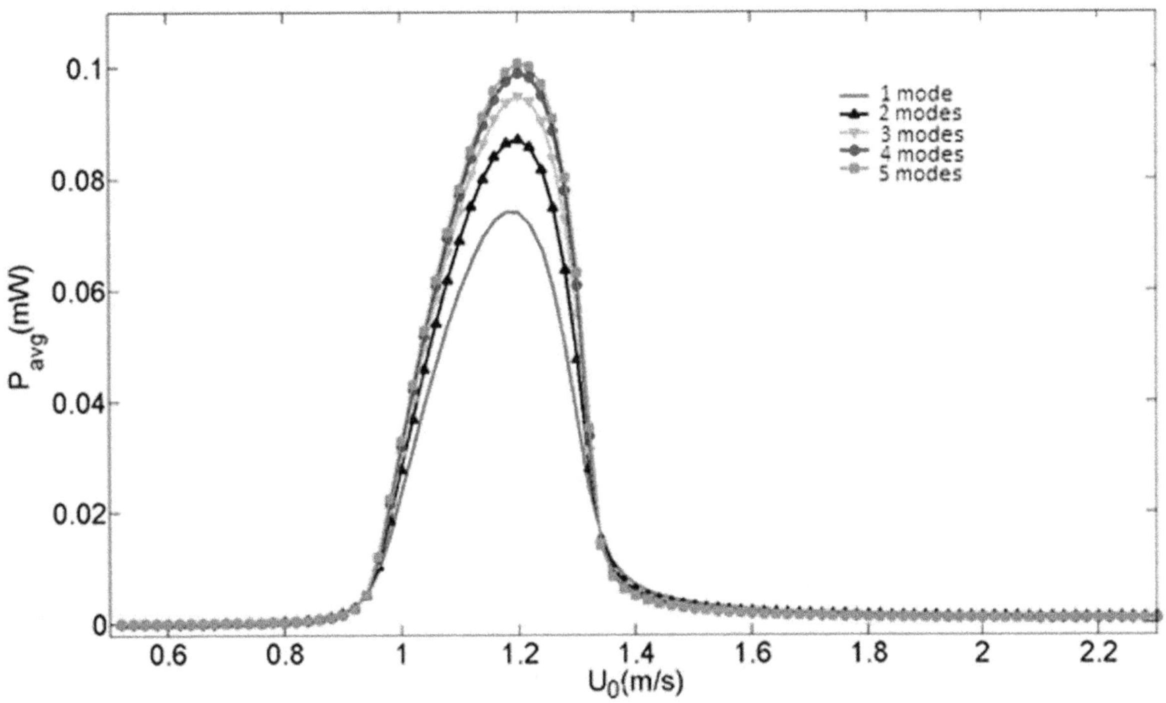

Clearly, a very good agreement is obtained between the numerical predictions of the derived model and the experimental measurements. In fact, in the experiments of Akaydin et al. (2012), the maximum average power is 0.1 *mW* when the freestream velocity is $U_0$=1.192 *m/s* whereas the maximum average power obtained by the present theoretical model is 0.1007 *mW* at $U_0$=1.20 *m/s*. In addition, the synchronization region is well-predicted by the nonlinear electroaeroelastic model and particularly its starting and its peak. On the other hand, an underestimation is obtained in the average harvested power when considering the derived model for freestream velocities larger than 1.4 *m/s* (post-synchronization region). The presence of a sudden drop in the simulated results is probably due to the accuracy of the wake oscillator model to the starting of the synchronization region and its peak.

## System's Performance: Effects of the Cylinder's Tip Mass

It is well-known that the general focus in designing these energy harvesters is on low freestream velocities and hence low shedding frequencies. One of the impediment of harvesting energy from vortex-induced vibrations is that maximum harvested power occurs only for a well-defined range of freestream velocities which are defined by the synchronization region (shedding frequency is near one of the natural frequencies of the system). To this end, the effects of the cylinder's tip mass on the synchronization region and the harvester's performance are investigated in order to design efficient piezoaeroelatsic energy harvesters depending on the operating freestream velocity.

*Figure 13. Comparisons of the variations of the average harvested power as a function of the freestream velocity when using the derived model (solid line) and the experimental measurements of Akaydin et al. (2012) (squares) when the electrical load resistance is set equal to 2.46 $10^6$ Ω*

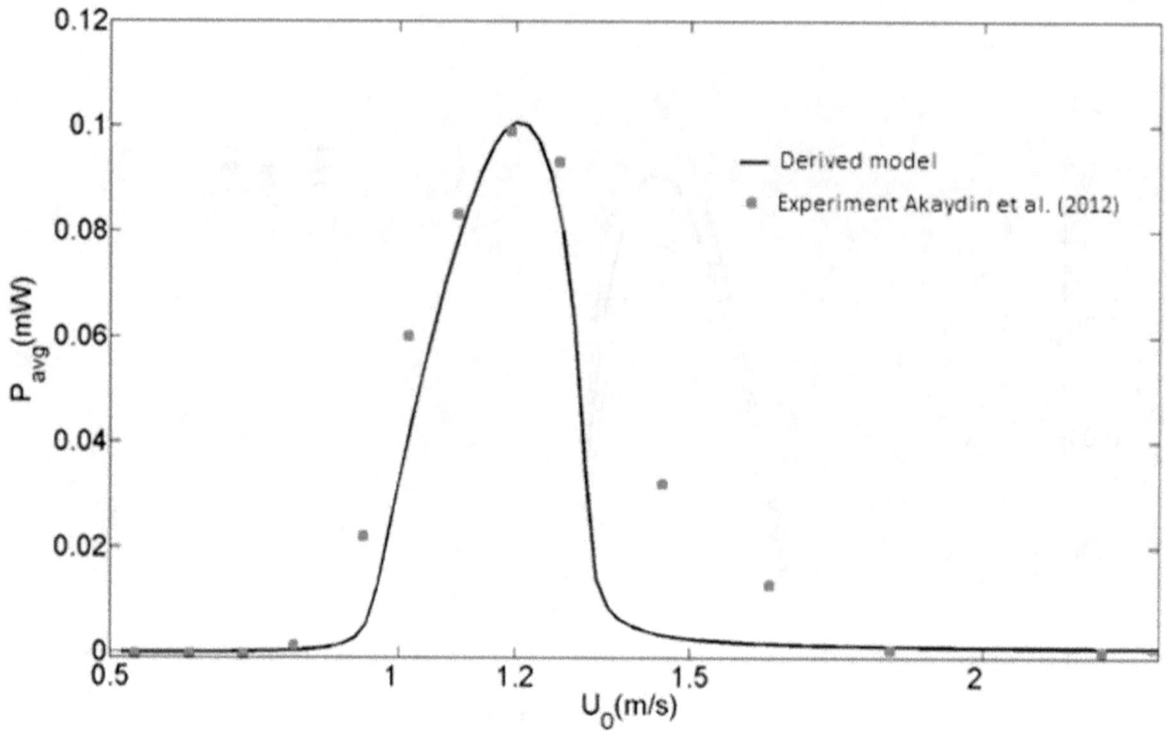

Using our validated model, Figure 14 shows the variation of the first natural frequency of the harvester as a function of the cylinder's tip mass. It follows form this plot that an increase in the cylinder's tip mass is accompanied by a decrease in the first natural frequency of the harvester. Therefore, it is expected that this decrease in the first natural frequency of the harvester is followed by a decrease in the resonant shedding frequency and hence a decrease in the resonant freestream velocity. This is very clear in Figure 15 which presents the variation of the harvested power as a function of the wind speed for different values of the cylinder's tip mass when the electrical load resistance is equal to 2.46 $10^6$ $\Omega$. Clearly, a decrease in the cylinder's tip mass is accompanied by a wider synchronization region and higher values of the average harvested power.

To determine the effects of the load resistance on the performance of the harvester, the variations of the resonant average harvested power as a function of the load resistance for different tip masses of the cylinder are plotted in Figure 16. It follows from this plot that there are two optimum values of the electrical load resistance that maximize the average harvested power for all tip mass values. It is also noted that the values of these optimum load resistance values depend on the cylinder's tip mass because the natural frequency of the harvester is changed. Clearly, there is a compromise between the load resistance and cylinder's tip mass to design enhanced harvesters at different speeds.

A nonlinear electroaeroelastic model for VIV-based energy harvesters has been derived. It was demonstrated that minimum four modes in the Galerkin discretization have to be considered to correctly evaluate the performance of the harvester. It was also shown that there is a very good agreement between our derived model and experimental measurements. The results showed that there is a compromise between the load resistance and cylinder's tip mass to design an enhanced piezoelectric energy harvester from vortex-induced vibrations at different operating wind speeds.

*Figure 14. Variation of the first natural frequency of the harvester as a function of the cylinder's tip mass*

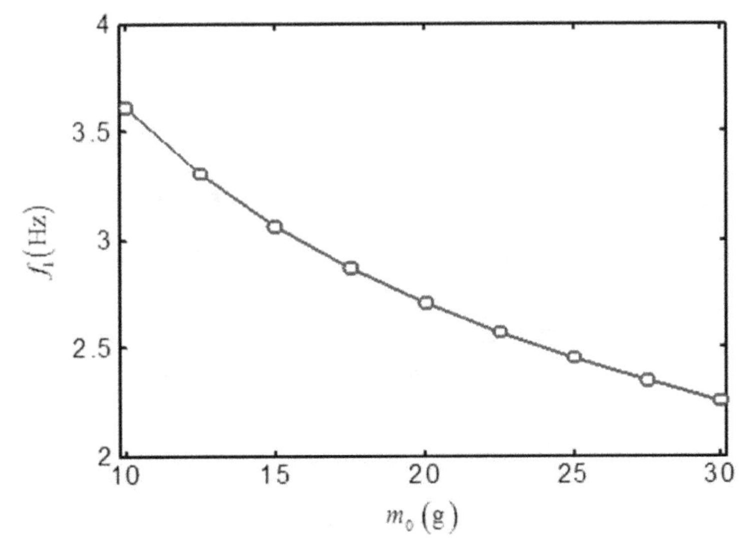

*Figure 15. Variation of the average harvested power as a function of the freestream velocity for different tip masses of the cylinder when the electrical load resistance is set equal to 2.46 $10^6$ Ω*

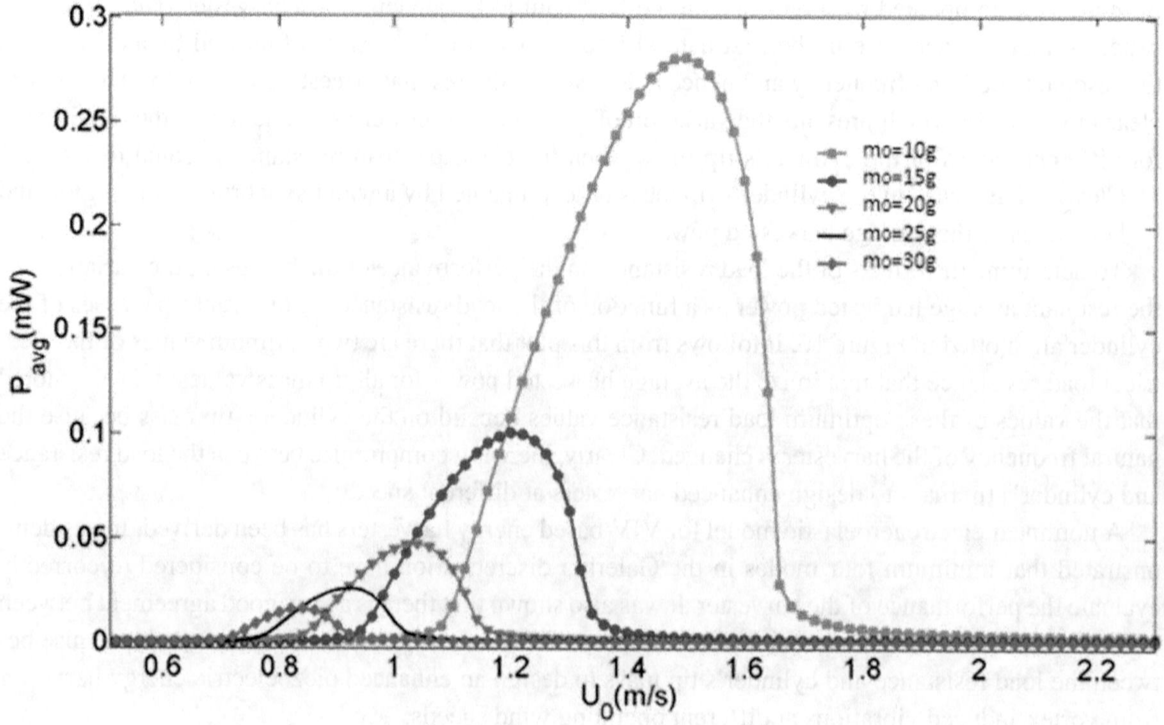


*Figure 16. Variation of the resonant average harvested power as a function of the load resistance for different values of the cylinder's tip mass*

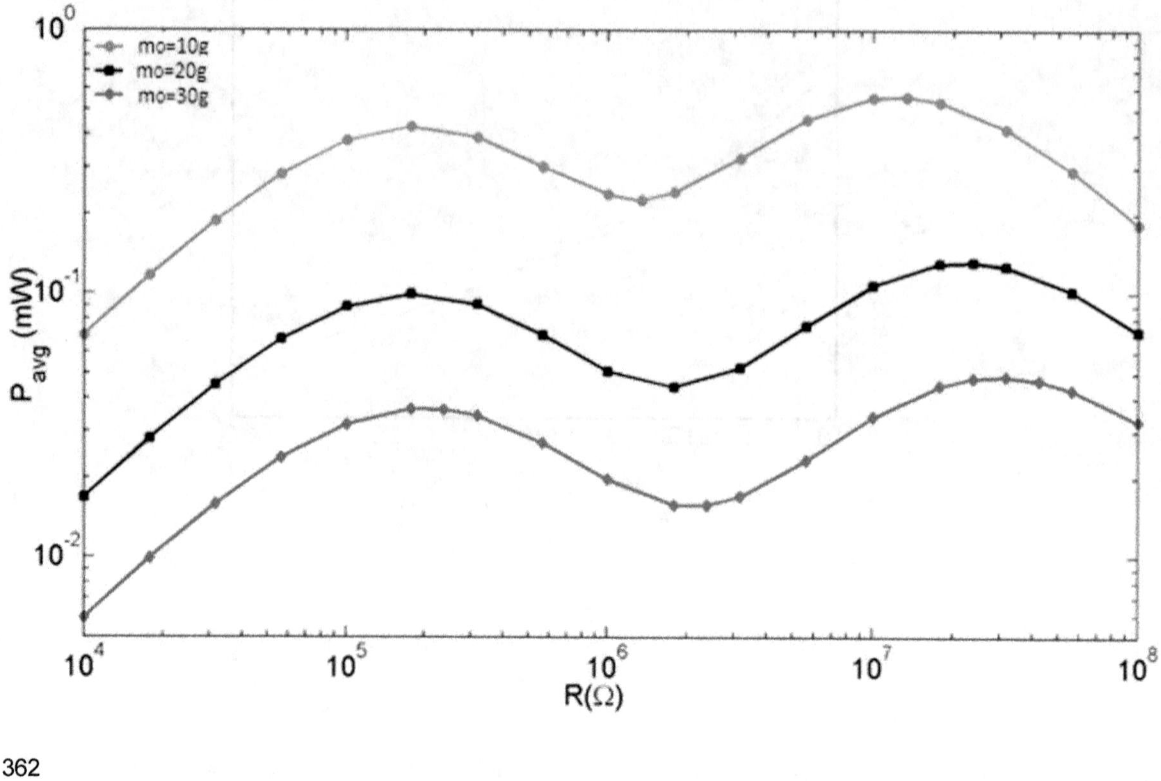

# GALLOPING-BASED AEROELASTIC ENERGY HARVESTER

## System's Modeling

The harvester under investigation consists of a bar with an equilateral triangle cross section attached to two multi-layered cantilever beams. As shown in Figure 17(a), this tip mass bar undergoes galloping in the transverse direction when subjected to an incoming flow. The cantilever beams are composed of aluminum and piezoelectric layers. The piezoelectric sheets are bounded by two-in-plane electrodes of negligible thickness connected in parallel with opposite polarity to an electrical load resistance. Our objective, in this chapter, is to develop a nonlinear electroaeroelastic model to predict the system's response and validate it with the experimental results of Sirohi and Mahadik (2011). Assuming the beam thickness is small compared to its length so that shear deformation and rotary inertia can be neglected, Euler–Bernoulli beam theory is used to model this harvester.

In order to minimize the blockage, a low frontal area was considered in the experiments when clamping the beams. Sirohi and Mahadik (2011) considered in their developed model a linear torsional spring of stiffness $k_\theta(\theta)$ at the root of the beam to model this flexibility, as shown in Figure 17(b). Furthermore, they considered a polynomial representation for the mode shapes when using the Galerkin discretization. It is well-known that the existence of a low frontal area or a freeplay in the system results in the presence of nonlinearity in the structural stiffness of the system. In addition, because the piezoelectric layers do not cover the whole beam, the mode shapes of the system are different than their counterparts when the piezoelectric layers cover the whole substrate beam. Consequently, in the current model, the exact mode shapes of the system are first determined. Furthermore, a more general representation of the stiffness of the torsional spring is considered by assuming a nonlinear spring with its stiffness given by

$$k_\theta(\theta) = k_{\theta 0} + k_{\theta 2}\theta^2 \tag{34}$$

Using the Euler-Bernoulli beam assumptions, the equation of motion of each of the two beams can be written as (Abdelkefi, Yan, & Hajj, 2013e)

*Figure 17. Schematic of the galloping-based energy harvester (a) three-dimensional view and (b) top view*

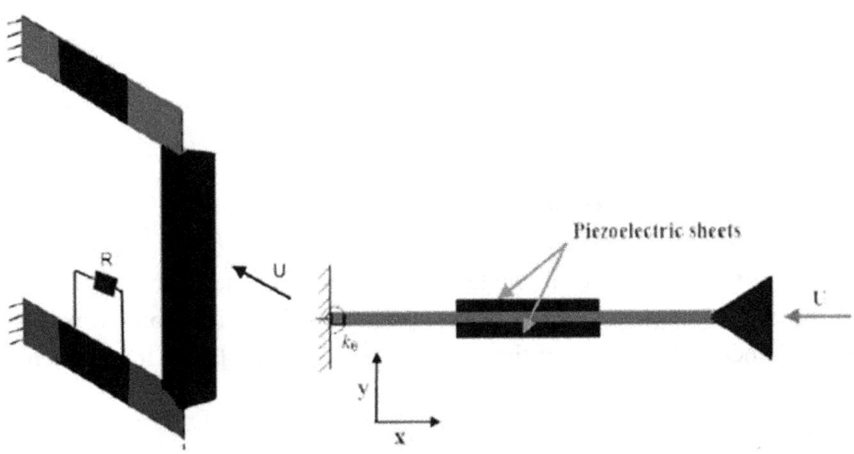

$$EI\frac{\partial^4 v(x,t)}{\partial x^4}+c_s I\frac{\partial^5 v(x,t)}{\partial x^4 \partial t}+c_a\frac{\partial v(x,t)}{\partial t}+m\frac{\partial^2 v(x,t)}{\partial t^2}+(\frac{d\delta(x-L_1)}{dx}-\frac{d\delta(x-L_2)}{dx})\vartheta_p V(t)$$
$$-\frac{d}{dx}(\delta(x)k_\theta v'(0,t))=F_{tip}\delta(x-L) \tag{35}$$

where $v(x,t)$ is the transverse displacement, $\delta(x)$ is the Dirac delta function, $F_{tip}$ is the galloping aerodynamic force at the tip of the beam as caused by the structure's oscillations, $L$ is the length of the beam, $c_a$ is the viscous air damping coefficient, $m$ is the mass of the beam per unit length, $V(t)$ is the generated voltage, $L_1$ is the distance from the left end of cantilever beam to the starting location of the piezoelectric sheet, $L_2$ is the distance from the left end of cantilever beam to the ending location of the piezoelectric sheet and $\vartheta_p$ is the piezoelectric coupling term which is given by

$$\vartheta_p=-2e_{31}b_2(h_p+h_s) \tag{36}$$

where $e_{31}=E_p d_{31}$ is the piezoelectric stress coefficient, $b_2$ is the width of the piezoelectric layer and $h_s$ and $h_p$ are the thicknesses of the aluminum and piezoelectric layers, respectively.

In equation (35), the stiffness $EI$ and mass of the beam per unit length $m$ are given by

$$EI=EI_1=\frac{1}{12}b_1E_sh_s^3 \text{ and } m=m_1=b_1\rho_sh_s \text{ for } 0\le x<L_1 \text{ or } L_2<x\le L$$

$$EI=EI_2=\frac{1}{12}b_1E_sh_s^3+\frac{2}{3}b_2E_p[(h_p+\frac{h_s}{2})^3-\frac{h_s^3}{8}] \text{ and } m=m_2=b_1\rho_sh_s+2b_2\rho_ph_p \text{ for } L_1\le x\le L_2$$

where $E_s$ and $E_p$ are the Young's Modulus of the aluminum and piezoelectric layers, respectively, and $\rho_s$ and $\rho_p$ are the respective densities of these layers.

To complete the problem formulation, the Gauss law is used to relate the mechanical and electrical part variables. Following Abdelkefi et al. (2013e), the equation of motion that governs the strain-voltage relation is

$$-2e_{31}(h_p+h_s)b_2\int_{L_1}^{L_2}\frac{\partial^3 v(x,t)}{\partial t\partial x^2}dx-\frac{4\epsilon_{33}^s b_2(L_2-L_1)}{h_p}\frac{dV(t)}{dt}=\frac{V(t)}{R} \tag{37}$$

## Galloping Force Representation

The quasi-steady approximation is used to model the aerodynamic loads. This approximation is justified by the fact that the characteristic time scale of the oscillations, which is approximately equal to $2\pi/\omega$, where $\omega$ is the fundamental frequency of the system, is, in general, larger than the characteristic time scale of the flow, which is of the order $b_{tip}/U$.

The lift force $F_L$ and the drag force $F_D$ are generally written as

$$F_L = \frac{1}{2}\rho_{air}V^2 b_{tip} C_L L_{tip}$$
$$F_D = \frac{1}{2}\rho_{air}V^2 b_{tip} C_D L_{tip} \quad (38)$$

In these equations, $\rho_{air}$ is the density of air, $L_{tip}$ is the length of the tip body, $V$ is the relative velocity of the wind with respect to tip body. $C_L$ and $C_D$ are, respectively, the lift and drag coefficients. These aerodynamic coefficients depend on the angle of attack which is given by

$$\alpha = -tan^{-1}(\frac{\dot{v}(L,t)}{U}) + v'(L,t) \quad (39)$$

where $U$ is the incoming wind speed.

The lift and drag coefficients of an equilateral cross-section geometry as measured experimentally by Alonso and Meseguer (2006) are plotted in Figure 18. These experimental data are fitted to model the aerodynamic coefficients as a function of the angle of attack as

$$C_L = 2.76215 + 2.63765\alpha \text{ for } \alpha < -0.6109$$
$$C_L = -2.2577\alpha + 1.0019\alpha^3 \text{ for } -0.6109 < \alpha < 0.6109 \quad (40)$$
$$C_L = -2.76215 + 2.63765\alpha \text{ for } \alpha > 0.6109$$

and

$$C_D = 1.6934 - 1.7373\alpha^2 \text{ for } -0.676 < \alpha < 0.676$$
$$C_D = 0.9 \text{ for } \alpha < -0.676 \text{ or } \alpha > 0.676 \quad (41)$$

## Eigenvalue Problem Analysis

To characterize the performance of the harvester, linear and nonlinear analyses have to be performed. To this end, the equations of motion are discretized using the Galerkin procedure, which requires the exact mode shapes of the structure. Hence, an eigenvalue problem analysis is performed to determine these mode shapes and associated natural frequencies. First, the damping, forcing, polarization, and nonlinear torsional terms are dropped from equation (35). Second, the transverse displacement is assumed to be written as $v(x,t)=\phi(x)e^{i\omega t}$. Third, because the piezoelectric layers are not covering the whole cantilever beam, the mode shape of this structure is divided into three different regions ($\phi_1(x)$, $\phi_2(x)$, and $\phi_3(x)$). The resulting eigenvalue problem is then given by

$$EI_1\phi_1^{iv} - m_1\omega^2\phi_1 = 0$$
$$EI_2\phi_2^{iv} - m_2\omega^2\phi_2 = 0 \quad (42)$$
$$EI_1\phi_3^{iv} - m_1\omega^2\phi_3 = 0$$

*Figure 18. Aerodynamic coefficients of an equilateral section: Analytical representations (solid lines) and experimental data (Alonso & Meseguer, 2006) (dots for the lift coefficient and stars for the drag coefficient)*

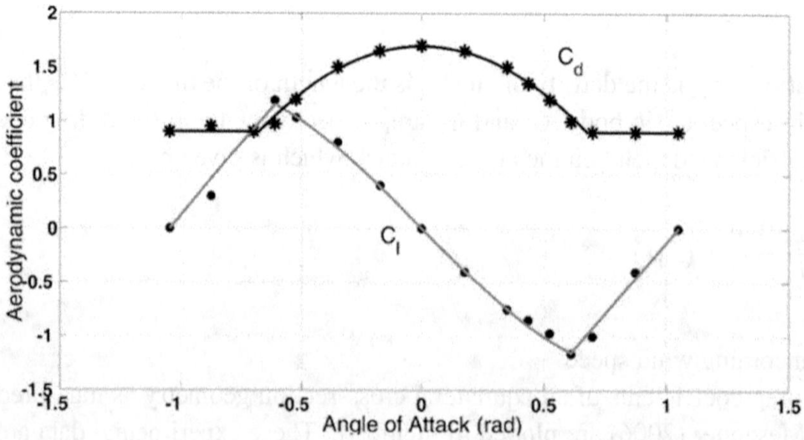


with the following boundary conditions:

$$\begin{aligned}&\phi_1(0)=0\,,\,EI\phi_1''(0)-k_{\theta 0}\phi_1'(0)=0,\,\phi_1(L_1)=\phi_2(L_1),\,\phi_1'(L_1)=\phi_2'(L_1)\,;\\&EI_1\phi_1''(L_1)=EI_2\phi_2''(L_1),\,EI_1\phi_1'''(L_1)=EI_2\phi_2'''(L_1),\,\phi_2(L_2)=\phi_3(L_2);\\&\phi_2'(L_2)=\phi_3'(L_2),\,EI_2\phi_2''(L_2)=EI_1\phi_3''(L_2),\,EI_2\phi_2'''(L_2)=EI_1\phi_3'''(L_2);\\&EI_1\phi_3''(L)-\omega^2 I_t\phi_3'(L)=0,\,EI_1\phi_3'''(L)+\omega^2 M_t\phi_3(L)=0\end{aligned} \tag{43}$$

where $I_t$ is the rotary inertia of the tip mass $M_t$.

The mode shapes for the three different regions are then written as

$$\begin{aligned}\phi_1(x)&=A_1\sin\beta_1 x+B_1\cos\beta_1 x+C_1\sinh\beta_1 x+K_1\cosh\beta_1 x\\\phi_2(x)&=A_2\sin\beta_2 x+B_2\cos\beta_2 x+C_2\sinh\beta_2 x+K_2\cosh\beta_2 x\\\phi_3(x)&=A_3\sin\beta_1 x+B_3\cos\beta_1 x+C_3\sinh\beta_1 x+K_3\cosh\beta_1 x\end{aligned} \tag{44}$$

where the coefficients of $\beta_1$ and $\beta_2$ are related by: $\beta_1=(m_1\,EI_2/\,m_2\,EI_1)^{1/4}\,\beta_2$.

Then, the mode shapes are normalized using the following orthogonality conditions:

$$\begin{aligned}&\int_0^{L_1}\phi_{1s}(x)m_1\phi_{1r}(x)dx+\int_{L_1}^{L_2}\phi_{2s}(x)m_2\phi_{2r}(x)dx+\int_{L_2}^{L_3}\phi_{3s}(x)m_1\phi_{3r}(x)dx\\&+\phi_{3s}(L)M_t\phi_{3r}(L)+\phi_{3s'}(L)I_t\phi_{3r'}(L)=\delta_{rs}\\&\frac{d\phi_{1s}(0)}{dx}k_{\theta 0}\frac{d\phi_{1r}(0)}{dx}+\int_0^{L_1}\frac{d^2\phi_{1s}(x)}{dx^2}EI_1\frac{d^2\phi_{1r}(x)}{dx^2}dx\\&+\int_{L_1}^{L_2}\frac{d^2\phi_{2s}(x)}{dx^2}EI_2\frac{d^2\phi_{2r}(x)}{dx^2}dx+\int_{L_2}^{L}\frac{d^2\phi_{3s}(x)}{dx^2}EI_1\frac{d^2\phi_{3r}(x)}{dx^2}dx=\delta_{rs}\omega_r^2\end{aligned} \tag{45}$$

where $s$ and $r$ are used to represent the vibration modes and $\delta_{rs}$ is the Kronecker delta, defined as unity when $s$ is equal $r$ and zero otherwise.

## Reduced-Order Model

To compute a reduced-order model of the system, the displacement $v(x,t)$ is expressed using the Galerkin procedure in the following form

$$v(x,t) = \sum_{i=1}^{\infty} \phi_i(x) q_i(t) \tag{46}$$

where $q_i(t)$ are the modal coordinates and $\phi_i(x)$ are the mode shapes of a cantilever beam with an equilateral tip mass.

Considering only one mode in the Galerkin discretization, one obtains

$$\begin{aligned} &\ddot{q}(t) + 2\xi\omega\dot{q}(t) + \omega^2 q(t) + \chi V(t) + k_{\theta 2}(\phi'(0))^4 q^3(t) = f(t) \\ &C_p \dot{V}(t) + \frac{V(t)}{R} - \chi \dot{q}(t) = 0 \end{aligned} \tag{47}$$

where $\xi$ is the mechanical damping ratio, $\omega$ is the fundamental natural frequency of the structure, $\chi$ is the piezoelectric coupling, $C_p$ is the capacitance of the harvester, and $f(t)$ is the galloping force which is given by (Abdelkefi, Yan, & Hajj, 2013f)

$$f(t) = \phi(L) F_{tip} = \frac{1}{2}\rho U^2 b_{tip} L_{tip} [a_1 \frac{\dot{q}}{U} + b_1 q] + \frac{1}{2}\rho b_{tip} L_{tip} [a_2 \frac{\dot{q}^3}{U} + b_2 U^2 q^3 + c_2 q \dot{q}^2 + e_2 U q^2 \dot{q}]$$

where $a_1$, $b_1$, $a_2$, $b_2$, $c_2$, and $e_2$ are functions of $\phi(L)$ and $\phi'(L)$.

It is noted that the galloping force is a combination of linear and cubic nonlinear terms. The Den Hartog stability criterion (Den Hartog, 1956) states that a section of a bluff body on a flexible support is susceptible to galloping when the linear term associated with the velocity of the aerodynamic galloping force is positive. The galloping phenomenon is a self-excited response which is initiated when the coupled damping of the system changes from positive to negative. As for the nonlinear terms of the aerodynamic galloping force, they affect the type of instability (supercritical or subcritical) and the amplitude of the ensuing limit-cycle oscillations.

To investigate the dependence of the onset speed of galloping on the electrical load resistance and to characterize the type of Hopf bifurcation of this harvester, linear and nonlinear analyses of the coupled electromechanical system are performed. To this end, the following state variables are defined as

$$\mathbf{X} = \begin{bmatrix} X_1 \\ X_2 \\ X_3 \end{bmatrix} = \begin{bmatrix} q \\ \dot{q} \\ V \end{bmatrix} \tag{48}$$

In terms of these state variables, the representative model of this galloping-based energy harvester is rewritten as

$$\begin{aligned}
\dot{X}_1 &= X_2 \\
\dot{X}_2 &= -(2\xi\omega - \frac{\rho U b_{tip} L_{tip} a_1}{2})X_2 - (\omega^2 - \frac{\rho U^2 b_{tip} L_{tip} b_1}{2})X_1 - \chi X_3 \\
&+\frac{1}{2}\rho b_{tip} L_{tip}[a_2 \frac{X_2^3}{U} + b_2 U^2 X_1^3 + c_2 X_1 X_2^2 + e_2 U X_1^2 X_2] - (\phi'(0))^4 k_{\theta 2} X_1^3 \\
\dot{X}_3 &= -\frac{1}{RC_p} X_3 + \frac{\chi}{C_p} X_2
\end{aligned} \tag{49}$$

Clearly, these equations can be expressed in a vector form as

$$\dot{\mathbf{X}} = B\mathbf{X} + \mathbf{C}(\mathbf{X},\mathbf{X},\mathbf{X}) \tag{50}$$

where C(X,X,X) is a cubic vector function of the state variables and

$$B = \begin{bmatrix} 0 & 1 & 0 \\ -\omega^2 + \dfrac{\rho U^2 b_{tip} L_{tip} b_1}{2} & -\left(2\xi\omega - \dfrac{\rho U b_{tip} L_{tip} a_1}{2}\right) & -\chi \\ 0 & \dfrac{\chi}{C_p} & -\dfrac{1}{RC_p} \end{bmatrix}$$

## Model Validation

To validate our model, two different steps are considered. First, the onset speed of galloping of the experiments of Sirohi and Mahadik (2011) is validated. Second, a nonlinear analysis is performed to identify the nonlinear torsional spring and validate the harvester's response.

As for the onset speed of galloping, a linear analysis is performed. At this speed, $U_g$, the coupled damping of the system changes sign from positive to negative. The matrix $B$ represents the dependence of the onset speed of galloping on the load resistance. This matrix has a set of three eigenvalues. The first two eigenvalues are similar to those of a pure galloping problem in the absence of the piezoelectricity effect. These two eigenvalues are complex-conjugate. The absolute value of the real part of these eigenvalues represents the coupled damping coefficient and the positive imaginary part corresponds to the coupled frequency of the harvester. Because the third eigenvalue is a result of the electromechanical coupling and is always real and negative, the stability of the trivial solution depends only on the real part of the first two eigenvalues. The solution of the linear part is asymptotically stable if the real part of the first eigenvalue is negative. On the other hand, if the real part of this eigenvalue is positive, the solution of the linearized system is unstable. At the transition point, the onset speed of galloping takes place.

The experimental measurements of Sirohi and Mahadik (2011) showed that the onset of galloping of this harvester took place at a speed $U_g = 3.57$ *m/s* when the load resistance was set equal to 3.7 $10^4\Omega$. In Figure 19, the variation of the onset speed of galloping as a function of the load resistance is plotted. When $R$=3.7 $10^4\Omega$, the onset speed of galloping is almost equal to 3.57 *m/s* which is in a very good agreement with the experimental value obtained by Sirohi and Mahadik (2011).

Inspecting Figure 19, it is clear that the onset speed of galloping strongly depends on the load resistance. In the lower (between $10^2\Omega$ and $10^3\Omega$) and higher (between $10^6\Omega$ and $10^8\Omega$) ranges, the variation of $U_g$ with the electrical load resistance is small. Between $10^3\Omega$ and $10^6\Omega$, a significant change in the onset speed of galloping is obtained with a peak value of about 3.7 *m/s* for a load resistance value of 3 $10^4\Omega$.

Next, the effects of the nonlinear torsional spring on the level of the harvested power are investigated, as shown in Figure 20. It follows from this plot that the nonlinear torsional spring affects the amplitude of the harvested power for all considered values of the load resistance. The plotted curves in Figure 21 show the variations of the harvested power as a function of the load resistance for the numerical and experimental data of Sirohi and Mahadik (2011) and two other curves when considering the nonlinear torsional stiffness equals to 0 *N.m* and 10 *N.m*, respectively. Although the linear representation of the torsional stiffness yields acceptable results, it is clear that including the nonlinear stiffness terms yields a better agreement with the experimental measurements. Consequently, including the effects of the mode shapes of a partially-covered piezoelectric cantilever beam and considering a nonlinear representation for the torsional spring are very beneficial and give excellent agreement with experiments.

*Figure 19. Variation of the onset speed of galloping as a function of the load resistance*

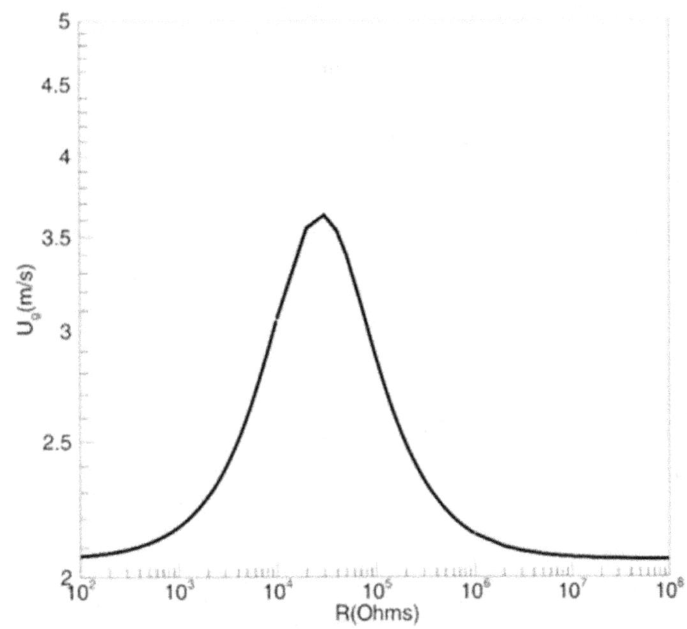

*Figure 20. Variation of the harvested power as a function of the nonlinear torsional spring coefficient for different values of the load resistance*

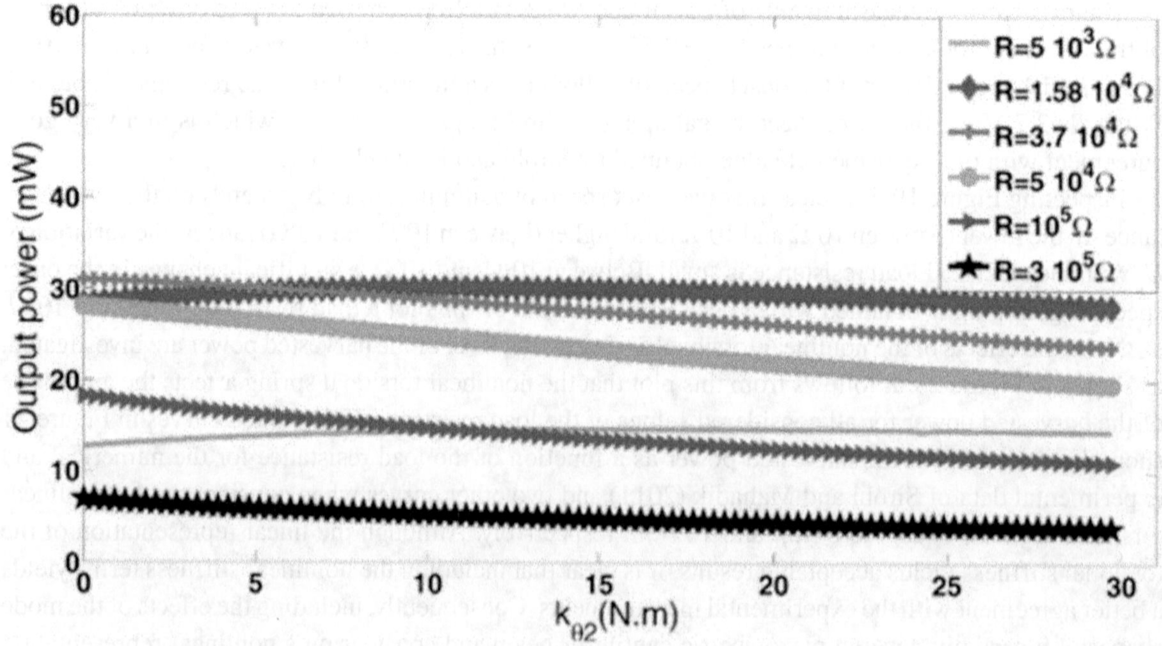


*Figure 21. Comparison of our derived model with the numerical and experimental measurements of Sirohi and Mahadik (2011)*

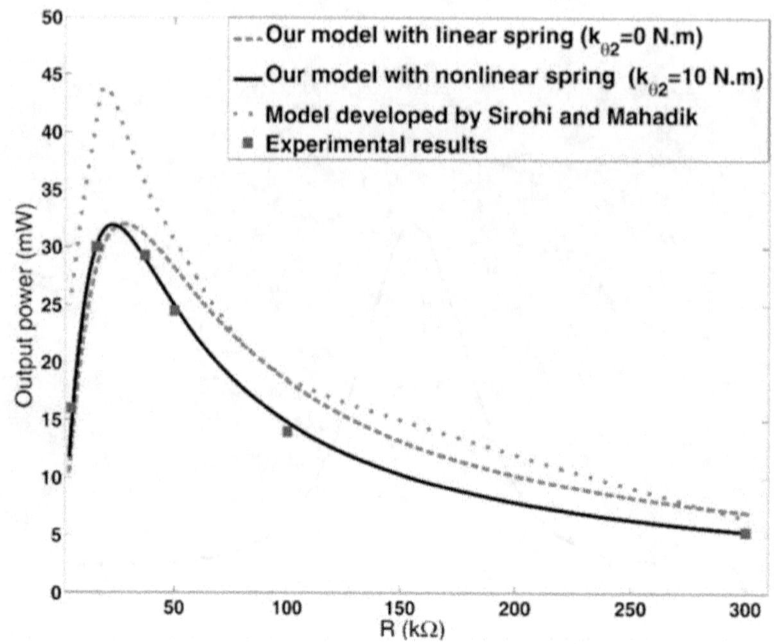

## Harvester's Performance

The plotted curves in Figure 22 show the bifurcation diagrams of the harvested power and tip displacement for five distinct values of the load resistance when considering $k_{\theta 2}$=10 *N.m*. It follows from these plots that the onset speed of galloping is affected by the load resistance as expected from the linear analysis results. Clearly, an increase in the value of the load resistance is not accompanied by an increase in the level of the harvested power. In fact, the highest level of harvested power is obtained when the load resistance is set equal to 3.7 $10^4\Omega$ for wind speeds above 3.5 *m/s*, as shown in Figure 22 (a). As for the transverse displacement, minimum displacement values are obtained when $R$=3.7 $10^4\Omega$ which corresponds to maximum onset speed of galloping and hence maximum electromechanical damping, as shown in Figure 19. It is interesting to contrast the facts that this resistance yields the minimum transverse displacement and maximum harvested power. It should be mentioned that a sudden jump takes place near the bifurcation. This is a characteristic of a subcritical instability. This observation is demonstrated based on the normal form of the Hopf bifurcation (Abdelkefi et al., 2013f).

To investigate more the effects of the load resistance on the performance of the harvester, the plotted curves, in Figures 23(a) and (b), show the variations of the harvested power and transverse displacement as function of the load resistance for three distinct wind speed values, namely, 2.8, 4, and 6 *m/s*. It follows from these figures that, at $U$=2.8 *m/s,* the harvester does not oscillate for load resistance values between 6 $10^3\Omega$ and $10^5\Omega$. This result is predicted because the onset speed of galloping in this region of load resistances is larger than 2.8 *m/s*. At $U$= 4 *m/s* and 6 *m/s*, there is a range of load resistances (6 $10^3\Omega$ and $10^5\Omega$) for which the transverse displacement is minimum for both wind speeds. This is due to the fact that the electromechanical damping is maximum over this band of electrical load resistance, as shown in Figure 23(b). For the same speeds, we note that there are optimum values of the load resistance for which the harvested power is a maximum, as shown in Figure 23(a). The region of load resistances over which the harvested power is maximized corresponds to the region of load resistances over which the transverse displacement is minimized for both wind speeds. This result is explained by the resistive shunt damping effect (Agneni, Mastroddi, & Polli, 2003). This realization is very useful for the design of enhanced galloping-based energy harvesters.

*Figure 22. Bifurcation diagrams of (a) the harvested power and (b) the transverse displacement for different values of the load resistance*

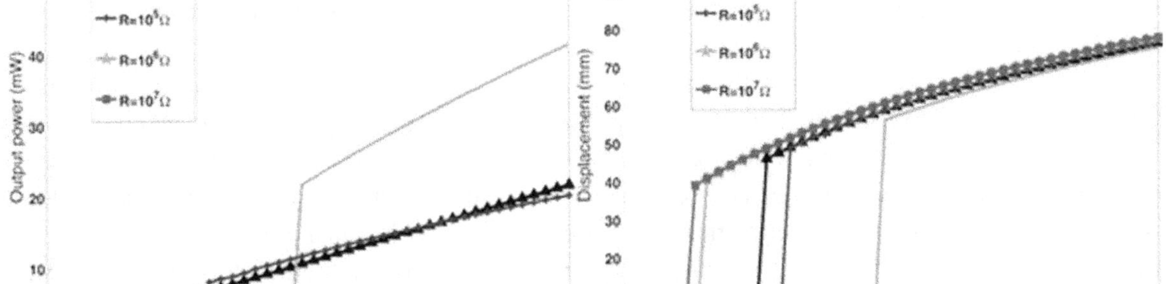

*Figure 23. Variations of the (a) harvested power and (b) transverse displacement as function of the load resistance for three values of the wind speed*

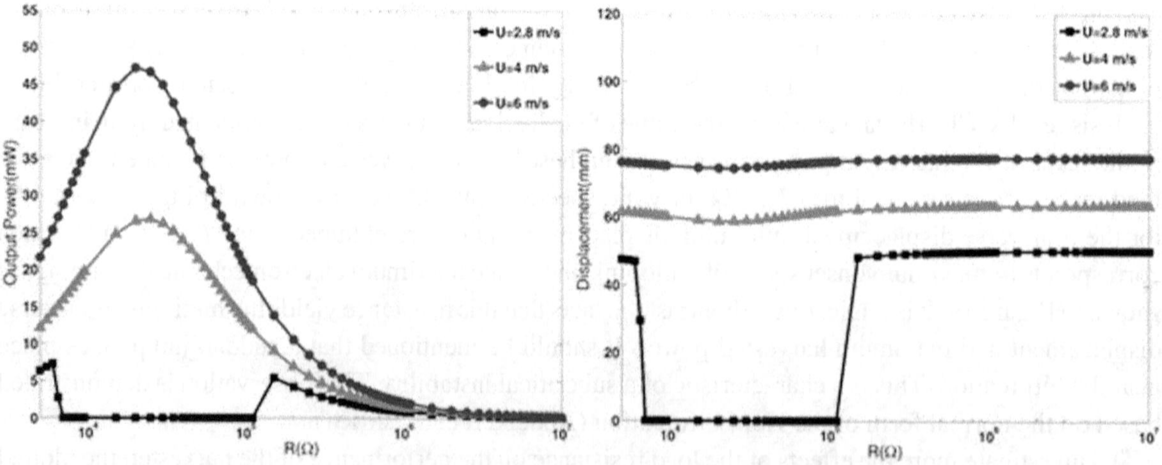


A coupled nonlinear electroaeroelastic model has been developed and validated with previous experimental results. The results showed that the onset speed of galloping is strongly affected by the electrical load resistance. It was also demonstrated that considering the mode shapes of a partially-covered piezoelectric cantilever beam and the nonlinear contribution of the torsional spring are very important to develop accurate analytical model. It was shown that maximum levels of the harvested power are accompanied by minimum transverse displacement values when varying the electrical load resistance which is associated to the shunt damping effect.

## QUALITATIVE AND QUANTITATIVE COMPARISONS

Based on the previous analyses and comparisons with experimental measurements, qualitative and quantitative comparisons are performed. First, it should be mentioned that these studied aeroelastic phenomena (flutter, VIV, and galloping) were tackled by many investigators in order to avoid possible damages in large scale systems, such as aircrafts, pipelines, bridges, transmission lines, etc. On the other hand, in the last few years, these aerodynamic instability phenomena have been exploited to harvest energy in small scale systems. The research findings in this topic were performed in the following chronological order: flutter from airfoil sections, VIV from circular cylinders, then galloping from prismatic structures (D-section, triangle, square, etc.), which reveals an implicit qualitative comparison between these types of harvesters.

Researchers first started by investigating the concept of energy harvesting from airfoil sections. In this approach, an airfoil section is attached at the tip of a multi-layered piezoelectric cantilever beam which is allowed to plunge and to pitch about an elastic axis. They studied the effects of the system's parameters of a flutter-based energy harvester on its linear flutter speed as well as its performance. Based on the performed analyses, it is clear that this designed harvester has many issues. In fact, there are many parameters in these two- or three- degree of freedom aeroelastic systems which make the design of efficient harvesters very difficult and complex. These parameters are related and a change in one of them results in a big change in the flutter speed as well as the system's performance. Furthermore, it is demonstrated that a decrease in the eccentricity (distance between the elastic axis and gravity axis) results in an enhancement in the level of the harvested power with a significant decrease in the pitch angle. However, this decrease in the eccentricity results in an increase in the flutter speed to higher and higher values. In addition, the presence of the pitch and/or flap degrees of freedom in the flutter case results in a reduction in the harvested power because there is an aerodynamic power is lost due to these rotational degrees of freedom. This is explained by the fact that, in the case of piezoelectric transduction, oscillations in the rotational modes are lost and not converted into useful electrical power. Consequently, considering a single-degree-of-freedom that allows harvesting energy from transverse displacement is expected to be more beneficial in terms of system's design parameters and performance.

Then, a particular focus was paid to the phenomenon of vortex-induced vibrations from circular cylinders. In this case, a circular cylinder is attached at the tip of a piezoelectric cantilever beam or a cantilever beam is placed in the wake of a circular cylinder such that the Karman vortex street takes place which allows the harvester to oscillate in a periodic manner. Resonant motions are obtained when the shedding frequency matches one of the natural frequencies of the harvester. Several studies focused on the possibility of harvesting energy from vortex-induced vibrations. This one-degree-of-freedom harvester is very beneficial in terms of system's design. However, its main drawback is the well-defined region of wind speed (synchronization region) in which the system can oscillate and then generate energy. Therefore, any change in the incoming wind speed can lead to a sudden drop in the level of the harvested power. This system is only good for the cases where the wind speed is uniform which is not realistic. Since the VIV-based energy harvesting system works only when the wind speed is in the synchronization region, it is considered inefficient for a wide range of wind speeds.

Afterwards, researchers focused on a novel aeroelastic energy harvester based on the galloping phenomenon. In this case, a prismatic structure (D-section, square, triangle, etc.) is attached to the free end of a piezoelectric cantilever beam which is allowed to translate vertically (single-degree-of-freedom). In this approach, when the wind speed is larger than the onset speed of galloping, large-amplitude oscillations are obtained which gives the opportunity to enhance the level of the harvested power. As such, this one-degree-of-freedom harvester is the easiest in terms of system's design.

In Table 2, quantitative comparisons in terms of efficiency, power density, and specific power between different considered aeroelastic energy harvesters are performed. It should be mentioned that these presented values are based on experimental measurements. The efficiency, power density, and specific power are calculated, respectively, by dividing the harvested power by the input aerodynamic power, the harvested power by the volume of the harvester, and the harvested power by the total mass of the harvester. Clearly, the efficiency of the flutter-based energy harvester is very low compared to the VIV- and galloping-based energy harvesters. In addition, the galloping-based energy harvester has a larger efficiency than the VIV-based energy harvester. Furthermore, the galloping-based energy harvester has a value of power density larger than 100 times its counterpart for the VIV- and flutter-based energy harvesters. Similarly, the galloping harvester has a very large specific power compared to the flutter and VIV harvesters. Therefore, based on qualitative and quantitative comparisons, it is concluded that the galloping-based energy harvester is the easiest to design and the most efficient to harvest energy at low wind speed.

*Table 2. Efficiency, power density, and specific power comparisons between different aeroelastic energy harvesters*

| | Flutter (Abdelkefi & Hajj, 2013) | VIV (Akaydin et al., 2012) | Galloping (Sirohi & Mahadik, 2011) |
|---|---|---|---|
| Efficiency | $6\ 10^{-5}$ | 0.026 | 0.07 |
| Power density (W/m$^3$) | 2.08 | 0.782 | 271 |
| Specific power (W/Kg) | $6.95\ 10^{-4}$ | $5.9\ 10^{-3}$ | 0.748 |

## CONCLUSION

In this chapter, a summary of different aeroelastic energy harvesters has been performed. Nonlinear aeroelectroelastic models have been developed to predict the responses of flutter-, VIV-, and galloping-based aeroelastic energy harvesters. Various aerodynamic loads representations have been discussed and used. Linear and nonlinear analyses were performed to investigate the effects of the system's parameters on the performance of these harvesters. The normal form of the Hopf bifurcation was derived to characterize the type of instability near bifurcation and to determine the importance of structural and aerodynamic nonlinearities on the performance of these energy harvesters. It was demonstrated that these analyses are very beneficial and can be used to breach traditional barriers in the design and performance enhancement of these aeroelastic energy harvesters, characterization of their behaviors, and identification of the contribution of different types of nonlinearities. Qualitative and quantitative comparisons were presented and discussed showing the potential of galloping-based aeroelastic energy harvesters in terms of design and effectiveness.

## REFERENCES

Abdelkefi, A. (2012). *Global nonlinear analysis of piezoelectric energy harvesting from ambient and aeroelastic vibrations* (PhD Dissertation). Virginia Tech, VA.

Abdelkefi, A., & Barsallo, N. (2014). Comparative modeling of low-frequency piezomagnetoelastic energy harvesters. *Journal of Intelligent Material Systems and Structures*. doi:10.1177/1045389X14523860

Abdelkefi, A., & Ghommem, M. (2013). Piezoelectric energy harvesting from morphing wing motions for micro air vehicles. *Theoretical & Applied Mechanics Letters*, *3*(5), 052001. doi:10.1063/2.1305204

Abdelkefi, A., Ghommem, M., Nuhait, A. O., & Hajj, M. R. (2014c). Nonlinear analysis and enhancement of wing-based piezoaeroelastic energy harvesters. *Journal of Sound and Vibration*, *333*(1), 166–177. doi:10.1016/j.jsv.2013.08.032

Abdelkefi, A., & Hajj, M. R. (2013). Performance enhancement of wing-based piezoaeroelastic energy harvesting through freeplay nonlinearity. *Theoretical & Applied Mechanics Letters*, *3*(4), 041001. doi:10.1063/2.1304101

Abdelkefi, A., Hajj, M. R., & Nayfeh, A. H. (2012a). Phenomena and modeling of piezoelectric energy harvesting from freely oscillating cylinders. *Nonlinear Dynamics*, *70*(2), 1377–1388. doi:10.1007/s11071-012-0540-x

Abdelkefi, A., Hajj, M. R., & Nayfeh, A. H. (2013a). Piezoelectric energy harvesting from transverse galloping of bluff bodies. *Smart Materials and Structures*, *22*(1), 015014. doi:10.1088/0964-1726/22/1/015014

Abdelkefi, A., Najar, F., Nayfeh, A. H., & Ben Ayed, S. (2011). An energy harvester using piezoelectric cantilever beams undergoing coupled bending–torsion vibrations. *Smart Materials and Structures*, *20*(11), 115007. doi:10.1088/0964-1726/20/11/115007

Abdelkefi, A., Nayfeh, A. H., & Hajj, M. R. (2012b). Design of piezoaeroelastic energy harvesters. *Nonlinear Dynamics*, *68*(4), 519–530. doi:10.1007/s11071-011-0233-x

Abdelkefi, A., Nayfeh, A. H., & Hajj, M. R. (2012c). Modeling and analysis of piezoaeroelastic energy harvesters. *Nonlinear Dynamics*, *67*(2), 925–939. doi:10.1007/s11071-011-0035-1

Abdelkefi, A., Nayfeh, A. H., & Hajj, M. R. (2012d). Enhancement of power harvesting from piezoaeroelastic systems. *Nonlinear Dynamics*, *68*(4), 531–541. doi:10.1007/s11071-011-0234-9

Abdelkefi, A., & Nuhait, A. O. (2013). Modeling and performance analysis of cambered wing-based piezoaeroelastic energy harvesters. *Smart Materials and Structures*, *22*(9), 095029. doi:10.1088/0964-1726/22/9/095029

Abdelkefi, A., Scanlon, J. M., McDowell, E., & Hajj, M. R. (2013b). Performance enhancement of piezoelectric energy harvesters from wake galloping. *Applied Physics Letters*, *103*(3), 033903. doi:10.1063/1.4816075

Abdelkefi, A., Vasconcellos, R., Nayfeh, A. H., & Hajj, M. R. (2013d). An analytical and experimental investigation into limit-cycle oscillations of an aeroelastic system. *Nonlinear Dynamics*, *71*(1-2), 159–173. doi:10.1007/s11071-012-0648-z

Abdelkefi, A., Yan, Z., & Hajj, M. R. (2013c). Temperature impact on the performance of galloping-based piezoaeroelastic energy harvesters. *Smart Materials and Structures*, *22*(5), 055026. doi:10.1088/0964-1726/22/5/055026

Abdelkefi, A., Yan, Z., & Hajj, M. R. (2013e). Modeling and nonlinear analysis of piezoelectric energy harvesting from transverse galloping. *Smart Materials and Structures*, *22*(2), 025016. doi:10.1088/0964-1726/22/2/025016

Abdelkefi, A., Yan, Z., & Hajj, M. R. (2013f). Nonlinear dynamics of galloping-based piezoaeroelastic energy harvesters. *European Physical Journal-Special Issues*, *222*(7), 1483–1501. doi:10.1140/epjst/e2013-01940-x

Agneni, H. D., Mastroddi, F., & Polli, G. M. (2003). Shunted piezoelectric patches in elastic and aeroelastic vibrations. *Computers & Structures*, *81*(2), 91–105. doi:10.1016/S0045-7949(02)00392-9

Akaydin, H. D., Elvin, N., & Andreopoulos, Y. (2010). Energy harvesting from highly unsteady fluid flow using piezoelectric materials. *Journal of Intelligent Material Systems and Structures*, *21*(13), 1263–1278. doi:10.1177/1045389X10366317

Akaydin, H. D., Elvin, N., & Andreopoulos, Y. (2012). The performance of a self-excited fluidic energy harvester. *Smart Materials and Structures*, *21*(2), 025007. doi:10.1088/0964-1726/21/2/025007

Akhtar, I. (2008). *Parallel simulations, reduced-order modeling, and feedback control of vortex shedding using fluidic actuators* (PhD Dissertation). Virginia Tech, VA.

Anton, S. R., & Sodano, H. A. (2007). A review of power harvesting using piezoelectric materials (2003-2006). *Smart Materials and Structures*, *16*(3), 1–21. doi:10.1088/0964-1726/16/3/R01

Beeby, S. P., Tudor, M. J., & White, N. M. (2006). Energy harvesting vibration sources for micro-systems applications. *Measurement Science & Technology*, *17*(12), 175–195. doi:10.1088/0957-0233/17/12/R01

Ben Ayed, S., Abdelkefi, A., Najar, F., & Hajj, M. R. (2014). Design and performance of variable-shaped piezoelectric energy harvesters. *Journal of Intelligent Material Systems and Structures*, *25*(2), 174–186. doi:10.1177/1045389X13489365

Bibo, A., & Daqaq, M. F. (2014). On the optimal performance and universal design curves of galloping energy harvesters. *Applied Physics Letters*, *104*(2), 023901. doi:10.1063/1.4861599

Birkhoff, G., & Zarantanello, E. H. (1957). *Jets wakes and cavities*. New York: Academic Press.

Bryant, M., & Garcia, E. (2009). Energy harvesting: A key to wireless sensor nodes. *Proceedings of the Society for Photo-Instrumentation Engineers*, *7493*, 74931W, 74931W-8. doi:10.1117/12.845784

Carnahan, B., Luther, H. A., & Wilkes, J. O. (1969). *Applied numerical methods*. New York: John Wiley & Sons, Inc.

Chen, S. S. (1987). *Flow-induced vibration of circular cylindrical structures. DC*. Washington: Hemisphere Publishing Corporation.

Cook-Chennault, K. A., Thambi, N., & Sastry, A. M. (2008). Powering MEMS portable devices-a review of non-regenerative and regenerative power supply systems with emphasis on piezoelectric energy harvesting systems. *Smart Materials and Structures*, *17*(4), 043001. doi:10.1088/0964-1726/17/4/043001

Dai, H. L., Abdelkefi, A., & Wang, L. (2014). Theoretical modeling and nonlinear analysis of piezoelectric energy harvesting from vortex-induced vibrations. *Journal of Intelligent Material Systems and Structures*, *25*(14), 1861–1874. doi:10.1177/1045389X14538329

De Marqui, C., Erturk, A., & Inman, D. J. (2011). Modeling and analysis of piezoelectric energy harvesting from aeroelastic vibrations using the doublet-lattice method. *Journal of Vibration and Acoustics*, *133*(1), 011003. doi:10.1115/1.4002785

Den Hartog, J. P. (1956). *Mechanical Vibrations*. New York: McGraw-Hill.

Dowell, E. H., & Tang, D. (2002). Nonlinear aeroelasticity and unsteady aerodynamics. *AIAA Journal*, *40*(9), 1697–1707. doi:10.2514/2.1853

Erturk, A., Vieira, W. G. R., De Marqui, C., & Inman, D. J. (2010). On the energy harvesting potential of piezoaeroelastic systems. *Applied Physics Letters*, *96*(18), 184103. doi:10.1063/1.3427405

Facchinetti, M. L., de Langre, E., & Biolley, F. (2004). Coupling of structure and wake oscillators in vortex-induced vibrations. *Journal of Fluids and Structures*, *19*(2), 123–140. doi:10.1016/j.jfluidstructs.2003.12.004

Ghommem, M., Abdelkefi, A., Nuhait, A. O., & Hajj, M. R. (2012). Aeroelastic analysis and nonlinear dynamics of an elastically mounted wing. *Journal of Sound and Vibration*, *331*(26), 5774–5787. doi:10.1016/j.jsv.2012.07.040

Gilliat, H. C., Strganac, T. W., & Kurdila, A. J. (2003). An investigation of internal resonance in aeroelastic systems. *Nonlinear Dynamics*, *31*(1), 1–22. doi:10.1023/A:1022174909705

Goldschmidtboeing, F., & Woias, P. (2008). Characterization of different beam shapes for piezoelectric energy harvesting. *Journal of Micromechanics and Microengineering*, *18*(10), 104013. doi:10.1088/0960-1317/18/10/104013

Grouthier, C., Michelin, S., & de Langre, E. (2012). Optimal energy harvesting by vortex-induced vibrations in cables. *10th International Conference on Flow-Induced Vibration (& Flow-Induced Noise).*

Hartlen, R. T., & Currie, I. G. (1970). Lift-oscillator model of vortex-induced vibration. *Journal of the Engineering Mechanics Division*, *96*, 577–591.

Jung, H. J., & Lee, S. W. (2011). The experimental validation of a new energy harvesting system based on the wake galloping phenomenon. *Smart Materials and Structures*, *20*(5), 055022. doi:10.1088/0964-1726/20/5/055022

Karami, A., & Inman, D. J. (2011). Equivalent damping and frequency change for linear and nonlinear hybrid vibrational energy harvesting systems. *Journal of Sound and Vibration*, *330*(23), 5583–5597. doi:10.1016/j.jsv.2011.06.021

Mackowski, A. W., & Williamson, C. H. K. (2013). An experimental investigation of vortex-induced vibration with nonlinear restoring forces. *Physics of Fluids*, *25*(8), 087101. doi:10.1063/1.4819082

Magoteaux, K. C., Sanders, B., & Sodano, H. A. (2008). Investigation of energy harvesting small unmanned air vehicle. Smart Materials and Structures: Active and PassiveSmart Structures and Integrated Systems II, *Proceedings of the SPIE*. San Diego, CA, 6928. doi:10.1117/12.775851

Masana, R., & Daqaq, M. F. (2011). Relative performance of a vibratory energy harvester in mono- and bi-stable potentials. *Journal of Sound and Vibration*, *330*(24), 6036–6052. doi:10.1016/j.jsv.2011.07.031

Mehmood, A., Abdelkefi, A., Akhtar, I., Nayfeh, A.H., Nuhait, A.O., & Hajj, M.R. (2013). Linear and nonlinear active feedback controls for vortex-induced vibrations of circular cylinders. *Journal of Vibration and Control,* doi: .10.1177/1077546312469425

Mehmood, A., Abdelkefi, A., Hajj, M. R., Nayfeh, A. H., Akhtar, I., & Nuhait, A. O. (2013). Piezoelectric energy harvesting from vortex-induced vibrations of circular cylinder. *Journal of Sound and Vibration*, *332*(19), 4656–4667. doi:10.1016/j.jsv.2013.03.033

Nayfeh, A. H. (1993). *Method of Normal Forms*. New York: Wiley Interscience.

Nuhait, A. O., & Mook, T. D. (2010). Aeroelastic behavior of flat plates moving near the ground. *Journal of Aircraft*, *47*(2), 464–474. doi:10.2514/1.41186

Rosa, M., & De Marqui, C. (2014). Modeling and analysis of a piezoelectric energy harvester with varying cross-sectional area. *Journal of Shock and Vibration*, *2014*, 1–9. doi:10.1155/2014/930503

Schulz, K. W., & Kallinderis, Y. (1998). Unsteady flow structure interaction for incompressible flows using deformable hybrid grids. *Journal of Computational Physics*, *143*(2), 569–597. doi:10.1006/jcph.1998.5969

Sirohi, J., & Mahadik, M. (2011). Piezoelectric wind energy harvester for low-power sensors. *Journal of Intelligent Material Systems and Structures*, *22*(18), 2215–2228. doi:10.1177/1045389X11428366

Skop, R. A., & Balasubramanian, S. (1997). A new twist on an old model for vortex-excited vibrations. *Journal of Fluids and Structures*, *11*(4), 395–412. doi:10.1006/jfls.1997.0085

Skop, R. A., & Griffin, O. M. (1975). On a theory for the vortex-exited oscillations of flexible cylindrical structures. *Journal of Sound and Vibration*, *41*(3), 263–274. doi:10.1016/S0022-460X(75)80173-8

Sodano, H., Park, G., & Inman, D. J. (2004). A review of power harvesting from vibration using piezoelectric materials. *Journal of Shock Vibration Digest*, *36*(3), 197–205. doi:10.1177/0583102404043275

Sousa, V. C., de M Anicézio, M., De Marqui, C. Jr, & Erturk, A. (2011). Enhanced aeroelastic energy harvesting by exploiting combined nonlinearities: Theory and experiment. *Smart Materials and Structures*, *20*(9), 094007. doi:10.1088/0964-1726/20/9/094007

Stanton, S. C., McGehee, C. C., & Mann, B. P. (2009). Reversible hysteresis for broadband magnetopiezoelastic energy harvesting. *Applied Physics Letters*, *95*(17), 174103. doi:10.1063/1.3253710

Tang, L. H., & Yang, Y. (2012). A nonlinear piezoelectric energy harvester with magnetic oscillator. *Applied Physics Letters*, *101*(9), 094102. doi:10.1063/1.4748794

Theodorsen, T. (1935). *General theory of aerodynamic instability and the mechanism of flutter* (Technical Report NACA 496).

Vasconcellos, R., & Abdelkefi, A. (2014). (in press). Nonlinear dynamical analysis of an aeroelastic system with multi-segmented moment in the pitch degree-of-freedom. *Communications in Nonlinear Science and Numerical Simulation*.

Vasconcellos, R., Abdelkefi, A., Marques, F. D., & Hajj, M. R. (2012). Representation and analysis of control surface freeplay nonlinearity. *Journal of Fluids and Structures*, *31*, 79–91. doi:10.1016/j.jfluidstructs.2012.02.003

Violette, R., de Langre, E., & Szydlowski, J. (2007). Computation of vortex-induced vibrations of long structures using a wake oscillator model: Comparison with DNS and experiments. *Computers & Structures*, *85*(11-14), 1134–1141. doi:10.1016/j.compstruc.2006.08.005

Yang, J., Preidikman, S., & Balaras, E. (2008). A strongly coupled, embedded-boundary method for fluid–structure interactions of elastically mounted rigid bodies. *Journal of Fluids and Structures*, *24*(2), 167–182. doi:10.1016/j.jfluidstructs.2007.08.002

Yang, Y. W., Zhao, L. Y., & Tang, L. H. (2013). Comparative study of tip cross-sections for efficient galloping energy harvesting. *Applied Physics Letters*, *102*(6), 064105. doi:10.1063/1.4792737

Zhao, L. Y., Tang, L. H., & Yang, Y. W. (2013). Comparison of modeling methods and parametric study for a piezoelectric wind energy harvester. *Smart Materials and Structures*, *22*(12), 125003. doi:10.1088/0964-1726/22/12/125003

## KEY TERMS AND DEFINITIONS

**Electromechanical Modeling:** Is the modeling of a dynamical system taking into account both electrical and mechanical aspects.

**Energy Harvesting:** Is the concept of generating energy from unused mechanical vibrations.

**Flutter:** Is an aeroelastic instability which takes place in wings. It is characterized by a critical speed, named flutter speed, at which self-excited motions take place.

**Nonlinear Dynamics:** Is the study of dynamical systems having nonlinear effects including aerodynamic and/or structural nonlinearities.

**Piezoelectricity:** Is the ability of certain materials to generate an AC voltage when subjected to mechanical stress.

**Transverse Galloping:** Is an aeroelastic instability which takes place in prismatic structures having one-degree-of-freedom.

**Vortex-Induced Vibrations:** Are resonant transverse vibrations that occur in a circular cylinder when the vortices are shed at a frequency near one of its natural frequencies.

**Wake Galloping:** Is the concept of having two or more bluff bodies placed in a parallel manner when subjected to an incoming flow.

# Chapter 13
# Energy Harvesting from Wastewater Using Nanofluidic Reverse Electrodialysis

**Kilsung Kwon**
*Sogang University, South Korea*

**Longnan Li**
*Sogang University, South Korea*

**Daejoong Kim**
*Sogang University, South Korea*

## ABSTRACT

*Energy is one of the most important issues in the world owing to the steady growth of energy demands, serious concern about fossil fuel exhaustion, and environmental problems. Reverse electrodialysis (RED), which is a technique to capture the useful energy from mixing the solutions of different concentration, is recently recognized as an efficient method to alleviate the current issue associated with the energy. This chapter reviews the RED studies using nanoporous membranes (named as nanofluidic reverse electordialysis, NRED) in place of conventional polymer membranes having no physical pore. This chapter is made up six parts: the mechanism, the fabrication of nanostructures, the influence of operating parameters (concentration gradient and pore size), the materials, the efforts on enhancing performance (surface modification technique and large area platform), the numerical modeling, and, lastly, the feasible applications.*

## INTRODUCTION

Global energy demand has been consistently increasing. According to the report prepared by the U.S. Energy Information Administration (EIA), world energy consumption will grow by 56 percent from 2010 (524 quadrillion British thermal units (Btu)) to 2040 (820 quadrillion Btu). Most of the enlargement in energy consumption occurs in non-organization for economic co-operation and development (non-OECD) countries due to their expanding economy, as shown in Figure 1 (U. S. Energy Information Administration (EIA), 2013).

DOI: 10.4018/978-1-4666-8254-2.ch013

*Figure 1. Global total energy consumption. The blue bar represents the OECD countries and the red bar is for the non-OECD countries (U.S. EIA, 2013).*

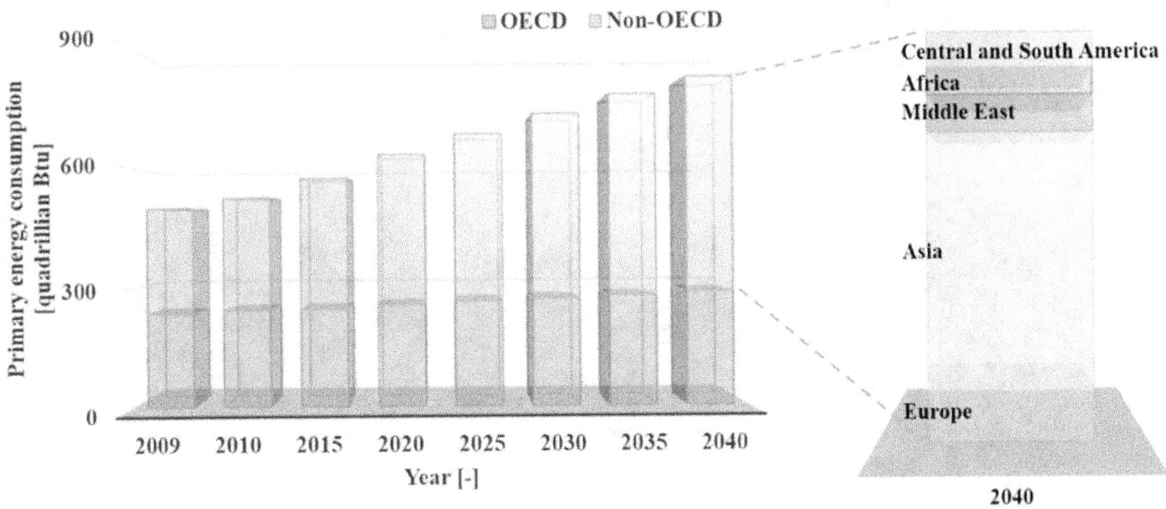


Fossil fuels like oil, natural gas, and coal are currently the main energy resources. It is expected that these will be still responsible for more than 80% of the world energy consumption in the future. The fossil fuels, however, are limited resources albeit with the increase in discoverable reserves due to technological advances. Concerns about their depletion (peak oil theory) have been continually raised by many experts during several decades (Mohr & Evans, 2008; Bardi, 2009). Environmental impacts such as climate change issues, resulting from fossil fuel-produced emissions, is also considered as a serious problem. Many countries implement emission regulations to protect the environment (Christoph et al., 2005).

The best solution for such a problems would be to use an energy resource which is sustainable and nonpolluting. Hence, various renewable energy resources including solar, wind, geothermal, hydraulic and ocean energy, have been actively studied (Lund, 2010; Panwar et al., 2011; Shi & Chew, 2012; Esteban & Leary, 2012; Kim et al., 2012; Alkan et al., 2013; Islam et al. 2013; Mahian et al., 2013; Redweik et al, 2013; Nolan et al. 2014). All the techniques are based on energy conversion to obtain a useful energy from intrinsic energy forms like potential, kinetic, and heat having renewable energy resources.

Renewable energy can also be captured with mixing two solutions having different concentrations (Vermaas, 2014). This is relatively an unconventional energy source, which many people have ignored. There however is a huge amount of potential energy. The most promising place for this concentration gradient energy is estuaries where river meets sea. The global river runoff is presently about 37300 $km^3$/year and therefore the theoretical potential energy is estimated to be 1.4~2.6 TW, which is comparable to the current generation of electricity (Vermaas, 2014). Figure 2 shows the potential energy at the major rivers of the world.

According to specific energy conversion methods, the concentration gradient energy can be mostly classified into three types: pressure retarded osmosis (PRO), reverse electrodialysis (RED), and capacitive mixing (CAPMIX). The PRO is an inverse process of reverse osmosis (RO), which is nowadays the most popular desalination process (Helfer et al., 2014). It is proposed by Sidney Loeb who is the professor at the Ben-Gurion University in 1973. The driving force is hydraulic pressure induced by water osmosis from a dilute solution (e.g. river water) to a concentrated solution (e.g. seawater) and power is

*Figure 2. Potential energy at the globally major rivers*

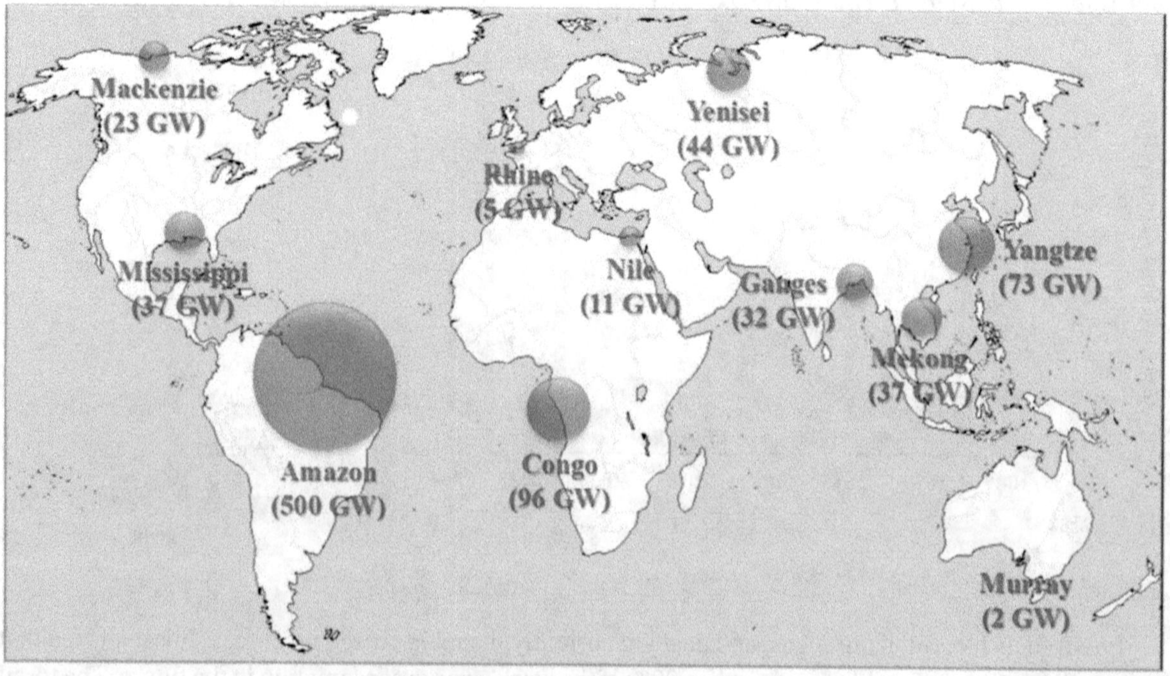


generated by depressurizing the permeation through the hydroturbine. This technique has reached the higher technological lever than the two other methods. The 4 kW pilot plant constructed by Stactkraft has been operated in Tofre, Norway. For RED, a proof-of-concept was reported by Pattle (1954) but it has not efficiently evolved due to several technological barriers. It was revives in mid 2000s due to its simple structure without moving parts (high reliability) and its direct energy conversion process (high efficiency). The number of published papers have steadily increasing, as presented in Figure 3. Ion movement through ion exchange membranes (IEMs) is the key to this method. Electricity is obtained by

*Figure 3. Increase in publications on RED between 2006 and 2013. The number of academic publications is based on the web of science and the google scholar databases.*

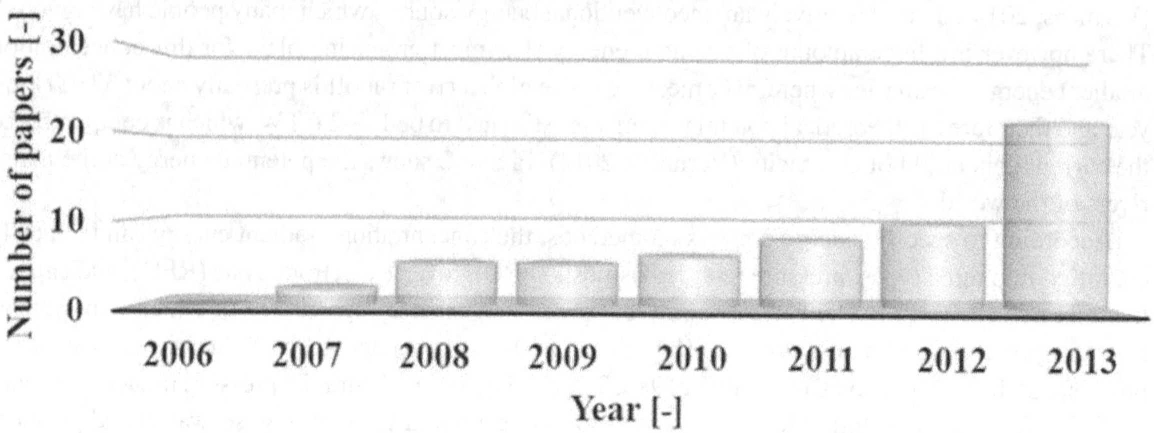

flowing electrons, generated by an oxidation reaction on electrode surfaces, through an external circuit integrated with an electrical load. The CAPMIX was first proposed in late 2000s (Brogioli, 2009). It can produces the power from a four-step cycle in contrast to the other two techniques. According to the method separating between cations and anions, there are two types of the CAPMIX: one is the capacitive double layer expansion (CDLE) using an external power supply; the other is capacitive donnan potential (CDP) employing IEMs (Rica et al., 2013).

The topic of this chapter is the review of novel RED development, one of strong candidates in concentration gradient energy. The typical RED system is composed of pumps, gaskets, spacers, electrodes, and IEMs. Figure 4 shows the schematic diagram of the typical RED system. The pumps is used to supply the diluted and concentrated solutions and circulate a so-called electrode rinse solution (redox couple) that reduces the activation loss. The gaskets are employed for preventing the leakage. The spacers located in the gaskets play three roles: one is supporting the IEMs; another is a flow path; the other is to alleviate the concentration polarization at interfaces between the IEMs and the solutions. The electrodes serve as exchanging the electrons on the surfaces and transporting them through the external circuit. The IEM is a key part in the RED because it selectively permeates the cations and anions. There are two IEM types: cation exchange membrane (CEM) and anion exchange membrane (AEM). The CEM and AEM are

*Figure 4. Schematic drawing of the RED system*

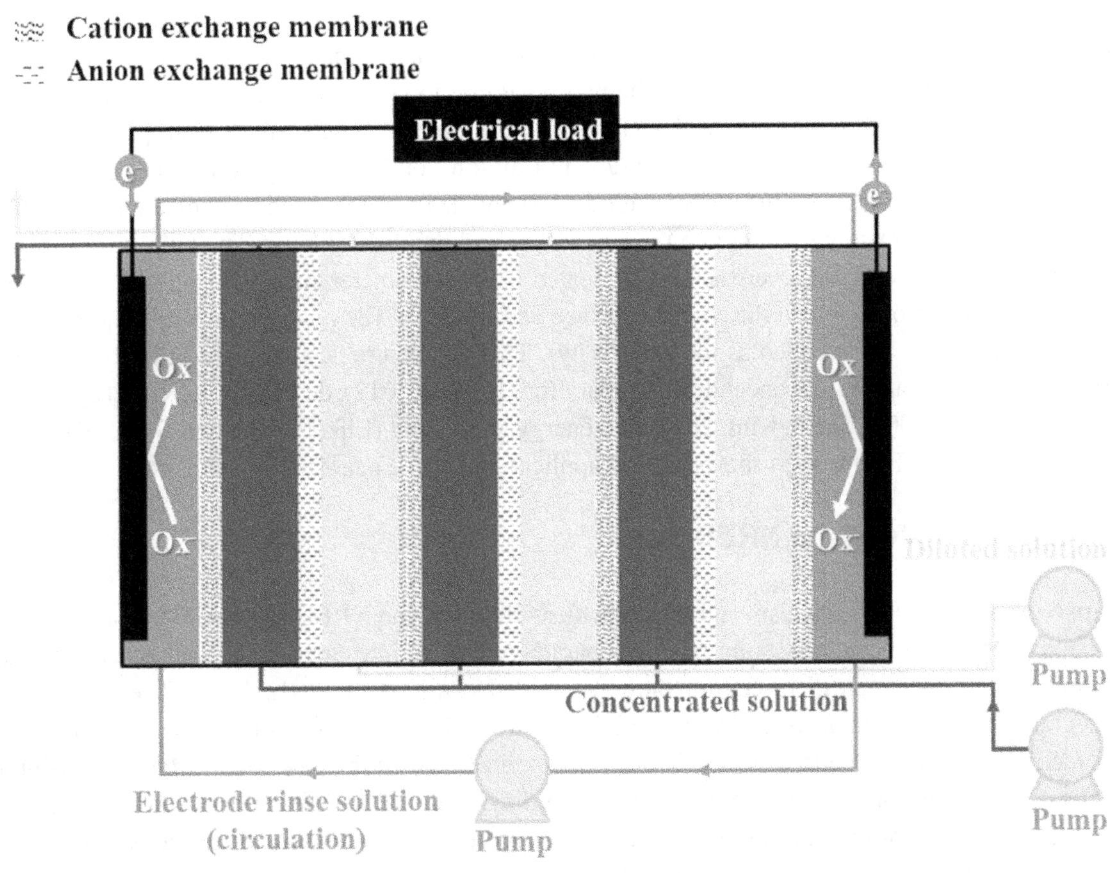

alternately placed and therefore the migrating direction of anions and cations becomes opposite, which enables the net flow of ions and electrons. Polymer membranes without a physical pore has been commonly utilized as IEMs for the RED system (Turek & Bandura, 2007; Veerman et al, 2009a; Veerman et al, 2009b). The high costs of these IEMs, however, may result in the difficulty in commercialization. A nanopore membrane is newly recognized as an alternative to replace the existing IEM (Kim & Duan et al, 2010; Cao et al, 2011; Kim et al., 2013; Kwon et al., 2014; Choi et al., 2014).

This chapter deals with the research on the RED with nanopore membranes, known as nanofluidic reverse electrodialysis (NRED). The following chapters on the NRED issues are organized as follows. Section 2 introduces the basic principle for the NRED system and effect of various parameters on the NRED performance. Section 3 summarizes existing studies on the nanoslit array, anodic aluminum oxide (AAO) membrane, track-etch membrane, and nanocrystal structure. Section 4 explains numerical approaches for system optimization. Section 5 is about suitable applications, followed lastly by the summary and the future research direction in Section 6.

## NRED SYSTEM

The appearance of novel technologies like micro electromechanical systems (MEMS) in the manufacturing sector allows the fabrication of nanometer-sized structures. New research fields have emerged in this regard. Nanofluidics, which is the study on fluid flows in the channel or pore with at least one characteristic dimension below 100 nm, is one of them. Several unique features caused by the extremely high surface to volume ratio have been found by several researchers in nanofluidics (Zangle et al, 2010; Kim et al, 2010). Figure 5 (a) presents a schematic diagram to compare a microchannel and a nanochannel. When a channel is contacted with an aqueous solution, the channel surface is typically charged due to surface ionization, ion adsorption, and/or ion dissolution. Counter-ions are dragged onto the channel surface and co-ions are repelled. An electric double layer (EDL) is thus formed on the channel surface. It can be intimately divided into the stern layer and the diffuse layer. The bulk layer exists in the micrometer-sized channel whereas it is eliminated in the nanometer-sized channel owing to strong electrostatic interaction between the charged surface and the ions. The nanochannel consequently may have perm-selectivity (also called as ion selectivity). This characteristics has been utilized in various applications such as sample preconcentration/separation (Ko et al. 2012), desalination (Kim et al. 2010b), mixing (Kim et al. 2008; Lee & Kim, 2012) and energy harvesting (Eijkel & van den Berg, 2010; Kim & Duan et al, 2010). Figure 5 (b) shows several applications of the nanofluidic platform.

### Working Principle of the NRED

Figure 6 (a) depicts the mechanism and the equivalent circuit of the NRED system. When a more concentrated and a less concentrated solutions are separately supplied into each end of the nanochannel, the counter-ions and co-ions are diffused from the concentrated solution to diluted solution. The counter-ions are migrated more easily than co-ions because of the electrostatic interaction inside the nanochannel. This asymmetric ion transport causes electrochemical redox reactions on the electrode surfaces to maintain the electro-neutrality of the solution. Electrons accordingly can be transferred to an external electric circuit, and electric power is produced at an electric load. The equivalent circuit of the NRED system is presented in Figure 6 (b). $E_{redox}$ is the potential on the electrodes resulting from the unequal voltage

*Figure 5. (a) Schematic of the microchannel (non-overlapped EDLs) and the nanochannel (overlapped EDLs) and (b) various application fields of the nanofluidic chip. These includes sample preconcentration, micromixer, portable desalination system, and energy harvesting using pressure and concentration gradient.*

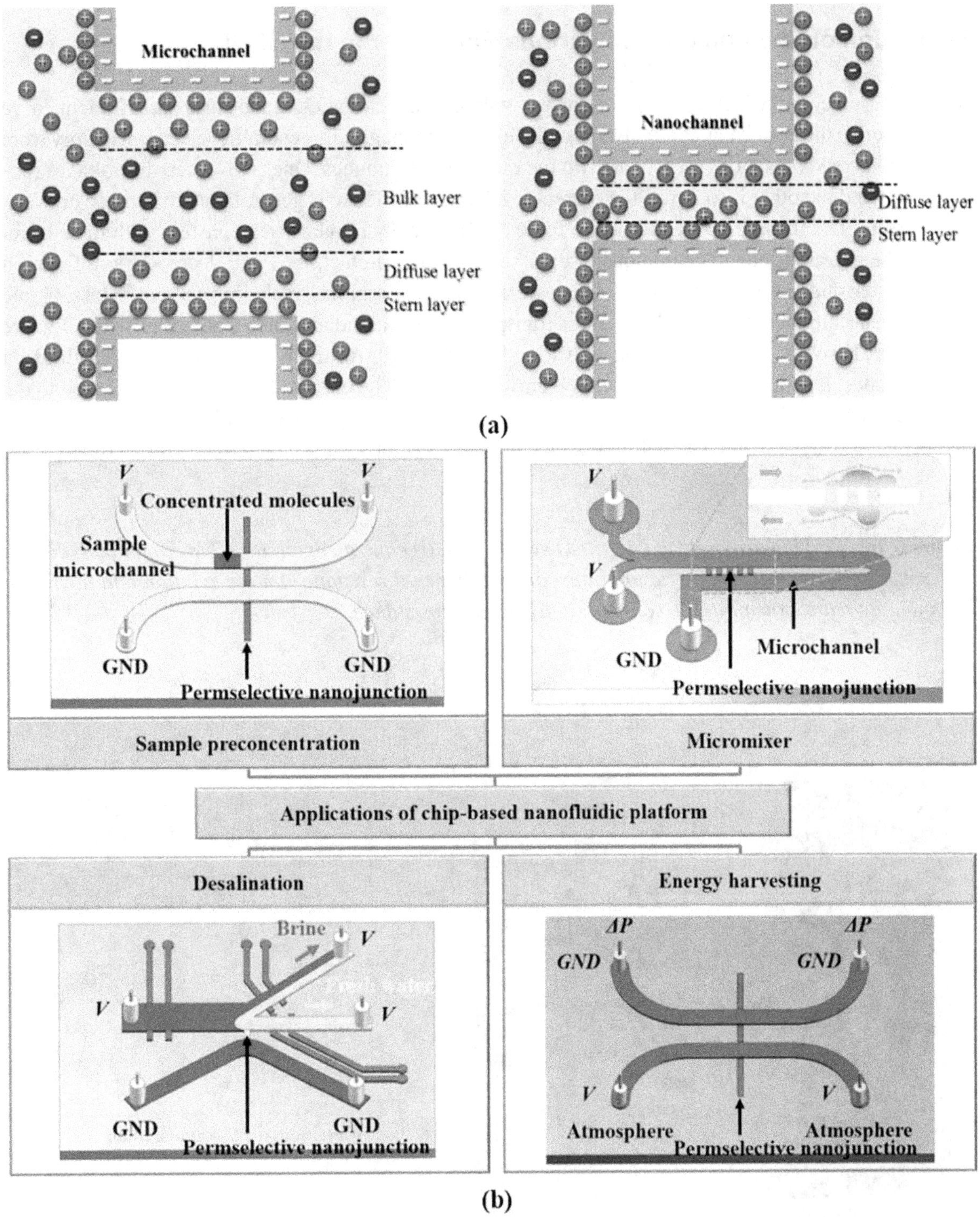

drops in different electrolyte concentrations and $E_{EMF}$ is the potential generated from the concentration gradient between both ends of the nanochannel. $I$ represents the current in NRED system. $R_{external}$ and $R_{internal}$ are electrical resistance of load and NRED system, respectively. $R_{internal}$ includes electrical resistance caused by nanochannel and solutions.

## Evaluation on the Influence of Various Parameter in the NRED

A polarization curve, which refers to the current-voltage characteristics, is the basic and efficient curve to characterize the performance of galvanic cells (electrochemical cells generating electrical energy from spontaneous redox reactions). There are two measurement techniques. One is to sweep the voltages (potentiostat) and the other is to sweep the currents (galvanostat). There is generally no difference between their results. The theoretical electromotive force (EMF) ideally remains as a constant with the current flow in the galvanic cells to continuously supply enough fuel. In fact, the EMF decreases as a function of the current due to irreversible losses. There are usually three regimes, including activation loss, ohmic loss, concentration loss (often called as mass transport loss), that induces such a potential drop (O'hayre et al. 2009; Kwon & Kim, 2013). The activation loss can result from the electrochemical reactions on the electrodes. It is dominant mainly at a low current regime. The ohmic loss is mainly attributed to the intrinsic internal resistances, which hamper the ion migration through the electrolyte. The concentration loss occurs mostly at a high current regime, where a large amount of fuel and oxidizer is fed.

*Figure 6. (a) Working principle of the NRED system and (b) equivalent circuit. This circuit includes the electromotive force by asymmetric ion transport and internal resistance on the solution and the nanochannel, the redox potential at electrodes, and the load resistance.*

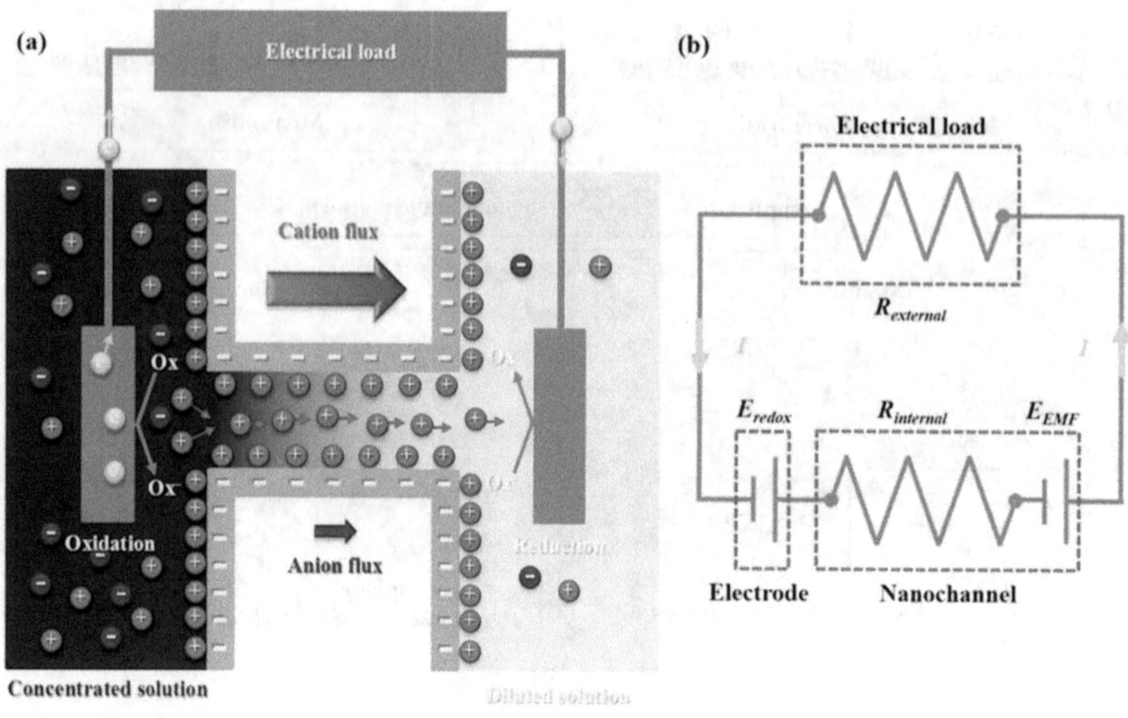

Most of NRED studies have generally reported the polarization curves in evaluating their NRED performance. Figure 7 (a) shows one example of the polarization curve in the NRED device. This plot is the experimental results from the NRED device with commercial track-etch polycarbonate membranes whose average pore size and active area is 15 nm and 0.79 $cm^2$, respectively. 300 mM and 1 M KCl solutions were used as the concentrated and diluted solutions. The voltage shows a linear trend with the increasing current density because the activation loss is largely diminished by the use of active electrodes (*e.g.* Ag/AgCl electrode). It is found that the major reason of the power generation drop is the ohmic resistance occurring at the working solutions and the nanochannel in the NRED system.

Open circuit voltage (OCV) and short circuit current (SCC) are important factors to determine the NRED performance. The OCV, the *y*-abscissa in the polarization curve, is defined as obtaining the theoretical maximum electromotive force under a given concentration difference. It can be represented through the Nernst equation (Veerman et al, 2011; Kwon et al, 2014).

$$V_{OC} = \alpha_{nanochannel} \frac{R_g T}{zF} \ln \frac{a_H C_H}{a_L C_L}. \quad (1)$$

*Figure 7. (a) Polarization curve and (b) power density curve of one example NRED system*

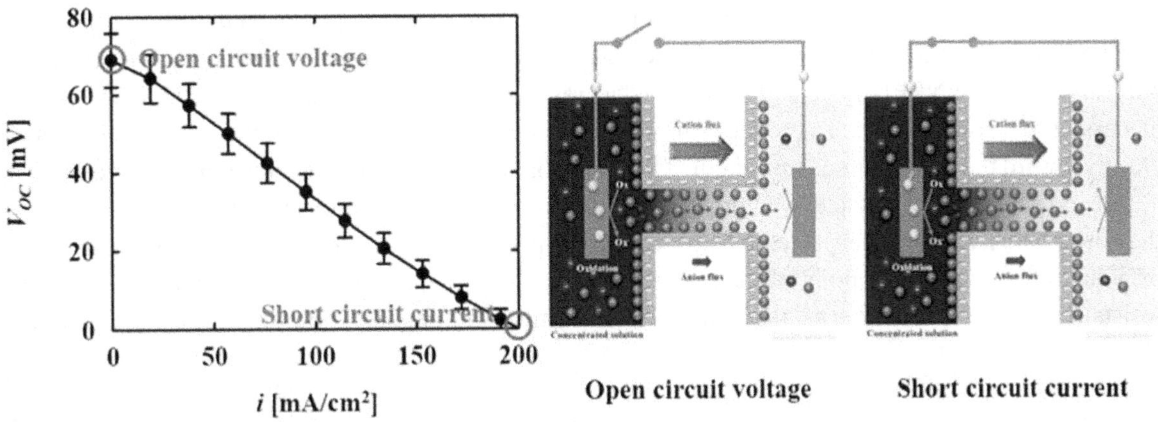


(a)

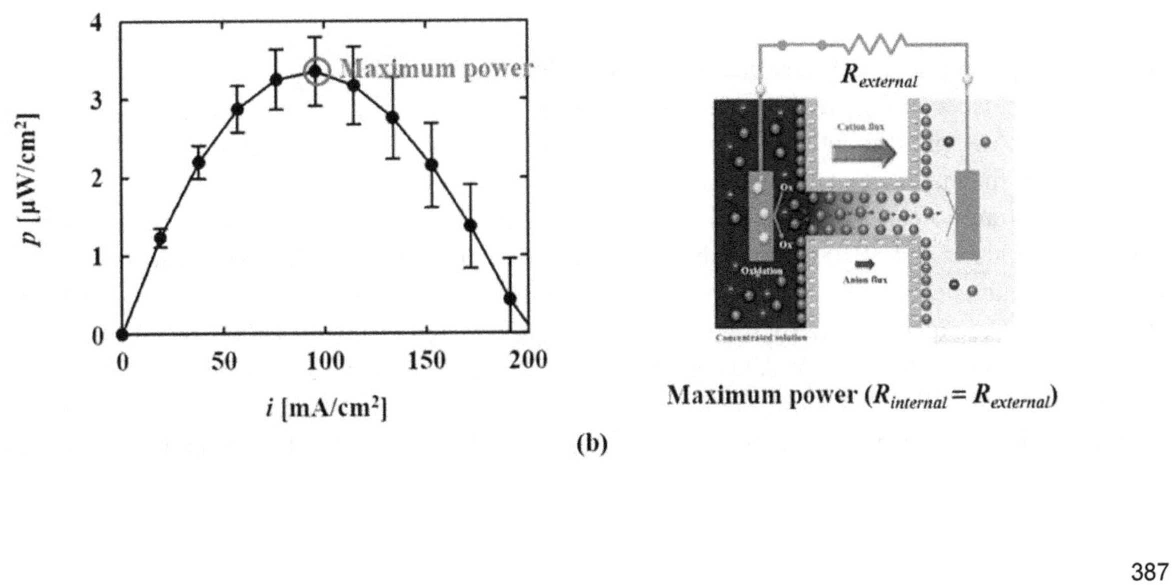


(b)

Here, $\alpha$ is the transference number which is one indicator of the permselectivity, $R$ is the universal gas constant (8.314 J/mol·K), $T$ is the operating temperature, $z$ is the ion valence number, $F$ is the Faraday constant (96485 C/mol), $a$ is the activity coefficient, and $C$ is the concentration of solution. The subscripts of $H$ and $L$ represent the concentrated and diluted solutions, respectively.

The OCV can become increasing with the concentration gradient in an ideal case where the perfect permselectivity (transference number of 1) of the nanofluidic channel and an ideal solution (no dependency of the activity coefficient on the concentration) are assumed. In reality, the permselectivity of the nanofluidic channel depends on various parameters. The concentration is one of them and it is directly related to the thickness of the EDL (also known as Debye length), which can determine the permselectivity of the nanofluidic channel. The EDL thickness can be expressed as (Probstein, 1994; Hunter, 1981)

$$\lambda_D = \sqrt{\frac{\varepsilon R_g T}{2F^2 z^2 C}}, \tag{2}$$

where ε is the permittivity of the working solution.

The EDL thickness is in inverse proportion to the square root of the concentration. The permselectivity decreases as the concentration of the working solution increases. The pH of the solutions, the intrinsic surface property of materials, and the pore size also affect on the permselectivity of the nanofluidic channel. The details about the influence of all these factors are discussed at the end of this section.

The SCC is the maximum current out of the NRED system. This can be changed along with the internal resistance. The NRED system generally has the resistance induced by the solutions located between electrodes and nanochannel. To produce a higher current, it is advantageous that the electrode is located closely toward the nanochannel, the length of the nanochannel is shortened, and the pore size is enlarged.

The power can be calculated by multiplying a terminal voltage by a circuit current. Figure 7 (b) shows the power density curve. The power density curve has a parabolic shape due to the linear ohmic behavior of polarization curve. The point of the maximum power is located at half the SCC where the internal resistance is equal to the external resistance. The maximum power generation, subsequently, is given as

$$P_{\max} = \frac{V_{OC}^2}{R_{internal}}, \tag{3}$$

where the $R_{internal}$ represents the internal resistance of the NRED system.

The performance of the NRED system is influenced by various parameters. These parameters are classified into two groups: on the working solution and on the nanochannel. Figure 8 shows the summarization of the major parameters related to NRED performance.

It is important to perform the experiments with a single nanopore device in understaning NRED phenomena. Guo et al. (2010) reported the parametric study using the NRED device made of a single nanopore membrane. The single nanopore is fabricated in the polyimide film by using the ion-track-etch method. The process is briefly summarized as follows. They first conducted the irradiation of the swift heavy ions of energy 11.4 MeV/u onto the polyimide membranes that are thickness of 12 μm; The membrane samples were contained in the single latent ion track; The etching process was conducted from one side of the membrane samples with sodium hypochlorite (active chlorine content of ~13%)

*Figure 8. Summarization of variety issues of the NRED system*

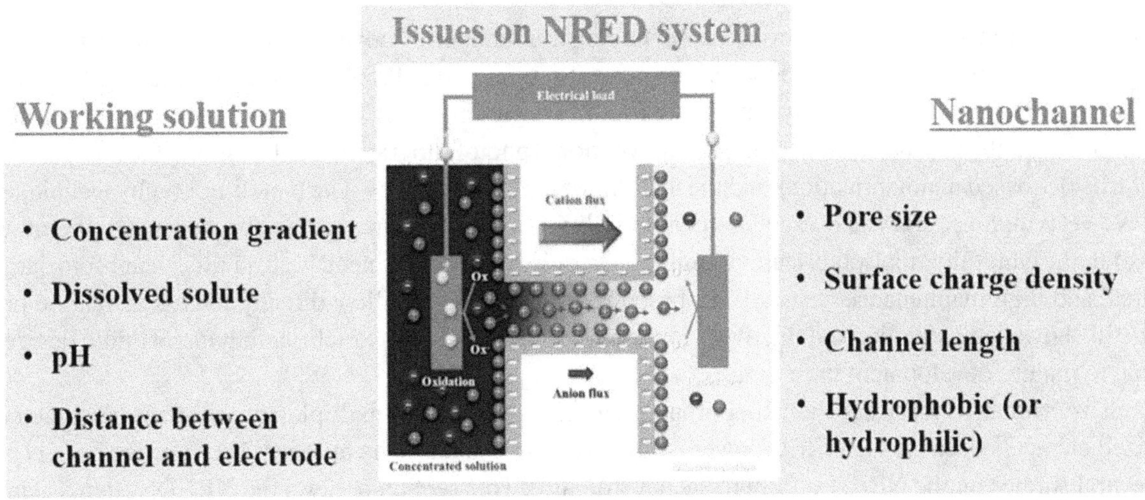


at the constant temperature of 50°C. The detailed fabrication can be found in Siwy et al. (2003). This nanopore was reported to have the conical internal shape. This study is about the evaluation on the influence of the pore diameter (four tip diameters of 23, 49, 72 and 269 nm), concentration gradient, and pH of working solutions. The potassium chloride (KCl) is used as the working solution. Similar to the case mentioned above, the OCV monotonously increases due to the decreasing permselectivity induced by the reduction of the electrostactic impact between the channel surface and the solutions. The SCC increases because of the declining resistance of the nanofluidic channel as a function of the pore size. The maximum power of 26 pW was obtained with the nanopore diameter of 72 nm in this study. The concentration gradient changed the OCV and the SCC. The OCV had the maximum value at the point between the 100 and 300 mM with the concentration of the diluted solution fixed at 1 mM while the SCC shows the dramatically growing trend. It was found that the pH enhances the permselectivity and ion conductance owing to increasing the number of deprotonated surface hydroxyl ($SiO^-$) groups (Schoch & Renaud, 2005). As a follow-up study of Guo et al., the effect of the diffusivity on the NRED performance was reported by Cao et al. (2011). They experimentally evaluated three different monovalent inorganic electrolytes: potassium chloride, sodium chloride (NaCl), and lithium chloride (LiCl). They used the bare polyimide membranes having the carboxyl groups (-COOH) on the surface of the nanopore wall and chemically modified membranes having the surface charges opposite to the bare membranes. The modification process to convert the carboxyl groups into the amino ($-NH^2$) terminated groups is as follows. The first step is the activation of the carboxyl groups into pentafluorophenyl esters by using the pentafluorophenol (PFP) and *N*-(3-dimethylaminopropyl)- *N''*-ethylacarbodiimide hydrochloride (EDC) coupling chemistry. The membranes were immersed in an ethanol solution containing 100 mM EDC and 200 mM PFP for 60 minutes at the room temperature. The second step is the covalent coupling of ethylenediamine (EDA). The membranes were immersed into a reaction vessel containing 100 mM EDA overnight. The final step is the cleaning of the membranes with the ethanol and deionized (DI) water. The harvested power was maximized with the KCl (45 pW) in the cation selective pore, while maximum power generation was obtained with the LiCl (22 pW) in the anion selective pore. It is found that the diffusion coefficient of ions is an important factor to decide the NRED performance.

## MATERIALS FOR NRED

With the advancement of the fabrication technology, a number of novel methods for nanometer-sized channels have being developed in the past decades (Sejsyan, 2011; Pimpin & Srituravanich, 2012; Duan et al., 2013). These are typically categorized into two groups: a MEMS-based nanofabrication and a nanolithography approaches excluding the conventional photolithography. Table 1 shows the comparison of MEMS-based nanofabrications and the nanolithography techniques. The nanolithography techniques have very high accuracy and excellent compatibility in fabricating nano-size structures. On the other hands, the nanolithography fabrication equipment requires high investment, high quality cleanroom, large area, and high maintenance cost. MEMS based technology cannot allow directly nano-size feature but it still shows high stability and relatively low performance/cost ratio in fabricating nanofluidic devices due to mature development these years.

It is critical to fabricate the nanoporous membrane having the multiple arrays of the nanometer-sized channels because the NRED power is linearly proportional to its number. There hence have been several studies on the NRED system in such a structure. This section reviews the NRED systems using the multiple nanopore membranes. There are four kinds of materials in these studies. These include the work on a 2D planar nanoslit, a commercial track-etch polycarbonate membrane, an anodic aluminum oxide membrane, and a nanocrystal membrane, as shown in Figure 9.

### 2D Planar Nanoslit

2D planar nanochannels are usually employed in most of nanofluidic studies. It can be fabricated in the conventional MEMS process. Figure 10 shows the schematic of the general process for planar nanochannel fabrication based on photolithography, etching, and bonding methods. This fabrication is performed in the order of the nanochannel, the microchannel, and the reservoir (Lee & Kim, 2012). The photolithography process is as follows: spin coating of a photoresist (PR), mask alignment, development, and soft/hard bake process. The nanochannel is etched by slow reactive ion etching (RIE) to obtain smooth nanochannel surfaces. The conventional Bosch deep reactive ion etching (DRIE) is normally used for the

*Table 1. Comparison of the MEMS-based nanofabrication and the nanolithography*

| | MEMS Based Nanofabrication | | Nanolithography | | |
|---|---|---|---|---|---|
| | Etching and bonding | Etching and deposition | Electron beam (EB) | Focused ion beam (FIB) | Nanoimprint (NI) |
| Feature size | Around 2 nm or larger | 30 nm or larger | Sub 10 nm or larger | Around 10 nm or larger | 20 nm or larger |
| Size controllability | Excellent | Good | Excellent | Excellent | Excellent |
| Dimension | 2D | 2D | 1D/2D | 0D/1D | 1D/2D |
| Running cost | Low | Medium | High | High | High |
| Facility | MEMS | MEMS | EBL system with MEMS | FIB system with MEMS | NI system with MEMS |
| Time | Short (60-80 wafers/hr) | Long | Long (8 hrs to write a chip pattern) | Long | Medium (> 5 wafers/hr) |

*Figure 9. Summarization of the previous studies on the NRED device using the nanoporous structures.*

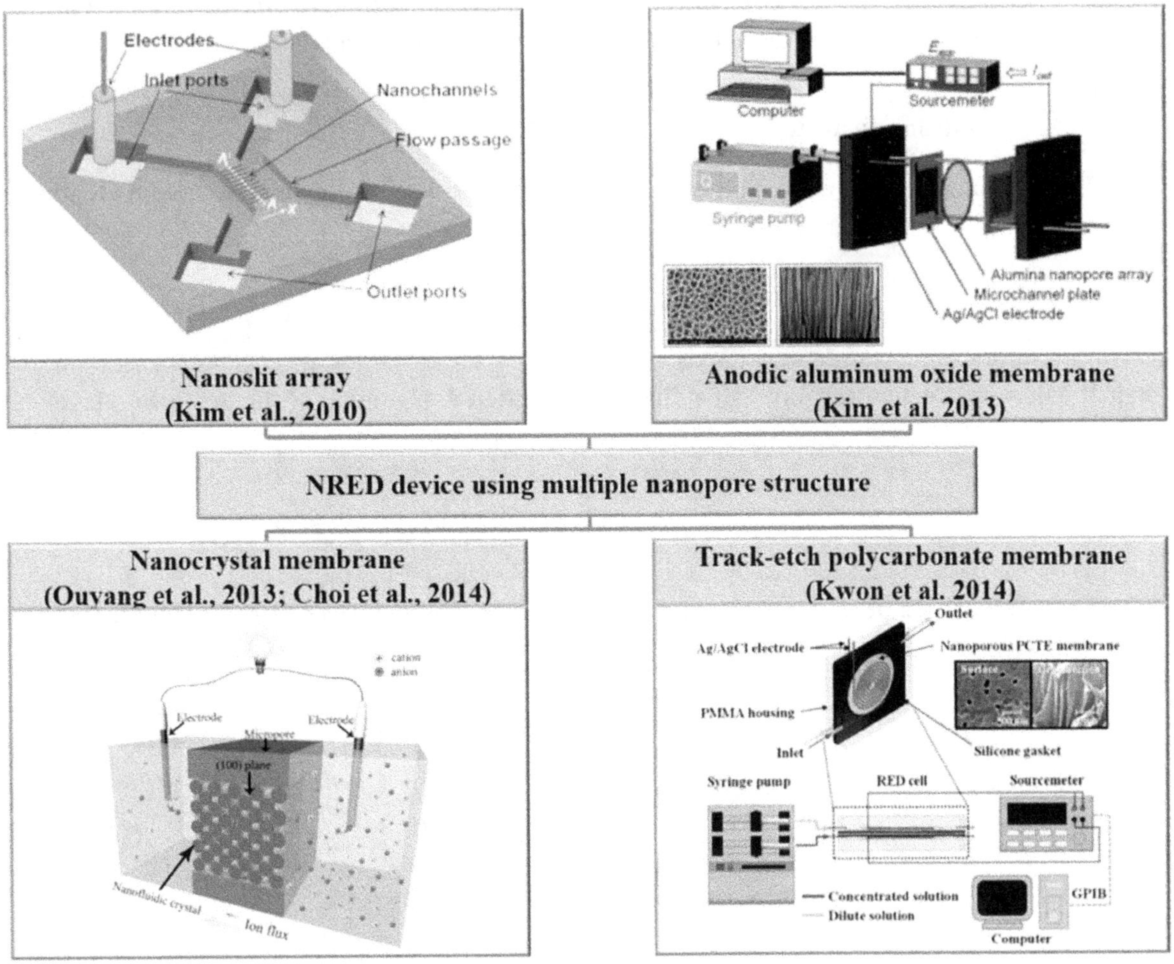


formation of microchannels. The deposition of insulation layer is an important process in applications requiring electrical connections. Thermal oxidation and chemical vapour deposition (CVD) are mainly used for its deposition. Anodic bonding, which is the last step, may be worth high attentions due to the concern about the collapse of the nanochannel. The nanochannels down to 2 nm was recently fabricated by using this approach (Duan & Majumdar, 2010).

Kim and Duan et al. (2010) studied the NRED device using the nanoslit array with the potassium chloride solutions. They evaluated the transference number, power generation, and energy conversion efficiency with the change in the concentration gradient and the nanochannel height (4, 26, and 80 nm). As consistent with the study mentioned above, the transference number decreases with the increasing nanochannel height and the increasingly concentrated solution. The power reached the near maximum with the concentrated and diluted solutions of 0.0001 and 0.1 M, respectively (Figure 11 (a)). The power density (power per effective area) increased five times that of the conventional RED system with existing IEMs. The energy conversion efficiency, which is defined as the ratio of the output energy (electrical power) to the input energy (Gibbs free energy of mixing), decreases monotonously with the concentration gradient (Figure 11 (b)).

*Figure 10. Schematic of the typical fabrication process for the 2D planar nanochannel. This process consists of seven steps about the fabrication of the nanochannel, microchannel, and reservoir, the formation of silicon dioxide, and anodic bonding between silicon and glass substrates.*

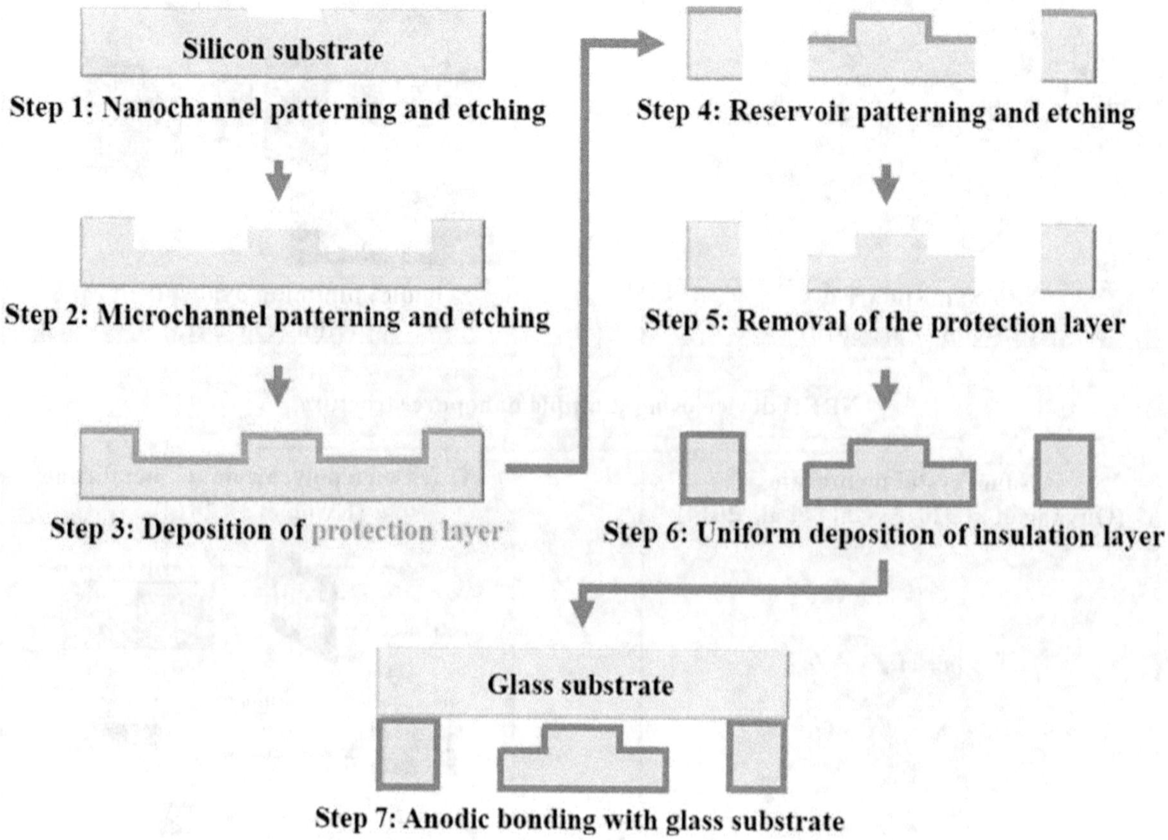


The 2D planar design is spatially ineffective to house many number of nanofluidic channels and therefore an alternate design may be needed. A 2D vertical design (with a high aspect ratio) could be an efficient possible alternative. This vertical design can be achieved with the combined etching and deposition processes (Ji et al, 2007; Wong et al, 2007; Mao & Han, 2009). Figure 12 shows the fabrication steps for the vertical nanochannel. First, a thin layer of silicon nitride was deposited on a silicon substrate as an etch mask for the following etching process. Standard photolithography was performed to define channel patterns. The narrow channel was formed by anisotropic wet etching (KOH) or DRIE. After the formation of the narrow channel, plasma-enhanced chemical vapor deposition (PECVD) oxide and low-pressure chemical vapor deposition (LPCVD) polysilicon were sequentially deposited on the trenches. Then, the thermal oxidation was conducted to reduce the gap size. Although the current technical level of the vertical nanochannel fabrication is somewhat low due to the difficulty in size controllability, this design may still be an answer for the efficient MEMS-based NRED system.

*Figure 11. (a) Maximum power output and (b) energy conversion efficiency (Kim and Duan et al., 2010)*

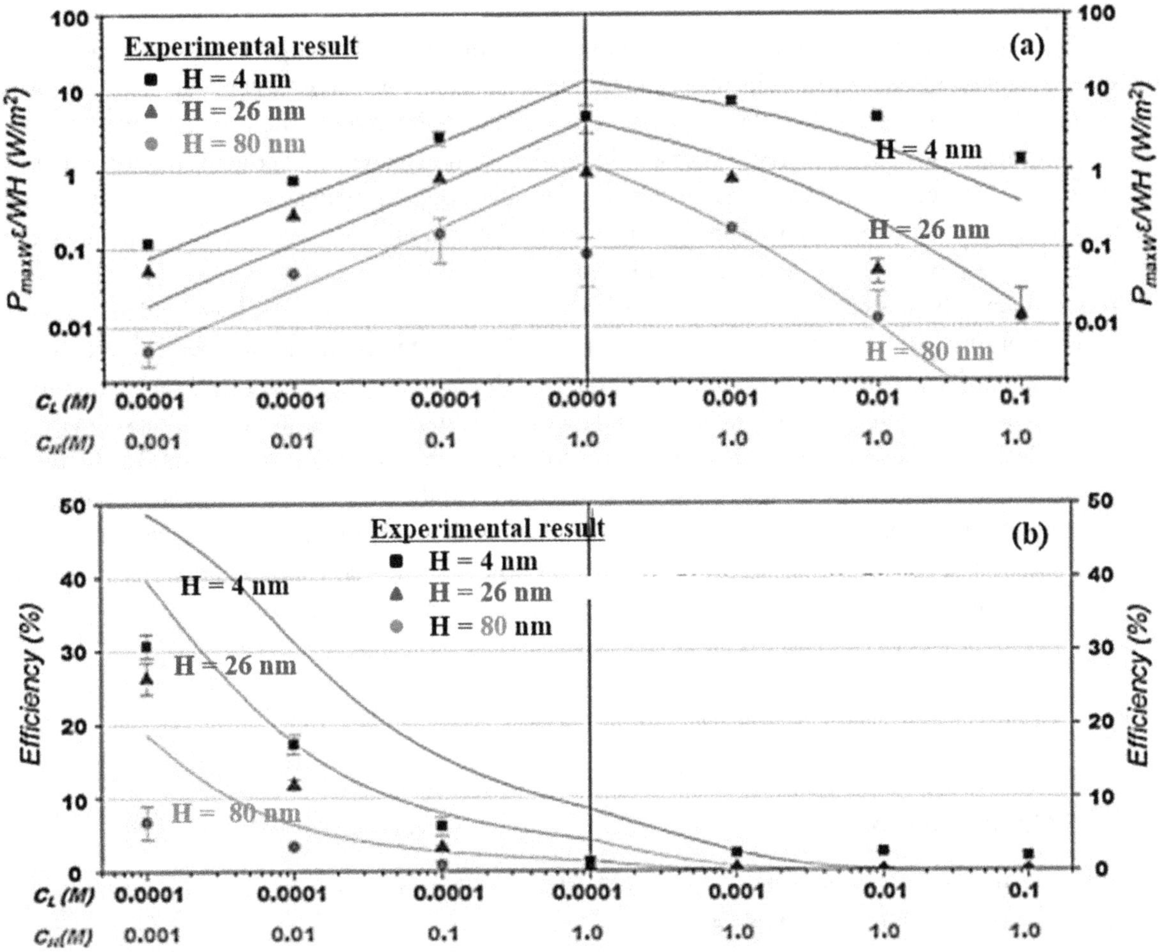


*Figure 12. Schematic of the fabrication process for the 2D vertical nanochannel*

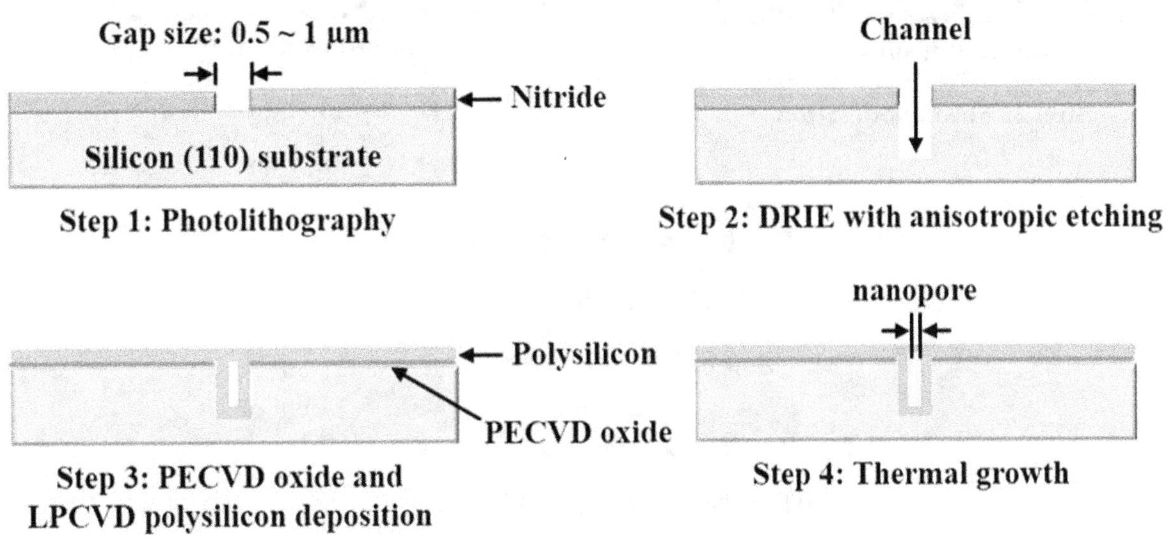

## Anodic Aluminum Oxide (AAO) Membranes

The anodization process of aluminum is known to produce highly ordered arrays of cylindrical shaped nanopores with controllable pore diameters ranging between 5 nm and 10 μm, periodicity, and density distribution (Gultepe et al., 2010; Poinern et al., 2011; Duan et al., 2013). The porous structures have several advantages such as high electrochemically insulation, transparency, chemical stability, and bio-compatibility. Consequently, various nanoscale devices have adopted them as the nanotemplate for the fabrication of the nanowire, nanotube, nanorod, and nanofilter to mitigate the undesirable matters, and as the cell culture substrate for biological applications.

The AAO membranes can be fabricated by a specialized anodization process. It is divided into three methods according to the number of steps. Among them, the two-step anodization process developed by Masuda et al. (1997) is the most popular method for the fabrication of the AAO membranes. Figure 13 shows the schematic of this two-step anodization process. The detailed process is as follows: (1) the Al sample is electropolished in perchloric acid-ethanol solution at the current density of ~500 mA/cm$^2$ for 1 min and then washed with warm and cold deionized water; (2) the polished Al sheets are anodized for 5-10 min to eliminate large ridges and to "texture" the Al surface; (3) the oxide film is removed using a mixed solution of 0.2 M $H_2C_rO_4$ and 0.4 M $H_3PO_4$ at 60°C; (4) second anodization is performed for 0.5-12 h to create long-range ordering.; (5) the aluminum substrate is eliminated.

*Figure 13. Schematic of the two-step anodization process*

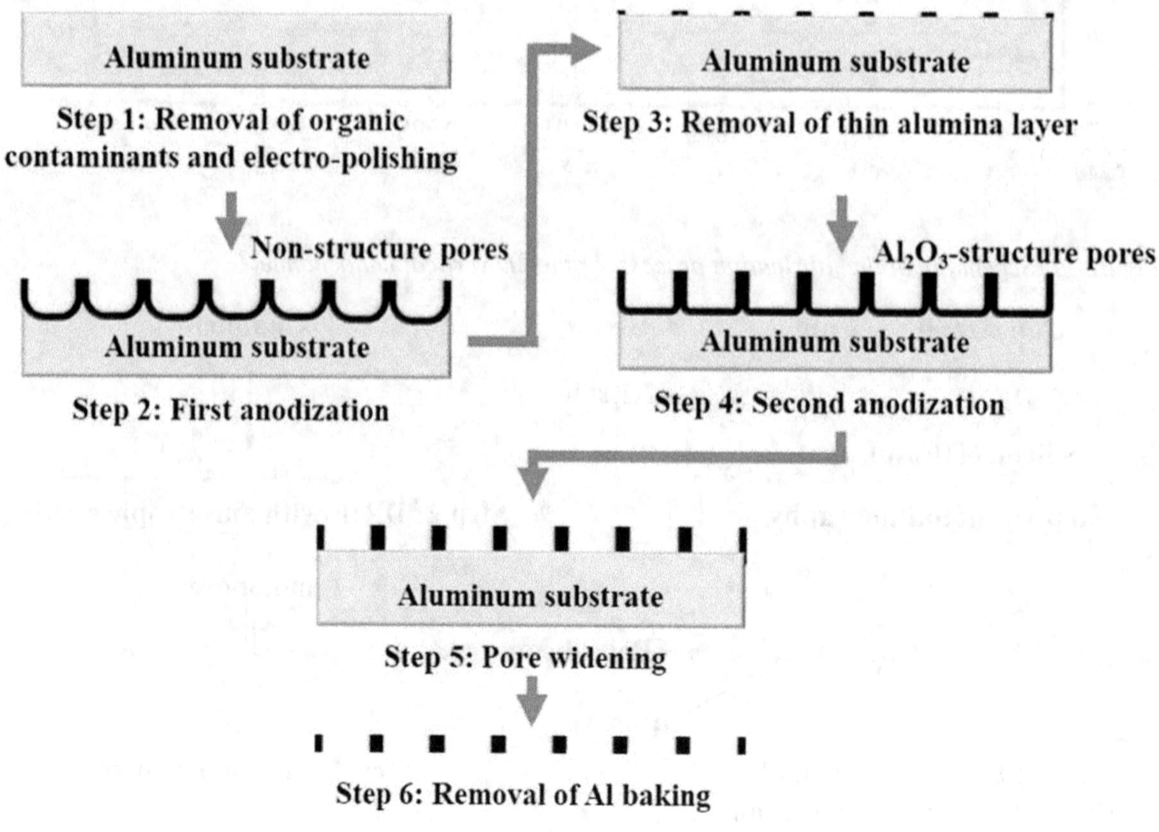

Kim et al. (2013) demonstrated the NRED device using the commercial AAO membranes (Anodisk 47, Whatman, Inc.) whose norminal pore radius and density are 10 nm and 1.18 × $10^{11}$ /$cm^2$, respectively. The permeation of the anions are far more preferred due to the positive surface charge near the neutral pH in the AAO membranes. The transference number of them is about 0.3. This value is nearly independent of the concentration below 200 mM. They also performed the evaluation on the AAO membranes coated with the silica layer. The silica layer was formed by the adsorption of $SiCl_4$ molecules followed by hydrolysis. The coated membranes allow the transport of the cations in contrast to the bare AAO membranes. The transference number was found to be 0.72 under the similar condition. The maximum power (542 nW) was greatly improved in comparison with the previous studies using the nanoslit array and the single track-etched membranes (Figure 14).

## Nanocrystal Structure

Nanocrystals (also called as photonic crystal or synthetic opal) refer to the well-organized structure by the self-assembly of nanoparticles (Duan et al., 2013). This material has being rapidly adopted with the advancement of nanoparticle synthesis. One is a nanofluidic pore system utilizing the interstices in a self-assembled nanoparticle crystal. These interstices can be generated by forming the nanoparticle crystal for example in a face-centered cubic (FCC) structure. The equivalent diameter can be estimated as 23.38% of the packed particle diameter (Chen et al., 2009).

Two approaches of nanocrystal formation was proposed for the NRED device: one is directly packing the porous membranes (Ouyang et al., 2013); the other is the chip-based fabrication (Choi et al., 2014). Ouyang et al. (2013) demonstrated the NRED device by packing the monodispersed nanoparticles with silica diameters in the micropore (40 × 40 $\mu m^2$) fabricated through the silicon microfabrication process. The effect of the nanoparticle size and concentration difference was studied in this research. The maximum output power and efficiency were 1.17 nW and 42.3%, respectively (Figure 15). Choi et al. (2014) demonstrated the chip-based NRED device using the nanocrystal structure. Its fabrication process is as follows: the PDMS mold having the shallow and deep channels was firstly fabricated from a SU-8 master; the solutions integrated with nanoparticles were supplied into the deep channels and then migrated from the deep channel to shallow channel by the pressure differences (it could not move over the another deep channel owing to the capillary stop valve); At the same time, the nanoparticles are dragged to the shallow channel and crystallized within the shallow channel by evaporation. They systematically investigated the dependence of the OCV, SCC, power generation, and energy conversion efficiency on the nanoparticle size and materials (silica and carboxylate polystyrene), and channel dimension. They achieved the maximum power density of 42.38 pW and the efficiency about 20%.

## Track-Etch Polycarbonate Membranes

Nanopore polycarbonate membranes fabricated by the track-etch method is a 1D nanoporous membrane widely used in various applications. It is commercially produced by many major companies (General Electronics (GE), Millipore, and Whataman) originally with the targeted application of filtration. There have been a number of studies on the nanofluidic systems integrated with the track-etch polycarbonate membranes because it has the good mechanical strength, thin thickness, and excellent bonding characteristics. The NRED device using the track-etch polycarbonate membrane was reported by Kwon et al. (2014). They investigated its performance according to the concentration difference (Figure 16), the pore

*Figure 14. Power generation versus electrical load (Kim et al., 2013)*

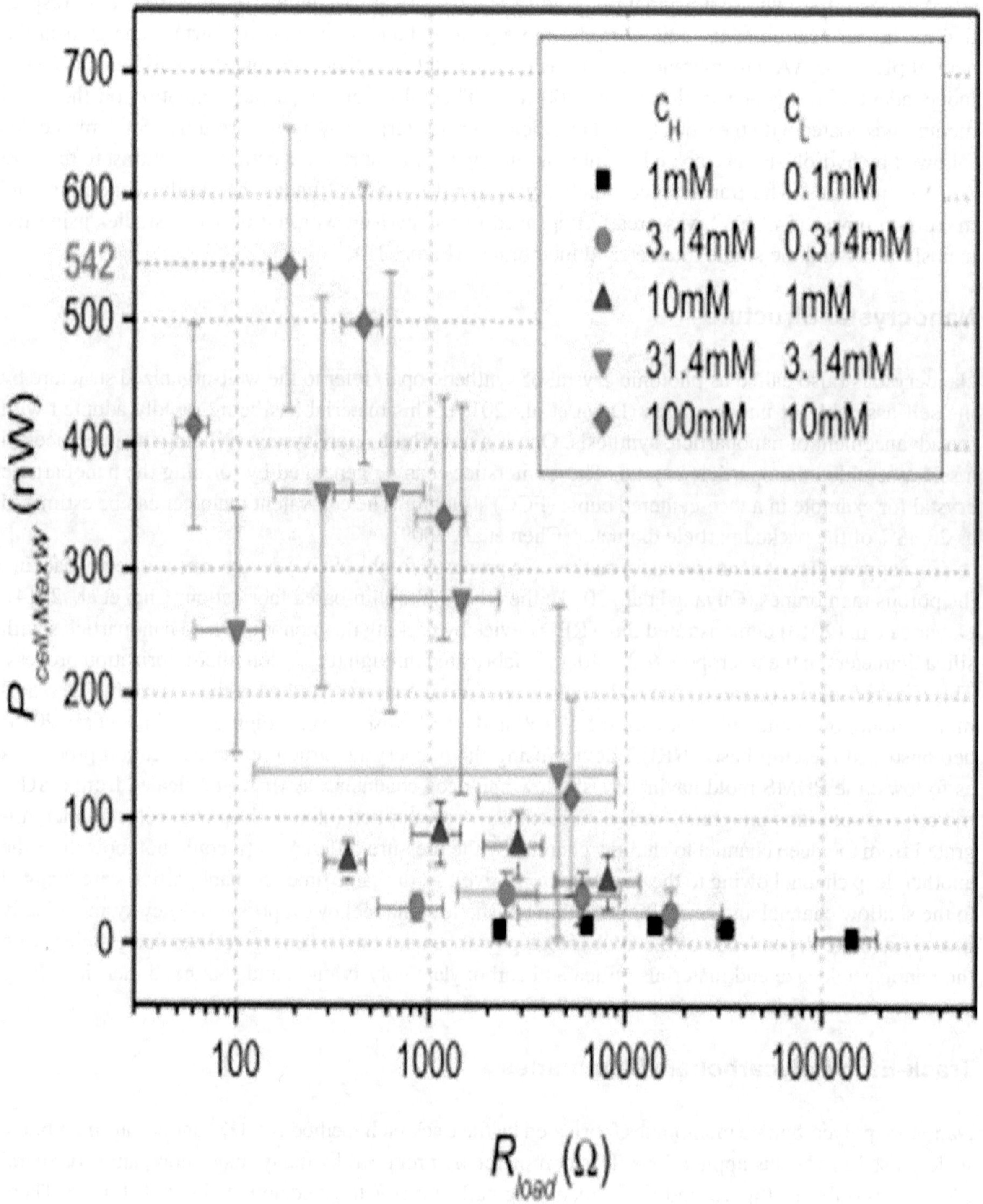

*Figure 15. (a) Maximum power density and (b) energy conversion efficiency with nanocrystal membranes (Ouyang et al, 2013)*

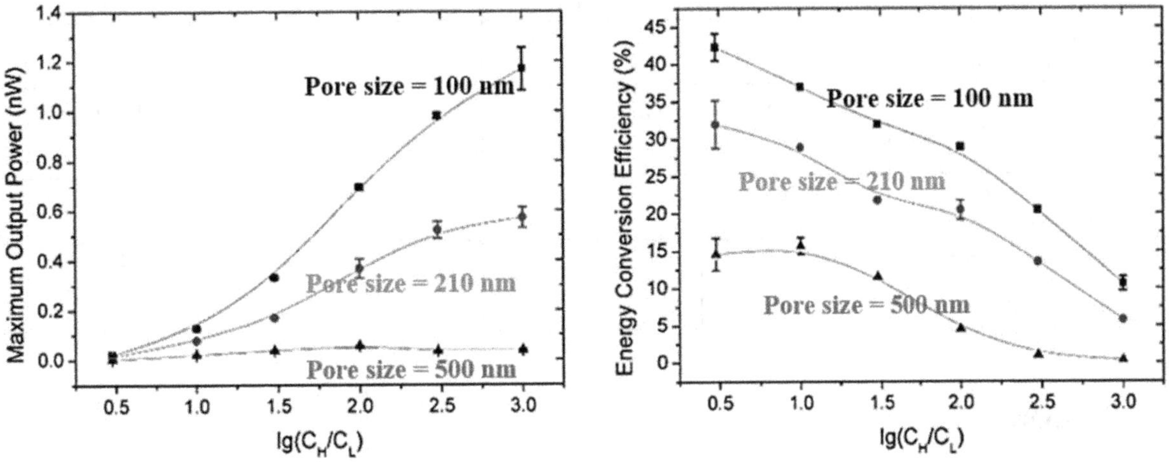


*Figure 16. Maximum power against the KCl concentration (Kwon et al., 2014)*

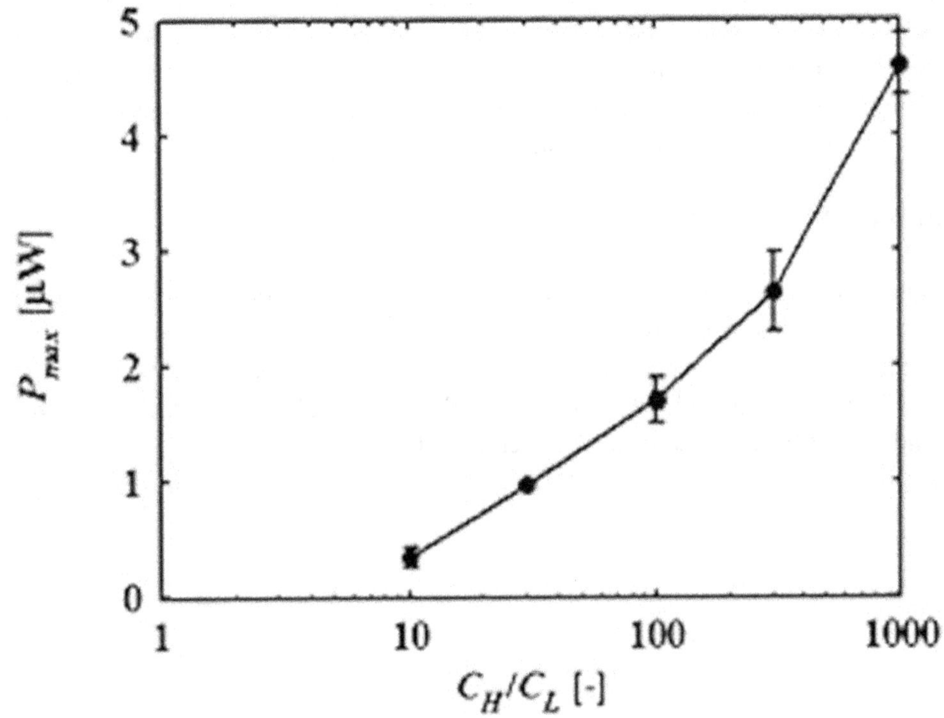


size of the membranes, and the types of electrolyte solutions. It is found that the NRED performance increases as the ion mobility increases and as the valence number decreases. In this study, the maximum power (4.8 μW) is greater than the power in NRED system with the single nanaopore membrane (Gue et al., 2010) and the nanoslits (Kim and Duan et al., 2010). Table 2 shows the performance comparison of different NRED systems.

*Table 2. Comparison of NRED system*

| NRED System | Maximum Power | Power Density [W/m²] | Efficiency[%] | Material |
|---|---|---|---|---|
| Nanoslit array (Kim et al., 2010) | 2.9 pW | 7.7 W/m² | 31% | Inorganic |
| Anodic aluminum oxide membrane (Kim et al., 2013) | 542 nW | - | - | Inorganic |
| Nanocrystal membrane (Ouyang et al, 2013) | 1170 pW | $2.82^{\pm 1.84}$ W/m² | $42.32^{\pm 1.84}$% | Inorganic |
| Track-etch polycarbonate membrane (Kwon et al. 2014) | 4.6 μW | 5.8 W/m² | - | Organic |

## MODELING

This section reports the fundamental theory of ion transport in NRED systems and the numerical method of NRED modeling. It is difficult to experimentally optimize all the major engineering parameters for the efficient ionic transport processes due to the complex connection among various physical parameters. A numerical simulation is useful to describe the ionic transport process theoretically and to establish the guidelines for the design of NRED systems. With these general theory and numerical method of NRED modeling, the NRED systems based on two kinds of nanofluidic channels (straight and conical nanochannels) are separately described below.

### Theory and Numerical Method

The continuum theory can be applied in the NRED systems as the characteristic dimension of the nanochannels is often larger than 5 nm (Bocquet & Barrat, 2007). The ionic flux through the nanochannel is contributed from ion diffusion and electromigration. The electroosmotic flow can be neglected in this case. Figure 17 shows a schematic of a NRED model. The flux of the ion species is governed by the continuity and Nernst-Planck equations (Probstein, 1994).

$$\nabla \cdot J_i = 0 \tag{4}$$

$$J_i = -D_i \nabla c_i - z_i c_i u_i F \nabla \phi \tag{5}$$

Here, $D$ is the diffusivity, $c$ is the concentration, $z$ is the valence number, $F$ is the Faraday constant, $R$ is the universal gas constant, and $T$ is the operating temperature. The subscript $i$ denotes the ion species and $u$ is the electric mobility.

The electric mobility can be expressed by Nernst-Einstein equation (Probstein, 1994).

*Figure 17. (a) Schematic of a NRED model and boundary conditions in a numerical simulation and (b) an example grid system of the NRED model*

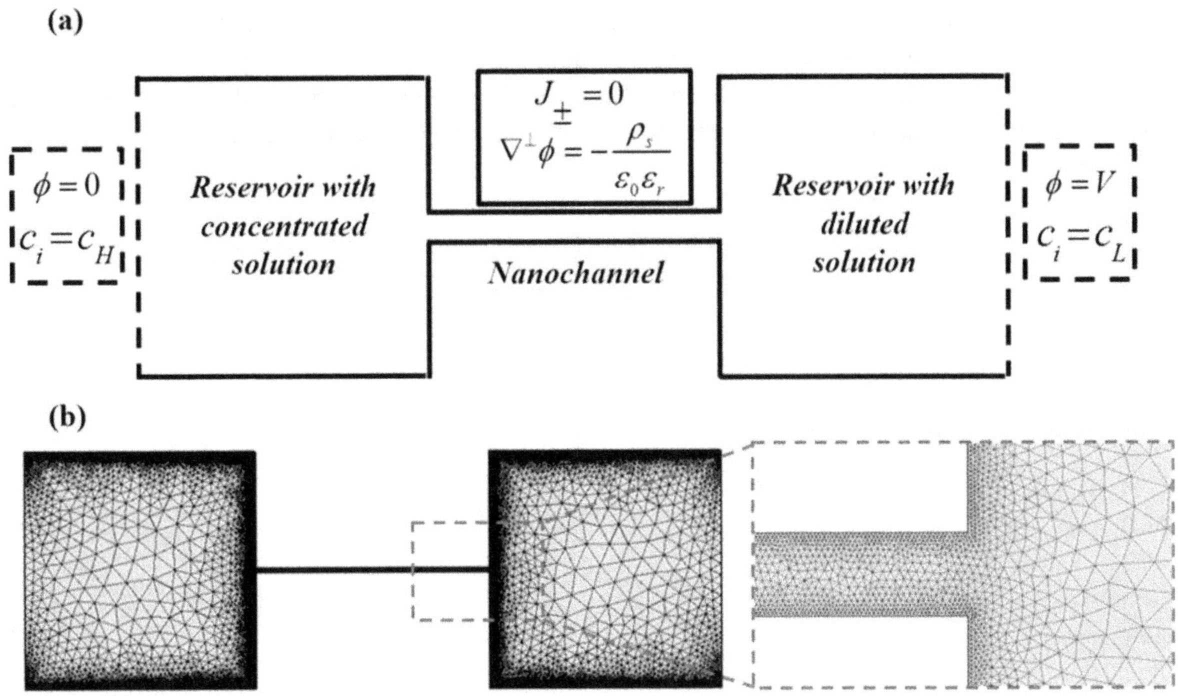


$$u_i = \frac{D_i}{RT} \tag{6}$$

The Poisson equation describes the electrical potential distribution in the nanofluidic system (Probstein, 1994).

$$\nabla^2 \phi = -\frac{\rho_e}{\varepsilon_0 \varepsilon_r} \tag{7}$$

Here, $\varepsilon_0$ is the permittivity of vacuum, $\varepsilon_r$ is the dielectric constant of the electrolyte solution. $\rho_e$ is the net electric charge density [C/m$^3$] and it can be expressed as (Probstein, 1994):

$$\rho_e = F \sum_i z_i c_i \tag{8}$$

The total electrical current through the nanochannel can be calculated by integrating the ion flux density across the cross section of nanochannel (Probstein, 1994):

$$I = z_i F \int_A (J_+ - J_-) dA \quad (9)$$

Here, $J$ is the ion flux through the nanochannel and $A$ is the cross section area of the nanochannel.

In order to describe the ionic transport in this multi-physical nature, the coupled governing equations, which is often called as Poisson-Nernst-Planck (PNP) equations (4-6), need to be solved. These coupled partial differential equations can be numerically solved under appropriate boundary conditions (Kim, 2011; Yeh et al, 2014):

$$J_{\pm}=0 \quad (10)$$

$$\nabla^{\perp}\phi = -\frac{\rho_s}{\varepsilon_0 \varepsilon_r} \quad (11)$$

$$\phi=0,\ c_i=c_H \quad (12)$$

$$\phi=V,\ c_i=c_L \quad (13)$$

Here, $^{\perp}$ denotes the wall normal component on the nanochannel and reservoir. $\rho_s$ is the surface charge density of the nanochannel surface and reservoir surface. The polarity of the surface charge is determined by the cation or anion selective nanochannel. Equation (8) corresponds to the flux of ion species through the nanochannel and reservoir wall being zero. Equation (9) corresponds to the boundary condition of the potential $\phi$ for the homogeneously charged nanochannel surface. Equation (10) and (11) are the boundary conditions at the end of the reservoir. Subscript $H$ and $L$ denote the concentrated and diluted solutions, respectively (Figure 18 (a)).

Various numerical simulation methods can be applied, for example, a finite element method (FEM) (Cao et al, 2011; Yeh et al, 2014), a finite volume method (FVM) (Mathur and Murthy, 2009; Oevermann and Klein, 2006) and a finite difference method (FDM) (Kim, 2011), in solving the PNP equations simultaneously. In order to capture the right physics, very fine grid system is needed at the nanochannel entrance and exit and near the surface of the charged nanochannel. (Figure 18 (b))

## Straight Channel

A straight (or cylindrical) nanochannel is the simplest geometry of the NRED system (Figure 14a). This kind of geometry is beneficial for directly finding the effects of basic parameters on NRED system performance due to its simplicity of geometry and the reduced computational time.

Cao et al. (2011) numerically calculated a cylindrical nanopore-based NRED model and investigated the effect of the electrolyte (*e.g.*, KCl, NaCl, LiCl and KF) on the performance of the NRED system (Figure 19). There result show that the two important parameters strongly influence the performance of the NRED system: the inherent asymmetric ion mobility and the charge properties at the nanopore

*Figure 18. (a) Schematic of a straight nanochannel based NRED system and (b) schematic of a conical nanochannel based NRED system*

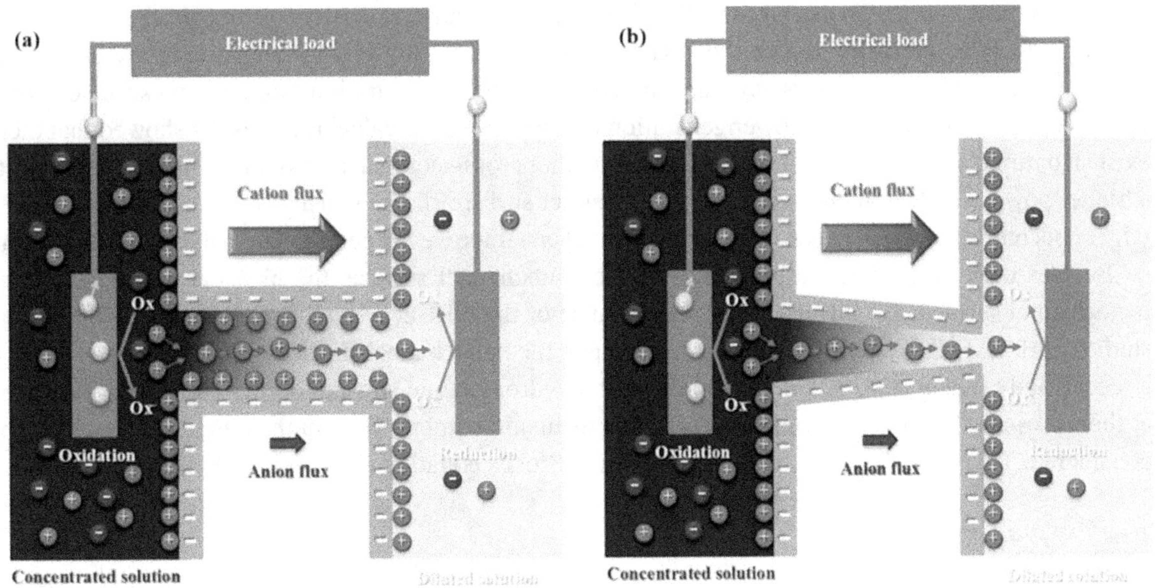


*Figure 19. Calculation results for ion concentration distribution (a, d) (identical for four types of electrolyte), local net current density (b, e) and total diffusion current through cation-(a-c) or anion-selective (d-f) nanopore. Symbols in (b-c) and (e-f) represent four types of electrolyte: KCl (square), NaCl (circle), LiCl (up triangle), and KF (down triangle) (Cao et al. (2011))*

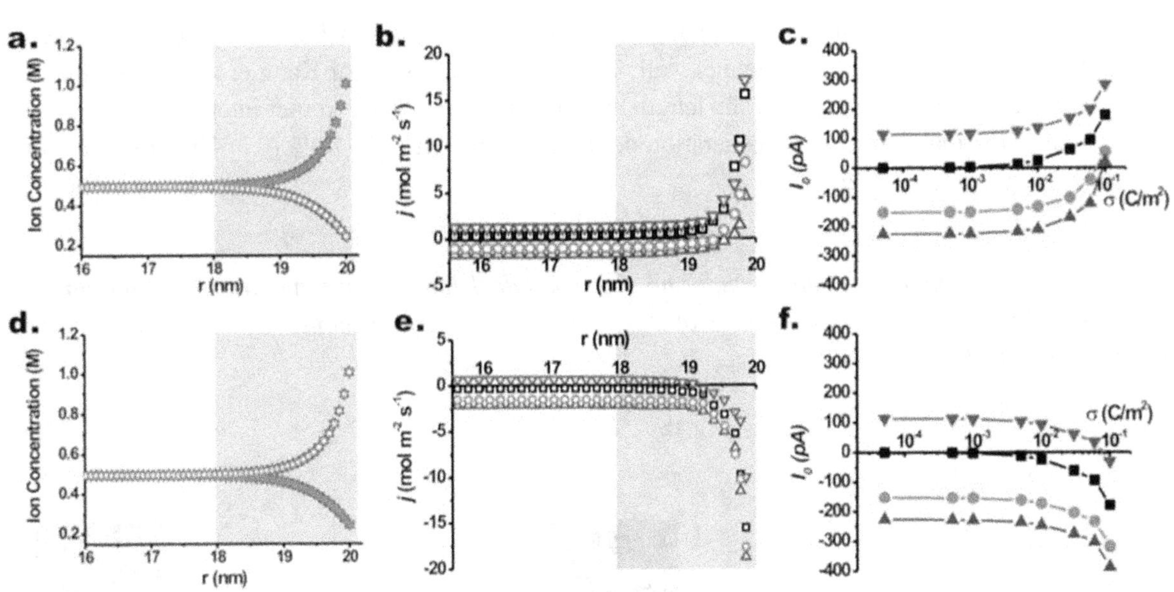

walls, which govern the charge separation and transport process. They suggested that a well-matched combination of electrolyte types and nanopore surface properties can further enhance the output power and energy conversion efficiency of the NRED system by nearly two orders of magnitude.

Kim (2011) numerically studied the NRED system with the straight nanochannel to investigate the effect of engineering parameters, *e.g.* the nanochannel size, concentration gradient and surface charge density of the nanochannel, on power generation performance. His calculation results showed that there exist an optimal nanochannel length and that the smaller cross-sectional area of the nanochannel exhibits a higher power generation density. Moreover, a higher surface charge density of the nanochannel and a larger concentration gradient across the nanochannel can improve the performance of the NRED system.

Besides the electrochemical properties of the nanochannel surface, the physical properties of the nanochannel surface also influence the performance of the NRED system. In the preceding numerical studies of the NRED system, a hydrodynamic slip at the nanochannel surface was neglected. However, a recent study (Kang et al, 2013) pointed out that a hydrodynamic slip can occurs along a solid surface at the nanoscale (Figure 20 (a)). In this case, the diffusion-osmotic flow in the nanochannel should be considered. This fluid flow can be described by the Stokes equation (Probstein, 1994):

$$0 = -\nabla P + \mu \nabla^2 \overline{u} + zF(c_+ - c_-)\nabla \phi \quad (14)$$

Here, $P$ and μ are the pressure and the viscosity of the fluid. For the boundary condition at the nanochannel surface, the slip length is considered for a partial slip condition (Kang et al, 2013):

$$u_x = b\nabla^{\perp} u_x \quad (15)$$

Here, $u_x$ is the slip velocity and $b$ (Figure 20 (b)) is the slip length, which is the vertical intercept along the axis normal to the nanochannel wall. The calculation results of Kang et al. showed that the electric conductance increases as the slip length increases (Figure 16b). Another interesting result is the peak value of the maximum power generation per unit area increases by 44% (b = 100 nm) compared to no slip case (Kang et al, 2013).

*Figure 20. (a) Schematic of comparison for no slip (solid dark line) condition and slip (solid light line) condition; (b) Counter ion concentration and velocity profile in nanochannel with slip fluid condition (Kang et al., 2011).*

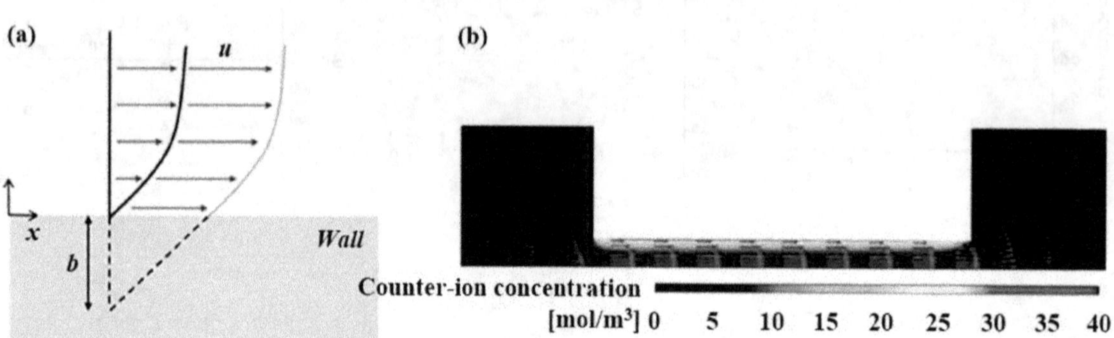

## Conical Channel

Conical nanochannels can be obtained from a track-etching process on a polymeric sheet. This kind of synthetic charged conical nanochannel has the properties of asymmetric ion transport (Figure 14b). To understand the asymmetric transport phenomena can be beneficial in designing a conical nanochannel based NRED system (Cao et al, 2011).

The mechanism of asymmetric ion transport through the conical nanochannel is originated from the interaction between the charged groups fixed on the nanochannel surface and the electrolyte ions (Cervera et al., 2005). Therefore, the conical nanochannel based NRED system exhibits different performances is from the base side to the tip side and from the tip side to the base side with the fixed concentration gradient.

Yeh et al. (2013) investigated the influence of a conical nanochannel on the performance of a NRED system through their numerical calculation. There results showed that the conical nanochannel has the direction dependence of the salt concentration gradient on their NRED performance (Figure 21). When the concentration gradient is from the base side to the tip side (Figure 21 (a)), the EDLs at the tip position are completely overlapped. In this case, the permselectivity of the conical nanochannel is increased as the transference number is increased. Usually, the large transference number indicates the higher conversion efficiency and the concentration gradient from the base side to the tip side would be beneficial to increase the efficiency of the NRED system. Other parameters, such as the nanochannel surface charge density, the magnitude of the concentration gradient across the nanochannel, showed a consistent trend with the straight nanochannel case.

## ENERGY HARVESTING APPLICATIONS

The power of the NRED system can be further enhanced by increasing the number of nanochannels, changing the channel geometry, and improving the surface properties. A NRED system contains no mechanical moving parts (such as turbines or motors) and it can use endless seawater or wastewater as a fuel. A NRED system is scalable and can be suitable for energy harvesting applications.

With a proper scaling up and assembly, a NRED system can directly convert a salinity energy from wastewater and servers as a portable energy harvester, *e.g.* for battlefield survival. A NRED system also can directly extract salinity gradient energy from a physiological environment and serve as an implantable biomedical devices that harvest electricity *in vivo*. By connecting the small sized NRED systems in series (to build up a voltage) and in parallel (to build up a current), the hierarchical integrated system may serve as a building block for a practical clean-energy-recovery plant (Guo et al, 2010; Cao et al, 2011)

An emerging application field for NRED system is a power source in micro total analysis system (μTAS) (Esquivel et al., 2012). The achievement of a higher degree of integration of the component on μTAS system requires a more stable and efficient miniature power source. A fuel cell and microbial fuel cell powered microfluidic platform for μTAS is studied in recent years (Qian et al., 2009; Esquivel et al., 2012) but all of these applications need an additional fuels such as hydrogen and bacteria for additional redox medium. A NRED system merely requires non-treated electrolytes for power generation and the NRED system can be simply integrated into the μTAS system as a power source.

*Figure 21. (a) Schematic for the power generation in base to tip diffusion NRED system and the power generation results of KCl, NaCl and LiCl solution in 5 – 55 nm and 5 – 110 nm size conical nanopore; (b) Schematic for the power generation in tip to base diffusion NRED system and the power generation results of KCl, NaCl and LiCl solution in 5 – 55 nm and 5 – 110 nm size conical nanopore (Yeh et al., 2013).*

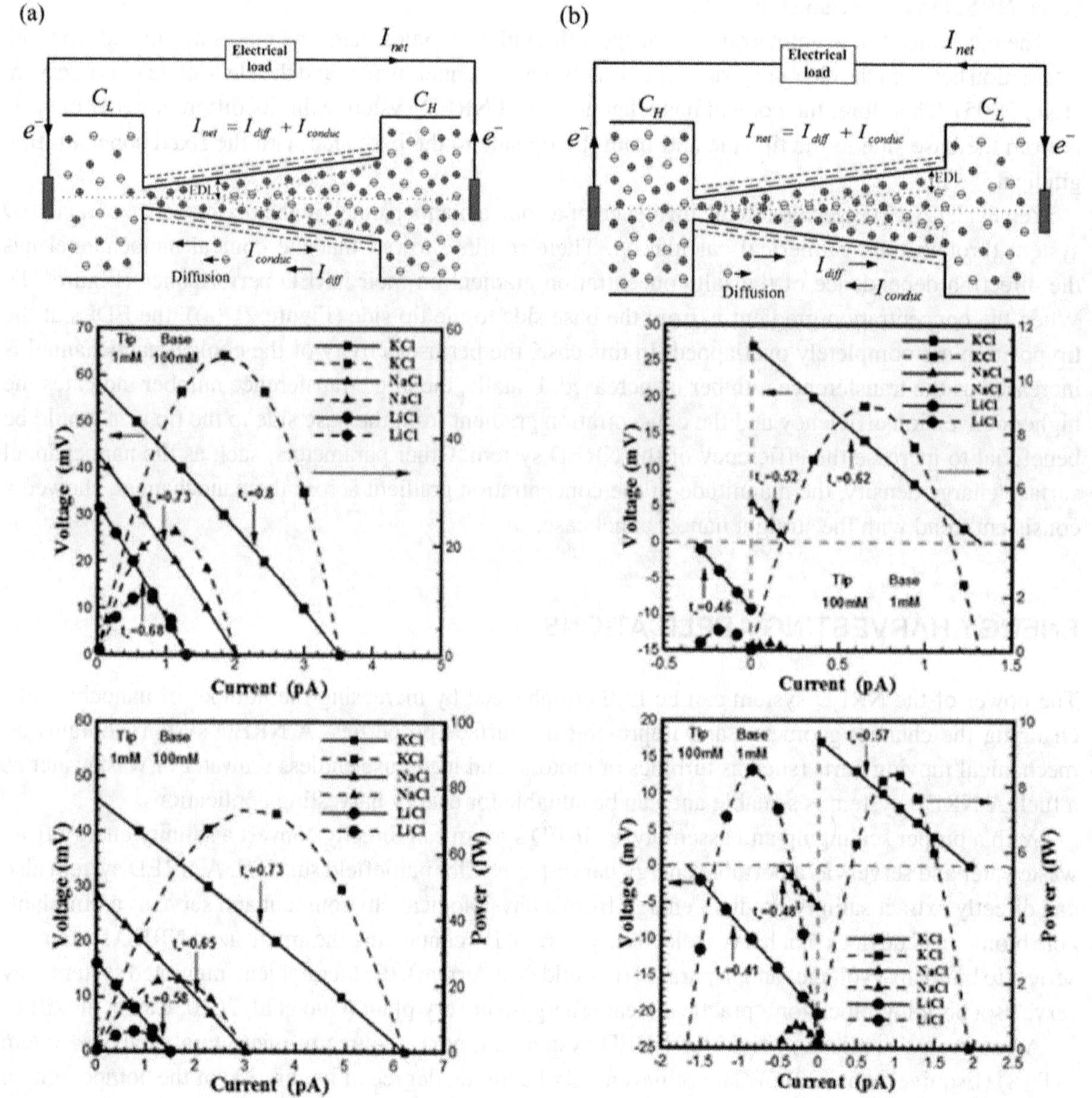

## CONCLUSION

The energy harvesting from environmental energy sources is recognized as the vital research field due to the diverse energy concerns like growing global energy demand and the depletion of fossil fuels these days. Various energy harvesting devices therefore have been developed recently. The NRED technology, one of such systems, is based on mixing two solutions of different concentrations with the utilization of nano-meter sized channels. This chapter summarizes on current research about the NRED technology. It is found that the NRED performance can be influenced by various parameters like the surface property, the pore size, the ion mobility, and the concentration gradient, as reported in a number of references. Especially, the concentration is a critical parameter because the EDL thickness determining the perm-selectivty of the nanochannel is significantly affected by the concentration. The power generation have steadily increased through adopting efficient nanoporous structures. The numerical modeling have been continuously evolved to optimize the NRED system. It is found that the NRED system can be enhanced through the optimization of not only the operating conditions but also the nanochannel shapes.

Although the NRED performance has been increasingly improved, the power generation is still low for most the target applications. Much efforts accordingly are in need for the enhancement of the power generation in the future. To achieve this goal, various surface modification techniques should be applied for the betterment of the nanopore properties and the structure should be converted from the current half-cell to the serial stack for the effective growth of the generating power.

## REFERENCES

Alkan, M. A., Keçebas, A., & Yamankaradeniz, N. (2013). Exergoeconomic analysis of a district heating system for geothermal energy using specific exergy cost method. *Energy*, *60*(1), 426–434. doi:10.1016/j.energy.2013.08.017

Bardi, U. (2009). Peak oil: The four stages of a new idea. *Energy*, *34*(3), 323–326. doi:10.1016/j.energy.2008.08.015

Bocquet, L., & Barrat, J. L. (2007). Flow boundary conditions from nano- to micro-scales. *Soft Matter*, *3*(6), 685–693. doi:10.1039/b616490k

Brogioli, D. (2009). Extracting renewable energy from a salinity difference using a capacitor. *Physical Review Letters*, *103*(5), 058501. doi:10.1103/PhysRevLett.103.058501 PMID:19792539

Cao, L., Guo, W., Ma, W., Wang, L., Xia, F., Wang, Y., & Zhu, D. et al. (2011). Towards understanding the nanofluidic reverse electrodialysis system: Well matched charge selectivity and ionic composition. *Energy & Environmental Science*, *4*(6), 2259–2266. doi:10.1039/c1ee01088c

Cervera, J., Schiedt, B., & Ramírez, P. (2005). A Poisson/Nernst-Planck model for ionic transport through synthetic conical nanopores. *Europhysics Letters*, *71*(1), 35–41. doi:10.1209/epl/i2005-10054-x

Chen, Z., Wang, Y., Wang, W., & Li, Z. (2009). Nanofluidic electrokinetics in nanoparticle crystal. *Applied Physics Letters*, *95*(10), 102105. doi:10.1063/1.3223774

Choi, E., Kwon, K., Kim, D., & Park, J. (2013). *Characterization of depletion layer formation using self-assembled nanoparticles in microchannels*. Paper presented at 2013 Transducers & Eurosensors XXVII: The 17th International Conference on Solid-State Sensors, Actuators and Microsystems, Barcelona. doi:10.1109/Transducers.2013.6627354

Choi, E., Kwon, K., Kim, D., & Park, J. (2014y). *Nanofluidic reverse electrodialysis platform using controlled assembly of nanoparticles for high power energy generation*. Paper presented at 2014 IEEE 27th International Conference on Micro Electro Mechanical Systems (MEMS), San Francisco, CA. doi:10.1109/MEMSYS.2014.6765666

Christoph, B., Tim, H., Andreas, L., Andreas, L., & Ulf, M. (2005). Assessing emission regulation in Europe: An interactive simulation approach. *The Energy Journal (Cambridge, Mass.)*, *26*(4), 1–21.

Chung, S., Yun, H., & Kamm, R. D. (2009). Nanointerstice-driven microflow. *Small*, *5*(5), 609–613. doi:10.1002/smll.200800748 PMID:19226594

Duan, C., & Majumdar, A. (2010). Anomalous ion transport in 2-nm hydrophilic nanochannels. *Nature Nanotechnology*, *5*(12), 848–852. doi:10.1038/nnano.2010.233 PMID:21113159

Duan, C., Wang, W., & Xie, Q. (2013). Review article: Fabrication of nanofluidic devices. *Biomicrofluidics*, *7*(2), 26501. doi:10.1063/1.4794973 PMID:23573176

Eijkel, J. C. T., & van der Berg, A. (2010). Nanofluidics and the chemical potential applied to solvent and solute transport. *Chemical Society Reviews*, *39*(3), 957–973. doi:10.1039/b913776a PMID:20179818

Esquivel, P. J., Castellarnau, M., Senn, T., Lochel, B., Samitier, J., & Sabate, N. (2012). Fuel cell-powered microfluidic platform for lab-on-a-chip applications. *Lab on a Chip*, *12*(1), 74–79. doi:10.1039/C1LC20426B PMID:22072241

Esteban, M., & Leary, D. (2012). Current developments and future prospects of offshore wind and ocean energy. *Applied Energy*, *90*(1), 128–136. doi:10.1016/j.apenergy.2011.06.011

Gultepe, E., Nagesha, D., Sridhar, S., & Amiji, M. (2010). Nanoporous inorganic membranes or coatings for sustained drug delivery in implantable devices. *Advanced Drug Delivery Reviews*, *62*(3), 305–315. doi:10.1016/j.addr.2009.11.003 PMID:19922749

Guo, W., Cao, L., Xia, J., Nie, F. Q., Ma, W., Xue, J., & Jiang, L. et al. (2010). Energy harvesting with single-ion selective nanopores: A concentration-gradient-driven nanofluidic power source. *Advanced Functional Materials*, *20*(8), 1339–1344. doi:10.1002/adfm.200902312

Helfer, F., Lemckert, C., & Anissimov, Y. G. (2014). Osmotic power with pressure retarded osmosis: Theory, performance, and trends – A review. *Journal of Membrane Science*, *453*(1), 337–358. doi:10.1016/j.memsci.2013.10.053

Hunter, R. J. (1981). *Zeta potential in colloid science: principles and applications*. London: Academic press.

Islam, M. R., Mekhilef, S., & Saidur, R. (2013). Progress and recent trends of wind energy technology. *Renewable & Sustainable Energy Reviews*, *21*, 456–468. doi:10.1016/j.rser.2013.01.007

Ji, L., Kim, J. K., Ji, Q., Leung, K. N., Chen, Y., & Gough, R. A. (2007). Conformal metal thin-film coatings in high-aspect ratio trenches using a self-sputtered rf-driven plasma source. *Journal of Vacuum Science & Technology B Microelectronics and Nanometer Structures*, *25*(4), 1227–1230. doi:10.1116/1.2749527

Kang, B., Kim, M., Joo, H., Kim, H. J., & Kim, D. K. (2013). *Experimental study on energy harvesting from salinity gradient by reverse electrodialysis in Ag/AgCl deposited AAO nanochannel array.* Presented at ASME 2013 International Mechanical Engineering Congress and Exposition, San Diego. doi:10.1115/IMECE2013-64364

Kang, B., Yoo, J., Kim, H. J., & Kim, D. K. (2013). Slip-enhanced reverse electrodialytic power generation in ion-selective nanochannels. *Journal of Thermal Science*, *22*(1), 36–41. doi:10.1007/s11630-013-0589-y

Kim, D. (2011). Numerical study of power generation by reverse electrodialysis in ion-selective nanochannels. *Journal of Mechanical Science and Technology*, *25*(1), 5–10. doi:10.1007/s12206-010-1113-x

Kim, D., Raj, A., Zhu, L., Masel, R. I., & Shannon, M. A. (2008). Non-equilibrium electrokinetic micro/nano fluidic mixer. *Lab on a Chip*, *8*(4), 625–628. doi:10.1039/b717268k PMID:18369520

Kim, D. K., Duan, C., Chen, Y. F., & Majumdar, A. (2010). Power generation from concentration gradient by reverse electrodialysis in ion-selective nanochannels. *Microfluidics and Nanofluidics*, *9*(6), 1215–1224. doi:10.1007/s10404-010-0641-0

Kim, G., Lee, M. E., Lee, K. S., Park, J. S., Jeong, W. M., Kang, S. K., & Kim, H. et al. (2012). An overview of ocean renewable energy resources in Korea. *Renewable & Sustainable Energy Reviews*, *16*(4), 2278–2288. doi:10.1016/j.rser.2012.01.040

Kim, J., Kim, S. J., & Kim, D. K. (2013). Energy harvesting from salinity gradient by reverse electrodialysis with anodic alumina nanopores. *Energy*, *51*(1), 413–421. doi:10.1016/j.energy.2013.01.019

Kim, S. J., Ko, S. H., Kang, K. H., & Han, J. (2010b). Direct seawater desalination by ion concentration polarization. *Nature Nanotechnology*, *5*(4), 297–301. doi:10.1038/nnano.2010.34 PMID:20305644

Kim, S. J., Song, Y. A., & Han, J. (2010a). Nanofluidic concentration devices for biomolecules utilizing ion concentration polarization: Theory, fabrication and applications. *Chemical Society Reviews*, *39*(3), 912–922. doi:10.1039/b822556g PMID:20179814

Ko, S. H., Song, Y. A., Kim, S. J., Kim, M., Han, J., & Han, J. (2012). Nanofluidic preconcentration device in a straight microchannel using ion concentration polarization. *Lab on a Chip*, *12*(21), 4472–4482. doi:10.1039/c2lc21238b PMID:22907316

Kwon, K., & Kim, D. (2013). Efficient water recirculation for portable direct methanol fuel cells using electroosmotic pumps. *Journal of Power Sources*, *221*(1), 172–176. doi:10.1016/j.jpowsour.2012.07.113

Kwon, K., Lee, S. J., Li, L., Han, C., & Kim, D. (2014). Energy harvesting system using reverse electrodialysis with nanoporous polycarbonate track-etch membranes. *International Journal of Energy Research*, *38*(4), 530–537. doi:10.1002/er.3111

Lee, S. J., & Kim, D. (2012). Millisecond-order rapid micromixing with non-equilibrium electrokinetic phenomena. *Microfluidics and Nanofluidics*, *12*(6), 897–906. doi:10.1007/s10404-011-0918-y

Liu, Q., Wang, Y., Guo, W., Ji, H., Xue, J., & Ouyang, Q. (2007). Asymmetric properties of ion transport in a charged conical. *Physical Review E: Statistical, Nonlinear, and Soft Matter Physics*, *75*(5), 051201. doi:10.1103/PhysRevE.75.051201 PMID:17677044

Loeb, S. (1976). Production of energy from concentrated brines by pressure-retarded osmosis: I. Preliminary technical and economic correlations. *Journal of Membrane Science*, *1*, 49–63. doi:10.1016/S0376-7388(00)82257-7

Lund, J. W. (2010). Direct utilization of geothermal energy. *Energies*, *3*(8), 1443–1471. doi:10.3390/en3081443

Mahian, O., Kianifar, A., Kalogirou, S. A., Pop, L., & Wongwises, S. (2013). A review of the applications of nanofluids in solar energy. *International Journal of Heat and Mass Transfer*, *57*(2), 582–594. doi:10.1016/j.ijheatmasstransfer.2012.10.037

Mao, P., & Han, J. (2009). Massively-parallel ultra-high-aspect-ratio nanochannels as mesoporous membranes. *Lab on a Chip*, *9*(4), 586–591. doi:10.1039/B809370A PMID:19190794

Masuda, H., Yamada, H., Satoh, M., Asoh, H., Nakao, M., & Tamamura, T. (1997). Highly ordered nanochannel-array architecture in anodic alumina. *Applied Physics Letters*, *71*(19), 2770. doi:10.1063/1.120128

Mathur, S., & Murthy, J. (2009). A multigrid method for the Poisson–Nernst–Planck equations. *International Journal of Heat and Mass Transfer*, *52*(17-18), 4031–4039. doi:10.1016/j.ijheatmasstransfer.2009.03.040

Mohr, S. H., & Evans, G. M. (2008). Peak oil: Testing Hubbert's curve via theoretical modeling. *Natural Resources Research*, *17*(1), 1–11. doi:10.1007/s11053-008-9059-8

Nolan, P., Lynch, P., & Sweeney, C. (2014). Simulating the future wind energy resource of Ireland using the COSMO-CLM model. *Wind Energy (Chichester, England)*, *17*(1), 19–37. doi:10.1002/we.1554

O'hayre, R., Cha, S.-W., Colella, W., & Prinz, F. B. (2009). *Fuel cell fundamentals* (2nd ed.). Hoboken, New Jersey: John Wiley & Sons, Inc.

Oevermann, M., & Klein, R. (2006). A Cartesian grid finite volume method for elliptic equations with variable coefficients and embedded interfaces. *Journal of Computational Physics*, *219*(2), 749–769. doi:10.1016/j.jcp.2006.04.010

Ouyang, W., Wei, W., Zhang, H., Wu, W., & Li, Z. (2013, January). *A low-cost, high-efficiency and high output-power nanofluidic energy harvester*. Paper presented at 2013 IEEE 26th International Confernece on Micro Electro Mechanical Systems (MEMS). Taipei. doi:10.1109/MEMSYS.2013.6474367

Ouyang, W., Wei, W., Zhang, H., Wu, W., & Li, Z. (2013). Nanofluidic crystal: A facile, high efficiency and high power density scaling up scheme for energy harvesting based on nanofluidic reverse electrodialysis. *Nanotechnology*, *24*(34), 345401. doi:10.1088/0957-4484/24/34/345401 PMID:23899953

Panqar, N. L., Kaushik, S. C., & Kothari, S. (2011). Role of renewable energy sources in environmental protection: A review. *Renewable & Sustainable Energy Reviews*, *15*(3), 1513–1524. doi:10.1016/j.rser.2010.11.037

Pattle, R. E. (1954). Production of electric power by mixing fresh and salt water in the hydroelectric pile. *Nature*, *174*(4431), 660. doi:10.1038/174660a0

Pimpin, A., & Srituravanich, W. (2012). Review on micro- and nanolithography techniques and their applications. *English Journal*, *16*(1), 37–55.

Poinern, G. E., Ali, N., & Fawcett, D. (2011). Progress in nano-engineered anodic aluminum oxide membrane development. *Materals*, *4*(3), 487–526. doi:10.3390/ma4030487

Probstein, R. F. (1994). *Physicochemical hydrodynamics*. New York: Wiley. doi:10.1002/0471725137

Qian, F., Baum, M., Gu, Q., & Morse, F. D. (2009). A 1.5 μL microbial fuel cell for on-chip bioelectricity generation. *Lab on a Chip*, *9*(21), 3076–3081. doi:10.1039/b910586g PMID:19823722

Redweik, P., Catita, C., & Brito, M. (2013). Solar energy potential on roofs and facades in an urban landscape. *Solar Energy*, *97*, 332–341. doi:10.1016/j.solener.2013.08.036

Revil, A., Pezard, P. A., & Glover, P. W. J. (1999). Streaming potential in porous media 1. Theory of the zeta potential. *Journal of Geophysical Research*, *104*(B9), 20021–20031. doi:10.1029/1999JB900089

Rica, R. A., Ziano, R., Salerno, D., Mantegazza, F., van Roij, R., & Brogioli, D. (2013). Capacitive mixing for harvesting the free energy of solutions at different concentrations. *Entropy*, *15*(4), 1388–1407. doi:10.3390/e15041388

Schoch, R. B., & Renaud, P. (2005). Ion transport through nanoslits dominated by the effective surface charge. *Applied Physics Letters*, *86*(25), 253111. doi:10.1063/1.1954899

Seisyan, R. P. (2011). Nanolithography in microelectronics: A review. *Technical Physics*, *56*(8), 1–14. doi:10.1134/S1063784211080214

Shi, L., & Chew, M. Y. L. (2012). A review on sustainable design of renewable energy systems. *Renewable & Sustainable Energy Reviews*, *16*(1), 192–207. doi:10.1016/j.rser.2011.07.147

Siwy, Z., Dobrev, D., Neumann, R., Trautmann, C., & Voss, K. (2003). Electro-responsive asymmetric nanopores in polyimide with stable ion-current signal. *Applied Physics. A, Materials Science & Processing*, *76*(5), 781–785. doi:10.1007/s00339-002-1982-7

Turek, M., & Bandura, B. (2007). Renewable energy by reverse electrodialysis. *Desalination*, *205*(1-3), 67–74. doi:10.1016/j.desal.2006.04.041

U. S. Energy Information Administration (EIA). (2013). *Annual energy outlook 2013*. Retrieved from http:www.eia.gov/forecasts/aeo/pdf/0383(2013).pdf

Veerman, J., de Jong, R. M., Saakes, M., Metz, S. J., & Harmsen, G. J. (2009a). Reverse electrodialysis: Comparison of six commercial membrane pairs on the thermodynamic efficiency and power density. *Journal of Membrane Science*, *343*(1-2), 7–15. doi:10.1016/j.memsci.2009.05.047

Veerman, J., Saakes, M., Metz, S. J., & Harmsen, G. J. (2009b). Reverse electrodialysis: Performance of a stack with 50 cells on the mixing of sea and river water. *Journal of Membrane Science*, *327*(1-2), 136–144. doi:10.1016/j.memsci.2008.11.015

Veerman, J., Saakes, M., Metz, S. J., & Harmsen, G. J. (2011). Reverse electrodialysis: A validated process model for design and optimization. *Chemical Engineering Journal*, *166*(1), 256–268. doi:10.1016/j.cej.2010.10.071

Vermaas, D. A. (2014). *Energy generation from mixing salt water and fresh water: Smart flow strategies for reverse electrodialysis.* (Unpublished doctoral dissertation). University of Twente.

Wong, C. C., Agarwai, A., Balasubramanian, N., & Kwong, D. L. (2007). Fabrication of self-sealed circular nano/microfluidic channels in glass substrates. *Nanotechnology*, *18*(13), 135304. doi:10.1088/0957-4484/18/13/135304 PMID:21730378

Yeh, H. C., Chang, C. C., & Yang, R. J. (2014). Reverse electrodialysis in conical-shaped nanopores: Salinity gradient-driven power generation. *RSC Advances*, *4*(6), 2705–2714. doi:10.1039/C3RA45392H

Zangle, T. A., Mani, A., & Santiago, J. G. (2010). Theory and experiments of concentration polarization and ion focusing at microchannel and nanochannel interfaces. *Chemical Society Reviews*, *39*(3), 1014–1035. doi:10.1039/b902074h PMID:20179822

## ADDITIONAL READING

Abdu, S., Marti-Calatayud, M. C., Wong, J. E., Garcia-Gabaldon, M., & Wessling, M. (2014). Layer-by-layer modification of cation exchange membranes controls ion selectivity and water splitting. *ACS Applied Materials & Interfaces*, *6*(3), 1843–1854. doi:10.1021/am4048317 PMID:24401042

Choi, E., Chang, H. K., Lim, C. Y., Kim, T., & Park, J. (2012). Concentration gradient generation of multiple chemicals using spatially controlled self-assembly of particles in microcahnnels. *Lab on a Chip*, *12*(20), 3968–3975. doi:10.1039/c2lc40450h PMID:22907568

Guo, W., Tian, Y., & Jiang, L. (2013). Asymmetric ion transport through ion-channel-mimetic solid-state nanopores. *Accounts of Chemical Research*, *46*(12), 2834–2846. doi:10.1021/ar400024p PMID:23713693

Kim, D., Posner, J. D., & Santiago, J. G. (2008). High flow rate per power electroosmotic pumping using low ion density solvents. *Sensors and Actuators. A, Physical*, *141*(1), 201–212. doi:10.1016/j.sna.2007.07.023

Kwon, K., Park, C.-W., & Kim, D. (2011). High-flowrate, compact electroosmotic pumps with porous polymer track-etch memrbanes. *Sensors and Actuators. A, Physical*, *175*(1-2), 108–115.

Ouyang, W., Wang, W., Zhang, H., Wu, W., & Li, Z. (2013, April). *Effect of crystal defects on the electrokinetics of nanofluidic crystal.* Paper presented at 2013 8th IEEE International Conference on Nano/Micro Engineered and Molecular Systems (NEMS), Suzhou. doi:10.1109/NEMS.2013.6559931

Siwy, Z., Apel, P., Dobrev, D., Neumann, R., Spohr, R., Trautmann, C., & Voss, K. (2003). Ion transport through asymmetric nanopores prepared by ion track etching. *Nuclear Instruments & Methods in Physics Research. Section B, Beam Interactions with Materials and Atoms*, *208*, 143–148. doi:10.1016/S0168-583X(03)00884-X

Vermaas, D. A., Bajracharya, S., Sales, B. B., Saakes, M., Hamelers, B., & Nijmeijer, K. (2013). Clean energy generation using capacitive electrodes in reverse electrodialysis. *Energy & Environmental Science*, *6*(2), 643–651. doi:10.1039/C2EE23562E

Wang, J., Zhang, M., Zhai, J., & Jiang, L. (2014). Theoretical simulation of the ion current rectification (ICR) in nano-pores based on the Poisson-Nerst-Plank (PNP) model. *Physical Chemistry Chemical Physics*, *16*(1), 23–32. doi:10.1039/C3CP51712H PMID:24253284

Yip, N. Y., Vermaas, D. A., Nijmeijer, K., & Elimelech, M. (2014). Thermodynamic, energy efficiency, and power density analysis of reverse electrodialysis power generation with natural salinity gradients. *Environmental Science & Technology*, *48*(9), 4925–4936. doi:10.1021/es5005413 PMID:24697542

## KEY TERMS AND DEFINITIONS

**Energy Harvesting:** The process to capture energy derived from external sources such as solar, thermal, kinetic, and wind energy.

**Nanofabrication:** A technique that fabricates the nano-sized structure.

**Nanofluidic:** The filed to study on the fluid flow in the nano-sized structure.

**Renewable Energy:** Energy resources that are no danger to depletion, including solar energy, wind energy, tidal energy, and geothermal energy.

**Reverse Electrodialysis:** Power generation by the ion transport from the concentrated solution to the diluted solution through ion exchange membranes.

**Salinity Gradient Power:** All of methods to generate electricity from the concentration gradient.

**Wastewater:** Any water that is polluted by human.

# Chapter 14
# Functionalization of Specific Electrostrictive Polymers for High Power Harvesting

**Bouchta Sahraoui**
*LUNAM Université, Université d'Angers, France*

**Daniel Guyomar**
*Institut National des Sciences Appliquées de Lyon (INSA), France*

**Adil Eddiai**
*Hassan II University, Morocco*

**Abdelowahed Hajjaji**
*National School of Applied Sciences El Jadida, Morocco*

**Mounir Meddad**
*Bachir El Ibrahimi University, Algeria*

**Yahia Boughaleb**
*Hassan II University, Morocco*

**Khalil Benkhouja**
*Chouaib Doukkali University, Morocco*

## ABSTRACT

*The energy harvesting based on electrostrictive polymers has great potential for remote applications such as in vivo sensors, embedded micro-electro-mechanical systems devices. The harvested energy and action are controlled by the permittivity, the Young's modulus and their dependence on frequency and level of stress. One should use a model which takes into account mechanical losses in order to obtain efficient devices. This chapter provides a brief overview of the methods for harvesting mechanical to electrical energy using elestrostrictive polymer. The second paragraph presents two new techniques which enable the improvement of the electromechanical performance of electrostrictive polymers in order to demonstrate their potential for the vibrational energy recovery. Based on the strong relationship between the frequency of operation and the mechanical strain from one could improve the electromechanical conversion. The development of a new prototype based on electrostrictive polymers for increasing the conversion AC–DC power is discussed.*

DOI: 10.4018/978-1-4666-8254-2.ch014

## 1. INTRODUCTION

Over the past 30 years, technological advancements have driven great changes in the communications, consumer electronics, medical, automotive as well as the leisure industry. During that time, sensor technology experienced substantial growth, both in use and development, thanks in large part to the implementation and large-scale commercialization of Micro Electro-Mechanical Systems or MEMs.

The invention of the transistor in 1947 by John Bardeen and Walter Brattain of Bell Labs marked the beginning of a revolution in electrical engineering that eventually led to the birth of Micro-electromechanical systems (MEMS). Soon after this pivotal discovery, a research ensued at a frenzied pace in the development of microelectronics resulting in the introduction of commercial silicon transistors in 1954 and the invention of the first integrated circuit (IC) by Jack Kilby of Texas Instruments in 1958. One of these early pioneers, Gordon Moore, made an astute observation in 1965 that the number of components per IC would double every 2 years. Moore's prediction, now popularly known as Moore's Law, is not completely correct and actually follows an 18-month doubling trend. Even so, this phenomenal growth rate means that we are now able to enjoy personal computers that run on the computing power of tens of millions of transistors in centimeter scale CPU packages (Nguyen & Werely, 2002).

As semiconductor technology developed, Micro-electric mechanical Systems (MEMS) emerged, combining mechanical structures and electronics on the micrometer scale. MEMS stops just short of being nano technology with feature sizes generally in the range of 1 to 100 micrometers. Examples of MEMS device applications include inkjet-printer cartridges, accelerometers, miniature robots, micro engines, locks, inertial sensors, micro transmissions, micro mirrors, micro actuators, optical scanners, fluid pumps, transducers, and chemical, pressure and flow sensors. New applications are emerging as the existing technology is applied to the miniaturization and integration of conventional devices. These systems can sense, control, and activate mechanical processes on the micro scale, and function individually or in arrays to generate effects on the macro scale. The micro fabrication technology enables fabrication of large arrays of devices, which individually perform simple tasks, but in combination can accomplish complicated functions. MEMS are not about any one application or device, nor are they defined by a single fabrication process or limited to a few materials. They are a fabrication approach that conveys the advantages of miniaturization, multiple components, and microelectronics to the design and construction of integrated electromechanical systems.

The MEMS industry has an estimated \$10 billion market, and with a projected 10-20% annual growth rate, it is estimated to have a \$34 billion market in 2002 (Micromachine Devices, 1997). The 2009 economic crisis has hit hard the semiconductor industry with a fall in turnover of 23%. Over the same period, the market for MEMS related to consumer applications, get out pretty well, since it will have helped generate \$1.2 billion, with an upward trend. The study by ISuppli expected by 2013 that this market will represent \$2.5 billion, driven by the development of devices (microphones, accelerometers, gyroscopes...) and telecommunications (BAW filters, RF Switch,...) (HIS Technology, 2012). The year 2012 was also a key date for the MEMS industry. This date corresponds in effect to the implementation of European laws on compulsory measure tire pressure in real time (TPMS), a measure that is likely to be achieved using MEMS (in force in the United States since 2007). Currently, the largest MEMS manufacturers are Hewlett Packard and Texas Instruments. MEMS have created numerous start-ups and it is possible to counttoday300 companies working in this sector.However, only10%of them is responsible for 90% of world-wide turn over.

Typically, electronics are used to interface MEMS devices from its functional domain (i.e., Physical, Chemical, or Biological) to the electrical domain for signal transduction and/or recording. It should be pointed out that the term MEMS was originally coined to refer to miniature sensors and actuators operating between electrical and mechanical domains. Gradually, the term MEMS has evolved to encompass a wide variety of other micro devices fabricated by micromachining. For example, a micro machined electrochemical sensor is referred to as a MEMS device even though there is no functional role played by this device in the mechanical domain. Similarly, the term "BioMEMS" is used to refer to the science and technology of micro devices fabricated by micromachining for biological and medical applications. BioMEMS may or may not include any electrical or mechanical functions. BioMEMS application areas include biomedical transducers, microfluidics, medical implants, microsurgical tools, and tissue engineering. The global BioMEMS market is expected to almost triple in size, from $1.9 billion in 2012 to $6.6 billion in 2018 (Yole Developpment, 2013).

Gradually, MEMS technology has demonstrated unique solutions and delivered innovative products in chemical, biological and medical domains as well. MEMS have penetrated into consumer electronics, home appliances, automotive industry, aerospace industry, biomedical industry, recreation and sports.

MEMs are a process technology that enables the combination of sensors, actuators, mechanical components, and to a certain extent, electronics on one silicon or germanium base. Typically, micro-fabrication is used to apply the various elements to a silicon or germanium wafer. MEMs offer various advantages over traditional electro-mechanical sensing technologies including: Smaller package size, Lower cost, Large-scale integration, Lighter weight, Less power consumption, Wider operating temperature, Higher output signal and Greater complexity.

MEMS devices can be made from many materials, but most are made from silicon due to the existing capacity and compatibility with existing semiconductor manufacturing equipment. Other common materials in MEMS include polymers, metals, and ceramics. Single crystal silicon offers very good mechanical properties, even better than many metals at micrometer scales, and is often used rather than metal. Ceramic materials offer material properties with the ability to create microscopic piezoelectric sensors, bio reactive sensors, and electrostatic sensors.

In recent years a growing interest in the field of autonomous wireless microsystems can be noticed. An autonomous system can be installed in a building, a machine or the human body. Generally, is difficult to access and it must have its own source of energy. Normally, this electrochemical battery provides energy needed, but its capacity is limited and its size dominates over the entire system. The considerable reduction of the energy consumption of electronic devices, and the development of MEMS technologies allow to consider supplying these systems from ambient energy. The advantages of this solution are obvious: the operating time of the device is limited only by the existence of an external source and in addition, the device size can be reduced.

## 2. ENERGY HARVESTING: CONTEXT, CHOICES AND COMPROMISES

Before going into detail of energy harvesting phenomena, we present in Table 1. The different energy sourceswith the average energy that can be obtained.

The major advantage of energy harvesting is the realization of completely autonomous microsystems requires no human intervention during the entire life of the system. This helps fill a major flaw batteries: the need to recharge or change them. The first limitation in the use of energy recovery systems from

*Table 1. Comparison of different sources ambient energy (HIS Technology, 2012)*

| Source | Power [μW/cm³] |
|---|---|
| Light (outdoors) | 15000 |
| Light (indoor) | 10 |
| Temperature gradient | 40 |
| human energy | 330 |
| Air flow | 380 |
| Pressure variation | 17 |
| Vibrations | 300 |

physics: the recoverable power depends on the size of the system. To stay in the dimensions of the order of 1 $cm^3$, we will retain the power output of the energy harvesting system will be of the order of 10-100 μW. The microstructures of energy recovery is therefore not intended to continuously feed a laptop or a mobile phone, by cons, they are ideal to give the energy needed to run simple sensors (measuring pressure, acceleration, temperature...). For such applications, energy harvesting becomes a viable solution capable of operating for several decades. Several sources of ambient energy have been considered to make autonomous electronic devices. Among these sources the vibration and mechanical stresses.

The idea of recovering mechanical energy in the environment is not new. By against the idea of using vibrations to power electronic devices are emerging. There are some commercial examples of powered movement or shaking devices, such as a flashlight from Applied Innovative Technologies or the Seiko Kinetics. More intensive research seeks to miniaturize the generators. According to the study made by Roundy et al. (Roundy, Wright, & Rabaey, 2003) vibrations are very interesting sources of energy densities of between 50 and 250 μW/$cm^3$. Another advantage of this type of source is that it requires only a simple mechanical connection with the vibrating object (as opposed to the difference in temperature and radiation or coupling with the generator source is often very difficult).

Most applications of wireless sensor networks provide mechanical vibration rich environments (industrial, transportation). The operating ambient vibrational energy is therefore a great way to replace the use of batteries which limits the lifetime of these networks. For this purpose, several types of micro generators have been developed which convert mechanical energy into electrical energy. We are interested in the first recovery vibration. At present, three basic principles (piezoelectricity, electromagnetism, electrostatic) can transform mechanical vibrations into electricity.

## 2.1. Electromagnetic Conversion

Many systems of vibrational energy recovery are based on the movement of a permanent magnet within a coil. This movement creates a current in the coil proportional to the variation of the magnetic flux in the coil, therefore proportional to the velocity of the magnet, through the magnetic field generated by the magnet in the coil and the surface of the coils. Faraday's law, according to Equation 1, determines the value of the generated voltage.

$$E = -N\frac{\partial \Phi}{\partial t} \tag{1}$$

*E*: Electromotive force.
*N*: Number of turns of the inductor.
**Φ**: Magnetic field through the inductor

These systems are capable of converting up to 30% of the energy supplied. However, they are difficult to miniaturize, firstly because the volume energy density decreases with the size of the system, and secondly because it is difficult to have a stable system with strong magnetic fields. In the case of energy recovery ambient, another problem arises: the frequencies involved are below 100 Hz and then coils tend to be stronger than inductive, leading to heavy losses by Joule effect. A lot of work has been done on this subject. Some embodiments relate to powers between 1μW and 100μW and correspond to the scale of MEMS (micro-electromechanical system MEMS in Anglo-Saxon literature). Other embodiments using larger structures reaching powers of a few hundred mW.

The device developed by Li et al (Li, Ho, Chan, Leong, & Wong, 2000), is used for supplying an infrared transmitter. It occupies an area of about 1cm$^3$ and helps provide 70μW. Since the power required to transmit is 3μW for 140ms, energy is first stored on a capacitor until it is sufficient for transmission.

## 2.2. Electrostatic Conversion

Electrostatic micro-generators are constituted by a charged capacitor, the electrodes may move. Unlike electromagnetic and piezoelectric systems, electrostatic generators need to be pre-loaded before being able to produce the power (Meninger, Mur-Miranda, Amirtharajah, Chandrakasan, & Lang, 2001).

Capacity is biased at a voltage *U* with a battery ($T_1$) via control electronics. This tension will be maintained throughout the cycle of decline in capacity. The capacity decreases with a constant voltage: the load decreases and thus there is a current flow charging the battery ($T_2$). When the capacitance reaches its minimum value, still present in the load capacity is entirely transferred to the battery in a short time before the mechanical movement ($T_3$).

## 2.3. Piezoelectric Conversion

Although the first observation of the piezoelectric effect was made by the French mineralogist A. Haüy in 1817, it is the Curie brothers that attribute the discovery in 1880 following their work on the relationship between crystal structure and pyroelectricity. A piezoelectricmaterial deforms when a voltage is applied to its terminals (the direct effect is that used by the generators), and reciprocally a voltage appears across its terminals when the material is constrained (that is the opposite effect). Piezoelectric micro-generators are based on the direct effect of piezoelectricity that the application of stress on a piezoelectric material electrodeposition leads to the appearance of a voltage between the electrodes. This type of converter uses the mechanical stresses in these quasi-static conditions is that the operating frequency is much lower than the natural resonance frequency of the recovery system. The energy delivered by the generator is proportional to the frequency of repetition of constraints used (Poulin, 2004).

*Figure 1. Variation of electrostatic converter-cycle constant voltage*

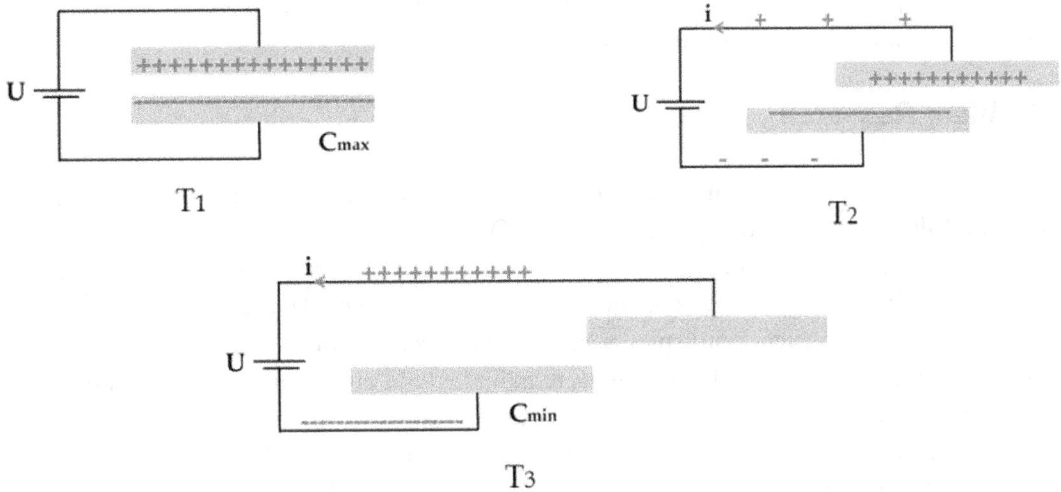


## 2.3.1. Operating Speed of the Piezoelectric Generator

The first step in designing a micro piezoelectric generator is the choice of operating conditions, which can be dynamic or quasi-static. The distinction between these two systems is not new. It is derived from the two types of piezoelectric ignition systems. The first type operates from an extremely brief compression (shock), the second from a slow compression, called (SQUEEZE). Following a shock, the piezoelectric system will vibrate at its resonant frequency. Under the effect of a slow compression, piezoelectric system does not vibrate. Of course, the equivalent model of piezoelectric system is different in the two cases: in the dynamic mode, a source associated with a *RLC* circuit; static scheme, a source associated with capacity.

## 2.3.2. Constraint Operation

Compliance with certain field boundaries, and restraint is necessary otherwise, the ceramic may deteriorate:

- **Dielectric Strength:** Like any insulating material, the ceramic has a dielectric strength which naturally leads to a limit value of the applied voltage.
- **Coercive Field:** On either their ferroelectric properties, ceramics have a hysteresis cycle *P*(*E*). This shows a boundary value (coercive field $E_C$ = 1500*V*/*mm*), beyond which the material is depolarized.
- **Maximum Stress Accepted:** Ceramics can withstand extremely high stress values up to 20*MPa*. However, they are less resistant to shocks.
- **Maximum Relative Deformation:** Superior strain $S_{max} = 3*10^{-4}$can cause cracks and breaks in ceramics. Moreover, these powders compacted and sintered less tolerant than the expansion compression. That is why there is often a preload applied in case of alternative stress.

- **Warm-Up:** When operating at high frequency, dielectric and mechanical losses causing heating of the structure. To avoid depolarization, it is often advisable not to exceed half of the Curie temperature, even if it leaves a substantial margin.

## 2.4. Problem Research

The aim of our study in the following section aims to provide a state of the art generators, based on an innovative and alternative technology to conventional solutions for known (piezoelectricity, electromagnetic and electrostatic). Indeed, the major criterion for this study is the nature of the work undertaken relates to the ductility. Most existing actuators and generators have made based on metals such as static structures or those based on very brittle piezoelectric ceramic as specified in the previous paragraph rigid structures. In addition, the coils and the magnets remain the same rigid elements it is possible to achieve on supports more flexible. Thus, the electro-active polymers, flexible materials with specific physical properties, are the best candidates to develop flexible generators and flexible actuators. The use of these polymers is expanding but little energy harvesting structures exploit their potential. Many technological advances are yet to be realized: the improvement of the material to use as a generator through its reliable characterization and modeling. More scientific teams are positioned in these field electro-active polymers, but none comprehensive comparative study generator and actuator fashion has been made.

# 3. IMPROVEMENT OF PHYSICAL PROPERTIES OF ELECTRO-ACTIVE POLYMERS FOR ENERGY HARVESTING

The intrinsic properties of electro-active polymers (electric permittivity, electrical losses, Young's modulus,...) play a crucial role in their performance in order to improve the response of actuator mode, but also on the density of power harvested of generator mode.The aim of the present section is to provide an overview of theelectrical and mechanical behavior of an electrostrictve polymer. As well as different methods currently utilized to increase their electromechanical properties in order to increase the energy harvesting of these polymers, it is principally playing on the permittivity of the material as well as in its Young's modulus.

## 3.1. Reinforcement of the Intrinsic Parameters of Electro-Active Polymers

The electromechanical transduction properties of any electrostrictive polymer are intrinsically regulated by the dielectric permittivity of the material.It has been pointed out that a high permittivity is required to increase energy harvesting and also to increase the actuation capability if the Young modulus is kept low.

Previous studies conducted by Eury et al. (Eury,et al., 1999) and by Guillot and Balizer (Guillot, & Balizer, 2003) have shown that the coefficient of electrostriction is proportional to the permittivity and Young's modulus. This dependence clearly demonstrates that increased permittivity material causes an increase in performance of electrostrictive polymers and dielectric elastomers. It is therefore essential to develop methods capable of increasing the dielectric properties while ensuring an optimal Young's modulus to maintain the flexible polymers. This increase in the permittivity generates a substantial increase in performance. Modifying a polymer matrix in order to increase its dielectric permittivity means acting on the dipolar moments of the material and, therefore, on its polarization.Currently, a variety of methods are available in order to increase the dielectric permittivity of polymer materials.Elaborate a composite by incorporating the conductive particles is an experimental approach for increasing the dielectric constant. It should be placed just before the percolation threshold where the insulator-conductor transition occurs. However, the Young's modulus is also slightly increased by the addition of these particles which mechanically strengthen the flexible matrix.

Smart materials are primary elements in energy harvesting because they represent the first stage of converting ambient vibrations into electrical energy. Consequently, the optimization of such a material is very important. According to the amazing physical properties of nanofillers, e.g., carbon nanopowders, random composites can be realized in the laboratory. The Matrices of bases chosen are the electroactive polymers: those are flexible and light organic compounds and capable to respond to an electrical stimulation by a change in size and shape. Among these polymers include the electrostrictive elastomers that have the particularity of being composed of a flexible matrix and a crystalline matrix (semi-crystalline). During the elaboration of a nanocomposite material, there are two principal difficulties to overcome: theassurance of a uniform repartition and the lack of aggregation of the nanoparticles within the material. Various manufacturing processes have been developed for different types of nanocomposites in order to obtain an ideal repartition of the nanofillers. Both methods are very distinct: the ex-situ fabrication by direct mixing of nano-reinforcements with the starting material previously melted and in-situ fabrication, with direct growth nanorenforts even within the matrix by chemical reactions.Three principal techniques for the nanocomposites to matrix polymer and nano reinforcements (particles, fibers, etc.) are referenced: the dispersion by solvent way, the dispersion by molten way and the in-situ polymerization.

According to a study in the laboratory LGEF, the dispersion of conductive nanofillers by melted way in a polymer matrix showed a good increase in the dielectric constant with a low rate of nano-objects and thus a conservation of Young's modulus.Among the commercially available polymers was employed in this work, polyurethane and P(VDF-TrFE-CFE). Moreover, these composites have been synthesized specifically for this study.

These polyurethane composites were preparedin the laboratory using thermoplastic polyurethane and the 58887 TPU elastomer (Estane) as the matrix. The neat polyurethane (PU) films as well as their filled counterparts were prepared by solution casting (You et al., 2011; Yuse et al., 2009; Su, Zhang, Wang, MacDiarmid, & Wynne, 1998).

PU is the general name of polymer consisting of a chain of organic units joined by burethane links. PU polymers are formed through step-growth polymerization to react a monomer containing at least two isocyanate functional groups with another monomer containing at least two hydroxyl (alcohol) groups in the presence of a catalyst. Polyurethane or urethane linkage is produced by reacting an isocyanate group, - N=C=O, with a hydroxyl (alcohol) group, -OH, Polyurethanes are produced by the polyaddition reaction of a polyisocyanate with a polyalcohol (polyol) in the presence of a catalyst and other additives. A polyisocyanate is a molecule with two or more isocyanate functional groups. The first essential component of a PU polymer is the isocyanate, The second essential component of a PU polymer is the polyol. Important characteristics of polyols are their molecular backbone, initiator, molecular weight, percent of primary hydroxyl groups, functionality, and viscosity.

The PU granules were dissolved in N,N dimethylformamide(DMF) at 80°C for 45 min and then carbon nanopowder (Aldrich, average particle size 30 nm) was added to the stirred solution. The volume content of carbon in the composite was fixed to 1% in order to ensure a homogeneous distribution of the filler. The nanopowder was ultrasonically dispersed in the mixture polyurethane (PU) (UP400S HIELSHER, probe H7; duty cycle 0.7%, 70% power) and DMF during 12 min. The mixture was cooled with ice after which the viscous mixture was poured onto a glass plate and cured at 60°C for 12h, and heated at 80°C during 6 h to remove most of the solvent. The obtained sample is rectangular (55x20 $mm^2$) and 60μm thick (Zhang, Neese, Ren, Chu, & Zhang, 2006). The dimensions of the sputtered gold electrode area on the obtained sample are 40x16 $mm^2$.The process presented below is described for the preparation of a composite using polyurethane, but it can be generalized other types of matrices.

## 3.2. Electrical and Mechanical Characterization of Electroactive Polymers

The purpose of this part is to present the principle of characterization of dielectric and mechanical properties as well as the interpretation of experimental results carried out to analyze the role of carbon nanoparticles and their influence on the electromechanical performance of electrostrictive polymers for energy recovery.

### 3.2.1. Dielectric Characterization

The dielectric characterization at low frequency will be focalized on the study of the following variables: the dielectric permittivity and dielectric losses of the two types of polymers and composites. The principle of dielectric characterizationwas based on the measurement of the capacitance of capacitor realized with the material to be studied as a dielectric element. For the dielectric characterization, gold was sputtered on both sides of the films to form electrodes (diameter of electrodes of 20 mm)deposited by cathode sputtering (Cressington208 HR).In order to evaluate the contribution of space charge, the dielectric constant of the composites loaded with fillers was measured using an HP 4284A LCR meter over a broad range of frequencies (from 0.01Hz to 1MHz) at room temperature.The dielectric constant and dielectric loss of all specimens were obtained using a lock-in amplifier (SR 830) combined with a current amplifier (Keithley428). The lock-in amplifier measures both amplitude and phase of the current of the specimen under low field dielectric properties.

The variation of dielectric constants for a pure PU composite, a pure P(VDF-TrFE-CFE) composite and filled composites versus frequency is shown in Figure 2. A large decrease in the dielectric constant was observed at around 1Hz for both composites when the frequency increased. Such a behavior is known to be due to the loss of one of the polarization contributions (interfacial polarization, orientation polarization, etc) of the dielectric constant value (Mitchell, 2004). Considering the value of the frequency, this decrease can be unambiguously attributed to the loss of the space-charge-induced interfacial polarization contribution. It can moreover been seen that the contribution of the space charge can be neglected for frequencies below 4Hz.

For the band of frequencies between 10Hz and $10^3$Hz frequency, the relative permittivity of the polymer is almost constant. This corresponds to the plateau of the orientation polarization. This can be explained by the orientation of the molecules that maintain a permanent dipole moment. As shown on Figure 2, the dielectric constant of the C nanofilled composites is always higher than this one of the pure composite.As expected, at higher frequencies, the gap between the values of the dielectric constant for pure composites and nanofilled composites is not so high, which confirms the fact that the fillers content is low compared to the threshold value.The incorporation of conductive charge has also probably increased the space charge density in addition to these intrinsically induced by the existence of soft and hard segments within the matrix.Similar observations have been reported by Z M Dang et al., who assumed an additional contribution to the quantity of accumulated charge when fillers are used (Dang, Zhang, & Tjong, 2004).

*Figure 2. Variation of the dielectric constant versus frequency for different filled composites based on polyurethane and de P(VDF-TrFE-CFE) matrix*

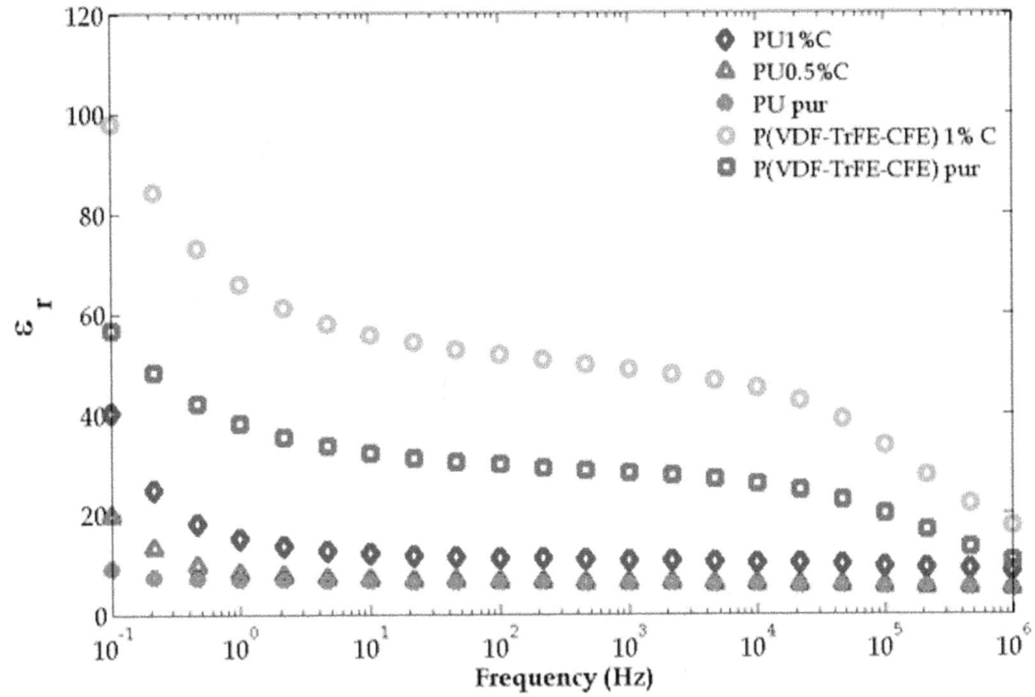


Figure 3 shows the variation of dielectric loss for a pure PU composite, a pure P(VDF-TrFE-CFE) composite and filled composites versus frequency. Wherever, the value of dielectric loss for the composites filled with nano-charges is closely equal the value of the pure composite.Despite the disadvantage that the permittivity increases in parallel with the conduction therefore the losses.Despite the drawback that the permittivity increases in parallel with the conduction, encouraging results suggest that this method of load is very promising in order to increase the electromechanical properties of electroactive polymers. The major effect of the presence of fillers in the dielectric is the modification of the internal distribution of electric field; this phenomenon is largely amplified when the conductive particles are dispersed in the dielectric matrix because they are equivalent to dipoles.

### 3.2.2. Mechanical Characterization

The study of the mechanical properties of polymers reveals itself as complex by reason of the diversity of observed behavior versus the temperature, deformation or frequencyof solicitation. The proposed approach involves measuring the Young's modulus for the different materials and composites studied based on polyurethane and the (P (VDF-TrFE-CFE)) as well as to see the influence of the incorporation of carbon black in the matrix on the variation of the elastic modulus.

*Figure 3. Variation of the dielectric loss versus frequency for different filled composites based on polyurethane and de P(VDF-TrFE-CFE) matrix*

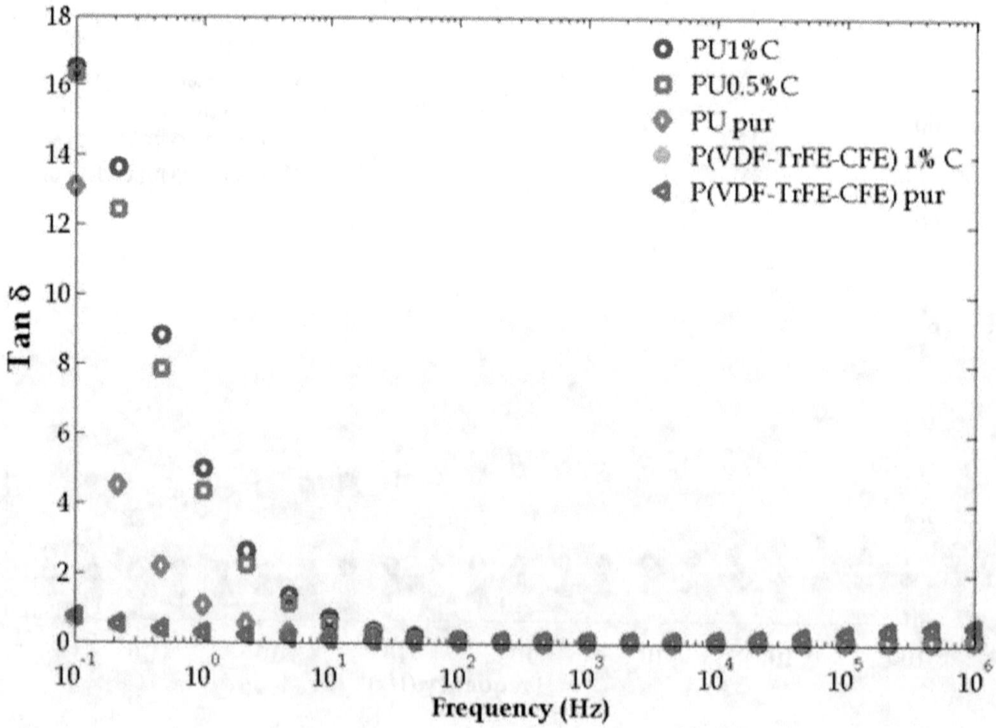


### 3.2.2.1. Principle of Measurement

Depending on the equipment, different modes of solicitation may be utilized (tensile, shearing, simple bending ...). Monitoring can be done either by fixing the applied force and measuring the deformation or strain imposed by working and by measuring the force.

For our study, the tensile testing has been chosen for ease of implementation. These measurements are made using the table with one degree of freedom Newport. Figure 4 illustrates the operating principle. The test sample is clamped between two jaws, one of said mobile as connected to the table with one degree of freedom, the other fixed as connected to the transducer. Sheets of specimen between 20and 100µm in thickness were cut into rectangular, 9.5mm for width and 25mm for length. Newport table is controlled using a function generator connected to the controller. The signals delivered by the force sensor and displacement are then displayed on an oscilloscope. It is possible to obtain a wide range of strain over a large frequency band.

### 3.2.2.2. Theoretical Model

This study allows us to present a model of hysteretic behavior in electrostrictive polymers and to determine the relationship between the strain S and stress T by introducing the mechanical losses. A great number of laboratory tests on polymers shows that the cyclic stress-strain curves are highly nonlinear and constitute a closed hysteresis loops. The results of these tests seem to indicate that the damping properties are especially of hysteretic type, and not viscous as those corresponding to the Kelvin-Voigt model (Bratosin, & Sireteanu, 2002).

*Figure 4. Schematic illustration of the experimental setup for measuring the stress-strain behavior in the electrostrictive polymers*

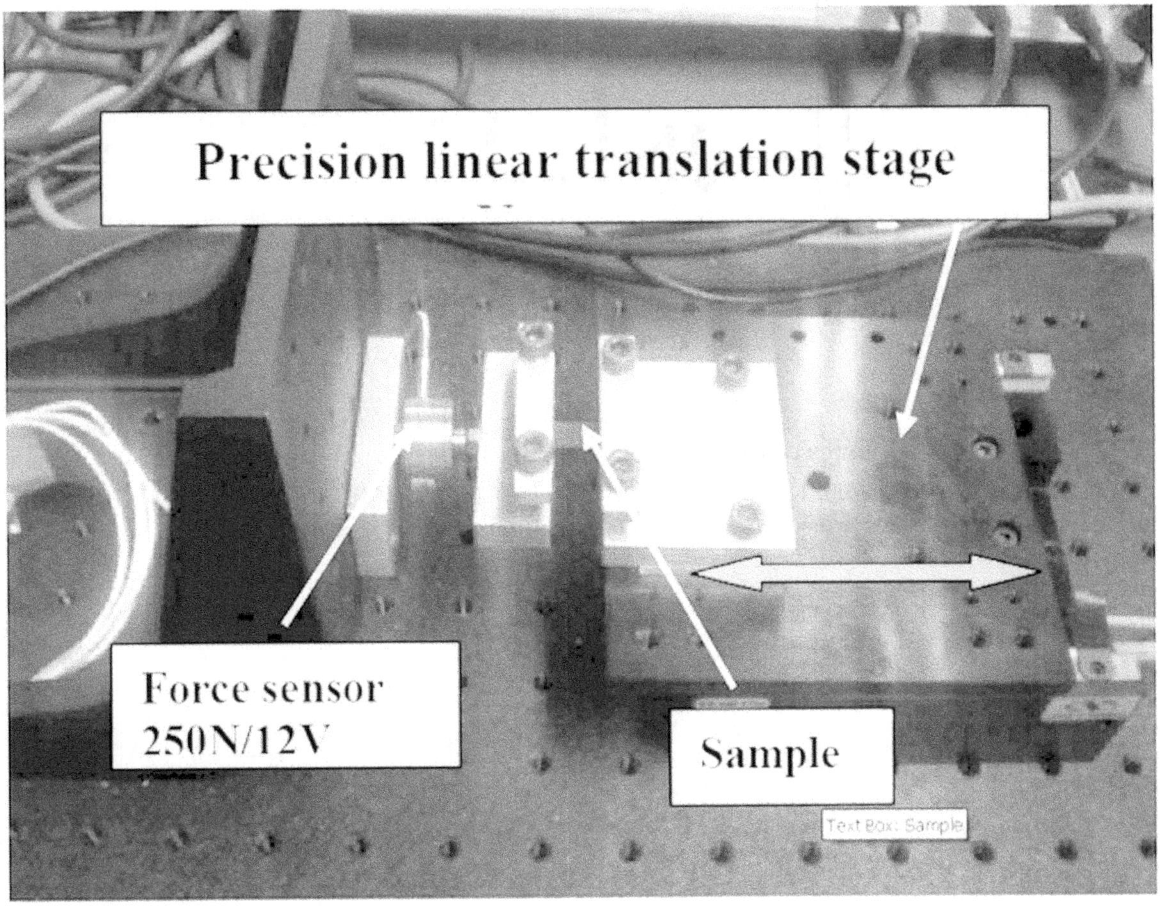


The generalized Hooke model (also known as the spring model) is the simplest model for describing the variation of the deformation S of a sample with the mechanical stress T. The model is linear; predicting that the displacement $U$ applied to the sample is proportional to the force F detected by force sensor. The mathematical representation of this behavior is given with the following equation:

$$T=YS \tag{2}$$

The previous model has allowed us to understand the phenomenon of non-linearity observed in the curves of stress as a function of deformation. By analogy, the application of a displacement on the material may be equivalent to a spring in series with a damper as shown in Figure 5, the spring characteristic of the polymer film and the damper allow us to describe the mechanical losses in the film, where (1) is the damper, (2) is the spring, and (3) is the direction of longitudinal displacement.

The basic principle is to introduce mechanical losses to the Equation (2) by the variation of Young's modulus versus times, and then the expression of stress becomes (Eddiai et al., 2012a):

*Figure 5. Equivalent structure of the polymer*

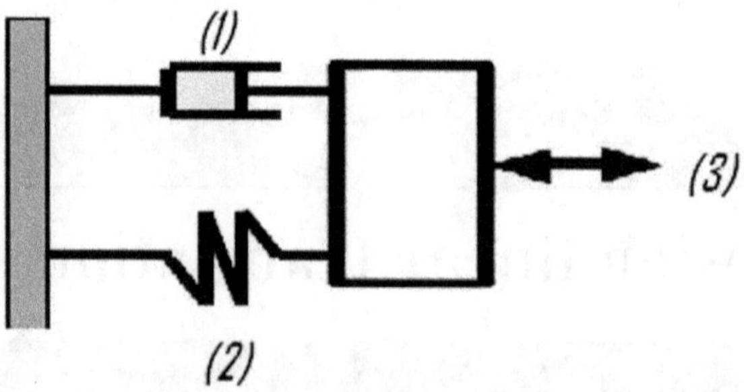


$$T(t) = Y_0 h(t) \otimes S(t) \quad (3)$$

Here, $\otimes$ corresponds to the convolution operation.

Where $Y_0$ is the Young's modulus of the electrostrictive polymer and $h(t)$ is a function to introduce mechanical losses through a mechanical characteristic time $\tau_{mec}$, and it is expressed as

$$h(t) = e^{\left(-\frac{t}{\tau_{mec}}\right)} \quad (4)$$

With

$$T(t) = \frac{F(t)}{eL} \quad (5)$$

and

$$S(t) = \frac{U(t)}{l} \quad (6)$$

where e is the thickness, $l$ is the length, and $L$ is the width.

So the final equation of force harvested as a function of imposed displacement becomes as follows:

$$F(t) = \left(\frac{eLY_0}{l}\right) e^{\left(-\frac{t}{\tau_{mec}}\right)} \otimes U(t) \quad (7)$$

In this experiment, a sinusoidal displacement which can be experimentally implemented and harvested sinusoidal signal strength was used. This displacement is given by

$$U(t) = U_{dc} + U_{ac}\sin(\omega t) \tag{8}$$

where $U_{dc}$, ω, and $U_{ac}$ correspond to the pre-displacement, the pulsation of the mechanical excitation, and the amplitude of displacement, respectively.

### 3.2.2.3. Validation of the model

The stress-strain behavior, in which contains the uniaxial tests, of polyurethane and P(VDF-TrFE-CFE) with carbon black powders (1wt.%C) was examined. The specimen was stretched withdifferent amplitude displacements 35, 70, and 100 mVpp in a quasi-static frequency range (100mHz up to 10Hz).

Figure 6 shows the comparison between the model and the experimental result for a sample type PU 1%C at displacement amplitude of 100 mVpp, a frequency of 10 Hzand a thickness of 57 μm. An excellent agreement has been found between the proposed modelling and the experimental data measured for the electrostrictive polymers. The behavior of the material has a clear hysteresis curve. This suggests that the character of nonlinear strain with stress is due to the existence of mechanical losses within the material.These losses become a form of energy dissipated within the material, so using this model; we can minimize these losses by reducing the surface of the stress-strain hysteresis by adjusting the amplitude of the applied deformation and frequency.

To obtain Figure 7, we used two different samples with the same amplitude of displacement 100 mVpp: (a) one is a P(VDF-TrFE-CFE)% C with a thickness of 49 μm, and (b) the other one is a PU 1% C with a thickness of 138 μm. Concerning the variation of hysteresis loop, the stress-strain is given as a function of frequency (quasi-static field). Figure 7 shows clearly that the area of the hysteresis loop remains almost unchanged with the change of frequency; therefore, when we increase the frequency, the slope of the curve increases which means the increase of the magnitude of Young's modulus. For frequenciesbelow 20 Hz, it is interesting to note that the mechanical losses in the polymer are low. Against them to become more important later with an increase in Young's modulus, effective for the frequency band where the maximum loss occurs, the effectiveness of our polymers, such as energy harvesting, will not be optimal as much energy mechanics will be dissipated as losses(heat). This maximum occurring in high frequencies is not a primary constraint to target applications, which the most of mechanical vibrations occur at lower frequencies.

On the other side, the variation in the magnitude of Young's modulus depends on type of material. For PU 1%C, this variation is unchanged from the increased hysteresis. So in this case, the magnitude of Young's modulus remains almost unchanged(Table 2.), but the increase in the hysteresis loop corresponds to the energy dissipated in each cycle.But in the case of P(VDF-TrFE-CFE) 1%C, we observed that the increase in the hysteresis loop with the variation of the displacement amplitude, resulted in a decreased in the amplitude of the module Young (Table 2.), can be explained by the nature of our polymer (thermoplastic), this material has a high strain rate to obtain a peak (maximum stress) which coincides with the constriction; locally in a band, the deformation can reach a high value (100%) and a significant softening of the polymer occurs. As a result, the stress required to maintain the constant strain rate decreases (plasticity) (Xu & Li, 2011a; Xu & Li, 2011b).

Into view of the results obtained, this model has allowed us to see more precisely the dependence of mechanical losses with the variation of the amplitude strain and mechanical frequency. As well as the control of the mechanical energy dissipated in the electro-active polymers in order to optimize the operating range of these materials for the actuators and energy harvesting.

*Figure 6. Strain-stress curve for PU 1% C*

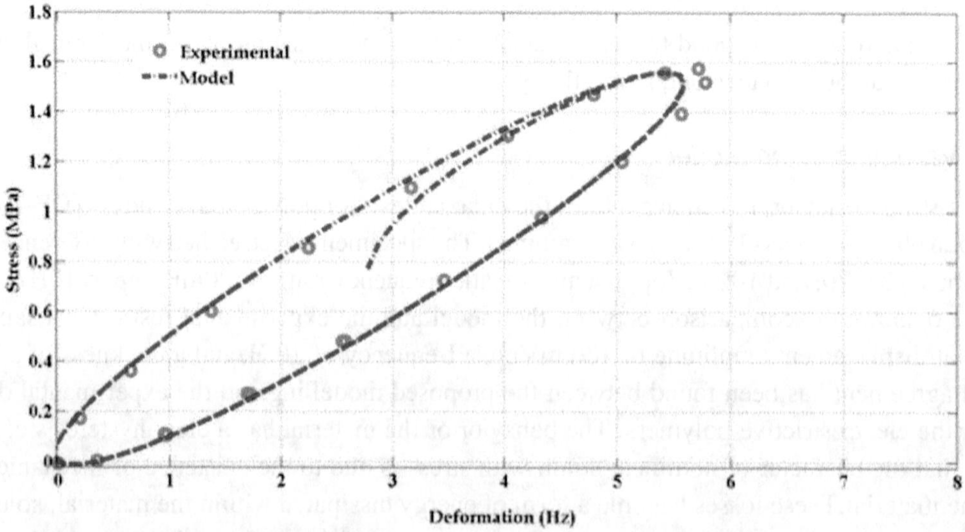


*Figure 7. Strain-stress curve*

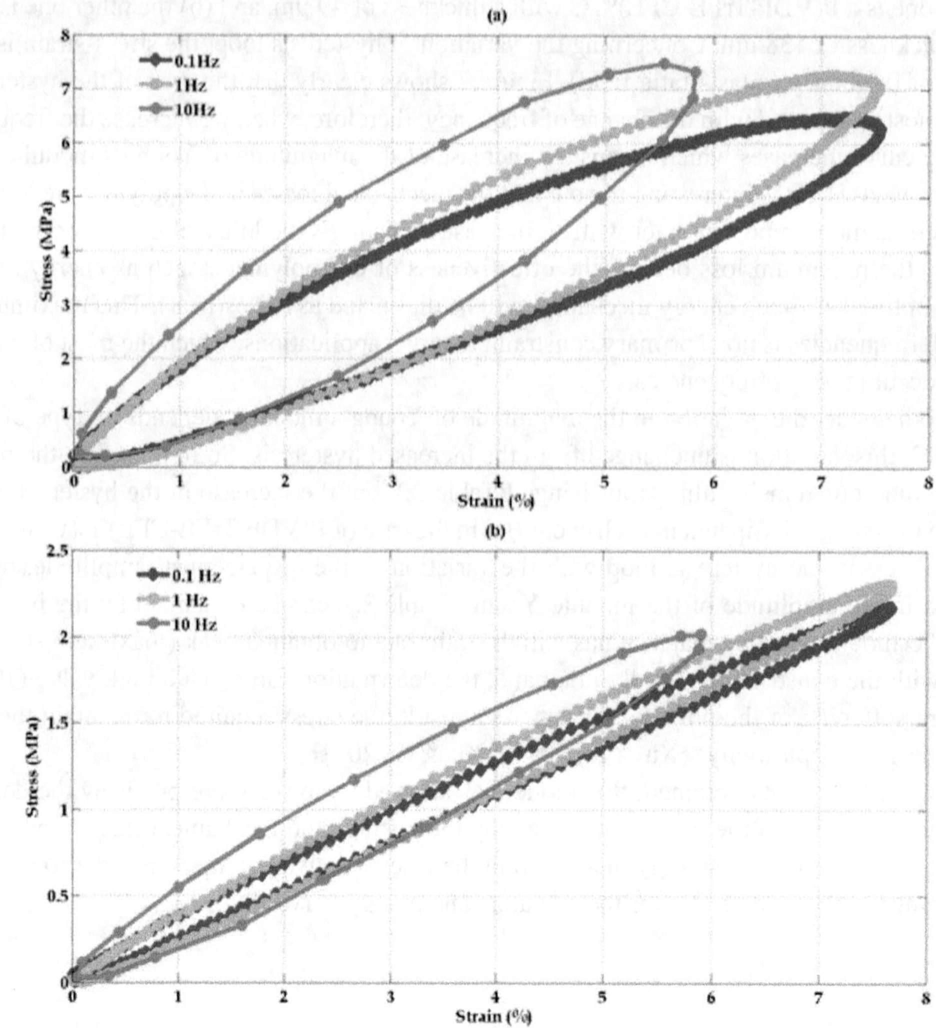

*Table 2. The amplitude of Young's modulus for PU 1%C and P(VDFTrFE-CFE) 1%C at different amplitude of displacement*

| | **Amplitude of Young's Modulus (MPa)** | |
|---|---|---|
| **Materials** | **PU 1%C at 1Hz** | **P(VDF-TrFE-CFE)1%C at 0.1Hz** |
| 35 mVpp | 30.02 | 156 |
| 70 mVpp | 30.18 | 148 |
| 100 mVpp | 29.22 | 121 |

## 3.3. Synthesis

The table below recapitulates the main results obtained during the electrical and mechanical characterization for different polymers and composites.This synthesis allows us to know the influence of nanoparticles incorporated into the starting matrix on the intrinsic parameters of polymers.It has also allowed us to examine the dependence of these parameters with the amplitude of mechanical displacement and the mechanical frequencyin order to optimize the operating interval for a better application in the actuating and energy recovery area.The whole of these characteristics portends an increase in the electromechanical activity in our polymers.Indeed, it is crucial to produce composites with a very high dielectric permittivity and a low Young's modulus in order to increase coefficient of electrostriction.

# 4. MODELING OF MICRO GENERATORS BASED ON ELECTROSTRICTIVE POLYMERS

The aim of this section is to present a model capable of predicting the harvested power using the electrostrictive polymers. Generally, the electrostriction phenomenon is defined as a quadratic coupling between strain ($S_{ij}$) and polarization ($P_m$):

$$\begin{cases} S_{ij} = s_{ijkl}^{D} T_{ij} + Q_{ijkl} P_l P_m \\ E_i = \beta_{ij}^{T} P_j + 2Q_{klij} T_{kl} P_j \end{cases} \tag{9}$$

*Table 3. Synthesis of results obtained in the characterization of polymers and composites*

| **Materials** | | **PU pur** | **PU 0.5%C** | **PU 1%C** | **P(VDF-TrFE-CFE) pur** | **P(VDF-TrFE-CFE)1%C** |
|---|---|---|---|---|---|---|
| Y(MPa) | 1 Hz | 35 | 36 | 38 | 232 | 240 |
| | 10 Hz | 36 | 39 | 43 | 246 | 259 |
| $\varepsilon_r$ | 1 Hz | 6.95 | 8.4 | 15 | 38 | 65.9 |
| | 10 Hz | 6.5 | 7.2 | 12 | 32.3 | 56 |
| Tanδ | 1 Hz | 1.066 | 4.35 | 4.99 | 0.26 | 0.27 |
| | 10 Hz | 0.14 | 0.56 | 0.57 | 0.094 | 0.095 |

where $T_{ij}$ is the stress, $s^D_{ijkl}$ is the elastic compliance, $Q_{ijkl}$ is the polarization-related electrostriction coefficient, and $\beta^T_{ij}$ is dielectric permeability. By assuming a linear relationship between the polarization and the electric field, the strain $S_{ij}$ and electric flux density $D_i$ are expressed as independent variables of the electric field intensity $E_k$, $E_l$ and the stress $T_{kl}$ by the constitutive relations as (Zhang, & Scheinbeim, 2004):

$$\begin{cases} S_{ij} = s^E_{ijkl} T_{kl} + M_{ijkl} E_k E_l \\ D_i = \varepsilon^T_{ijkl} E_k + 2M_{ijkl} E_l T_{kl} \end{cases} \tag{10}$$

where $M_{ijkl}$ is the electric-field-related electrostriction coefficient, and $\varepsilon^T_{ijkl}$ is the linear dielectric permittivity.

One way of harvesting energy using electrostrictive polymers is to operate in the pseudo-piezoelectric mode. For this, the electrostrictive polymer was subjected to a dc-biased electric field. Because the polymer was not piezoelectric, it was necessary to induce polarization with a dc bias to obtain pseudo-piezoelectric behavior (Guyomar et al., 2009; Lallart, Cottinet, Lebrun, Guiffard, & Guyomar, 2010). An isotropic electrostrictive polymer film contracts along the thickness direction (the electric field direction) and expands along the film direction when an electric field is applied across the thickness, assuming that only a nonzero stress is applied along the length of the film. The constitutive relation (10) can then be simplified as (Liu, Ren, Hofmann, & Zhang, 2005):

$$\begin{cases} S_1 = M_{31} E_3^2 + s^E_{11} T_1 \\ D_3 = \varepsilon^T_{33} E_3 + 2M_{31} E_3 T_1 \end{cases} \tag{11}$$

The electric displacement $D_3$ flowing through the polymer film can be calculated from Equation (12) and from the expressions for $S_1$ and $E_3$ by replacing $T_1$ with $T_1 = \dfrac{(S_1 - M_{31}E_3^2)}{s^E_{11}}$:

$$D_3 = \varepsilon^T_{33} E_3 + \frac{2M_{31}}{s^E_{11}} E_3 S_1 - 2\frac{M^2_{31} E^3_3}{s^E_{11}} \tag{12}$$

The current induced by the transverse vibration can be measured as $I = \int_A \frac{\partial D_3}{\partial t} dA$, where A is the area of the electrostrictive polymer. The current produced by the polymer can thus be related to the strain and electric field by:

$$I = \int_A \left[ \frac{\partial E_3}{\partial t} \left( \varepsilon^T_{33} + \frac{2M_{31}S_1 - 6M^2_{31}E^2_3}{s^E_{11}} \right) + \frac{2M_{31} \frac{\partial S_1}{\partial t} E_3}{s^E_{11}} \right] dA \tag{13}$$

Here, $\frac{\partial E_3}{\partial t}$ and $\frac{\partial S_1}{\partial t}$ are the time derivatives of the electrical field and strain. Since a dc electric field ($E_{dc}$) was applied to the sample so that $\frac{\partial E_3}{\partial t} = 0$, the short-circuit current can be expressed as

$$I = 2M_{31}^{*}YE_{dc}\int_{A}\frac{\partial S_1}{\partial t}dA \tag{14}$$

with $\frac{1}{s_{11}^{E}} = Y$:

Here, Y is the Young modulus. Here, $M_{31}^{*}$ is the apparent electrostrictive coefficient used todescribe the experimental dependence of the current response ofthe material on the applied strain and electric field. In fact, whenan electric field is applied to any material, it determines the chargedisplacement that leads to the field induced strains. If the strain isproportional to the electric field, the phenomenon is called "piezoelectricity". If the strain depends quadratically on the electric field, then two phenomena "electrostriction" and "electrostatic"(Maxwell effect) can be involved (Diaconu, Ddorohoi, & Policeanu, 2006). The experimental measured of $M_{31}^{*}$ should thus be:

$$M_{31}^{*} = M_{electrostriction} + M_{\max well} \tag{15}$$

assuming a constant strain ($S_1$), the relationship between the mea-sured displacement and the strain in the polymer can be expressed according to the following equation:

$$S_1 = \frac{\Delta l}{l_0} \tag{16}$$

where, $\Delta l$ and $l_0$ represent respectively the variation of the length.0and the initial length.

The model for the polymer was thus simplified by the circuit ($R_p$, $C_p$) in parallel with the motional branch. Figure 8 displays the most commonly adopted form of an electrical scheme (Eddiai et al., 2012b) where $C_p$ is the capacitance of the clamped polymer and $R_p$ is a resistance representing the dielectric losses.

This approach rendered it possible to determine the power harvested by the EPC in a pseudo piezo-electric state. The third branch is the motional branch modelled by the current source $I_h$ (Equation (14)) that can model the harvestedcurrent from vibrations.

For the active energy harvesting system investigated within thescope of this study, there was electric energy consumed by the polarization of the polymer. No direct measurement on the amountof electric energy consumed was performed. Based on the results of $R_p$ and the equivalent electric scheme of the setup, it was possibleto estimate the efficiency of the conversion.

Consequently, the power harvested power could be calculated from the harvested current using Equation (17):

$$P_{harvested} \propto R_C I^2 \tag{17}$$

where $R_C$ is the resistor load.

*Figure 8. The equivalent electric circuit of an electrostrictive polymer*

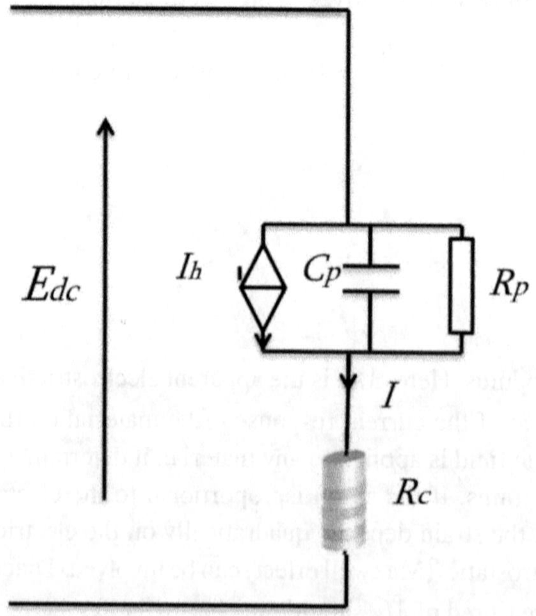


## 5. ENERGY HARVESTING

Recent trends in both industrial and research fields have focused on electrostrictive polymers that possess many advantages such as high productivity, flexibility, and high electromechanical response (Eddiai et al., 2012c; Zhang, Neese, Ren, Chu, & Zhang, 2006; Bobnar, Levstik, Huang, & Zhang, 2007; Meddad et al., 2012a; Liu, Zhang, & Cross, 1999; Eury, Yimniriun, Sundar, & Moses, 1999). A lot of effort has been devoted recently in developing device configurations and concepts in the energy harvesting with electro-active materials with the objective of achieving high electric power output (Jean-Mistral, Basour, & Chaillout, 2008; Ren, liu, Hofmann, & Zhang, 2007). The aim of this section consists of proposing new techniques in order to improve the power recovered. Thus, a study of the effect of stain amplitude and operating frequency on the electrostrictive polymer composite in order to optimize the optimal range of the current recovered by these two variables to obtain a better electromechanical conversion.

### 5.1. New Techniques for the Maximization of Power Recovered

In this section, two techniques will be presented in order to improve the power recovered using the electrostrictive polymers. The first will be based on reversal polarization in the half period and a phase shift between the electrical and mechanical excitations will be considered in order to maximize the density of energy harvested. For the second technique, we will propose a new structure that is able to predict mechanical frequency excitation in order to increase power-harvesting capabilities of electrostrictive polymers.

### 5.1.1. The Maximization of the Power Harvested by the Reversal of Polarization

Energy harvesting closely depends on the mechanical behavior of the device as well as the electromechanical coupling coefficient of the polymer.This technique investigates the effects of different signals of electrical field Ein order to develop a more in-depth understanding of the changes in electrostrictive polymers composites (EPCs) response for increased current and energy harvesting (Eddiai et al., 2014).

#### 5.1.1.1. Modeling and Abilities Analysis

The goal of this section is to expose a theoretical derivation of the maximal output power that can be obtained using electrostrictive materials. In this configuration, the current generated by the polymer was used to measure the harvested power.Based on the Equation 9, the approach for modeling the currentcan be written for an electrostrictive material as:

$$\begin{aligned} T_{ij} &= Y_{ijkl}{}^{P}.S_{kl} + \alpha_{ijmn}.P_m.P_n \\ P_n &= \varepsilon_{mn}{}^{S}.E_{sat}th\left(E_m/E_{sat}\right) + 2.\alpha_{klmn}.\varepsilon_{mn}{}^{S}.P_n.S_{kl} \end{aligned} \tag{18}$$

where $E_{sat}$ is the saturation field and $\alpha_{ijmn} = Q_{ijkl} \bullet Y_{klmn}$ is a coefficient with $Q_{ijkl}$ representing the electrostrictive coefficient. An isotropic electrostrictive polymer film contracts along the thickness direction and expands along the film direction when an electric field is applied across the thickness, assuming that only a non-zero stress is employed along the length of the film. The constitutive relation is then simplified as:

$$\begin{cases} T_1 = Y_{11}.S_1 + \alpha_{31}.P_3^2 \\ P_3 = \varepsilon_{33}.E_{sat}th\left(E_3/E_{sat}\right) + 2.\alpha_{31}.\varepsilon_{33}.P_3.S_1 \end{cases} \tag{19}$$

By assuming uniform strain and electric field over the sample, the macroscopic equation of the current $I = A\dfrac{\partial P}{\partial t}$ flowing out of the polymer can be expressed as:

$$\frac{\partial P_3}{\partial t} = \frac{\varepsilon_{33}.(1 - th^2(E_3/E_{sat})).\dfrac{\partial E_3}{\partial t} + 2.\alpha_{31}.\varepsilon_{33}.P_3.\dfrac{\partial S_1}{\partial t}}{1 - 2.\alpha_{31}.\varepsilon_{33}.S_1} \tag{20}$$

The current induced by the transverse vibration is then given by the equation below where A represents the electrode surface:

$$I = A.\left[\frac{\varepsilon_{33}.(1 - th^2(E_3/E_{sat})).\dfrac{\partial E_3}{\partial t}}{1 - 2.\alpha_{31}.\varepsilon_{33}.S_1} + \frac{2.\alpha_{31}.\varepsilon^2{}_{33}.E_{sat}.th(E_3/E_{sat}).\dfrac{\partial S_1}{\partial t}}{(1 - 2.\alpha_{31}.\varepsilon_{33}.S_1)^2}\right] \tag{21}$$

The total electric field is equal to the sum of a static field with reversal $E_{dc}$ and an alternating electric field $E_{ac}$, across the resistor $R_C$; ($E_3=E_{dc}+E_{ac}$).

According to Figure8 which presents the electrical scheme of the electrostrictive sample, the ac component of the electric field is generated by the load such as

$$E_{ac} = \frac{V_{ac}}{e} = \frac{R.I}{e} \tag{22}$$

where is the thickness of sample and the current expression thus becomes:

$$I = A.\left[\frac{\varepsilon_{33}.(1 - th^2(\frac{E_{ac} + E_{dc}}{E_{sat}})).(\frac{R}{e}.\frac{\partial I}{\partial t})}{1 - 2.\alpha_{31}.\varepsilon_{33}.S_1} + \frac{2.\alpha_{31}.\varepsilon^2{}_{33}.E_{sat}.th(\frac{E_{ac} + E_{dc}}{E_{sat}}).\frac{\partial S_1}{\partial t}}{(1 - 2.\alpha_{31}.\varepsilon_{33}.S_1)^2}\right] \tag{23}$$

In this experiment, a sinusoidal strain $S_1$ and electric field $E_3$ were used. They are given by the following equations:

$$E_3 = E_{dc} + E_0.\sin(\omega.t) \tag{24}$$

$$S_1 = S_{dc} + S_0.\sin(\omega.t + \varphi) \tag{25}$$

where $\omega=2.\pi.f$ is the angular frequency, $S_{dc}$ is the static component of the strain, $S_0$ denotes the strain magnitude and $\varphi$ is the phase shift between electrical field and strain. In our case the dynamic component of the electric field is low and can be neglected compared to static component ()Equation (23) can therefore be simplified as follows:

$$I = A\left[\frac{2.\alpha_{31}.\varepsilon^2{}_{33}.E_{sat}.th(E_{dc}/E_{sat}).\frac{\partial S_1}{\partial t}}{(1-2.\alpha_{31}.\varepsilon_{33}.S_1)^2}\right]_2 \tag{26}$$

The relationship (26) shows clearly that if we reverse the polarity in the half-period with the application of a phase shift between the two excitations, one can gain more electric power compared to a conventional system based the application of a static field $E_{dc}$. The harvested energy density for one period of the sinusoidal excitation is given by (Ren, liu, Hofmann, & Zhang, 2007),

$$W = \int_0^T E_3.I.dt \tag{27}$$

Replacing the current by its expression (Equation (I.23)), the harvested energy density becomes as follows,

$$W = \int_0^T E_3.A.\left[\frac{\varepsilon_{33}.(1 - th^2(E_3/E_{sat})).\frac{\partial E_3}{\partial t}}{1 - 2.\alpha_{31}.\varepsilon_{33}.S_1} + \frac{2.\alpha_{31}.\varepsilon^2{}_{33}.E_{sat}.th(E_3/E_{sat}).\frac{\partial S_1}{\partial t}}{(1 - 2.\alpha_{31}.\varepsilon_{33}.S_1)^2}\right].dt \tag{28}$$

thus,

$$W = 2.\pi.\alpha_{31}.\varepsilon_{33}^2.A.E_{sat}.th(\frac{E_{dc}}{E_{sat}}).E_{dc}.S_0.\sin(\varphi) \tag{29}$$

It can be seen from this latter equation, that the harvested density energy reaches its maximal value when sin(φ)=1.

### 5.1.1.2. Experimental Setup

Electrostrictive polymers were subjected to a dc-biased electric field in order to induce polarization, because the polymer is not piezoelectric. The setup developed for characterizing the electrostrictive coefficient of the polymer film is shown schematically in Figure 9.The polymer film is mounted in sample holder composedof two parts: one fixed and another that can be moved in 1 direction with the help of an XM550 Ironless linear motor. The film is stretched with 0.5% maximum amplitude of strain at 15 Hz. It is assumed that the sample strained along the 1 direction, by applying a phase shift with slot electric field using a generator function (Agilent 33220A) connected to the high voltage amplifier (Trek 609D-6). The sample was connected to an electrical load $R_C$, and the current was monitored by a current amplifier (Stanford SR570). Then, the power harvested on the load is derived using equation. (17).

### 5.1.1.3. Validation of the Model

In order to access the validity of the model of the harvested power presented in section 5.1.1.1., various measurements were carried out. Figure 10 shows the experimental results which globally validate the theoretical model for the evaluation of the harvested power. A possible reason for the slight discrepancy between both results (for $R_C \rangle 7M\Omega$) could be unmodeled loss mechanisms in the electrostrictive device, which represents the signal of the strain and the electric field as a function of time. The experimental data also shows the existence of an optimal load resistance $R_C$=*7MΩ* which correspond to a maximum harvested power (2.7 μW) in quite good agreement with the theoretical model. In Figure 11, we present the harvested power with two different polarization signals. It is clear that the harvested power with this technique is found to be 7 times more important than the one obtained by classical techniques (polarization fields with a constant signal).This value demonstrated the excellent potential of this technique for energy harvesting.

*Figure 9. Schema view of the experimental setup*

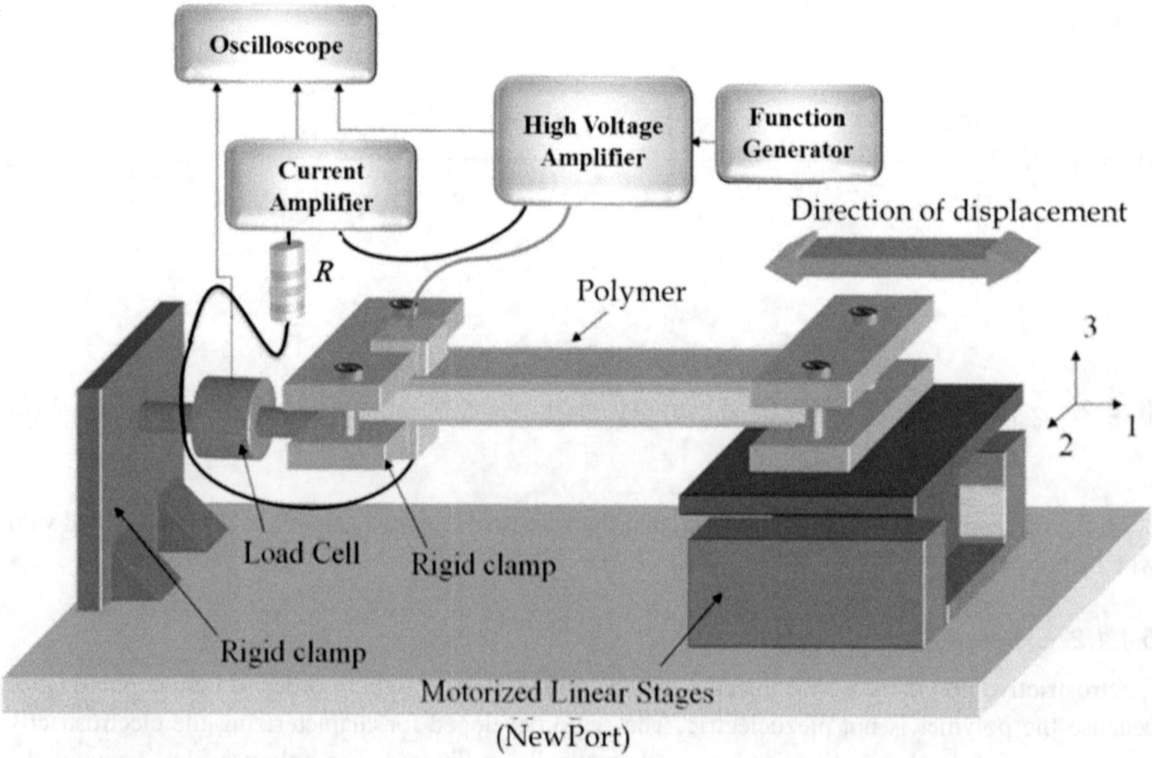


*Figure 10. Harvested power under proposed electrical field signals of 10 V/μm at 15Hz for a strain 0.5%*

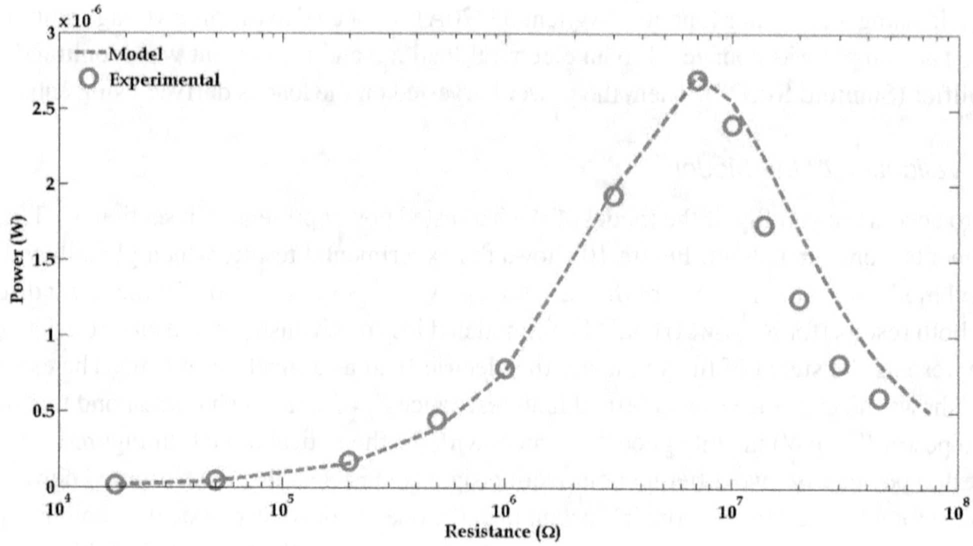

*Figure 11. The experimental harvested Power under different electrical field signals of10 V/µm at 15 Hz for a strain 0.5%*

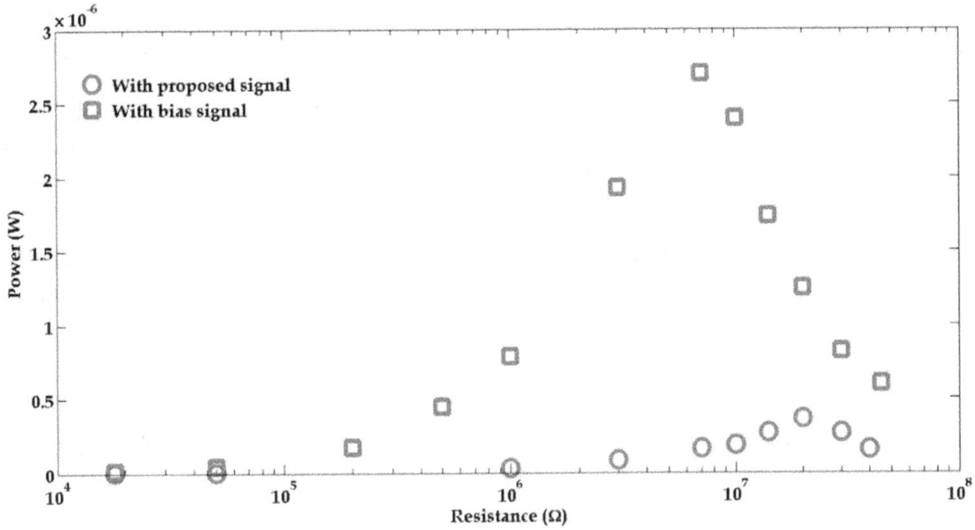


Electrostrictive polymer composites are thus promising candidate for replacing piezoelectric (PZT: Lead Zirconat-Titanate) materials power harvesting. The technique developed in this paper could be a good tool for estimating and then improving energy harvesting capabilities in the future.

### 5.1.2. An Adaptive Prototype Design to Maximize Power Harvesting

The energy harvesting with electrostrictive polymers has great potential for remote applications such as in vivo sensors, embedded micro-electro-mechanical systems devices, and distributed network instruments. The use of such polymers for energy harvesting is a growing field, which has great potential from an energy density viewpoint. The output power is inversely proportional to the harvester's frequency bandwidth. Consequently, it is much harder to efficiently harvest power from low-frequency sources with a large frequency band response and with a very small system size than from a stabilized high-frequency vibration source.

Many investigations have been performed on electrostrictive polymers to predict their energy harvesting capabilities by increasing the permittivity or incorporating conductive nano-particles (Putson et al., 2011). Other studies, such as those carried out by Ren et al. (Ren, liu, Hofmann, & Zhang, 2007) and Cottinet et al., (Eddiai, 2013) show the effect of the strain frequency in the current generated by thepolymer in order to increase the harvested power. The objectiveof this technique was to demonstrate the feasibility of the changes in system response by a prototype design for an increased current and energy harvesting by a frequency doubler.

#### *5.1.2.1. Principle of Measurement of the Harvested Power*

Electrostriction is generally defined as a quadratic coupling between the strain and the polarization. Here, an electrostrictive polymer was subjected to a dc bias electric field. This was necessary in order to obtain a pseudo piezoelectric behavior due to the polymer not being piezoelectric. Two configurations of an electrical scheme, illustrated in Figure 12,were compared in this study. In configuration (a) following the application of a mechanical excitation with a frequency $f_m$, it was stretched with the same frequency. Consequently, when the polymer was excited electrically, the harvested current was related to $f_m$. Assuming a uniform strain and electric field over the sample, we can express the macroscopic equation of the current $I = A\dot{P}$ (where A refers to the electrode surface) flowing out of the polymer as a function of the strain and electric field. The relation between the polarization-electric field was linear when the electrical E was small, $P=\varepsilon E$, and the power harvested on the load was derived using the Equation (17).

For the second configuration (b), the principle part was a metal on which the polymer film was attached. Consequently, the mechanical excitation of the cycle gave rise to two cycles of deformation of the polymer with a frequency of $2f_m$. The main advantage of this configuration was the simplicity of increasing the conversion abilities of the electrostrictive materials. The electromechanical properties and energy-harvesting capacities of electrostrictive polymer are important both from the viewpoints of application and fundamental understanding. This new configuration was simple, convenient to use, and capable of increasing the harvested power. The configuration of the mechanical system, shown in Figure 12, is the following: the polymer film was attached in one end to the metal and its other end was attached at the solid (fixed) base. By applying an excitation mechanism by a shaker, the polymer sample was stretched in two directions.

From Equation 14, it is clear that the current generated in the first configuration depended on $f_m$, whereas in the second configuration, involving stretching for a single cycle of mechanical excitation, the generated current depended on $2f_m$.

Therefore, the power harvested ($P_{harvested1}$) with the first configuration can be calculated as

$$P_{harvested1} = \frac{(\alpha^2.E_{dc}^2.S_0^2)}{4.C_p}.f \tag{30}$$

where $\alpha = \dfrac{2M_{31}A}{s_{11}}$.For the second configuration, the harvested power ($P_{harvested2}$) can be determined with the following equation:

$$P_{harvested2} \propto 4R_c.M_{31}^{*\,2}.Y^2.E_{dc}^{\ 2}.\left[\int_A \frac{\partial S_1}{\partial t}\,dA\right]^2 \propto R_c.\alpha^2.E_{dc}^2.S_0^2\left(2f_m\right) \tag{31}$$

### *5.1.2.2. Modeling of the Harvested Power*

It is clear from Equations (30) and (31) that the power harvested in the second configuration was twice that of the first configuration. The equivalent electric scheme is shown in Figure 8, in which $R_C$ is the electric load.

*Figure 12. Diagram of the prototype circuit*

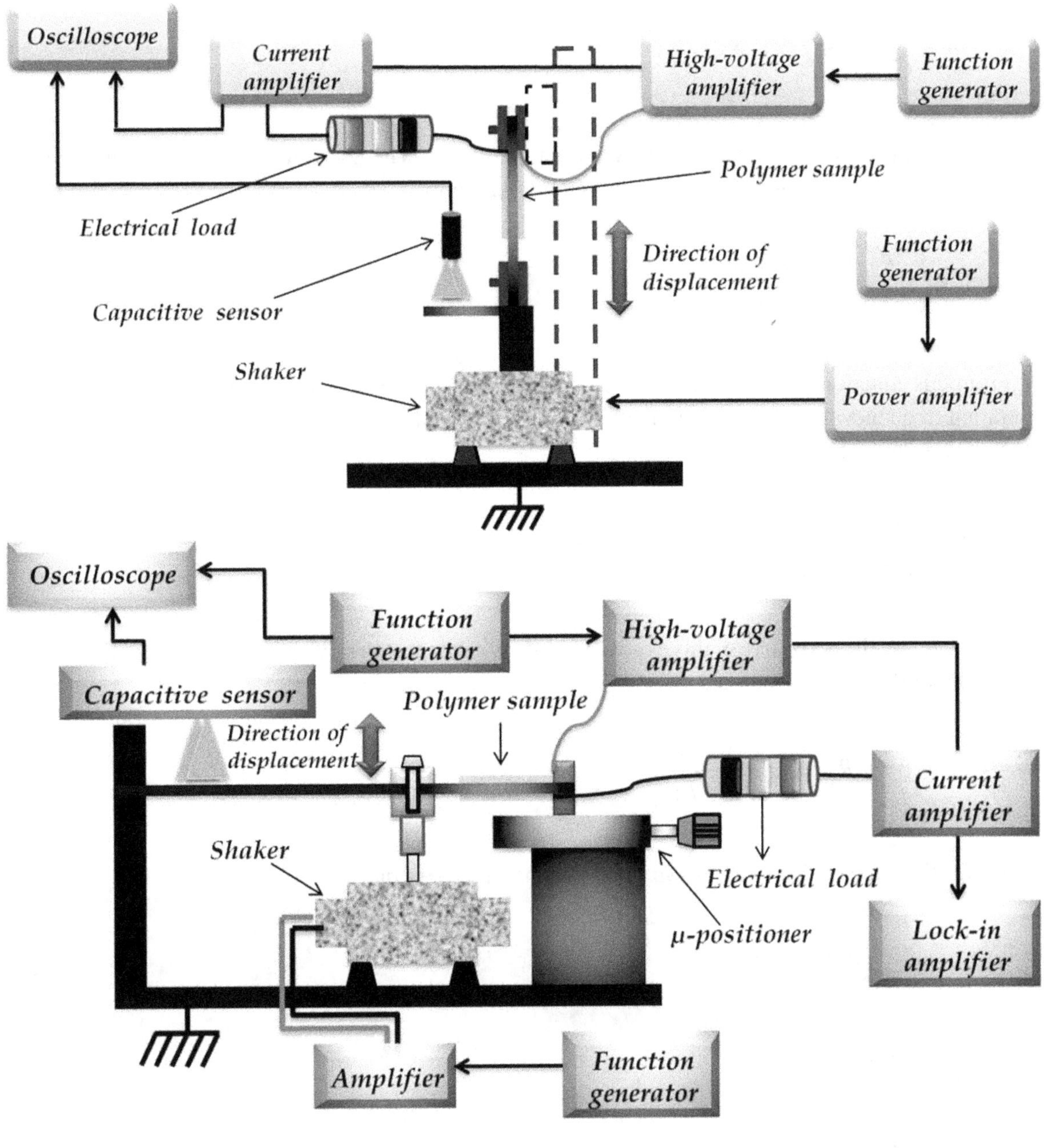


A purely resistive load directly connected to the electrostrictive elements (Lebrun, Guyomar, Guiffard, Cottinet, & Putson, 2009) demonstrates that when the polymer operates in the pseudo-piezoelectric mode, the alternative component on the load resistance may be neglected compared with the dc-bias electric field

$$I = \alpha.E_{dc}.\frac{\partial S_1}{\partial t} - C_p.\frac{\partial V}{\partial t} \qquad (32)$$

The dynamic voltage on the electrostrictive element can be expressed in the frequency domain as:

$$\tilde{V} = \frac{\alpha.E_{dc}.R_c}{1+jR_c.C_p.2\pi f}.j2\pi f.\tilde{S}_1 \tag{33}$$

where j is the imaginary unit and $f$ is the vibration frequency. The harvested power can be written as a function of the strain amplitude $S_0$ by the following relation:

$$P_{harvested} = \frac{\alpha^2.E_{dc}^2.R_c.(2\pi f)^2 S_0^2}{1+(R_c C_p 2\pi f)^2} \tag{34}$$

From this equation, it is clear that there is a strong relationship between the harvested power and the operating frequency. The power is optimal for a given load and demonstrates by deriving the expression of the power with respect to R:

$$\frac{\partial P_{harvested}}{\partial t} = \frac{1-(R_c.C_p.2\pi.f)^2}{(1+(R_c.C_p.2\pi f)^2)^2}.\frac{\alpha^2 E_{dc}^2.(2\pi.f)^2.S_0^2}{2} \tag{35}$$

By canceling the derivative of this expression (35) with respect to R, we, therefore, obtain the optimal load and maximum power output:

$$\frac{\partial P_{harvested}}{\partial R_c} = 0 \Rightarrow R_{opt} = \frac{1}{C_p.(2\pi.f)} \tag{36}$$

So, for the matched load, the maximum power harvested is equal to

$$P_{harvested\_max} = \frac{(\alpha^2.E_{dc}^2.S_0^2)}{4.C_p}.f \tag{37}$$

For $\beta = \frac{(\alpha^2.E_{dc}^2.S_0^2)}{4.C_p}$ we get

$$P_{harvested\_max} = \beta.f \tag{38}$$

It can be shown that the maximal power depends not only on the material properties but also on external condition and dimensions.

Figure 13 shows that the relative permittivity was nearly constant in the 1–10 Hz range with the relationship (37) and (38). Moreover, in the two configurations mentioned above, it is clear that the frequency $f$ of operation is the only parameter that intervenes in the change of the power harvested.

*Figure 13. Current and displacement vs. time*

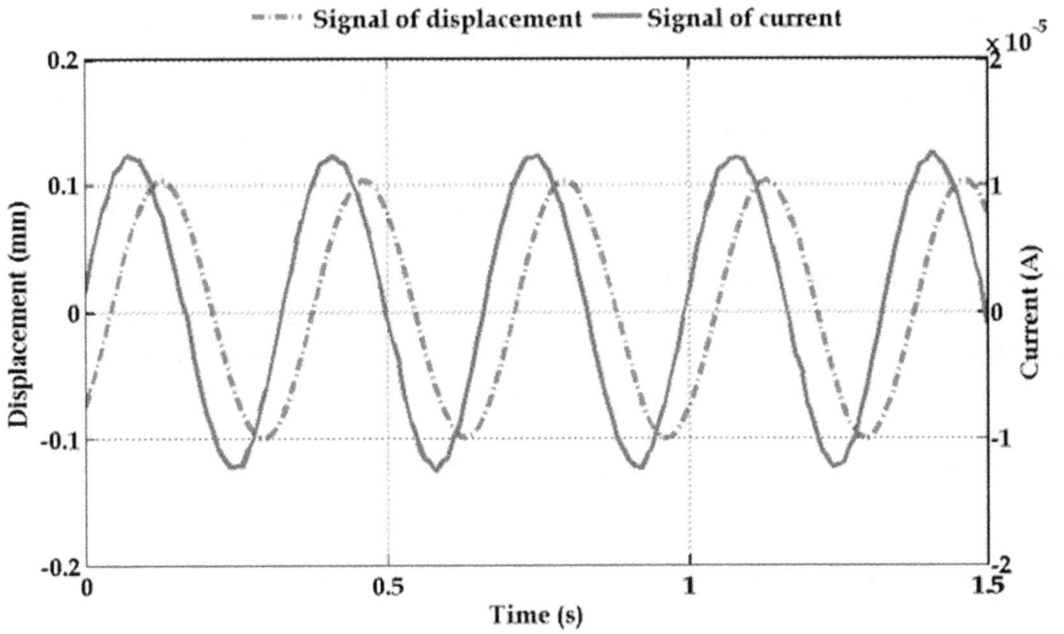


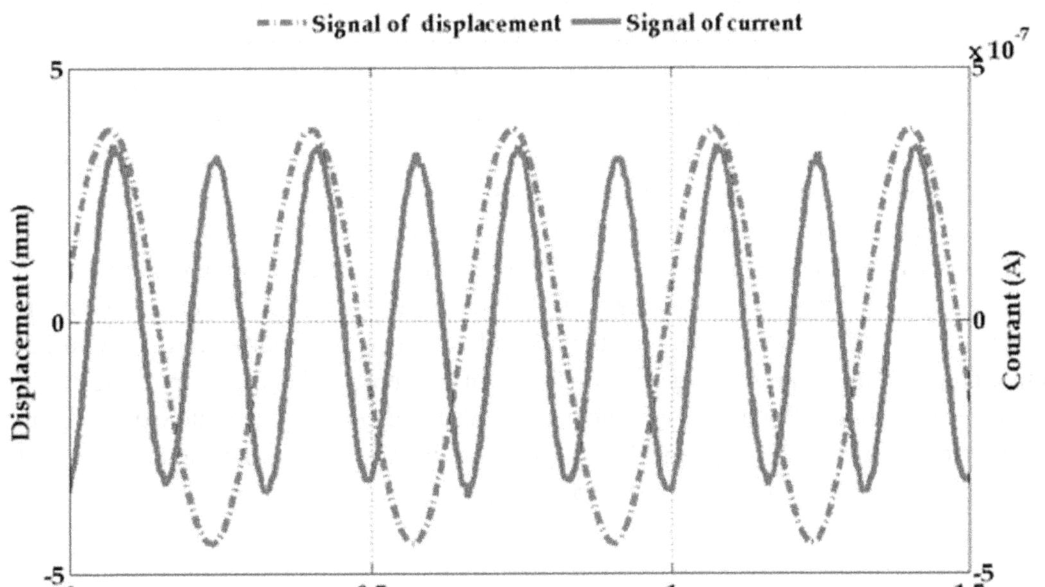

### 5.1.2.3. Validation of the model

The following section judges the accuracy of the measured output current, as well as the effect of the proposed configuration. The film was stretched with a maximum strain amplitude of 0.5% at $3Hz$ in both configurations. The parameters for the polymers used in this experiment are given in Table 4.

The modeling and experiments were first realized as displayed in the analysis of the short-circuit current. This was done to illustrate the theory presented in section 5.1.2.2 and to demonstrate the effect of this system with regard to the performance of electrostictive polymers for energy harvesting (Meddad et al., 2012b).

The variation in harvested current and displacement versus time is given in Figure 13a and Figure 13b. As can be seen when analyzing this current, the curve of the current in the first configuration is in phase with the displacement, while in the second configuration, there is an increase of the harvested current with a double frequency as compared tothat of the displacement.Various solutions exist to improve the energy harvesting. One can for instance increase the area of the polymer, the bias electric field, or the strain, but also the frequency. The latter parameter was an important factor in the conversion.

The next experiment was performed to validate the increase in power. According to the theory presented in section 5.1.2.2, the difference between the last experiments resides in an added electrical resistance for measuring the harvested power. Figures14 and 15 present the power versus electric load for a given electric field of 10 V/µm and strain of 0.5%. These data display the existence of an optimal load resistance of 39MΩ and a maximum harvested power of 0.285 µw in the classical configuration.

For the proposed configuration, on the other hand, the maximum power reached a value of 0.5 µwwith an optimal load resistance of 20MΩ, as theoretically expected according to Equation (36) for the two configurations. A good agreement between the modeling and the experimental results was observed. The magnification made it possible to see that the proposed configuration clearly demonstrated an output power increase that was 80% greater as compared to the standard technique and this without the use of an electronic circuit. The second advantage in the proposed structure was an increase of the harvested power associated with a decrease in the optimal resistance of 20MΩ.

This technique proposes an application approach for increasing the conversion abilities of electrostrictive polymers without the use of an electronic circuit. Based on a mechanism of doubling the frequency of displacement of the polymer through a proposed mechanical structure, we can also obtain a higher dynamic current. This effect renders possible a significant increase in terms of converted power. The structure proposed thus represents a good tool for estimating potential energy harvesting candidates for technologies that collect vibrations from the environment and convert them into electricity in order to power a variety of sensors. Examples include applications with low-frequency vibration (smart textiles), recovering for instance microseismic or human energy.

*Table 4. Parameters of polymers (PU 1 wt. % C)*

| **Polymer** | $\varepsilon_r$ | $Y(MPa)$ | $A(cm^2)$ | $E(\mu m)$ |
|---|---|---|---|---|
| **PU 1%C** | 11 | 34 | 4×1.6 | 53 |

*Figure 14. Harvested power under different electrical load values at 6Hz and 3Hz with S=0.5%*

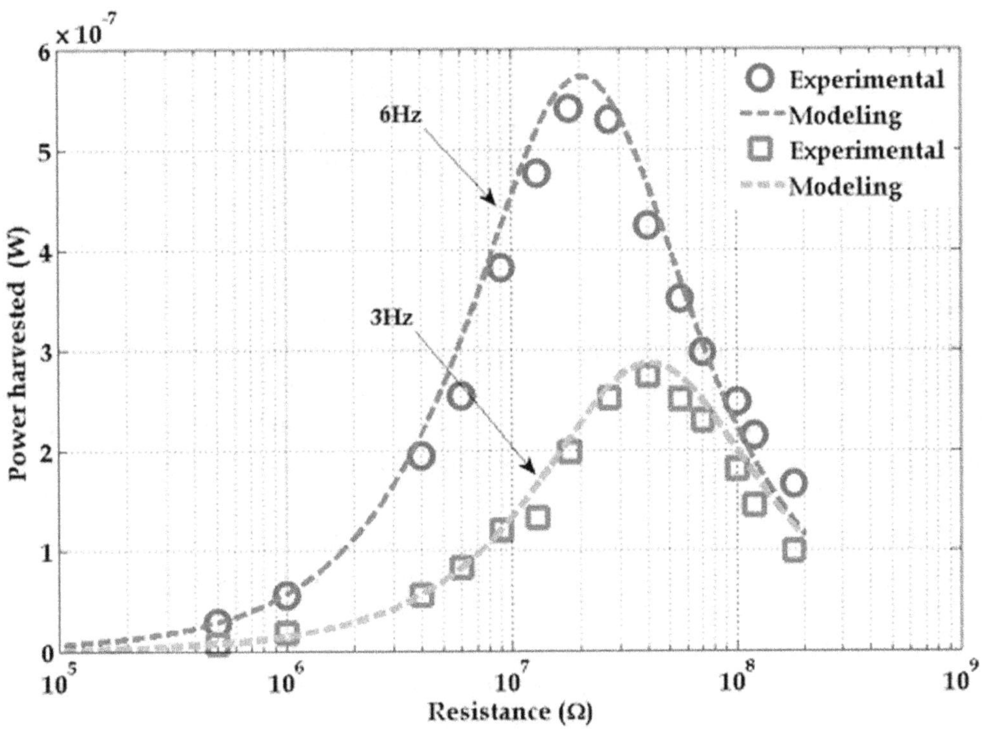


*Figure 15. Harvested power as a function of frequency and electrical load*

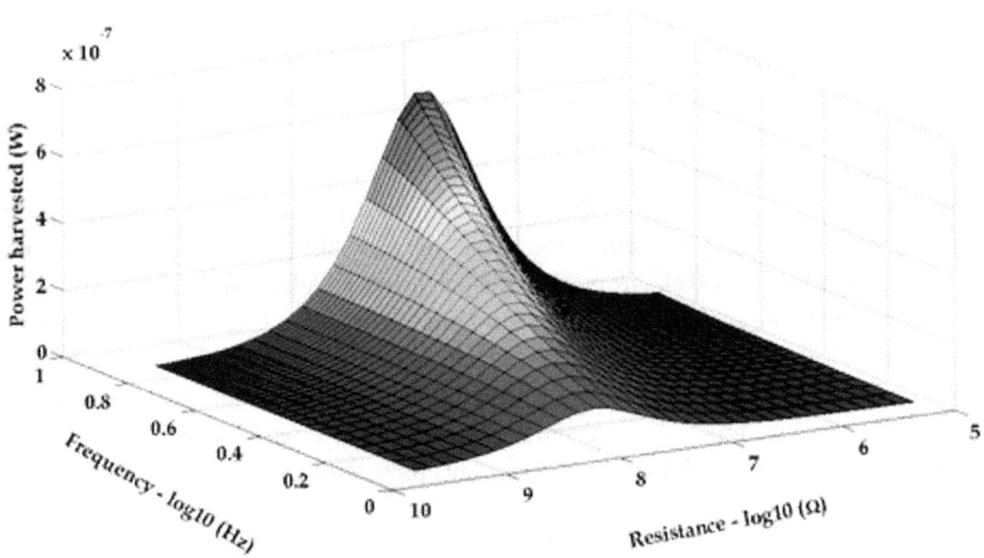

## 5.2. Evaluation of the Electromechanical Conversion

At present, the investigation of using electrostrictive polymers for energy harvesting (a conversion of mechanical toelectrical energy) is beginning to show potential for this application. Basically, the relative energy gaindepends on the current induced by the mechanical strain and frequency. The purpose of this part was todetermine the optimum operating range for improved electro-mechanical conversion efficiency of electrostrictive polymer composite by the effect of mechanical parameters (mechanical frequency and the amplitude strain) that leads to an increase in the generated current and improve the output power in relation to the injected power.

Previous studies have been performed on electrostrictive polymers to predict their energy harvesting capabilities by increasing the permittivity or incorporating conductive nanoparticles (Meddad et al., 2014a; Lallart, Cottinet, Lebrun, Guiffard, & Guyomar, 2010). Other investigations, Guyomar et al. have demonstrated in their work (Guyomar, Cottinet, Lebrun, & Sebald, 2011a; Meddad et al., 2012c; Guyomar, Lallart, & Cottinet, 2011b) that there is strong relationship between the frequency ofoperation and the mechanical deformation from fast Fourier trans-forms analysis (FFT) of the current harvested by electrostrictives polymers. Therefore, our study is focused on the analysisof the effect of stainamplitude and operating frequency on the electrostrictive polymer composite in order to optimize the optimal range of the current recovered by these two variables in order to obtain a better electromechanical conversion (Meddad et al., 2014b).

The choice of frequency and amplitude of strain are not trivial tostudy the effectiveness of electrostrictive polymers. However, the optimization of the two parameters is very important in order toimprove the electromechanical conversion of the electrostrictive polymer for harvesting electrical energy. Current measurements under short-circuit Equation (14) are based on the optimization of the frequency electric field, mechanical frequency, and the amplitude of mechanical excitation which lead to an increase of the current generated and enhancing the output power.Tables 5 and 6 presentthe electrical efficiency as a function of mechanical frequency and transverse deformation for a dc electric field of 13 V/µmat 1 Hz, respectively. Since the electrical efficiency of the polymer calculated by the ratio between the input power and that harvested increases with transverse strain (also shown in Table 6, likewise the efficiency increases by the increasing of mechanical frequency (Table 5). It is noteworthy that these experimental measurements of electrical efficiency clearly show the strong dependence on mechanical parameters.

This dependence allows us to know in advance the mechanical frequency range and transverse deformation in order to achieve good electromechanical conversion of electrostrictive polymers for energy harvesting. For example, in Table 6, above a deformation of 4% at 6*Hz* and for the conditions of the used electric field (13 V/µm, 1 Hz), the electrical efficiency becomes positive. Finally, we will present a comparison in terms of converted power and energy density of our results with those recently published. In our study, we have found a power of 0.3 µW converted using PU 1 wt% C (6.4 $cm^2$and 52 µm thickness) under 10 $V.\mu m^{-1}$with 0.5% strain at 3 Hz. This represents an energy density of 0.0082 $mJ.g^{-1}$

*Table 5. The electrical efficiency as a function of mechanical frequency for a constant strain of 5% and a static field of 13V/µm at 1Hz*

| **Frequency (Hz)** | 3 | 6 | 9 |
|---|---|---|---|
| **Electrical Efficiency (%)** | 37.14 | 52.32 | 62.34 |

*Table 6. The electrical efficiency as a function of transverse strain for a mechanical frequency of 6Hz and a static field of 13 V/μm at 1Hz*

| Strain (%) | 0.75 | 2 | 4 | 6.5 |
|---|---|---|---|---|
| Electrical Efficiency (%) | 33.3 | 37.6 | 50.6 | 53.3 |

(the specific mass of PU 1 wt.% Cis 1.1 g.cm$^{-3}$). Comparison with the works realized with regard to electrostrictive polymers, Lallart et al. (Lallart, Cottinet, Lebrun, Guiffard, & Guyomar, 2010) showed an energy density of about 0.056 mJ.g$^{-1}$ scavenged by their device (5 cm$^2$and 50 μm thickness) made by P(VDF–TrFE–CFE)–1%C under 10 MV.μm$^{-1}$ and 0.5%strain at 100 Hz. Finally, we note that the energy density recovered by our system is very large compared to that obtained by Lallart et al., also knowing that the electrical losses in the P(VDF–TrFE–CFE)is higher than that in PU.

Finally, we conclude that the combination of the two parameters (the trans-verse strain amplitude and the operating frequency) has a strong influence on the electrical efficiency conversion of electrostrictive polymers for mechanical energy harvesting.

## 6. SYSTEM MANAGEMENT AND TRANSFER ENERGY TO THE STORAGE BATTERY

### 6.1. Transfer of Electrical Energy from a Source of Electrostrictive Polymer

The recovery system associated with the piezoelectric system is important because the energy generated is generally not used directly by the target portable device. The conversion of electrical energy delivered by a micro power generator a problem that has been raised in many articles (Guyomar, Lefeuvre, Richard, & Badel, 2004). The electric conversion is most often simply discharge the piezoelectric system capacity, via a rectifier bridge circuit. This Simple is inefficient in terms of energy efficiency: the micro power generator having essentially capacitive internal impedance, the transfer of energy in another capacity is always partial. And will lead to a decrease in energy to 25% of the maximum power delivered by a piezoelectric generator that can be shown in Figure (16). Suppose, a capacitor of capacitance $C_0$, which is the initial power$W_0$, and a capacitor $C_1$ initially discharged, the two capacitors are connected together through a circuit resistance R. (Meddad et al., 2014a)

For the initial state with the initial open circuit and power $W_0 = \dfrac{1}{2.C_0}.V_0^2$ with $C_0$=$C_l$ after charging $V_f = \dfrac{q_0}{2.C_0}$ with $V_f = \dfrac{V_0}{2}$ where the energy capacitor is $W_l = \dfrac{1}{8.C_0}V_0^2 = \dfrac{W_0}{4}$. During the energy transfer, $\dfrac{W_0}{2}$ was dissipated by the Joule effect in the circuit. This simple calculation shows the inefficiency of the direct transfer of energy between these capacitors for this reason several system (Burns, Taylor, & Welsh, 2003), have been realized to increase the proportion of energy transferred to the energy available without try to increase it. However, the piezoelectric generator is primarily an electromechanical system, which offers four degrees of freedom, two electrical and two mechanical quantities sizes, linked together. Therefore, it is possible, by acting on the voltage and the current, changing the force and speed, to increase the power converted.

Figure 16. Energy transfer between two capacitors

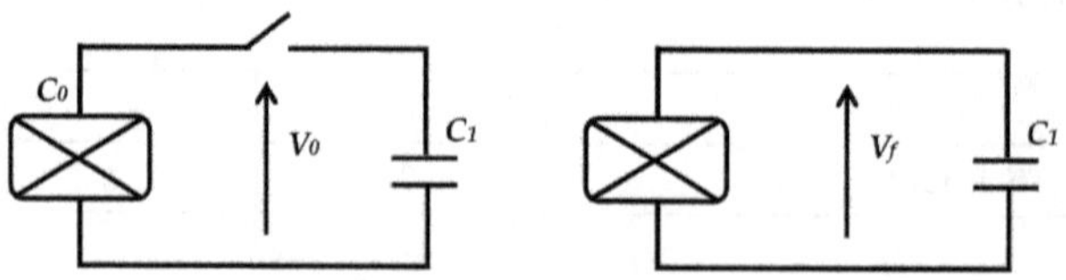


Unlike the conventional case of piezoelectricity example, where the rectifier bridge is connected directly in parallel with the piezoelectric element, it is necessary in the case of electro-active operating system in pseudo-piezoelectric materials, adapt the architecture assembly Figure 17 (Meddad, 2012c). We must add two components. A resistor $R_s$. to impose that the current produced by the conversion electromechanically flows mainly in the loop where the diode bridge is connected. $C_b$ capacity eliminates the high voltage component across the AC-DC converter followed by a filtering capacitor $C_f$.

Expression of the voltage across the resistor is expressed as:

$$\tilde{V} = \frac{c_b}{c_b + c_p} \frac{\alpha.E_{dc}.R}{1 + \frac{jR.c_b.c_p.\omega}{c_b + c_p} j.\omega.\tilde{S}_1} j\omega\tilde{S}_1 \quad (39)$$

The power in the resistor R is equal to:

$$P_{harvested\_AC} = \frac{\tilde{V}.\tilde{V}^*}{2.R} = \left(\frac{c_b}{c_b + c_p}\right)^2 \frac{\alpha^2.R}{1 + \left(\frac{R.c_b.c_p.\omega}{c_b + c_p}\right)^2} \frac{\omega^2 E_{dc}^2 S_0^2}{2} \quad (40)$$

Figure 17. Electrical schematic used with AC conversion

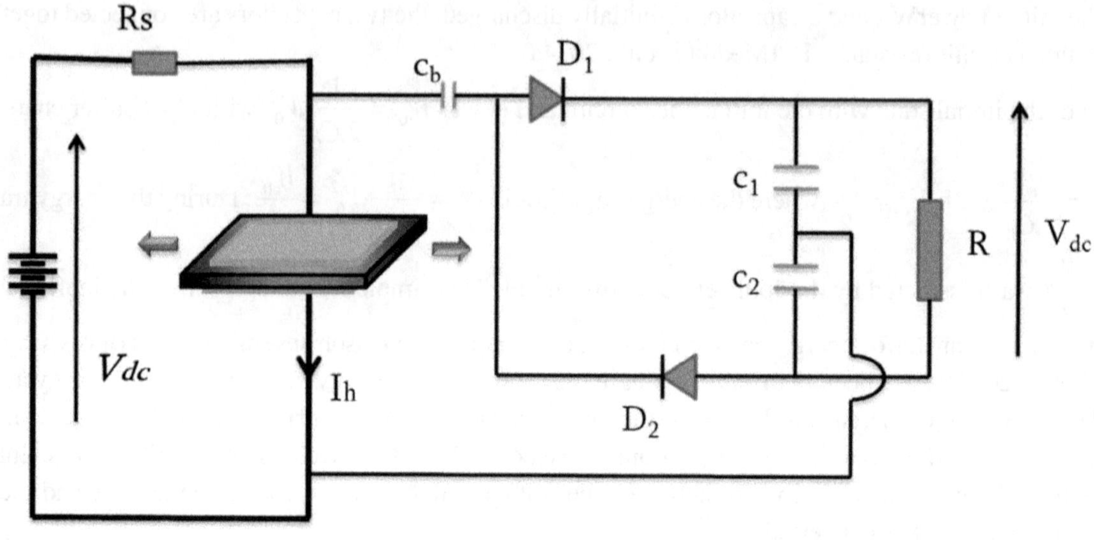

where S represents the magnitude of strain

To overcome the problem related to the failure of the conventional structure, a solution has been proposed is to use a frequency doubler (Figure 12b). Indeed this type of architecture has only two alternating strain for a single mechanical excitation alternately. From a theoretical point of view the maximum power is obtained for a greater resistance. Ultimately, result obtained in Figure 18 recovered a rectified power supply by the best conventional system.

## 6.2. Optimization Based on the Storage Capacity

The choice of storage capacity is an important parameter in the design and performance of a micro power generator. The Figure 19 shows a simplified circuit of a rectifier bridge with a storage pattern.

From this circuit:

$$\begin{cases} V(t) = I_q.R_d.\left(1-e^{-\left(\frac{t}{R_d.C_a}\right)}\right) & 0 \le t < t_d \\ V_1(t) = \langle V_1 \rangle & \end{cases} \tag{41}$$

$\langle V_1 \rangle$ represents the average value of the voltage across the capacitor accumulate $t_d$, represents the conduction time for the two diodes. $R_d$ represents the dielectric losses. The energy transfer occurs between the locked and the capacity storage capacity only when $V(t)=V_1(t)+2V_{dd}$. So (41) can be written as:

$$2V_{dd} + \langle V_1 \rangle = I_q.R_d.\left(1-e^{-\left(\frac{t}{R_d.C_a}\right)}\right) \tag{42}$$

Found the following equation:

$$W_1 = \frac{1}{R_1}\int_0^T V_1(t)^2 dt \tag{43}$$

Therefore

$$W_1 = \frac{1}{R_1}.\left[\int_{t_d}^{a}\left(\left(I_q.R_e - 2.V_{dd}.\frac{R_e}{R_d}\right).\left(1-e^{\frac{t-t_d}{R_e.C_e}}\right)+\left\langle V_1\right\rangle.e^{\frac{t-t_d}{R_e.C_e}}\right)^2.dt + \int_a^T\left(V_1(a).e^{\frac{t-a}{R_1.C_d}}\right)^2.dt\right] \tag{44}$$

Then the energy density becomes in the following form:

*Figure 18. Power recovered according to the load resistor an DC in AC mode, for a deformation of 0.5% and an electric field 7V/μm using the proposed structure*

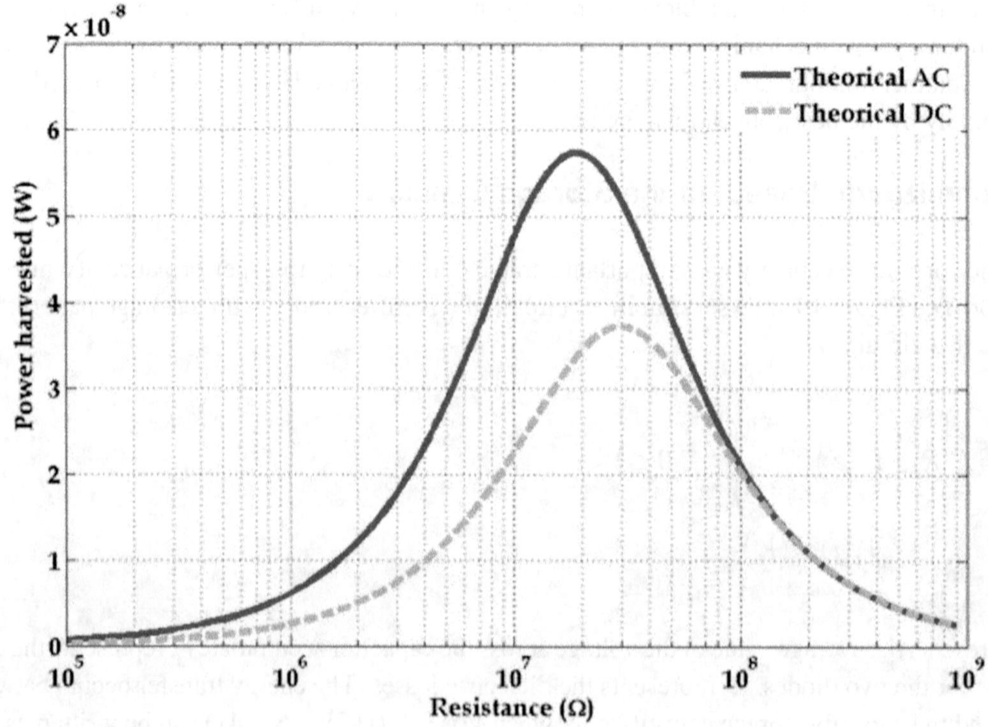


*Figure 19. Equivalent circuit for charging a capacitor*

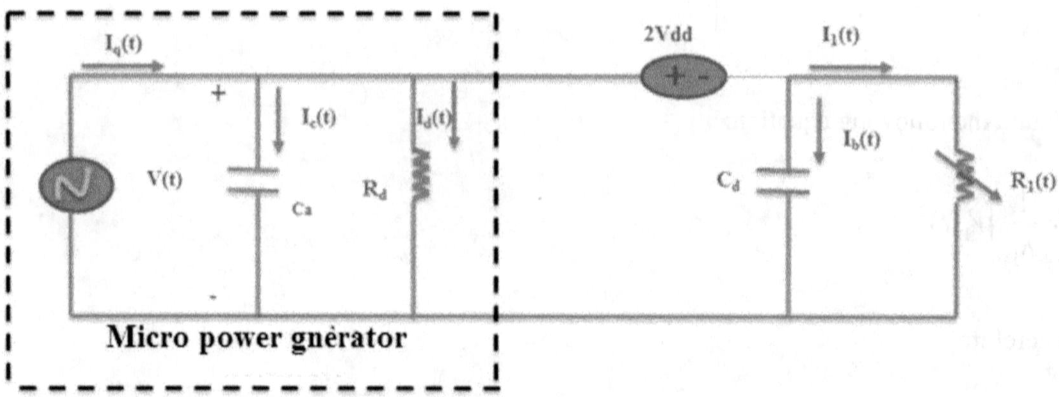


$$\begin{aligned} W_1 = {} & \frac{1}{R_1}\cdot\left(I_q.R_e - 2.V_{dd}.\frac{R_e}{R_d}\right)^2\cdot\left((a-t_d) - 2.R_eC_e.\left(1-e^{\frac{-(a-t_d)}{R_e.C_e}}\right) + \frac{R_e.C_e}{2}.\left(1-e^{\frac{-2(a-t_d)}{R_e.C_e}}\right)\right) \\ & + \frac{2}{R_1}\cdot\langle V_1\rangle\cdot\left(I_q.R_e - 2.V_{dd}.\frac{R_e}{R_d}\right)\cdot\left(R_e.C_e.\left(1-e^{\frac{-(a-t_d)}{R_e.C_e}}\right) - \frac{R_e.C_e}{2}.\left(1-e^{\frac{-2(a-t_d)}{R_e.C_e}}\right)\right) \end{aligned} \quad (42)$$

With a=0.075s, T=0.55s, we represent the simulation results of the power generated based on the load and the storage capacity represented in Figure 20. The graphic representation of the generated power and the storage capacity according to the load resistance has allowed us to follow, now, the impact of the parameter $C_1$ (storage capacity) of the power generated. It should be noted that the characteristics of energy transfer at startup are almost the same for different loads that support relatively low voltages.

In all cases the maximum energy in the load reaches its maximum when the capacitor $C_1$ is less than some pF. Beyond this value of the storage capacity of the power is always constant. However, a small capacity $C_1$ accumulation does not store a lot of energy and will have little effect at the time constant, by cons for large values of $C_1$ yield decreases and the power delivered to the load is minimum. With an average value of the storage capacity of the order of a few nF one corresponds to function optimally and in accordance with the results described in Figure 19.Based on a mechanism of doubling the frequency of displacement of the polymer through a proposed mechanical structure, we can also obtain a higher dynamic current. This effect renders possible a significant increase in terms of converted power; it was found that the maximum resistive output power could reach 0.058 μW, which is 80% more than in classical techniques. This value demonstrated the excellent potential of this technique for energy harvesting. The structure proposed in this section thus represents a good tool for estimating potential energy harvesting candidates for technologies that collect vibrations from the environment and convert them into.

## 7. CONCLUSION

The work presented in this chapter relates to the generation of electrical energy to power portable electronics (mobile) based on the recovery of human mechanical energy. The electro-active polymers were studied as a transducer for converting mechanical energy provided by human movement into electrical

*Figure 20. Power harvested according to the load and the storage capacity*

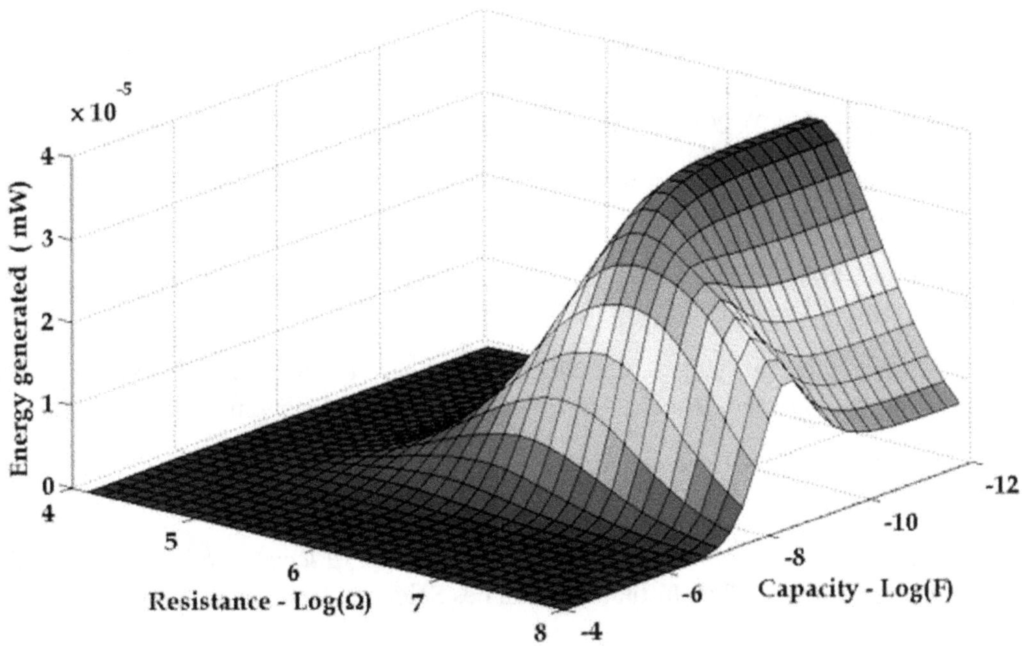

energy. Four different themes complementary and necessary were treated. After a state of the art materials used in MEMS and performance, two distinct objectives were defined.

At first, the study was focused on the synthesis process materials and types of filler to be added to polymers to improve properties in operation and energy recovery. These properties were analyzed in the second part. Electrical analysis was conducted on the various composites. In this context, it has been shown that the permittivity depends closely on the operating frequency. The influence of the carbon black on the improvement of dielectric parameters were shown notably increasing the dielectric permittivity and the reduction of Young's modulus. The result of these two parameters as these polymers promising candidates for potential energy collector's small low frequency operating range which fits with the resonance frequency of the micro-generators. Three samples of polymers PU 0.5% C. PU 1% Cand P (VDF-TrFE-CFE)1% C were therefore developed and characterized.

A test bench with the measurement protocols was introduced characterize these three polymers. The results obtained were satisfactory: increased permittivity when adding nanoparticles and low base frequency losses in the frequency range (10Hzto 100kHz). All these features portend a particular interest in these polymers in this area particularly PU 1% C.

In the next section, a mathematical model was established for maximizing energy. Using the reverse bias to the half period of displacement, for its simplicity, the recovered power was multiplied by a factor of 4. Characterization by FFT spectral analysis and polarization corona allowed us to low level admissible and to analyze the effect of the electric field dependent. The experimental results compared to theory were satisfactory as to the energy recovery. Finally, thanks to the specific circuits energy management, developed a system with frequency doubler and other circuits, we were able to test a complete system of energy generation. In conclusion, the electro-active polymer materials are flexible expansion having specific physical properties, are better candidates for the future development of flexible micro-generators and flexible actuators.

## REFERENCES

Bobnar, V., Levstik, A., Huang, C., & Zhang, Q. M. (2007). Enhanced dielectric response in all-organic polyaniline–poly (vinylidene fluoride-trifluoroethylene-chlorotrifluoroethylene) composite. *Journal of Non-Crystalline Solids*, *353*(2), 205–209. doi:10.1016/j.jnoncrysol.2006.10.003

Bratosin, D., & Sireteanu, T. (2002). A nonlinear Kelvin-Voigt model for soils. In *Proceedings of the Romanian Academy*, Series A. 3.

Burns, R., Taylor, G. W., & Welsh,T. R. (2003, 18 February). Sensors for power conversion systems. *PatentUS*, 6 522 048.

Dang, Z. M., Zhang, Y. H., & Tjong, S. C. (2004). Dependence of dielectric behavior on the physical property of composites. *Synthetic Metals*, *146*, 79–84. doi:10.1016/j.synthmet.2004.06.011

Diaconu, I., Ddorohoi, D. O., & Policeanu, F. (2006, July). Electrostriction of a polyurethane elastomer-based polyester. *Sensors Journal, IEEE*, *6*(4), 876–880. doi:10.1109/JSEN.2006.877978

Eddiai, A. (2013). *Caractérisation et Modélisation des Polymères Electro-actifs: Application à la Récupération d'Energie*. France: PhD. Institut National des Sciences Appliquées de Lyon.

Eddiai, A., Meddad, M., Guyomar, D., Hajjaji, A., Boughaleb, Y., Yuse, K., & Sahraoui, B. et al. (2012b). Enhancement of electrostrictive polymer efficiency for energy harvesting with cellular polypropylene electrets. *Synthetic Metals*, *162*(21-22), 1948–1953. doi:10.1016/j.synthmet.2012.08.012

Eddiai, A., Meddad, M., Hajjaji, A., Guyomar, D., Boughaleb, Y., &Sahraoui, B. (2012c). Mechanical effect on the performance of electrostrictive polymers for energy harvesting. *Xplore IEEE. Org.*, 6253823.

Eddiai, A., Meddad, M., Sbiaai, K., Boughaleb, Y., Hajjaji, A., & Guyomar, D. (2014). A new technique for maximize the energy harvested using electrostrictive polymer composite. *Optical Materials*, *36*(1), 13–17. doi:10.1016/j.optmat.2013.07.014

Eddiai, A., Meddad, M., Touhtouh, S., Hajjaji, A., Boughaleb, Y., Guyomar, D., & Sahraoui, B. et al. (2012a). The mechanical characterization an electrostrictive polymer for actuation and energy harvesting. *Journal of Applied Physics*, *111*(12), 124115. doi:10.1063/1.4729532

Eury, S., Yimniriun, R., Sundar, V., & Moses, P. J. (1999). Converse electrostriction in polymers and composites. *Materials Chemistry and Physics*, *61*(1), 18–23. doi:10.1016/S0254-0584(99)00107-8

Eury, S., Yimnirun, R., Sundar, V., Moses, P. J., Jang, S.-J., & Newnham, R. E. (1999). Converse electrostriction in polymers and composites. *Materials Chemistry and Physics*, *61*(1), 18–23. doi:10.1016/S0254-0584(99)00107-8

Guillot, F. M., & Balizer, E. (2003). Electrostrictive effect in polyurethanes. *Journal of Applied Polymer Science*, *89*(2), 399–404. doi:10.1002/app.12096

Guyomar, D., Cottinet, P.-J., Lebrun, L., & Sebald, G. (2011a). Characterization of an electroactive polymer simultaneously driven by an electrical field and a mechanical excitation: An easy means of measuring the dielectric constant, the Young modulus and the electrostrictive coefficients. *Physics Letters. [Part A]*, *375*(16), 1699–1702. doi:10.1016/j.physleta.2011.03.011

Guyomar, D., Lallart, M., & Cottinet, P.-J. (2011b). Electrostrictive conversion enhancement of polymer composites using a nonlinear approach. *Physics Letters. [Part A]*, *375*(3), 260–264. doi:10.1016/j.physleta.2010.12.026

Guyomar, D., Lebrun, L., Putson, C., Cottinet, P. J., Guiffard, B., & Muensit, S. (2009). Electrostrictive energy conversion in polyurethane nanocomposites. *Journal of Applied Physics*, *106*(1), 014910. doi:10.1063/1.3159900

Guyomar, D., Lefeuvre, E., Richard, C., & Badel, A. (2004, May). *Piezoelectric energy harvesting circuit using a synchronized switch technique.* Paper presented at SMEBA'03, Suzhou, China.

Jean-Mistral, C., Basour, S., & Chaillout, J.-J. (2008, March). Smart Structures and Materials. In *Proc. SPIE Conference on Electroactive Polymer Actuators and Devices, San Diego*, 6927, 692716.

Lallart, M., Cottinet, P.-J., Lebrun, L., Guiffard, B., & Guyomar, D. (2010). Evaluation of energy harvesting performance of electrostrictive polymer and carbon-filled terpolymer composites. *Journal of Applied Physics*, *108*(3), 034901. doi:10.1063/1.3456084

Lebrun, L., Guyomar, D., Guiffard, B., Cottinet, P. J., & Putson, C. (2009). The Characterization of the harvesting capabilities of an electrostrictive polymer composite. *Sensors and Actuators. A, Physical, 153*(2), 251–257. doi:10.1016/j.sna.2009.05.009

Li, W. J., Ho, T. C. H., Chan, G. M. H., Leong, P. H. W., & Wong, H. Y. (2000). Infrared signal transmission by a laser-micromachined, vibration-induced power generator. In *Proceedings of the 43rd IEEE Midwest Symposium on Circuits and Systems* (pp. 236-239). doi:10.1109/MWSCAS.2000.951628

Liu, R., Zhang, Q., & Cross, L. E. (1999). Experimental investigation of electrostrictive polarization biased direct apparent piezoelectric properties in polyurethane elastomer under quasistatic conditions. *Journal of Applied Polymer Science, 73*(13), 2603–2609. doi:10.1002/(SICI)1097-4628(19990923)73:13<2603::AID-APP7>3.0.CO;2-R

Liu, Y., Ren, K. L., Hofmann, H. F., & Zhang, Q. (2005). Investigation of electrostrictive polymers for energy harvesting. *IEEE Transactions on Ultrasonics, Ferroelectrics, and Frequency Control, 52*(12), 2411–2417. doi:10.1109/TUFFC.2005.1563285 PMID:16463508

Meddad, M. (2012c). *Etude de l'énergie renouvelable dans l'environnement humain pour alimenter les systèmes électroniques portables.* Algeria: PhD. Université Ferhat Abbas of Sétif.

Meddad, M., Eddiai, A., Guyomar, D., Belkhiat, S., Cherif, A., Yuse, K., & Hajjaji, A. (2012b). An adaptive prototype design to maximize power harvesting using electrostrictive polymers. *Journal of Applied Physics, 112*(5), 054109. doi:10.1063/1.4751456

Meddad, M., Eddiai, A., Guyomar, D., Belkhiat, S., Hajjaji, A., Cherif, A., & Boughaleb, Y. (2012a). Study of the behaviour of electrostrictive polymers for energy harvesting with FFT analysis. *Journal of Optoelectronics and Advanced Materials, 14*, 55–60.

Meddad, M., Eddiai, A., Guyomar, D., Belkhiat, S., Hajjaji, A., Yuse, K., & Boughaleb, Y. (2012c). Evaluation by fast Fourier transforms analysis of energy harvesting in electrostrictive polymers driven by an electric field and a mechanical excitation. *Journal of Intelligent Material Systems and Structures, 24*(4), 411–420. doi:10.1177/1045389X12461077

Meddad, M., Eddiai, A., Hajjaji, A., Boughaleb, Y., Guyomar, D., & Fliyou, M. (2014b). Optimization of the energy harvested by the effect of strain and frequency on an electrostrictive polymer composite. *Synthetic Metals, 188*, 72–76. doi:10.1016/j.synthmet.2013.11.022

Meddad, M., Eddiai, A., Hajjaji, A., Guyomar, D., Belkhiat, S., Boughaleb, Y., & Chérif, A. (2014a). Lowest of AC–DC power output for electrostrictive polymers energy harvesting systems. *Optical Materials, 36*(1), 80–85. doi:10.1016/j.optmat.2013.05.008

Meninger, S., Mur-Miranda, J. O., Amirtharajah, R., Chandrakasan, A., & Lang, J. (2001). Vibration-to-electric energy conversion. *IEEE Transactions on Very Large Scale Integration Systems, 9*, 64–76.

Micromachine Devices. (1997, May). *European Study Sees MEMS Market at More Than $34 billion by 02.*

Mitchell, B. S. (2004). *An introduction to materials engineering and science for chemical and materials engineers.* New York: Wiley.

Nguyen, N. T., & Werely, S. T. (2002). *Fundamentals and Applications of Microfluidics.* Norwod, MA: Artech House, Inc.

Poulin, G. (2004). *Contribution to the development of a piezoelectric generator for mobile applications* (Ph.D. Thesis). University Paris XI Orsay, France.

Putson, C., Lebrun, L., Guyomar, D., Muensit, N., Cottinet, P. J., Seveyrat, L., & Guiffard, B. (2011). Effects of copper filler sizes on the dielectric properties and the energy harvesting capability of nonpercolated polyurethane composites. *Journal of Applied Physics*, *109*(2), 024104. doi:10.1063/1.3534000

Ren, K., Liu, Y., Hofmann, H., Zhang, Q. M., & Blottman, J. (2007). An active energy harvesting scheme with an electroactive polymer. *Applied Physics Letters*, *91*(13), 132910. doi:10.1063/1.2793172

Roundy, S., Wright, P. K., & Rabaey, J. (2003). A study of low level vibrations as a power source for wireless sensor nodes. *Computer Communications*, *26*(11), 1131–1144. doi:10.1016/S0140-3664(02)00248-7

Su, J., Zhang, Q., Wang, P.-C., MacDiarmid, A. G., & Wynne, K. J. (1998). Preparation and characterization of electrostrictive polyurethane films with conductive polymer electrodes. *Polymers for Advanced Technologies*, *9*(6), 317–321. doi:10.1002/(SICI)1099-1581(199806)9:6<317::AID-PAT758>3.0.CO;2-Y

Xu, T., & Li, G. (2011a). Cyclic stress-strain behavior of shape memory polymer based syntactic foam programmed by 2-D stress condition. *Polymer*, *52*(20), 4571–4580. doi:10.1016/j.polymer.2011.08.005

Xu, T., & Li, G. (2011b). A shape memory polymer based syntactic foam with negative Poisson's ratio. *Materials Science and Engineering A-structural Materials*, *528*(22-23), 6804–6811. doi:10.1016/j.msea.2011.05.034

Yole Developpment. (2013). BioMEMS: Microsystem Device Market for Healthcare Applications. Author: France.

You, K. M., Park, S. S., Lee, C. S., Kim, J. M., Park, G. P., & Kim, W. N. (2011). Preparation and characterization of conductive carbon nanotube-polyurethane foam composites. *Journal of Materials Science*, *46*(21), 6850–6855. doi:10.1007/s10853-011-5645-y

Yuse, K., Guiffard, B., Belouadah, R., Petit, L., Seveyrat, L., & Guyomar, D. (2009). Polymer nanocomposites for microactuation and magneto-electric transduction. *Frontiers of Mechanical Engineering in China*, *4*, 350–354.

Zhang, Q. M., & Scheinbeim, J. (2004). Electric EAP. In Y. Bar-Cohen, (Ed.), Electroactive Polymer (EAP) Actuators as Artificial Muscles: Reality, Potential, and Challenges (pp. 95 – 100). Bellingham: SPIE Press.

Zhang, S., Neese, B., Ren, K., Chu, B., & Zhang, Q. M. (2006). Microstructure and electromechanical responses in semicrystalline ferroelectric relaxor polymer blends. *Journal of Applied Physics*, *100*(4), 044113. doi:10.1063/1.2335778

Zhang, S., Neese, B., Ren, K., Chu, B., & Zhang, Q. M. (2006). Microstructure and electromechanical responses in semicrystalline ferroelectric relaxor polymer blends. *Journal of Applied Physics*, *100*(4), 044113. doi:10.1063/1.2335778

# Compilation of References


Abdelkefi, A. (2012). *Global nonlinear analysis of piezoelectric energy harvesting from ambient and aeroelastic vibrations* (PhD Dissertation). Virginia Tech, VA.

Abdelkefi, A., & Barsallo, N. (2014). Comparative modeling of low-frequency piezomagnetoelastic energy harvesters. *Journal of Intelligent Material Systems and Structures*. doi:10.1177/1045389X14523860

Abdelkefi, A., & Ghommem, M. (2013). Piezoelectric energy harvesting from morphing wing motions for micro air vehicles. *Theoretical & Applied Mechanics Letters*, *3*(5), 052001. doi:10.1063/2.1305204

Abdelkefi, A., Ghommem, M., Nuhait, A. O., & Hajj, M. R. (2014c). Nonlinear analysis and enhancement of wing-based piezoaeroelastic energy harvesters. *Journal of Sound and Vibration*, *333*(1), 166–177. doi:10.1016/j.jsv.2013.08.032

Abdelkefi, A., & Hajj, M. R. (2013). Performance enhancement of wing-based piezoaeroelastic energy harvesting through freeplay nonlinearity. *Theoretical & Applied Mechanics Letters*, *3*(4), 041001. doi:10.1063/2.1304101

Abdelkefi, A., Hajj, M. R., & Nayfeh, A. H. (2012a). Phenomena and modeling of piezoelectric energy harvesting from freely oscillating cylinders. *Nonlinear Dynamics*, *70*(2), 1377–1388. doi:10.1007/s11071-012-0540-x

Abdelkefi, A., Hajj, M. R., & Nayfeh, A. H. (2013a). Piezoelectric energy harvesting from transverse galloping of bluff bodies. *Smart Materials and Structures*, *22*(1), 015014. doi:10.1088/0964-1726/22/1/015014

Abdelkefi, A., Najar, F., Nayfeh, A. H., & Ben Ayed, S. (2011). An energy harvester using piezoelectric cantilever beams undergoing coupled bending–torsion vibrations. *Smart Materials and Structures*, *20*(11), 115007. doi:10.1088/0964-1726/20/11/115007

Abdelkefi, A., Nayfeh, A. H., & Hajj, M. R. (2012a). Modeling and analysis of piezoaeroelastic energy harvesters. *Nonlinear Dynamics*, *67*(2), 925–939. doi:10.1007/s11071-011-0035-1

Abdelkefi, A., Nayfeh, A. H., & Hajj, M. R. (2012b). Design of piezoaeroelastic energy harvesters. *Nonlinear Dynamics*, *68*(4), 519–530. doi:10.1007/s11071-011-0233-x

Abdelkefi, A., Nayfeh, A. H., & Hajj, M. R. (2012b). Global nonlinear distributed-parameter model of parametrically excited piezoelectric energy harvesters. *Nonlinear Dynamics*, *67*(2), 1147–1160. doi:10.1007/s11071-011-0059-6

Abdelkefi, A., Nayfeh, A. H., & Hajj, M. R. (2012c). Effects of nonlinear piezoelectric coupling on energy harvesters under direct excitation. *Nonlinear Dynamics*, *67*(2), 1221–1232. doi:10.1007/s11071-011-0064-9

Abdelkefi, A., Nayfeh, A. H., & Hajj, M. R. (2012d). Enhancement of power harvesting from piezoaeroelastic systems. *Nonlinear Dynamics*, *68*(4), 531–541. doi:10.1007/s11071-011-0234-9

Abdelkefi, A., & Nuhait, A. O. (2013). Modeling and performance analysis of cambered wing-based piezoaeroelastic energy harvesters. *Smart Materials and Structures*, *22*(9), 095029. doi:10.1088/0964-1726/22/9/095029

Abdelkefi, A., Scanlon, J. M., McDowell, E., & Hajj, M. R. (2013b). Performance enhancement of piezoelectric energy harvesters from wake galloping. *Applied Physics Letters*, *103*(3), 033903. doi:10.1063/1.4816075

Abdelkefi, A., Vasconcellos, R., Nayfeh, A. H., & Hajj, M. R. (2013d). An analytical and experimental investigation into limit-cycle oscillations of an aeroelastic system. *Nonlinear Dynamics*, *71*(1-2), 159–173. doi:10.1007/s11071-012-0648-z

Abdelkefi, A., Yan, Z., & Hajj, M. R. (2013c). Temperature impact on the performance of galloping-based piezoaeroelastic energy harvesters. *Smart Materials and Structures*, *22*(5), 055026. doi:10.1088/0964-1726/22/5/055026

Abdelkefi, A., Yan, Z., & Hajj, M. R. (2013e). Modeling and nonlinear analysis of piezoelectric energy harvesting from transverse galloping. *Smart Materials and Structures*, *22*(2), 025016. doi:10.1088/0964-1726/22/2/025016

Abdelkefi, A., Yan, Z., & Hajj, M. R. (2013f). Nonlinear dynamics of galloping-based piezoaeroelastic energy harvesters. *European Physical Journal-Special Issues*, *222*(7), 1483–1501. doi:10.1140/epjst/e2013-01940-x

Abdel-Rahman, M., González, F., & Boreman, G. (2004). Antenna-coupled metal-oxide-metal diodes for dual-band detection at 92.5 GHz and 28 THz. *Journal of Electronics Letters*, *40*(2), 116–118. doi:10.1049/el:20040105

Abdina, Z., Alim, M. A., Saidur, R., Islam, M. R., Rashmi, W., Mekhilef, S., & Wadi, A. (2013). Solar energy harvesting with the application of nanotechnology. *Renewable & Sustainable Energy Reviews*, *26*, 837–852. doi:10.1016/j.rser.2013.06.023

Adhikari, S., Friswell, M., & Inman, D. (2009). Piezoelectric energy harvesting from broadband random vibrations. *Smart Materials and Structures*, *18*(11), 115005. doi:10.1088/0964-1726/18/11/115005

Agneni, H. D., Mastroddi, F., & Polli, G. M. (2003). Shunted piezoelectric patches in elastic and aeroelastic vibrations. *Computers & Structures*, *81*(2), 91–105. doi:10.1016/S0045-7949(02)00392-9

Ajitasaria, J., Choe, S. Y., Shen, D., & Kim, D. J. (2007). Modeling and analysis of a bimorph piezoelectric cantilever beam for voltage generation. *Smart Materials and Structures*, *16*(2), 447–454. doi:10.1088/0964-1726/16/2/024

Akaydin, H. D., Elvin, N., & Andreopoulos, Y. (2010). Energy harvesting from highly unsteady fluid flow using piezoelectric materials. *Journal of Intelligent Material Systems and Structures*, *21*(13), 1263–1278. doi:10.1177/1045389X10366317

Akaydin, H. D., Elvin, N., & Andreopoulos, Y. (2012). The performance of a self-excited fluidic energy harvester. *Smart Materials and Structures*, *21*(2), 025007. doi:10.1088/0964-1726/21/2/025007

Akhtar, I. (2008). *Parallel simulations, reduced-order modeling, and feedback control of vortex shedding using fluidic actuators* (PhD Dissertation). Virginia Tech, VA.

Akkermans, J. A. G., van Beurden, M. C., Doodeman, G. J. N., & Visser, H. J. (2005). Analytical models for low-power rectenna design. *IEEE Antennas and Wireless Propagation Letters*, *4*(1), 187–190. doi:10.1109/LAWP.2005.850798

Al Agha, K., Bertin, M.-H., Dang, T., Guitton, A., Minet, P., Val, T., & Viollet, J.-B. (2009, October). Which wireless technology for industrial wireless sensor networks? The development of OCARI technology. *IEEE Transactions on Industrial Electronics*, *56*(10), 4266–4278. doi:10.1109/TIE.2009.2027253

Alamin Dow, A. B., Schneider, M., Koo, D., Rubaye, H. A., Bittner, A., Schmid, U., & Kherani, N. (2012). Modeling the performance of a micromachined piezoelectric energy harvester. *Microsystem Technologies*, *18*(7-8), 1035–1043. doi:10.1007/s00542-012-1436-x

Alda, J., Rico-Garcia, J. M., Lopez-Alonso, J. M., & Boreman, G. (2005). Optical antennas for nano-photonic applications. *Nanotechnology*, *16*(5), S230–S234. doi:10.1088/0957-4484/16/5/017

Alkan, M. A., Keçebas, A., & Yamankaradeniz, N. (2013). Exergoeconomic analysis of a district heating system for geothermal energy using specific exergy cost method. *Energy*, *60*(1), 426–434. doi:10.1016/j.energy.2013.08.017

Allen, C., Karam, K. Z., Cam, P., Hill, M., & Tindle, T. (1995). Application of virtual reality devices to the quantitative assessment of manual assembly forces in a factory environment. In *Proceedings of IEEE International Conference on Industrial Electronics, Control and Instrumentation*, *2*, 1048-1053. doi:10.1109/IECON.1995.483940

Anton, S. R., & Sodano, H. A. (2007). A review of power harvesting using piezoelectric materials (2003–2006). *Smart Materials and Structures*, *16*(3), R1–R21. doi:10.1088/0964-1726/16/3/R01

Anvar, A., & Li, D. Y. (2012). Ocean wave kinetic energy harvesting system for automated sub sea sensors. *World Academy of Science. Engineering and Technology*, *6*, 9–13.

Arigliano, A., Caricato, P., Grieco, A., & Guerriero, E. (2014). Producing, storing, using and selling renewable energy: The best mix for the small medium industry. *Computers in Industry*, *65*, 408–418.

Arnold, D. (2007). Review of microscale magnetic power generation. *IEEE Transactions on Magnetics*, *43*(11), 3940–3951. doi:10.1109/TMAG.2007.906150

Arrieta, A. F., Hagedorn, P., Erturk, A., & Inman, D. J. (2010). A piezoelectric bistable plate for nonlinear broadband energy harvesting. *Applied Physics Letters*, *97*(10), 104102. doi:10.1063/1.3487780

Bailey, R. L. (1972). A proposed new concepts for solar energy converter. *Journal of Engineering for Power*, *94*(2), 73–77. doi:10.1115/1.3445660

Baird, T. J., Aulbach, A., O'Donohue, K. P., Dalton, J. A., & Gauvin, D. V. (2014). Background variability in standard clinical pathology biomarkers in beagle dogs instrumented with chronic indwelling telemetry devices. *Journal of Pharmacological and Toxicological Methods*, *69*(3), 223–228. doi:10.1016/j.vascn.2014.01.005 PMID:24508734

Ballas, R. G. (2007). Piezoelectric multilayer beam bending actuators: Static and dynamic behavior and aspects of sensor integration. London: Springer London, Limited.

Balpande, S. S., Lande, S. B., & Rungta, S. (2009). Modeling of PZT based power harvester and power aware control approach for pervasive node BSN. In *Proceeding of International Advance Computing Conference* (IACC 2009). Patiala, India: IEEE.

Bao, H., & Ruan, X. (2010). Optical absorption enhancement in disordered vertical silicon nanowire arrays for photovoltaic applications. *Optics Letters*, *35*(20), 3378–3380. doi:10.1364/OL.35.003378 PMID:20967072

Barak, A., Hen, L., Boniel-Nissim, M., & Shapira, N. (2008). A comprehensive review and a meta-analysis of the effectiveness of internet-based psychotherapeutic interventions. *Journal of Technology in Human Services*, *26*(2-4), 109–114. doi:10.1080/15228830802094429

Bardi, U. (2009). Peak oil: The four stages of a new idea. *Energy*, *34*(3), 323–326. doi:10.1016/j.energy.2008.08.015

Bareiß, M., Tiwari, B. N., Hochmeister, A., Jegert, G., Zschieschang, U., Klauk, H., & Lugli, P. et al. (2011). Nano antenna array for terahertz detection. *IEEE Transactions on Microwave Theory and Techniques*, *59*(10), 2751–2757. doi:10.1109/TMTT.2011.2160200

Barlev, D., Vidu, R., & Stroeve, P. (2011). Innovation in concentrated solar power. *Solar Energy Materials and Solar Cells*, *95*(10), 2703–2725. doi:10.1016/j.solmat.2011.05.020

Barton, D., Burrow, S., & Clare, L. (2010). Energy harvesting from vibrations with a nonlinear oscillator. *Journal of Vibration and Acoustics*, *132*(2), 0210091. doi:10.1115/1.4000809

Becker, C., Amkreutz, D., Sontheimer, T., Preidel, V., Lockau, D., Haschke, J., & Rech, B. et al. (2013). Polycrystalline silicon thin-film solar cells: Status and perspectives. *Solar Energy Materials and Solar Cells*, *119*, 112–123. doi:10.1016/j.solmat.2013.05.043

Beeby, S. P., Tudor, M. J., & White, N. M. (2006). Energy harvesting vibration sources for micro-systems applications. *Measurement Science & Technology*, *17*(12), 175–195. doi:10.1088/0957-0233/17/12/R01

Behar, O., Khellaf, A., & Mohammedi, K. (2013). A review of studies on central receiver solar thermal power plants. *Renewable & Sustainable Energy Reviews*, *23*, 12–39. doi:10.1016/j.rser.2013.02.017

Ben Ayed, S., Abdelkefi, A., Najar, F., & Hajj, M. R. (2014). Design and performance of variable-shaped piezoelectric energy harvesters. *Journal of Intelligent Material Systems and Structures*, *25*(2), 174–186. doi:10.1177/1045389X13489365

Benasciutti, D., & Moro, L. (2010). Harvested power and sensitivity analysis of vibrating shoe-mounted piezoelectric cantilevers. *Smart Materials and Structures*, *19*(11), 115011. doi:10.1088/0964-1726/19/11/115011

Benasciutti, D., Moro, L., & Gallina, M. (2013). On the optimal bending deflection of piezoelectric scavengers. *Journal of Intelligent Material Systems and Structures*, *24*(5), 627–639. doi:10.1177/1045389X12469451

Benasciutti, D., Moro, L., Zelenika, S., & Brusa, E. (2010). Vibration energy scavenging via piezoelectric bimorphs of optimized shapes. *Microsystem Technologies*, *16*(5), 657–668. doi:10.1007/s00542-009-1000-5

Berland, B. (2003). *Photovoltaic technologies beyond the horizon: Optical rectenna solar cell*. Retrieved from http://www.nrel.gov/docs/fy03osti/33263.pdf

Berland, B., Simpson, L., Nuebel, G., Collins, T., & Lanning, B. (2001). Optical rectenna for direct conversion of sunlight to electricity. In *Proceedings of the Conference National Center for Photovoltaics Program Review Meeting (NREL)*. Shaffer Parkway Littleton, Colorado.

Bharadwaj, P., Deutsch, B., & Novotny, L. (2009). Optical antennas. *Advances Optics and Photonics*, *1*(3), 438–483. doi:10.1364/AOP.1.000438

Biagioni, P., Huang, J., & Hecht, B. (2012). Nanoantennas for visible and infrared radiation. *Journal of Progress in Physics*, *75*, 1–10. PMID:22790344

Bibo, A., & Daqaq, M. F. (2014). On the optimal performance and universal design curves of galloping energy harvesters. *Applied Physics Letters*, *104*(2), 023901. doi:10.1063/1.4861599

Bionic-Power. (n. d.). *Website*. Retrieved from www.bionic-power.com

Birkhoff, G., & Zarantanello, E. H. (1957). *Jets wakes and cavities*. New York: Academic Press.

Bitschi, A. (2009). *Modelling of thermoelectric devices for electric power generation* (Ph.D. dissertation). Swiss Federal Institute of Technology Zurich, ETH Zurich.

Blankevoort, L., Huiskes, R., & de Lange, A. (1991). The envelope of passive knee-joint motion. In L. Blankevoort, Passive motion characteristics of the human knee joint: experiments and computer simulations. Katholieke Universiteit te Nijmegen.

Bobnar, V., Levstik, A., Huang, C., & Zhang, Q. M. (2007). Enhanced dielectric response in all-organic polyaniline–poly (vinylidene fluoride-trifluoroethylene-chlorotrifluoroethylene) composite. *Journal of Non-Crystalline Solids*, *353*(2), 205–209. doi:10.1016/j.jnoncrysol.2006.10.003

Bocquet, L., & Barrat, J. L. (2007). Flow boundary conditions from nano- to micro-scales. *Soft Matter*, *3*(6), 685–693. doi:10.1039/b616490k

Böttner, H., Nurnus, J., Gavrikov, A., Schubert, A., & Schlereth, K. H. (2004). New thermoelectric components using microsystems technologies. *Journal of Microelectromechanical Systems*, *13*(3), 414–420. doi:10.1109/JMEMS.2004.828740

Bouchouicha, D., Latrach, M., Dupont, F., & Ventura, L. (2010, Sept). An experimental evaluation of surrounding RF energy harvesting devices. *The 40th Microwave Conference*, 1381-1384.

Boukai, A., Bunimovich, Y., Tahir-Kheli, J., Yu, J., Goddard, W. III, & Heath, J. (2008). Si nanowires as efficient thermoelectric materials. *Nature*, *451*(7175), 168–171. doi:10.1038/nature06458 PMID:18185583

BP. (June 2012). *Statistical Review of World Energy*. Retrieved November 2012 from http://www.bp.com/statisticalreview

Bratosin, D., & Sireteanu, T. (2002). A nonlinear Kelvin-Voigt model for soils. In *Proceedings of the Romanian Academy*, Series A. 3.

Briones, E., Alda, J., & González, F. J. (2013). Conversion efficiency of broad-band rectennas for solar energy harvesting applications. *Optics Express*, *14*(S3), A412–A418. doi:10.1364/OE.21.00A412 PMID:24104428

Brogioli, D. (2009). Extracting renewable energy from a salinity difference using a capacitor. *Physical Review Letters*, *103*(5), 058501. doi:10.1103/PhysRevLett.103.058501 PMID:19792539

Brown, W. C. (1984, September). The history of power transmission by radio waves. *IEEE Transactions on Microwave Theory and Techniques*, *32*(9), 1230–1242. doi:10.1109/TMTT.1984.1132833

Brown, W. C. (1991). An experimental low power density rectenna.*IEEE MTT-S International Microwave Symposium Digest*, 197-200.

Bryant, M., & Garcia, E. (2009). Energy harvesting: A key to wireless sensor nodes. *Proceedings of the Society for Photo-Instrumentation Engineers*, *7493*, 74931W, 74931W-8. doi:10.1117/12.845784

Bryson, S., & Levit, C. (1992). The virtual wind tunnel. *IEEE Computer Graphics and Applications*, *12*(4), 25–34. doi:10.1109/38.144824

Burdea, G. C., Weiss, P. L. T., & Thalmann, D. (2007). Guest editorial special theme on virtual rehabilitation. *IEEE Transactions on Neural Systems and Rehabilitation Engineering*, *15*(1), 1. doi:10.1109/TNSRE.2007.891367

Burns, R., Taylor, G. W., & Welsh,T. R. (2003, 18 February). Sensors for power conversion systems. *PatentUS*, 6 522 048.

Byrnes S. J., Blanchard R., & Capasso F. (2014). Harvesting renewable energy from earth's mid-infrared emissions. *PNAS Early Edition*, 1-6.

Cao, L. Y., Fan, P. Y., Vasudev, A. P., White, J. S., Yu, Z., Cai, W., & Brongersma, M. L. et al. (2010). Semiconductor nanowire optical antenna solar absorbers. *Journal of Nano Letters*, *10*(2), 439–445. doi:10.1021/nl9036627 PMID:20078065

Cao, L., Guo, W., Ma, W., Wang, L., Xia, F., Wang, Y., & Zhu, D. et al. (2011). Towards understanding the nanofluidic reverse electrodialysis system: Well matched charge selectivity and ionic composition. *Energy & Environmental Science*, *4*(6), 2259–2266. doi:10.1039/c1ee01088c

Cao, L., White, J. S., Park, J. S., Schuller, J. A., Clemens, B. M., & Brongersma, M. L. (2009). Engineering light absorption in semiconductor nanowire devices. *Journal of Nature Material*, *8*(8), 643–647. doi:10.1038/nmat2477 PMID:19578337

Carnahan, B., Luther, H. A., & Wilkes, J. O. (1969). *Applied numerical methods*. New York: John Wiley & Sons, Inc.

Cervera, J., Schiedt, B., & Ramírez, P. (2005). A Poisson/Nernst-Planck model for ionic transport through synthetic conical nanopores. *Europhysics Letters*, *71*(1), 35–41. doi:10.1209/epl/i2005-10054-x

Challa, V., Prasad, M., Shi, Y., & Fisher, F. (2008). A vibration energy harvesting device with bidirectional resonance frequency tunability. *Smart Materials and Structures*, *75*, 1–10.

Chan, C. M., Cao, G. Z., Fong, H., & Sarikaya, M. (2000). *Coatings on polyester*, 148–154.

Chaves, A. A., Keller, W. J., O'Sullivan, S., Mixson, L., & Briscoe, R. J. (2006). Cardiovascular monkey telemetry: Sensitivity to detect QT interval prolongation. *Journal of Pharmacological and Toxicological Methods*, *54*(2), 150–158. doi:10.1016/j.vascn.2006.03.004 PMID:16679034

Chen, H., Xi, N., Lai, K. W. C., Chen, L., Fung, C. K. M., & Lou, J. (2012). Plasmonic-resonant bowtie antenna for carbon nanotube photodetectors. *International Journal of Optics*, *2012*, 1–9. doi:10.1155/2012/318104

Chen, S. K., Gonella, S., Chen, W., & Liu, W. K. (2010). A level set approach for optimal design of smart energy harvesters. *Computer Methods in Applied Mechanics and Engineering*, *199*(37-40), 2532–2543. doi:10.1016/j.cma.2010.04.008

Chen, S. S. (1987). *Flow-induced vibration of circular cylindrical structures. DC*. Washington: Hemisphere Publishing Corporation.

Chen, X., Xu, S., Yao, N., & Shi, Y. (2010). 1.6 V nanogenerator for mechanical energy harvesting using PZT nanofibers. *Nano Letters*, *10*(6), 2133–2137. doi:10.1021/nl100812k PMID:20499906

Chen, Z., Wang, Y., Wang, W., & Li, Z. (2009). Nanofluidic electrokinetics in nanoparticle crystal. *Applied Physics Letters*, *95*(10), 102105. doi:10.1063/1.3223774

Cherednichenko, S., Hammar, A., Bevilacqua, S., Drakinskiy, V., Stake, J., & Kalabukhov, A. (2011). A room temperature bolometer for terahertz coherent and incoherent detection. *IEEE Transaction Terahertz Science Technology*, *1*(2), 395–402. doi:10.1109/TTHZ.2011.2164654

Choi, E., Kwon, K., Kim, D., & Park, J. (2013). *Characterization of depletion layer formation using self-assembled nanoparticles in microchannels*. Paper presented at 2013 Transducers & Eurosensors XXVII: The 17th International Conference on Solid-State Sensors, Actuators and Microsystems, Barcelona. doi:10.1109/Transducers.2013.6627354

Christoph, B., Tim, H., Andreas, L., Andreas, L., & Ulf, M. (2005). Assessing emission regulation in Europe: An interactive simulation approach. *The Energy Journal (Cambridge, Mass.)*, *26*(4), 1–21.

Chung, S., Yun, H., & Kamm, R. D. (2009). Nanointerstice-driven microflow. *Small*, *5*(5), 609–613. doi:10.1002/smll.200800748 PMID:19226594

Chutinan, A., & John, S. (2008). Light trapping and absorption optimization in certain thin-film photonic crystal architectures. *Journal of Physical Review A*, *78*(2), 1–15. doi:10.1103/PhysRevA.78.023825

Clair, D., Bibo, A., Sennakesavababu, V., & Daqaq, M. F. (2010). A scalable concept for micropower generation using flow-induced self-excited oscillations. *Applied Physics Letters*, *96*, 144103. doi:10.1063/1.3385780

Clanaman, S. (2013). MEMS medical gadgets for Obamacare impress CES visitors. *MEMS Journal*. Retrieved from www.memsjournal.com

Cook-Chennault, K. A., Thambi, N., & Sastry, A. M. (2008). Powering MEMS portable devices-a review of non-regenerative and regenerative power supply systems with emphasis on piezoelectric energy harvesting systems. *Smart Materials and Structures*, *17*(4), 043001. doi:10.1088/0964-1726/17/4/043001

Cottone, F., Vocca, H., & Gammaitoni, L. (2009). Nonlinear energy harvesting. *Physical Review Letters*, *102*(8), 080601. doi:10.1103/PhysRevLett.102.080601 PMID:19257728

Crane, D. T. (May 2012). *Thermoelectric waste heat recovery program for passenger vehicles,*. U.S. Department Energy Efficiency & Renewable Energy (EERE). Retrieved August 2014 from http://www1.eere.energy.gov

Dagenais, M., Choi, K., Yesilkoy, F., Chryssis, A. N., & Peckerar, M. C. (2010). Solar spectrum rectification using nano-antennas and tunneling diodes. In *Proceedings of the conference on Optoelectronic Integrated Circuits XII*. San Francisco, California: SPIE.

Dai, H. L., Abdelkefi, A., & Wang, L. (2014). Theoretical modeling and nonlinear analysis of piezoelectric energy harvesting from vortex-induced vibrations. *Journal of Intelligent Material Systems and Structures*, *25*(14), 1861–1874. doi:10.1177/1045389X14538329

Dang, Z. M., Zhang, Y. H., & Tjong, S. C. (2004). Dependence of dielectric behavior on the physical property of composites. *Synthetic Metals*, *146*, 79–84. doi:10.1016/j.synthmet.2004.06.011

Daqaq, M. F. (2012). On intentional introduction of stiffness nonlinearities for energy harvesting under white Gaussian excitations. *Nonlinear Dynamics*, *69*(3), 1063–1079. doi:10.1007/s11071-012-0327-0

*Datasheet of linear technology LT3588*. (n. d.). Retrieved from http://www.linear.com

*Datasheet of MEAS LDT1-028K*. (n. d.). Retrieved from http://www.meas-spec.com

*Datasheet of Mide Volture V22B*. (n. d.). Retrieved from http://www.mide.com/index.php

De Marqui Junior, C., Erturk, A., & Inman, D. J. (2009). An electromechanical finite element model for piezoelectric energy harvester plates. *Journal of Sound and Vibration*, *327*(1), 9–25. doi:10.1016/j.jsv.2009.05.015

De Marqui, C., Erturk, A., & Inman, D. J. (2011). Modeling and analysis of piezoelectric energy harvesting from aeroelastic vibrations using the doublet-lattice method. *Journal of Vibration and Acoustics*, *133*(1), 011003. doi:10.1115/1.4002785

De Pasquale, G., & Somà, A. (2011). *Device for diagnosing railway bogies by applying an energy-autonomous measuring and transmitting bolt, and corresponding control method.* PCT Patent Number WO2011/117718.

De Pasquale, G. (2013). Energy harvesters for powering wireless systems. In D. Uttamchandani (Ed.), *Handbook of MEMS for wireless and mobile applications* (pp. 345–400). Cambridge, UK: Woodhead Publishing. doi:10.1533/9780857098610.2.345

De Pasquale, G., & Somà, A. (2013). Energy harvesting from human motion with piezo fibers for the body monitoring by MEMS sensors. In *Proceedings of Symposium on Design, Test, Integration and Packaging of MEMS/MOEMS* (pp. 79-84). Barcelona, Spain.

De Pasquale, G., Somà, A., & Fraccarollo, F. (2012). Piezoelectric energy harvesting for autonomous sensors network on safety-improved railway vehicles. *Proceedings of the Institution of Mechanical Engineers. Part C, Journal of Mechanical Engineering Science*, *226*(4), 1107–1117. doi:10.1177/0954406211418158

De Pasquale, G., Somà, A., & Fraccarollo, F. (2013). Comparison between piezoelectric and magnetic strategies for wearable energy harvesting. *Journal of Physics: Conference Series*, *476*, 012097. doi:10.1088/1742-6596/476/1/012097

De Pasquale, G., Somà, A., & Zampieri, N. (2012). Design, simulation, and testing of energy harvesters with magnetic suspensions for the generation of electricity from freight trains vibrations. *Journal of Computational and Nonlinear Dynamics*, *7*(4), 041011. doi:10.1115/1.4006920

Debije, M. G., & Verbunt, P. P. C. (2012). *Thirty years of luminescent solar concentrator research: Solar energy for the built environment.* Advanced Energy Materials; doi:10.1002/aenm.201100554

Den Hartog, J. P. (1956). *Mechanical Vibrations*. New York: McGraw-Hill.

Diaconu, I., Ddorohoi, D. O., & Policeanu, F. (2006, July). Electrostriction of a polyurethane elastomer-based polyester. *Sensors Journal, IEEE*, *6*(4), 876–880. doi:10.1109/JSEN.2006.877978

Diftler, M. A., Culbert, C. J., Platt, R., & Bluethmann, W. J. (2003). Evolution of the NASA/DARPA robonaut control system. In *Proceedings of IEEE International Conference on Robotics and Automation*, *2*,2543-2548. doi:10.1109/ROBOT.2003.1241975

*Diminution de l'expositionaux ondes électromgnétiques émises par les antennes relais de téléphonie mobile.* (2013) Technical report. Retrieved from http://www.developpement-durable.gouv.fr/IMG/pdf/rapport_COPIC_31_juillet_2013.pdf

Dmitruk, N. L., Malynych, S. Z., & Moroz, I. E., & Kurlyak, V. Y. (2010). Optical efficiency of Ag and Au nanoparticles. *Journal of Semiconductor Physics, Quantum Electronics & Optoelectronics*, *13*, 369–373.

Doerr, K. U., Rademacher, H., Huesgen, S., & Kubbat, W. (2007). Evaluation of a low-cost 3D sound system for immersive virtual reality training system. *IEEE Transactions on Visualization and Computer Graphics*, *13*(2), 204–212. doi:10.1109/TVCG.2007.37 PMID:17218739

Dolgov, A., Zane, R., & Papovic, Z. (2010). Power management system for online low power RF energy harvesting optimization. *Circuits and Systems I: Regular Papers, IEEE Transactions on, 57(7),* 1802- 1811.

Douyère, A., Alicalapa, F., & Lan Sun Luk, J.-D. (2008). High efficiency microwave rectenna circuit: Modeling and design. *Electronics Letters*, *44*(24), 1409–1410. doi:10.1049/el:20081794

Dowell, E. H., & Tang, D. (2002). Nonlinear aeroelasticity and unsteady aerodynamics. *AIAA Journal*, *40*(9), 1697–1707. doi:10.2514/2.1853

Duan, C., & Majumdar, A. (2010). Anomalous ion transport in 2-nm hydrophilic nanochannels. *Nature Nanotechnology*, *5*(12), 848–852. doi:10.1038/nnano.2010.233 PMID:21113159

Duan, C., Wang, W., & Xie, Q. (2013). Review article: Fabrication of nanofluidic devices. *Biomicrofluidics*, *7*(2), 26501. doi:10.1063/1.4794973 PMID:23573176

Duché, D., Escoubas, L., Simon, J. J., Torchio, P., Vervisch, W., & Flory, F. (2008). Slow bloch modes for enhancing the absorption of light in thin films for photovoltaic cells. *Journal of Applied Physics Letter*, *92*(19), 193310–193313. doi:10.1063/1.2929747

DuToil, N. E., Wardle, B. L., & Kim, S. G. (2005). Design considerations for MEMS-scale piezoelectric mechanical vibration energy harvesters. *Integrated Ferroelectrics*, *71*(1), 121–160. doi:10.1080/10584580590964574

Dutoit, N. E., & Wardle, B. L. (2007). Experimental verification of models for microfabricated piezoelectric vibration energy harvesters. *AIAA Journal, 45*(5), 1126–1137. doi:10.2514/1.25047

Eddiai, A., Meddad, M., Hajjaji, A., Guyomar, D., Boughaleb, Y., &Sahraoui, B. (2012c). Mechanical effect on the performance of electrostrictive polymers for energy harvesting. *Xplore IEEE. Org.,* 6253823.

Eddiai, A. (2013). *Caractérisation et Modélisation des Polymères Electro-actifs: Application à la Récupération d'Energie*. France: PhD. Institut National des Sciences Appliquées de Lyon.

Eddiai, A., Meddad, M., Guyomar, D., Hajjaji, A., Boughaleb, Y., Yuse, K., & Sahraoui, B. et al. (2012b). Enhancement of electrostrictive polymer efficiency for energy harvesting with cellular polypropylene electrets. *Synthetic Metals, 162*(21-22), 1948–1953. doi:10.1016/j.synthmet.2012.08.012

Eddiai, A., Meddad, M., Sbiaai, K., Boughaleb, Y., Hajjaji, A., & Guyomar, D. (2014). A new technique for maximize the energy harvested using electrostrictive polymer composite. *Optical Materials, 36*(1), 13–17. doi:10.1016/j.optmat.2013.07.014

Eddiai, A., Meddad, M., Touhtouh, S., Hajjaji, A., Boughaleb, Y., Guyomar, D., & Sahraoui, B. et al. (2012a). The mechanical characterization an electrostrictive polymer for actuation and energy harvesting. *Journal of Applied Physics, 111*(12), 124115. doi:10.1063/1.4729532

Eder, A., & Linde, M. (2011). *Efficient and Dynamic – The BMW Group Roadmap for the Application of Thermoelectric Generators*. Paper presented at 2nd Thermoelectrics Applications Workshop, San Diego. Retrieved August 2014 from http://www1.eere.energy.gov

Edmison, J., Jones, M., Nakad, Z., & Martin, T. (2002). Using piezoelectric materials for wearable electronic textiles. In *Proceedings of IEEE International Symposium on Wearable Computers*, (41-48). doi:10.1109/ISWC.2002.1167217

Edwards, C. (1947, Nov). Microwave converters.*Proc. IRE*, 35(11), 1181-1191.

Edwards, H., Debord, J., Tran, T., Freeman, D., & Maggio, K. (2013), Performance metrics for thermoelectric energy harvesting studied using a novel planar 65 nm silicon CMOS-based thermopile.*Proceedings of the 2013 IEEE Sensors Conference*, Baltimore, MD, Nov 3-6, 2013. doi:10.1109/ICSENS.2013.6688611

Edwards, H., Niu, Q., & de Lozanne, A. (1993). A quantum dot refrigerator. *Applied Physics Letters, 63*(12), 1815–1817. doi:10.1063/1.110672

Edwards, H., Niu, Q., & de Lozanne, A. (1999). Thermoelectric conversion. In *Wiley encyclopedia of electrical and electronics engineering* (Vol. 22, pp. 80–90). New York: John Wiley and Sons. doi:10.1002/047134608X.W3044

Edwards, H., Niu, Q., Georgakis, G., & de Lozanne, A. (1995). Cryogenic cooling using tunneling structures with sharp energy features. *Physical Review B: Condensed Matter and Materials Physics, 52*(8), 5714–5736. doi:10.1103/PhysRevB.52.5714 PMID:9981758

Eijkel, J. C. T., & van der Berg, A. (2010). Nanofluidics and the chemical potential applied to solvent and solute transport. *Chemical Society Reviews, 39*(3), 957–973. doi:10.1039/b913776a PMID:20179818

Elvin, N. G., & Elvin, A. A. (2011). Effects of axial forces on cantilever piezoelectric resonators for structural energy harvesting. *Strain, 47*, 153–157. doi:10.1111/j.1475-1305.2010.00751.x

Elvin, N. G., Lajnef, N., & Elvin, A. A. (2006). Feasibility of structural monitoring with vibration powered sensors. *Smart Materials and Structures, 15*(4), 977–986. doi:10.1088/0964-1726/15/4/011

Elvin, N., Elvin, A., & Choi, D. H. (2003). A self-powered damage detection sensor. *Journal of Strain Analysis for Engineering Design, 38*(2), 115–124. doi:10.1243/030932403321163640

Energizer Battery Manufacturing Inc. (2008). *Energizer charger handbook*. Retrieved September 2014 from http://data.energizer.com/PDFs/charger_appman.pdf

Epp, L. W., Khan, A. R., Smith, H. K., & Smith, R. P. (2000, January). A compact dual polarized 8.51 GHz rectenna for high voltage (50V) actuator applications. *IEEE Transactions on Microwave Theory and Techniques, 48*(1), 111–120. doi:10.1109/22.817479

Erickson, R. W. (2001). *Fundamentals of Power Electronics* (2nd ed., pp. 440–482). New York: Kluwer Academic Publishers. doi:10.1007/b100747

Erturk, A. (2009). *Electromechanical modeling of piezoelectric energy harvesters* (PhD Dissertation). Virginia Tech, VA.

Erturk, A., & Inman, D. J. (2008). On mechanical modeling of cantilevered piezoelectric vibration energy harvesters. *Journal of Intelligent Material Systems and Structures*, *19*(11), 1311–1325. doi:10.1177/1045389X07085639

Erturk, A., & Inman, D. J. (2009). An experimentally validated bimorph cantilever model for piezoelectric energy harvesting from base excitations. *Smart Materials and Structures*, *18*(2), 025009. doi:10.1088/0964-1726/18/2/025009

Erturk, A., & Inman, D. J. (2011). Broadband piezoelectric power generation on high-energy orbits of the bistable Duffing oscillator with electromechanical coupling. *Journal of Sound and Vibration*, *330*(10), 2339–2353. doi:10.1016/j.jsv.2010.11.018

Erturk, A., & Inman, D. J. (2011). *Piezoelectric energy harvesting*. USA: John Wiley & Sons. doi:10.1002/9781119991151

Erturk, A., Vieira, W. G. R., De Marqui, C., & Inman, D. J. (2010). On the energy harvesting potential of piezoaeroelastic systems. *Applied Physics Letters*, *96*(18), 184103. doi:10.1063/1.3427405

Esquivel, P. J., Castellarnau, M., Senn, T., Lochel, B., Samitier, J., & Sabate, N. (2012). Fuel cell-powered microfluidic platform for lab-on-a-chip applications. *Lab on a Chip*, *12*(1), 74–79. doi:10.1039/C1LC20426B PMID:22072241

Esteban, M., & Leary, D. (2012). Current developments and future prospects of offshore wind and ocean energy. *Applied Energy*, *90*(1), 128–136. doi:10.1016/j.apenergy.2011.06.011

Eury, S., Yimniriun, R., Sundar, V., & Moses, P. J. (1999). Converse electrostriction in polymers and composites. *Materials Chemistry and Physics*, *61*(1), 18–23. doi:10.1016/S0254-0584(99)00107-8

Everredtronics. (2012). *TEG127-40A Technical Info*. Retrieved September 2014 from http://www.everredtronics.com/TEG.html

Facchinetti, M. L., de Langre, E., & Biolley, F. (2004). Coupling of structure and wake oscillators in vortex-induced vibrations. *Journal of Fluids and Structures*, *19*(2), 123–140. doi:10.1016/j.jfluidstructs.2003.12.004

Falkenstein, E. A. (2011). *Characterization and design of a low-power wireless power delivery system* (PhD dissertation). University of Colorado at Boulder, CO.

Falkenstein, E. A., Roberg, M., & Papovic, Z. (2012, July). Low-power wireless power delivery. *IEEE Transactions on Microwave Theory and Techniques*, *60*(7), 2277–2286. doi:10.1109/TMTT.2012.2193594

Fang, H., Liu, J. Q., Zu, Z. Y., Dong, L., Wang, L., Chen, D., & Liu, Y. et al. (2006). Fabrication and performance of MEMS-based piezoelectric power generator for vibration energy harvesting. *Microelectronics Journal*, *37*(11), 1280–1284. doi:10.1016/j.mejo.2006.07.023

Fang, H., Li, X., Song, S., Xu, Y., & Zhu, J. (2008). Fabrication of slantingly aligned silicon nanowire arrays for solar cell applications. *Journal of Nanotechnology*, *19*(255703), 1–10. PMID:21828663

Fan, S. H., & Joannopoulos, J. D. (2002). Analysis of guided resonances in photonic crystal slabs. *Journal of Physical Review B*, *65*(235112), 1–8.

Feichtner, T., Selig, O., Kiunke, M., & Hecht, B. (2012). Evolutionary optimization of optical antennas. *Journal of Physical Review Letters*, *109*(127701), 1–8. PMID:23005987

Fend, T., Hoffschmidt, B., Jorgensen, G., Küster, H., Krüger, D., Pitz-Paal, R., & Riffelmann, K. J. et al. (2003). Comparative assessment of solar concentrator materials. *Solar Energy*, *74*(2), 149–155. doi:10.1016/S0038-092X(03)00116-6

Fend, T., Jorgensen, G., & Küster, H. (2000). Applicability of highly reflective aluminium coil for solar concentrators. *Solar Energy*, *68*(4), 361–370. doi:10.1016/S0038-092X(00)00027-X

Ferrari, M., Ferrari, V., Guizzetti, M., Andò, B., Baglio, S., & Trigona, C. (2010). Improved energy harvesting from wideband vibrations by nonlinear piezoelectric converters. *Sensors and Actuators. A, Physical*, *162*(2), 425–431. doi:10.1016/j.sna.2010.05.022

Föll H., Leisner M., Cojocaru A. & Carstensen J., (2010), Macroporous semiconductors. *Journal of Materials, 3*, 3006-3076.

Forni, I., & Cappellini, O. (1974). *Compendio di meccanica articolare*. Urbino, Italy: Agraria Editore.

Fox, M. (2012). *Optical Properties of Solids*. Oxford: Oxford University Press.

Fujino, Y., Ito, T., Fujita, M., Kaya, N., Matsumoto, H., Kawabata, K., & Onodera, T. et al. (1994, April). A driving test of a small DC motor with rectenna array. *IEICE Transactions on Communications*, *E77-B*(4), 526–528.

Gadalla, M. N. (2013). *Nano antenna integrated diode (Rectenna) for infrared energy harvesting*. (Master dissertation). Retrieved from Kaust Repository.

Gadalla, M. N., Abdel-Rahman, M., & Shamim, A. (2014). Design, optimization and fabrication of a 28.3 THz nanorectenna for infrared detection and rectification. *Scientific Reports*, *4*, 1–9. doi:10.1038/srep04270 PMID:24599374

Gallo, M., Mescia, L., Losito, O., Bozzetti, M., & Prudenzano, F. (2012). Design of optical antenna for solar energy collection. *Energy*, *39*(1), 27–32. doi:10.1016/j.energy.2011.02.026

Garnett, E. C., & Yang, P. (2010). Light trapping in silicon nanowire solar cells. *Journal of Nano*, *3*, 1082–1087. PMID:20108969

Georgiadis, A., Andia, G., & Collado, A. (2010). Rectenna deisgn and optimization using reciprocity theory and harmonic balance analysis for electromagnetic (EM) energy harvesting. *IEEE Antennas and Wireless Propagation Letters*, *9*, 444–446. doi:10.1109/LAWP.2010.2050131

Ghommem, M., Abdelkefi, A., Nuhait, A. O., & Hajj, M. R. (2012). Aeroelastic analysis and nonlinear dynamics of an elastically mounted wing. *Journal of Sound and Vibration*, *331*(26), 5774–5787. doi:10.1016/j.jsv.2012.07.040

Gilliat, H. C., Strganac, T. W., & Kurdila, A. J. (2003). An investigation of internal resonance in aeroelastic systems. *Nonlinear Dynamics*, *31*(1), 1–22. doi:10.1023/A:1022174909705

Glaser, P. E. (1968). Power from the sun. *Science*, *162*(3856), 857–886. doi:10.1126/science.162.3856.857 PMID:17769070

Glenn, J. (1994). *The complete patents of Nikola Tesla*. New York: Barnes and Noble Books.

Glynne-Jones, P., Beeby, S. P., & White, N. M. (2001). Towards a piezoelectric vibration-powered microgenerator. *IEE Proceedings. Science Measurement and Technology*, *148*(2), 68–72. doi:10.1049/ip-smt:20010323

Goetzberger, A., & Hoffmann, V. U. (2005). *Photovoltaic solar energy generation*. Berlin, Heidelberg: Springer-Verlag.

Goh, J., & Shoji, M. (2011). *Electric power generator for a bicycle*. US Patent Number US 8053912.

Goldschmidtboeing, F., & Woias, P. (2008). Characterization of different beam shapes for piezoelectric energy harvesting. *Journal of Micromechanics and Microengineering*, *18*(10), 104013. doi:10.1088/0960-1317/18/10/104013

Goldsmid, H., & Douglas, R. (1954). The use of semiconductors in thermoelectric refrigeration. *British Journal of Applied Physics*, *5*(11), 386–390. doi:10.1088/0508-3443/5/11/303

Gonzalez, F. J., & Boreman, G. D. (2005). Comparison of dipole, bowtie, spiral and log-periodic IR antennas. *Infrared Physics & Technology*, *46*(5), 418–428. doi:10.1016/j.infrared.2004.09.002

Granstrom, J., Feenstra, J., Sodano, H. A., & Farinholt, K. (2007). Energy harvesting from a backpack instrumented with piezoelectric shoulder straps. *Smart Materials and Structures*, *16*(5), 1810–1820. doi:10.1088/0964-1726/16/5/036

Gritz, M. A., Puscasu, I., Spencer, D., & Boreman, G. D. (2003). Fabrication of an infrared antenna-coupled microbolometer linear array using chrome as a mask. *Journal of Vacuum Science & Technology B Microelectronics and Nanometer Structures*, *21*(6), 2608–2611. doi:10.1116/1.1629714

Grouthier, C., Michelin, S., & de Langre, E. (2012). Optimal energy harvesting by vortex-induced vibrations in cables.*10th International Conference on Flow-Induced Vibration (& Flow-Induced Noise)*.

Grover, S., & Moddel, G. (2011). Applicability of metal/insulator/metal (MIM) diodes to solar rectennas. *IEEE Journal of Photovoltic*, *1*(1), 78–83. doi:10.1109/JPHOTOV.2011.2160489

Guan, M., & Liao, W. (2007). On the efficiencies of piezoelectric energy harvesting circuits towards storage device voltages. *Smart Materials and Structures*, *16*(2), 498–505. doi:10.1088/0964-1726/16/2/031

Guillot, F. M., & Balizer, E. (2003). Electrostrictive effect in polyurethanes. *Journal of Applied Polymer Science*, *89*(2), 399–404. doi:10.1002/app.12096

Guillot, F. M., Beckham, H. W., & Leisen, J. (2007). *Piezoelectric fabrics for energy harvesting*. National Textile Center.

Gultepe, E., Nagesha, D., Sridhar, S., & Amiji, M. (2010). Nanoporous inorganic membranes or coatings for sustained drug delivery in implantable devices. *Advanced Drug Delivery Reviews*, *62*(3), 305–315. doi:10.1016/j.addr.2009.11.003 PMID:19922749

Gungor, V. C., & Hancke, G. P. (2009, Oct). Industrial wireless sensor networks: Challenges, design principles and technical approaches. *IEEE Transactions on Industrial Electronics*, *56*(10), 4258–4265. doi:10.1109/TIE.2009.2015754

Guo, W., Cao, L., Xia, J., Nie, F. Q., Ma, W., Xue, J., & Jiang, L. et al. (2010). Energy harvesting with single-ion selective nanopores: A concentration-gradient-driven nanofluidic power source. *Advanced Functional Materials*, *20*(8), 1339–1344. doi:10.1002/adfm.200902312

Gustafsson, S., Karawacki, E., & Khan, M. (1979). Transient hot-strip method for simultaneously measuring thermal conductivity and thermal diffusivity of solids and fluids. *Journal of Physics. D, Applied Physics*, *12*(9), 1411–1421. doi:10.1088/0022-3727/12/9/003

Guyomar, D., Lefeuvre, E., Richard, C., & Badel, A. (2004, May). *Piezoelectric energy harvesting circuit using a synchronized switch technique.* Paper presented at SMEBA'03, Suzhou, China.

Guyomar, D., Cottinet, P.-J., Lebrun, L., & Sebald, G. (2011a). Characterization of an electroactive polymer simultaneously driven by an electrical field and a mechanical excitation: An easy means of measuring the dielectric constant, the Young modulus and the electrostrictive coefficients. *Physics Letters. [Part A]*, *375*(16), 1699–1702. doi:10.1016/j.physleta.2011.03.011

Guyomar, D., Lallart, M., & Cottinet, P.-J. (2011b). Electrostrictive conversion enhancement of polymer composites using a nonlinear approach. *Physics Letters. [Part A]*, *375*(3), 260–264. doi:10.1016/j.physleta.2010.12.026

Guyomar, D., Lebrun, L., Putson, C., Cottinet, P. J., Guiffard, B., & Muensit, S. (2009). Electrostrictive energy conversion in polyurethane nanocomposites. *Journal of Applied Physics*, *106*(1), 014910. doi:10.1063/1.3159900

Hagedorn, F., Leicht, J., Sanchez, D., Hebn, T., & Manoli, Y. (2013). Impedance matching for broadband piezoelectric harvesting. *Journal of Physics: Conference Series*, *476*, 0120283. doi:10.1088/1742-6596/476/1/012083

Hagerty, J., Helmbrecht, F., McCalpin, W., Zane, R., & Papovic, Z. (2004, March). Recycling ambient microwave energy with broadband rectenna arrays. *IEEE Transactions on Microwave Theory and Techniques*, *52*(3), 1014–1024. doi:10.1109/TMTT.2004.823585

Hagglund, C., Zach, M., Petersson, G., & Kasemo, B. (2008). Electromagnetic coupling of light into a silicon solar cell by nano disk plasmons. *Journal of Applied Physics Letters, 92*, 053110 1-3.

Halvorsen, E. (2008). Energy harvesters driven by broadband random vibrations. *Journal of Microelectromechanical Systems*, *17*(5), 1061–1071. doi:10.1109/JMEMS.2008.928709

Harne, R. L., & Wang, K. W. (2013). A review of the recent research on vibration energy harvesting via bistable systems. *Smart Materials and Structures*, *22*(2), 023001. doi:10.1088/0964-1726/22/2/023001

Hartlen, R. T., & Currie, I. G. (1970). Lift-oscillator model of vortex-induced vibration. *Journal of the Engineering Mechanics Division*, *96*, 577–591.

Hartschuh, A., Beversluis, M. R., Bouhelier, A., & Novotny, L. (2004). Tip-enhanced optical spectroscopy. *Journal of Philosophical Transactions of the Royal Society of London, Series A: Mathematical. Physical and Engineering Sciences*, *362*(1817), 807–819. doi:10.1098/rsta.2003.1348

Helfer, F., Lemckert, C., & Anissimov, Y. G. (2014). Osmotic power with pressure retarded osmosis: Theory, performance, and trends – A review. *Journal of Membrane Science*, *453*(1), 337–358. doi:10.1016/j.memsci.2013.10.053

He, S., Cui, Y., Ye, Y., Zhang, P., & Jin, Y. (2009). Optical nano-antennas and metamaterials. *Materialtoday*, *12*, 16–24.

He, X.-F., & Gao, J. (2013). Wind energy harvesting based on flow-induced-vibration and impact. *Microelectronic Engineering*, *111*, 82–86. doi:10.1016/j.mee.2013.02.009

Hicks, L., & Dresselhaus, M. (1993). Effect of quantum-well structures on the thermoelectric figure of merit. *Physical Review B: Condensed Matter and Materials Physics*, *47*(19), 12727–12731. doi:10.1103/PhysRevB.47.12727 PMID:10005469

Hobeck, J. D., & Inman, D. J. (2012). Artificial piezoelectric grass for energy harvesting from turbulence-induced vibration. *Smart Materials and Structures*, *21*(10), 105024. doi:10.1088/0964-1726/21/10/105024

Ho, C. J., Sing, T. K., & Li, P. C. (2004). Preferred orientation control and characterization of AlN thin films using reactive sputtering. *Tamkang Journal of Science and Engineering*, *7*, 1–4.

Hochbaum, A., Chen, R., Delgado, R., Liang, W., Garnett, E., Najarian, M., & Yang, P. et al. (2008). Enhanced thermoelectric performance of rough silicon nanowires. *Nature*, *451*(7175), 163–167. doi:10.1038/nature06381 PMID:18185582

Hoffschmidt, B., Alexopoulos, S., Rau, C., Sattler, J., Anthrakidis, A., Boura, C., & Hilger, P. et al. (2012). Concentrating solar power. *Comprehensive Renewable Energy*, *3*(2), 595–636. doi:10.1016/B978-0-08-087872-0.00319-X

Holland, R., & EerNisse, E. P. (1969). *Design of resonant piezoelectric devices*. Cambridge, MA: MIT Press.

Hong, L., Rusli, Wang, X., Zheng H., Wang, H., & Yu, H. (2013). Design guideline of Si Nanohole/PEDOT:PSS hybrid structure for solar cell application. *Journal of Nanotechnology*, *24*, 1–6. PMID:23940111

Hong, L., Wang, R. X., Zheng, H., & Wang, H. (2013). Design guidelines for slanting silicon nanowire arrays for solar cell application. *Journal of Applied Physics*, *114*(8), 084301–0843016. doi:10.1063/1.4819175

Huang, Z., Fang, H., & Zhu, J. (2007). Fabrication of silicon nanowire arrays with controlled diameter, length, and density. *Journal of Advanced Materials*, *19*(5), 744–748. doi:10.1002/adma.200600892

Huang, Z., Geyer, N., Werner, P., Boor, J. D., & Gosele, U. (2011). Metal assisted chemical etching of silicon: A review. *Journal of Advanced Materials*, *23*(2), 285–308. doi:10.1002/adma.201001784 PMID:20859941

Huang, Z., Shimizu, T., Senz, S., Zhang, Z., Geyer, N., & Gosele, U. (2010). Oxidation rate effect on the direction of metal-assisted chemical and electrochemical etching of silicon. *The Journal of Physical Chemistry C*, *114*(24), 10683–10690. doi:10.1021/jp911121q

Hudak, N., & Amatucci, G. (2008). Small-scale energy harvesting through thermoelectric, vibration, and radiofrequency power conversion. *Journal of Applied Physics*, *103*(10), 101301–101324. doi:10.1063/1.2918987

Hu, L., & Chen, G. (2007). Analysis of optical absorption in silicon nanowire arrays for photovoltaic applications. *Journal of Nano Letters*, *7*(11), 3249–3252. doi:10.1021/nl071018b PMID:17927257

Hunter, R. J. (1981). *Zeta potential in colloid science: principles and applications*. London: Academic press.

Huson, A. (1974). Biomechanische problem des kniegelenks. *Ortopädie*, *3*, 119–126.

Hussein M., Hameed, M. F. O., Areed, N. F. F., & Obayya, S. S. A. (2014b). Ultra high efficient solar cell based on decagonal arrays of silicon nanowires. *Journal of Optical Engineering, 53*, 117105 1-7.

Hussein, M., Areed, N. F. F., Hameed, M. F. O., & Obayya, S. S. A. (2013). Flower-shaped dipole based nano-antenna for energy harvesting. In *Proceedings of the Conference on Progress in electromagnetics research symposium (PIERS)*. Stockholm, Sweden: PIER.

Hussein, M., Areed, N. F. F., Hameed, M. F. O., & Obayya, S. S. A. (2014c). Hybrid core semiconductor nanowires for solar cell applications. *In the proceeding of the 14th International Conference on Numerical Simulation of Optoelectronic Devices (NUSOD)*. Alma de Mallorca, Spain.

Hussein, M., Areed, N. F. F., Hameed, M. F. O., & Obayya, S. S. A. (2014a). Design of flower-shaped dipole nano-antenna for energy harvesting. *Journal of IET, 8*, 167–173.

Iba, S., Weghe, J. M. V., Paredis, C. J. J., & Khosla, P. K. (1999). An architecture for gesture-based control of mobile robots. In *Proceedings International Conference On Intelligent Robots and Systems*, (pp. 851-857). doi:10.1109/IROS.1999.812786

Ikeda, T. (1996). *Fundamentals of piezoelectricity*. Oxford University Press.

Imai, S., Tamaru, S., Fujimori, K., Sanagi, M., & Nogi, S. (2011, May). Efficiency and harmonics generation in microwave to dc conversion circuits of half-wave and full-wave rectifier types. In *IEEE MTT- S International Microwave Workshop Series on Innovative Wireless Power Transmission: Technologies, Systems, and Applications (IMWS)*, 15 –18. doi:10.1109/IMWS.2011.5877081

Inman, D. J. (2000). *Engineering vibration*. Prentice Hall.

International Commission on Non-Ionizing Radiation Protection. (1998). Guidelines for limiting exposure to time-varying electric, magnetic, and electromagnetic fields (up to 300 GHz). *Health Physics, 74*(4), 494–52. PMID:9525427

International Standard ISO 7250. (n. d.). *Basic human body measurements for technological design. Part 1 (Body measurement definitions and landmarks) and Part 2 (Statistical summaries of body measurements from individual ISO populations)*.

Isarakorn, D., Briand, D., Gariglio, S., Sambri, A., Stucki, N., Triscone, J. M., & De Rooji, N. F. et al. (2008). A study on epitaxial piezoelectric thin films grown on silicon for energy scavenging applications. In *Proceedings of PowerMEMS*. Sendai, Japan: IOP.

Islam, M. R., Mekhilef, S., & Saidur, R. (2013). Progress and recent trends of wind energy technology. *Renewable & Sustainable Energy Reviews, 21*, 456–468. doi:10.1016/j.rser.2013.01.007

Jaana, M., Paré, G., & Sicotte, C. (2009). Home telemonitoring for respiratory conditions: A systematic review. *The American Journal of Managed Care, 15*, 313–320. PMID:19435399

Jadidian, B., Janas, V., Safari, A., French, J., Weitz, G., Luke, J., & Cass, R. (1996). Development of fine scale piezoelectric ceramic/polymer composites via incorporation of fine PZT fibers. In *IEEE International Symposium on Application of Ferroelectrics*, 1, 31-34. doi:10.1109/ISAF.1996.602705

Jager, T., Guigon, R., Chaillout, J. J., & Despesse, G. (2008). Harvesting raindrop energy: Experimental study. *Smart Materials and Structures, 17*(1).

Jean-Mistral, C., Basour, S., & Chaillout, J.-J. (2008, March). Smart Structures and Materials. In *Proc.SPIE Conference on Electroactive Polymer Actuators and Devices, San Diego*, 6927, 692716.

Ji, L., Kim, J. K., Ji, Q., Leung, K. N., Chen, Y., & Gough, R. A. (2007). Conformal metal thin-film coatings in high-aspect ratio trenches using a self-sputtered rf-driven plasma source. *Journal of Vacuum Science & Technology B Microelectronics and Nanometer Structures, 25*(4), 1227–1230. doi:10.1116/1.2749527

Joannopoulos, J. D., Meade, R. D., & Winn, J. N. (2008). *Photonic crystals: Molding the flow of light*. New Jersey: Princeton University Press.

Jovanovic, V., Ghamaty, S., & Bass, J. (2006). Design, fabrication and testing of a novel energy harvesting thermoelectric power supply for wireless sensors. In *Proceedings of PWR2006 ASME Power, Atlanta, GA, USA* (pp. 643-651). doi:10.1115/POWER2006-88150

Jung, B., Cho, C., Yoon, H., Yoon, H., Youn, B. D., & Kim, Y. (2012). *Statistical Model Calibration for Energy Harvesting Skin Analysis and Design.* Paper presented at the 53rd AIAA/ASME/ASCE/AHS/ASC Structures, Structural Dynamics and Materials. doi:10.2514/6.2012-1833

Jung, H. J., & Lee, S. W. (2011). The experimental validation of a new energy harvesting system based on the wake galloping phenomenon. *Smart Materials and Structures, 20*(5), 055022. doi:10.1088/0964-1726/20/5/055022

Kandala, A., Betti, T., & Fontcubertai, A. M. (2008). General theoretical considerations on nanowire solar cell designs. *Journal of Physica Status Solidi, 206*(1), 173–178. doi:10.1002/pssa.200723672

Kang, B., Kim, M., Joo, H., Kim, H. J., & Kim, D. K. (2013). *Experimental study on energy harvesting from salinity gradient by reverse electrodialysis in Ag/AgCl deposited AAO nanochannel array.* Presented at ASME 2013 International Mechanical Engineering Congress and Exposition, San Diego. doi:10.1115/IMECE2013-64364

Kang, B., Yoo, J., Kim, H. J., & Kim, D. K. (2013). Slip-enhanced reverse electrodialytic power generation in ion-selective nanochannels. *Journal of Thermal Science, 22*(1), 36–41. doi:10.1007/s11630-013-0589-y

Karami, A. (2011). *Micro-scale and nonlinear vibrational energy harvesting* (PhD Dissertation). Virginia Tech, VA.

Karami, A., & Inman, D. J. (2011). Equivalent damping and frequency change for linear and nonlinear hybrid vibrational energy harvesting systems. *Journal of Sound and Vibration, 330*(23), 5583–5597. doi:10.1016/j.jsv.2011.06.021

Karami, A., & Inman, D. J. (2012). Powering pacemakers from heartbeat vibrations using linear and nonlinear energy harvesters. *Applied Physics Letters, 100*(4), 042901. doi:10.1063/1.3679102

Karthaus, U., & Fisher, M. (2003, October). Fully integrated passive UHF RFID transpoder IC with 16.7-μW minimum RF input power. *IEEE Journal of Solid-State Circuits, 38*(10), 1602–1608. doi:10.1109/JSSC.2003.817249

Kasyap, A., Lim, J., Johnson, D., Horowitz, S., Nishida, T., Ngo, K., . . . Cattafesta, L. (2002). *Energy reclamation from a vibrating piezoceramic composite beam.* Paper presented at the International Congress on Sound and Vibration.

Keefe, D. F., Feliz, D. A., Moscovich, T., Laidlaw, D. H., & LaViola, J. J. (2001). Cave painting: A fully immersive 3D artistic medium and interactive experience. In *Proceedings of Symposium on Interactive 3D Graphics and Games* (pp. 85-93).

Kelzenberg, M. D., Boettcher, S. W., Petykiewicz, J. A., Turner, D. B., Putnam, M. C, & Warren, E. L. Spurgeon, J. M., Briggs, R. M., Lewis, N. S., & Atwater, H. A. (2010). Enhanced absorption and carrier collection in Si wire arrays for photovoltaic applications. *Journal of Nature Materials, 9*, 239–244. PMID:20154692

Kennedy, C. E., & Terwilliger, K. (2005). Optical durability of candidate solar reflectors. *Journal of Solar Energy Engineering, 127*(May), 262. doi:10.1115/1.1861926

Khaligh, A., & Onar, O. (2010). *Energy harvesting: solar, wind, and ocean energy conversion systems.* CRC Press.

Kielkowski, R. M. (1994). *Inside SPICE - overcoming the obstacles of circuit simulation, chapter Understanding circuit simulation.* New York: McGraw-Hill.

Kim, H. (2014). *Segmented piezoelectric skin energy harvesters* (Unpublished doctoral dissertation). Seoul National University, South Korea.

Kim, H. S., Kim, J.-H., & Kim, J. (2011). A review of piezoelectric energy harvesting based on vibration. *International Journal of Precision Engineering and Manufacturing, 12*(6), 1129-1141.

Kim, D. (2011). Numerical study of power generation by reverse electrodialysis in ion-selective nanochannels. *Journal of Mechanical Science and Technology, 25*(1), 5–10. doi:10.1007/s12206-010-1113-x

Kim, D. K., Duan, C., Chen, Y. F., & Majumdar, A. (2010). Power generation from concentration gradient by reverse electrodialysis in ion-selective nanochannels. *Microfluidics and Nanofluidics, 9*(6), 1215–1224. doi:10.1007/s10404-010-0641-0

Kim, D., Raj, A., Zhu, L., Masel, R. I., & Shannon, M. A. (2008). Non-equilibrium electrokinetic micro/nano fluidic mixer. *Lab on a Chip*, *8*(4), 625–628. doi:10.1039/b717268k PMID:18369520

Kim, G., Lee, M. E., Lee, K. S., Park, J. S., Jeong, W. M., Kang, S. K., & Kim, H. et al. (2012). An overview of ocean renewable energy resources in Korea. *Renewable & Sustainable Energy Reviews*, *16*(4), 2278–2288. doi:10.1016/j.rser.2012.01.040

Kim, H.-U., Bedekar, V., Islam, R. A., Lee, W., Leo, D., & Priya, S. (2008). Laser-machined piezoelectric cantilevers for mechanical energy harvesting. *Transaction on Ultrasonics, Ferroelectics, and Frequency Control*, *55*(9), 1900–1905. doi:10.1109/TUFFC.881 PMID:18986886

Kim, J., Kim, S. J., & Kim, D. K. (2013). Energy harvesting from salinity gradient by reverse electrodialysis with anodic alumina nanopores. *Energy*, *51*(1), 413–421. doi:10.1016/j.energy.2013.01.019

Kim, M., Dugundji, J., & Wardle, B. L. (2013). Optimal design of piezoelectric materials and devices for energy harvesting. *Journal of the Korean Physical Society*, *62*(11), 1689–1695. doi:10.3938/jkps.62.1689

Kim, M., Hoegen, M., Dugundji, J., & Wardle, B. L. (2010). Modeling and experimental verification of proof mass effects on vibration energy harvester performance. *Smart Materials and Structures*, *19*(4), 045023. doi:10.1088/0964-1726/19/4/045023

Kim, M., Hong, S., Miller, D. J., Dugundji, J., & Wardle, B. L. (2011). Size effect of flexible proof mass on the mechanical behavior of micron-scale cantilevers for energy harvesting applications. *Applied Physics Letters*, *99*(24), 243506, 243506–243504. doi:10.1063/1.3663858

Kim, S. J., Ko, S. H., Kang, K. H., & Han, J. (2010b). Direct seawater desalination by ion concentration polarization. *Nature Nanotechnology*, *5*(4), 297–301. doi:10.1038/nnano.2010.34 PMID:20305644

Kim, S. J., Song, Y. A., & Han, J. (2010a). Nanofluidic concentration devices for biomolecules utilizing ion concentration polarization: Theory, fabrication and applications. *Chemical Society Reviews*, *39*(3), 912–922. doi:10.1039/b822556g PMID:20179814

Klimiec, E., Zaraska, W., Zaraska, K., Gąsiorski, K. P., Sadowski, T., & Pajda, M. (2008). Piezoelectric polymer films as power converters for human powered electronics. *Microelectronics and Reliability*, *48*(6), 897–901. doi:10.1016/j.microrel.2008.04.001

Kocakarin, I., & Yegin, K. (2013). Glass superstrate nanoantennas for infrared energy harvesting applications. *International Journal of Antennas and Propagation*, *2013*, 1–7. doi:10.1155/2013/245960

Korolak, D., Taris, T., Deval, Y., Begueret, J. B., & Mariano, A. (2013). Bulk biased RF energy harvesting dedicated to 900MHz/2.4GHz ISM bands. *Workshop on Circuits and System design*, 176- 179.

Ko, S. H., Song, Y. A., Kim, S. J., Kim, M., Han, J., & Han, J. (2012). Nanofluidic preconcentration device in a straight microchannel using ion concentration polarization. *Lab on a Chip*, *12*(21), 4472–4482. doi:10.1039/c2lc21238b PMID:22907316

Kostiučenko, O. (2011). *Nano-structure-induced field enhanced nonlinear optics of plasmonic substrates.* (Master dissertation). University of Southern, Denmark.

Kotani, K., & Ito, T. (2007, Nov). High efficiency CMOS rectifier circuit with self-Vth-cancellation and power regulation functions for UHF RFIDs. *IEEE Asian Solid-State Circuits Conference*, (pp. 119-122). doi:10.1109/ASSCC.2007.4425746

Kotani, K., Sazaki, A., & Ito, T. (2009, November). High efficiency differential drive CMOS rectifier for UHF RFIDs. *IEEE Journal of Solid-State Circuits*, *44*(11), 3011–3018. doi:10.1109/JSSC.2009.2028955

Kotter D. K., Novack S. D., Slafer W. D., & Pinhero P. J. (2010). Theory and manufacturing processes of solar nanoantenna electromagnetic collectors. *Journal of Solar Energy Engineering*, *132*(1).

Kotter, D. K., Novack, S. D., Slafer, W. D., & Pinhero, P. J. (2008). Nanatenna electromagnetic collector. In the Proceedings of Conference on Energy Sustainability (ASME). Florida, USA.

Kottmann, J. O., Martin, D. S., & Schultz, S. (2001). Non-regularly shaped plasmon resonant nanoparticle as localized light source for near-field microscopy. *Journal of Microscopy*, *202*(1), 60–65. doi:10.1046/j.1365-2818.2001.00866.x PMID:11298871

Krenz, P. M., Tiwari, B., Szakmany, G. P., Orlov, A. O., González, F. J., Boreman, G. D., & Porod, W. (2012). Response increase of IR antenna-coupled thermocouple using impedance matching. *IEEE Journal of Quantum Electronics*, *48*(5), 659–664. doi:10.1109/JQE.2012.2189758

Kuiqing, P., Mingliang, Z., Aijiang, L., Ning, W., Ruiqin, Z., & Shuit, L. (2007). Ordered silicon nanowire arrays via nanosphere lithography and metal-induced etching. *Journal of Applied Physics Letter*, *90*(16), 163123. doi:10.1063/1.2724897

Kulah, H., & Najafi, K. (2008). Energy scavenging from low-frequency vibrations by using frequency up-conversion for wireless sensor applications. *IEEE Sensors Journal*, *8*(3), 261–268. doi:10.1109/JSEN.2008.917125

Kulkarni, V., Ben-Mrad, R., Eswar Prasad, S., & Nemana, S. (2013). A shear-mode energy harvesting device based on torsional stresses. *Transaction on Mechatronics*, *19*(3), 801–807. doi:10.1109/TMECH.2013.2259635

Kwon, K., & Kim, D. (2013). Efficient water recirculation for portable direct methanol fuel cells using electroosmotic pumps. *Journal of Power Sources*, *221*(1), 172–176. doi:10.1016/j.jpowsour.2012.07.113

Kwon, K., Lee, S. J., Li, L., Han, C., & Kim, D. (2014). Energy harvesting system using reverse electrodialysis with nanoporous polycarbonate track-etch membranes. *International Journal of Energy Research*, *38*(4), 530–537. doi:10.1002/er.3111

Kymissis, J., Kendall, C., Paradiso, J., & Gershenfeld, N. (1998). Parasitic power harvesting in shoes. In *Proceedings of IEEE International Symposium on Wearable Computers* (pp. 132-139).

Ladan, S., Ghassemi, N., Ghiotto, A., & Wu, K. (2013). Highly efficient compact rectenna for wireless energy harvesting Application. *IEEE Microwave Magazine*, *14*(1), 117–122. doi:10.1109/MMM.2012.2226629

Lallart, M., Cottinet, P.-J., Lebrun, L., Guiffard, B., & Guyomar, D. (2010). Evaluation of energy harvesting performance of electrostrictive polymer and carbon-filled terpolymer composites. *Journal of Applied Physics*, *108*(3), 034901. doi:10.1063/1.3456084

Lebrun, L., Guyomar, D., Guiffard, B., Cottinet, P. J., & Putson, C. (2009). The Characterization of the harvesting capabilities of an electrostrictive polymer composite. *Sensors and Actuators. A, Physical*, *153*(2), 251–257. doi:10.1016/j.sna.2009.05.009

Lee, R.-G., Chen, K.-C., Chiang, S.-S., Lai, C.-C., Liu, H.-S., & Wei, M.-S. (2006, May). A backup routing with wireless sensor network for bridge monitoring system. *The 4th Annual Communication Networks and Services Research Conference (CNSR)*, (pp. 161-165).

Lee, S., Youn, B. D., & Giraud, M. (2010). *Designing energy harvesting skin structure utilizing outdoor unit vibration.*

Lee, M., Chen, C. Y., Wang, S., Cha, S. N., Park, Y. J., Kim, J. M., & Wang, Z. L. et al. (2012). A hybrid piezoelectric structure for wearable nanogenerators. *Advanced Materials*, *24*(13), 1759–1764. doi:10.1002/adma.201200150 PMID:22396355

Lee, S. J., & Kim, D. (2012). Millisecond-order rapid micromixing with non-equilibrium electrokinetic phenomena. *Microfluidics and Nanofluidics*, *12*(6), 897–906. doi:10.1007/s10404-011-0918-y

Lee, S., & Youn, B. D. (2011a). A design and experimental verification methodology for an energy harvester skin structure. *Smart Materials and Structures*, *20*(5), 057001. doi:10.1088/0964-1726/20/5/057001

Lee, S., & Youn, B. D. (2011b). A new piezoelectric energy harvesting design concept: Multimodal energy harvesting skin. *IEEE Transactions on Ultrasonics, Ferroelectrics and Frequency Control*, *58*(3), 629–645.

Lee, S., Youn, B. D., & Jung, B. C. (2009). Robust segment-type energy harvester and its application to a wireless sensor. *Smart Materials and Structures*, *18*(9), 095021. doi:10.1088/0964-1726/18/9/095021

Leland, E. S., Lai, E. M., & Wright, P. K. (2004). *A self-powered wireless sensor for indoor environmental monitoring*. Paper presented at the Wireless Networking Symposium, University of Texas at Austin Department of Electrical & Computer Engineering.

Leland, E. S., & Wright, P. K. (2006). Resonance tuning of piezoelectric vibration energy scavenging generators using compressive axial preload. *Smart Materials and Structures*, *15*(5), 1413–1420. doi:10.1088/0964-1726/15/5/030

Leonov, V. (2011). Energy harvesting for self-powered wearable devices. In A. Bonfiglio & D. De Rossi, (Eds.), Wearable monitoring systems. New York: Springer. doi:10.1007/978-1-4419-7384-9_2

Le, T., Mayaram, K., & Fiez, T. (2008, May). Efficient far field radio frequency energy harvesting for passively powered sensor networks. *IEEE Journal of Solid-State Circuits*, *43*(5), 1278–1302. doi:10.1109/JSSC.2008.920318

Li, W. J., Ho, T. C. H., Chan, G. M. H., Leong, P. H. W., & Wong, H. Y. (2000). Infrared signal transmission by a laser-micromachined, vibration-induced power generator. In *Proceedings of the 43rd IEEE Midwest Symposium on Circuits and Systems* (pp. 236-239). doi:10.1109/MWSCAS.2000.951628

Liang, J., & Liao, W. S. (2011). Energy flow in piezoelectric energy harvesting systems. *Smart Materials and Structures*, *20*(1), 015005. doi:10.1088/0964-1726/20/1/015005

Lien, I. C., & Shu, Y. C. (2012). Array of piezoelectric energy harvesting by the equivalent Impedance approach. *Smart Materials and Structures*, *21*(8), 082001. doi:10.1088/0964-1726/21/8/082001

Li, J., Yu, H., Wong, S. M., Li, X., Zhang, G., & Lo, P. G. (2009). Design guidelines of periodic Si nanowire arrays for solar cell application. *Journal of Applied Physics Letter*, *95*, 243113-1–4.

Lin, C., & Povinelli, M. L. (2010). Optical absorption enhancement in silicon nanowire and nanohole arrays for photovoltaic applications. In the Proceedings of Conference on Next Generation (Nano) Photonic and Cell Technologies for Solar Energy Conversion, (SPIE). California.

Lin, G. H., Abdu, R., & Bockris, J. O. M. (1996). Investigation of resonance light absorption and rectification by sub nanostructures. *Journal of Applied Physics*, *80*(1), 565–568. doi:10.1063/1.362762

Li, Q., Naing, V., & Donelan, J. M. (2009). Development of a biomechanical energy harvester. *Journal of Neuroengineering and Rehabilitation*, *6*. PMID:19549313

Liu, Q., Wang, Y., Guo, W., Ji, H., Xue, J., & Ouyang, Q. (2007). Asymmetric properties of ion transport in a charged conical. *Physical Review E: Statistical, Nonlinear, and Soft Matter Physics*, *75*(5), 051201. doi:10.1103/PhysRevE.75.051201 PMID:17677044

Liu, R., Zhang, Q., & Cross, L. E. (1999). Experimental investigation of electrostrictive polarization biased direct apparent piezoelectric properties in polyurethane elastomer under quasistatic conditions. *Journal of Applied Polymer Science*, *73*(13), 2603–2609. doi:10.1002/(SICI)1097-4628(19990923)73:13<2603::AID-APP7>3.0.CO;2-R

Liu, Y., Ren, K. L., Hofmann, H. F., & Zhang, Q. (2005). Investigation of electrostrictive polymers for energy harvesting. *IEEE Transactions on Ultrasonics, Ferroelectrics, and Frequency Control*, *52*(12), 2411–2417. doi:10.1109/TUFFC.2005.1563285 PMID:16463508

Li, X. (2012). Metal assisted chemical etching for high aspect ratio nanostructures: A review of characteristics and applications in photovoltaics. *Journal of Current Opinion in Solid State and Materials Science*, *16*(2), 71–81. doi:10.1016/j.cossms.2011.11.002

Locatelli, A. (2010). Analysis of the optical properties of wire antennas with displaced terminals. *Optics Express*, *18*(9), 9504–9510. doi:10.1364/OE.18.009504 PMID:20588796

Loeb, S. (1976). Production of energy from concentrated brines by pressure-retarded osmosis: I. Preliminary technical and economic correlations. *Journal of Membrane Science*, *1*, 49–63. doi:10.1016/S0376-7388(00)82257-7

Lu, C., & Tsui, C. Y. Ki, & W.H (2011). Vibration energy scavenging system with maximum power tracking for micropower applications. In Proceedings of Very Large Scale Integration Systems (pp. 2109-2119).

Lui, H., Lee, C., Kobayashi, T., Tay, C. J., & Quan, C. (2012). Investigation of a MEMS piezoelectric energy harvester system with a frequency-widened-bandwidth mechanism introduced by mechanical stoppers. *Smart Materials and Structures*, *21*(3), 035005. doi:10.1088/0964-1726/21/3/035005

Lund, J. W. (2010). Direct utilization of geothermal energy. *Energies*, *3*(8), 1443–1471. doi:10.3390/en3081443

Mabrouki, A., Latrach, M., & Lorrain, V. (2014, May). High efficiency low power design using zero bias schottky diodes. *13th Edition of IEEE Faible Tension Faibe Consommation*, (pp. 1-4). Monaco, France.

Mabrouki, A., Latrach, M., & Sayegh, Z. (2013, Sept). Design and experiment of RF rectifiers for wireless power transmission. In *13th Mediterranean Microwave Symposium*, (pp. 1-4). Saida, Lebanon. doi:10.1109/MMS.2013.6663079

Mackowski, A. W., & Williamson, C. H. K. (2013). An experimental investigation of vortex-induced vibration with nonlinear restoring forces. *Physics of Fluids*, *25*(8), 087101. doi:10.1063/1.4819082

Magoteaux, K. C., Sanders, B., & Sodano, H. A. (2008). Investigation of energy harvesting small unmanned air vehicle. Smart Materials and Structures: Active and PassiveSmart Structures and Integrated Systems II, *Proceedings of the SPIE*. San Diego, CA, 6928. doi:10.1117/12.775851

Mahian, O., Kianifar, A., Kalogirou, S. A., Pop, L., & Wongwises, S. (2013). A review of the applications of nanofluids in solar energy. *International Journal of Heat and Mass Transfer*, *57*(2), 582–594. doi:10.1016/j.ijheatmasstransfer.2012.10.037

Mainwaring, A., Culler, D., Polastre, J., Szewczyk, R., & Anderson, J. (2002). Wireless sensor networks for habitat monitoring. In *Proceedings of the 1st ACM International Workshop on Wireless sensor networks and applications, WSNA'02*, (pp. 88-97). New York, USA.

Malik, N., & Ray, M. C. (2003). Effective coefficients of piezoelectric fiber-reinforced composites. *AIAA Journal*, *41*(4), 704–710. doi:10.2514/2.2001

Malinovska, D. D. (2010). The state-of-the-art and future development of the photovoltaic technologies–the route from crystalline to nanostructured and new emerging materials. *Journal of Physics: Conference Series*, 253.

Malkawi, A., & Srinivasan, R. (2004). Multimodal human-computer interaction for immersive visualization: Integrating speech-gesture recognitions and augmented reality for indoor environments. In *Proceedings International Conference on Computer Graphics* (pp. 171-176).

Mallick, S. B., Agrawal, M., & Peumans, P. (2010). Optimal light trapping in ultra-thin photonic crystal crystalline silicon solar cells. *Journal of Optics Express*, *18*(6), 5691–5706. doi:10.1364/OE.18.005691 PMID:20389585

Mann, B. P., & Sims, N. D. (2009). Energy harvesting from the nonlinear oscillations of magnetic levitation. *Journal of Sound and Vibration*, *319*(1-2), 515–530. doi:10.1016/j.jsv.2008.06.011

Manzano-Agugliaro, F., Alcayde, A., Montoya, F. G., Zapata-Sierra, A., & Gil, C. (2013). Scientific production of renewable energies worldwide: An overview. *Renewable & Sustainable Energy Reviews*, *18*, 134–143. doi:10.1016/j.rser.2012.10.020

Mao, P., & Han, J. (2009). Massively-parallel ultra-high-aspect-ratio nanochannels as mesoporous membranes. *Lab on a Chip*, *9*(4), 586–591. doi:10.1039/B809370A PMID:19190794

Marian, V. M., Vollaire, C., Verdier, J., & Allard, B. (2012). Low power rectenna systems for wireless energy transfer. In J. Agbinya (Ed.), *Wireless Power Transfer* (pp. 67–107). Melbourne, Australia: River Publishers.

Marian, V., Menudier, C., Thevenot, M., Vollaire, C., Verdier, J., & Allard, B. (2011, June). Efficient design of rectifying antennas for low power detection.*MTT –S International Microwave Symposium Digest*, 1-4.

Marks, A. M. (1984). *Device for conversion of light power to electrical power.* U.S. patent 4445050.

Marzencki, M., Basrour, S., Charlot, B., Spirkovich, S., & Colin, M. (2005). A MEMS piezoelectric vibration energy harvesting device. In Proceedings of PowerMEMS, (pp. 45-48). Tokyo, Japan: IOP.

Masana, R., & Daqaq, M. F. (2011a). Electromechanical modeling and nonlinear analysis of axially-loaded energy harvesters. *Journal of Vibration and Acoustics*, *133*(1), 011007. doi:10.1115/1.4002786

Masana, R., & Daqaq, M. F. (2011b). Relative performance of a vibratory energy harvester in mono- and bi-stable potentials. *Journal of Sound and Vibration*, *330*(24), 6036–6052. doi:10.1016/j.jsv.2011.07.031

Massaro, A., & Cingolani, R. (2011c). Parametric analysis of piezoelectric resonances of AlN layouts for vibrating energy harvesting. *Journal of Computational and Theoretical Nanoscience*, *8*(11), 1–8. doi:10.1166/jctn.2011.1955

Massaro, A., Cingolani, R., & Passaseo, A. (2011b). 3D FEM modelling and technology of piezoelectric ring MEMS antenna. *Progress in Electromagnetics Research, C*, 23, 123–135.

Massaro, A., De Guido, S., Ingrosso, I., Cingolani, R., De Vittorio, M., Cori, M., & Passaseo, A. et al. (2011a). Freestanding piezoelectric rings for high efficiency energy harvesting at low frequency. *Applied Physics Letters*, *98*(5), 053502. doi:10.1063/1.3551725

Massaro, A., Ingrosso, I., Giordano, C., Todaro, M. T., Cingolani, R., De Vittorio, M., & Passaseo, A. (2009). Micromachining Mo/ALN/Mo piezoelectric ring resonators. In *Proceedings of European Microwave Conference (EuMA)*. Rome, Italy: IEEE.

Masuda, H., Yamada, H., Satoh, M., Asoh, H., Nakao, M., & Tamamura, T. (1997). Highly ordered nanochannel-array architecture in anodic alumina. *Applied Physics Letters*, *71*(19), 2770. doi:10.1063/1.120128

Mateu, L., Codrea, C., Lucas, N., Pollak, M., & Spies, P. (2006). Energy harvesting for wireless communication systems using thermogenerators. In *Proceedings of the XXI Conference on Design of Circuits and Integrated Systems (DCIS)*, Barcelona, Spain.

Mateu, L., & Moll, F. (2005). Optimum piezoelectric bending beam structures for energy harvesting using shoe inserts. *Journal of Intelligent Material Systems and Structures*, *16*(10), 835–845. doi:10.1177/1045389X05055280

Mathur, S., & Murthy, J. (2009). A multigrid method for the Poisson–Nernst–Planck equations. *International Journal of Heat and Mass Transfer*, *52*(17-18), 4031–4039. doi:10.1016/j.ijheatmasstransfer.2009.03.040

Maxim. (2014). *MAX17710 Energy-Harvesting Charger and Protector*. Retrieved September 2014 from http://www.maximintegrated.com/en/products/power/battery-management/MAX17710.html

Ma, Z., & Vandenbosch, G. A. E. (2013). Optimal solar energy harvesting efficiency of nano-rectenna systems. *Solar Energy*, *88*, 163–174. doi:10.1016/j.solener.2012.11.023

Mazar, B. (2012). *State of the art prototype vehicle with a thermoelectric generator*. Paper presented at the TE Application Workshop, Baltimore. Retrieved August 2014 from http://www1.eere.energy.gov

McCoy, J. (Jan 2012). *Thermoelectric technology: Materials, processes, devices and systems*. ASM San Diego Chapter. Retrieved August 2014 from http://www.hi-z.com

McKnight, W. H. (2002). *Energy-harvesting device using electrostrictive polymers*. US Patent Number US 6433465B1. Milkline. Retrieved from http://www.milkline.com

McMahon, J. M., Gray, S. K., & Schatz, G. C. (2010). Optical properties of nanowire dimers with a spatially nonlocal dielectric function. *Journal of Nano Letters*, *10*(9), 3473–3481. doi:10.1021/nl101606j PMID:20715807

McMahon, J. M., Henry, A. I., Wustholz, K. L., Natan, M. J., Freeman, R. G., & Van Duyne, R. P. (2009). Gold nanoparticle dimer plasmonics: Finite element method calculations of the electromagnetic enhancement to surface-enhanced Raman spectroscopy. *Journal of Analytical and Bioanalytical Chemistry*, *394*, 1819–1825.

McSpadden, J. O., & Chang, K. (1994). A dual polarized circular patch rectifying antenna at 2.45GHz for microwave power conversion and detection. In *IEEE MTT-S International Microwave Symposium Digest*, (pp. 1749-1752).

Meddad, M. (2012c). *Etude de l'énergie renouvelable dans l'environnement humain pour alimenter les systèmes électroniques portables*. Algeria: PhD. Université Ferhat Abbas of Sétif.

Meddad, M., Eddiai, A., Guyomar, D., Belkhiat, S., Cherif, A., Yuse, K., & Hajjaji, A. (2012b). An adaptive prototype design to maximize power harvesting using electrostrictive polymers. *Journal of Applied Physics, 112*(5), 054109. doi:10.1063/1.4751456

Meddad, M., Eddiai, A., Guyomar, D., Belkhiat, S., Hajjaji, A., Cherif, A., & Boughaleb, Y. (2012a). Study of the behaviour of electrostrictive polymers for energy harvesting with FFT analysis. *Journal of Optoelectronics and Advanced Materials, 14*, 55–60.

Meddad, M., Eddiai, A., Guyomar, D., Belkhiat, S., Hajjaji, A., Yuse, K., & Boughaleb, Y. (2012c). Evaluation by fast Fourier transforms analysis of energy harvesting in electrostrictive polymers driven by an electric field and a mechanical excitation. *Journal of Intelligent Material Systems and Structures, 24*(4), 411–420. doi:10.1177/1045389X12461077

Meddad, M., Eddiai, A., Hajjaji, A., Boughaleb, Y., Guyomar, D., & Fliyou, M. (2014b). Optimization of the energy harvested by the effect of strain and frequency on an electrostrictive polymer composite. *Synthetic Metals, 188*, 72–76. doi:10.1016/j.synthmet.2013.11.022

Meddad, M., Eddiai, A., Hajjaji, A., Guyomar, D., Belkhiat, S., Boughaleb, Y., & Chérif, A. (2014a). Lowest of AC–DC power output for electrostrictive polymers energy harvesting systems. *Optical Materials, 36*(1), 80–85. doi:10.1016/j.optmat.2013.05.008

Mehmood, A., Abdelkefi, A., Akhtar, I., Nayfeh, A.H., Nuhait, A.O., & Hajj, M.R. (2013). Linear and nonlinear active feedback controls for vortex-induced vibrations of circular cylinders. *Journal of Vibration and Control,* doi: .10.1177/1077546312469425

Mehmood, A., Abdelkefi, A., Hajj, M. R., Nayfeh, A. H., Akhtar, I., & Nuhait, A. O. (2013). Piezoelectric energy harvesting from vortex-induced vibrations of circular cylinder. *Journal of Sound and Vibration, 332*(19), 4656–4667. doi:10.1016/j.jsv.2013.03.033

Mein, W. S., Yu, Y. H., Yali, L., Junshuai, L., Wei, S. X., Singh, N., Lo, P. G. Q., & Dim, K. (2011). Boosting short-circuit current with rationally designed periodic Si nanopillar surface texturing for solar cells. *Journal of IEEE Transaction on Electron Devices, 58*, 3224–3229.

Meirovitch, L. (2001). *Fundamentals of vibrations.* McGraw-Hill New York.

Meisner, G. P. (2011). *Advanced thermoelectric materials and generator technology for automotive waste heat at GM.* Paper presented at Thermoelectrics Applications Workshop, San Diego.

Melcor. (2012). *CP1.0-127-05L-RTV Heat Pump.* Retrieved November 2012 from http://www.datasheetpro.com/part-info/CP1.0-127-05L-RTV/2981405

Meninger, S., Mur-Miranda, J. O., Amirtharajah, R., Chandrakasan, A. P., & Lang, J. H. (2001). Vibration-to-electric energy conversion. *IEEE Transactions on Very Large Scale Integration (VLSI) Systems, 9*(1), 64–76.

Merabet, B., Cirio, L., Takhedmit, H., Costa, F., Vollaire, C., Allard, B., & Picon, O. (2009). Low-cost converter for harvesting of microwave electromagnetic energy. In *Proceedings of the IEEE Energy Conversion Congress and Exposition,* (pp. 2592-2599). doi:10.1109/ECCE.2009.5316093

Micromachine Devices. (1997, May). *European Study Sees MEMS Market at More Than $34 billion by 02.*

Midrio, M., Romagnoli, M., Boscolo, S., Angelis, C., Locatelli, A., Modotto, D., & Capobianco, A. (2011). Flared monopole antennas for 10 μm radiation. *Journal of Quantum Electronics, 47*(1), 84–91. doi:10.1109/JQE.2010.2057501

Mikeka, C., & Arai, H. (2011). Design Issues in Radio Frequency Energy Harvesting System. In T. Y. Kheng (Ed.), Sustainable energy harvesting technologies – Past, present and future, (pp. 235-256). Intech. doi:10.5772/25348

Ming-Liang, Z., Kui-Qing, P., Xia, F., Jian-Sheng, J., Rui-Qin, Z., Shuit-Tong, L., & Ning-Bew, W. (2008). Preparation of large-area uniform silicon nanowires arrays through metal-assisted chemical etching. *Journal of Physical Chemistry, 112*, 4444–4450. PMID:18433167

Minnich, A. (2008). *Modeling the Thermoelectric Properties of Bulk and Nanocomposite Thermoelectric Materials* (MSc dissertation). MIT, USA.

Mitchell, B. S. (2004). *An introduction to materials engineering and science for chemical and materials engineers.* New York: Wiley.

Mitcheson, P. D., Yeatman, E. Y., Rao, G. K., Holmes, A. S., & Green, T. C. (2008). Energy harvesting from human and machine motion from wireless electronic devices. In Proceedings of IEEE (pp. 1457-1486).

Mitcheson, P. D., Green, T. C., & Yeatman, E. M. (2007). Power processing circuits for electromagnetic, electrostatic and piezoelectric inertial energy scavengers. *Microsystem Technologies-Micro-and Nanosystems-Information Storage and Processing Systems*, *13*, 1629–1635.

Moddel, G., & Grover, S. (2013). *Rectenna solar cells.* New York: Springer. doi:10.1007/978-1-4614-3716-1

Mohr, S. H., & Evans, G. M. (2008). Peak oil: Testing Hubbert's curve via theoretical modeling. *Natural Resources Research*, *17*(1), 1–11. doi:10.1007/s11053-008-9059-8

Mossi, K., Mouhli, M., Smith, B. F., Mane, P. P., & Bryant, R. G. (2006). Shape modeling and validation of stress-biased piezoelectric actuators. *Smart Materials and Structures*, *15*(6), 1785–1793. doi:10.1088/0964-1726/15/6/033

Motter, D., Lavarda, J. V., Dias, F. A., & Da Silva, S. (2012). Vibration energy harvesting using piezoelectric transducer and non-controlled rectifiers circuits. *Journal of the Brazilian Society of Mechanical Sciences and Engineering*, *34*(spe), 378–385. doi:10.1590/S1678-58782012000500006

Muskens, O. L., Rivas, J. G., Algra, R. E., Bakkers, E. P. A. A., & Lagendijk, A. (2008). Design of light scattering in nanowire materials for photovoltaic applications. *Journal of Nano Letter*, *8*, 2638–2642.

Myung, S. K., McDonnell, D. D., Kazinets, G., Seo, H. G., & Moskowitz, J. M. (2009). Concerns about a meta-analysis of computer smoking cessation programs – Reply. *Archives of Internal Medicine*, *169*, 929–937. doi:10.1001/archinternmed.2009.109 PMID:19468084

Nahum, M., Eiles, T., & Martinis, J. (1994). Electronic microrefrigerator based on a normal-insulator-superconductor tunnel junction. *Applied Physics Letters*, *65*(24), 3123–3125. doi:10.1063/1.112456

Nakamoto, H., Yamazaki, D., Yamamoto, T., Kurata, H., Yamada, S., Mukaida, K., & Gotoh, K. et al. (2007, January). A passice UHF RF identification CMOS tag IC using ferroelectric RAM in 0,35um technology. *IEEE Journal of Solid-State Circuits*, *42*(1), 101–110. doi:10.1109/JSSC.2006.886523

Nayfeh, A. H. (1993). *Method of Normal Forms.* New York: Wiley Interscience.

Nelson, J. (2003). *The physics of solar cells.* London: Imperial College. doi:10.1142/p276

Neubeck, L., Redfern, J., Fernandez, R., Briffa, T., Bauman, A., & Freedman, S. B. (2009). Telehealth interventions for the secondary prevention of coronary heart disease: A systematic review. *European Journal of Cardiovascular Prevention and Rehabilitation*, *16*(3), 281–289. doi:10.1097/HJR.0b013e32832a4e7a PMID:19407659

Nguyen, N. T., & Werely, S. T. (2002). *Fundamentals and Applications of Microfluidics.* Norwod, MA: Artech House, Inc.

Nolan, P., Lynch, P., & Sweeney, C. (2014). Simulating the future wind energy resource of Ireland using the COSMO-CLM model. *Wind Energy (Chichester, England)*, *17*(1), 19–37. doi:10.1002/we.1554

Novotny, L. (2007). Effective wavelength scaling for optical antennas. *Physical Review Letters*, *98*(26), 266802-1, 266802–266804. doi:10.1103/PhysRevLett.98.266802 PMID:17678117

Novotny, L., & van Hulst, N. (2011). Antennas for light. *Nature Photonics*, *5*(2), 83–90. doi:10.1038/nphoton.2010.237

Nuffer, J., & Bein, T. (2006). *Applications of piezoelectric materials in transportation industry.* Paper presented at the Global Symposium on Innovative Solutions for the Advancement of the Transport Industry.

Nuhait, A. O., & Mook, T. D. (2010). Aeroelastic behavior of flat plates moving near the ground. *Journal of Aircraft*, *47*(2), 464–474. doi:10.2514/1.41186

O'hayre, R., Cha, S.-W., Colella, W., & Prinz, F. B. (2009). *Fuel cell fundamentals* (2nd ed.). Hoboken, New Jersey: John Wiley & Sons, Inc.

Oevermann, M., & Klein, R. (2006). A Cartesian grid finite volume method for elliptic equations with variable coefficients and embedded interfaces. *Journal of Computational Physics*, *219*(2), 749–769. doi:10.1016/j.jcp.2006.04.010

Orlo, J., Utlaut, M., & Swanson, L. (2002). *High resolution focused ion beams: FIB and applications*. Kluwer Academic/Plenum Publishers.

Ottman, G. K., Hofmann, H. F., Bhatt, A. C., & Lesieutre, G. A. (2002). Adaptive piezoelectric energy harvesting circuit for wireless remote power supply. *Power Electronics. IEEE Transactions on*, *17*(5), 669–676.

Ouyang, W., Wei, W., Zhang, H., Wu, W., & Li, Z. (2013, January). *A low-cost, high-efficiency and high output-power nanofluidic energy harvester*. Paper presented at 2013 IEEE 26th International Confernece on Micro Electro Mechanical Systems (MEMS). Taipei. doi:10.1109/MEMSYS.2013.6474367

Ouyang, W., Wei, W., Zhang, H., Wu, W., & Li, Z. (2013). Nanofluidic crystal: A facile, high efficiency and high power density scaling up scheme for energy harvesting based on nanofluidic reverse electrodialysis. *Nanotechnology*, *24*(34), 345401. doi:10.1088/0957-4484/24/34/345401 PMID:23899953

Ovejas, V. J., & Cuadras, A. (2011). Multimodal piezoelectric wind energy harvesters. *Smart Materials and Structures*, *20*(8), 085030. doi:10.1088/0964-1726/20/8/085030

Paing, T., Morroni, J., Dolgov, A., Shin, J., Brannan, J., Zane, R., & Papovic, Z. (2007). Wirelessly powered wireless sensor platform.*European Microwave Conference*, (pp. 999-1002).

Palik, E. D. (1997). *Handbook of optical constants of solids*. San Diego: Academic Press.

Panqar, N. L., Kaushik, S. C., & Kothari, S. (2011). Role of renewable energy sources in environmental protection: A review. *Renewable & Sustainable Energy Reviews*, *15*(3), 1513–1524. doi:10.1016/j.rser.2010.11.037

Papovic, Z. B. (2006, November). Wireless powering for low-power distributed sensors. *Serbian Journal of Electrical Engineering*, *3*(2), 149–162. doi:10.2298/SJEE0603149P

Pattle, R. E. (1954). Production of electric power by mixing fresh and salt water in the hydroelectric pile. *Nature*, *174*(4431), 660. doi:10.1038/174660a0

Pellegrini, S. P., Tolou, N., Schenk, M., & Herder, J. L. (2013). Bistable vibration energy harvesters: A review. *Journal of Intelligent Material Systems and Structures*, *24*(11), 1303–1312. doi:10.1177/1045389X12444940

Peng, K., Hu, J., Yan, Y., Wu, Y., Fang, H., Xu, Y., & Zhu, J. et al. (2006). Fabrication of single-crystalline silicon nanowires by scratching a silicon surface with catalytic metal particles. *Journal of Advanced Functional Materials*, *16*(3), 387–394. doi:10.1002/adfm.200500392

Perera, K. C. R., Sampath, B.G., Dassanayake, V.P.C. & Hapuwatte, B. M. (2014). Harvesting of kinetic energy of the raindrops. *International Journal of Mathematical, Computational, Physical and Quantum Engineering, 8*(2).

PhysOrg.com. (2011). *Energy harvesting skin generates power from air conditioners*. Retrieved May 21, 2014, from http://phys.org/news/2011-04-energy-harvesting-skin-power-air.html

Pimpin, A., & Srituravanich, W. (2012). Review on micro- and nanolithography techniques and their applications. *English Journal*, *16*(1), 37–55.

Pinna, L., Dahiya, R., & Valle, M. (2009). *SPICE model for piezoelectric bender generators*. Paper presented at 16th IEEE International Conference on Electronics, Circuits, and Systems, (ICECS). Hammamet, Tunisia.

Pirola, V. (2003). *Cinesiologia*. Milano, Italy: Edi.Ermes Editors.

Planck, M. (1901). Über das Gesetz der Energieverteilung im Normalspektrum. *Annalen der Physik*, *4*(3), 553–558. doi:10.1002/andp.19013090310

Poinern, G. E., Ali, N., & Fawcett, D. (2011). Progress in nano-engineered anodic aluminum oxide membrane development. *Materals*, *4*(3), 487–526. doi:10.3390/ma4030487

Pop, V., van de Molengraft, J., Schnitzler, F., Penders, J., van Schaijk, R., & Vullers, R. (2008). Power optimization for wireless autonomous transducer solutions. In Proceedings of PowerMEMS (pp. 141-144).

Poulin, G. (2004). *Contribution to the development of a piezoelectric generator for mobile applications* (Ph.D. Thesis). University Paris XI Orsay, France.

Pozzi, M., & Zhu, M. (2011). *Pizzicato excitation for wearable energy harvesters. Solar & Alternative Energy*. SPIE Newsroom.

Pozzi, M., & Zhu, M. (2012). Characterization of a rotary piezoelectric energy harvester based on plucking excitation for knee-joint wearable applications. *Smart Materials and Structures*, *21*(5), 055004. doi:10.1088/0964-1726/21/5/055004

Prasad, T., Colvin, V. L., & Mittleman, D. M. (2008). Dependence of guided resonances on the structural parameters of terahertz photonic crystal slabs. *Journal of Applied Physics*, *25*, 633–644.

Priya, S. (2007). Advances in energy harvesting using low profile piezoelectric transducers. *Journal of Electroceramics*, *19*(1), 167–182. doi:10.1007/s10832-007-9043-4

Probstein, R. F. (1994). *Physicochemical hydrodynamics*. New York: Wiley. doi:10.1002/0471725137

Putson, C., Lebrun, L., Guyomar, D., Muensit, N., Cottinet, P. J., Seveyrat, L., & Guiffard, B. (2011). Effects of copper filler sizes on the dielectric properties and the energy harvesting capability of nonpercolated polyurethane composites. *Journal of Applied Physics*, *109*(2), 024104. doi:10.1063/1.3534000

Qian, F., Baum, M., Gu, Q., & Morse, F. D. (2009). A 1.5 µL microbial fuel cell for on-chip bioelectricity generation. *Lab on a Chip*, *9*(21), 3076–3081. doi:10.1039/b910586g PMID:19823722

Qi, S., Shuttleworth, R., Olutunde Oyadiji, S., & Wright, J. (2010). Design of a multiresonant beam for broadband piezoelectric energy harvesting. *Smart Materials and Structures*, *19*(9), 094009. doi:10.1088/0964-1726/19/9/094009

Qi, Y., Jafferis, N. T., Lyons, K. Jr, Lee, C. M., Ahmad, H., & McAlpine, M. C. (2010). Piezoelectric ribbons printed onto rubber for flexible energy conversion. *Nano Letters*, *10*(2), 524–528. doi:10.1021/nl903377u PMID:20102189

Qi, Y., Kim, J., Nguyen, T. D., Lisko, B., Purohit, P. K., & McAlpine, M. C. (2011). Enhanced piezoelectricity and stretchability in energy harvesting devices fabricated from buckled PZT ribbons. *Nano Letters*, *11*(3), 1331–1336. doi:10.1021/nl104412b PMID:21322604

Quinn, D., Triplett, L., Vakakis, D., & Bergman, L. (2011). Comparing linear and essentially nonlinear vibration-based energy harvesting. *Journal of Vibration and Acoustics*, *133*(1), 011001. doi:10.1115/1.4002782

Rai-Choudhury, P. (1997). *SPIE handbook of microlithography, micromachining and microfabrication*. Washington: SPIE Press.

Rakic, A. D., Djurisic, A. B., Elazar, J. M., & Majewski, M. L. (1998). Optical properties of metallic films for vertical-cavity optoelectronic devices. *Applied Optics*, *37*(22), 5271–5283. doi:10.1364/AO.37.005271 PMID:18286006

Ramadan, K. S., Sameoto, D., & Evoy, S. (2014). A review of piezoelectric polymers as functional materials for electromechanical transducers. *Smart Materials and Structures*, *23*(3), 033001. doi:10.1088/0964-1726/23/3/033001

Ramadass, Y. K., & Chandrakasan, A. P. (2010). An efficient piezoelectric energy harvesting interface circuit using a bias-flip rectifier and shared inductor. *Journal of Solid-State Circuits*, *45*(1), 189–204. doi:10.1109/JSSC.2009.2034442

Ramanandraibe, E., Latrach, M., Abdouni, W., & Sharaiha, A. (2013, Sept). A half loop antenna associated with one SRR cell. *International Conference on Electromagnetics in Advanced applications*, (pp. 1442-1445). Torino, Italy. doi:10.1109/ICEAA.2013.6632488

Rashid, M. H. (2007). *Power Electronics Handbook* (2nd ed.). Burlington, MA: Butterworth-Heinemann.

Raytheon Company. (1977, Aug). *Electronic and mechanical improvement of the receiving terminal of a free-space microwave power transmission system* (Technical Report, PT – 4964, NASA Rep. CR-135194). Wayland MA: Raytheon Company.

Redweik, P., Catita, C., & Brito, M. (2013). Solar energy potential on roofs and facades in an urban landscape. *Solar Energy*, *97*, 332–341. doi:10.1016/j.solener.2013.08.036

REN21. (2011). *Renewables 2011 Global Status Report*. Retrieved November 2012 from http://www.ren21.net

REN21. (2012). *Renewables 2012 Global Status Report*. Retrieved November 2012 from http://www.ren21.net

Renaud, M., Karakaya, K., Sterken, T., Fiorini, P., Hoof, C. V., & Puers, R. (2008). Fabrication, modelling and characterization of MEMS piezoelectric vibration harvesters. *Sensors and Actuators. A, Physical, 145-146*, 380–386. doi:10.1016/j.sna.2007.11.005

Renaud, M., Sterken, T., Fiorini, P., Puers, R., Baert, K., & van Hoof, C. (2005). Scavenging energy form human body, design of a piezoelectric transducer. In *Proceedings of 13th International Conference on Solid-State Sensors, Actuators and Microsystems* (pp. 784-787). Seoul- Korea.

Renaud, M., Sterken, T., Schmitz, A., Fiorini, P., Van Hoof, C., & Puers, R., R. (2007). Piezoelectric harvesters and MEMS technology: fabrication, modeling and measurements. In *Proceeding of International Conference of Solid State Sensors, Actuators and Microsystems*. New York: IEEE. doi:10.1109/SENSOR.2007.4300274

Ren, K., Liu, Y., Hofmann, H., Zhang, Q. M., & Blottman, J. (2007). An active energy harvesting scheme with an electroactive polymer. *Applied Physics Letters*, *91*(13), 132910. doi:10.1063/1.2793172

Report, S. (2006). *Science of Sustainability 2006 - A view from Japan*. Retrieved 1st November 2012 from http://www.sos2006.jp

Revil, A., Pezard, P. A., & Glover, P. W. J. (1999). Streaming potential in porous media 1. Theory of the zeta potential. *Journal of Geophysical Research*, *104*(B9), 20021–20031. doi:10.1029/1999JB900089

Rica, R. A., Ziano, R., Salerno, D., Mantegazza, F., van Roij, R., & Brogioli, D. (2013). Capacitive mixing for harvesting the free energy of solutions at different concentrations. *Entropy*, *15*(4), 1388–1407. doi:10.3390/e15041388

Riemer, R., & Shapiro, A. (2011). Biomechanical energy harvesting from human motion: Theory, state of the art, design guidelines, and future directions. *Journal of Neuroengineering and Rehabilitation*, *8*. PMID:21521509

Roberg, M., Reveyrand, T., Ramos, I., Falkenstein, E. A., & Papović, Z. (2012, December). High-Efficiency Harmonically Terminated Diode and Transistor Rectifiers. *IEEE Transactions on Microwave Theory and Techniques*, *60*(12), 4043–4052. doi:10.1109/TMTT.2012.2222919

Rosa, M., & De Marqui, C. (2014). Modeling and analysis of a piezoelectric energy harvester with varying cross-sectional area. *Journal of Shock and Vibration*, *2014*, 1–9. doi:10.1155/2014/930503

Rosenberg, R., & Slater, M. (1999). The chording glove: A glove-based text input device. *IEEE Transactions on Systems, Man, and Cybernetics - Part C*, *29*(2), 186–191. doi:10.1109/5326.760563

Roundy, S., & Wright, P. K. (2004). A piezoelectric vibration based generator for wireless electronics. *Smart Materials and Structures*, *22*(13), 1131–1142.

Roundy, S., Wright, P. K., & Rabaey, J. M. (2003). *Energy scavenging for wireless sensor networks*.

Roundy, S., Leland, E. S., Baker, J., Carleton, E., Reilly, E., Lai, E., & Sundararajan, V. et al. (2005). Improving power output for vibration-based energy scavengers. *Pervasive Computing, IEEE*, *4*(1), 28–36. doi:10.1109/MPRV.2005.14

Roundy, S., & Wright, P. K. (2004). A piezoelectric vibration based generator for wireless electronics. *Smart Materials and Structures*, *13*(5), 1131–1142. doi:10.1088/0964-1726/13/5/018

Roundy, S., & Wright, P. K. (2005). A piezoelectric vibration based generator for wireless electronics. *Journal of Intelligent Material Systems and Structures*, *16*, 809–823. doi:10.1177/1045389X05054042

Roundy, S., Wright, P. K., & Rabaey, J. (2003). A study of low level vibrations as a power source for wireless sensor nodes. *Computer Communications*, *26*(11), 1131–1144. doi:10.1016/S0140-3664(02)00248-7

Rowe, D. (Ed.). (2010). *Thermoelectrics handbook: macro to nano*. New York: CRC Press, Taylor & Francis.

Rowe, D. M. (2006). *Thermoelectrics Handbook, Macro to Nano*. Boca Raton, FL: CRC Press, Taylor & Francis Group.

Ruan, X. L., & Kaviany, M. (2005). Enhanced nonradiative relaxation and photoluminescence quenching in random, doped nanocrystalline powders. *Journal of Applied Physics, 97*(10), 104331–104335. doi:10.1063/1.1900937

Ruan, X. L., & Kaviany, M. (2006). Enhanced laser cooling of rare-earth-ion-doped nanocrystalline powders. *Journal of Physical Review B, 73*(15), 155422-1–15. doi:10.1103/PhysRevB.73.155422

Rupp, C. J., Evgrafov, A., Maute, K., & Dunn, M. L. (2009). Design of piezoelectric energy harvesting systems: A topology optimization approach based on multilayer plates and shells. *Journal of Intelligent Material Systems and Structures, 20*(16), 1923–1939. doi:10.1177/1045389X09341200

Russel, K. (2011). *A Hybrid Thermoelectric Cooler Thermal Management System for Electronic Packaging* (MSc dissertation). McMaster University, Canada.

Russell, P. (2003). Photonic crystal fibers. *Journal of Science, 299*(5605), 358–362. http://www.fgglass.com/light_and_thermal_performance.html doi:10.1126/science.1079280 PMID:12532007

Sabaawi, A. M., Tsimenidis, C., & Sharif, B. S. (2013). *Overview of nanoantennas for solar rectennas*. Retrieved from http://link.springer.com/chapter/10.1007%2F978-1-4614-3716-1_11

Sabaawi, A. M. A., Tsimenidis, C. C., & Sharif, B. S. (2013). Analysis and modeling of infrared solar rectennas. *IEEE Journal of Selected Topics in Quantum Electronics, 19*(3), 9000208. doi:10.1109/JSTQE.2012.2227686

Sakamoto, W. K., Souza, E., & Das-Gupta, D. K. (2011). Electroactive properties of flexible piezoelectric composites. *Materials Research, 4*(3), 201–2014. doi:10.1590/S1516-14392001000300010

Schissel, P., Kennedy, C., & Goggin, R. (1995). Role of inorganic oxide interlayers in improving the adhesion of sputtered silver film on PMMA. *Journal of Adhesion Science and Technology, 9*(4), 413–424. doi:10.1163/156856195X00356

Schoch, R. B., & Renaud, P. (2005). Ion transport through nanoslits dominated by the effective surface charge. *Applied Physics Letters, 86*(25), 253111. doi:10.1063/1.1954899

Schock, H., Caillet, T., Case, E., Fleurial, J. P., Hogan, T., Lyle, M., & Zhu, G. (Jan 2011). *Thermoelectric conversion of waste heat to electricity in an IC engine powered vehicle*. US Department of Energy, U.S. Department Energy Efficiency & Renewable Energy (EERE). Retrieved November 2012 from http://www.eere.energy.gov

Schulz, K. W., & Kallinderis, Y. (1998). Unsteady flow structure interaction for incompressible flows using deformable hybrid grids. *Journal of Computational Physics, 143*(2), 569–597. doi:10.1006/jcph.1998.5969

Seisyan, R. P. (2011). Nanolithography in microelectronics: A review. *Technical Physics, 56*(8), 1–14. doi:10.1134/S1063784211080214

Selit. (2012). *Selitherm Thermal Insulation Board*. Info Sheet. Retrieved November 2012 from http://www.selit.com

Semenov, A. D., Richter, H., Hübers, H. W., Günther, B., Smirnov, A., Il'in, K. S., & Karamarkovic, J. P. et al. (2007). Terahertz performance of integrated lens antennas with a hot-electron bolometer. *IEEE Transactions on Microwave Theory and Techniques, 55*(2), 239–247. doi:10.1109/TMTT.2006.889153

Shahruz, S. M. (2006). Design of mechanical band-pass filters for energy scavenging. *Journal of Sound and Vibration, 292*(3-5), 987–998. doi:10.1016/j.jsv.2005.08.018

Shenck, N. S., & Paradiso, J. A. (2001). Energy scavenging with shoe-mounted piezoelectrics. *IEEE Micro, 21*(3), 30–42. doi:10.1109/40.928763

Shen, D., Park, J.-H., Ajitsaria, J., Choe, S.-Y., Wikle, H. C. III, & Kim, D.-J. (2008). The design, fabrication and evaluation of a MEMS PZT cantilever with an integrated Si proof mass for vibration energy harvesting. *Journal of Micromechanics and Microengineering, 18*(5), 055017. doi:10.1088/0960-1317/18/5/055017

Shi, L., & Chew, M. Y. L. (2012). A review on sustainable design of renewable energy systems. *Renewable & Sustainable Energy Reviews, 16*(1), 192–207. doi:10.1016/j.rser.2011.07.147

Shin, H., & Song, J. T. (2010). Piezoelectric coefficient measurement of AlN thin films at the nanometer scale using piezoresponse force microscopy. *Journal of the Korean Physical Society*, *56*(2), 580–585. doi:10.3938/jkps.56.580

Shinohara, N., & Matsumoto, H. (1998, March). Experimental study of large rectenna array for microwave energy transmisison. *IEEE Transactions on Microwave Theory and Techniques*, *46*(3), 261–267. doi:10.1109/22.661713

Shrestha, S., Noh, S. K., & Choi, D. Y. (2013). Comparative study of antenna designs for RF energy harvesting. *International Journal of Antennas and Propagation*, *2013*, 1–7. doi:10.1155/2013/385260

Shu, Y. C., & Lien, I. C. (2006). Analysis of power output for piezoelectric energy harvesting systems. *Smart Materials and Structures*, *15*(6), 1499–1512. doi:10.1088/0964-1726/15/6/001

Sirohi, J., & Mahadik, M. (2011). Piezoelectric wind energy harvester for low-power sensors. *Journal of Intelligent Material Systems and Structures*, *22*(18), 2215–2228. doi:10.1177/1045389X11428366

Siwy, Z., Dobrev, D., Neumann, R., Trautmann, C., & Voss, K. (2003). Electro-responsive asymmetric nanopores in polyimide with stable ion-current signal. *Applied Physics. A, Materials Science & Processing*, *76*(5), 781–785. doi:10.1007/s00339-002-1982-7

Skop, R. A., & Balasubramanian, S. (1997). A new twist on an old model for vortex-excited vibrations. *Journal of Fluids and Structures*, *11*(4), 395–412. doi:10.1006/jfls.1997.0085

Skop, R. A., & Griffin, O. M. (1975). On a theory for the vortex-exited oscillations of flexible cylindrical structures. *Journal of Sound and Vibration*, *41*(3), 263–274. doi:10.1016/S0022-460X(75)80173-8

Snyder, J. (2008). Small thermoelectric generators. *The Electrochemical Society Interface*, *17*(3), 54-56. Retrieved from http://www.electrochem.org/dl/interface/fal/fal08/fal08_p54-56.pdf

Sodano, H. A., Park, G., & Inman, D. (2004). Estimation of electric charge output for piezoelectric energy harvesting. *Strain*, *40*(2), 49–58. doi:10.1111/j.1475-1305.2004.00120.x

Sodano, H., Park, G., & Inman, D. J. (2004). A review of power harvesting from vibration using piezoelectric materials. *Journal of Shock Vibration Digest*, *36*(3), 197–205. doi:10.1177/0583102404043275

Somà, A., De Pasquale, G., & Fraccarollo, F. (2011). *Ski- or trekking-racket equipped with magnetic-inductive harvester device for electric generation.* PCT Patent Number PCT/IT1012/000280.

Song, H. J., Choi, Y.-T., Purekar, A. S., & Wereley, N. M. (2009). Performance evaluation of multi-tier energy harvesters using macro-fiber composite patches. *Journal of Intelligent Material Systems and Structures*, *20*(17), 2077–2088. doi:10.1177/1045389X09347017

Song, H. J., Choi, Y.-T., Wereley, N. M., & Purekar, A. S. (2010). Energy harvesting devices using macro-fiber composite materials. *Journal of Intelligent Material Systems and Structures*, *21*(6), 647–658. doi:10.1177/1045389X10361633

Sousa, V. C., de M Anicézio, M., De Marqui, C. Jr, & Erturk, A. (2011). Enhanced aeroelastic energy harvesting by exploiting combined nonlinearities: Theory and experiment. *Smart Materials and Structures*, *20*(9), 094007. doi:10.1088/0964-1726/20/9/094007

Standards Committee of the IEEE Ultrasonics, Ferroelectrics, & Frequency Control Society. (1988). IEEE Standard on Piezoelectricity: An American National Standard. New York: Institute of Electrical and Electronics Engineers, Inc.

Stanton, S. C., McGehee, C. C., & Mann, B. P. (2009). Reversible hysteresis for broadband magnetopiezoelastic energy harvesting. *Applied Physics Letters*, *95*(17), 174103. doi:10.1063/1.3253710

Stanton, S. C., McGehee, C. C., & Mann, B. P. (2010). Nonlinear dynamics for broadband energy harvesting: Investigation of a bistable piezoelectric inertial generator. *Physica D. Nonlinear Phenomena*, *239*(10), 640–653. doi:10.1016/j.physd.2010.01.019

Stark, I. (2006). Thermal energy harvesting with Thermo Life. In *Proceedings of the International Workshop on Wearable and Implantable Body Sensor Networks*. doi:10.1109/BSN.2006.37

Steffan, R., Schull, U., & Kuhlen, T. (1998). Integration of virtual reality based assembly simulation into CAD/CAM environments. In *Proceedings of Annual Conference of the IEEE Industrial Electronics Society* (pp. 2535-2537). doi:10.1109/IECON.1998.724128

Stokes, J. L., Yu, Y., Yuan, Z. H., Pugh, J. R., Lopez-Garcia, M., Ahmad, N., & Cryan, M. J. (2014). Analysis and design of a cross dipole nanoantenna for fluorescence-sensing applications. *Journal of the Optical Society of America. B, Optical Physics*, *31*(2), 302–310. doi:10.1364/JOSAB.31.000302

Street, R. A., Wong, W. S., & Paulson, C. (2009). Analytic model for diffuse reflectivity of silicon nanowire. *Journal of Nano Letters*, *9*(10), 3494–3497. doi:10.1021/nl901683y PMID:19588905

Sturman, D. J., & Zeltzer, D. (1994). A survey of glove-based input. *IEEE Computer Graphics and Applications*, *14*(1), 30–39. doi:10.1109/38.250916

Su, J., Vullers, R. J. M., Goedbloed, M., van Andel, Y., Leonov, V., & Wang, Z. (2010). Thermoelectric energy harvester fabricated by Stepper. *Microelectronic Engineering*, *87*(5-8), 1242–1244. doi:10.1016/j.mee.2009.11.135

Su, J., Zhang, Q., Wang, P.-C., MacDiarmid, A. G., & Wynne, K. J. (1998). Preparation and characterization of electrostrictive polyurethane films with conductive polymer electrodes. *Polymers for Advanced Technologies*, *9*(6), 317–321. doi:10.1002/(SICI)1099-1581(199806)9:6<317::AID-PAT758>3.0.CO;2-Y

Sundaramurthy, A., Crozier, K., Kino, G., Fromm, D., Schuck, P., & Moerner, W., (2005). Field enhancement and gap-dependent resonance in a system of two opposing tip-to-tip Au nano triangles. *Journal of Physical Review B, 72*, 165409 1-6.

Taflove, A., & Hagness, S. (2005). *Computational Electrodynamics: The Finite Difference Time-Domain Method* (3rd ed.). London: Artech House.

Takahashi, T., & Kishino, F. (1991). Hand gesture coding based on experiments using a hand gesture interface device. *SIGCHI Bullettin*, *23*(2), 67–74. doi:10.1145/122488.122499

Takhedmit, H., Merabet, B., Cirio, L., Allard, B., Costa, F., Vollaire, C., & Picon, O. (2010). A 2.45-ghz low cost and efficient rectenna. *The Fourth European Conference Antennas and Propagation (EuCAP)*, (pp. 1-5). Barcelona, Spain.

Tang, L. H., & Yang, Y. (2012). A nonlinear piezoelectric energy harvester with magnetic oscillator. *Applied Physics Letters*, *101*(9), 094102. doi:10.1063/1.4748794

Tani, B., Maia, R. F., & von Wangenheim, A. (2007). A gesture interface for radiological workstation. In *Proceedings of International Symposium on Computer-Based Medical Systems* (pp. 27-32).

Taranovich, S. (2012). The Rover's Energy. *EDN*. Retrieved from http://www.edn.com/electronics-blogs/mission-to-mars--nasa-engineering-and-the-red-planet-/4392098/The-rover-s-energy

Techet, A. H., Allen, J. J., & Smits, A. J. (2002). Piezoelectric eels for energy harvesting in the ocean. In *Proceedings of The Twelfth (2002) International Offshore and Polar Engineering Conference Kitakyushu*, Japan: ISOPE.

Technology, L. (2014) *LTC 3105 Step-Up DC/DC Converter*. Retrieved September 2014 from http://www.linear.com/product/LTC3105, 2011

Teng, X. F., Zheng, Y. T., Poon, C. C., & Bonato, P. (2008). Wearable medical systems for p-health. *IEEE Reviews in Biomedical Engineering*, *1*, 62–74. doi:10.1109/RBME.2008.2008248 PMID:22274900

Theodorsen, T. (1935). *General theory of aerodynamic instability and the mechanism of flutter* (Technical Report NACA 496).

Thums, M., Whiting, S. D., Reisser, J. W., McMahon, C. R., & Meekan, M. G. (2013). Tracking sea turtle hatchlings - A pilot study using acoustic telemetry. *Journal of Experimental Marine Biology and Ecology*, *440*, 156–163. doi:10.1016/j.jembe.2012.12.006

Tian, B., Zheng, X., Kempa, T. J., Fang, Y., Yu, N., Yu, G., & Lieber, C. M. et al. (2007). Coaxial silicon nanowires as solar cells and nanoelectronic power sources. *Journal of Nature*, *449*(7164), 885–889. doi:10.1038/nature06181 PMID:17943126

Tikhodeev, S. G., Yablonskii, A. L., Muljarov, E. A., Gippius, N. A., & Ishihara, T. (2002). Quasi guided modes and optical properties of photonic crystal slabs. *Journal of Physical Review, 66*.

Tiwari, G. N., & Dubey, S. (2010). *Fundamentals of photovoltaic modules and their applications*. RSC Publishing.

Tomasiello, S. (1998). Differential quadrature method: Application to initial-boundary-value problems. *Journal of Sound and Vibration, 218*(4), 573–585. doi:10.1006/jsvi.1998.1833

Tran, K., Polisena, J., Coyle, D., Coyle, K., Kluge, E., Cimon, K., . . . Scott, R. (2008). Home telehealth for chronic disease management (Technology report n. 113). Ottawa, Canada: Canadian Agency for Drugs and Technologies in Health.

Trenberth, K. E., Fasullo, J., & Kiehl, J. (2009). Earth's global energy budget. *Bulletin of the American Meteorological Society, 90*(3), 311–323. doi:10.1175/2008BAMS2634.1

Tsakalakos, L. (2008). Nanostructures for photovoltaics. *Journal of Materials Science and Engineering, 62*, 175–189.

Tumbleston, J. R., Ko, D. H., Samulski, E. T., & Lopez, R. (2009). Absorption and quasi guided mode analysis of organic solar cells with photonic crystal photoactive layers. *Journal of Optics Express, 17*(9), 7670–7681. doi:10.1364/OE.17.007670 PMID:19399146

Turek, M., & Bandura, B. (2007). Renewable energy by reverse electrodialysis. *Desalination, 205*(1-3), 67–74. doi:10.1016/j.desal.2006.04.041

Tyagi, V. V., Rahim, A. A., Rahim, N. A., & Selvaraj, L. (2013). Progress in solar PV technology: Research and achievement. *Renewable & Sustainable Energy Reviews, 20*, 443–461. doi:10.1016/j.rser.2012.09.028

U. S. Energy Information Administration (EIA). (2013). *Annual energy outlook 2013*. Retrieved from http:www.eia.gov/forecasts/aeo/pdf/0383(2013).pdf

Umeda, T., Toshida, H., Sekine, S., Fujita, Y., Suzuki, T., & Otaka, S. (2006, January). A 950-MHz rectifier circuit for sensor network tags with 10-m distance. *IEEE Journal of Solid-State Circuits, 41*(1), 35–41. doi:10.1109/JSSC.2005.858620

Ungan, T., Le Polozec, X., Walker, W., & Reindl, L. (2009, Sept). RF energy harvesting design using high Q resonators.*IEEE MTT-S International Microwave Workshop on Wireless Sensing, Local Positioning, and RFID*, 1-4. doi:10.1109/IMWS2.2009.5307869

Van den Ende, D. A., van de Wiel, H. J., Groen, W. A., & van der Zwaag, S. (2012). Direct strain energy harvesting in automobile tires using piezoelectric PZT–polymer composites. *Smart Materials and Structures, 21*(1), 015011. doi:10.1088/0964-1726/21/1/015011

van Dorp, W. F., & Hagen, C. W. (2008). A critical literature review of focused electron beam induced deposition. *Journal of Applied Physics, 104*(8), 081301. doi:10.1063/1.2977587

Vandenbosch, G. A. E., & Ma, Z. (2012). Upper bounds for the solar energy harvesting efficiency of nano-antennas. *Journal of Nano Energy, 1*(3), 494–502. doi:10.1016/j.nanoen.2012.03.002

Vasconcellos, R., & Abdelkefi, A. (2014). (in press). Nonlinear dynamical analysis of an aeroelastic system with multi-segmented moment in the pitch degree-of-freedom. *Communications in Nonlinear Science and Numerical Simulation*.

Vasconcellos, R., Abdelkefi, A., Marques, F. D., & Hajj, M. R. (2012). Representation and analysis of control surface freeplay nonlinearity. *Journal of Fluids and Structures, 31*, 79–91. doi:10.1016/j.jfluidstructs.2012.02.003

Vatansever, D., Hadimani, R. L., Shah, T., & Siores, E. (2011). An investigation of energy harvesting from renewable sources with PVDF and PZT. *Smart Materials and Structures, 20*(5), 055019. doi:10.1088/0964-1726/20/5/055019

Veerman, J., de Jong, R. M., Saakes, M., Metz, S. J., & Harmsen, G. J. (2009a). Reverse electrodialysis: Comparison of six commercial membrane pairs on the thermodynamic efficiency and power density. *Journal of Membrane Science, 343*(1-2), 7–15. doi:10.1016/j.memsci.2009.05.047

Veerman, J., Saakes, M., Metz, S. J., & Harmsen, G. J. (2009b). Reverse electrodialysis: Performance of a stack with 50 cells on the mixing of sea and river water. *Journal of Membrane Science, 327*(1-2), 136–144. doi:10.1016/j.memsci.2008.11.015

Veerman, J., Saakes, M., Metz, S. J., & Harmsen, G. J. (2011). Reverse electrodialysis: A validated process model for design and optimization. *Chemical Engineering Journal, 166*(1), 256–268. doi:10.1016/j.cej.2010.10.071

Vermaas, D. A. (2014). *Energy generation from mixing salt water and fresh water: Smart flow strategies for reverse electrodialysis.* (Unpublished doctoral dissertation). University of Twente.

Viola, F., Romano, P., & Miceli, R., Acciari, G., & Spataro, C. (2014). *Piezoelectric model of rainfall energy harvester.* Paper presented at 2014 Ninth International Conference on Ecological Vehicles and Renewable Energies (EVER), Monaco.

Viola, F., Romano, P., Miceli, R., & Acciari, G. (2013). *Harvesting rainfall energy by means of piezoelectric transducer.* Paper presented at International Conference on CLEAN ELECTRICAL POWER Renewable Energy Resources Impact 2013, Alghero, Italy doi:10.1109/ICCEP.2013.6586952

Violette, R., de Langre, E., & Szydlowski, J. (2007). Computation of vortex-induced vibrations of long structures using a wake oscillator model: Comparison with DNS and experiments. *Computers & Structures, 85*(11-14), 1134–1141. doi:10.1016/j.compstruc.2006.08.005

Walsh, C., Rondineau, S., Jankovic, M., Zhao, G., & Papovic, Z. (2005). A conformal 10ghz rectenna for wireless powering of piezoelectric sensor electronics.*IEEE MTT-S International Microwave Symposium Digest*, 4-12.

Wang, Z. (2010). Prospectives for China's solar thermal power technology development. *Energy, 35*(11), 4417–4420. doi:10.1016/j.energy.2009.04.004

Weber-Bargioni, A., Schwartzberg, A., Schmidt, M., Harteneck, B., Ogletree, D. F., Schuck, P. J., & Cabrini, S. (2010). Functional plasmonic antenna scanning probes fabricated by induced-deposition mask lithography. *Nanotechnology, 21*(6), 065306. doi:10.1088/0957-4484/21/6/065306 PMID:20061594

Wong, C.-H., Dahari, Z., Abd Manaf, A., Sidek, O., Miskam, M.A., & Mohamed, J. J. (2013). Simulation of piezoelectric raindrop energy harvester. In TENCON Spring Conference, 2013 IEEE, (pp. 465 – 469). doi: 10.1109/TENCONSpring.2013.6584488

Wong, C. C., Agarwai, A., Balasubramanian, N., & Kwong, D. L. (2007). Fabrication of self-sealed circular nano/microfluidic channels in glass substrates. *Nanotechnology, 18*(13), 135304. doi:10.1088/0957-4484/18/13/135304 PMID:21730378

World Health Organization. (1999). *What are electromagnetic fields? Technical report.* WHO Regional Office for Europe.

Xiang, H. J., Wang, J. J., Shi, Z. F., & Zhang, Z. W. (2013). Theoretical analysis of piezoelectric energy harvesting from traffic induced deformation of pavements. *Smart Materials and Structures, 22*(9), 095024. doi:10.1088/0964-1726/22/9/095024

Xie, J., Lee, C., & Feng, H. (2010). Design, fabrication, and characterization of CMOS MEMS-Based thermoelectric power generators. *Journal of Microelectromechanical Systems, 19*(2), 317–324. doi:10.1109/JMEMS.2010.2041035

Xie, L., & Du, R. (2012). Harvest human kinetic energy to power portable electronics. *Journal of Mechanical Science and Technology, 26*(7), 2005–2008. doi:10.1007/s12206-012-0503-7

Xu, T., & Li, G. (2011a). Cyclic stress-strain behavior of shape memory polymer based syntactic foam programmed by 2-D stress condition. *Polymer, 52*(20), 4571–4580. doi:10.1016/j.polymer.2011.08.005

Xu, T., & Li, G. (2011b). A shape memory polymer based syntactic foam with negative Poisson's ratio. *Materials Science and Engineering A-structural Materials, 528*(22-23), 6804–6811. doi:10.1016/j.msea.2011.05.034

Yang, J., Preidikman, S., & Balaras, E. (2008). A strongly coupled, embedded-boundary method for fluid–structure interactions of elastically mounted rigid bodies. *Journal of Fluids and Structures, 24*(2), 167–182. doi:10.1016/j.jfluidstructs.2007.08.002

Yang, Y. W., Zhao, L. Y., & Tang, L. H. (2013). Comparative study of tip cross-sections for efficient galloping energy harvesting. *Applied Physics Letters, 102*(6), 064105. doi:10.1063/1.4792737

Yang, Y., & Tang, L. (2009). Equivalent circuit modeling of piezoelectric energy harvester. *Journal of Intelligent Material Systems and Structures*, *20*(18), 2223–2235. doi:10.1177/1045389X09351757

Yan, Z., & Abdelkefi, A. (2014). Nonlinear characterization of concurrent energy harvesting from galloping and base excitations. *Nonlinear Dynamics*, *77*(4), 1171–1189. doi:10.1007/s11071-014-1369-2

Yeh, H. C., Chang, C. C., & Yang, R. J. (2014). Reverse electrodialysis in conical-shaped nanopores: Salinity gradient-driven power generation. *RSC Advances*, *4*(6), 2705–2714. doi:10.1039/C3RA45392H

Yole Developpment. (2013). BioMEMS: Microsystem Device Market for Healthcare Applications. Author: France.

Yoo, J., Yan, L., Lee, S., Kim, Y., & Yoo, H. J. (2010). A 5.2 mW self-configured wearable body sensor network controller and a 12 μW wirelessly powered sensor for a continuous health monitoring system. *IEEE Journal of Solid-State Circuits*, *45*(1), 178–188. doi:10.1109/JSSC.2009.2034440

Yoon, H. (2013). *Analytical model for electric power prediction of piezoelectric energy harvesting* (Unpublished master's thesis). Seoul National University, South Korea.

Yoon, H., Youn, B. D., & Kim, H. S. (2014). *Analysis of electromechanical performance of energy harvesting skin based on the Kirchhoff plate theory*. Paper presented at the ASME 2014 International Design Engineering Technical Conferences & Computers and Information in Engineering Conference. doi:10.1115/DETC2014-35433

Yoon, H., & Youn, B. D. (2014). Stochastic quantification of the electric power generated by a piezoelectric energy harvester using a time–frequency analysis under non-stationary random vibrations. *Smart Materials and Structures*, *23*(4), 045035. doi:10.1088/0964-1726/23/4/045035

Yoo, T., & Chang, K. (1992, June). Theoretical and experimental development of 10 and 35 GHz rectennas. *IEEE Transactions on Microwave Theory and Techniques*, *40*(6), 1259–1266. doi:10.1109/22.141359

Yoshio, M., Brodd, R. J., & Kozawa, A. (Eds.). (2009). Lithium-ion batteries science and technologies. New York: Springer Science + Business Media. doi:10.1007/978-0-387-34445-4

You, K. M., Park, S. S., Lee, C. S., Kim, J. M., Park, G. P., & Kim, W. N. (2011). Preparation and characterization of conductive carbon nanotube-polyurethane foam composites. *Journal of Materials Science*, *46*(21), 6850–6855. doi:10.1007/s10853-011-5645-y

Youn, B. D., Jung, B. C., Xi, Z., Kim, S. B., & Lee, W. (2011). A hierarchical framework for statistical model calibration in engineering product development. *Computer Methods in Applied Mechanics and Engineering*, *200*(13), 1421–1431. doi:10.1016/j.cma.2010.12.012

Yu, J., & Zhao, H. (2007). A numerical model for thermoelectric generator with the parallel-plate heat exchanger, Science Direct. *Journal of Power Sources*, *172*(1), 428–434. doi:10.1016/j.jpowsour.2007.07.045

Yuse, K., Guiffard, B., Belouadah, R., Petit, L., Seveyrat, L., & Guyomar, D. (2009). Polymer nanocomposites for microactuation and magneto-electric transduction. *Frontiers of Mechanical Engineering in China*, *4*, 350–354.

Zangle, T. A., Mani, A., & Santiago, J. G. (2010). Theory and experiments of concentration polarization and ion focusing at microchannel and nanochannel interfaces. *Chemical Society Reviews*, *39*(3), 1014–1035. doi:10.1039/b902074h PMID:20179822

Zhang, Q. M., & Scheinbeim, J. (2004). Electric EAP. In Y. Bar-Cohen, (Ed.), Electroactive Polymer (EAP) Actuators as Artificial Muscles: Reality, Potential, and Challenges (pp. 95 – 100). Bellingham: SPIE Press.

Zhang, H. L., Baeyens, J., Degrève, J., & Cacères, G. (2013). Concentrated solar power plants: Review and design methodology. *Renewable & Sustainable Energy Reviews*, *22*, 466–481. doi:10.1016/j.rser.2013.01.032

Zhang, S., Neese, B., Ren, K., Chu, B., & Zhang, Q. M. (2006). Microstructure and electromechanical responses in semicrystalline ferroelectric relaxor polymer blends. *Journal of Applied Physics*, *100*(4), 044113. doi:10.1063/1.2335778

Zhao, L. Y., Tang, L. H., & Yang, Y. W. (2013). Comparison of modeling methods and parametric study for a piezoelectric wind energy harvester. *Smart Materials and Structures*, *22*(12), 125003. doi:10.1088/0964-1726/22/12/125003

Zhao, S., & Erturk, A. (2012). Electroelastic modeling and experimental validations of piezoelectric energy harvesting from broadband random vibrations of cantilevered bimorphs. *Smart Materials and Structures*, *22*(1), 015002. doi:10.1088/0964-1726/22/1/015002

Zheng, B., Chang, C. J., & Gea, H. C. (2009). Topology optimization of energy harvesting devices using piezoelectric materials. *Structural and Multidisciplinary Optimization*, *38*(1), 17–23. doi:10.1007/s00158-008-0265-0

Zhou, Y. (2013). *Contribution à la récupération d'énergie ambiante pour les objets communicants autonomes* (PhD dissertation). Université de nantes, France.

# About the Contributors

**Luciano Mescia** received the Master degree in Electronic Engineering in 2000 and the Ph.D. degree in Electromagnetic Fields in 2003. In January 2005, he joined Politecnico of Bari as Assistant Professor. His research interests include theoretical aspects for the development of artificial neural networks, genetic algorithm, and swarm intelligence applied to rare earth doped fiber lasers and amplifiers. Moreover, he is performing studies regarding the design of innovative antenna array for energy harvesting applications, the analysis and synthesis of novel dielectric lens antennas operating in the microwave and millimeter frequency range, the development of novel FDTD schemes based on fractional calculus. He has cooperated with many national and international research institutions and he joined several research projects with academic and industrial partners. His research work has resulted in more than 100 publications in high-impact ISI scientific and engineering peer-reviewed journal, leading international conferences, lectures and invited papers. He is a member of Italian Society of Optics and Photonics (SIOF-EOS) and Italian Society of Electromagnetism (SIEm).

**Onofrio Losito** took his degree in Electronic Engineering, in 2002 from Politecnico di Bari, and in the September 2007, a PhD degree in Information Engineering at the University of Lecce. Since 1999 he has cooperated as an outside expert with high schools, as activities on the techniques of installation for systems of telecommunications and measures of electromagnetic fields focusing the research activity, to the analysis and design of one dimensional and periodic leaky wave antennas, antennas array, microwave filters, and electromagnetic shielding, for RF. Since 2007 he has joined at the ITEL telecommunication s.r.l., as researcher coordinator in the ITEL Laboratory of electromagnetic compatibility. Since 2012 he is a fellow research in the Electromagnetic Fields Group of the Dipartimento di Ingegneria Elettrica e dell'Informazione at Politecnico di Bari. He was executive scientific manager for some national project (PON01_02238, PON02_00576_33297, LECJBT5) and a reviewer of some international journals as IEEE AP. Dr. Losito is a member of IEEE AP-Society, IEEE EMC Society, EurAAP, SIEm and CNIT.

**Francesco Prudenzano**, PhD, was born in Manduria (Italy) in 1964, he graduated in Electronic Engineering from the University of Bari in April 1990. Since 2003 he has been Associate Professor in Electromagnetic Fields. Since 1998 he was responsible of the Laboratories and of Integrated Optics and Technology and since 2003 of Campi Elettromagnetici of Politecnico di Bari. From 2006 to 2012 he was Coordinator of the Teacher Council of the Information Engineering of II FacoltÃ di Ingegneria of Politecnico di Bari. His research activity regards the modeling and the experimental characterization of integrated optical devices, photonic crystal fibers, photonic band gap devices; design and fabrication of lasers and optical amplifiers, in both photonic crystal fibers (PCF) and planar optics, for near (NIR) and medium (MID-IR) infrared wavelength application. He has co-authored over 300 publications, 220 of which published on journals and international conferences, lectures and invited papers. He is involved in several national and international research projects and co-operations.

* * *

**Abdessattar Abdelkefi** is an Assistant Professor in the Department of Mechanical and Aerospace Engineering at New Mexico State University. Prior to this appointment, he was a Postdoctoral Research Fellow at Virginia Tech between August 2012 and July 2014. He earned his Ph.D. in Engineering Mechanics from Virginia Tech in July 2012 and a Master of Science in Computational Mechanics from Tunisia Polytechnic School in collaboration with Virginia Tech in December 2009. His main research interests are in the fields of nonlinear dynamics, energy harvesting, smart structures, renewable energy, vibration and control, aeroelasticity, fluid- structure interactions, and biomimetic locomotion. Over the past 5 years, Dr. Abdelkefi has one book which is a product of his research work in the period between 2010 and 2012. He also has a filed US patent in low-frequency energy harvesting systems with Samsung and one invention disclosure. Dr. Abdelkefi has published more than forty archival journal articles, and more than fifty conference presentations at national and international conferences. During the past few years, Dr. Abdelkefi has served as a reviewer in more than twenty journals, such as, *Journal of Applied Physics, Applied Physics Letters, Smart Materials and Structures, Intelligent Material Systems and Structures, Nonlinear Dynamics, Journal of Sound and Vibration*, etc.

**Nihal F. F. Areed** was born in Mansoura, Egypt, in 1979. She received the B.Sc., M.Sc., and Ph.D. degrees in electrical communications engineering from Mansoura University (MU), Mansoura, Egypt, in 2000, 2003, and 2008, respectively. In 2000, she joined the Department of Electronics and communications Engineering, Faculty of Engineering, MU, Mansoura, Egypt as a research assistant. Since May 2008, she was an Assistant Professor and became an Associate Professor in September 2013. Also, since September 2012, she was with the Centre for Photonics and Smart Materials, Zewail City of Science and Technology, Giza, Egypt. Her current research interests include modeling, design, and numerical simulation of nanophotonic devices. She has 15 journal publications, mostly in applied physics, PIER, IEEE, and 19 conference presentations. Additionally, she has authored one book published by the world-leading scientific publisher Wiley.

**Khalil Benkhouja** has held a doctorate in Solid Chemistry and especially in magneto-chemistry since 1988 from University Louis Pascal Strasbourg, France. Khalil Benkhouja is Research Professor at the sciences Faculty of the Chouaib Doukkali University. His research covers a variety of subjects related to materials: Elaboration of the powder materials, preliminary characterizations and determination of the Physic properties (magnetism, ferroelectrics, batteries and so on) with their Correlation to crystalline structures by fit using a theoretical model. He has 24 years of research experience and important papers in scientific journals and several Congress Communications in the field of solid materials principally the phosphate compounds.

**Pietro Bia** received the Bachelor's and Master's Degree in 2007 and 2010, respectively. In 2008 he received the best Italian geoscience and remote sensing thesis prize form IEEE GRSS Italian Chapter. During 2011 he received a Scholarship for post graduate research activities for the topic: Design of innovative cladding-pumped fiber lasers, nonlinear effects induced by high optical power density and microsphere laser. In 2011 he attended a Master II level course from Università degli studi di Bari on Technologies for space remote sensing. Since 2012 he is a PhD student of Politecnico di Bari and his research activity is focused on the analysis of dielectric lens antennas for microwave application and FDTD modeling for PEF propagation in complex dispersive media. He is a member of Italian Society of Optics and Photonics (SIOF-EOS), Italian Society of Electromagnetism (SIEm) and student member of IEEE.

**Yahia Boughaleb** received his MSc and his PhD in solid state physics from University Paris-Sud, France. He has been a research associate in the department of chemistry and material research Center in Northwestern University USA from 1984 to 1986 and in Ecole Polytechnique, Paris in 1987. He is actually full professor of physics in University Hassan II of Casablanca. He is also since 2006 a member of Hassan II Academy of Science and Technology. He was elected in 2006 vice president of Moroccan society of statistical physics and in 2012 president of Moroccan society of physics of advanced materials and their applications. His main research interests include surface growth, diffusion in solids, fractals and fracture phenomena in composites materials, polymers for energy harvesting and computational physics.

**Mihail Cernaianu** has a Bachelor degree in electronics engineering and finished his PhD in the field of thermoelectric materials for energy harvesting systems in 2013. Since then, he is working at ELI-NP in Romania, where is leading the ELI-NP team dealing with the monitoring and control systems for experimental set-ups and the two-major instruments (the High Power Laser System of 10PW and the Gamma beam system).

**Chulmin Cho** received his B.S. degree from the Department of Mechanical and Aerospace Engineering at Seoul National University, Republic of Korea, in 2012 and the M.S. degree from the Department of Mechanical and Aerospace Engineering at Seoul National University, Republic of Korea, in 2014. He is currently working at Mechatronics R&D Center, Samsung Electronics.

**Giorgio De Pasquale** received his M.S. degree ('Dynamic behavior of microsystems under squeeze film damping') in Mechanical Engineering in 2006 at Politecnico di Torino. From 2006 to 2007 he was designer in the fields of heat generators and injectors for polyurethane deposition. In 2009 he was visiting student at University of South Florida (Tampa, USA); in the same year he received the 'Sapio' Award for research from the President of Italian Deputies' Chamber. In 2010 he received his Ph.D. in Mechanical Engineering ('Mechanical design of microsystems: dynamic and fatigue behavior') at Politecnico di Torino. Until 2011 he was postdoctoral researcher, when he became Assistant Professor; in the same year he received a special mention from the ASME Italian Section. In 2013 he was visiting researcher at MIT (Boston, USA). He participated to national and international financed projects in the fields of MEMS design and testing, energy harvesting and vehicle monitoring systems. He is author of scientific publications in international journals and conferences and he owns national and international patents.

**Jeff DeBord** is a Distinguished Member of Technical Staff at Texas Instruments with over 19 years of experience in silicide and BEOL integration on technologies from 0.25 um to 65 nm. He is currently the Process Integration Manager in the DMOS6 wafer fab in Dallas, Texas. He has four US patents and has authored over 25 publications in the areas of semiconductor development and manufacturing, organometallic synthesis, organically templated transition metal phosphates and organically templated metal oxides. His educational background includes a Ph.D. in Inorganic Chemistry from the University of Nebraska-Lincoln and postdoctoral work at the NEC Research Institute in Princeton, New Jersey.

**Adil Eddiai** received his master's degree in condensed matter physics from Chouaib El Doukkali University in El Jadida, Morocco in 2008 and his Ph.D. at Institut National des Sciences Appliquées (INSA) de Lyon, France and Chouaib El Doukkali University in El Jadida, Morocco in 2013. He is currently Associate Professor at the Faculty of Science Ben M'sik of Casablanca, with research interests concerning electroactive materials characterization, smart structures, actuation performance and environmental energy scavenging.

**Hal Edwards** is a Distinguished Member of Technical Staff at Texas Instruments in Dallas, Texas, where he has held a variety of roles in research, development, and lab management. At present Hal works in TI's Analog Technology Development organization, where he develops high-voltage power transistors for new smart power technologies. He also spends some time developing integrated sensors in smart power technologies, for example the Ambient Light Sensor photodiode in TI's new OPT3001 product. He has over 40 refereed publications and holds over 20 US patents. Prior to joining TI in 1995, Hal completed his Ph. D. in Physics at the University of Texas at Austin, where he held the John Wheeler Graduate Fellowship in Physics and carried out research in superconductivity and quantum devices. He graduated with his Bachelors of Science in Physics with Honors from Stanford University in 1990. In his spare time, Hal volunteers as a Scoutmaster with Boy Scouts of America and spends about 10% of his nights in a tent on campouts with his four sons. He and his wife Susie enjoy traveling with (and without!) their sons and are active in furthering their sons' education through homeschooling.

**Mohamed Farhat** was born in Egypt in 1979. He received the B.Sc. degree in electronics and control engineering and the M.Sc. degree in engineering physics from Mansoura University, Mansoura, Egypt, in 2000 and 2005, respectively, and the Ph.D. degree from a channel scheme between Mansoura University and University of South Wales, U.K., in 2010. He was a recipient of the Best Ph.D. Thesis Prize, and Incentive Award from Mansoura University in 2012 and 2014, respectively. He is currently with Faculty of Engineering, Mansoura University, Egypt. Also, he is with Centre for Photonics and Smart Materials (CPSM), Zewail City of Science and Technology, Egypt. He is involved in computational electromagnetics, with good track records of publications. Based on his research, 43 journal papers, mostly in the IEEE and the IEE journals, have been published, and 40 conference papers have been presented in the best national and international meetings. Additionally, he has authored one book published by the world-leading scientific publisher Wiley.

**Dave Freeman** is a Texas Instruments Fellow and CTO for the High Voltage Power Solutions business unit at TI. Dave has expertise in the areas of battery management ranging from charging to capacity estimation. Dave joined TI in 1999 through the Unitrode acquisition. In the areas of power management, he covers low power DC/DC, high frequency power conversion and digitally controlled power. Other areas of focus for Dave are renewable energy systems and low power energy harvesting. Addition interest includes sensors and analytical methods used to evaluate physical properties of materials. Dave works closely with the power research group inside TI's Kilby Labs. Dave has a BS degree in Physics from Midwestern State University.

**Aurel Gontean** graduated in 1986 the Electronics and Telecommunications Faculty, Politehnica Univesity Timisoara, Romania. He has a PhD Degree in Electronics (1998). He has over 140 papers, 10 books, 2 patents. Currently, he is professor at the Applied Electronics Department. His current interest is energy harvesting. An IEEE Member since 1994, he is involved in COST Actions 1301 and 1401. He enjoys history and classical music.

**Ernesto González De León** is a student of the Instituto Tecnológico de Tepic. He has participated in two investigation summers in the Centro de Investigación y Desarrollo Tecnológico en Electroquímica (Cideteq).

**M. en C. Monserrat Gutiérrez** obtained her master degree in science of chemical engineering from the Universidad Autónoma Metropolitana (UAM-I, México). She is a doctoral student of electrochemistry from Centro De Investigación Y Desarrollo Tecnológico En Electroquímica (CIDETEQ) and, at the same time is a doctoral student of Materials science from Ecole Centrale de Lyon (ECL, Francia).

**Daniel Guyomar** received a degree in Physics from Amiens University, Amiens, France, engineering and a doctor–engineer degree in acoustics from Compiègne University, France, as well as a PhD degree in Physics from Paris VII University, Paris, France. In 1982–1983 he worked as a research associate in fluid dynamics at the University of Southern California, Los Angeles, CA. He was a National Research Council Awardee (1983–1984) detached at the Monterey Naval Postgraduate School, California, to develop transientwave propagation modelling. He was hired by Schlumberger in 1984 to lead several projects dealing with borehole imaging; and then moved to Thomson Submarine activities in the field of underwater acoustics. In 1992, Dr. Guyomar co-created the Techsonic company, which is involved in research, development, and production of piezoelectric and ultrasonic devices. He is presently Full-time University Professor at Institut National des Sciences Appliquées de Lyon (INSA), Lyon, France, where he manages the Laboratoire de Génie Electrique et Ferroélectricité (LGEF). He also works as a consultant for several companies. His present research interests include the field of piezo-material characterization, piezoactuators, acoustics, power ultrasonics, vibration control, and energy harvesting.

**Abdelowahed Hajjaji** was born in Morocco. He received his Master's degree in Materials Sciences from National Institute for Applied Sciences, Lyon, in 2004 and his PhD degree in materials behaviours, vibration control and energy harvesting in 2007. He is currently Associate Professor at National School of Applied Sciences (ENSAJ), El Jadida, Morocco. His current research activities include piezoelectric systems, energy harvesting, vibration control, and noise reduction.

**Mohamed Hussein** was born in Egypt in 1984. He received the B.Sc. degree in electronics-physics and the M.Sc. degree in electronics-physics from the Faculty of Science, Ain Shams University, Cairo, Egypt, in 2006 and 2010, respectively. He is currently working toward the Ph.D. degree in physics between

Ain Shams University and Centre for Photonics and Smart Materials (CPSM), Zewail City of Science and Technology, Egypt. His research interests include modelling and simulation of energy harvesting devices such as nanoantennas and nanowires. He has 5 journal papers and 5 conference papers based on his research.

**Byung C. Jung** is a Senior Researcher in the Department of System Dynamics at the Korea Institute of Machinery and Materials. He received his B.S. in mechanical engineering from Hanyang University, his M.S. degree in mechanical engineering from KAIST, Korea and his Ph.D. in reliability engineering from the University of Maryland - College Park, USA. His research interests are model verification & validation, energy harvester design, underwater shock analysis and total ship survivability.

**Daejoong Kim** received the PhD degree in mechanical engineering at Stanford University, Stanford, California, in 2007. Before that, he earned the BS and MS degrees in mechanical engineering at Seoul National University, Seoul, Korea, in 1999 and 2001, respectively. He was a postdoctoral research associate at the University of Illinois, Urbana-Champaign (UIUC). He is currently an associate professor at Sogang University in Department of Mechanical Engineering since 2008. He specializes in fluid mechanics and transport phenomena at micro- and nanoscale. His research includes micropumps, micromixers, micro power sources, reverse electrodialysis, seawater desalination, and molecular dynamics simulation. He served as a reviewer for many archived journals including Sensors and Actuators A: Physical and Microfluidics and Nanofluidics. He also served as a secretary general, a session organizer, or a session chair in many conferences sponsored by American Society of Mechanical Engineers (ASME) and Korean Society of Mechanical Engineers (KSME).

**Hongjin Kim** received his B.S., M.S., and Ph.D. degrees from the Department of Mechanical and Aerospace Engineering at Seoul National University, Republic of Korea, in 2006, 2009, and 2014, respectively.

**Yoon Young Kim** obtained B.S. and M.S. degrees from Seoul National University, in 1981 and 1983. From Stanford University, he received a PhD degree in Mechanical Engineering in 1989. He joined Seoul National University in 1991 as a faculty member. He is the director of the Institute of Advanced Machinery and Design at Seoul National University, a fellow of KAST (The Korean Academy of Science and Technology, 2009) and also a member of NAEK (The National Academy of Engineering of Korea, 2009). His research has been on Integrated Analysis and Design of Multiphysics Systems. In particular, his main research areas are Topology Design Optimization of Multiphysics Systems, Magnetostrictive Transducers for Guided Wave Transduction, and Elastic/Acoustic Meta-materials. He has published over 150 international journal papers and 20 patents. His research has been recognized with many awards including Research Excellence Award, Seoul National University (2010), Prime Minister Award for 2008 Korea's Patent and Invention Exhibition (2008), Research Excellence Award, Seoul National University (2008), Emerald-COMPEL Award for Best Paper, in Symposium on Electromagnetic Fields in Electrical Engineering (ISEF2007) (2007), KSME (Korean Society of Mechanical Engineers) Award (2005), Korea Minister of Commerce, Industry and Energy's Award for Korea's New Technology (Nondestructive Pipe Evaluation System Using Tailed-Patch Magnetostrictive Ultrasonic Transducers) (2005), Best Research Project Award in Power Electricity Basic Research (2001), and Research Excellence Award from Korean Society for Noise and Vibration Engineers Award (2001).

**Kilsung Kwon** is currently a doctoral student in Department of Mechanical Engineering at Sogang University. He received BS degree and master degree in Mechanical Engineering at Seoul National University of Technology in 2009 and Sogang University at 2011. His research work focuses on the development of efficient fuel and air supply module for portable fuel cells.

**Mohamed Latrach** (IEEE member) received the Ph.D. degree in electronics from the University of Limoges, Limoges, France. He is currently Professor of microwave engineering in the Ecole Supérieure d'Electronique de l'Ouest (ESEO), Angers, France, where he is Head of the Radio-Frequency & Microwave research group. His research interests include: design of hybrid & monolithic active and passive microwave circuits, metamaterials, LH materials, antennas, rectennas and their applications in wireless communications and wireless power transmission (WPT). Mohamed Latrach has supervised several Ph.D students, postdoctoral students and master/engineer students, with many publications in national and international conferences and journals. He has also filed 3 patents. He has delivered numerous invited presentations and has participated in several projects. He is currently the leader for ESEO in the SPINNAKER project, which is an important collaborative project between academic and industrial partners, which deals with the design of new antennas structures for RFID Applications.

**Longnan Li** is currently a Ph.D. student in Department of Mechanical Engineering at Sogang University. He received BS degree in Civil Engineering at NEPU in 2007. His research work focuses on the water and ion transport in nanofluidic system and its desalination applications.

**Aya Mabrouki** received her Master and Ph.D. degrees in Microelectronics in 2007 and 2010 respectively, both from the University of Bordeaux, France. Her research interests include RF circuit design under low power constraint and microwave energy harvesting techniques for wireless sensor networks. She joined ESEO (Ecole Supérieur d'électronique de l'Ouest, Angers) in 2011 as an associate professor. She has published several communications in international and national conferences and journals (Springer, RFIT, ICECS, NEWCAS, PRIME…). In 2009, she received the IEEE NEWCAS Best Student Paper Award, in Toulouse, France.

**Kenneth Maggio** graduated from Oklahoma State University and joined Texas Instruments, Inc., Dallas, in 1989, in the Defense and Electronics Group. He has worked on a wide range of applications including phase array radar, and radar jammers, cellular radio, motor control, and battery management. He joined the Hard Disk Drive Products Group where he was a Key Instigator and Design Manager of a new design group for read/write preamplifiers which captured a majority market share. In 1999, he moved to the Wireless Terminals Group holding management positions including Chief Technical Officer of the Digital RF Processor (DRP) Group, Texas Instruments, Inc. He has been involved in various research and design in Power electronics including work in Kilby labs for thermoelectrics, with continued interest in quantum effect devices.

**Alessandro Massaro** received the Laurea degree in electronic engineering and the Ph.D. degree in telecommunication engineering from the Università Politecnica delle Marche, Ancona, Italy, in 2001 and 2004, respectively. From 2004 to 2006 he worked as Research Scientist (post-doc) in the Department of Electromagnetism and Bioengineering at Università Politecnica delle Marche. In 2006, he spent one year in Research and Development at medical and industrial optics industry (endoscope design and optical systems). He worked for two years with National Nanotechnology Laboratory of CNR-INFM, Università del Salento, as principal investigator. He was team leader in Robotics Lab. platform of the Center for Bio-Molecular Nanotechnology of Italian Institute Technology (IIT), Arnesano, Lecce, Italy. He was team leader in Smart Materials platform (during the submission of the proposed chapter) for the Center for Bio-Molecular Nanotechnology of Italian Institute Technology (IIT), Arnesano, Lecce, Italy. His research interests are in the design and modeling of photonic band gap circuits, in the development of computer aided design (CAD) tools inthe area of integrated optics, MEMS technology and systems, and smart material implementation.

**Yunny Meas Vong** obtained his engineering degree in electrochemistry from the National Polytechnique Institute of Grenoble (INPG-ENSEEG-France) and his doctorate in 1978. He is a member (level 3) of the National System of Researchers (SNI) and the Mexican Academy of Sciences. In 1991, as General Director (1991-2000), founded the Center of Research and Technological Development in Electrochemistry (CIDETEQ, CONACYT), and currently Director of Technology. His main areas of research are related to electrocatalytic materials, surface functionalization and environmental electrochemistry. Dr. Meas has published over 135 articles in indexed journals, several book chapters and several books. He is co-author of 19 patents and has received several awards. He is a member of editorial boards of several journals. He was the founder and president twice (1983-1985, 1998-2000) of the Mexican Society of Electrochemistry, president of the Iberoamerican Society of Electrochemistry (2002-2004) and is currently vice-president of the International Society of Electrochemistry.

**Meddad Mounir** received his engineer's degree in electrical engineering from setif University in Algeria in 1999 and his Ph.D. at Institut National des Sciences Appliquées (INSA) de Lyon, France and setif University Algeria in 2012. He is currently Associate professor at Bachir El Ibrahimi University, BBA Algeria, with research interests concerning electroactive materials characterization, smart structures, actuation performance and environmental energy scavenging.

**Rosario Miceli** is Associate Professor at the Electrical, Electronic and Telecommunication Engineering Department of the University of Palermo where he teaches "Electrical Systems for the Automatization" and "Electrical Drives for the Automatization". He has carried out and nowadays carries out his activity of research mainly at the Unit of Palermo of the National Section "Converters, Electrical Machines and Drives". As researcher he has worked and works on various programs of research funded by European Community, by MURST, by CNR, by the University of Palermo, by the Sicilian Region and by some private firms and state-owned enterprises. He has been secretary of the Program Technical Committee for the Unit of Palermo, regarding the all-important working Program of Scientific Research: "Integrated System with Electrical Vehicles for Multiple and Single Mobility of the Disable People in Urban Environment" funded by MURST in the years 1999/2001. He has been person in charge of the Project of Bilateral Scientific Research "Innovative control and diagnosis algorithms for electrical drives for the industrial processes automatization" between the University of Palermo and the Universidad Central de Venezuela. He is person in charge of the University laboratory "Laboratory for the Sustainable Development and the Energetic Saving" funded by the Sicilian Region with the 3.15 Sicily POR action. He is person in charge of the Unit of University Research of Palermo inside the "ENERGO" project funded by the Sicilian Region with the 3.14 Sicily POR action. He is person in charge of the Unit of University Research of Palermo inside the european project "BEYWATCH" (Building Energy Watcher) funded by the European Community with the FP7 action. He is also member of the Technical Committee TC2 "Rotating Electrical Machines" of CEI (Comitato Elettrotecnico Italiano). He is "probo viro" of the Associazione Italiana Cultura Qualità (AICQ Sicily). During his activity of research Dr. Rosario Miceli has been involved in the following subjects of research: automatic design of electrical network, improvement of the reliability and of the continuity of the electric network of distribution, dynamic and stationary behaviour of the induction motors and of the linear induction motors, behaviour of electrical drives with variable speed, diagnosis of electrical drives, applications of hard-computing and soft-computing AI techniques to the control and diagnosis of electrical drives, electromagnetic compatibility of electrical drives, pre-standardization activity on electrical drives, fuel cell in road electrical vehicles and in distributed generation.

**Salah Obayya** received the B.Sc. and M.Sc. degrees in Electronics and Communications Engineering from Mansoura University, Egypt, in 1991 and 1994, respectively. In addition, he got his Ph.D. under "joint supervision scheme" between City University London and Mansoura University, in December 1999. Professor Obayya is now Full Professor of Photonics and Vice Chair and Academic President of Zewail City of Science and Technology, Egypt. He has a track record of 121 journal publications mostly in IEEE and IET/IEE, 2 authored books published by the world-leading scientific publisher Wiley, and 154 conference presentations with many of them invited talks, presentations and keynote lectures. He led the team that developed one of the world's best comprehensive numerical packages for the analysis, design and optimization of nano-photonic devices, and subsystems for low-carbon sustainable energy issues, where a number of "world-first" numerical approaches have been developed.

**José de Jesús Pérez Bueno** was born in 1973 in Zacatecas, Mexico. He has been developing both basic and applied research, at the CIDETEQ (Centro de Investigación y Desarrollo Tecnológico en Electroquímica, S.C.), in the following areas: Hybrid organic-inorganic coatings, sol-gel, nanotechnology, water remediation, semiconductors, electrodeposits, and others. Dr. Perez has been advisor in several concluded theses: 9 Bachelor, 9 Master Degree, 3 PhDs, and hosted 2 Postdoctorals. His published papers are currently 34, which up to date have 250 citations. In addition, he had published 5 papers in national journals, 11 book chapters and 3 books. His work includes a national patent granted in 2010, and 6 pending patent applications. He received thirteen awards, such as: three Alejandrina Awards granted by the Autonomous University of Querétaro – Mexico; 2010 TECNOS Award; and the prize of the Institute of Materials Research of the UNAM, both to his doctoral thesis in 2003 as well as co-advisor in 2010.

**Pietro Romano** was born in Palermo, Italy on 15 May 1967. He received the MSc. and Ph.D. degrees in Electrical Engineering from the University of Palermo, Italy, in 1993 and 1998, respectively. He is now a researcher at the Department of Electrical Engineering and teaches Basic Electrical Engineering and Electrotechnics. His research activity is mainly in the field of insulating systems diagnosis, partial discharge measurements, multifactor stress effects and electric field simulations, photovoltaic array reconfiguration, energy harvesting.

**Bouchta Sahraoui** was born in Morocco. He received the M.Sc. degree in Physics from the Nicolaus Copernicus University (Torun, Poland) in 1992 and received the Ph.D. degree in Physics in 1996 from the University of Angers (France) and also from the Nicolaus Copernicus University (in 1998). His Ph.D. thesis's were supervised by Professors: Jaroslaw Zaremba (Torun, Poland) and Genevieve Rivoire (Angers, France) and involved the third order nonlinear optical properties of new TTF derivatives. Professor Sahraoui joined the Laboratory of Optical Properties of Materials and their Applications (POMA) at the University of Angers in 1996 and since 2003 has held the position of Professor of Physics. Since the 1st of January 2010 Professor Bouchta Sahraoui became a member of the Institute of Sciences and Molecular Technologies of Angers (MOLTECH Anjou)- UMR CNRS 6200 Professor Bouchta SAHRAOUI has authored over 292 peer reviewed publications (h-index 26), 2 books moreover he was nominated for many Editorial Activities. Professor Sahraoui was nominated as distinguished professor in September 2011for more details please visit his home page at the address: http://sb-lerm-foibnoa.univ-angers.fr/.

**Guy Stremsdoerfer** was entitled to supervise research in 1992. - Recipient of the training allowance of Research of the Ministry: 1994, 1998, 2003,2007. - Chairman of several international congresses. - Author in many international journals: "Journal of Electrochemical Society, Journal of Material Science, Coating, ... " and project REI of the DGA. - International Distinctions: 2nd prize Brenner at USA in 1999, 1er prize of Innovation technology at Mexico in 2006, 2010 Award for best thesis in Materials Engineering (joint supervision Lorena Magallon), Mexico 2010. - National Distinctions (France): 1er prize of Technological Innovation, in 1997 France. - Opening to the international to create a Franco-Lebanese cooperation 1997-2003 (2 Programs MIRA Region) - Confirmation to the international for creating Franco-Mexican collaboration 2001-2013. -Responsible and host of four Post Doctoral Programmes (PCP) with the Ministry of Foreign Affairs (France) and the Mexican (CONACYT). - French and European Expert in coatings in liquid medium (about 140 expertise) -Impliqué In the competence center "Chemistry and Environment" (Duramat project) and the National Action of Research "INFIME". Scientific Production 1. 4 Book Chapters. 2. 33 publications in national and international journals. 3. 20 conference proceedings publications including 15 international. 4. 72communications including 38 international communications. 5. 17 invited conferences including four international. Action research development - 5 patents (1993,1997,1997, 2005, 2008); 2 envelopes "Soleau" one trademark application in 1999. - Creating a Business "JetMetal Technology" (2007), winner of the national competition of business creation in 2006, 3 prices technological innovation (2006, TOTAL, Contractors, Rhone-Alpes Undertaking). 1er levé bottom of € 2.5 million in 2009. -45 Contract research programs. -158 presentations of Research and expertise. - 56 Contract reports, 7 patent licenses. - Cited more than 40 times in newspaper articles from major newspapers or broadcast TV, radio ... Scientific activities and research supervision. - Responsible of the theme "Chemical Deposition" from 1991 to 1994 in the laboratory IFOS UMR 5621. -Responsible of the "Development of multifunctional surface structures" operation IFOS laboratory 1994-2003. - Advisor of 9 PhD, 20 Master students, 13 TFE, 12 Engineers in Training. - Reviewer in 4 Theses (2005-2010) -Agrement Scientific EXPERT by the Directorate General for Research and Innovation for 2011-2012-2013. Collaboration with laboratories from academia and industry research - About 30 French and foreign laboratories.

**Toan Tran** is a Process Integration Engineer at Texas Instruments with over 25 years' experience in new Process Technology Development ranging from 1.0um to 65nm for CMOS and Embedded Flash. He graduated from UT at Arlington with BSEE degree.

**Fabio Viola** received the Laurea degree in electrical engineering and the Ph.D. degree from the Università degli Studi di Palermo, Palermo, Italy, in 2002 and 2006, respectively. He is currently assistant professor in University of Palermo. He is the holder of the course of "Electrotechnics" at the Engineering Faculty of Palermo. His research interests are in the field of electromagnetic compatibility, where he works on developing and employing numerical methods for the study of linear and nonlinear aspects of electromagnetic compatibility. Other interests are in the field of renewable energy.

**Heonjun Yoon** received his B.S. degree from the Department of Mechanical Engineering at Inha University, Republic of Korea, in 2011 and the M.S. degree from the Department of Mechanical and Aerospace Engineering at Seoul National University, Republic of Korea, in 2013. Currently, he is pursuing the Ph.D. degree in the Department of Mechanical and Aerospace Engineering at Seoul National University, Republic of Korea. He is exploring two research avenues: piezoelectric energy harvesting (PEH) and statistical model verification and validation (V&V). Mr. Yoon's research has led to one featured article from the Smart Materials and Structures journal, two best paper awards, and best M.S. thesis award in the area of piezoelectric energy harvesting.

**Byeng D. Youn** received his Ph.D. degree from the department of Mechanical Engineering at the University of Iowa, Iowa City, IA, in 2001. He was a research associate at the University of Iowa (until 2004), an assistant professor in Michigan Technical University (until 2007), and an assistant professor in the University of Maryland College Park (until 2010). Currently, he is an associate professor at the School of Mechanical and Aerospace Engineering at Seoul National University, Republic of Korea. He is currently exploring three research avenues: system risk-based design, prognostics and health management (PHM), and energy harvester design. Dr. Youn's research and educational dedication has led to: six notable awards, including the ISSMO/Springer Prize for the Best Young Scientist in 2005 from the International Society of Structural and Multidisciplinary Optimization (ISSMO), and more than 100 publications in the area of system-risk-based design and PHM and energy harvester design.

# Index

## R



## S



## T



## U



## V



## W



## Z

www.ingramcontent.com/pod-product-compliance
Lightning Source LLC
LaVergne TN
LVHW060350300926
849162LV00006B/1242